1986

513

COLLECTED ACCOUNTS OF

TRANSITION METAL CHEMISTRY

VOLUME II

Comprising articles reprinted
from Volumes 5-8, **Accounts of Chemical Research**

EDITORS

Fred Basolo, *Northwestern University*
Joseph F. Bunnett, *University of California, Santa Cruz*
Jack Halpern, *University of Chicago*

An ACS Reprint Collection

American Chemical Society
Washington, D.C. 1977

Library of Congress CIP Data

Basolo, Fred, 1920- comp.
Collected accounts of transition metal chemistry.
(An ACS reprint collection)
"Comprising articles reprinted from volumes 1—Accounts of chemical research."

Includes bibliographical references.

1. Transition metals—Collected works.
I. Bunnett, Joseph F., joint comp. II. Halpern, Jack, joint comp. III. Accounts of chemical research. IV. Title. V. Title: Transition metal chemistry.

QD172.T6B37 546'.6 72-95642
ISBN 0-8412-0348-2 (hard); 0-8412-0356-3 (paper)
(v. 2) 1-365

PRINTED IN THE UNITED STATES OF AMERICA

Contents

Synthesis, Structure, and Bonding

Redox Chemistry

Organometallic Reactions and Catalytic Applications

Applications to Organic Synthesis

Transition Metal Chemistry of Biological Interest

Preface

This volume presents 32 papers from Volumes 5–8 of *Accounts of Chemical Research.* In character it is similar to "Collected Accounts of Transition Metal Chemistry, Volume I, which reprinted papers from Volumes 1–4 of the journal. It is our hope and belief that this volume will be useful to students of transition metal chemistry, to persons doing research in this field, and to others who may wish to read about the subject.

Editors of a volume such as this are always faced with difficult and somewhat arbitrary choices as to which papers to include. Some of high quality and considerable relevance inevitably are excluded, and we regret any errors in editorial judgement.

Authors of papers included in this volume were offered the opportunity of contributing short addenda in which important recent developments could be mentioned. Most chose to contribute such addenda, feeling that the updates would add to the book's usefulness.

In the organization of this volume, as well as in the editing of the journal from which it is derived, we have been greatly assisted by Helga Koivisto, and we wish to express our deep appreciation. We also thank members of the American Chemical Society Books and Journals staff in Washington, D. C. and Easton, Penn. for their support and cooperation.

FRED BASOLO

JOSEPH F. BUNNETT

September 1976 JACK HALPERN

Some Recent Studies on Poly(tertiary phosphines) and Their Metal Complexes

R. Bruce King

Department of Chemistry, University of Georgia, Athens, Georgia 30601

Received June 18, 1971

Reprinted from Accounts of Chemical Research, *5, 177* (*1972*)

During the past 15 years tertiary phosphines have become important ligands in coordination chemistry.[1] Certain metal complexes of tertiary phosphines have been shown to have unusual chemical properties of importance in such diverse areas as homogenous hydrogenation,[2] oxygen transfer,[3] and nitrogen fixation.[4] Such interesting applications have provided impetus to the further development of this relatively young field.

The chemical properties of metal complexes of tertiary phosphines vary with the electronic and steric properties of the ligands. For this reason further development of the coordination chemistry of tertiary phosphines is highly dependent upon the concurrent development of synthetic organophosphorus chemistry.

These considerations prompted efforts to develop new methods for the syntheses of tertiary phosphines which would possess novel coordination chemistry. One of the early objectives was synthesis of a hexa(tertiary phosphine) with the six donor phosphorus atoms situated so as to bond to a single metal atom much as do the six donor nitrogen and oxygen atoms in ethylenediaminetetraacetic acid (EDTA). The strong affinity of EDTA for a great variety of metal atoms, that in the early 1950's provided the basis for the development of complexometric analytical techniques,[5] suggested that such a chelating hexa(tertiary phosphine) might form unusual transition metal complexes with useful properties.

This research program in synthetic organophosphorus chemistry was started in late 1968 in collaboration with Dr. Pramesh Kapoor, who had had considerable prior experience in synthesizing and handling the often air-sensitive, malodorous, and toxic organophosphorus compounds. At that time the only generally useful method for the preparation of poly(tertiary phosphines) employed reactions of alkali metal dialkylphosphides and diarylphosphides with organic polyhalides. Alkali metal diphenylphosphides, $MP(C_6H_5)_2$,[6] were convenient reagents for this type of synthesis since they are readily obtained by reactions of the alkali metal with the readily available triphenylphosphine in an appropriate coordinating solvent such as tetrahydrofuran (THF) (M = Li), liquid ammonia (M = Na or Li), or dioxane (M = K). This synthetic method is exemplified by the preparation of $(C_6H_5)_2PCH_2CH_2P(C_6H_5)_2$ (I) from the reaction of lithium diphenylphosphide with 1,2-dichloroethane according to[6]

$$2LiP(C_6H_5)_2 + ClCH_2CH_2Cl \xrightarrow{THF} 2LiCl + \underset{\text{I}}{(C_6H_5)_2PCH_2CH_2P(C_6H_5)_2}$$

Compound I is frequently known by the trivial name "diphos," but this term is becoming confusing, since other di(tertiary phosphines) are beginning to receive considerable attention from coordination chemists. The analogous reactions between lithium diphenylphosphide and the cis and trans isomers of 1,2-dichloroethylene give stereospecifically the cis and trans isomers of the olefinic di(tertiary phosphine) $(C_6H_5)_2PCH{=}CHP(C_6H_5)_2$ according to[7]

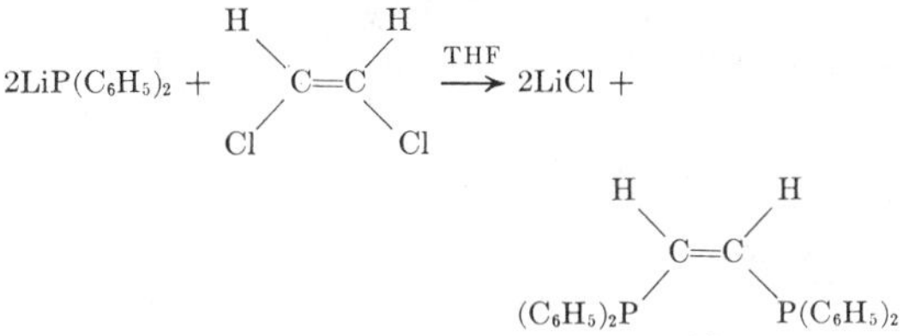

II

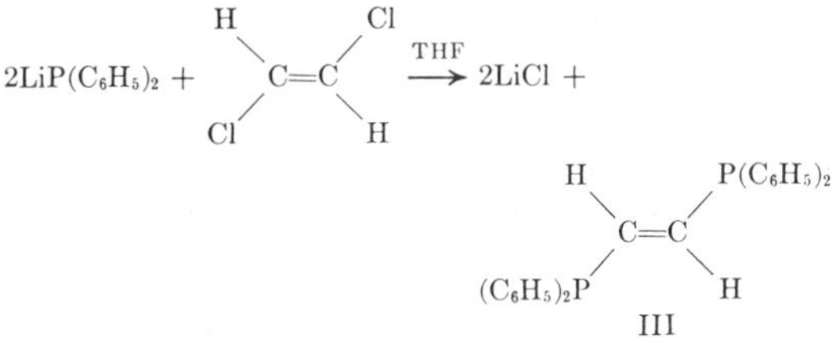

III

In a few cases analogous reactions of alkali metal diphenylphosphides with organic polyhalides have been

R. Bruce King was born in Rochester, N. H., on Feb 27, 1938. He did undergraduate work at Oberlin College and received his Ph.D. degree from Harvard. After a year with E. I. du Pont de Nemours and Company, he joined the Independent Research Division of the Mellon Institute as a Fellow. In 1966 Dr. King moved to the University of Georgia, where he is now Research Professor.

Dr. King's research interests have encompassed a wide range with emphasis on preparative transition-metal organometallic chemistry. Recently he has also become interested in cyanocarbon chemistry, the synthesis of unusual organophosphorus ligands, and chemical applications to topology and group theory. He is the 1971 recipient of the American Chemical Society award in Pure Chemistry sponsored by Alpha Chi Sigma.

(1) G. Booth, *Advan. Inorg. Chem. Radiochem.*, **6**, 1 (1964); T. A. Manuel, *Advan. Organometal. Chem.*, **3**, 181 (1965).

(2) J. A. Osborn, F. H. Jardine, J. F. Young, and G. Wilkinson, *J. Chem. Soc. A*, 1171 (1966).

(3) L. Vaska, *Science*, **140**, 809 (1963); *Accounts Chem. Res.*, **1**, 335 (1968).

(4) A. Yamamoto, S. Kitazume, L. S. Pu, and S. Ikeda, *Chem. Comm.*, 79 (1967).

(5) F. J. Welcher, "The Analytical Uses of Ethylenediamine Tetraacetic Acid," Van Nostrand, Princeton, N. J., 1958.

(6) J. Chatt and F. A. Hart, *J. Chem. Soc.*, 1378 (1960).

(7) A. M. Aguiar and D. Daigle, *J. Amer. Chem. Soc.*, **86**, 2299 (1964).

useful for the synthesis of certain tri- and tetra(tertiary phosphines), *e.g.*, eq 1–3. These syntheses of tri- and

$$3NaP(C_6H_5)_2 + CH_3C(CH_2Cl)_3 \xrightarrow[\text{2. THF}]{\text{1. } NH_3} 3NaCl + CH_3C[CH_2P(C_6H_5)_2]_3 \quad (1)^7$$

IV

$$2NaP(C_6H_5)_2 + C_6H_5P(CH_2CH_2Br)_2 \xrightarrow{NH_3} 2NaBr + C_6H_5P[CH_2CH_2P(C_6H_5)_2]_2 \quad (2)^8$$

V

$$4NaP(C_6H_5)_2 + C(CH_2Cl)_4 \longrightarrow 4NaCl + [(C_6H_5)_2PCH_2]_4C \quad (3)^9$$

tetra(tertiary phosphines), however, have various disadvantages. In the synthesis of the tri(tertiary phosphine) IV, the mole ratio of sodium diphenylphosphide to the halide $CH_3C(CH_2Cl)_3$ is critical. If a deficiency of sodium diphenylphosphide is used, the reaction proceeds differently to form the phosphetanium salt VII.[10]

$$2NaP(C_6H_5)_2 + CH_3C(CH_2Cl)_3 \longrightarrow 2NaCl +$$

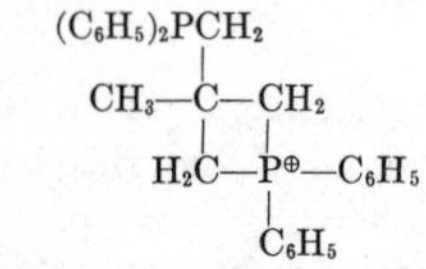

VII

The synthesis of the tri(tertiary phosphine) V is reported[8] to give only a 16% yield and uses the rather unstable and difficulty accessible halide $C_6H_5P(CH_2CH_2Br)_2$[11] as a starting material. Steric considerations prevent the tetra(tertiary phosphine) VI from acting as a tetradentate chelating agent. Only three phosphorus atoms in VI at most can bond to a single metal atom. An analogous synthetic method is not applicable for the syntheses of other tetra(tertiary phosphines) which might act as tetradentate ligands by having all four phosphorus atoms bond to a single metal atom.

These and other limitations in the reactions of alkali metal dialkylphosphides with organic polyhalides as methods for the preparations of poly(tertiary phosphines) indicate the need for a new general method for poly(tertiary phosphine) synthesis. We were particularly interested in the synthesis of poly(tertiary phosphines) with multidentate chelating properties. Compounds containing the PCH_2CH_2P and *cis*-PCH=CHP units were, therefore, selected as synthetic objectives since both of these types of units should be bidentate. The unit PCH_2CH_2P can act either as a bidentate chelating unit as in VIIIa or as a bridging unit as in VIIIb because of free rotation around the carbon–carbon single bond.

Synthesis of Poly(tertiary phosphines)

One possible method for constructing a PCH_2CH_2P

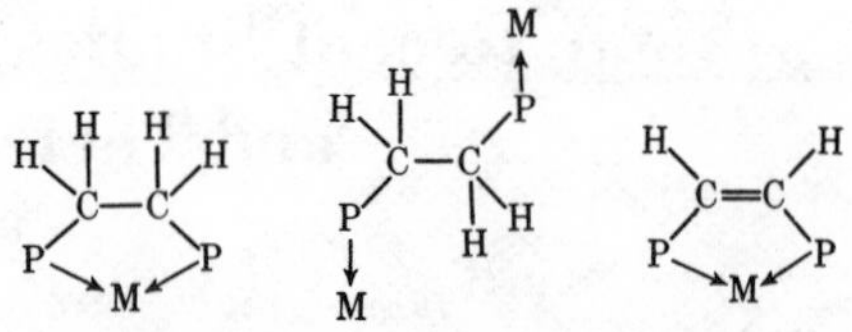

VIIIa VIIIb IX

unit in a poly(tertiary phosphine) appeared to be the addition of a phosphorus–hydrogen bond across the vinyl double bond of a vinylphosphine as shown in eq 4.

$$P{-}H + CH_2{=}CHP \longrightarrow PCH_2CH_2P \quad (4)$$

Previous work on the addition of phosphorus–hydrogen bonds to other unsaturated compounds such as acrylonitrile[12] or 2,4-hexadiyne[13] suggested that a basic catalyst such as phenyllithium or potassium hydroxide might be required. The necessary vinylphosphines were generally prepared by reactions of vinylmagnesium bromide in tetrahydrofuran with the corresponding halophosphines[14,15] or alkoxyphosphines.[16]

In order to evaluate this proposed new method, the reaction between diphenylphosphine and diphenylvinylphosphine in the presence of a phenyllithium catalyst was investigated. An ~80% yield of the desired phosphine $(C_6H_5)_2PCH_2CH_2P(C_6H_5)_2$ (I) was obtained according to eq 5.[17,18] The ditertiary phosphine was

$$(C_6H_5)_2PH + (C_6H_5)_2PCH{=}CH_2 \longrightarrow I \quad (5)$$

identified by comparison of its physical and spectroscopic properties with those of authentic material prepared from lithium diphenylphosphide and 1,2-dichloroethane as described above. The synthesis of I from diphenylphosphine and diphenylvinylphosphine does not offer any advantages in terms of efficiency and convenience over the previously discussed synthesis of I from lithium diphenylphosphide and 1,2-dichloroethane[6] when allowance is made for the lesser availability of both diphenylphosphine and diphenylvinylphosphine relative to triphenylphosphine, which is easily converted to lithium diphenylphosphide. However, the synthesis of I from diphenylphosphine and diphenylvinylphosphine provided the first unequivocal demonstration that the PCH_2CH_2P unit could be synthesized efficiently by the base-catalyzed addition of a phosphorus–hydrogen bond across the vinyl double bond of a vinylphosphine.

We next investigated the applicability of this new synthetic technique for the syntheses of more complex poly(tertiary phosphines) containing PCH_2CH_2P units. The tri(tertiary phosphine) $C_6H_5P[CH_2CH_2P(C_6H_5)_2]_2$,

(8) W. Hewertson and H. R. Watson, *J. Chem. Soc.*, 1490 (1962).
(9) J. Ellmann and K. Dorn, *Ber.*, **99**, 653 (1966).
(10) D. L. Berglund and D. W. Meek, *J. Amer. Chem. Soc.*, **90**, 518 (1968); *Inorg. Chem.*, **8**, 2602 (1969).
(11) C. H. S. Hitchock and F. G. Mann, *J. Chem. Soc.*, 2081 (1958).

(12) M. M. Rauhut, I. Hechenbleikner, H. A. Currier, F. C. Schaefer, and V. P. Wystrach, *J. Amer. Chem. Soc.*, **81**, 1103 (1959).
(13) G. Märkl and R. Potthast, *Angew. Chem., Int. Ed. Engl.*, **6**, 86 (1967).
(14) L. Maier, D. Seyferth, F. G. A. Stone, and E. G. Rochow, *J. Amer. Chem. Soc.*, **79**, 5884 (1957).
(15) K. D. Berlin and G. B. Butler, *J. Org. Chem.*, **26**, 2537 (1961); R. Rabinowitz and J. Pellon, *ibid.*, **26**, 4623 (1961).
(16) M. I. Kabachnik, C. Y. Chang, and E. N. Tsvetkov, *Dokl. Akad. Nauk SSSR*, **135**, 603 (1960); *Chem. Abstr.*, **55**, 12272 (1961).
(17) R. B. King and P. N. Kapoor, *J. Amer. Chem. Soc.*, **91**, 5191 (1969).
(18) R. B. King and P. N. Kapoor, *ibid.*, **93**, 4158 (1971).

which was previously[8] obtained in only 16% yield by the reaction of sodium diphenylphosphide with the halide $C_6H_5P(CH_2CH_2Br)_2$, could now be obtained in much better yield (50–90%) and much more conveniently by the method of either eq 6 or eq 7.[17,18] Suitable basic catalysts for these reactions are phenyllithium and potassium *tert*-butoxide. Suitable solvents for these reactions are benzene and tetrahydrofuran.

$$C_6H_5PH_2 + 2(C_6H_5)_2PCH{=}CH_2 \longrightarrow V \quad (6)$$

$$2(C_6H_5)_2PH + C_6H_5P(CH{=}CH_2)_2 \longrightarrow V \quad (7)$$

This new synthetic approach could be used to prepare two isomeric tetra(tertiary phosphines) of the empirical composition $C_{42}H_{42}P_4$ and containing PCH_2CH_2P structural units. The tripod tetra(tertiary phosphine) $P[CH_2CH_2P(C_6H_5)_2]_3$ (X) could be prepared in good yield by the methods of either eq 8 or eq 9.[17,18]

$$PH_3 + 3(C_6H_5)_2PCH{=}CH_2 \longrightarrow \underset{X}{P[CH_2CH_2P(C_6H_5)_2]_3} \quad (8)$$

$$3(C_6H_5)_2PH + P(CH{=}CH_2)_3 \longrightarrow X \quad (9)$$

The reaction between PH_3 and $(C_6H_5)_2PCH{=}CH_2$ is especially convenient for the preparation of the tripod tetra(tertiary phosphine) X since this reaction proceeds in nearly quantitative yield in a few hours' time simply by bubbling the PH_3 generated by hydrolysis of commercial aluminum phosphide into a boiling tetrahydrofuran solution of diphenylvinylphosphine containing some potassium *tert*-butoxide catalyst.[19] The linear tetra(tertiary phosphine) $(C_6H_5)_2PCH_2CH_2P(C_6H_5)CH_2CH_2P(C_6H_5)CH_2CH_2P(C_6H_5)_2$ (XI) can be prepared by eq 10.[17,18] The preparation of the required

$$(C_6H_5)(H)PCH_2CH_2P(H)(C_6H_5) + 2(C_6H_5)_2PCH{=}CH_2 \longrightarrow \underset{XI}{(C_6H_5)_2PCH_2CH_2P(C_6H_5)CH_2CH_2P(C_6H_5)CH_2CH_2P(C_6H_5)_2} \quad (10)$$

$C_6H_5P(H)CH_2CH_2P(H)C_6H_5$ starting material for the synthesis of the linear tetra(tertiary phosphine) XI is somewhat troublesome, but can be accomplished either by reaction of $C_6H_5P(OR)_2$ (R = isopropyl, etc.) with 1,2-dibromoethane followed by lithium aluminum hydride reduction of the intermediate $C_6H_5P(O)(OR)CH_2CH_2P(O)(OR)C_6H_5$[20] or by cleavage of $(C_6H_5)_2PCH_2CH_2P(C_6H_5)_2$ (I) with sodium in liquid ammonia followed by protonation with ammonium chloride;[21] the latter method is less preferable because of the concurrent formation of $(C_6H_5)_2PH$ which must be removed by a careful fractional vacuum distillation.

The successful preparation of the tri(tertiary phosphine) V and the tetra(tertiary phosphines) X and XI containing PCH_2CH_2P structural units encouraged us to investigate the similar preparation of a hexa(tertiary phosphine) with PCH_2CH_2P structural units. We found that base-catalyzed addition of the four phosphorus–hydrogen bonds in $H_2PCH_2CH_2PH_2$[22] to 4 equiv of diphenylvinylphosphine gave the hexa(tertiary phosphine) $[(C_6H_5)_2PCH_2CH_2]_2PCH_2CH_2P[CH_2CH_2P(C_6H_5)_2]_2$ (XII) according to eq 11.[17,18] This approach

$$H_2PCH_2CH_2PH_2 + 4(C_6H_5)_2PCH{=}CH_2 \longrightarrow \underset{XII}{[(C_6H_5)_2PCH_2CH_2]_2PCH_2CH_2P[CH_2CH_2P(C_6H_5)_2]_2} \quad (11)$$

provided the first synthetic route to a hexa(tertiary phosphine) but had the disadvantage of requiring the use of the malodorous and pyrophoric $H_2PCH_2CH_2PH_2$. This disadvantage could be partially alleviated by generating the $H_2PCH_2CH_2PH_2$ by reduction of $(C_2H_5O)_2P(O)CH_2CH_2P(O)(OC_2H_5)_2$ with lithium aluminum hydride in diethyl ether followed by dilute aqueous acid treatment to remove the lithium and aluminum salts. The resulting diethyl ether solution of $H_2PCH_2CH_2PH_2$ could be used directly for the preparation of the hexa(tertiary phosphine) XII without isolation of pure $H_2PCH_2CH_2PH_2$.[23] By means of this technique, 50-g quantities of the hexa(tertiary phosphine) XII were easily obtained in a single preparation, thereby making this compound available for use as a ligand in transition metal chemistry.

We also investigated the possibility of preparing mixed arsine–phosphines by the base-catalyzed addition of arsenic–hydrogen bonds across the vinyl double bonds of vinylphosphines. The base-catalyzed addition of the arsenic–hydrogen bond in diphenylarsine, $(C_6H_5)_2AsH$, across the vinyl double bond in diphenylvinylphosphine gave the mixed arsine–phosphine $(C_6H_5)_2AsCH_2CH_2P(C_6H_5)_2$ (XIII).[17,18] Similarly, the

$$(C_6H_5)_2AsH + (C_6H_5)_2PCH{=}CH_2 \longrightarrow \underset{XIII}{(C_6H_5)_2AsCH_2CH_2P(C_6H_5)_2}$$

base-catalyzed addition of the arsenic–hydrogen bonds in 2 equiv of $(C_6H_5)_2AsH$ across the vinyl double bonds in $C_6H_5P(CH{=}CH_2)_2$ gave the mixed diarsine–phosphine $C_6H_5P[CH_2CH_2As(C_6H_5)_2]_2$ (XIV).[17,18] Potas-

$$2(C_6H_5)_2AsH + C_6H_5P(CH{=}CH_2)_2 \longrightarrow \underset{XIV}{C_6H_5P[CH_2CH_2As(C_6H_5)_2]_2}$$

sium *tert*-butoxide was used as a catalyst for these reactions; phenyllithium appeared to be unsuitable.

The successful synthesis of poly(tertiary phosphines) with PCH_2CH_2P structural units by base-catalyzed addition of phosphorus–hydrogen bonds across the vinyl double bonds of vinylphosphines suggested that olefinic poly(tertiary phosphines) with $PCH{=}CHP$ structural units might be analogously prepared by base-catalyzed addition of phosphorus–hydrogen bonds across the carbon–carbon triple bonds of ethynylphosphines according to eq 12. In this reaction there is an ambiguity as to whether the two phosphorus atoms in the adduct will be in cis positions (XVc) or in trans positions (XVt). In order to elucidate the stereochem-

(19) R. B. King, R. N. Kapoor, M. S. Saran, and P. N. Kapoor, *Inorg. Chem.*, **10**, 1851 (1971).

(20) K. Issleib and H. Weichmann, *Ber.*, **101**, 2197 (1968).

(21) K. Sommer, *Z. Anorg. Allgem. Chem.*, **376**, 37 (1970).

(22) L. Maier, *Helv. Chim. Acta*, **49**, 842 (1966).

(23) R. B. King and M. S. Saran, *Inorg. Chem.*, **10**, 1861 (1971).

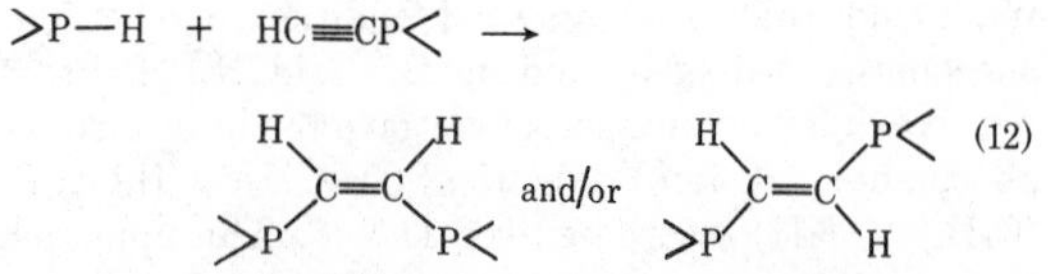

$$>P\!-\!H \ + \ HC\!\equiv\!CP< \ \longrightarrow$$

$$\text{XVc} \qquad \text{and/or} \qquad \text{XVt} \qquad (12)$$

istry, the reaction between diphenylphosphine and diphenylethynylphosphine was investigated because both the cis (II) and the trans (III) isomers of the olefinic di(tertiary phosphine) $(C_6H_5)_2PCH\!=\!CHP(C_6H_5)_2$ are well-characterized compounds.[7] The reaction between diphenylphosphine and diphenylethynylphosphine afforded a product, $(C_6H_5)_2PCH\!=\!CHP(C_6H_5)_2$, which was shown to be the pure trans isomer III on the basis of infrared and proton nmr spectra as well as its reaction with $CH_3Mo(CO)_3C_5H_5$ in acetonitrile solution to give a bridged product $(C_6H_5)_2PCH\!=\!CHP(C_6H_5)_2\text{-}[Mo(CO)_2(COCH_3)(C_5H_5)]_2$ rather than a chelate $[(C_6H_5)_2PCH\!=\!CHP(C_6H_5)_2]_2Mo(CO)_2$.[24] Thus the reaction between diphenylphosphine and diphenylethynylphosphine occurs in a syn fashion to give a trans product.

Poly(tertiary phosphines) with trans-PCH=CHP bridges are of less value to coordination chemists because both phosphorus atoms of the trans-PCH=CHP moiety cannot bond to the same metal atom (i.e., to form chelates). This reaction was therefore not extended to other examples. However, an olefinic mixed phosphine–arsine, trans-$(C_6H_5)_2PCH\!=\!CHAs(C_6H_5)_2$ (XVI), was prepared[17,18]

$$(C_6H_5)_2AsH \ + \ (C_6H_5)_2PC\!\equiv\!CH \ \longrightarrow$$

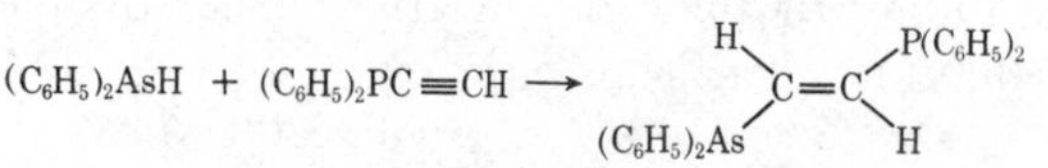

XVI

This reaction demonstrates that arsenic–hydrogen bonds can add to the carbon–carbon triple bonds of ethynylphosphines just as they can add to the vinyl carbon–carbon double bonds of vinylphosphines as discussed above.

Synthesis of Other Trivalent Phosphorus and Arsenic Derivatives

The successful base-catalyzed addition of phosphorus–hydrogen and arsenic–hydrogen bonds across the vinyl bonds of vinylphosphines suggested investigation of base-catalyzed additions of phosphorus–hydrogen and arsenic–hydrogen bonds across multiple bonds in other types of compounds in order to synthesize other organophosphorus and organoarsenic compounds of potential value as ligands. Vinyl isocyanide, $CH_2\!=\!CHNC$,[25] was first investigated because of the usefulness of isocyanides as ligands in coordination chemistry.

The base-catalyzed additions of $(C_6H_5)_2EH$ (E = P or As) to vinyl isocyanide gave the expected addition

products $(C_6H_5)_2ECH_2CH_2NC$ (XVII, E = P or As) according to eq 13.[26] The presence of an isocyanide

$$(C_6H_5)_2EH + CH_2\!=\!CHNC \longrightarrow (C_6H_5)_2ECH_2CH_2NC \quad (13)$$
$$E = \text{P or As} \qquad\qquad \text{XVII, E = P or As}$$

group in the products was demonstrated by a strong infrared absorption frequency around 2150 cm^{-1}. The reactions of $(C_6H_5)_2EH$ (E = P or As) with vinyl isocyanide thus proceed by 1,2 addition of the phosphorus–hydrogen or arsenic–hydrogen bond across the vinyl double bond of the vinyl isocyanide.

In an attempt to prepare the phosphine diisocyanide $C_6H_5P(CH_2CH_2NC)_2$, the base-catalyzed reaction of phenylphosphine with vinyl isocyanide was investigated. This reaction instead gave the 3-azaphosphole derivative XVIII[26] (eq 14). The absence of an isocy-

$$C_6H_5PH_2 \ + \ CH_2\!=\!CHNC \ \longrightarrow$$

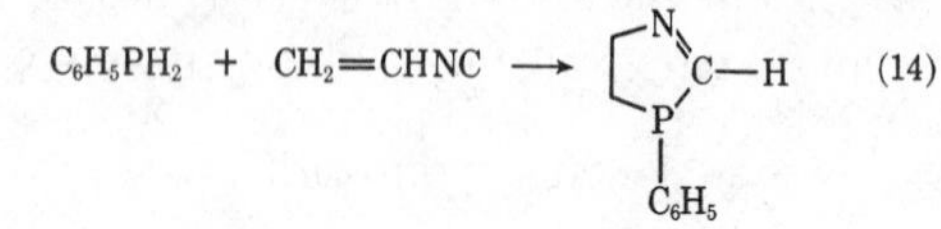

XVIII

anide group in the 3-azaphosphole product XVIII is demonstrated by the absence of an isocyanide infrared frequency around 2150 cm^{-1}. The reaction of phenylphosphine with vinyl isocyanide thus proceeds by 1,4 addition of the phosphorus–hydrogen bonds across the multiple bonds of the conjugated vinyl and isocyanide groups. Compound XVIII is of interest as the first simple derivative of the 3-azaphosphole system to be prepared. It is potentially a very reactive molecule because of the presence of both a trivalent phosphorus atom and a carbon–nitrogen multiple bond; however, its chemistry has not yet been investigated.

We also investigated the base-catalyzed addition of diphenylphosphine to the carbon–carbon double bond of diethyl vinylphosphonate, $CH_2\!=\!CHP(O)(OC_2H_5)_2$. This reaction gave a viscous pasty residue which was not characterized but which instead was reduced with lithium aluminum hydride to give the colorless, malodorous, air-sensitive compound $(C_6H_5)_2PCH_2CH_2PH_2$ (XIX), apparently according to reaction sequence 15.[27]

$$(C_6H_5)_2PH + CH_2\!=\!CHP\overset{\displaystyle O}{\underset{\displaystyle OC_2H_5}{|}}\!-\!OC_2H_5 \longrightarrow$$

$$(C_6H_5)_2PCH_2CH_2P\overset{\displaystyle O}{(OC_2H_5)_2} \quad (15)$$

$$\Big\downarrow \text{LiAlH}_4$$

$$(C_6H_5)_2PCH_2CH_2PH_2$$
$$\text{XIX}$$

$(C_6H_5)_2PCH_2CH_2PH_2$ is the first known compound incorporating both the primary phosphine and the tertiary phosphine functional groups. Its ^{31}P nmr spectrum exhibits a singlet and a triplet of approximately equal intensities. The triplet arises from the phos-

(24) R. B. King, L. W. Houk, and P. N. Kapoor, *Inorg. Chem.*, **8**, 1792 (1969).

(25) D. S. Matteson and R. A. Bailey, *J. Amer. Chem. Soc.*, **90**, 3761 (1968).

(26) R. B. King and A. Efraty, *ibid.*, **93**, 564 (1971).

(27) R. B. King and P. N. Kapoor, *Angew. Chem., Int. Ed. Engl.*, **10**, 734 (1971).

phorus of the PH_2 group, split by the two hydrogen atoms directly bonded to it. The singlet arises from the tertiary phosphorus atom.

The conversion of $(C_6H_5)_2PH$ to $(C_6H_5)_2PCH_2CH_2PH_2$ as discussed above is the first useful method for converting a P–H bond to a $PCH_2CH_2PH_2$ unit. Further applications of this synthetic technique for the preparation of compounds containing $PCH_2CH_2PH_2$ units have been hampered by the disagreeable, pervasive, and persistent odors of most primary phosphines, including compounds of this type. The value of compounds containing the $PCH_2CH_2PH_2$ unit as building blocks for the syntheses of poly(tertiary phosphines) is demonstrated by the base-catalyzed addition of the two phosphorus–hydrogen bonds in $(C_6H_5)_2PCH_2CH_2PH_2$ to the vinylic moieties in 2 equiv of diphenylvinylphosphine to give the tripod tetra(tertiary phosphine) $P[CH_2CH_2P(C_6H_5)_2]_3$ (X) (eq 16).[27] X

$$(C_6H_5)_2PCH_2CH_2PH_2 + 2(C_6H_5)_2PCH{=}CH_2 \longrightarrow \underset{\text{X}}{P[CH_2CH_2P(C_6H_5)_2]_3} \quad (16)$$

prepared by this method was shown by its infrared spectrum, melting point, and complexing properties with nickel chloride to be identical with authentic tetra(tertiary phosphine) X prepared by the base-catalyzed reaction of phosphine with 3 equiv of diphenylvinylphosphine as described above.

Metal Complexes of Poly(tertiary phosphines)

The new synthetic techniques in organophosphorus (and organoarsenic) chemistry discussed in this Account provide a wide range of new ligands of considerable interest to coordination chemists. We have made a survey of the reactions of several of the new poly(tertiary phosphines) with transition metal halides, metal carbonyls, and cyclopentadienylmetal carbonyls. The ligands studied include the tri(tertiary phosphine) $C_6H_5P[CH_2CH_2P(C_6H_5)_2]_2$ (V), the isomeric tetra(tertiary phosphines) $P[CH_2CH_2P(C_6H_5)_2]_3$ (X) and $(C_6H_5)_2PCH_2CH_2P(C_6H_5)CH_2CH_2P(C_6H_5)CH_2CH_2P(C_6H_5)_2$ (XI), and the hexa(tertiary phosphine) $[(C_6H_5)_2PCH_2CH_2]_2PCH_2CH_2P[CH_2CH_2P(C_6H_5)_2]_2$ (XII).

The Tri(tertiary phosphine) V. The tri(tertiary phosphine) $C_6H_5P[CH_2CH_2P(C_6H_5)_2]_2$ (V: abbreviated as Pf–Pf–Pf) can bond to transition metals in the following six fundamentally different ways:[28] monoligate monometallic (XXa or XXb), biligate monometallic (XXc), triligate monometallic (XXd), biligate

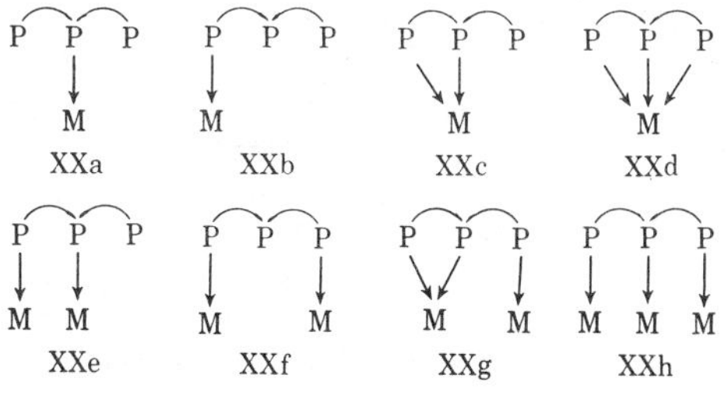
XXa XXb XXc XXd XXe XXf XXg XXh

(28) R. B. King, P. N. Kapoor, and R. N. Kapoor, *Inorg. Chem.*, **10**, 1841 (1971).

bimetallic (XXe or XXf), triligate bimetallic (XXg), and triligate trimetallic (XXh). (The phenyl rings are omitted in these and the following structures and the CH_2CH_2 bridges are represented by a simple arc.) We have been able to prepare complexes which exemplify each of these six fundamentally different types of bonding of the tri(tertiary phosphine) V to transition metals. However, currently available physical and/or chemical techniques have not yet made possible a distinction between the two different types of monoligate monometallic complexes XXa and XXb and between the two different types of biligate bimetallic complexes XXe and XXf. Factors affecting the types of complexes (XXa through XXb) formed in a reaction with a particular transition metal derivative include the ease of generating vacant coordination positions on the transition metal and the number of vacant positions so generated.

Preparation of a monoligate monometallic complex (XXa or XXb) of the tri(tertiary phosphine) Pf–Pf–Pf (V) requires a transition metal system in which exactly one vacant coordination position is readily generated and where the rate of reaction with tertiary phosphines is sufficiently sluggish that only one phosphorus atom will be complexed. These conditions seem to be satisfied by the methyliron complex $CH_3Fe(CO)_2C_5H_5$ which reacts with the tri(tertiary phosphine) Pf–Pf–Pf (V) in boiling acetonitrile to form the monoligate monometallic acetyliron derivative $CH_3COFe(CO)(\text{Pf–Pf–Pf})(C_5H_5)$[28] (eq 17). This reaction corresponds to the re-

$$CH_3Fe(CO)_2C_5H_5 + \text{Pf–Pf–Pf} \xrightarrow[\Delta]{CH_3CN} CH_3COFe(CO)(\text{Pf–Pf–Pf})(C_5H_5) \quad (17)$$

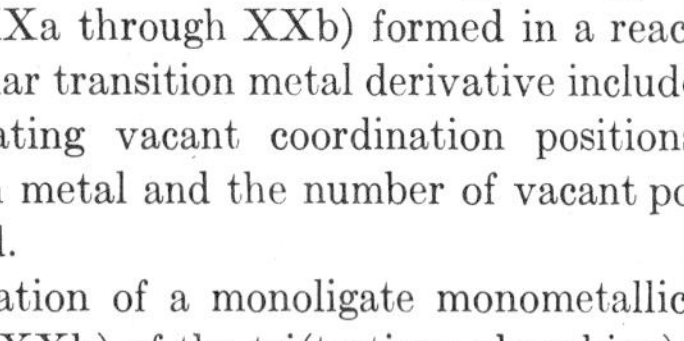

XXI

action of $CH_3Fe(CO)_2C_5H_5$ with mono(tertiary phosphines) (*e.g.*, triphenylphosphine) to form similar acetyl derivatives of the type $CH_3COFe(CO)(PR_3)(C_5H_5)$.[29] The availability of exactly one coordination position in the methyliron derivative $CH_3Fe(CO)_2C_5H_5$ for replacement by a tertiary phosphine appears to be a consequence of the availability of exactly one methyl–iron bond to remove one carbonyl ligand by insertion of carbon monoxide to form an acetyl–iron bond.

Biligate monometallic complexes (XXc) of the tri(tertiary phosphine) Pf–Pf–Pf (V) are considerably more common than monoligate monometallic complexes (XXa or XXb) because there are several types of metal carbonyl derivatives in which exactly two reactive coordination positions are readily generated by loss of carbonyl groups and/or other ligands. Specific

(29) J. P. Bibler and A. Wojcicki, *ibid.*, **5**, 889 (1966); P. M. Treichel, R. L. Shubkin, K. W. Barnett, and D. Reichard, *ibid.*, **5**, 1177 (1966).

examples of biligate monometallic complexes of the tri(tertiary phosphine) Pf–Pf–Pf (V) include reactions 18–20.

$$\text{Pf–Pf–Pf} + C_7H_8Cr(CO)_4 \xrightarrow[5.5\ \text{hr}]{\text{hexane, }\Delta} \underset{\text{XXII}}{(\text{Pf–Pf–Pf})Cr(CO)_4} + C_7H_8 \quad (18)$$

$$\text{Pf–Pf–Pf} + CH_3Mn(CO)_5 \xrightarrow[67\ \text{hr}]{\text{benzene, }\Delta} \underset{\text{XXIII}}{CH_3Mn(CO)_3(\text{Pf–Pf–Pf})} + 2CO \quad (19)$$

$$\text{Pf–Pf–Pf} + C_5H_5Mo(CO)_3Cl \xrightarrow[28\ \text{hr}]{\text{benzene, }25°} \underset{\text{XXIV}}{[C_5H_5Mo(CO)_2(\text{Pf–Pf–Pf})]Cl} + CO \quad (20)$$

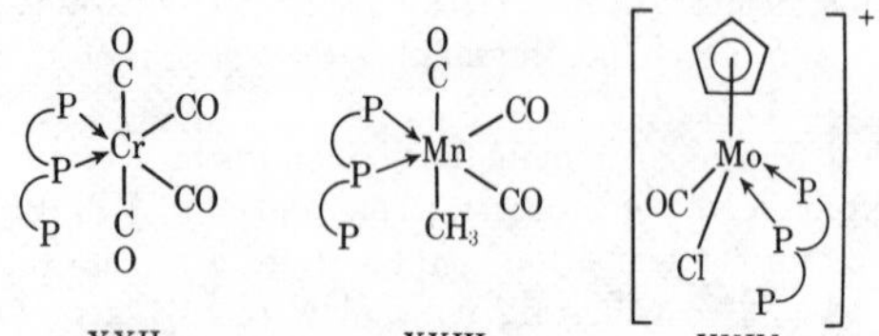

XXII XXIII XXIV

Triligate monometallic complexes (XXd) of the tri(tertiary phosphine) Pf–Pf–Pf (V) are formed if the ligand acts as a tridentate chelating agent. Examples of complexes of this type include reactions 21–26. Formation of a triligate monometallic complex (XXd) from the tri(tertiary phosphine) Pf–Pf–Pf (V) and a transition metal derivative only requires that three reactive coordination positions on the transition metal atom can be readily generated. This occurs in reactions of the tri(tertiary phosphine) Pf–Pf–Pf (V) with most metal halide derivatives but less frequently with metal carbonyl derivatives, since, when carbonyl groups in metal carbonyls are successively replaced with poorer π acceptors such as tertiary phosphines, the remaining carbonyl groups become more firmly bonded to the central transition metal atom because of increased back-bonding using filled metal d orbitals and antibonding carbon monoxide π^* orbitals.

$$\underset{M = \text{Pd or Pt}}{\text{Pf–Pf–Pf} + MCl_4^{2-}} \xrightarrow{H_2O\text{–EtOH}} \underset{\text{converted to } PF_6^- \text{ salt}}{[(\text{Pf–Pf–Pf})MCl]Cl + 2Cl^-} \quad (21)$$

$$\text{Pf–Pf–Pf} + RhCl_3 \cdot 3H_2O \xrightarrow{\text{EtOH}} (\text{Pf–Pf–Pf})RhCl_3 + 3H_2O \quad (22)$$

$$\underset{M = \text{Cr, Mo, and W}}{\text{Pf–Pf–Pf} + M(CO)_6} \xrightarrow{\text{xylene, }\Delta} (\text{Pf–Pf–Pf})M(CO)_3 + 3CO \quad (23)$$

$$\text{Pf–Pf–Pf} + Mn(CO)_5Br \xrightarrow[41\ \text{hr}]{\text{benzene, }\Delta} (\text{Pf–Pf–Pf})Mn(CO)_2Br + 3CO \quad (24)$$

$$\text{Pf–Pf–Pf} + C_5H_5Mo(CO)_3Cl \xrightarrow[\text{uv, 41 hr}]{\text{benzene}} C_5H_5Mo(\text{Pf–Pf–Pf})Cl + 3CO \quad (25)$$

$$\text{Pf–Pf–Pf} + C_5H_5Fe(CO)_2Br \xrightarrow[\text{uv, 15 hr}]{\text{benzene}} \underset{\text{converted to } PF_6 \text{ salt}}{[C_5H_5Fe(\text{Pf–Pf–Pf})]Br} + 2CO \quad (26)$$

The formation of transition metal complexes of the tri(tertiary phosphine) Pf–Pf–Pf (V) with two or more transition metals attached to a single Pf–Pf–Pf (V) ligand requires more unusual transition metal systems since the following conditions must be satisfied. (1) The number of vacant coordination positions generated on the transition metal must be limited. Otherwise the tritertiary phosphine Pf–Pf–Pf (V) will form a triligate monometallic complex (XXd). (2) The reactivity of the transition metal system must be sufficiently high that it reacts rapidly with trivalent phosphorus ligands. Otherwise only one transition metal will be attached to the Pf–Pf–Pf (V) ligand.

The one suitable reaction for preparing a biligate bimetallic complex of the tri(tertiary phosphine) Pf–Pf–Pf (V) uses the bimetallic cyclopentadienyliron carbonyl $[C_5H_5Fe(CO)_2]_2$ according to eq 27. (Pf–Pf–Pf)$Fe_2(CO)_2(C_5H_5)_2$ probably has structure XXV because of the potential stability of the unusual six-membered chelate ring containing two iron atoms, two phosphorus atoms, and two carbon atoms. The infrared spectrum of this complex exhibits only a single bridging ν(CO) frequency at 1672 cm^{-1}; no terminal ν(CO) frequencies were observed. The green color and other spectroscopic properties of (Pf–Pf–Pf)$Fe_2(CO)_2(C_5H_5)_2$ resemble those of recently reported[30] (diphos)$Fe_2(CO)_2(C_5H_5)_2$ complexes in accord with the proposed bonding of only two of the three phosphorus atoms of the tri(tertiary phosphine) Pf–Pf–Pf (V) in XXV.

$$\text{Pf–Pf–Pf} + [C_5H_5Fe(CO)_2]_2 \longrightarrow (\text{Pf–Pf–Pf})Fe_2(CO)_2(C_5H_5)_2 + 2CO \quad (27)$$

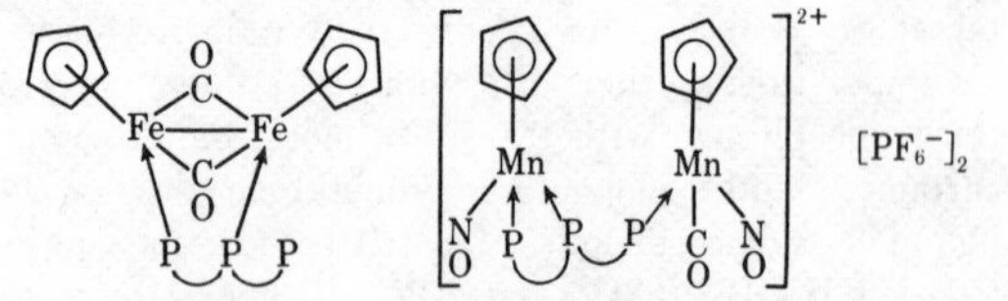

XXV XXVI

The one suitable reaction for preparing a triligate bimetallic complex of the tri(tertiary phosphine) Pf–Pf–Pf (V) uses the cation $[C_5H_5Mn(CO)_2NO]^+$ according to eq 28. The two carbonyl groups in the cation $[C_5H_5Mn(CO)_2NO]^+$ are known[31] to be easily displaced by tertiary phosphines and other related ligands by mild heating; however, both the nitrosyl and cyclopentadienyl groups in this cation are resistant to substitution with tertiary phosphines. The reactivity of the cation $[C_5H_5Mn(CO)_2NO]^+$ in boiling methanol or acetone is so high that it reacts with all available trivalent phosphorus atoms in the system. The availability of two reactive coordination positions in $[C_5H_5$-

$$\text{Pf–Pf–Pf} + 2[C_5H_5Mn(CO)_2NO][PF_6] \xrightarrow[21\ \text{hr}]{\text{MeOH, }\Delta} \underset{\text{XXVI}}{[(C_5H_5)_2Mn_2(CO)(NO)_2(\text{Pf–Pf–Pf})][PF_6]_2} + 3CO \quad (28)$$

(30) R. J. Haines and A. L. Du Preez, *J. Organometal. Chem.*, **21**, 181 (1970).

(31) R. B. King and A. Efraty, *Inorg. Chem.*, **8**, 2374 (1969).

$Mn(CO)_2NO]^+$ causes $[C_5H_5Mn(CO)_2NO]^+$ to form the triligate bimetallic (XXg) complex $[(C_5H_5)_2Mn_2(CO)(NO)_2$(Pf–Pf–Pf)]$[PF_6]_2$ (XXVI) upon reaction with the tritertiary phosphine Pf–Pf–Pf (V).

A suitable reaction for preparing a triligate trimetallic complex of the tri(tertiary phosphine) Pf–Pf–Pf (V) uses the methylmolybdenum derivative $CH_3Mo(CO)_3C_5H_5$ in acetonitrile solution according to eq 29. The

$$\text{Pf–Pf–Pf} + 3CH_3Mo(CO)_3C_5H_5 \xrightarrow{CH_3CN,\ 25°} (\text{Pf–Pf–Pf})[Mo(CO)_2(COCH_3)(C_5H_5)]_3 \quad (29)$$

methylmolybdenum derivative $CH_3Mo(CO)_3C_5H_5$ in acetonitrile solution easily provides one coordination position (possibly through an acyl intermediate of the type $CH_3COMo(CO)_2(NCCH_3)(C_5H_5)$) and thus reacts very readily with any and all available trivalent phosphorus atoms. However, only one coordination position is easily generated, and therefore each $CH_3Mo(CO)_3C_5H_5$ unit can only complex with a single trivalent phosphorus atom. This means that $CH_3Mo(CO)_3C_5H_5$ in acetonitrile solution has the unusual ability of complexing individually with all available trivalent phosphorus atoms to produce a complex containing one metal atom for each trivalent phosphorus atom.

The Two Tetra(tertiary phosphines) X and XI. The observed coordination chemistry of the tri(tertiary phosphine) Pf–Pf–Pf (V) demonstrates that complexes with any of the six fundamentally different modes of bonding XXa through XXh of the ligand to metal atoms can be obtained if appropriately selected transition metal systems are used. The primary factors affecting the type of complex formed are the ease of providing reactive coordination positions and the number of such positions that can be readily generated on each metal atom. The tetra(tertiary phosphines) X (abbreviated as $P(\text{–Pf})_3$) and XI (abbreviated as Pf–Pf–Pf–Pf) can bond to one or more transition metal atoms in the following ten fundamentally different ways: (a) monoligate monometallic; (b) biligate monometallic; (c) triligate monometallic; (d) tetraligate monometallic; (e) biligate bimetallic; (f) triligate bimetallic; (g) tetraligate bimetallic; (h) triligate trimetallic; (i) tetraligate trimetallic; (j) tetraligate tetrametallic. Because of the complexity of the systems, complexes with not all of the ten possible ways of bonding either tetra(tertiary phosphine) (X or XI) to metal atoms have been prepared. Four of the ten possible ways of bonding the tripod tetra(tertiary phosphine) $P(\text{–Pf})_3$ (X) to metal atoms are known as exemplified by the following complexes:[19] (1) biligate monometallic: $P(\text{–Pf})_3PtCl_2$, $P(\text{–Pf})_3M(CO)_4$ (M = Cr and Mo), $CH_3COMn(CO)_3P(\text{–Pf})_3$, $[C_5H_5Mn(NO)P(\text{–Pf})_3][PF_6]$, and $[C_5H_5Fe(CO)P(\text{–Pf})_3]I$; (2) triligate monometallic: $P(\text{–Pf})_3MCl_3$ (M = Rh and Re), $P(\text{–Pf})_3M(CO)_3$ (M = Cr and Mo), $CH_3Mn(CO)_2P(\text{–Pf})_3$, and $P(\text{–Pf})_3Mn(CO)_2Br$; (3) tetraligate monometallic: $[P(\text{–Pf})_3MCl]^+$ (M = Ni, Co, and Fe) and $P(\text{–Pf})_3M(CO)_2$ (M = Mo and W); (4) tetraligate tetrametallic: $P(\text{–Pf})_3[Mo(CO)_2(COCH_3)(C_5H_5)]_4$ and $P(\text{–Pf})_3[Fe_2(CO)_2(C_5H_5)_2]_2$.

Six of the ten possible ways of bonding the linear tetra(tertiary phosphine) Pf–Pf–Pf–Pf (XI) to metal atoms are known as exemplified by the following complexes:[19] (1) monoligate monometallic: $CH_3COFe(CO)$(Pf–Pf–Pf–Pf)(C_5H_5); (2) biligate monometallic: (Pf–Pf–Pf–Pf)$M(CO)_4$ (M = Cr and Mo), $CH_3Mn(CO)_3$(Pf–Pf–Pf–Pf), $[C_5H_5Mo(CO)_2$(Pf–Pf–Pf–Pf)]Cl, $C_5H_5Mn(CO)$(Pf–Pf–Pf–Pf); (3) triligate monometallic: (Pf–Pf–Pf–Pf)MCl_3 (M = Rh and Re), (Pf–Pf–Pf–Pf)$M(CO)_3$ (M = Cr and Mo), (Pf–Pf–Pf–Pf)$Mn(CO)_2Br$, and $[C_5H_5Fe$(Pf–Pf–Pf–Pf)$]^+$; (4) tetraligate monometallic: [(Pf–Pf–Pf–Pf)M$]^{2+}$ (M = Ni, Pd, and Pt), [(Pf–Pf–Pf–Pf)CoCl$]^+$, and [(Pf–Pf–Pf–Pf)Rh$]^+$; (5) triligate bimetallic: $[(C_5H_5)_2Mn_2(CO)(NO)_2$(Pf–Pf–Pf–Pf)]$[PF_6]_2$; (6) tetraligate tetrametallic: (Pf–Pf–Pf–Pf)$[Mo(CO)_2(COCH_3)(C_5H_5)]_4$ and (Pf–Pf–Pf–Pf)$[Fe_2(CO)_2(C_5H_5)_2]_2$.

Most of the gaps in the known ways of bonding of the tetra(tertiary phosphines) X and XI to metal atoms occur in the cases where one tetra(tertiary phosphine) unit is bonded to two or three metal atoms. In many of these cases it is difficult to conceive of a transition metal system that will possess the correct number of readily generated coordination positions so that two or three rather than one or four metal atoms will bond to each tetra(tertiary phosphine) unit.

The two tetra(tertiary phosphines) X and XI represent an unusual isomeric pair of ligands with the same types of bridges (CH_2CH_2) between each pair of trivalent phosphorus atoms and with the same groups (C_6H_5) to fill each nonbridging position of the trivalent phosphorus atoms. They differ only in the relative arrangements of the phosphorus atoms and the bridging CH_2CH_2 groups. Indeed, the relationship between the linear tetra(tertiary phosphine) $(C_6H_5)_2PCH_2CH_2P(C_6H_5)CH_2CH_2P(C_6H_5)CH_2CH_2P(C_6H_5)_2$ (Pf–Pf–Pf–Pf, XI) and the tripod tetra(tertiary phosphine) $P[CH_2CH_2P(C_6H_5)_2]_3$ ($P(\text{–Pf})_3$, X) resembles the relationship between the isomeric hydrocarbons *n*-butane, $CH_3CH_2CH_2CH_3$, and isobutane, $HC(CH_3)_3$. For these reasons, a comparison of the complexing behaviors of the linear tetra(tertiary phosphine) Pf–Pf–Pf–Pf (XI) and the tripod tetra(tertiary phosphine) $P(\text{–Pf})_3$ (X) was of particular interest.

The following transition metal systems were found to form different types of metal complexes with the linear tetra(tertiary phosphine) Pf–Pf–Pf–Pf (XI) than with the tripod tetra(tertiary phosphine) $P(\text{–Pf})_3$ (X).[19] (1) The linear tetra(tertiary phosphine) Pf–Pf–Pf–Pf (XI) reacts with nickel(II) chloride in boiling ethanol to form brown to yellow square-planar nickel(II) derivatives (*e.g.*, (Pf–Pf–Pf–Pf)$NiCl_2$ and [(Pf–Pf–Pf–Pf)Ni]$[PF_6]_2$, whereas the tripod tetra(tertiary phosphine) $P(\text{–Pf})_3$ (X) reacts with nickel(II) chloride under similar conditions to give blue to violet five-coordinate nickel(II) derivatives (*e.g.*, $[P(\text{–Pf})_3NiCl][PF_6]$). (2) The linear tetra(tertiary phosphine) Pf–Pf–Pf–Pf (XI) does not react with ferrous chloride in boiling ethanol. However, the tripod tetra(tertiary phosphine) $P(\text{–Pf})_3$ (X) reacts with ferrous chloride under these conditions to give the violet five-coordinate derivative $[P(-Pf)_3FeCl]^+$. (3) The linear tetra(tertiary phosphine) Pf–Pf–Pf–Pf (I) reacts with $[C_5H_5Mn(CO)_2NO][PF_6]$ to form the triligate bimetallic derivative $[(C_5H_5)_2Mn_2(CO)(NO)_2(Pf–Pf–Pf–Pf)][PF_6]$ whereas the tripod tetra(tertiary phosphine) $P(-Pf)_3$ reacts with $[C_5H_5Mn(CO)_2NO][PF_6]$ under the same conditions, including mole ratios of reactants, to give the biligate monometallic derivative $[C_5H_5Mn(NO)P(-Pf)_3][PF_6]$.

These results suggest that among the two tetra(tertiary phosphine) ligands X and XI the tripod tetra(tertiary phosphine) $P(-Pf)_3$ (X) is the better "chelating agent" (*i.e.*, better at forming complexes with several phosphorus atoms all bonded to a single metal atom). However, the linear tetra(tertiary phosphine) Pf–Pf–Pf–Pf (XI) is better at forming polymetallic complexes (*i.e.*, complexes with two or more metal atoms bonded to a single tetra(tertiary phosphine) ligand) presumably because of the greater possible maximum phosphorus–phosphorus distance in Pf–Pf–Pf–Pf (XI) than in $P(-Pf)_3$ (X). The greater possible maximum phosphorus–phosphorus distance in the linear tetra(tertiary phosphine) ligand Pf–Pf–Pf–Pf (XI) relative to the tripod tetra(tertiary phosphine) ligand $P(-Pf)_3$ (X) arises from the fact that the two end phosphorus atoms in Pf–Pf–Pf–Pf (XI) are separated by three CH_2CH_2 bridges and two middle phosphorus atoms, whereas no pair of phosphorus atoms in $P(-Pf)_3$ (X) is separated by more than two CH_2CH_2 bridges and one middle phosphorus atom.

The Hexa(tertiary phosphine) XII. The coordination chemistry of the hexa(tertiary phosphine) $[(C_6H_5)_2PCH_2CH_2]_2PCH_2CH_2P[CH_2CH_2P(C_6H_5)_2]_2$ (XII; abbreviated as $P_2(-Pf)_4$) is of interest because of the general similarity of the relative positions of the six donor phosphorus atoms in XII as compared with those of the six donor oxygen and nitrogen atoms in the analytically important ethylenediaminetetraacetic acid (EDTA).[5] However, investigation of the reactions of the hexa(tertiary phosphine) $P_2(-Pf)_4$ (XII) with a wide variety of transition metal systems failed to reveal any cases where the hexa(tertiary phosphine) unambiguously acted as a hexadentate (hexaligate monometallic) ligand (XXVII).[23] This contrasts with the behavior of EDTA which forms hexadentate metal complexes very easily.[5] Apparently, it is difficult to bend the five CH_2CH_2 bridges in the hexa(tertiary phosphine) $P_2(-Pf)_4$ (XII) around so that all six phosphorus atoms in $P_2(-Pf)_4$ can bond to a single metal atom as in structure XXVII.

The hexa(tertiary phosphine) $P_2(-Pf)_4$ (XII) reacts with metal(II) chloride derivatives of the nickel, palladium, and platinum triad to form the diamagnetic metal(II) cations $[P_2(-Pf)_4M]^{2+}$ where the hexa(tertiary phosphine) acts as a tetradentate (tetraligate monometallic) ligand as in structure XXVIII. The complexes of the type $[P_2(-Pf)_4M]^{2+}$ resemble the corresponding tetraligate monometallic complexes $[(Pf–Pf–Pf–Pf)M]^{2+}$ obtained from the same metal chloride derivatives and the linear tetra(tertiary phosphine) Pf–Pf–Pf–Pf (XI). The hexa(tertiary phosphine) $P_2(-Pf)_4$ (XII) reacts with metal chloride derivatives of cobalt, rhodium, and ruthenium to form the metal(III) cations $[P_2(-Pf)_4MCl_2]^+$ in which the hexa(tertiary phosphine) XII acts as a tetradentate (tetraligate monometallic) ligand as in structure XXIX. The cobalt and rhodium derivatives of type XXIX exhibit the expected diamagnetism for these metals in the +3 oxidation state whereas the ruthenium derivative of type XXIX exhibits the expected paramagnetism corresponding to one unpaired electron. The cobalt(III) derivative $[P_2(-Pf)_4CoCl_2]Cl$ is formed by reaction of cobalt(II) chloride with the hexa(tertiary phosphine) $P_2(-Pf)_4$ (XII); since this reaction was carried out in a nitrogen atmosphere, the source of the oxidation from cobalt(II) to cobalt(III) is not clear.

Our work with the hexa(tertiary phosphine) $P_2(-Pf)_4$ (XII) has failed to yield any complexes with all six phosphorus atoms bonded to a single metal atom. However, bimetallic complexes of $P_2(-Pf)_4$ were prepared in which all six phosphorus atoms are bonded to metal atoms.[23] The most distinctive complexes of this type are the derivatives $P_2(-Pf)_4[M(CO)_3]_2$ (XXX, M = Cr, Mo, and W), which are readily prepared by heating the hexa(tertiary phosphine) $P_2(-Pf)_4$ (XII) with excess of the metal hexacarbonyl in boiling xylene. A similar complex, $P_2(-Pf)_4[Mn(CO)_2CH_3]_2$ (XXXI), can be prepared by heating $CH_3Mn(CO)_5$ with the hexa(tertiary phosphine) $P_2(-Pf)_4$ (XII) in at least a 2:1 metal:ligand ratio in boiling mesitylene. However, if the mole ratio of $CH_3Mn(CO)_5$ to $P_2(-Pf)_4$ (XII) is only 1:1 and the reaction is carried out in boiling xylene rather than boiling mesitylene, the product is the

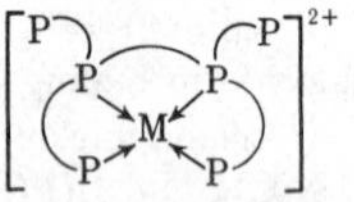
XXVIII

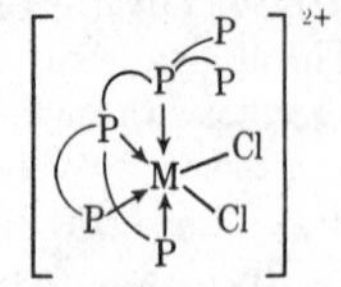
XXIX

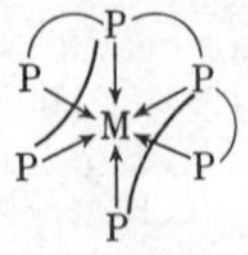
XXVII

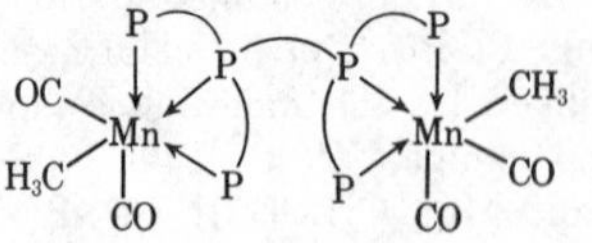
XXXI

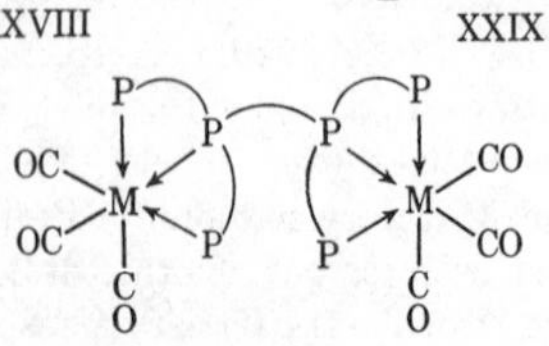
XXX

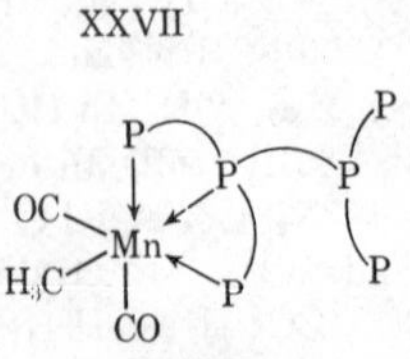
XXXII

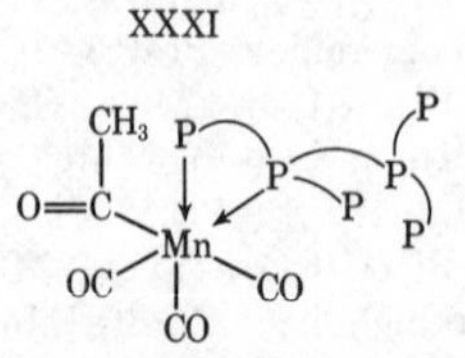
XXXIII

triligate monometallic derivative $P_2(-Pf)_4Mn(CO)_2CH_3$ (XXXII) in which only half of the hexa(tertiary phosphine) XII is involved in bonding to the metal atom. If the reaction between $CH_3Mn(CO)_5$ and $P_2(-Pf)_4$ (XII) is carried out under milder conditions in the polar solvent tetrahydrofuran, one of the liberated molecules of carbon monoxide inserts into the methyl–manganese bond to form the biligate monometallic acetyl derivative $CH_3COMn(CO)_3P_2(-Pf)_4$ (XXXIII).

One of the most unusual metal complexes of the hexa(tertiary phosphine) $P_2(-Pf)_4$ (XII) is the hexaligate hexametallic derivative $P_2(-Pf)_4[Mo(CO)_2(COCH_3)(C_5H_5)]_6$ which is formed as a yellow precipitate upon reaction of $CH_3Mo(CO)_3C_5H_5$ with the hexa(tertiary phosphine) $P_2(-Pf)_4$ (XII) in acetonitrile solution. This reaction demonstrates dramatically the great ease of generating from $CH_3Mo(CO)_3C_5H_5$ in acetonitrile solution one vacant coordination position which is so reactive that it exhaustively seeks out all available uncomplexed trivalent phosphorus atoms.

Conclusion

This work has shown that numerous poly(tertiary phosphines, arsines, and phosphine–arsines) can be prepared by the base-catalyzed additions of phosphorus–hydrogen bonds or arsenic–hydrogen bonds to the vinyl bonds of vinylphosphines. Other trivalent phosphorus derivatives of interest can be prepared by similar base-catalyzed additions of phosphorus–hydrogen bonds to other unsaturated compounds such as vinyl isocyanide and diethyl vinylphosphonate. The potential for the synthesis of additional organophosphorus compounds by techniques related to those summarized in this Account is very large for the following reasons. (1) Phosphorus–hydrogen bonds can be added to a large variety of appropriate unsaturated compounds particularly those known to undergo Michael addition[32] reactions. (2) Compounds with "active hydrogen" other than those with phosphorus–hydrogen or arsenic–hydrogen bonds can be added to the carbon–carbon multiple bonds in vinylphosphorus or ethynylphosphorus derivatives. (3) Compounds can be prepared where the "extra" coordination positions of the trivalent phosphorus atoms are occupied by groups other than phenyl. Phenylphosphorus derivatives were used in the work summarized in this article for two reasons: (a) they are readily available as compared with analogous organophosphorus derivatives in which the phenyl groups are replaced by other organic groups; (b) related aliphatic alkylphosphorus derivatives are more toxic and air-sensitive and also are more likely to form less readily characterized liquid products rather than crystalline poly(tertiary phosphines) when subjected to reactions analogous to those summarized in this Account.

The new poly(tertiary phosphines) prepared during the course of this research are interesting to coordination chemists because they form a wide variety of metal complexes. Thus complexes of the tri(tertiary phosphine) $C_6H_5P[CH_2CH_2P(C_6H_5)_2]_2$ were prepared in which the tri(tertiary phosphine) ligand is bonded to metal atoms in each of the six possible fundamentally different ways. The variety of unusual poly(tertiary phosphine) types that can be prepared by these new synthetic methods and the variety of transition metal systems which form tertiary phosphine complexes make the coordination chemistry of the poly(tertiary phosphines) and related compounds a nearly limitless field. Furthermore, studies on the coordination chemistry of poly(tertiary phosphines) are likely to lead eventually to major contributions to areas of practical importance such as nitrogen fixation, fuel cell technology, homogenous hydrogenation, and new catalytic processes for the synthesis of useful organic chemicals.

I am indebted to the Air Force Office of Scientific Research for support of much of the research discussed in this article. I also acknowledge the expert collaboration of Dr. Pramesh N. Kapoor in the experimental work which led to the development of the new poly(tertiary phosphine) synthesis. Other postdoctoral research associates who made major contributions to the work described in this article include Dr. Avi Efraty, Dr. Mohan S. Saran, and Dr. Ramesh N. Kapoor.

Addendum (May 1976)

Since this Account was published in 1972, these techniques have been adapted and extended to the preparation of extensive series of polyphosphines with terminal groups other than phenyl such as hydrogen,[33] methyl,[34] and neopentyl.[35] Much of this newer work uses base-catalyzed additions of phosphorus–hydrogen bonds to vinylphosphorus sulfides rather than to vinyl derivatives containing trivalent phosphorus such as discussed above. If desired, the sulfur can be removed from the phosphine sulfide units after building up the polyphosphorus system by using a strong reducing agent, of which $LiAlH_4$ in boiling dioxane has proved to be the most generally useful.[34] The base-catalyzed addition of a primary phosphine RPH_2 to a vinyl derivative of tetracoordinate (but not tricoordinate) phosphorus can be readily controlled to give the 1:1 adduct still containing a phosphorus–hydrogen bond.[36] This has made accessible for the first time unusual trivalent phosphorus derivatives of potential interest as ligands such as $C_6H_5P(H)CH_2CH_2P(CH_3)_2$, $(CH_3)_2PCH_2CH_2P(H)CH_2CH_2PH_2$, and $C_6H_5P(CH_2CH_2PH_2)[CH_2CH_2P(CH_3)_2]$.[36] Polyphosphines with terminal phosphorus–hydrogen bonds which we have prepared since 1972 include $(C_6H_5)_2PCH_2CH_2P(H)C_6H_5$, $C_6H_5P(H)CH_2CH_2PH_2$, $C_6H_5P(CH_2CH_2PH_2)_2$, and $P(CH_2CH_2PH_2)_3$.[33] New polytertiary phosphines with terminal methyl groups include the unsymmetrical

(32) E. D. Bergmann, D. Ginsburg, and R. Pappo, *Org. React.*, **10**, 179 (1959).

(33) R. B. King, J. C. Cloyd, Jr., and P. N. Kapoor, *J. Chem. Soc. Perkin Trans.*, I, 2226 (1973).

(34) R. B. King and J. C. Cloyd, Jr., *J. Am. Chem. Soc.*, **97**, 53 (1975).

(35) R. B. King, J. C. Cloyd, Jr., and R. H. Reimann, *J. Org. Chem.*, **41**, 972 (1976).

(36) R. B. King and J. C. Cloyd, Jr., *J. Am. Chem. Soc.*, **97**, 46 (1975).

ditertiary phosphine $(C_6H_5)_2PCH_2CH_2P(CH_3)_2$, the tritertiary phosphines $R'P(CH_2CH_2PR_2)_2$ ($R = CH_3$, $R' = CH_3$ and C_6H_5; $R = C_6H_5$, $R' = CH_3$), the tripod

tetratertiary phosphine $P[CH_2CH_2P(CH_3)_2]_3$, and the hexatertiary phosphine $[(CH_3)_2PCH_2CH_2]_2$-$PCH_2CH_2P[CH_2CH_2P(CH_3)_2]_2$.[34]

Carbamoyl and Alkoxycarbonyl Complexes of Transition Metals

ROBERT J. ANGELICI

Department of Chemistry, Iowa State University, Ames, Iowa 50010

Received March 20, 1972

Reprinted from ACCOUNTS OF CHEMICAL RESEARCH, *5, 335 (1972)*

Although the carbamoyl and alkoxycarbonyl functional groups have been known in organic chemistry for over a century, only in the past 9 years have inorganic analogs been reported. It is the intent of

$$L_nM-C\overset{O}{\underset{NRR'}{\diagup}}$$ carbamoyl (or carboxamido)

$$L_nM-C\overset{O}{\underset{OR}{\diagup}}$$ alkoxycarbonyl (or carboalkoxy)

(L_nM is a transition metal with other ligands, L)

this Account to summarize the methods used in their preparation, their structural and spectral features, and their reactions, and to speculate about their role in synthetic organic and metal-catalyzed reactions.

Since these compounds sit directly on the borderline between inorganic and organic chemistry, it is hoped that chemists in both areas will see applications to their own research.

Preparative Methods

Metal Carbonyl Complexes with Amines or Alkoxides. The reaction

Robert J. Angelici was born in Minnesota, attended St. Olaf College, and received his Ph.D. from Northwestern University in 1962. He spent the following year working with E. O. Fischer at the University of Munich. He returned to this country and began his teaching career at Iowa State University where he is now Professor. In 1970–1972, he was an Alfred P. Sloan Foundation Fellow. His research interests are in two broad areas: (1) the synthesis of organometallic compounds and mechanistic studies of their reactions, and (2) transition metal complex catalysis of reactions of biological interest.

$$L_nMC\equiv O^+ + 2HNRR' \rightleftharpoons L_nMC\overset{O}{\underset{NRR'}{\diagup}} + H_2NRR'^+ \quad (1)$$

represents by far the most general method of preparing carbamoyl complexes. Usually the reactions are rapid and essentially quantitative at room temperature. Although NH_3 and primary and secondary alkylamines react smoothly, aniline and other aromatic amines have not yet been found to undergo a similar reaction with any metal carbonyl complex. In addition to those designated in Table I, complexes such as (C_5H_5)-$W(CO)_4^+$,[1] $(C_5H_5)Mo(CO)_4^+$,[1] $(C_5H_5)W(CO)_3(NH_3)^+$,[2] and $(C_5H_5)Mo(CO)_3(NH_3)^+$,[2] also yield carbamoyl complexes. A related reaction[3] of the carbonyl–thiocarbonyl complex, $(C_5H_5)Fe(CO)_2(CS)^+$, with methylamine gives the thiocarbamoyl derivative $(C_5H_5)Fe(CO)_2CS$-$NHCH_3$. Amine attack on a coordinated CO ligand is presumably also involved at an intermediate stage in the formation of $Mn(CO)_4(NH_3)CONH_2$,[2] $Re(CO)_4$-$(NH_3)CONH_2$,[4] $Mn(CO)_3(NH_3)[P(C_6H_5)_3]CONH_2$,[2] Co-$(CO)_3[P(C_6H_5)_3]CONC_5H_{10}$,[5] and $Hg(CONR_2)_2$.[6]

(1) W. Jetz and R. J. Angelici, *J. Amer. Chem. Soc.*, **94**, 3799 (1972).

(2) H. Behrens, H. Krohberger, R. J. Lampe, J. Langer, D. Maertens, and P. Pässler, Proceedings of the 13th International Conference on Coordination Chemistry, Cracow-Zakopane, Poland, Sept 1970, Vol. II, p 339.

(3) L. Busetto, M. Graziani, and U. Belluco, *Inorg. Chem.*, **10**, 78 (1971).

(4) H. Behrens, E. Lindner, and P. Pässler, *Z. Anorg. Allg. Chem.*, **365**, 137 (1969).

Table I
Correlation of C–O Force Constants with Metal Carbonyl–Amine Reactions (Equation 1)

Compound	Force constant, mdyn/Å	Amine reaction
$\{[(C_6H_5)_3P]_2PtCl(CO)\}BF_4$[a]	18.2	Yes[b]
$[(C_5H_5)Ru(CO)_3]PF_6$	17.6	Yes[c]
$[(C_5H_5)Fe(CO)_3]PF_6$	17.6	Yes[d]
$[(CH_3NH_2)Mn(CO)_5]PF_6$	17.5 (eq)	Yes[e]
	16.8 (ax)	
$[(CH_3NH_2)Re(CO)_5]PF_6$	17.5 (eq)	Yes[f]
	16.5 (ax)	
$Fe(CO)_5$	17.0 (ax)	Equil[g]
	16.4 (eq)	
$\{[(C_6H_5)_3P]_2Mn(CO)_4\}PF_6$[a]	16.9	Equil[h]
$[(toluene)Mn(CO)_3]PF_6$	16.9	Equil[i]
$[C_6(CH_3)_6Mn(CO)_3]PF_6$	16.5	Equil[i]
$[(C_6H_{11}NH_2)_3Mn(CO)_3]PF_6$	15.7	No[i]
$(C_5H_5)Mn(CO)_3$	15.6	No[i]
$\{[(CH_3)_2CHNH_2]_3Mn(CO)_3\}PF_6$	15.5	No[i]
$\{[(CH_3)_2CHNH_2]_3Re(CO)_3\}PF_6$	15.5	No[i]
$[(dien)Mn(CO)_3]I$	15.3	No[i]
$[(NH_3)_3Re(CO)_3]B(C_6H_5)_4$	15.2	No[i]

[a] Trans geometry. [b] C. R. Green and R. J. Angelici, *Inorg. Chem.*, submitted for publication. [c] A. E. Kruse and R. J. Angelici, *J. Organometal. Chem.*, **24**, 231 (1970). [d] L. Busetto and R. J. Angelici, *Inorg. Chim. Acta*, **2**, 391 (1968). [e] R. J. Angelici and D. L. Denton, *ibid.*, **2**, 3 (1968). [f] R. J. Angelici and A. E. Kruse, *J. Organometal. Chem.*, **22**, 461 (1970). [g] Reference 9. [h] R. Brink and R. J. Angelici, to be published. [i] Reference 7.

It should be emphasized that not all metal carbonyl complexes react according to eq 1. In Table I,[7] the complexes are divided into three groups which reflect their tendency to react with amines: *yes*, which means a carbamoyl complex was isolated and there was no evidence for the reverse reaction; *equil*, which means that the carbonyl compound was not completely converted to the carbamoyl derivative or that it could readily be regenerated from the carbamoyl complex (*e.g.*, a CH_2Cl_2 solution of (mesitylene)$Mn(CO)_3^+$ and excess cyclohexylamine contains (mesitylene)$Mn(CO)_2CONHC_6H_{11}$; yet on evaporation (mesitylene)$Mn(CO)_3^+$ is recovered quantitatively);[7] or *no*, which means that there was no evidence for reaction even when the carbonyl complex was dissolved in pure alkylamine.

In an attempt to understand this wide range of reactivity we have correlated the tendency to react with the C–O stretching force constant of the coordinated CO group. Darensbourg and Darensbourg[8] pointed out that these force constants are a measure of the positive charge on the carbon of the CO group—*e.g.*, the higher the force constant, the higher the positive charge on the carbon. A high positive charge would favor, both kinetically and thermodynamically, the formation of a carbamoyl complex. However, since these reactions are so rapid, it appears that the observed reactivity is controlled by thermodynamics.

Indeed, there does appear to be a rather good correlation between force constants and reactivity which may be summarized as follows: *CO ligands with force constants greater than 17.2 mdyn/Å readily form carbamoyl complexes; those with force constants between 16.0 and 17.0 give reversible equilibrium mixtures; and those with constants below 16.0 show no evidence for carbamoyl formation.*[7] In terms of C–O stretching frequencies, it appears that carbonyl complexes with C–O stretching absorptions below approximately 2000 cm^{-1} do not yield carbamoyl complexes.[7]

The force constant correlation not only aids in predicting what metal carbonyl complexes will form carbamoyl derivatives but also in complexes where there are two chemically different kinds of CO groups, as in $Fe(CO)_5$, $(CH_3NH_2)Mn(CO)_5^+$, or $(CH_3NH_2)Re(CO)_5^+$, it correctly predicts that the CO group with the highest C–O force constant is that which is converted to the carbamoyl ligand.

With the exception of $Fe(CO)_5$,[9] neutral metal carbonyl complexes have not been treated with amines for purposes of generating carbamoyl complexes. Fischer and coworkers, however, have isolated $Cr(CO)_5CON(C_2H_5)_2^-$ from the reaction of $Cr(CO)_6$ and $(C_2H_5)_2N^-$. Also, spectroscopic evidence has been obtained for $Ni(CO)_3CON(CH_3)_2^-$ produced in the reaction of $Ni(CO)_4$ with $(CH_3)_2N^-$,[10] while $Fe(CO)_4CON(C_2H_5)_2^-$ has been postulated as an intermediate in the reaction of $Fe(CO)_5$ with $(C_2H_5)_2N^-$.[11,12]

Closely related to the amine reactions is that of hydrazine which proceeds as follows

$$L_nMC{\equiv}O^+ + 2NH_2NH_2 \longrightarrow N_2H_5^+ + MC({=}O)NHNH_2 \longrightarrow MN{=}C{=}O + NH_3 \quad (2)$$

The carbazoyl intermediates have not been isolated in pure form because of their tendency to decompose to the isocyanate complexes. The reaction is a useful method of converting a terminal CO group into an isocyanate ligand and has been used with $(C_5H_5)Fe(CO)_3^+$,[13] $(C_5H_5)Fe(CO)_2(CS)^+$[3] (to give the isothiocyanate complex), $(C_5H_5)Ru(CO)_3^+$,[14] $Re(CO)_6^+$,[15] $(CH_3NH_2)Re(CO)_5^+$,[16] $Re(CO)_4[P(C_6H_5)_3]Br$,[17] $Mn(CO)_3[P(CH_3)_2(C_6H_5)]_2Br$,[18] and (arene)$Mn(CO)_3^+$.[7]

(5) J. Palágyi and L. Markó, *J. Organometal. Chem.*, **17**, 453 (1969).
(6) U. Schöllkopf and F. Gerhart, *Angew. Chem.*, *Int. Ed. Engl.*, **5**, 664 (1966).
(7) R. J. Angelici and L. Blacik, *Inorg. Chem.*, **11**, 1754 (1972).
(8) D. J. Darensbourg and M. Y. Darensbourg, *ibid.*, **9**, 1691 (1970).
(9) W. F. Edgell, M. T. Yang, B. J. Bulkin, R. Bayer, and N. Koizumi, *J. Amer. Chem. Soc.*, **87**, 3080 (1965); W. F. Edgell and B. J. Bulkin, *ibid.*, **88**, 4839 (1966).
(10) S. Fukuoka, M. Ryang, and S. Tsutsumi, *J. Org. Chem.*, **33**, 2973 (1968).
(11) E. O. Fischer and H. J. Kollmeier, *Angew. Chem.*, *Int. Ed. Engl.*, **9**, 309 (1970); E. O. Fischer, E. Winkler, C. G. Kreiter, G. Huttner, and B. Krieg, *ibid.*, **10**, 922 (1971); G. Huttner and K. Bernhard, *Chem. Ber.*, **105**, 67 (1972); E. O. Fischer, H. Beck, C. G. Kreiter, J. Lynch, J. Müller, and E. Winkler, *ibid.*, **105**, 162 (1972).
(12) E. O. Fischer and V. Kiener, *J. Organometal. Chem.*, **27**, C56 (1971).
(13) R. J. Angelici and L. Busetto, *J. Amer. Chem. Soc.*, **91**, 3197 (1969).
(14) A. E. Kruse and R. J. Angelici, *J. Organometal. Chem.*, **24**, 231 (1970).
(15) R. J. Angelici and G. C. Faber, *Inorg. Chem.*, **10**, 514 (1971).

Alkoxycarbonyl complexes may be prepared by the analogous reaction, eq 3. Some cationic complexes

$$L_nMC{\equiv}O^+ + OR^- \longrightarrow L_nMCOOR \quad (3)$$

which are known to react in this manner are $(C_5H_5)Fe(CO)_3^+$,[19] $(C_5H_5)Ru(CO)_3^+$,[14] $Os(CO)_3[P(C_6H_5)_3]_2Cl^+$,[20] $Mn(CO)_4[P(C_6H_5)_3]_2^+$,[21] $(C_5H_5)Mn(CO)_2(NO)^+$,[22] $Re(CO)_4[P(C_6H_5)_3]_2^+$,[21] $Re(CO)_6^+$,[23] $Co(CO)_3[P(C_6H_5)_3]_2^+$,[24] $Ir(CO)_3[P(C_6H_5)_3]_2^+$,[25] $Ir(CO)_2[Sb(C_6H_5)_3]_3^+$,[26] $Rh(CO)_2[Sb(C_6H_5)_3]_3^+$,[26] and $Ir(CO)(H)[(CH_3)_2PCH_2CH_2P(CH_3)_2]_2^+$.[27] Complexes such as $Ir(CO)_2[P(C_6H_5)_3]_2I_2^+$ [28] and $Pt(CO)Cl[P(C_6H_5)_3]_2^+$ [29] react directly with alcohols to give alkoxycarbonyl derivatives. Although the reactions of $Pt[P(C_6H_5)_3]_2Cl_2$,[30] $Pd[P(C_6H_5)_3]_2Cl_2$,[30] and $Pt[P(C_6H_5)_3]_2(N_3)(NCO)$[31] with CO and alkoxides (or alcohols) are slightly more complex, they presumably proceed through the cationic carbonyl intermediates, $[(C_6H_5)_3P]_2M(X)(CO)^+$. The alkoxycarbonyl complexes, $Rh_6(CO)_{15}CO_2CH_3^-$ [32] and $Ir(CO)[P(CH_3)_2(C_6H_5)]_3(HgCl)CO_2CH_3^+$,[33a] result from still more complicated reactions. Methoxycarbonyl derivatives of corrinoids have been obtained from the reaction of CO and CH_3OH with the Co(III) corrinoids.[33b]

In complexes containing both CO and CS ligands, such as $(C_5H_5)Fe(CO)_2(CS)^+$ and $Ir[P(C_6H_5)_3]_2(CO)_2(CS)^+$, methoxide attack occurs at both CO and CS of the former,[3] depending on the conditions, whereas the iridium complex[34] yields only $Ir[P(C_6H_5)_3]_2(CO)(CS)CO_2CH_3$.

Like the alcohol reactions, water adds to a CO group in $IrCl_2[P(CH_3)_2(C_6H_5)]_2(CO)_2^+$ to give the carboxylic acid derivative $IrCl_2[P(CH_3)_2(C_6H_5)]_2(CO)CO_2H$.[35] Although CO_2H ligands have been postulated in a variety of reactions, this seems to be the only isolated member of this family.

Metal Isocyanide Complexes with Hydroxide. Although OH^- attack on a coordinated isocyanide ligand followed by proton migration to the nitrogen would appear to be a general method of carbamoyl synthesis, it has been reported for only one complex[36] (eq 4).

$$trans\text{-}[(C_6H_5)_3P]_2Pt(CNCH_3)_2^{2+} + OH^- \longrightarrow trans\text{-}[(C_6H_5)_3P]_2Pt(CNCH_3)CONHCH_3^+ \quad (4)$$

Reaction with SH^- yields the analogous thiocarbamoyl complex, *trans*-$[(C_6H_5)_3P]_2Pt(CNCH_3)CSNHCH_3$.[36] The reaction of $[(C_6H_5)_3P]_2Rh(CNCH_3)Cl$ with water to give $[(C_6H_5)_3P]_2Rh(CO)Cl$ presumably proceeds *via* a carbamoyl intermediate.[37]

Oxidative Addition of Carbamoyl Chlorides and Chloroformates. Metal complexes which are known to undergo oxidative addition reactions[38] also yield carbamoyl and alkoxycarbonyl derivatives by this route. For example[30,38]

$$Pt[P(C_6H_5)_3]_4 + ClCON(CH_3)_2 \longrightarrow [(C_6H_5)_3P]_2Pt(Cl)CON(CH_3)_2 \quad (5)$$

$$Pt[P(C_6H_5)_3]_4 + ClCOOCH_3 \longrightarrow [(C_6H_5)_3P]_2Pt(Cl)CO_2CH_3 \quad (6)$$

The analogous reactions of $Pd[P(C_6H_5)_3]_4$ have also been reported.[39,40a] The thiocarbamoyl derivatives $[(C_6H_5)_3P]_2Pt(Cl)CSN(CH_3)_2$, $[(C_6H_5)_3P]_2Pd(Cl)CSN(CH_3)_2$,[39] and $[(C_6H_5)_2PCH_2CH_2P(C_6H_5)_2]Ni(Cl)CSN(CH_3)_2$[40b] and the dithioethoxycarbonyl complex $[(C_6H_5)_3P]_2Pt(Cl)CS_2C_2H_5$[41] have been prepared similarly. Oxidative additions of $ClC(S)SC_2H_5$ to $RhCl[P(C_6H_5)_3]_3$ and $IrCl(CO)[P(C_6H_5)_3]_2$ also give dithioethoxycarbonyl complexes.[41] The analogous reaction of $ClCSN(CH_3)_2$ with $RhCl[P(C_6H_5)_3]_3$ or $Rh(CO)Cl[P(C_6H_5)_3]_2$ gives the dimeric compound, $\{[(C_6H_5)_3P]_2Rh(Cl)_2CSN(CH_3)_2\}_2$ which presumably contains bridging Cl groups.[40b] Chloroformates also react with $IrCl(CO)[P(CH_3)_2(C_6H_5)]_2$ to yield alkoxycarbonyl complexes, $IrCl_2(CO)[P(CH_3)_2(C_6H_5)]_2CO_2R$.[35]

Nucleophilic Attack on Chloroformates and Carbamoyl Chlorides. The first known carbamoyl complex was prepared by King[22,42] using this route

$$(C_5H_5)Fe(CO)_2^- + ClC(O)N(CH_3)_2 \longrightarrow (C_5H_5)Fe(CO)_2CON(CH_3)_2 + Cl^- \quad (7)$$

Since then transition metal nucleophiles also have been used to prepare $(C_5H_5)Fe(CO)_2CO_2R$,[19] $(C_5H_5)Ru(CO)_2CO_2R$,[14] $Mn(CO)_5CO_2R$,[21] $Re(CO)_5CO_2R$,[43] and $Co(CO)_3[P(C_6H_5)_3]CO_2R$[24,44] from the chloroformates, ClC(O)OR. Moreover, several derivatives

(16) R. J. Angelici and A. E. Kruse, *J. Organometal. Chem.*, **22**, 461 (1970).
(17) J. T. Moelwyn-Hughes, A. W. B. Garner, and A. S. Howard, *J. Chem. Soc. A*, 2361 (1971).
(18) J. T. Moelwyn-Hughes, A. W. B. Garner and A. S. Howard, *ibid.*, 2370 (1971).
(19) L. Busetto and R. J. Angelici, *Inorg. Chim. Acta*, **2**, 391 (1968).
(20) W. Hieber, V. Frey, and P. John, *Chem. Ber.*, **100**, 1961 (1967).
(21) T. Kruck and M. Noack, *ibid.*, **97**, 1693 (1964).
(22) R. B. King, M. B. Bisnette, and A. Fronzaglia, *J. Organometal. Chem.*, **5**, 341 (1966).
(23) A. M. Brodie, G. Hulley, B. F. G. Johnson, and J. Lewis, *ibid.*, **24**, 201 (1970).
(24) W. Hieber and H. Duchatsch, *Chem. Ber.*, **98**, 1744 (1965).
(25) L. Malatesta, G. Caglio, and M. Angoletta, *J. Chem. Soc.*, 6974 (1965).
(26) W. Hieber and V. Frey, *Chem. Ber.*, **99**, 2614 (1966).
(27) S. D. Ibekwe and K. A. Taylor, *J. Chem. Soc. A*, 1 (1970).
(28) L. Malatesta, M. Angoletta, and G. Caglio, *ibid.*, 1836 (1970).
(29) W. J. Cherwinski and H. C. Clark, *Inorg. Chem.*, **10**, 2263 (1971); W. J. Cherwinski and H. C. Clark, *Can. J. Chem.*, **47**, 2665 (1969); H. C. Clark, K. R. Dixon, and W. J. Jacobs, *J. Amer. Chem. Soc.*, **91**, 1346 (1969).
(30) E. Dobrzynski and R. J. Angelici, to be published.
(31) W. Beck, M. Bauder, G. La Monnica, S. Cenini, and R. Ugo, *J. Chem. Soc. A*, 113 (1971).
(32) P. Chini and S. Martinengo, Proceedings of the 13th International Conference on Coordination Chemistry, Cracow-Zakopane, Poland, Sept 1970, Vol. I, p 282.
(33) (a) A. J. Deeming and B. L. Shaw, *J. Chem. Soc. A*, 3356 (1970); (b) W. Friedrich and M. Moskophidis, *Z. Naturforsch. B*, **26**, 879 (1971).
(34) M. J. Mays and F. P. Stefanini, *J. Chem. Soc. A*, 2747 (1971).
(35) A. J. Deeming and B. L. Shaw, *ibid.*, 443 (1969).
(36) W. J. Knebel and P. M. Treichel, *Chem. Commun.*, 516 (1971); P. M. Treichel, W. J. Knebel, and R. W. Hess, *J. Amer. Chem. Soc.*, **93**, 5424 (1971).
(37) A. L. Balch and J. Miller, *J. Organometal. Chem.*, **32**, 263 (1971).
(38) J. Halpern, *Accounts Chem. Res.*, **3**, 386 (1970).
(39) C. R. Green and R. J. Angelici, *Inorg. Chem.*, in press.
(40) (a) P. Fitton, M. P. Johnson, and J. E. McKeon, *Chem. Commun.*, 6, (1968); (b) B. Corain and M. Martelli, *Inorg. Nucl. Chem. Lett.*, **8**, 39 (1972).
(41) D. Commereuc, I. Douek, and G. Wilkinson, *J. Chem. Soc. A*, 1771 (1970).
(42) R. B. King, *J. Amer. Chem. Soc.*, **85**, 1918 (1963).
(43) T. Kruck, M. Höfler, and M. Noack, *Chem. Ber.*, **99**, 1153 (1966).
(44) R. F. Heck, *J. Organometal. Chem.*, **2**, 195 (1964).

containing Co–CO_2R groups have been prepared from reduced forms of vitamin B_{12},[45] bis(dimethylglyoximato)cobalt(II),[45] and bis(salicylaldehyde)ethylenediiminatocobalt(II)[46] and chloroformates.

Alkyl Isocyanate Insertion into a Metal–Hydrogen Bond. The insertion reaction 8 in the presence of

$$(C_5H_5)W(CO)_3H + CH_3NCO \xrightarrow{(C_2H_5)_3N} (C_5H_5)W(CO)_3CONHCH_3 \quad (8)$$

catalytic amounts of triethylamine proceeds rapidly and in good yield to the carbamoyl product.[1] There is good evidence that the reaction proceeds by initial reaction of $(C_5H_5)W(CO)_3H$ and $(C_2H_5)_3N$ to give $(C_2H_5)_3NH^+$ and $(C_5H_5)W(CO)_3^-$, which attacks the isocyanate carbon. The resulting $(C_5H_5)W(CO)_3CO$-NCH_3^- then gains a proton to give the product. Reaction 8 is reversible[1] and will be discussed in detail later in this Account.

Although $(C_5H_5)Fe(CO)_2H$ has been reported[47] to react with *tert*-butyl isocyanate to give a low yield of the insertion product $(C_5H_5)Fe(CO)_2CONHC(CH_3)_3$, the major products were $(C_5H_5)Fe(CO)[CNC(CH_3)_3]$-$[CONHC(CH_3)_3]$ and $(C_5H_5)_2Fe_2(CO)_3[CNC(CH_3)_3]$. The simple insertion product probably results from the presence of catalytic amounts of $(C_5H_5)Fe(CO)_2^-$. In fact, we have now found[48] that the carbamoyl complexes, $(C_5H_5)Fe(CO)_2CONHR$, may be prepared according to eq 9. Similarly $Re(CO)_5^-$ reacts[49] with

$$(C_5H_5)Fe(CO)_2^- + RNCO + (C_2H_5)_3NH^+ \longrightarrow (C_5H_5)Fe(CO)_2CONHR + (C_2H_5)_3N \quad (9)$$

CH_3NCO and $(C_2H_5)_3NH^+$ to give $Re(CO)_5CONHCH_3$.

Structure and Bonding

X-Ray Studies. The compound *cis*-$Mn(CO)_4(NH_2$-$CH_3)CONHCH_3$ crystallizes in both tetragonal and monoclinic forms, yet the molecular structure in both[50] is

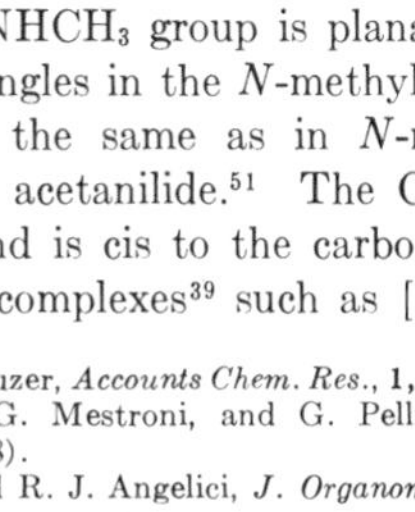

The Mn–C(O)$NHCH_3$ group is planar, and the bond distances and angles in the *N*-methylcarbamoyl group are essentially the same as in *N*-methylacetamide, acetamide, and acetanilide.[51] The CH_3 group in the carbamoyl ligand is cis to the carbonyl oxygen; this contrasts with complexes[39] such as $[(C_6H_5)_3P]_2Pt(Cl)$-$CONHCH_3$ where nmr results suggest that the CH_3 group is trans to the oxygen.

The Mn–C bond distance (2.07 Å) is considerably longer than the M–CO distances which range from 1.78 to 1.86 Å. This suggests that the Mn–C bond order is substantially lower for the carbamoyl group and may correspond to a simple σ bond. The Mn–C(sp^2) σ-bond distance might be calculated from the known C(sp^2) radius of 0.74 Å and a Mn radius. Although the latter value has been estimated[52] as 1.46 and 1.39 Å, perhaps the best value in this instance may be calculated from the Mn–N distance (2.11 Å) in *cis*-$Mn(CO)_4(NH_2CH_3)CONHCH_3$ by subtracting the N(sp^3) radius[53] of 0.70 Å. This gives an Mn radius of 1.41 Å and an expected Mn–C(sp^2) bond distance of 2.15 Å. The shorter observed distance suggests there is some, although not much, double bond character in the Mn–carbamoyl bond.

A dimeric thiocarboxamido complex, $\{[(CH_3O)_3P]$-$Pd(Cl)CSN(CH_3)_2\}_2$, results from the oxidative-addition reaction (analogous to eq 5) of $Pd[P(OCH_3)_3]_4$ with $ClC(S)N(CH_3)_2$ which presumably proceeds initially *via* formation of $[(CH_3O)_3P]_2Pd(Cl)CSN(CH_3)_2$. On losing $P(OCH_3)_3$ from this intermediate, the bridged dimer is formed. Its novel structure[54]

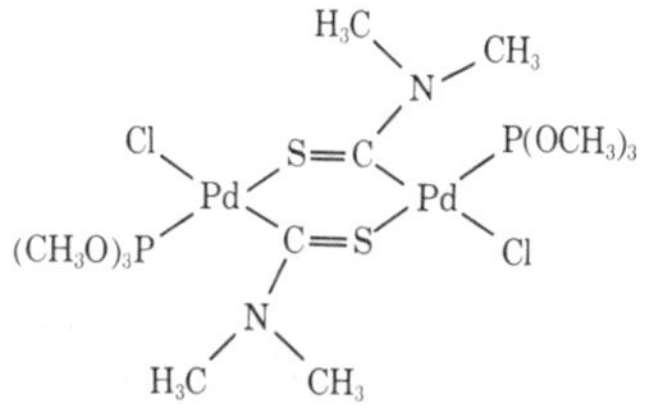

shows essentially square-planar coordination around each Pd with the plane of the thiocarbamoyl ligand roughly (~67°) perpendicular to the coordination plane. The sulfur atom of each thiocarbamoyl group is bound to the other Pd atom. As a result of the geometry of the bridging ligands, the square planes around the Pd atoms are oriented at an angle of 57° with respect to each other. The 2.00-Å Pd–C bond distance is that expected for σ bonding only, and the geometry of the thiocarbamoyl ligands is essentially that which is found in organic thioamides. The analogous Ni complex $\{[(C_6H_5O)_3P]Ni(Cl)CSN(CH_3)_2\}_2$ has been postulated to have the same structure.[40b] These structures indicate, as in organic thioamides, that the sulfur atom of thiocarbamoyl complexes may be a donor atom in complexing with other metals. In general, the complexing ability of the oxygen or sulfur atom in carbamoyl and thiocarbamoyl complexes has yet to be examined extensively.

(45) G. N. Schrauzer, *Accounts Chem. Res.*, **1**, 97 (1968).
(46) G. Costa, G. Mestroni, and G. Pellizer, *J. Organometal. Chem.*, **11**, 333 (1968).
(47) W. Jetz and R. J. Angelici, *J. Organometal. Chem.*, **35**, C37 (1972).
(48) B. D. Dombek and R. J. Angelici, to be published.
(49) R. Brink and R. J. Angelici, to be published.
(50) D. M. Chipman and R. A. Jacobson, *Inorg. Chim. Acta*, **1**, 393 (1967); G. L. Breneman, D. M. Chipman, C. J. Galles, and R. A. Jacobson, *ibid.*, **3**, 447 (1969).
(51) "Interatomic Distances," *Chem. Soc., Spec. Publ.*, **No. 11** (1958); M. B. Robin, F. A. Bovey, and H. Basch in "The Chemistry of Amides," J. Zabicky, Ed., Interscience, New York, N. Y., 1970, p 1.
(52) M. R. Churchill, *Perspect. Struct. Chem.*, **3**, 91 (1970).
(53) F. A. Cotton and D. C. Richardson, *Inorg. Chem.*, **5**, 1851 (1966).
(54) J. Clardy, S. Porter, H. White, C. R. Green, and R. J. Angelici, to be published.

Infrared and Nmr Studies. Infrared spectra of the *N*,*N*-dimethylcarbamoyl ligands show a characteristic C=O stretching absorption in the range 1565 to 1615 cm^{-1}. Similar assignments for the secondary carbamoyl groups, CONHR, are more difficult because of the occurrence of the N–H bending mode in the same region; generally a broad absorption which presumably includes both modes is observed. Carbonyl (C=O) stretching frequencies for the alkoxycarbonyl group are found in the 1610- to 1700-cm^{-1} range. Like their organic analogs, the carbonyl frequency range for the carbamoyl group is lower than that of the alkoxycarbonyl group.

The frequency ranges for both the carbamoyl and alkoxycarbonyl ligands are roughly 80 cm^{-1} lower than observed for their organic counterparts; this indicates that the C=O bond strength is decreased by metals as compared to organic groups and that metals in these complexes are better electron donors than are alkyl groups.

Some effort has been directed at studying the barrier to rotation around the C–N bond of *N*,*N*-dimethylcarbamoyl ligands. In proton nmr spectra of these

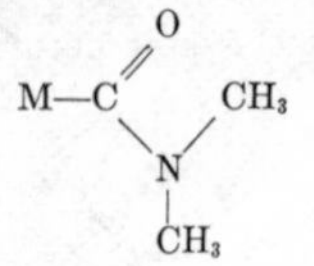

derivatives the CH_3 groups are observed in the region τ 7.1 to 8.5 and are nonequivalent. On warming, they broaden and coalesce into one resonance. The coalescence temperature (T_c) is a measure of the rate of rotation around the C–N bond. These values are recorded in Table II together with similar results for organic derivatives.[55] The coalescence temperatures have been interpreted in terms of the double bond character of the C–N bond and the bulkiness of groups near the rotating $N(CH_3)_2$ group. The high coalescence temperature of the platinum complexes appears to be related to the size of the phosphine (L) groups. This is supported by the decrease in T_c as phenyl groups are replaced by smaller CH_3 groups in these derivatives.

Table II
Coalescence Temperatures of Inorganic and Organic Carbamoyl Derivatives, $ZCON(CH_3)_2$

Z	Solvent	T_c, °C
OCH_3[a]	$CHCl_3$	−13
$(\pi\text{-}C_5H_5)Mo(CO)_3$[b]	$(CD_3)_2CO$	~18
$(\pi\text{-}C_5H_5)W(CO)_3$[b]	$(CD_3)_2CO$	~18
CH_3[a]	Neat	87
$[(CH_3)_2(C_6H_5)P]_2PtCl$[b]	*c*	120–130
$[CH_3(C_6H_5)_2P]_2PtCl$[b]	*c*	140–145
$[(C_6H_5)_3P]_2PdCl$[b]	*c*	155–165
$[(C_6H_5)_3P]_2PtCl$[b]	*c*	>200
$2,4,6\text{-}[(CH_3)_3C]_3C_6H_2$[a]	*d*	>200

[a] Reference 55. [b] Reference 39. [c] *o*-Dichlorobenzene. [d] Chloronaphthalene–benzotrichloride (1:1).

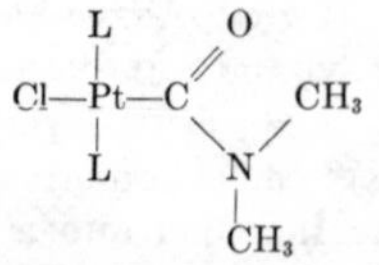

In the absence of steric effects, as in $(C_5H_5)M(CO)_3CON(CH_3)_2$ (where M = Mo or W), low coalescence temperatures are observed. Attempts to observe rotation in the monomethylcarbamoyl complex $(C_5H_5)W(CO)_3CONHCH_3$ were unsuccessful;[56] the CH_3 resonance remains an unchanged sharp doublet over the temperature range −25 to 40°.

The methyl protons of methoxycarbonyl complexes noted in this Account have been reported to range from τ 6.3 to 7.6. There is no evidence relating to the possibility of restricted rotation around the C–O bond in these derivatives.

Optically Active Alkoxycarbonyl Complexes. Brunner[57] and coworkers have prepared several optically active alkoxycarbonyl complexes with four different groups around the metal atom. For example, the optically active mentholate anion ($C_{10}H_{19}O^-$) reacts with racemic $(C_5H_5)Mn(NO)[P(C_6H_5)_3](CO)^+$ (eq 10)

$$(C_5H_5)Mn(NO)[P(C_6H_5)_3](CO)^+ + C_{10}H_{19}O^- \longrightarrow (C_5H_5)Mn(NO)[P(C_6H_5)_3]CO_2C_{10}H_{19} \quad (10)$$

to give the diastereomeric alkoxycarbonyl complexes which can be readily resolved by crystallization. These derivatives in solution racemize slowly (2–3-hr half-lives at room temperature) at the asymmetric Mn atom by a mechanism which is postulated to involve rate-determining dissociation of $P(C_6H_5)_3$. On reaction with HCl, each diastereomer is converted to one enantiomer of the original cation (eq 11). The con-

$$\underset{\text{diastereomer}}{(C_5H_5)Mn(NO)[P(C_6H_5)_3]CO_2C_{10}H_{19}} \xrightarrow{HCl} \underset{\text{optically active}}{(C_5H_5)Mn(NO)[P(C_6H_5)_3](CO)^+} + C_{10}H_{19}OH \quad (11)$$

figurational stability of these optically active cations is evidenced by the observation that specific rotations of their solutions at room temperature do not decrease on storage for several weeks. These enantiomers also react with CH_3O^- to give optically active alkoxycarbonyl complexes whose kinetics of racemization have also been studied.[57]

Similarly $(C_5H_5)Fe[P(C_6H_5)_3](CO)_2^+$ reacts with mentholate ion to create a complex with four different groups (eq 12). The resulting diastereomeric mixture

$$(C_5H_5)Fe[P(C_6H_5)_3](CO)_2^+ + C_{10}H_{19}O^- \longrightarrow (C_5H_5)Fe[P(C_6H_5)_3](CO)CO_2C_{10}H_{19} \quad (12)$$

can be resolved by crystallization. These diastereomers are much more stable to racemization than the alkoxycarbonyl derivatives of manganese discussed above. This lack of racemization indicates not only that $P(C_6H_5)_3$ dissociation is slow but also that the

(55) H. Kessler, *Angew. Chem., Int. Ed. Engl.*, **9**, 219 (1970).

(56) C. G. Green, W. Jetz, and R. J. Angelici, unpublished results.

(57) H. Brunner, *Angew. Chem., Int. Ed. Engl.*, **10**, 249 (1971), and references therein.

mentholate group cannot readily migrate from one carbonyl group to the other in the complex since this migration would also result in racemization. This is the only available evidence which suggests that alkoxy groups remain fixed on one carbonyl carbon atom and do not rapidly migrate to other CO ligands in the complex.

Reactions of Carbamoyl and Alkoxycarbonyl Ligands

Since these complexes have become available only recently, their chemistry has not been extensively investigated. Their potential utility in organic synthesis will almost certainly bring about more studies in this area. Those reactions which have been reported to date are summarized below.

With Acids. Carbamoyl and alkoxycarbonyl complexes react with acids to generate metal carbonyl complexes as shown in eq 13 and 14. With strong

$$MCONRR' + 2HA \longrightarrow MCO^+A^- + (H_2NRR')^+A^- \quad (13)$$

$$MCOOR + HA \longrightarrow MCO^+A^- + ROH \quad (14)$$

acids such as HCl, there are no reported examples in which reactions 13 and 14 do not take place. The reliability of these reactions has in fact made them a useful chemical means of establishing the presence of these groups in complexes.

A qualitative study[39] of *trans*-[(C₆H₅)₃P]₂Pd(Cl)-CON(CH₃)₂ with several acids shows that HCl, Cl₃CCO₂H (pK_a = 0.70), and BrCH₂CO₂H (pK_a = 2.69) do react, whereas weaker acids such as p-NO₂C₆H₄CO₂H (pK_a = 3.41), C₆H₅CO₂H (pK_a = 4.19), and CH₃CO₂H do not. Although a systematic study has not been made, the strength of the acid required to carry out these reactions must be related to the reverse of reactions 13 and 14. That is, it must be related to the tendency of the metal carbonyl complex to add amines (or alcohols) according to eq 1 (or 3). In cases where the tendency is low (Table I), as for (arene)Mn(CO)₃⁺, the corresponding carbamoyl complex (arene)Mn(CO)₂CONRR' reacts[7] with acids even as weak as H₂NRR'⁺. Where the tendency is high, as for [(C₆H₅)₃P]₂Pt(Cl)CO⁺, strong acids are required to remove the amine from the corresponding carbamoyl complex, [(C₆H₅)₃P]₂Pt(Cl)CONRR'.

A thiocarbonyl complex, (C₅H₅)Fe(CO)₂(CS)⁺, has been prepared[58] by making use of this general type of reaction (eq 15). A similar attempt[39] with the thio-

$$(C_5H_5)Fe(CO)_2CSOC_2H_5 + HCl \longrightarrow$$
$$(C_5H_5)Fe(CO)_2(CS)^+Cl^- + C_2H_5OH \quad (15)$$

carbamoyl complexes, [(C₆H₅)₃P]₂M(Cl)CSN(CH₃)₂ (where M = Pd or Pt), gave no reaction.

With R₃O⁺. In contrast to H⁺, the Lewis acid C₂H₅⁺ reacts with anionic carbamoyl complexes to give "carbene" complexes.[11] Similar reactions of W(CO)₆, Fe(CO)₅, and Ni(CO)₄ first with LiNR₂ (where R = CH₃ or C₂H₅) followed by (C₂H₅)₃O⁺ give analogous products, (CO)₅WC(OC₂H₅)NR₂, (CO)₄FeC-

$$(CO)_5CrC\begin{smallmatrix}O^-\\N(CH_3)_2\end{smallmatrix} + (C_2H_5)_3O^+ \longrightarrow$$

$$(CO)_5CrC\begin{smallmatrix}OC_2H_5\\N(CH_3)_2\end{smallmatrix} + (C_2H_5)_2O \quad (16)$$

(OC₂H₅)NR₂, and (CO)₃NiC(OC₂H₅)NR₂.[11,59] To date there have been no reports of (C₂H₅)₃O⁺ reactions with *uncharged* carbamoyl or alkoxycarbonyl complexes.

Interconversion of Carbamoyl and Alkoxycarbonyl Ligands. Like the reaction of organic esters with amines, Re(CO)₅CO₂CH₃ reacts rapidly at room temperature with 1 equiv of methylamine as in eq 17.[49]

$$Re(CO)_5CO_2CH_3 + CH_3NH_2 \longrightarrow$$
$$Re(CO)_5CONHCH_3 + CH_3OH \quad (17)$$

The reverse reaction, the conversion of a carbamoyl to an alkoxycarbonyl group, has been accomplished by refluxing (C₅H₅)Fe(CO)₂CON(CH₃)₂ in CH₃OH for 1 hr[22] (eq 18). In an ester interchange reaction, the

$$(C_5H_5)Fe(CO)_2CON(CH_3)_2 + CH_3OH \longrightarrow$$
$$(C_5H_5)Fe(CO)_2CO_2CH_3 + (CH_3)_2NH \quad (18)$$

ethoxycarbonyl complex, [(C₆H₅)₃P]₂Pt(Cl)CO₂C₂H₅, reacts with methanol at 63° (12 hr) to give the corresponding methoxycarbonyl derivative.[29] In related reactions, it has been suggested[60] that the alkoxycarbonyl complexes (C₅H₅)Fe(CO)₂CO₂CH₃ and Mn(CO)₅CO₂C₂H₅ react with H₂O to give carboxylic acid intermediates, (C₅H₅)Fe(CO)₂CO₂H and Mn(CO)₅CO₂H, which decompose with loss of CO₂ to the hydrides (C₅H₅)Fe(CO)₂H and Mn(CO)₅H.

With Bases. Triethylamine reacts rapidly at room temperature with the secondary carbamoyl complex (C₅H₅)W(CO)₃CONHCH₃ (eq 19). This is an equi-

$$(C_5H_5)W(CO)_3CONHCH_3 + (C_2H_5)_3N \rightleftharpoons$$
$$(C_5H_5)W(CO)_3^- + CH_3NCO + (C_2H_5)_3NH^+ \quad (19)$$

librium reaction which can be displaced to the right or left (see eq 8) depending upon the reactant quantities present. Attempts to carry out the same reaction on (C₅H₅)Fe(CO)₂CONHCH₃ and [(C₆H₅)₃P]₂Pt(Cl)-CONHCH₃ showed no reaction. The lack of reaction in these cases is probably related to the stability of the products. For example, the reaction of (C₅H₅)Fe(CO)₂CONHCH₃ (eq 20) would yield the strongly

$$(C_5H_5)Fe(CO)_2CONHCH_3 + (C_2H_5)_3N \rightleftharpoons$$
$$(C_5H_5)Fe(CO)_2^- + CH_3NCO + (C_2H_5)_3NH^+ \quad (20)$$

nucleophilic (C₅H₅)Fe(CO)₂⁻ which would favor the left side of eq 20. This is supported by the facility of the reverse reaction (eq 9).

If a primary amine such as CH₃NH₂ is used as the base in eq 19, the intermediate CH₃NCO reacts with excess amine to give the urea, and the overall reaction becomes[1]

(58) L. Busetto and R. J. Angelici, *J. Amer. Chem. Soc.*, **90**, 3283 (1968); L. Busetto, U. Belluco, and R. J. Angelici, *J. Organometal. Chem.*, **18**, 213 (1968).

(59) E. Winkler, Ph.D. Dissertation, Technische Universität München, 1971.
(60) H. C. Clark and W. J. Jacobs, *Inorg. Chem.*, **9**, 1229 (1970).

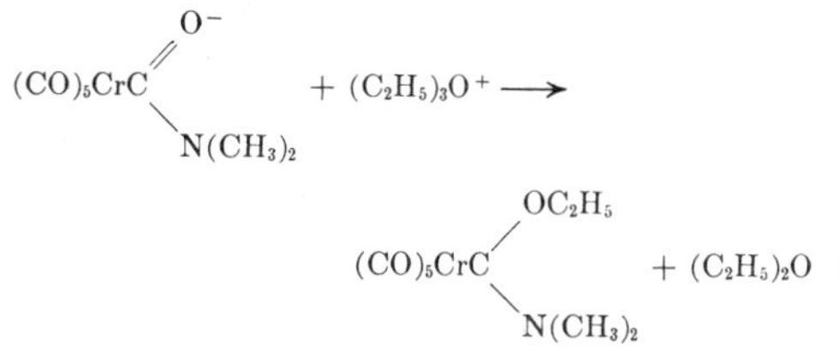

$$(C_5H_5)W(CO)_3CONHCH_3 + CH_3NH_2 \longrightarrow (C_5H_5)W(CO)_3^- + CH_3NH_3^+ + (CH_3NH)_2CO \quad (21)$$

Although the analogous Mo complex also undergoes this reaction, $C_5H_5Fe(CO)_2CONHCH_3$ does not, presumably for the same reasons that it does not react with $(C_2H_5)_3N$. The facile conversion of a carbamoyl group to an alkyl isocyanate or urea suggests that carbamoyl complexes may be intermediates in metal-catalyzed reactions of CO and alkylamines to form these organic compounds.

Dehydration to Cyano Complexes. Behrens and coworkers[61] have observed the dehydration of the $CONH_2$ ligand under widely varying temperature conditions. For example, the reaction

$$M(CO)_4(NH_3)CONH_2 + NH_3 \longrightarrow M(CO)_3(NH_3)_2CN + H_2O + CO \quad (22)$$

in liquid NH_3 proceeds at $-45°$ for M = Mn, but 120° is required for the analogous Re complex. In addition to the main product, the manganese complex reaction also yields some $Mn(CO)_3(NH_3)_2NCO$. The dehydration of $Mn(CO)_3[P(C_6H_5)_3](NH_3)CONH_2$ proceeds without the loss of CO to give $Mn(CO)_3[P(C_6H_5)_3](NH_3)CN$ in ether solution at $-33°$. Likewise $Mn(CO)_3[P(C_6H_5)_3]_2CONH_2$ is converted to $Mn(CO)_3(NH_3)_2CN$ in liquid ammonia at 60°.

With Halogens. Cleavage of the Hg–C bond is reported[6] in the reaction of $Hg[CON(C_2H_5)_2]_2$ with Br_2 (eq 23). The reaction of $(C_5H_5)Fe(CO)_2CONHCH_3$

$$Hg[CON(C_2H_5)_2]_2 + Br_2 \longrightarrow HgBr_2 + BrCON(C_2H_5)_2 \quad (23)$$

with Br_2 presumably also proceeds by initial bond cleavage[62] (eq 24). The resulting carbamoyl bromide

$$(C_5H_5)Fe(CO)_2CONHCH_3 + Br_2 \longrightarrow (C_5H_5)Fe(CO)_2Br + BrCONHCH_3 \quad (24)$$

is known to decompose to HBr and CH_3NCO. This HBr would rapidly react (eq 13) with some of the starting carbamoyl complex to give $(C_5H_5)Fe(CO)_3^+$ and $CH_3NH_3^+$. The overall reaction is

$$3(C_5H_5)Fe(CO)_2CONHCH_3 + 2Br_2 \longrightarrow 2(C_5H_5)Fe(CO)_2Br + [(C_5H_5)Fe(CO)_3]Br + 2CH_3NCO + [CH_3NH_3]Br \quad (25)$$

Thus the halogen reaction is another method (see eq 19) of converting carbamoyl complexes into alkyl isocyanates.

With Mercuric Chloride. Reaction of the cyclohexylcarbamoyl complex $(C_5H_5)Fe(CO)_2CONHC_6H_{11}$ with $HgCl_2$ appears to proceed[62] in much the same way as that with halogens. In this case, the Fe–C bond is cleaved with ClHg–Cl to give Fe–HgCl and Cl–C. Subsequent reaction of the carbamoyl chloride gives the overall reaction 26. In contrast, the reaction of

$$3(C_5H_5)Fe(CO)_2CONHC_6H_{11} + 2HgCl_2 \longrightarrow 2(C_5H_5)Fe(CO)_2HgCl + [(C_5H_5)Fe(CO)_3]Cl + C_6H_{11}NCO + [C_6H_{11}NH_3]Cl \quad (26)$$

(61) H. Behrens, E. Lindner, D. Maertens, P. Wild, and R. Lampe, *J. Organometal. Chem.*, **34**, 367 (1972).

the anionic carbamoyl complex $Ni(CO)_3CON(CH_3)_2^-$ with $HgCl_2$ proceeds quite differently[10] (eq 27) to give

$$2Ni(CO)_3CON(CH_3)_2^- + HgCl_2 \longrightarrow (CH_3)_2NC(O)C(O)N(CH_3)_2 + Hg + LiCl + 2(Ni + CO) \quad (27)$$

N,N,N′,N′-tetramethyloxamide as the main organic product.

With Organic Halides. Both $Hg[CON(C_2H_5)_2]_2$ and $Ni(CO)_3CON(CH_3)_2^-$ react with organic acid chlorides to give the simple amides $RCON(CH_3)_2$ resulting from removal of the $N(CH_3)_2$ group from the complex. With benzyl chloride the mercury complex[6] gives only $C_6H_5CH_2N(CH_3)_2$, whereas benzyl bromide with the nickel complex[63] gives both $C_6H_5CH_2N(CH_3)_2$ and $C_6H_5CH_2CON(CH_3)_2$. The nickel complex also reacts with *trans*-$C_6H_5CH{=}CHBr$, C_6H_5I, and 3-bromocyclohexene to give high yields of the product ($RCON(CH_3)_2$) containing the $CON(CH_3)_2$ group in place of the halogen.

Other Reactions. In addition to the above reactions which have been definitely shown to involve carbamoyl or alkoxycarbonyl complexes, there is a large number of metal-promoted reactions of carbon monoxide with amines or alcohols which very probably involve these types of complexes as intermediates. For example, Heck[64] has made extensive use of $Hg{-}CO_2CH_3$ complexes to generate very reactive $Pd{-}CO_2CH_3$ derivatives which on reaction with olefins yield methoxycarbonyl-substituted olefins $>C{=}C(CO_2CH_3){-}$. Similar intermediates are presumably involved in the reaction[65] of organic halides (RX) with $Ni(CO)_4$ and alkoxides ($R'O^-$) to give esters RCO_2R'.

There is an extensive literature of metal-catalyzed reactions of CO and amines to form organic isocyanates, formamides, and ureas.[66] For example, $Mn_2(CO)_{10}$ catalyzes the reaction of primary alkylamines with CO to give dialkylureas and H_2. Copper(II) catalyzes the reaction of amines and CO and oxygen to yield ureas and water.[67] The complex $(RNH_2)_2PdCl_2$ reacts with CO to give organic isocyanates (RNCO) and Pd metal.[68] The known existence of carbamoyl complexes of all of these metals suggests that carbamoyl complexes are probable intermediates in these reactions.

I am grateful to my coworkers for their major contributions to the ideas and results summarized here. I also very much appreciate the support of our research by the National Science Foundation, the Alfred P. Sloan Foundation, and the donors of the Petroleum Research Fund, administered by the American Chemical Society.

(62) W. Jetz and R. J. Angelici, *Inorg. Chem.*, **11**, 1960 (1972).
(63) S. Fukuoka, M. Ryang, and S. Tsutsumi, *J. Org. Chem.*, **36**, 2721 (1971).
(64) R. F. Heck, *J. Amer. Chem. Soc.*, **93**, 6896 (1971), and references therein.
(65) E. J. Corey and L. S. Hegedus, *ibid.*, **91**, 1232 (1969).
(66) A. Rosenthal and I. Wender in "Organic Syntheses Via Metal Carbonyls," I. Wender and P. Pino, Ed., Interscience, New York, N. Y., 1968, p 405.
(67) W. Brackman, *Discuss. Faraday Soc.*, **46**, 122 (1968).
(68) E. W. Stern and M. L. Spector, *J. Org. Chem.*, **31**, 596 (1966).

Mixed-Valence Ferrocene Chemistry[1]

Dwaine O. Cowan,* Carole LeVanda, Jongsei Park, and Frank Kaufman

Chemistry Department, The Johns Hopkins University, Baltimore, Maryland 21218

Davy Faraday Research Laboratory of The Royal Institution, London, W1X 4BS, United Kingdom

Received June 12, 1972

Reprinted from ACCOUNTS OF CHEMICAL RESEARCH, *6, 1 (1973)*

Mixed-valence compounds contain at least two atoms or two identical molecular moieties in different oxidation states.[2] Compounds **1–6** are examples

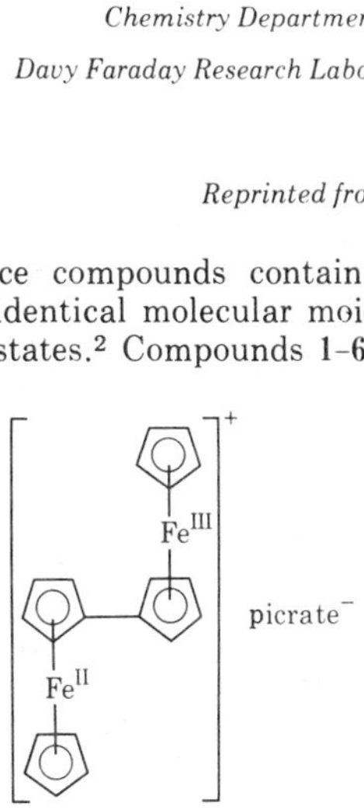

1

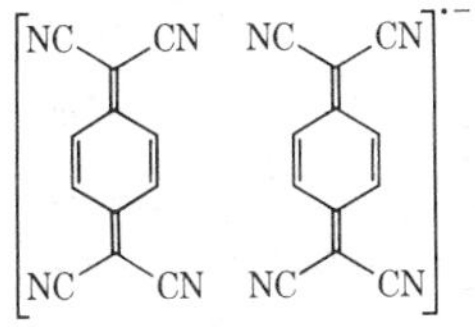

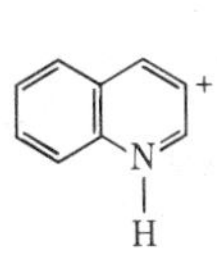

(FeO·Fe_2O_3)
Fe_3O_4

2 3

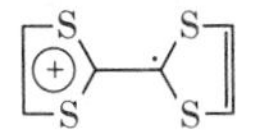

4 5

$K_2Pt(CN)_4Br_{0.3}\cdot xH_2O$

6

of intra- and intermolecular mixed-valence compounds.[2-7] These and other related mixed-valence compounds are of great interest because they form an important link in the study of inorganic, organic, and biological oxidation–reduction reactions. Moreover, data derived from the study of inorganic mixed-valence molecules suggest that the properties of mixed-valence compounds will rarely be just the sum of the properties of the component parts taken separately. For example, most Fe(II) and Fe(III) salts are colorless to pale green, while most Fe(II)-Fe(III) mixed-valence salts are deep blue to black. Likewise, the electrical conductivity of Fe_3O_4, a mixed-valence Fe(II)Fe(III) oxide, is 10^6 times larger than the trivalent Fe_2O_3.

The properties of these mixed-valence compounds will depend upon the amount of delocalization or the extent of interaction (α) between the two moieties. A classification now in common usage is[2a]

Class I	Little or no interaction Interaction parameter $\alpha \approx 0$ Properties = sum of components	Integral oxidation states, $\alpha > 0.25$
Class II	Interaction parameter $0.707 > \alpha > 0$ New optical and electronic properties + properties of constituent components	
Class III	Interaction parameter $\alpha = 0.707$ New optical and electronic properties	Fractional oxidation states, $\alpha > 0.25$

where α is defined for the ground state of the complex as

$$\psi_G = (1 - \alpha^2)^{1/2}\Phi_i + \alpha\Phi_j \quad (1)$$

Φ_i and Φ_j are wave functions for the donor–acceptor components of the mixed-valence system.

Dwaine O. Cowan, a native of Fresno, Calif., received the B.S. degree from Fresno State College in 1958, and the Ph.D. degree from Stanford University. After a postdoctoral year at the California Institute of Technology, he joined The Johns Hopkins University, where he is now Professor of Chemistry. His research interests include the organic solid state, organometallic chemistry, photochemistry, and electron transport in biological systems.

Frank Kaufman received the B.S. degree from the University of Rochester and the Ph.D. degree from The Johns Hopkins University and is currently an NIH Postdoctoral Fellow at The Royal Institution. Jongsei Park and Carole LeVanda are graduate students at The Johns Hopkins University.

(1) Paper VIII, The Organic Solid State. Paper VII: D. O. Cowan, J. Park, C. U. Pittman, Jr., Y. Sasaki, T. K. Mukheijee, and N. A. Diamond, *J. Amer. Chem. Soc.*, **94**, 5110 (1972).

(2) (a) M. B. Robin and P. Day, *Advan. Inorg. Chem. Radiochem.*, **10**, 247 (1967); (b) G. C. Allen and N. S. Hush, *Progr. Inorg. Chem.*, **8**, 357 (1967); (c) N. S. Hush, *ibid.*, **8**, 391 (1967).

(3) D. O. Cowan and F. Kaufman, *J. Amer. Chem. Soc.*, **92**, 219 (1970).

(4) (a) W. J. Siemons, P. E. Bierstedt, and R. G. Kepler, *J. Chem. Phys.*, **39**, 3523 (1963); (b) V. Walatka, Jr., and J. H. Perlstein, *Mol. Cryst. Liq. Cryst.*, **15**, 269 (1971); (c) L. I. Burovov, D. N. Fedutin, and I. F. Shchegolev, *Zh. Eksp. Teor. Fiz.*, **59**, 1125 (1971) [*Sov. Phys.-JETP*, **32**, 612 (1971)].

(5) F. Wudl, D. Wobschall, and E. J. Hufnagel, *J. Amer. Chem. Soc.*, **94**, 672 (1972).

(6) C. Creutz and H. Taube, *J. Amer. Chem. Soc.*, **91**, 3988 (1969).

(7) (a) K. Krogmann and H. D. Hausen, *Z. Anorg. Allg. Chem.*, **358**, 67 (1968); (b) K. Krogmann, *Angew. Chem., Int. Ed. Engl.*, **8**, 35 (1969); (c) M. J. Minot and J. H. Perlstein, *Phys. Rev. Lett.*, **26**, 371 (1971); (d) D. Kuse and H. R. Zeller, *Phys. Rev. Lett.*, **27**, 1060 (1971).

The optical and electronic properties of compounds **1**,[3] **3**,[2a] and **5**[6] suggest that they belong in class II, while compounds **2**,[4] **4**,[5] and **6**[7] exhibit greater delocalization and probably belong in class III. While the preparation and properties of mixed-valence inorganic compounds have been explored in some depth, relatively little work has been published regarding mixed-valence organic and organometallic compounds. This situation should rapidly change inasmuch as the following statement made by Robin and Day with specific reference to inorganic compounds is equally applicable to organic and organometallic compounds:

"For those interested in the interrelationships between electronic structure, molecular structure, electronic spectra, electronic conduction, and molecular magnetism, the mixed-valence systems offer a class of compounds unique in chemistry."[2a]

Furthermore it is our prejudice that mixed-valence organometallic compounds are more amenable to architectural tinkering than are their inorganic counterparts and thus provide the means with which to probe the variety and complexity of mixed-valence interactions.

Molecular and Electronic Structure of Ferrocene and Ferricenium Ion

The ferrocenyl unit is an attractive component in the study of mixed-valence compounds due to the ease with which it may be oxidized. A variety of ferrocene compounds[8] have been prepared (**7**–**10**). It

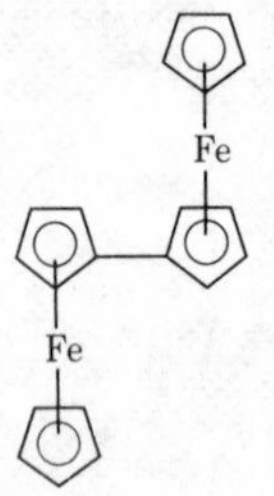

7

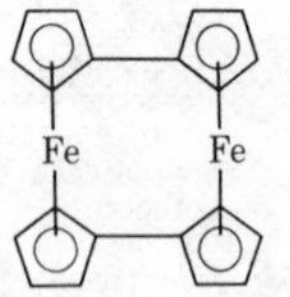

8

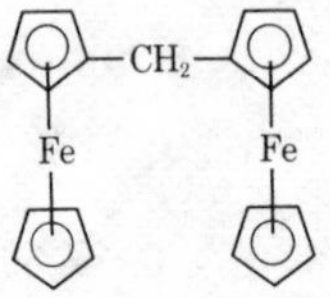

9

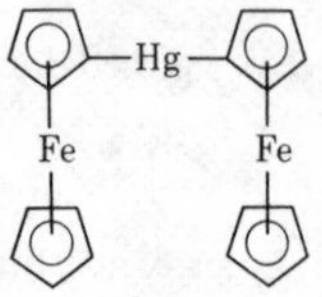

10

should be possible to convert these and similar compounds into mixed-valence compounds *via* chemical and electrochemical means. Compound **1** was readily prepared by oxidation of **7**, with an excess of benzoquinone in the presence of picric acid.[3,9] In the conformation shown, **1** has less opportunity for both through-bond and through-space interaction than the oxidized form of **8**, while compounds **9** and **10** when oxidized should exhibit less delocalization than **1**. Biferrocene [Fe(II)Fe(III)] picrate (**1**) and related

(8) (a) M. Rosenblum, "Chemistry of the Iron Group Metallocenes," Wiley, New York, N. Y., 1965; (b) D. E. Bublitz and K. L. Rinehart, Jr., *Org. React.*, **17**, 1 (1969).

(9) F. Kaufman and D. O. Cowan, *J. Amer. Chem. Soc.*, **92**, 6198 (1970).

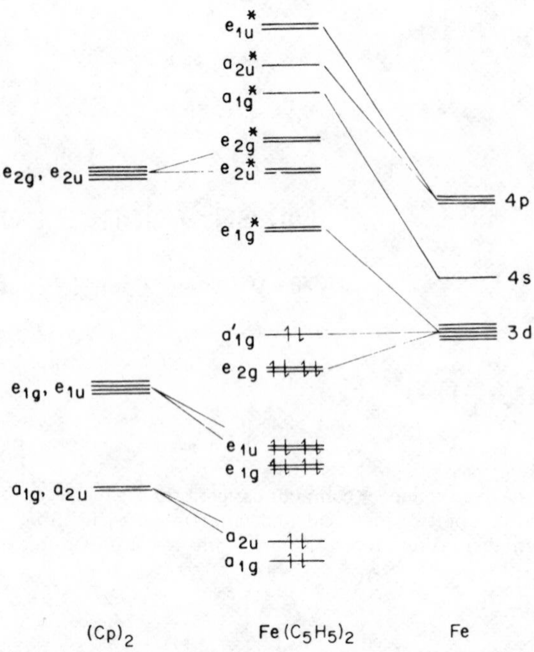

Figure 1. Molecular orbital energy level diagram of ferrocene (redrawn from R. Prins, *Mol. Phys.*, **19**, 603 (1970)).

mixed-valence organometallic compounds possess a number of interesting properties that will be considered after a short review of the current ideas regarding the electronic structure of ferrocene and ferrocenium cation.

Most semiempirical molecular orbital theory calculations for ferrocene[10] give a molecular orbital energy level scheme similar to the one shown in Figure 1. While there are some level-ordering differences among the various calculations, it is now generally accepted that the LUMO and the two HOMO are primarily metal d orbital in character, as shown by Mossbauer, optical, esr, and photoelectron spectroscopic studies.

Ferrocenium ion (Fc^+) has been assigned the orbitally degenerate $^2E_{2g}$ $[(e_{2g})^3(a_{1g})^2]$ configuration based on magnetic susceptibility[10a,11,12] and esr[10a,11a,12,13] studies which indicate that the e_{2g} orbital is not completely filled. For example, the magnetic moment of $Fc^+BF_4^-$ at room temperature is 2.49 μ_B; this value is much larger than the spin only moment of 1.73 μ_B. However, the e_{2g} and a_{1g} orbitals are close in energy, and it is difficult to decide which is the HOMO, $(e_{2g})^3(a_{1g})^2$ or $(a_{1g})^2(e_{2g})^3$.

The blind application of photoelectron spectroscopy and Koopmans' theorem ($\epsilon_{\text{ionization}} = -\epsilon_{\text{orbital}}^{\text{SCF}}$) to the study of ferrocene would suggest that the HOMO is the e_{2g} level since the intensity of the signal corresponding to the first ionization is

(10) (a) R. Prins, *Mol. Phys.*, **19**, 603 (1970); (b) I. H. Hillier and R. M. Canadine, *Discuss. Faraday Soc.*, **47**, 27 (1969); (c) Y. S. Sohn, D. N. Hendrickson and H. B. Gray, *J. Amer. Chem. Soc.*, **93**, 3603 (1971).

(11) (a) R. H. Maki and T. E. Berry, *J. Amer. Chem. Soc.*, **87**, 4437 (1965); (b) D. N. Hendrickson, Y. S. Sohn and H. B. Gray, *Inorg. Chem.*, **10**, 1559 (1971).

(12) D. O. Cowan, G. A. Candela, and F. Kaufman, *J. Amer. Chem. Soc.*, **93**, 3889 (1971).

(13) R. Prins and F. J. Reinders, *J. Amer. Chem. Soc.*, **91**, 4929 (1969).

twice as great as that of the second ionization. However, since the first two ionizations differ by only 0.35 eV (2800 cm^{-1}), all that can safely be concluded is that the two levels are close in energy.[10a,12,13,14] The success of Koopmans' theorem depends upon the approximate cancelation of (a) the reorganizational energy required by the ion after the electron has been ejected and (b) the electron–electron correlation energy difference between the neutral molecule and ion. It is therefore not surprising that, in contrast to PES, the optical spectrum of ferrocene is best interpreted in terms of a highest occupied a_{1g} orbital.[10c]

The 617-nm (1600-cm^{-1}) absorption of the ferricenium ion was originally attributed to the $^2E_{2g} \rightarrow {}^2A_{1g}$ transition in which an electron is excited from the a_{1g}(3d) to the e_{2g}(3d) orbital.[15] Prins[10a,16] and Gray[11] have argued that this assignment is improbable since the molar absorptivity (ϵ = 340) is larger than that observed for similar 3d–3d transitions in the absorption spectra of other metallocenes. They assign this band to the symmetry-allowed $^2E_{2g} \rightarrow {}^2E_{2u}$ transition, in which an electron is excited from the e_{1u} ligand orbital to the partly filled e_{2g} ion orbital. In favor of this assignment are the observations that substitution of an electron-donating methyl group on the ferrocenium ion produces a red shift of 700 cm^{-1}, while an electron-withdrawing acetyl group produces a blue shift of 2700 cm^{-1}.

Electronic Absorption Spectroscopy of Mixed-Valence Compounds

In order to discuss the electronic spectra of biferrocene [Fe(II)Fe(III)] picrate and related compounds, it is desirable to know more about the extent of delocalization (α) in these compounds. X-Ray photoelectron spectra (PES) of the following six ferrocene compounds were measured[17] in order to evaluate the extent of interaction between the two ferrocenyl units: ferrocene, biferrocene, ferrocene [Fe(III)] picrate, ferrocene [Fe(III)] fluoroborate, biferrocene [Fe(II)Fe(III)] picrate, and biferrocene [2Fe(III)] fluoroborate.

If there is only weak interaction between the two ferrocene units in biferrocene, then upon formation of biferrocene [Fe(II)Fe(III)] picrate the charge should be relatively localized on one ferrocene unit. In the closely related series of ferrocene compounds described, the binding energies of a specific iron core level are expected to vary approximately linearly with the charge on these atoms. If our postulate that biferrocene [Fe(II)Fe(III)] picrate can be described as a class II (α = 0–0.1) mixed-valence compound is correct, then we should observe two transitions for each Fe core level. This is in fact what was observed for biferrocene [Fe(II)Fe(III)] picrate. The transition at a binding energy of 707.7 eV corresponds in shape and position to Fe $^2p_{3/2}$ bands found in ferrocene and biferrocene, while the transition at a binding energy of 711.1 eV corresponds in position and shape to the Fe $^2p_{3/2}$ bands observed in ferrocene [Fe(III)] picrate, ferrocene [Fe(III)] fluoroborate, and biferrocene [2Fe(III)] fluoroborate. This latter ionization band is broadened by exchange interaction of the core electrons and unpaired valence electrons.

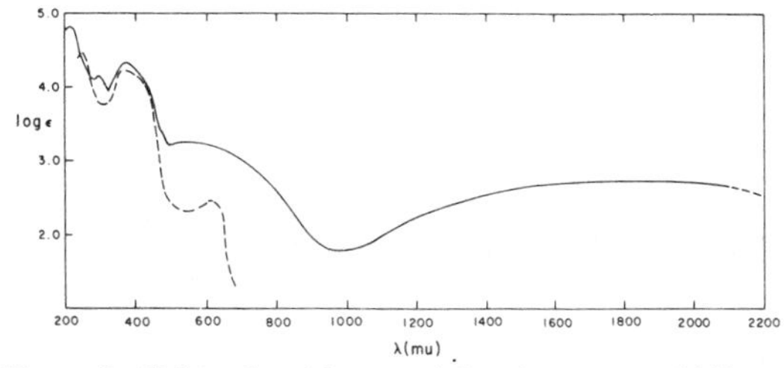

Figure 2. Visible–ultraviolet–near infrared spectrum of biferrocene [Fe(II)Fe(III)] picrate (1), solid line; spectrum of ferrocene [Fe(III)] picrate, broken line.

The X-ray PES results indicate that there is at best weak interaction between the two ferrocenyl rings in biferrocene [Fe(II)Fe(III)] picrate. This conclusion is confirmed by the Mossbauer spectrum of this compound.[18] Here again the Mossbauer spectrum appears to be a composite of the spectra obtained for ferrocene and ferrocene [Fe(III)] picrate.

There are several factors that could account for the weak interaction between the ferrocene and ferrocenium portions of biferrocene [Fe(II)Fe(III)] picrate. The most important is probably the fact that the HOMO is nonbonding in character.

Since α is small for biferrocene [Fe(II)Fe(III)] picrate, we expect to observe electronic transitions similar to those observed for ferrocene and ferrocenium picrate. The electronic spectrum of biferrocene [Fe(II)Fe(III)] picrate is shown in Figure 2 along with the spectrum of ferrocene [Fe(III)] picrate.[9] The presence of the ferrocene–ferrocenium transitions is consistent with the suggested weak interaction between the two portions of the molecule. There are, however, at least two important differences. The 600-nm ($^2E_{2g} \rightarrow {}^2E_{2u}$) band is broadened and shifted to higher energy, and a new near-infrared transition is observed at 1900 nm (ϵ = 550, acetonitrile) which is assigned to a mixed-valence electron-transfer transition.

A simple substitution effect cannot explain the broadening and change in energy of the 600-nm transition since ferrocene, like an alkyl group, is electron donating relative to the ferrocenium ion and should give rise to a red shift. The determination of the Hammett σ_p for the ferrocenyl group as a substituent on benzene shows it to be a stronger electron-donating ligand than the methyl group.[19]

It is possible that the broad 600-nm band is composed of two transitions (see Figure 3). If this is true, one would expect the $Fc(e_{1u}) \rightarrow Fc^+(e_{2g})$ transition to be blue shifted in relation to the known $Fc^+(e_{1u}) \rightarrow Fc^+(e_{2g})$ transition. Attempts to resolve this 600-nm band by spectroscopic measurements at low temperatures have only been partially successful to date.

All mixed-valence class II and III compounds prepared and studied thus far are characterized by a

(14) D. N. Hendrickson, *Inorg. Chem.*, **11**, 1161 (1972).
(15) D. R. Scott and R. S. Becker, *J. Phys. Chem.*, **69**, 3207 (1965).
(16) R. Prins, *Chem. Commun.*, **280** (1970).
(17) D. O. Cowan, J. Park, M. Barber, and P. Swift, *Chem. Commun.*, 1444 (1971).
(18) D. O. Cowan, R. L. Collins, and F. Kaufman, *J. Phys. Chem.*, **75**, 2026 (1971).
(19) A. N. Nesmenanov, E. G. Perevolova, S. P. Gubin, K. I. Grandberg, and A. G. Kozlovosky, *Tetrahedron Lett.*, 2381 (1966).

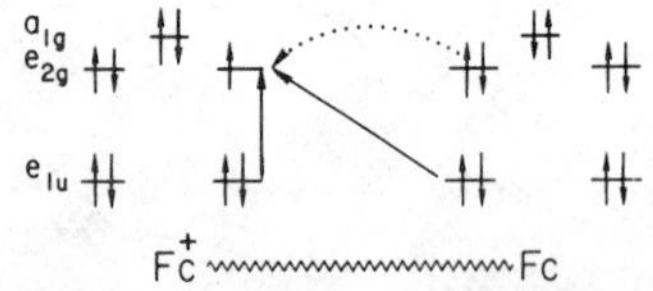

Figure 3. Possible mixed valence (electron transfer) transition in biferrocene [Fe(II)Fe(III)] picrate.

Configuration Change

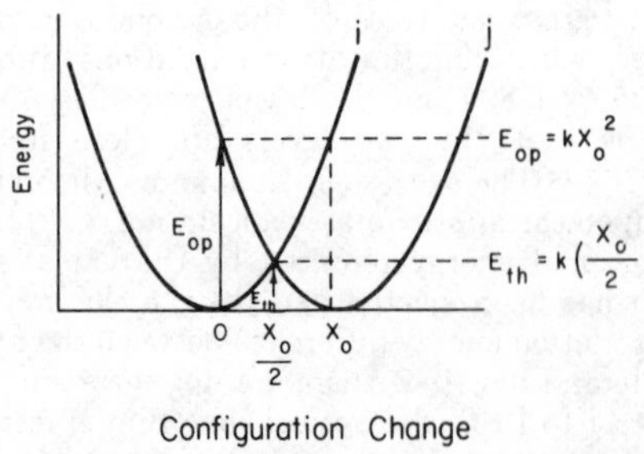

Figure 4. Potential energy–configuration diagram for a symmetrical one-electron transfer. The curves represent two weakly interacting states i and j, with different equilibrium configurations $(0, X_0)$. If these states are represented as harmonic oscillators $[E(X) = k \Delta X^2]$, then the maximum for the optical transition is equal to kX_0^2 and the maximum for the thermal transition (at $X_0/2$) is equal to $k(X_0/2)^2$; or the thermal activation energy is one-quarter of the optical transition energy ($E_{th} = E_{op}/4$) (redrawn from ref 9).

new electronic transition in the visible to near-infrared region of the spectrum. The intensity of this new transition, which is seen in the absorption spectrum of the mixed-valence species but not in the spectra of any of the equal valence precursor compounds, is an extremely sensitive gauge of the extent of interaction, even when α is very small.[2] As such, the appearance of this new absorption band is a simple and useful diagnostic tool to identify the presence of a compound which is mixed valent in character. Mixed-valence electronic transition data for a number of new compounds are given in Table I.[3,6,20-23] Recently mixed-valence transitions have also been observed for mixed oxidation state ferrocene compounds prepared from terferrocene, quaterferrocene, sexiferrocene, and polyferrocene.[24]

For biferrocene [Fe(II)Fe(III)] picrate this transition corresponds to the intramolecular electron transfer shown in Figure 3 (- - - - →)

$$[{}^1A_{1g}(e_{2g})^4(a_{1g})^2]_i \,\rule[0.5ex]{1em}{0.4pt}\, [{}^2E_{2g}(e_{2g})^3(a_{1g})^2]_j \xrightarrow{h\nu}$$

$$[{}^2E_{2g}(e_{2g})^3(a_{1g})^2]_i \,\rule[0.5ex]{1em}{0.4pt}\, [{}^1A_{1g}(e_{2g})^4(a_{1g})^2]_j{}^*$$

where * indicates a vibrationally excited ground state. As such, the transition involves an electron transfer between two metallike orbitals (e_{2g}) that interact only weakly with the π conjugated system of the molecule. From the following discussion it will be seen that the energy (λ) of the mixed-valence transition is dependent upon differences in geometry and bond lengths in the donor and acceptor portions of the molecule but is *not* very dependent upon the extent of delocalization in the ground state (α) or upon the distance separating the donor and acceptor moieties. However, the intensity of the allowed transition is proportional to the square of the probability of finding the electron on the acceptor portion in the ground state and upon the square of the distance between the two moieties. A simple model (Figure 4) first proposed by Hush[2b,c] provides qualitative insight into how the various molecular parameters alter the mixed-valence transition. If we have a mixed-valence compound where the mixed-valent moieties have oxidation states i and j, respectively, then the ground-state electron distribution is $[i,j]$. The new absorption band is assigned to the $[i,j] + h\nu$

(20) G. Emschwiller and C. K. Jorgensen, *Chem. Phys. Lett.*, **5**, 561 (1970).

(21) G. R. Eaton and R. H. Holm, *Inorg. Chem.*, **10**, 805 (1971).

(22) C. Sigwart, P. Hemmerich, and J. T. Spence, *Inorg. Chem.*, **7**, 2545 (1968).

(23) S. B. Saha and S. Basu, *J. Chim. Phys. Physicochim. Biol.*, **67**, 2069 (1970).

(24) (a) Unpublished observations of M. Rausch, F. Kaufman, J. Park, and D. O. Cowan. (b) Aharoni and Litt (S. M. Aharoni and M. H. Litt, *J. Organometal. Chem.*, **22**, 171, 179 (1970)) have prepared a ferrocene–ferricenium mixed crystal; no new near-ir band was reported.

→ $[j,i]^*$ transition in which the optical excitation results in intermolecular or intramolecular electron transfer. In the limit of $X_0 = 0$ in Figure 4, the equilibrium (configuration) normal coordinates of the i and j states would be equivalent and there would be no Franck–Condon barrier to optical electron transfer. This means that, as the electron-transfer transition in a series of mixed-valence compounds (Table I) tends toward higher energy, the atoms which are in different oxidation states find themselves in increasingly different ligand environments. This explains why the electron-transfer transition in biferrocene [Fe(II)Fe(III)] has such a low energy compared with other members of Table I. While the X-ray data[8a,25] are not as detailed as one would desire, we know that ferrocenium ion is structurally very similar to ferrocene. There is a very small difference in the cyclopentadienyl–iron–cyclopentadienyl distances between the two, and a small distortion from the D_5 symmetry of ferrocene has been noted for the ferrocenium ion.[11,12] The energy of the optical transition is dependent on the difference in configuration at the two sites because structural and/or electronic reorganization is required to accommodate the intramolecular electron transfer.

If the mixed-valence compound is unsymmetrical, *i.e.*, the structure of the two moieties is different, the interpretation of the electron-transfer bands is complicated by the fact that the zero point energy levels for the initial and final states may be vertically displaced relative to each other. A case in point is provided by the mixed-valence ruthenium complexes 11 (Table I) where substitution of pyrazine (**11b**) for

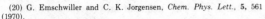

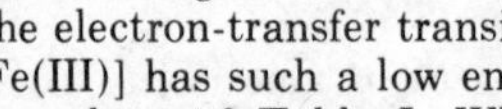

$$[(H_3N)_4(L)RuN\bigcirc NRu(NH_3)_5]^{5+}$$

11

NH₃ (**11a**) results in an electron-transfer band which is shifted to higher energy[6] (410-nm shift). In these complexes both configurational and energy differences between the $[i,j]$ and the $[j,i]$ states lead to the observed shift in wavelength maximum of the mixed-valence transition.

(25) (a) T. Bernstein and F. H. Herbstein, *Acta Crystallogr.*, *Sect. B*, **24**, 1640 (1968); (b) R. C. Petersen, Ph.D. Thesis, The University of California, Berkeley, Calif., 1966.

Table I
Spectral Data for Mixed Valence Compounds

Compd	Formula	λ_{max}, nm	ϵ	Ref
1	(In text)	1900	550	3
11a	L = NH_3	1570	7×10^3	6
11b	L = pyrazine	1160		6
12	$[(NC)_4Fe(CN)_2Fe(CN)_4]^{6-}$	1280	5.5×10^3	20
13	$[(R_4C_4S_4)_2Co(DPPE)Co(R_4S_4C_4)_2]^{-}$ [a]	1360		21
14	$[(X)_2Cu(OAc)_2Cu(X)_2]^{-}$ [a]	900		22
15	$[(H_3N)_2Cu(Br)_2Cu(Br)_2Cu(NH_3)_2]$	625	160	23

[a] DPPE = *trans*-1,2-bis(diphenylphosphino)ethylene; X = CH_3OH.

The intensity of the electronic transition depends upon the magnitude of the transition dipole moment (M_{12}) (eq 2) between the ground (1) and excited

$$M_{12} = \langle 1|\Sigma_i e_i X_i|2\rangle \quad (2)$$

state (2); see Figure 4. Since 1 and 2 are composed of mixed donor (Φ_i) and acceptor (Φ_j) wave functions it is possible to show[2c,26] *via* first order perturbation theory that for a simple isolated donor-acceptor system the magnitude of M_{12} is

$$|M_{12}| = e\alpha r \quad (3)$$

where r is the distance between the donor and acceptor, and the α is approximately defined in equation 1. The intensity (oscillator strength) will then depend upon the square of both the interaction parameter (α) and the distance (r) separating the interacting donor-acceptor moieties. Based on this type of perturbation treatment it is possible to estimate α^2 from spectral data using the following equation

$$\alpha^2 \cong (4.5 \times 10^{-4})\epsilon_{max}\Delta_{1/2}/(\bar{\nu}r^2) \quad (4)$$

In this equation, the band shape of the electron transfer transition is assumed to be Gaussian, ϵ_{max} is the molar absorptivity at the band maximum, $\Delta_{1/2}$ is the band half-width in wave numbers, $\bar{\nu}$ is the frequency in wave numbers and r is the donor-acceptor distance in Å. Using this equation, α^2 is calculated to be 1.610×10^{-3} for Prussian blue[2c] [$KFe^{II}Fe^{III}(CN)_6 \cdot H_2O$]; $0.9\text{–}1.4 \times 10^{-2}$ for biferrocene [Fe(II)Fe(III)] picrate (**1**); and 1.7×10^{-2} for Creutz and Taube's ruthenium complex (**5**). All three of these compounds meet the trapped valence requirement, $\alpha < 0.25$.[26]

Thermally Activated Electron Transfer

In addition to the optical Franck–Condon path for electron transfer, mixed-valence compounds can also undergo a thermally activated process which leads to electron exchange when the transition-state equilibrium geometry is achieved (see Figure 4). From this model it can be seen that the activation energy for the thermal electron-transfer process is equal to or greater than one-fourth the energy of the optical transition.[2c] This allows one to calculate a thermally activated unimolecular electron-transfer rate constant (k_{th}).

$$k_{th} \approx kT/h \exp[hc\bar{\nu}/(4RT)] \quad (5)$$

For biferrocene [Fe(II)Fe(III)] picrate the calculated rate constant is 1.3×10^{10} sec^{-1}. This is comparable to the unimolecular rate constant (4×10^9 sec^{-1}) estimated from the product of the experimental second-order rate constant for the exchange of ferrocene and ferrocenium ion and a concentration term calculated so that the average iron–iron distance would be the same as that found for compound **1**.[3,9]

A second example which involves fewer approximations in the rate calculation is the binuclear copper complex **14** (Table I) which has a mixed-valence absorption band at 900 nm.[23] An esr study of this compound indicates that intramolecular exchange rate is greater than 10^8 sec^{-1}, while the value calculated from the energy of the optical electron-transfer transition is 5×10^8 sec^{-1}.

It seems very probable that the relationship between electronic structure and electron-transfer rates in mixed-valence compounds will be useful in defining the important rate-determining parameters in oxidation–reduction reactions. This is especially true for bimolecular reactions which proceed through a short-lived activated complex, where the rate-determining step is electron transfer within the complex.[27,28] The importance of precursor complex formation has been discussed for electron-transfer reactions,[27] and kinetic and spectral data have been used as evidence for the occurrence of certain mixed-valent complexes in the reaction mechanism. Recently it has been suggested that mixed-valence species serve as intermediates in several interesting reactions. Esr measurements during the autocatalytic decomposition of alkylcopper(I) species[29a] have been interpreted in terms of complexes of the $RCu^{I}Cu^{0}$ type. Optical changes during the Mn(III)-catalyzed oxidative decarboxylation of acids[29b] are thought to be due to the formation of $Mn^{III}(RCO_2)_xMn^{II}$ complexes.

There are many components of living systems which contain two or more metal atoms in close proximity.[30] When the metal atoms of these com-

(26) B. Mayoh and P. Day, *J. Amer. Chem. Soc.*, **94**, 2885 (1972). When the resonance interaction between the i and j components of the mixed valence system is as large as E_{th} ($E_{th} = E_{op/4}$), then complete delocalization exists. Since the resonance interaction β between the i and j components is given by $\beta = \alpha(E_{op})$, then α must be less than ¼ to have trapped valences. This classification criterion is analogous to Mott's classification of the electronic structure of crystals (small polaron or band models). I. G. Austin and N. F. Mott, *Science*, **168**, 71 (1970).

(27) (a) R. H. Marcus, *Annu. Rev. Phys. Chem.*, **15**, 155 (1964); (b) A. G. Sykes, *Advan. Inorg. Chem. Radiochem.*, **10**, 153 (1967).

(28) (a) J. F. Endicott, *J. Phys. Chem.*, **73**, 2594 (1969); (b) R. C. Patel, R. E. Ball, J. F. Endicott, and R. G. Hughes, *Inorg. Chem.*, **9**, 23 (1970); (c) D. P. Rillema, J. F. Endicott, and R. C. Patel, *J. Amer. Chem. Soc.*, **94**, 394 (1972).

(29) (a) K. Wada, M. Tamura, and J. Kochi, *J. Amer. Chem. Soc.*, **92**, 6656 (1970); (b) J. M. Anderson and J. Kochi, *ibid.*, **92**, 2450 (1970).

pounds are involved in valence changes, mixed-valence effects may be important in determining physical parameters such as oxidation–reduction potentials and rates of chemical reactions. In addition, mixed-valence intermediates are involved as redox couples in the electron-transfer chains of chloroplasts and mitochondria, where a number of metal-containing proteins shuttle electrons between oxidizing and reducing substrates. In this regard, a model system for aerobic oxidation of cytochrome *c* has been prepared in which the electrons are thought to be transferred by tunneling between Fe(II)–Fe(III) potential wells.[31]

One group of biological compounds that can be considered as mixed valence in character is the non-heme iron proteins.[30a,32] These substances, which have active sites of the type $(Fe–S)_n$, where $n = 1–8$, have been examined by a variety of physical techniques which indicate that the iron atoms ($n > 2$) can be in close structural and electronic proximity.[33] The 2.5-Å resolution X-ray map of bacterial ferredoxin ($n = 8$) indicates that there are two clusters of tetrahedrally arranged 4Fe–4S units in the molecule.[34] Magnetic susceptibility and Mossbauer measurements on a number of these iron–sulfur proteins indicate that the oxidized form is diamagnetic, while the reduced form has one unpaired electron, which is delocalized over all the iron atoms in the active site.[33b] A model with antiferromagnetically coupled Fe(III) atoms (oxidized form) and Fe(II)–Fe(III) atoms (reduced form) has been suggested to account for these results.[35] Spinach ferredoxin and beef adrenodoxin, on the other hand, have Fe–S active sites which are not characterized by a high degree of metal–metal interaction. Circular dichroism and absorption spectra of these substances have transitions at about 11,000 cm^{-1} that have been interpreted as mixed-valence bands due to weak Fe(III)–Fe(II) orbital overlap.[32] The weak coupling hypothesis is consistent with the low mixed-valence band intensity and the presence of the individual component transitions in the Mossbauer and electronic absorption spectra.

Many of the iron–sulfur proteins have unique physical properties, such as extremely low oxidation–reduction potentials (near the value of the H_2 electrode) and *g* values of less than 2 as measured by esr experiments for the reduced (mixed-valence) forms. The reasons for the extraordinary properties of individual species, or the explanation for differences in the behavior of related iron–sulfur proteins, is not immediately apparent from the structural data on the active sites. For example, ferredoxin and HiPIP have structurally similar Fe–S clusters,[34] yet their E_0 values differ by 0.7 V.[36]

(30) (a) D. O. Cowan, G. Pasternak, and F. Kaufman, *Proc. Natl. Acad. Sci. U. S.*, **66**, 837 (1970); R. J. P. Williams in "Current Topics in Bioenergetics," D. R. Sanadi, Ed., Academic Press, New York, N. Y., 1969.

(31) J. H. Wang, *Accounts Chem. Res.*, **3**, 90 (1970).

(32) W. A. Eaton, G. Palmer, J. A. Fee, T. Kimura, and W. Lovenberg, *Proc. Natl. Acad. Sci. U. S.*, **68**, 3015 (1971).

(33) (a) B. B. Buchanan and D. J. Arnon, *Advan. Enzymol. Relat. Areas Mol. Biol.*, **33**, 119 (1970); (b) J. C. M. Tsibris and R. W. Woody, *Coord. Chem. Rev.*, **5**, 417 (1970).

(34) L. C. Sieker, E. Adman, and L. A. Jensen, *Nature (London)*, **235**, 40 (1972).

(35) K. K. Rao, R. Cammack, D. O. Hall, and C. E. Johnson, *Biochem. J.*, **122**, 257 (1971).

It is possible that geometrical distortions from the normal geometry may be important in determining the properties of these systems.

Structural Change

It is expected that interaction between the appropriate atoms in a mixed-valence compound [*i,j*] may manifest itself in discrete structural changes relative to the [*i,i*] or [*j,j*] parent species. These changes are expected to depend on the nature of the interaction between the HOMO levels and the structural rigidity of the molecular framework. A major distortion in structure may result in bonding changes which will be reflected in new bond lengths and bond angles that can be easily measured by X-ray crystallography. On the other hand, some subtle geometric distortions could elude X-ray detection.

Mixed-valence organometallic chalcogen complexes have been prepared and studied by X-ray crystallography in order to determine the stereochemical consequences of removing a valence electron. The tetramercapto-bridged Mo(III) dimer, $[Mo(h^5\text{-}C_5H_5)(SCH_3)_2]_2$, and the corresponding mixed-valence cation were prepared and found to have very similar bond distances and bond angles.[37] These results were in strong contrast to those found for the mixed-valence iron–sulfur dimer $[Fe(h^5\text{-}C_5H_5)(CO)(SCH_3)]_2^+$. The latter compound has an iron–iron distance of 2.925 Å which was a decrease of 0.46 Å relative to that found for the diamagnetic neutral dimer.[38] The decrease in the metal–metal distance was also reflected as a large distortion (16°) of the bridging Fe–S–Fe angle (98° neutral species).

The difference in sensitivity to mixed-valence interaction upon removal of one valence electron from the Mo *vs.* Fe dimer was ascribed to the presence of angular strain in the bridging ligands of the former compound which prevented any shortening of the Mo–Mo bond upon oxidation. This interpretation is consistent with the very small bridging Mo–S–Mo angles (64°) found in the neutral dimer.

The decrease in the Fe–Fe bond length of the dimercapto mixed-valence compound was attributed to the formation of a one-electron metal–metal bond upon removal of a single electron from the completely filled σ antibonding orbital by oxidation of the neutral dimer.

Geometrical changes have also been detected in ferrocene dimers[12] upon the removal of one or more valence electrons, even though the metal orbitals involved in the valence change are only weakly coupled and nonbonding in character. Biferrocene can exist in three stable oxidation states, the diamagnetic [2,2] and the two paramagnetic states [2,3] and [3,3]. The choice of one- *vs.* two-electron oxidation depends upon the oxidizing agent used.[9] The paramagnetic derivatives, which contain one [2,3] and two

(36) HiPIP also has unique esr properties since it is the oxidized form ($g_{\parallel} = 2.12$, $g_{\perp} = 2.04$) and not the reduced form which has an unpaired electron. See G. Palmer, H. Britzinger, R. W. Estabrook, and R. H. Sands in "Magnetic Resonance in Biological Systems," H. Ehrenberg, B. G. Malmstrom and T. Vanngard, Ed., Pergamon Press, London, 1967, p 159.

(37) N. G. Connelly and L. F. Dahl, *J. Amer. Chem. Soc.*, **92**, 7470 (1970).

(38) N. G. Connelly and L. F. Dahl, *J. Amer. Chem. Soc.*, **92**, 7472 (1970).

Table II
Electrical Conductivity at 298 K and Activation Energy (E_a)

Compd	σ, $(\Omega\ \mathrm{cm})^{-1}$	E_a, eV	Ref
Ferrocene	10^{-14}	0.90	9
Ferricenium picrate	10^{-13}	. . .	9
1	10^{-8}	0.43	3
2	100	0.023[b]	4b
3	250		2a
4	0.25	0.19	5
4 reduced	10^{-12}		5
6	0.3–120	0.085[b]	7c

[a] Conductivity data for **4** was determined on a pellet; all other data were from single crystal measurements. [b] Low-temperature activation energy.

[3,3] unpaired electrons respectively, can be studied by magnetic susceptibility and electron spin resonance. The differences in observed behavior indicate that geometric distortions in these compounds can lead to changes in electronic structure.

Electrical Conductivity

Table II gives electrical conductivity data for several mixed-valence compounds and their non-mixed-valence parents. The electrical conductivity of single crystals of the biferrocene [Fe(II)Fe(III)] picrate were found to be 10^6 times larger than that of ferrocene or ferrocenium picrate.[3,9] Measurement of the conductivity as a function of temperature indicated that compound **1** was an intrinsic semiconductor, with an activation energy for the conduction of 0.43 eV. This is in comparison to an activation energy of 0.9 eV found for ferrocene. The difference between the activation energy for electrical conduction (0.43 eV) and the thermal activation energy for intramolecular electron transfer (0.16 eV) indicates that the rate-limiting process in the conduction is electron hopping between the individual molecules in the crystal. The electrical conductivity of ferrocene polymers (**16**, **17**) can be dramatically increased when the polymers are converted into mixed-valence poly salts.[1]

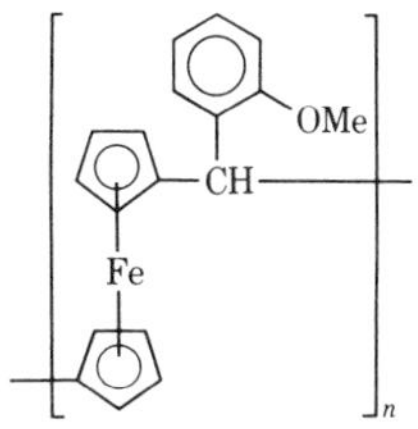

16

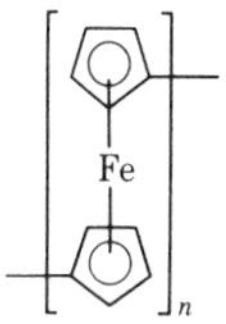

17

A simple electron-hopping model for the electrical conductivity is consistent with the observations that the conductivity is relatively insensitive to the anion structure, or to conjugation, but is very sensitive to the fraction of the ferrocene units oxidized. The higher conductivity for compounds **2**, **4**, and **6** which stack in one-dimensional chains is probably due to (a) stronger interaction between individual units and (b) appropriate molecular orientation within the crystal for interaction between more than two of the molecular units. However, recent experimental[39] and theoretical[40] work suggest that the probability of finding a very highly conducting one-dimensional system is quite low. This is a consequence of the fact that even small random potentials (caused, for example, by counterions) will result in the localization of an otherwise extensively delocalized system.

The organic compound with the highest known electrical conductivity combines the cation of compound **4**, tetrathiafulvalenium cation (TTF^+), with the tetracyano-*p*-quinodimethane anion ($TCNQ^-$).[41] Single crystals of this material have a conductivity of 650 $\Omega^{-1}\ \mathrm{cm}^{-1}$ at room temperature and a value of 15,000 $\Omega^{-1}\ \mathrm{cm}^{-1}$ at 77 K.

(39) J. H. Perlstein, M. J. Minot, and V. Walatka, *Mater. Res. Bull.*, **7**, 309 (1972).
(40) A. N. Bloch, R. B. Weisman, and C. M. Varma, *Phys. Rev. Lett.*, **28**, 753 (1972).
(41) Unpublished observations of D. O. Cowan, J. Ferraris, J. H. Perlstein, and V. Walatka.

Tricobalt Carbon, an Organometallic Cluster

Bruce R. Penfold

Chemistry Department, University of Canterbury, Christchurch, New Zealand

Brian H. Robinson*

Chemistry Department, University of Otago, Dunedin, New Zealand

Received April 17, 1972

Reprinted from ACCOUNTS OF CHEMICAL RESEARCH, *6, 73 (1973)*

The metal cluster is now a well-known structural feature among the compounds of transition metals in their lower valency states.[1-3] Metal cluster compounds occur especially among the carbonyls of iron, cobalt, nickel, and their congeners, mostly with the metal in a state of zero valence. There are authenticated examples which contain triangular, tetrahedral, square-pyramidal, and octahedral clusters of metal atoms.[2,3] In these compounds the metal–metal bonds which hold the clusters together are strong, and metal–metal bond formation is a dominating influence in their chemistry.[1,4]

Bruce Penfold was born in Christchurch, New Zealand, where he obtained an M.Sc. in Chemistry at the University of Canterbury. He obtained his training in X-ray crystallography with Wilham Cochran at the Cavendish Laboratory, Cambridge University, where he received his Ph.D. in 1952. He then returned to the Chemistry Department at Canterbury to establish an X-ray diffraction laboratory and now holds a personal chair in this department. He has applied crystallographic methods to a variety of chemical structural problems, especially in recent years to metal clusters and to transition metal fluorocarbon complexes.

Brian H. Robinson was born in Christchurch, New Zealand, and received his Ph.D. degree from the University of Canterbury in 1965. After 2 years of postdoctoral work, he joined the staff at the University of Otago where he is now Senior Lecturer. Dr. Robinson is interested in donor–acceptor relationships in organometallic chemistry and spectroscopic properties of biological complexes.

(1) F. A. Cotton, *Quart. Rev., Chem. Soc.*, **20**, 389 (1966).
(2) B. R. Penfold, *Perspect. Struct. Chem.*, **2**, 71 (1968).
(3) P. Chini, *Inorg. Chim. Acta Rev.*, **2**, 31 (1968).
(4) R. D. Johnston, *Advan. Inorg. Chem. Radiochem.*, **13**, 471 (1970).

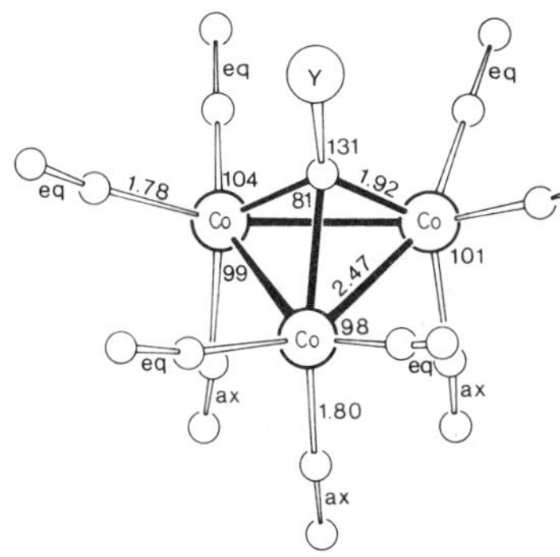

Figure 1. Molecular structure for compounds of general formula $YCCo_3(CO)_9$. Bond lengths (Å) and angles (degrees) are mean values over a number of independent structure determinations. Estimated standard deviations in these mean values are not greater than 0.01 Å or 1°.

In some of the metal cluster carbonyls, nonmetal atoms form symmetrical bridges across two or more atoms of the cluster. The triangular tricobalt cluster appears to be uniquely amenable to this type of bridging. For example, many carbonyls have been prepared in which a sulfur[5-9] (or one of its higher congeners)[10] or a carbon atom is symmetrically bridged to a Co_3 triangle to form a tetrahedral grouping. The tetrahedral Co_3C unit is the structural core of a whole series of methinyltricobalt enneacarbonyls of general composition $YCCo_3(CO)_9$ (Y = halogen, alkyl, aryl, carboxyl etc.) first prepared in 1958[11] and whose basic molecular structure, as illustrated in Figure 1, was elucidated by X-ray diffraction 8 years later.[12]

In the $YCCo_3(CO)_9$ compounds, three of the CO groups will be referred to as *axial* and the remaining six as *equatorial*, as indicated in Figure 1. The -C-$Co_3(CO)_9$ group possesses idealized C_{3v} symmetry and the essential molecular dimensions averaged over all appropriate compounds are given in the same figure. In all cases the equatorial CO groups are bent out of the Co_3 plane toward Y. For those compounds where the size of Y does not introduce additional constraints, the dihedral angle between the Co_3 triangle and a CoCO(eq)CO(eq) triangle is 30°, on the average. These dimensions are highly consistent. Significant departures from average values may be related to special features of individual compounds: these are noted below in the section on molecular structures.

The principles governing the reactivity and stereochemistry of metal cluster carbonyls in general are fairly well understood, but this cannot be said for the methinyltricobalt compounds which are in many ways unique. They are, without exception, unusually stable to oxidation and thermal degradation, relative to the other cobalt carbonyls and comparable tetrahedral metal cluster compounds. The tricobalt carbon group in these compounds is the only authenticated cluster which is truly organometallic in the sense that traditional inorganic and organic reactions may both be carried out on the $YCCo_3$ unit. Reactivity is determined by steric and electronic requirements of the cluster as a whole. (Other cluster compounds, such as $Ru_6(CO)_{18}C$[13,14] and $Fe_5(CO)_{15}C$,[15] which contain an isolated carbon atom have been prepared, but these are best considered as carbides.)

When we commenced our studies, we decided to focus on three basic questions. First, was the four-coordinate apical carbon atom indeed aliphatic (sp^3), or did the basal cobalt triangle alter the nature of this carbon atom? If the Co_3C fragment was an integrated unit, would the reactivity of the apical substituent in turn be affected? Second, would intramolecular steric interactions place any constraints on the coordination behavior of the basal cobalt triangle? Third, could a satisfactory description of the bonding be formulated which could explain the unusual stability of these compounds as well as the details of molecular stereochemistry which began to emerge from the accompanying sequence of X-ray diffraction studies? Answers at least to the first two of these questions are now appearing.

Preparation

A comprehensive review of the synthetic routes has appeared recently;[16] we shall consider only those aspects pertinent to this Account. The original preparation[11] from acetylene complexes $Co_2(CO)_6RCCH$ is restricted to terminal acetylenes, and a $-CH_2$ is necessarily introduced.

$$Co_2(CO)_6RC_2H \xrightarrow{H_2SO_4\text{-methanol}} RCH_2CCo_3(CO)_9 \quad (1)$$

A general method utilizing α,α,α-trihalo compounds was announced independently from two laboratories[17,18] in 1961, and clusters with a wide variety of apical substituents were described.

$$5Co_2(CO)_8 + 3CYX_3 \rightleftharpoons 2Co_3(CO)_9CY + 4CoX_2 + 22CO + [YCX] \quad (2)$$

Extraneous carbon is not necessary, however. In fact these clusters, easily recognizable by their distinctive purple or dark purple-brown color, turn up in many reactions of $Co_2(CO)_8$ and $Co_4(CO)_{12}$,[16] reflecting the high stability of the Co_3C unit. The source of carbon in these reactions is undoubtedly coordinated CO, although the actual intermediates have not been isolated.

A significant development is the recent preparation[19] of the anion $[OCCo_3(CO)_9]^-$ derived from the

(5) L. Marko, G. Bőr, E. Klumpp, B. Marko, and G. Almasy, *Chem. Ber.*, **96**, 955 (1963).
(6) C. H. Wei and L. F. Dahl, *Inorg. Chem.*, **6**, 1229 (1967).
(7) E. Klumpp, G. Bor, and L. Marko, *Chem. Ber.*, **100**, 1451 (1967).
(8) D. L. Stevenson, V. R. Magnuson, and L. F. Dahl, *J. Amer. Chem. Soc.*, **89**, 3727 (1967).
(9) C. E. Strouse and L. F. Dahl, *Discuss. Faraday Soc.*, **47**, 93 (1969).
(10) C. E. Strouse and L. F. Dahl, *J. Amer. Chem. Soc.*, **93**, 6032 (1971).
(11) R. Markby, J. Wender, R. A. Friedel, F. A. Cotton, and H. W. Sternberg, *J. Amer. Chem. Soc.*, **80**, 6529 (1958).
(12) P. W. Sutton and L. F. Dahl, *J. Amer. Chem. Soc.*, **89**, 261 (1967).
(13) B. F. G. Johnson, R. D. Johnston, and J. Lewis, *J. Chem. Soc. A*, 2865 (1968).
(14) A. Sirigu, M. Bianchi, and E. Benedetti, *Chem. Commun.*, 596 (1969).
(15) E. H. Braye, L. F. Dahl, W. Hubel, and D. L. Wampler, *J. Amer. Chem. Soc.*, **84**, 4633 (1962).
(16) G. Palyi, F. Piacenti, and L. Marko, *Inorg. Chim. Acta Rev.*, **4**, 109 (1970).
(17) G. Bor, B. Marko, and L. Marko, *Acta Chim. Acad. Sci. Hung.*, **27**, 395 (1961).
(18) W. T. Dent, L. A. Duncanson, R. G. Guy, H. W. B. Reed, and B. L. Shaw, *Proc. Chem. Soc., London*, 169 (1961).
(19) S. A. Fieldhouse, B. H. Freeland, C. D. M. Mann, and R. J. O'Brien, *Chem. Commun.*, 181 (1971).

Table I
Reactions of Methinyltricobalt Enneacarbonyls, $YCCo_3(CO)_9$, Involving Substitution of CO

Apical group, Y	Ligand	Product	Comments	Ref
Me, Ph, F, Cl, Br	Phosphines Arsines	$YCCo_3(CO)_8L$	Equatorial substitution or CO-bridged structure	28, 35
		$YCCo_3(CO)_7L_2$	Limited formation	
		$YCCo_3(CO)_6L_3$	Et_2PhP only	
Me, Ph, F	Arenes	$YCCo_3(CO)_6$arene	Very labile	40
	Cyclooctatetraene	$YCCo_3(CO)_6COT$	Labile, fluxional	38
	Trienes	$YCCo_3(CO)_6$triene	Static structure, stable	41
Me, Ph, F, Cl	Norbornadiene	$YCCo_3(CO)_7$norb	Nonrigid structure	40
Me, Ph, F, Cl	Cyclopentadiene	$YCCo_3(CO)_4Cp_2$	Stable, CO bridged	40, 41, 64
Me, Ph, F	Olefins	Brown insoluble powders, carbonylation products?		41
Me, Ph, F	Acetylenes	Decomposition; $Co_2(CO)_6RCCR$ etc.		41
Me, Cl	Phosphine oxides	Co(II) phosphine complexes		42
Me	Diarsines	$YCCo_3(CO)_7[(CH_3)_2As]_2C_2(CF_2)_2$	CO-bridged structure	43, 44
CF_3	Diarsines	Rearrangement to give tetrahedral Co_4 structure		43, 45

$$Co_2(CO)_8 \xrightarrow{\text{K or Na-ether}} [OCCo_3(CO)_9]^- \quad (3)$$

hydroxy cluster $HOCCo_3(CO)_9$. A CO-bridged configuration is adopted by the anion presumably as a means of dissipating the negative charge. At this point it is worth noting that $HOCCo_3(CO)_9$ is unknown. Our current ideas on the electronic structure of these clusters lead us to predict that it would be stable only in a strongly acidic medium. The anion could be the precursor for a myriad of compounds of the type $R'OCCo_3(CO)_9$ where R′ is a metallic or nonmetallic moiety. Some boron[20,21] and silicon[20-22] derivatives have been described.

$$LiCo_3(CO)_{10} + R_3SiCl \rightleftharpoons R_3SiOCCo_3(CO)_9 + LiCl \quad (4)$$

$$Co_2(CO)_8 + H_2BNMe_3 \longrightarrow Me_3NBH_2OCCo_3(CO)_9 \quad (5)$$

$$NaCo_3(CO)_{10} + Me_3NBH_2I \rightleftharpoons Me_3NBH_2OCCo_3(CO)_9 + NaI \quad (6)$$

In theory, the following reaction should provide methinyltricobalt enneacarbonyls and analogs in which carbon is replaced by a main group or transition element (eq 7). However, these reactions with

$$3NaCo(CO)_4 + MX_3 \rightleftharpoons 3NaX + MCo_3(CO)_9 + 3CO \quad (7)$$

heavier main group elements give open polymeric structures, *e.g.*, $RSn[Co(CO)_4]_3$,[23] or simpler carbonyls, *e.g.*, $AsCo(CO)_4$.[24] The major metallic products when boron or silicon is involved do not contain boron or silicon bonded to cobalt but are in fact CCo_3 cluster compounds,[25] and reports of silicon analogs have been refuted.[26] We believe that boron or silicon clusters may have a transient existence but that CO insertion or transfer is so facile that anionic intermediates (say $[OCCo_3(CO)_9]^-$) are produced which subsequently undergo electrophilic attack by the boron or silicon reagent, *e.g.*

$$NaCo(CO)_4 \longrightarrow NaCo_3(CO)_{10} + RSiX_3 \longrightarrow NaCl + RCCo_3(CO)_9 + [\text{siloxanes, etc.}] \quad (8)$$

The only proven analogs to the carbon clusters are the paramagnetic group VI derivatives in which the carbon has been replaced by S, SR, Se, and Te,[6,10] although there is strong indirect evidence for the germanium analog $PhGeCo_3(CO)_9$.[26]

Reactions of Methinyltricobalt Enneacarbonyls

The chemistry of these cluster compounds has two facets, (a) the coordination chemistry of the basal Co_3 triangle (summarized in Table I) and (b) reactions involving transformation of the apical group, Y. Both aspects are interrelated in the sense that it is the steric and electronic requirements of the Co_3C core which govern the overall reactivity.

Coordination Chemistry of the Basal Triangle. Several years ago Ercoli and coworkers[27] found that three CO groups (presumably axial) exchange with ^{14}CO faster than the other six and that the relative rates fall in the order Y = H < Br < Cl < F. When Y = $COOCH_3$, only one CO group appeared to exchange. It is difficult to reconcile these observations with the structures of the clusters, but the influence of the apical group is significant and is reminiscent of the trans effect in classical inorganic reactions. In fact, the thermodynamic stability of any derivative and the conformation adopted follow a definite pattern based on the type of apical group. As a general rule aryl derivatives are the most resistant to thermal and oxidative decomposition. Apical C–halogen bonds are very susceptible to nucleophilic attack and hetereolytic fission. For this reason many simple derivatives cannot be isolated; compounds with direct C–C linkages are produced instead. For instance, dehalogenation occurs on reaction with Ph_3As.[28]

$$2ClCCo_3(CO)_9 + Ph_3As \xrightarrow{373\ K} [CCo_3(CO)_9]_2 + Ph_3AsCl_2? \quad (9)$$

Even more intriguing are the acetylene com-

(20) C. D. M. Mann, A. J. Cleland, S. A. Fieldhouse, B. H. Freeland, and R. J. O'Brien, *J. Organometal. Chem.*, **24**, C61 (1970).
(21) F. Klanberg, W. B. Askew, and L. J. Guggenberger, *Inorg. Chem.*, **7**, 2265 (1968).
(22) S. A. Fieldhouse, A. J. Cleland, B. H. Freeland, C. D. M. Mann, and R. J. O'Brien, *J. Chem. Soc. A*, 2536 (1971).
(23) D. F. Patmore and W. A. G. Graham, *Inorg. Chem.*, **5**, 2222 (1966).
(24) A. S. Foust, M. S. Foster, and L. F. Dahl, *J. Amer. Chem. Soc.*, **91**, 5633 (1969).
(25) B. Nicholson and J. Simpson, University of Otago, personal communication.
(26) R. Ball, M. J. Bennett, E. H. Brooks, W. A. G. Graham, J. Hoyane, and S. M. Illingworth, *Chem. Commun.*, 592 (1970).
(27) G. Cetini, R. Ercoli, O. Gambino, and G. Vaglio, *Atti. Accad. Sci. Torino, Cl. Sci. Fis. Mat. Natur.*, **99**, 1123 (1964).
(28) B. H. Robinson and W. S. Tham, *J. Organometal. Chem.*, **16**, 45 (1969).

pounds, $[CCo_3(CO)_9]_2C_2$, $[CCo_3(CO)_9][Co_2(CO)_6C_2H]$, and $[CCo_3(CO)_9]_2C_2[Co_2(CO)_6C_2]$, formed in many reactions with organic molecules such as arenes[29-33] and also with nucleophiles. The clusters themselves appear, from deuteriation studies, to be the source of carbon, but the mechanism is unknown. Seyferth and coworkers[34] have shown that $[CCo_3(CO)_9]_2C_2[Co_2(CO)_6C_2]$ can also be prepared from hexachlorocyclopropane.

Lewis base derivatives illustrate how the coordination behavior and electronic structure of the cluster are interrelated. Generally solid-state structures correspond to simple substitution of L in an equatorial position for complexes $YCCo_3(CO)_8L$. However, when Y = CH_3, the cluster may reorganize to a CO-bridged configuration.[35] Both nonbridged and bridged isomers coexist in solution, irrespective of the group Y. A CO-bridged configuration is also adopted by the more highly substituted phosphine derivatives $YCCo_3(CO)_7L_2$ and $YCCo_3(CO)_6L_3$, probably because a bridging CO is more effective in dissipating the increased charge on the cluster from σ donation. Activation energies [E_A (Y): 105 (CH_3), 83 (C_6H_5), 63 (H), 51 (F)] derived in a kinetic study[36] of the reactions

$$YCCo_3(CO)_9 \underset{}{\overset{-CO}{\rightleftharpoons}} YCCo_3(CO)_8 \xrightarrow{L} YCCo_3(CO)_8L$$

which proceed by the usual dissociative SN1 mechanism are in accord with the concept of an electron-delocalized Co_3C core. The cobalt–CO bond is strongest with an electron-donating apical substituent and as expected the ν(CO) frequencies decrease from the fluoro to methyl derivative.[35]

Certain of the polyene and diene complexes have interesting spectral properties. The cyclooctatetraene derivatives, $YCCo_3(CO)_6COT$, where the polyene is coordinated to all three cobalt atoms,[36,37] are typical nonrigid (fluxional) molecules. An analysis of the pmr spectrum of $PhCCo_3(CO)_6COT$ strongly indicates that the predominant rearrangement pathway is a sequence of 1,2 shifts.[38] A different type of nonrigid behavior is exhibited by the complexes, $YCCo_3(CO)_7$norb.[39] From scale models it seems that the diene should be coordinated to one cobalt atom, but there are still two conformational possibilities: cis (adjacent to the apical group),cis or cis,trans. An analysis of the pmr spectra at low temperatures indicates that the instantaneous configurations are cis–trans (a 2:3:1:2 spectrum). As the temperature increases the individual resonances collapse and a time-averaged 4:2:2 spectrum is observed at ~223 K, while at room temperature the olefin and bridgehead protons become equivalent (6:2 spectrum).[40-45] A mechanism involving (cis,trans)-(cis,cis) isomerization is probably operating. At temperatures above ~60° the norbornadiene complexes also undergo a reverse Diels–Alder reaction giving cyclopentadienyl complexes[41,46] and at the same time function as catalysts in the formation of the norbornadiene dimers.[41]

A number of derivatives have a tremendous affinity for CO, due to the lability of the ligand, and thus are ideal substrates for catalytic reactions. In fact, where the cluster–ligand bond is weak (alkenes, cyclooctadiene, acetylenes) the organic products are usually those formed through CO incorporation or polymerization.[41] The possibility that the phosphine complexes may act as carbonylation catalysts is currently being explored. A patent has already been granted to Bor[47] for the application of the parent clusters as antiknock agents, and Bamford[48] found that $PhCH_2CCo_3(CO)_9$ is an adept initiator of vinyl polymerization.

Reactions Involving the Apical Carbon and Apical Substituent (Y). The chemistry of the apical linkage, C–Y, is influenced by electronic effects and the severe stereochemical constraints imposed on the clusters. Certain physical features of the clusters suggest that the apical linkage has properties inconsistent with a saturated (sp^3) carbon atom. In particular, a number of C–Y bond lengths are shorter than expected for a "single" bond[30,31,37,49,50] and C–halogen stretching frequencies are unusually high; for example ν(C–Cl) in $ClCCo_3(CO)_9$ appears at 906 cm^{-1}.[18] These observations may be explained using the concept of a delocalized electron-rich Co_3C core which is electron-withdrawing with respect to the apical group. The high electron density in C–Y bonds may be due to polarization of this core by an electronegative halogen or to π bonding,[16] although the availability of suitable orbitals for π bonding is open to question. On these arguments it is anticipated that the apical carbon atom should be activated toward nucleophilic attack and intermediate carbonium ions stabilized. Moreover, the C–Y bond is sterically protected from "backside" attack and a dissociative SN1 mechanism will be favored.

Reactions with Electrophiles. The Friedel–Crafts substitution of halo clusters is an important route to aryl clusters and illustrates the steric constraint on such reactions. These substitution reactions may be accomplished, often in high yields, using a nitrogen atmosphere,[51,52] *e.g.*

(29) B. H. Robinson, J. L. Spencer, and R. Hodges, *Chem. Commun.*, 1480 (1968).
(30) R. J. Dellaca, B. R. Penfold, B. H. Robinson, W. T. Robinson, and J. L. Spencer, *Inorg. Chem.*, **9**, 2197 (1970).
(31) R. J. Dellaca, B. R. Penfold, B. H. Robinson, W. T. Robinson, and J. L. Spencer, *Inorg. Chem.*, **9**, 2204 (1970).
(32) B. H. Robinson and J. L. Spencer, *J. Organometal. Chem.*, **30**, 267 (1971).
(33) M. D. Brice, B. R. Penfold, W. T. Robinson, and S. R. Taylor, *Inorg. Chem.*, **9**, 362 (1970).
(34) D. Seyferth, R. J. Spohn, M. R. Churchill, K. Gold, and F. Scholer, *J. Organometal. Chem.*, **23**, 237 (1970).
(35) T. W. Matheson, B. H. Robinson, and W. S. Tham, *J. Chem. Soc. A*, 1457 (1971).
(36) J. L. Spencer, Thesis, University of Otago, 1971.
(37) M. D. Brice, R. J. Dellaca, B. R. Penfold, and J. L. Spencer, *Chem. Commun.*, 72 (1971).
(38) B. H. Robinson and J. L. Spencer, *J. Organometal. Chem.*, **33**, 97 (1971).
(39) P. E. Elder and B. H. Robinson, *J. Organometal. Chem.*, **36**, C45 (1972).
(40) B. H. Robinson and J. L. Spencer, *J. Chem. Soc. A*, 2045 (1971).
(41) P. E. Elder and B. H. Robinson, unpublished results.
(42) T. W. Matheson and B. H. Robinson, unpublished results.
(43) W. R. Cullen, personal communication.
(44) F. W. B. Einstein and R. D. G. Jones, *Inorg. Chem.*, **11**, 395 (1972).
(45) F. W. B. Einstein and R. D. G. Jones, personal communication.
(46) I. U. Khand, G. R. Knox, P. I. Pauson, and W. E. Watts, *Chem. Commun.*, 36 (1971).
(47) G. Bor, L. Marko, and M. Freund, Hungarian Patent, 149692 (1961).
(48) C. H. Bamford, G. C. Eastmond, and W. K. Maltman, *Trans. Faraday Soc.*, **60**, 1432 (1964).
(49) R. J. Dellaca and B. R. Penfold, *Inorg. Chem.*, **10**, 1269 (1971).
(50) M. D. Brice and B. R. Penfold, *Inorg. Chem.*, **11**, 138 (1972).
(51) R. Dolby and B. H. Robinson, *Chem. Commun.*, 1058 (1970).

$$ClCCo_3(CO)_9 + C_6H_5(CH_3) \xrightarrow[N_2]{AlCl_3} (CH_3)C_6H_4CCo_3(CO)_9 \quad 60\% \qquad (10)$$

Substitution is almost wholly para because of interference from the equatorial carbonyls in the ortho product. With small substituents the isomer distribution is kinetically controlled; for example, in reaction 10 the isomer distribution is

Temp, °C	Solvent	Ortho, %	Para, %
40	CH_2Cl_2, PhMe	62	38
110	PhMe	36	64

Unstable green intermediates, $[HRCCo_3(CO)_9]^+AlCl_4^-$, have been isolated from some reactions. Other more versatile routes to aryl clusters, employing, for example, Grignard and mercury reagents,[53] are now available. The coordinated aromatic moieties in these aryl clusters also undergo typical aromatic Friedel–Crafts reactions.[54,55]

All clusters are protonated in strong acids, and the deep red solutions exhibit the characteristic high-field resonance ($\tau \sim 30$) of a coordinated hydrogen.[42] Only monoprotonation is achieved, an observation difficult to interpret on a structural basis although other clusters (*e.g.*, $Ru_3(CO)_{12}$[56]) exhibit similar behavior. A useful approach might be to describe this protonation in terms of acid–base interaction and as a logical extension to derive a basicity function for the Co_3CY unit.

Substituents with basic functions are protonated in weak acids also and a range of salts of the type shown in (11) have been studied.[55] Physicochemical

$$Me_2NC_6H_4CCo_3(CO)_9 + HX \text{ (or RX)} \rightleftharpoons [H(\text{or } R)Me_2NC_6H_4CCo_3(CO)_9]^+X^- \qquad (11)$$

measurements on these salts have provided Hammett functions for the tricobaltcarbon cluster which resemble those of a typical deactivating group, CO_2H. Analysis of pK_b (for the salts) and pK_a (for phenol derivatives prepared by route 12) values also suggest that the cluster is a strongly deactivating entity.[53]

$$MeOC_6H_4CCo_3(CO)_9 \xrightarrow[CH_2Cl_2]{BBr_3} HOC_6H_4CCo_3(CO)_9 \qquad (12)$$

Recently it has been demonstrated[57,58] that the acid, $HO_2CCCo_3(CO)_9$, and esters dissolve in concentrated H_2SO_4 to generate the acylium cation, $^+OCCCo_3(CO)_9$, a precursor of considerable importance. This reaction is a direct consequence of the sterically hindered environment of the apical group. The following examples show the proton is only one of several electrophiles which give the acylium cation in high yield. Since reactions with halo clusters involve CO insertion, these reactions must be carried out under CO for maximum yield.

$$(CO)_9Co_3C\text{-}Cl \xrightarrow[CH_2Cl_2]{AlCl_3\text{-}CO} 60\% \qquad (13)$$

$$(CO)_9Co_3C\text{-}Cl \xrightarrow[Et_2O]{BF_3\text{-}CO} 50\% \qquad {}^+OCCo_3(CO)_9 \qquad (14)$$

$$(CO)_9Co_3C\text{-}CO_2Et \xrightarrow[CH_2Cl_2]{BCl_3\text{-}N_2} 60\% \qquad (15)$$

The above reactions may be quenched with an appropriate nucleophile (see ref 57) and probably (13) is the most convenient route (quench with H_2O) to the acid from $Co_2(CO)_8$. For steric reasons it is difficult to esterify the carboxylic acid cluster by conventional procedures, but esterification may be accomplished using oxonium salts.[55]

$$(CO)_9Co_3C\text{-}CO_2H \xrightarrow{R_3O^+BF_4^-} (CO)_9Co_3C\text{-}CO_2R \quad 80\% \qquad (16)$$

Reactions with Nucleophiles. Nucleophiles such as Grignards, RMgX, attack the apical carbon atom of halo clusters, yielding the expected substituted cluster provided R is an aryl group, the reaction is carried out under nitrogen, and the cluster/nucleophile ratio is greater than 1:9[55] (eq 17). Yields from the

$$ClCCo_3(CO)_9 + RMgX \rightarrow RCCo_3(CO)_9 + MgXCl \qquad (17)$$

Grignard reactions are high (80%) and it is a stereospecific synthesis (*cf.* (10)). The reaction conditions suggest that there is initial attack on coordinated CO followed by attack by R^- on the apical carbon atom. Moreover, when R is a primary or secondary alkyl a coordinated CO is attacked by 2 mol of RMgX (eq 18) which on work-up leads to complete decomposition of the cluster (not substitution as originally thought[51]).

$$ClCCo_3(CO)_8CO + 2RMgX \longrightarrow ClCCo_3(CO)_8\text{—}\overset{(+)}{\underset{RMgX}{C}}\text{—}\overset{(-)}{\underset{R}{O}}MgX \qquad (18)$$

In contrast nucleophilic substitution *under CO* invariably gives compounds resulting from CO insertion.[42,55,59] Clearly, the initial step is again nucleophilic attack on coordinated CO, but the nature of the intermediates (possibly $ClOCCo_3(CO)_9$) is un-

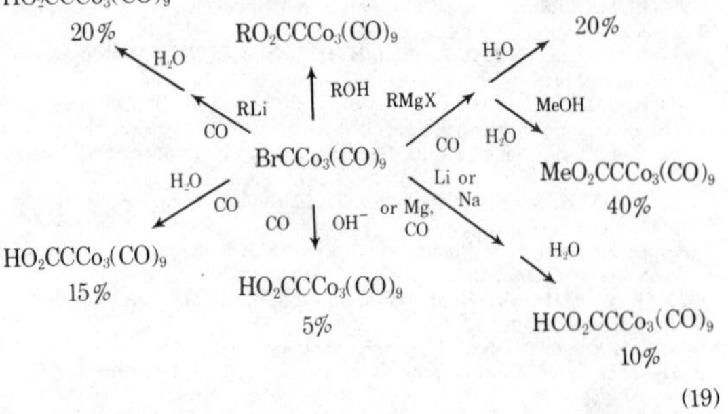

(19)

(52) R. Dolby and B. H. Robinson, *J. Chem. Soc., Dalton Trans.*, in press.

(53) D. Seyferth, J. E. Hallgren, and R. J. Spohn, *J. Organometal. Chem.*, **23**, C55 (1970).

(54) D. Seyferth and A. T. Wehman, *J. Amer. Chem. Soc.*, **92**, 5520 (1970).

(55) R. Dolby and B. H. Robinson, unpublished work.

(56) A. J. Deeming, B. F. G. Johnson, and J. Lewis, *J. Chem. Soc. A*, 2967 (1970).

(57) J. E. Hallgren, C. S. Eschbach, and D. Seyferth, *J. Amer. Chem. Soc.*, **94**, 2547 (1972).

(58) D. Seyferth, J. E. Hallgren, R. J. Spohn, A. T. Wehman, and E. H. Williams, Special Lectures, XXIIIrd Congress, IUPAC, Boston, Mass., 1971.

(59) R. Ercoli, E. Santambrigio, and E. Tattamanti-Cassagrande, *Chim. Ind. (Milan)*, **44**, 1344 (1962).

known. The crucial step is the intramolecular CO insertion which we postulate goes *via* a cyclic transition state. In fact, the close proximity of an equatorial CO to the apical group in these clusters is ideal for the formation of such a transition state. It seems likely that the majority of CO insertions in these compounds proceeds by this mechanism rather than the alternative intermolecular CO insertion into a strong C–Y bond, and this would account for the large number of insertion products (see ref 16). Significant yields of the derivatives $HCCo_3(CO)_9$,$[CCo_3(CO)_9]_2C_2$, $[CCo_3(CO)_9]_2CO$, and $[CCo_3(CO)_9][Co_2(CO)_6C_2H]$ are always obtained in the above reactions. It has been proved[42] by deuteriation experiments that the hydrogen in $HCCo_3(CO)_9$ is derived from H_2O, but the mechanism for its formation is still uncertain.

Conclusion. The unique electronic and steric properties of methinyltricobalt enneacarbonyls give rise to novel reactivity patterns, but there are large areas yet to be explored. Apart from an isolated report that the double bond in $HO_2CC{=}CCCo_3(CO)_9$ can be hydrogenated under mild conditions, little is known about the reactivity of coordinated apical aryl and alkyl groups;[16] profitable routes to a variety of organic substrates should result from work in this area (see, for example, the work reviewed in ref 58).

The coordination chemistry of the basal triangle is modified by the apical group, and it will be interesting to see if there is a synergic relationship. Finally, the concept of a delocalized electron-withdrawing Co_3C core adequately explains the known chemical and spectral data.

Molecular Structures

Detailed crystal structures have been determined for 13 compounds containing the Co_3C cluster. In none of these structures is there any pronounced intermolecular association, so that we may reasonably regard the solid-state molecular dimensions as applying also to the free molecules. The structures so far determined may be classified into five groups, each of which we shall discuss in turn relative to the basic $YCCo_3(CO)_9$ structure.

1. Substitution of Y only. Besides the methyl compound already referred to,[12] the only structure reported is that of $(C_2H_5)_3NBH_2OCCo_3(CO)_9$, where Y is triethylamine–oxyborane, arising from the reaction of triethylamine–borane with $Co_2(CO)_8$ in benzene.[21] This is the only structure reported in which an oxygen atom is directly bonded to the cluster carbon although compounds of general formula $R_3Si{-}OCCo_3(CO)_9$ presumably contain this feature. The C–O bond length of 1.28 (2) Å approximates to that of a double bond if the carbon atom is assumed to be sp^2 hybridized, but is also consistent with an interpretation in terms of a single bond involving an approximately sp-hybridized carbon atom.

2. Substitution of CO by σ Donor. The only structure reported[33] is that of $CH_3CCo_3(CO)_8P(C_6H_5)_3$ in which a single *equatorial* carbonyl has been replaced by triphenylphosphine following the reaction of $CH_3CCo_3(CO)_9$ with $P(C_6H_5)_3$ in hexane. There are no other major consequential structural changes, although infrared spectra show that in solution there

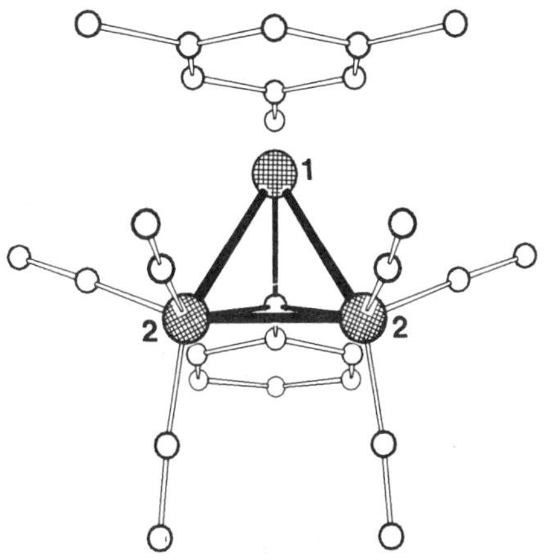

Figure 2. Molecular structure of the π-mesitylene complex $C_6H_5CCo_3(CO)_6{\cdot}(CH_3)_3C_6H_3$. The mean value of the distance from Co(1) to the mesitylene ring carbon atoms is 2.15 (3) Å.[60]

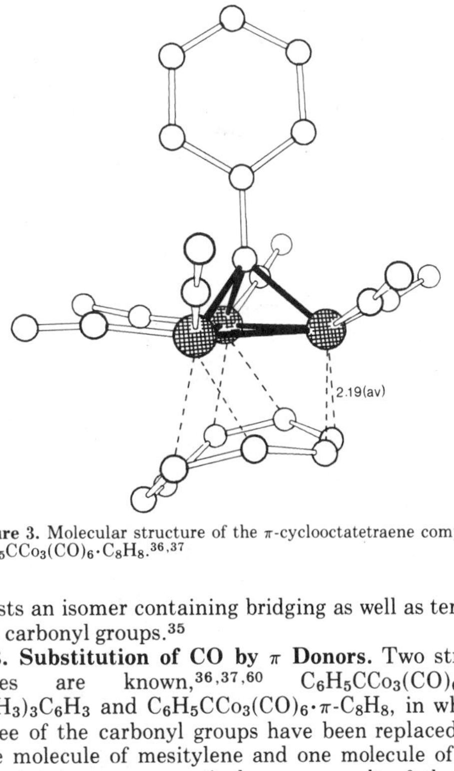

Figure 3. Molecular structure of the π-cyclooctatetraene complex $C_6H_5CCo_3(CO)_6{\cdot}C_8H_8$.[36,37]

exists an isomer containing bridging as well as terminal carbonyl groups.[35]

3. Substitution of CO by π Donors. Two structures are known,[36,37,60] $C_6H_5CCo_3(CO)_6{\cdot}\pi{-}(CH_3)_3C_6H_3$ and $C_6H_5CCo_3(CO)_6{\cdot}\pi{-}C_8H_8$, in which three of the carbonyl groups have been replaced by one molecule of mesitylene and one molecule of cyclooctatetraene, respectively, as a result of the direct reaction of $C_6H_5CCo_3(CO)_9$ with the π-donor compounds. In the mesitylene complex (Figure 2) all three carbonyls (two equatorial and one axial) associated with a single cobalt have been replaced and the unique Co atom is approximately equidistant (mean value 2.15 (3) Å) from all atoms of the mesitylene ring.[60] The Co–Co bonds involving the unique Co atom are 0.036 (4) Å shorter than the remaining Co–Co bond.

In the cyclooctatetraene complex (Figure 3) the three axial carbonyls have been replaced and the

(60) R. J. Dellaca and B. R. Penfold, *Inorg. Chem.*, 11, 1855 (1972).

Table II
Known Molecular Structures Containing the CCo_3 Cluster, Grouped as in the Text

1	2	3	4	5
$CH_3CCo_3(CO)_9$[12]	$CH_3CCo_3(CO)_8$-$P(C_6H_5)_3$[33]	$C_6H_5CCo_3(CO)_6$·π-$(CH_3)_3C_6H_3$[60]	$[CCo_3(CO)_9]_2$[50]	$CH_3CCo_3(CO)_8P$-$(C_6H_{11})_3$[63]
$(C_2H_5)_3NBH_2OCCo_3(CO)_9$[21]		$C_6H_5CCo_3(CO)_6$·π-C_8H_8[36,37]	$[CCo_3(CO)_9]_2CO$[61]	$CH_3CCo_3(CO)_7$-$[(CH_3)_2As]_2C_2(CF_2)_2$[44]
$R_3SiOCCo_3(CO)_9$[a,20]			$[CCCo_3(CO)_9]_2$[31]	$[OCCo_3(CO)_9]^-$ [a,19]
			$[Co_2(CO)_6]$-$HC_2CCo_3(CO)_9$[30]	$CH_3CCo_3(CO)_4$·$(\pi$-$C_5H_5)_2$[64]
			$[Co_2(CO)_6]$-$[C_2CCo_3(CO)_9]_2$[34,49]	

[a] Structure assumed from indirect evidence but not yet confirmed.

$CCo_3(CO)_6$ group retains idealized C_{3v} symmetry. The manner in which the cyclooctatetraene molecule (in the tub conformation) is attached to the Co_3 triangle is most interesting. Three of the four double bonds in the polyene ring are associated respectively with the three Co atoms, while the fourth double bond is bent away from the Co_3 triangle.[36,37] The mean value for the close Co–C(ring) contacts is 2.19 (2) Å. Notable also in this compound are the short Co–CO bonds which are in this case all equatorial. Their mean length is 1.69 (3) Å or 0.09 Å less than the mean value for all compounds in which no substitution of axial carbonyls has occurred. The indication is that axial–equatorial carbonyl repulsions may be a factor controlling the length of the Co–CO bonds, but further confirmation is required. The reason why the mesitylene molecule does not replace three axial carbonyls may be rationalized in a similar way—a model suggests that the nonbonded repulsions between equatorial carbonyls and a planar aromatic ring symmetrically placed below the Co_3 triangle would be prohibitive.[60]

4. Linkage of Two Clusters. There are five distinct examples of the linkage of one Co_3C cluster through its carbon atom to another cluster either directly or through an intermediate carbon chain.[30,31,34,49,50,61] These structures are compared in Figure 4 which shows the Co_3C clusters in corresponding orientations and which also indicates the bond lengths in the carbon chains. All compounds were formed in the reaction of $YCCo_3(CO)_9$ (Y = Cl, Br) with toluene, *m*-xylene, or mesitylene.[31] Compound A is a special case in that the Y group is not strictly a cluster but rather a dicobalt hexacarbonyl fragment coordinated by an acetylene.[30] However, it is properly to be included in this group of compounds because of its close structural relation to compound C.

In all three compounds, A, B, and C, an acetylene bond has been linked to the carbon atom of one Co_3C cluster. In A and C this bond is coordinated to $Co_2(CO)_6$. In compounds B and C the acetylene is linked to a second Co_3C cluster, in the case of C through a second acetylene bond. In compound D the two Co_3C clusters are linked through a carbonyl bridge, and in E the link is direct. In compound E, $[CCo_3(CO)_9]_2$, the Co–C bonds within the clusters are significantly longer [1.96 (1) Å] than the mean value (1.92) found for all of the other compounds. This lengthening may be associated with the minimization of nonbonded repulsions between equatorial carbonyls from the two halves of the molecule.[50] These equatorial carbonyls are already significantly bent back toward the Co_3 planes (mean dihedral angle between Co_3 and CoC(eq)C(eq) is 24°, compared with a mean value of 30° in compounds where there are no such intermolecular interactions), and a lengthening of the Co–C cluster bonds will still further lengthen the nonbonded contacts. As it is, there are O···O contacts as short as 3.08 Å.

This series of compounds provides quantitative data concerning the bonds formed by the carbon atom of the cluster. Compound E, the dimer $[CCo_3(CO)_9]_2$, is particularly important. Although

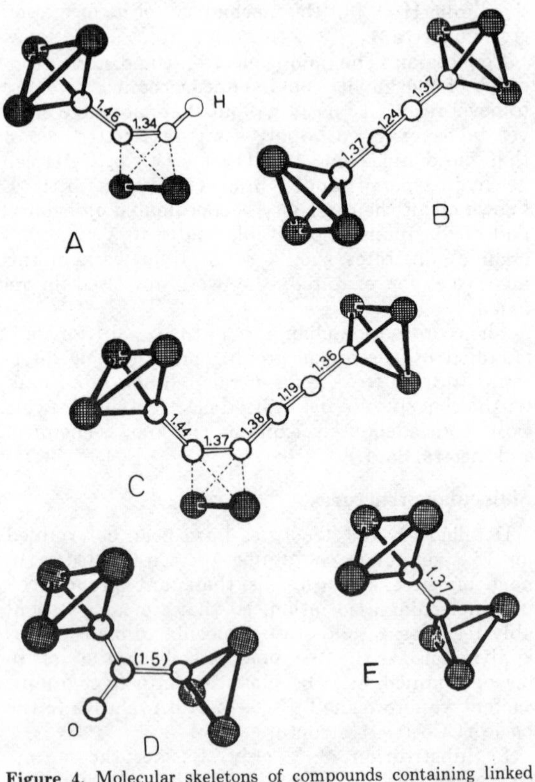

Figure 4. Molecular skeletons of compounds containing linked clusters. Carbonyl groups (except for the bridging carbonyl in compound D) have been omitted for clarity. A is $[Co_2(CO)_6]$-$HC_2CCo_3(CO)_9$; B is $[CCCo_3(CO)_9]_2$; C is $[Co_2(CO)_6][C_2CCo_3(CO)_9]_2$; D is $[CCo_3(CO)_9]_2CO$; E is $[CCo_3(CO)_9]_2$. (See ref 30, 31, 34, and 49, 61, 50 respectively.)

(61) G. Allegra and S. Valle, *Acta Crystallogr., Sect. B*, **25**, 107 (1969).

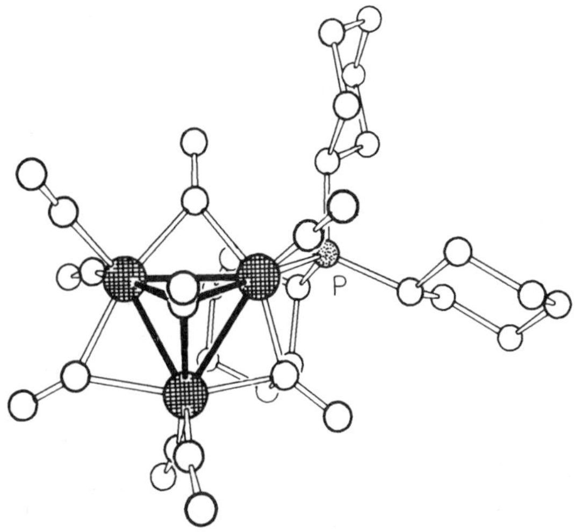

Figure 5. Molecular structure of $CH_3CCo_3(CO)_8{\cdot}P(C_6H_{11})_3$.[63]

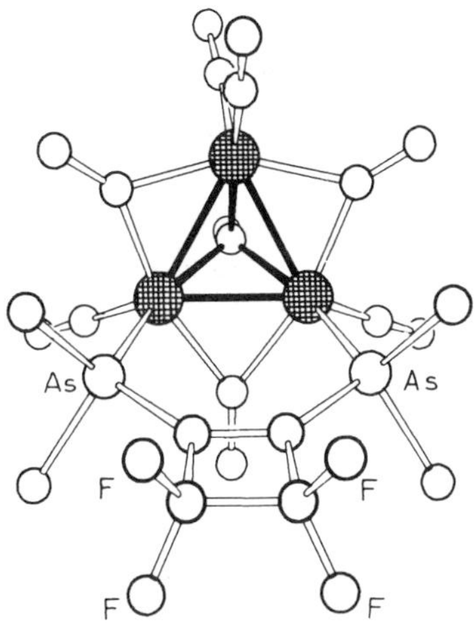

Figure 6. Molecular structure of $CH_3CCo_3(CO)_7.[(CH_3)_2As]_2$-$C_2(CF_2)_2$.[44]

there appears to be no way in which the central C–C bond may possess π character (see, however, Kettle and Khan[62]), the length of 1.37 (1) Å for this bond is very short and if it is a single bond there will be very much reduced p character in the orbitals of the carbon atoms which point to each other. Given this information, there is no need to attribute π contributions to any of the C–C bonds formed by the cluster carbon atoms in this series of compounds.[50] The observed lengths in the carbon chains of compounds A, B, C, and E are all consistent with a description in terms of formally single bonds from the cluster carbon atom, triple bonds for the uncoordinated acetylene and double bonds for the coordinated acetylene. The C–C bond lengths in compound D are unfortunately not of sufficiently high precision to be considered in detail in this context.

5. Rearrangement of Carbonyls to Form Metal Bridges. The compound $CH_3CCo_3(CO)_8P(C_6H_{11})_3$ (Figure 5) results from the replacement of one axial carbonyl group by a molecule of tricyclohexylphosphine[63] while the compound $CH_3CCo_3(CO)_7$-$[(CH_3)_2As]_2C_2(CF_2)_2$ (Figure 6) results from the replacement of two axial carbonyl groups by one molecule of 1,2-bis(dimethylarsino)tetrafluorocyclobutene.[44] In both cases there has been a rearrangement of the equatorial carbonyl groups so that three of them are no longer terminal to the cluster but form bridges across each of the sides of the Co_3 triangle. The CCo_3 cluster is, however, only slightly distorted as a result. The mean value of the Co–CO (bridge) bond lengths is 1.94 Å.

In the compound $CH_3CCo_3(CO)_4(\pi\text{-}C_5H_5)_2$ (Figure 7) a rather different rearrangement has occurred.[64] The two cyclopentadienyl groups have replaced five of the carbonyl groups associated with two cobalt atoms and the single bridging carbonyl is axial rather than equatorial. The mean value of the Co–CO (bridge) bond lengths is 1.84 (2) Å.

(62) S. F. A. Kettle and J. Khan, *J. Organometal. Chem.*, 5, 588 (1961).
(63) T. W. Matheson and B. R. Penfold, unpublished results.
(64) R. S. McCallum and B. R. Penfold, unpublished results.

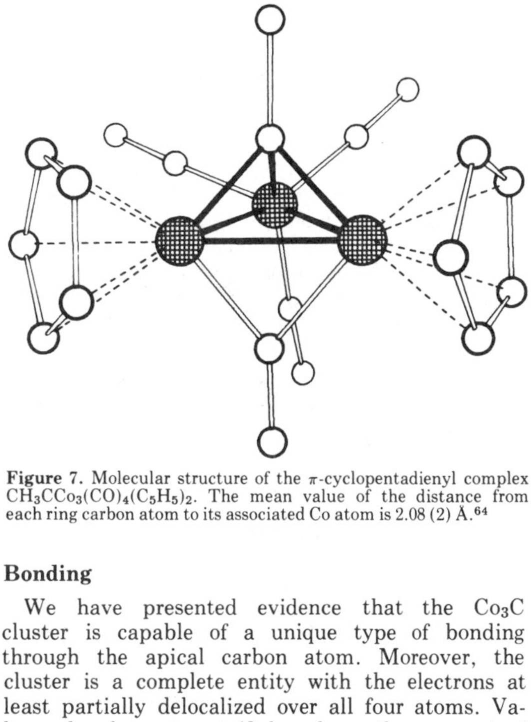

Figure 7. Molecular structure of the π-cyclopentadienyl complex $CH_3CCo_3(CO)_4(C_5H_5)_2$. The mean value of the distance from each ring carbon atom to its associated Co atom is 2.08 (2) Å.[64]

Bonding

We have presented evidence that the Co_3C cluster is capable of a unique type of bonding through the apical carbon atom. Moreover, the cluster is a complete entity with the electrons at least partially delocalized over all four atoms. Valence bond treatments[12] based on the concept of "bent bonds" and on sp^3-hybridized carbon atom are therefore unsatisfactory. Descriptions in terms of the carbon as sp^2 or sp offer no advantages. The theoretical problems are essentially the same as encountered in tackling simpler metal clusters; all orbitals must be considered. From topological arguments Kettle[62] concluded that all available orbitals in the Co_3C cluster were just filled by the 30 valence electrons but only if three orbitals of the carbon are used in bonding. A better approach would be to consider a Co_3CR rather than Co_3C unit as in this way a MO treatment would permit the calculation of accessible experimental parameters such as bond lengths, magnetic anisotropy, etc.

Addendum (May 1976)

Further evidence for the electronic structure of the cluster proposed in this Account and in earlier papers is provided by recent work. The exceptional stability of the carbonium ions $R_2\overset{+}{C}$-$CCo_3(CO)_9$ (c.f. the acylium cation $\overset{+}{O}C$-$CCo_3(CO)_9$), prepared by Et_3SiH reduction of the acyl derivatives, $RCOCco_3(CO)_9$, is attributed to electron donation to the positively charged center via σ–π conjugation.[65] Steric protection of the electrophilic center by the equatorial carbonyl groups will also hinder attack by nucleophiles; thus, the carbonium ions are weak electrophiles. Small shifts in the ^{13}C nmr resonance of the carbinyl carbon on going from an alcohol to the carbonium ion, and the deshielded resonance of the apical carbon atom (between 310 and 255 ppm) are consistent with this hypothesis.[66] The paramagnetic radical anions, $YCCo_3(CO)_9^-$, also show exceptional stability. For example, ether solutions of $PhCCo_3(CO)_9^-$ are stable under nitrogen at ambient temperatures for many hours.[67] Spectral and electrochemical studies on these species confirm that the LUMO in the neutral clusters is largely metal in character whereas the HOMO has contributions from the cobalt atoms, apical carbon atom, and possibly, apical substituent. There is also strong evidence for a σ–π interaction with an apical fluorine substituent.[67]

Full details for the preparation of a wide variety of cluster derivatives have appeared,[68–70] and the organic chemistry of the apical substituent has been extensively developed by Seyferth's group.[66,71] The non-rigid behavior of $YCCo_3(CO)_7$ norb compounds is now believed to result from a flipping motion of the ligand about the cobalt atom rather than (cis,trans)–(cis,cis) isomerization.[69] Halo-clusters act as catalysts for the formation of Binor-S from norbornadiene.[69,72] An interesting observation is that $HCCo_3(CO)_9$ adds to C=C and C≡C bonds;[69,75] for example, norbornadiene undergoes a retro Diels-Alder reaction.[72] Radical-initiated addition has been demonstrated for allyl compounds,[74] but the mechanism in most cases seems to involve prior coordination of the substrate to the cobalt atoms.[69,73]

Two further crystal structures of complexes resulting from the substitution of carbonyls by π donors have been determined. The cycloheptatriene derivative $C_6H_5CCo_3(CO)_6.\pi$-C_7H_8 (ref. 75) closely resembles the corresponding cyclooctatetraene derivative[37] in that the three axial carbonyls of the parent cluster have been replaced, and there is a one-to-one association between the three Co atoms and the three double bonds of the triene. In the norbornadiene derivative C_2H_5-$CCo_3(CO)_7$.norb (norb = C_7H_8) (ref. 76) the two equatorial carbonyls of one Co atom have been replaced by the diene, and the two double bonds of the ligand occupy positions close to the original locations of the replaced carbonyls.

(65) D. Seyferth, G. H. Williams, and J. E. Hallgren, *J. Am. Chem. Soc.*, **95**, 266 (1973).

(66) D. Seyferth, G. H. Williams, and D. D. Traficante, *J. Organomet. Chem.*, **97**, C11 (1975), and references therein.

(67) B. M. Peake, B. H. Robinson, J. Simpson, and D. J. Watson, *J. Chem. Soc. Chem. Commun.*, 945 (1974) and unpublished work.

(68) R. Dolby and B. H. Robinson, *J. Chem. Soc. Dalton Trans.*, 1794 (1974) and references therein.

(69) P. E. Elder, B. H. Robinson, and J. Simpson, *J. Chem. Soc. Dalton Trans.*, 1771 (1975) and references therein.

(70) D. Seyferth, J. E. Hallgren, and P. L. K. Hung, *J. Organomet. Chem.*, **50**, 265 (1973).

(71) D. Seyferth, G. H. Williams, A. T. Wehman, and M. O. Nestle, *J. Am. Chem. Soc.*, **97**, 2107 (1975) and references therein.

(72) T. Kitumura and T. Joh, *J. Organomet. Chem.*, **65**, 235 (1974).

(73) Unpublished work, University of Otago.

(74) D. Seyferth and J. E. Hallgren, *J. Organomet. Chem.*, **49**, C41 (1973).

(75) R. G. Holloway and B. R. Penfold, unpublished results.

(76) Ng Yew Sun and B. R. Penfold, unpublished results.

Stereochemical Studies of Metal Carbonyl–Phosphorus Trifluoride Complexes

Ronald J. Clark* and Marianna Anderson Busch

Department of Chemistry, Florida State University, Tallahassee, Florida

Received August 14, 1972

Reprinted from ACCOUNTS OF CHEMICAL RESEARCH, **6**, *246* (*1973*)

Transition metal carbonyl chemistry originated with the discovery of $Ni(CO)_4$ in 1890. Since that time many other carbonyl complexes have been prepared, as well as a large number of compounds containing a variety of Lewis bases substituted for one or more of the carbonyl groups. As a general rule, the bonding properties of these other bases are quite different from those of CO, and the physical properties of the species formed by substitution of these bases are quite different from those of the parent carbonyl. Phosphorus trifluoride and its complexes are a striking exception.

The similarity of PF_3 and CO as ligands was first noted by Chatt.[1] In 1950, he suggested that phosphorus trifluoride should be capable of forming a complex with nickel, analogous to $Ni(CO)_4$. One year later Wilkinson[2] carried out the reactions

$$Ni(CO)_4 + 4PCl_3 \longrightarrow Ni(PCl_3)_4 + 4CO$$

$$Ni(PCl_3)_4 + 4PF_3 \longrightarrow Ni(PF_3)_4 + 4PCl_3$$

in which the final product, $Ni(PF_3)_4$, had properties quite similar to the carbonyl.

Chatt and Williams[1] at about the same time prepared a material of the approximate composition $Ni(CO)_2(PF_3)_2$. The properties of this material were so similar to those of the tetracarbonyl that it was not then possible to prove that the composition represented an actual intermediate substitution compound, as opposed to a mixture of $Ni(CO)_4$ and $Ni(PF_3)_4$. More recent work, however, has shown that, in addition to the tetracarbonyl and tetraphosphine, all possible intermediates do exist.[3]

The similarity between the properties of metal carbonyls and their phosphorus trifluoride analogs has now been demonstrated for a large number of cases. This Account describes some of these studies, placing major emphasis on work with mixed carbonyl-phosphine complexes carried out in the authors' laboratory. The totally substituted species have been extensively reviewed by Kruck[4] and more recently by Nixon.[5]

Work with these complexes has yielded considerable information on such subjects as bonding, stereochemistry, and ligand site exchange. We believe that PF_3 can serve as an excellent model for CO, and, as such, produces compounds readily studied by fluorine nmr. Much of the information which this work has yielded would have been difficult, if not impossible, to obtain using ^{13}C nmr.

Preparation

Most Lewis base substitution products of metal carbonyls are prepared by direct ligand replacement of carbon monoxide in the parent carbonyl. As applied to phosphorus trifluoride, the process is called trifluorophosphinedecarbonylation. With most lig-

$$M(CO)_n + xPF_3 \longrightarrow M(CO)_{n-x}(PF_3)_x + xCO$$

ands the activation energy required to progress through successive stages of substitution increases markedly at each stage.[6] Thus, through judicious control of time and temperature in a thermal substitution it is usually possible to obtain a product essentially composed of a single compound. A simple purification procedure such as recrystallization is frequently sufficient to yield the pure complex. However, this state of affairs does not exist in metal carbonyl–trifluorophosphine chemistry.

Although definitive data are lacking, the activation energies for each stage of PF_3 substitution appear to be nearly the same. As a result, the substituted compounds react with PF_3 almost as readily as the parent carbonyl, and complex mixtures are produced which are difficult to separate. With most monometallic carbonyls, $M(CO)_n$, these mixtures usually contain all possible compositions of the type $M(PF_3)_x(CO)_{n-x}$. If more than one isomer can result from each degree of PF_3 substitution, all possible isomers are usually present as well. By controlling the PF_3:CO ratio and the time of reaction, only limited control of the system composition is possible.

Reactions of PF_3 with metal carbonyls containing such readily replaceable ligands as amines or certain cyclic polyenes sometimes produce a limited number of products. Examples include the reaction of PF_3 with norbornadienemolybdenum tetracarbonyl[7] to yield *cis*-$Mo(CO)_4(PF_3)_2$, with aminemolybdenum

Ronald J. Clark received his Ph.D. in 1958 with Dr. Jacob Kleinberg at the University of Kansas. He spent three and a half years as a research chemist with the Linde Division of Union Carbide. After a short period with Dr. John D. Corbett at The Iowa State University, he joined the staff of Florida State University where he is now Professor of Inorganic Chemistry. His interests are in the field of coordination chemistry with emphasis on metal carbonyl-phosphorus trifluoride chemistry.

Marianna A. Busch recieved her Ph.D. in chemistry at Florida State University in May 1972, while working on dieneiron carbonyl–trifluorophosphine complexes. She is now doing postdoctoral work at Cornell University with Dr. Earl L. Muetterties.

(1) J. Chatt, *Nature (London),* **165**, 637 (1950); J. Chatt and A. A. Williams, *J. Chem. Soc.*, 3061 (1951).
(2) G. Wilkinson, *J. Amer. Chem. Soc.*, **73**, 5501 (1951).
(3) R. J. Clark and E. O. Brimm, *Inorg. Chem.*, **4**, 651 (1965).
(4) T. Kruck, *Angew. Chem., Int. Ed. Eng.*, **6**, 53 (1967).
(5) J. F. Nixon, *Advan. Inorg. Chem. Radiochem.*, **13**, 363 (1970).
(6) F. Basolo and R. G. Pearson, "Mechanisms of Inorganic Reactions," 2nd ed, Wiley, New York, N. Y., 1967.
(7) C. G. Barlow, J. F. Nixon, and M. Webster, *J. Chem. Soc. A*, 2216 (1968); R. J. Clark, unpublished observations.

pentacarbonyl[8] to yield $Mo(CO)_5PF_3$, and with cycloheptatrienemolybdenum tricarbonyl to yield $Mo(CO)_3(PF_3)_3$. In the last case, the *mer*, or a mixture of the *mer* and *fac*, isomers can be obtained, depending on the technique.[7] (In the *mer* isomer three PF_3 ligands are planar, and in the *fac* isomer they are mutually perpendicular.) The factors controlling this isomer distribution have not been fully determined.

The speed of reaction between PF_3 and a metal carbonyl is variable, but, in general, the rate decreases as the coordination number of the metal increases. For $Ni(CO)_4$, $HCo(CO)_4$, and $Co(NO)(CO)_3$ reaction occurs at room temperature or below. For complexes of higher coordination number, either high-pressure–high-temperature or photochemical conditions are required.

Only limited work has been done on the substitution of PF_3 into polymetallic carbonyls.[8] However, there appears to be a marked reluctance for simple replacement to occur, especially with complexes of the first-row transition metals.

Iron pentacarbonyl readily undergoes photochemical reaction to yield $Fe_2(CO)_9$, but once PF_3 groups have been introduced, the $Fe(PF_3)_x(CO)_{5-x}$ species do not yield dimers, even after extensive irradiation.[9] The hydride $HCo(CO)_4$ loses H_2 at −20° to yield $Co_2(CO)_8$, but $HCo(PF_3)_4$ is stable to 200°. The $HCo(PF_3)_x(CO)_{4-x}$ species have intermediate stabilities, but we have not been able to demonstrate[10] simple loss of H_2. Dimanganese decacarbonyl can have four CO groups replaced sequentially, but any attempt at further substitution produces $Mn_2(PF_2)_2(PF_3)_x(CO)_{8-x}$ compounds with PF_2 bridges.[11] The $Ru_3(CO)_{12}$ trimer can have up to six CO groups replaced sequentially before breaking down to yield the $Ru(PF_3)_x(CO)_{5-x}$ species.[12]

Separation

The physical similarity of the $M(PF_3)_x(CO)_{n-x}$ species makes their separation difficult. The solubilities and volatilities are too much alike for such techniques as recrystallization, sublimation, or pot-to-pot distillation to be completely successful. However, Bigorgne[13] has claimed an enrichment of the $Fe(CO)_x(PF_3)_{5-x}$ compounds by careful pot-to-pot distillation.

Distillation of the $Mo(PF_3)_x(CO)_{6-x}$ species on a 24-in. spinning band column at reduced pressures has also yielded partial enrichment of the more highly substituted species, but only the monophosphine could be obtained pure. In general, however, fractional distillation is an unattractive separation technique. The compounds are quite toxic, some four-coordinate systems disproportionate readily, and, in general, the volatilities are simply too similar.

The separation technique that has proven most generally useful has been gas–liquid chromatography on a small preparative scale. Columns packed with various silicon oils, phthalates, or squalene on firebrick are used. For the $Ni(PF_3)_x(CO)_{4-x}$ and $Co(NO)(PF_3)_x(CO)_{3-x}$ species, disproportionation can still be troublesome. High column temperatures can also result in compound decomposition.

Elution chromatography from silica or alumina columns has not been particularly successful. The long retention times often result in extensive reaction with the basic alumina or with adsorbed water. However, a possible tool for the future, especially for systems with low volatility, is high-pressure liquid chromatography.

Characterization

The carbonyl–trifluorophosphine species can be characterized by a variety of techniques which have been found to be generally applicable from one group of complexes to another. *Sequence of formation* will generally provide the first clue in any new system. If replacement of CO by PF_3 is sequential, then the order in which the new species appear should be a function of the extent of PF_3 substitution. When prudently applied, an assignment based on this technique has never failed to be confirmed by other methods of characterization.

Sequence of elution is a consistent pattern in the glpc separations of these systems. The retention times of the species decrease as the degree of substitution increases. Thus, the parent carbonyl is always the last to be eluted, the monophosphine somewhat earlier, the diphosphine earlier yet, and so forth.

The matching of an experimental *infrared spectrum* with that predicted by group theory usually adds further confirmation to the identification of compounds containing one or more carbonyl groups.[14] Frequently this method has also been useful in detecting the presence of isomers, and in determining possible structures for these species.

Fluorine nmr yields readily recognizable patterns. The typical pattern for a monophosphine consists of a doublet with $^1J_{PF}$ equal to about 1300 Hz. The patterns become more complex as the number of PF_3 groups increases, but they can be readily identified.[5]

Mass spectroscopy provides final confirmation. The parent ion peak is usually of strong intensity, and the high mass region is characterized by a pattern resulting from the successive loss of CO and PF_3 from the parent ion. Since mass spectroscopy became available, we have ceased doing elemental analysis as the formula obtained from an accurate analysis always agreed with that determined by other methods of characterization.

Properties

The individual carbonyl–trifluorophosphine complexes are volatile and yellow, light-yellow, or colorless solids or liquids at room temperature. The thermal stability of these species seems to increase steadily as the degree of PF_3 substitution increases. Most are stable in nonpolar solvents, but undergo

(8) A. K. Wensky, Ph.D. Dissertation, Florida State University, 1970; D. J. Darensbourg and T. L. Brown, *Inorg. Chem.*, **7**, 1679 (1968).
(9) R. J. Clark, *Inorg. Chem.*, **3**, 1395 (1964).
(10) C. A. Udovich and R. J. Clark, *Inorg. Chem.*, **8**, 938 (1969).
(11) P. E. Brotman and R. J. Clark, unpublished observations.
(12) C. A. Udovich and R. J. Clark, *J. Organometal. Chem.*, **36**, 355 (1972).
(13) J. B. Tripathi and M. Bigorgne, *J. Organometal. Chem.*, **9**, 307 (1967).
(14) L. M. Haines and M. H. B. Stiddard, *Advan. Inorg. Chem. Radiochem.*, **12**, 53 (1969).

solvolysis rather rapidly in hydroxylic solvents like alcohols or wet THF. However, they are sufficiently hydrophobic that the reaction with pure water is usually slow. Most compounds are not particularly sensitive to dry air and, once purified, can be stored for long periods of time in a deep freeze without noticeable decomposition, especially in sealed tubes from which water and air have been removed.

These compounds must be assumed to be quite toxic and should be treated with the same respect due the metal carbonyls themselves.

Bonding

Phosphorus trifluoride readily forms more extensive series of substitution products with more carbonyls than any other ligand. This has been attributed to a high degree of similarity in the bonding of PF_3 and CO groups to transition metals. Carbon monoxide is a weak σ donor, but has empty antibonding orbitals of the appropriate energy and symmetry to act as a π acceptor of electrons from filled metal d orbitals. The presence of the highly electronegative fluorine atoms on PF_3 also makes phosphorus trifluoride a weak Lewis base. However, these same fluorines also lower the energy of the empty phosphorus 3d orbitals and enhance the ability of PF_3 to accept electrons from filled metal orbitals. The stability of a PF_3 complex can thus be attributed to strong $d\pi$–$d\pi$ bonding between the ligand and the metal, in contrast to $p\pi^*$–$d\pi$ bonding between CO and the metal.

A considerable amount of physical evidence has accumulated which indicates that PF_3 has essentially the same π-bonding tendencies as carbon monoxide. Mass spectroscopic studies on certain mixed carbonyl–trifluorophosphine complexes have shown that, within experimental error, the metal–carbonyl and metal–trifluorophosphine bond energies are equal.[15] Infrared and Raman studies also show that the CO stretching force constants change less on replacing CO by PF_3 than by any other ligand, except possibly NO which is not a conventional two-electron donor. There is little ^{19}F and ^{31}P chemical shift between the complexes as the number of PF_3 groups changes. It would also seem that only ligands of comparable bonding ability could be able to displace one another so readily. Additional evidence of this type has been reviewed by Nixon.[5]

Aside from bonding properties, the factor of size or cone angle[16] is probably also important in determining the ability of PF_3 to coordinate so readily. Cone angle is a measure of how much crowding or steric interaction a ligand is likely to cause. A study of a series of complexes containing a variety of ligands suggests that, with the probable exception of CO itself (plus PH_3 and a caged phosphine), phosphorus trifluoride has the smallest cone angle. Thus, PF_3 would be expected to replace CO more readily than most other ligands since the presence of this group causes relatively little steric crowding around the metal.

(15) F. E. Saalfield, M. V. McDowell, S. K. Gondon, and A. G. MacDiarmid, *J. Amer. Chem. Soc.*, **90**, 3684 (1968).
(16) C. A. Tolman, *J. Amer. Chem. Soc.*, **92**, 2956 (1970).

In most metal carbonyls, PF_3 displays little site preference, and when isomers are possible, the relative isomer populations can often be approximately predicted from a simple statistical distribution of CO and PF_3 groups over all possible coordination sites. The slightly larger ligand $P(OCH_3)_3$ reacts to yield only the more crowded cis form of $Mo(CO)_4L_2$. However, in $Mo(CO)_4(PF_3)_2$ the cis:trans ratio is about 2.5:1 (compared to a 4:1 statistical distribution), suggesting that PF_3 has a slight trans-directing ability. These isomer abundances cannot be predicted from purely steric arguments and should still be attributed to the electronic similarity of CO and PF_3.

A third factor which can also explain the large number of PF_3 derivatives of carbonyl complexes has to do with the physical properties imparted to the new compounds by PF_3 substitution. Generally, the introduction of some ligand other than CO results in a severe loss of volatility and solubility as compared to the parent. Frequently the materials are so intractable that only major products can be isolated and identified. In contrast, the presence of fluorine in the outer environment of PF_3 complexes yields materials of high volatility which can be subjected to gas–liquid chromatography, a technique ideally suited for the isolation of minor components and the separation of complex mixtures. It is interesting to speculate on the discovery of minor components in the carbonyl complexes of other Lewis bases, were these complexes also amenable to glpc separation.

Stereochemical Rearrangements

The preceding discussion of the similarities between the properties, bonding, and ligand sizes in the complexes of PF_3 and CO suggests why we consider phosphorus trifluoride to be such a good model for carbon monoxide in metal carbonyls. It therefore seems reasonable to assume that the presence or absence of ligand rearrangement in the carbonyl–trifluorophosphine complexes probably reflects analogous behavior in the parent carbonyl. We have carried out extensive work during recent years studying ligand rearrangement phenomena in many of these complexes. The rates of most of these processes fall conveniently within the "time scale" of high-resolution nuclear magnetic resonance. The importance of having a good model for CO can be appreciated by considering the difficulty of doing ^{13}CO nmr, especially when spectra must be examined at frequent intervals over large temperature ranges. However, nmr is ideally suited for use with carbonyl–trifluorophosphine systems because of the presence of the more easily studied fluorine nuclei.

Among the simple $M(PF_3)_x(CO)_{n-x}$ complexes, numerous cases of ligand rearrangement have been found. In most cases the processes appear to be intramolecular. However, some evidence for a process involving an intermolecular exchange has also been observed in several systems.

In general, six-coordinate systems appear to be completely rigid, while the five-coordinate systems frequently are not. Due to site equivalency in the tetrahedral four-coordinate complexes, it is quite difficult to find a technique to detect ligand rearrangements in these systems.

Six Coordination. Most six-coordinate systems like $Mo(PF_3)_x(CO)_{6-x}$ are stereochemically rigid. We have found no case in which the isomerization of a compound like *cis*-$Mo(CO)_4(PF_3)_2$ occurs without accompanying disproportionation,[17] implying a general scrambling scheme as shown below. Complete

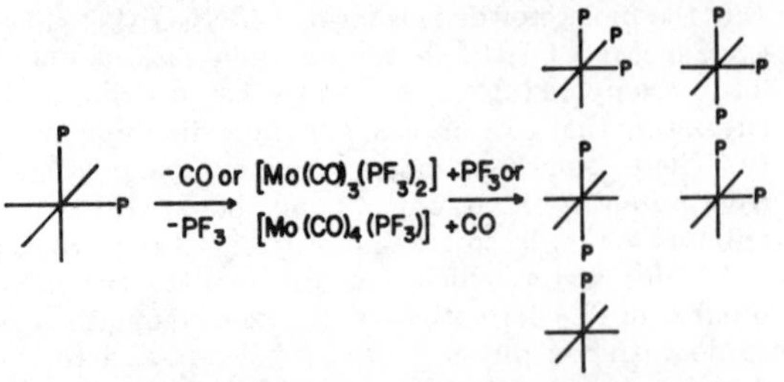

rigidity is also observed for such six-coordinate systems as $Cr(PF_3)_x(CO)_{6-x}$,[8] $W(PF_3)_x(CO)_{6-x}$,[8] $CF_3Mn(PF_3)_x(CO)_{5-x}$,[18] and $CF_2HCF_2Mn(PF_3)_x(CO)_{5-x}$.[18] Rigidity in these latter systems implies only that the isomers are stable enough to be isolated by glpc at elevated temperatures and exist for lengthy periods of time without any indication of isomerization or disproportionation. What might occur under more forcing conditions has not been determined.

The $HMn(PF_3)_x(CO)_{5-x}$ system is an exception.[18] Irradiation of the parent carbonyl and PF_3 with uv light yields a mixture of complexes representing all possible compositions, $x = 1$ to 5. These can be readily isolated by glpc, but the individual isomers for each degree of PF_3 substitution cannot be separated from one another. The fluorine spectrum of $HMn(CO)_4(PF_3)$ shown in Figure 1 indicates that both the cis and trans isomers are present, with relative populations of about 7:1.[19] It can be inferred that the interconversion of the isomers is faster than glpc time (several minutes), but slower than nmr time (10^{-3} to 10^{-5} sec).

In the $HRe(PF_3)_x(CO)_{5-x}$ system the individual isomers can be isolated if the column temperature is carefully controlled. The rate of the isomerization was found to be about the same as an intermolecular hydrogen–deuterium exchange reaction.

$$HRe(PF_3)(CO)_4 + DMn(CO)_5 \rightleftarrows DRe(PF_3)(CO)_4 + HMn(CO)_5$$

A similar, but faster exchange also occurs in the Mn system, suggesting that the isomerization of the CO and PF_3 ligands could be a result of this process.

Five Coordination. Within the last few years an ever-increasing number of molecules have been discovered which exhibit "nonrigid" or "fluxional" behavior. These molecules have more than one thermally accessible structure and pass rapidly from one of these structures to another by some kind of intramolecular rearrangement.[20]

Our interest in this area began with the $Fe(CO)_5$ problem. Both Cotton[21] and the late Sir Ronald Nyholm[21] with their coworkers attempted to determine

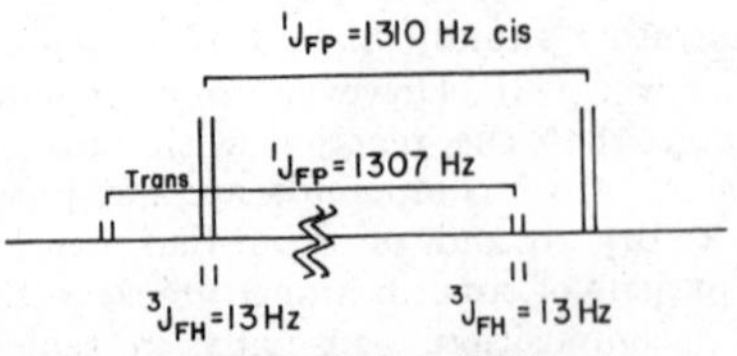

Figure 1. ^{19}F spectrum of $HMn(CO)_4(PF_3)$.

the solution structure of $Fe(CO)_5$ by ^{13}C (^{13}CO) nmr. A 3:2 or 4:1 ^{13}CO environment should distinguish between the trigonal-bipyramid or square-based pyramid, respectively. Each research group reported only a single ^{13}C frequency, unchanged to −60° (more recently to −110°).[22] Equivalence of CO groups was attributed to a rapid intramolecular process, postulated as the Berry pseudorotation.[23]

Due to the electronic similarity of PF_3 and CO, our group felt that the substitution of phosphorus trifluoride into $Fe(CO)_5$ should yield species similar to the parent with PF_3 distributed over all possible coordination sites. It also seemed likely that the much heavier PF_3 might slow the inversion process sufficiently to observe the fluorine nmr spectrum of the limiting structures. Although the first hope proved to be well founded,[9] the second was not.

All compositions of the general formula $Fe(PF_3)_x(CO)_{5-x}$ can be readily isolated by glpc.[9,24] However, efforts to isolate the individual isomers have failed. The infrared spectra of these stable compounds give clear evidence that, based on a trigonal-bipyramid structure, each composition contains significant amounts of every possible isomer. Complete assignments of the CO stretching frequencies have been made.[25] However, the nmr spectra of the different compositions show all PF_3 groups to be equivalent down to temperatures of −120° with no broadening of the spectral lines. Since intermolecular CO and PF_3 substitution is quite slow, except at much higher temperatures or under uv irradiation, the isomers of each composition are probably undergoing a rapid intramolecular rearrangement at speeds much faster than the nmr "time scale," but slower than the vibrational frequencies of the atoms within the molecule.

Some elegant studies by Udovich[24] gave additional evidence for intramolecular exchange in these compounds. Methanol causes the solvolysis of PF_3 ligands into $PF_x(OCH_3)_{3-x}$, with the latter preferring axial sites over equatorial sites. An isomerization occurs *via* an intramolecular process during the solvolysis, quantitatively converting equatorial $Fe(CO)_4PF_3$ into axial $Fe(CO)_4(PF_2OCH_3)$.[25] The $Mn(NO)(PF_3)_x(CO)_{4-x}$,[26] $Ru(PF_3)_x(CO)_{5-x}$,[12] and $[Mn(PF_3)_x(CO)_{5-x}]^-$ [27] systems show similar behavior.

(17) R. J. Clark and P. I. Hoberman, *Inorg. Chem.*, **4**, 1771 (1965).
(18) W. J. Miles, Jr., and R. J. Clark, *Inorg. Chem.*, **7**, 1801 (1968).
(19) W. J. Miles, Ph.D. Dissertation, Florida State University, 1970.
(20) E. L. Muetterties, *Accounts Chem. Res.*, **3**, 266 (1970).
(21) F. A. Cotton, A. Danti, J. S. Waugh, and R. W. Fessenden, *J. Chem. Phys.*, **29**, 1427 (1958); R. Bramley, B. M. Figgis, and R. S. Nyholm, *Trans. Faraday Soc.*, **58**, 1893 (1962).
(22) O. A. Gansow, A. R. Burke, and W. D. Vernon, *J. Amer. Chem. Soc.*, **94**, 2552 (1972).
(23) S. Berry, *J. Chem. Phys.*, **32**, 933 (1960).
(24) C. A. Udovich, R. J. Clark, and H. Haas, *Inorg. Chem.*, **8**, 1066 (1969).
(25) H. Haas and R. K. Sheline, *J. Chem. Phys.*, **47**, 2996 (1967).
(26) C. A. Udovich and R. J. Clark, *J. Organometal. Chem.*, **25**, 199 (1970).

A second type of nonrigid system was observed in the $R_fCo(PF_3)_x(CO)_{4-x}$ compounds[27,28] where R_f represents CF_3, C_2F_5, or C_3F_7, and $x = 1$ to 4. The species are expected to be trigonal bipyramids with the perfluoroalkyl ligand occupying an axial site. In the simplest system,[28] $CF_3Co(CO)_3(PF_3)$, both the averaged and limiting spectra could be studied.

At slightly above room temperature, fluorine nmr shows only one type of PF_3 group and one type of CF_3 group. On cooling, the lines broaden and then resharpen into a more complex pattern indicative of two different species, each with spectra qualitatively similar to the room temperature pattern. A comparison of the averaged coupling constants at low temperatures with those observed at high temperatures suggests that the room temperature pattern is an average of about equal concentrations of the two isomers.

The room temperature spectra of the remaining trifluoromethyl[27] species, as well as those of all the ethyl and propyl compounds,[27] show that all PF_3 groups are equivalent for each substitution value x. In the methyl compounds, cooling to low temperatures causes broadening of the spectrum followed by the appearance of a pattern indicative of the individual isomers. Assuming that the CF_3 group occupies the axial position in a trigonal-bipyramidal structure, the time-averaged value of the CF_3–P coupling constants suggested that roughly statistical amounts of each isomer are present.

For the ethyl and propyl compounds, cooling to temperatures as low as $-100°$ revealed no sign of spectral broadening. Since accidental degeneracy of all the resonances is hardly likely, this could indicate that either the molecular structure of each isomer is such that all PF_3 groups are naturally equivalent or else the rearrangement process is too fast to observe by nmr.

The first suggestion seems unlikely since the infrared spectra in the carbonyl region as well as the fluorine nmr spectra (particularly the $^3J_{CF_2\text{-}P}$ coupling constants) suggest that the methyl, ethyl, and propyl species have similar structures at room temperature. The second suggestion requires the exchange process to be faster in the ethyl and propyl compounds than in the methyl complexes. However, since the ethyl and propyl groups are bulkier than the methyl, this possibility also seems unlikely, especially if the process occurs by the Berry mechanism which must involve the axial groups. A different rearrangement mechanism, however, might explain this phenomenon.

Other systems which show stereochemical nonrigidity are the dieneiron carbonyl–trifluorophosphine complexes, of which butadieneiron tricarbonyl is the prototype. These complexes have proven to be especially interesting not only because the phosphine groups undergo rearrangement but also because the investigation has revealed some significant differences between the CO and PF_3 ligands.

For butadieneiron tricarbonyl in the solid state, X-ray diffraction studies[29] have shown that the coordination of the diene and carbonyl groups around the iron is in an approximately square-based pyramidal arrangement. The butadiene skeleton is cisoid and planar. Two carbonyl carbons and the two terminal carbons of the butadiene form the base and the third carbonyl group is at the apex.

For butadieneiron tricarbonyl in solution, three well-resolved ^{12}CO stretching vibrations are observed. Assuming a model requiring two of the three carbonyls to be equivalent, the close agreement of the observed ^{13}CO frequencies with those predicted from calculations based on a Cotton–Kraihanzel force field[14] indicated that the structure of the complex in solution is essentially the same as that in the crystalline state.[30] Because of the similarity in size and bonding of the PF_3 and CO ligands, it was assumed that PF_3 substitution should not alter this structure.

In the butadienetrifluorophosphineiron dicarbonyl complex, the number and location of the ^{13}CO stretching vibrations indicated that the PF_3 group prefers the apical position in the square-based pyramidal structure. This preference is so great that only one of the two possible isomers could be detected by fluorine nmr.

In the butadienebis(trifluorophosphine)iron carbonyl complex, the infrared spectrum again indicated that only one of the two possible isomers is present, but could not determine the identity of the preferred species. The room-temperature fluorine nmr spectrum shows equivalent PF_3 groups, suggesting that both phosphines occupy basal positions. However, a temperature-dependent study of the spectrum revealed that the apparent PF_3 equivalence is actually due to a rapid intramolecular exchange of the phosphine groups. The limiting spectrum showed that the only isomer present has one PF_3 apical and one PF_3 basal.

Phosphine exchange was also evident in the butadienetris(trifluorophosphine)iron complex. At room temperature, all fluorines are equivalent, while at $-120°$ the limiting pattern shows two resonances with a 2:1 ratio of intensities.

There are several possible descriptions of the motion which could make the PF_3 groups nmr equivalent in the bis and tris phosphine complexes. One possibility is simple position exchange of the two PF_3 ligands. A more reasonable possibility is for all three ligands (three PF_3 groups in the tris phosphine and two PF_3 groups and one carbonyl in the bis phosphine) to undergo a sort of concerted rotation around the iron atom. The latter possibility is also suggested by a study[31] of several diene$Fe(PF_3)_x(CO)_{3-x}$ systems in which the two basal coordination sites are made nonequivalent by the presence of a group other than hydrogen on one of the diene carbons. Because the three sites are not mutually perpendicular, there must be some bending in addition to a rotation about the metal.

An examination of the metal orbitals appropriate for π bonding with each of the ligands in these com-

(27) C. A. Udovich, Ph.D. Dissertation, Florida State University, 1969.
(28) C. A. Udovich and R. J. Clark, *J. Amer. Chem. Soc.*, **91**, 526 (1969).
(29) O. S. Mills and G. Robinson, *Acta Crystallogr.*, **16**, 758 (1963).
(30) J. D. Warren and R. J. Clark, *Inorg. Chem*, **9**, 373 (1970).
(31) M. A. Busch, Dissertation, Florida State University, 1972.

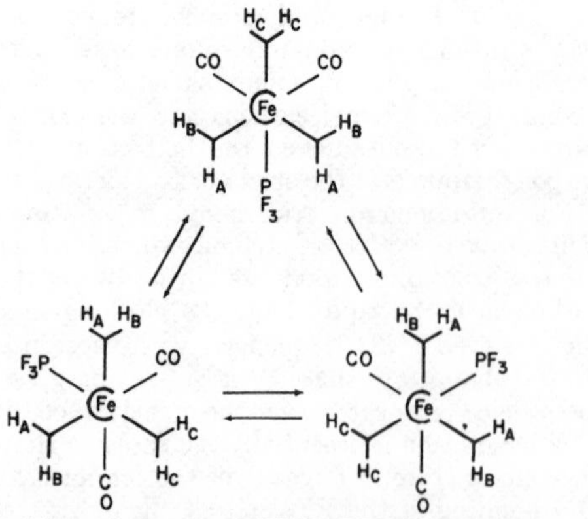

Figure 2. Schematic drawing of TMMFe(PF_3)($CO)_2$ rotamers.

plexes reveals a possible explanation for the site preference of PF_3 in the square-based pyramidal structure. The apical position appears to have a poorer potential for back-donation of electrons from the metal than the basal sites, resulting in a lower C–O bond order for the basal carbonyls than for the apical carbonyls. In agreement with this prediction, the carbonyl stretching force constant for the apical carbonyl in butadieneiron tricarbonyl was found to be at least 0.15 mdyn/Å larger than for the basal carbonyls. Therefore, site preference exhibited by PF_3 may be electronic in origin, with PF_3, a slightly weaker π-bonding group, being forced into the apical site by the better π-bonding CO ligand.

An argument based on bonding differences is supported by the study of a second group of structurally similar complexes, the (C_6H_8)Fe($PF_3)_x$($CO)_{3-x}$ (C_6H_8, cyclohexadiene) system.[32] In this case, the force constants for the two types of carbonyls in the parent compound are essentially equal, suggesting that the apical and basal sites are more nearly equivalent. All possible phosphine isomers are found to be present, although PF_3 substitution still favors the apical site. Temperature-dependent fluorine nmr shows that a rapid intramolecular process makes the PF_3 groups appear equivalent at room temperature.

When the two basal positions are made nonequivalent by an asymmetric diene, the PF_3 ligands exhibit additional site preferences.[31] The possibility that such preferences can be generally predicted from the relative values of the CO stretching force constants is being investigated.

A final type of ligand rearrangement was discovered in the PF_3-substituted complexes of trimethylenemethaneiron tricarbonyl, (TMM)Fe($CO)_3$. The TMM ligand is an unstable organic intermediate, thought to be a diradical, stabilized by the Fe($CO)_3$ group.[33] Phosphorus trifluoride substitutes into the complex photochemically to yield all possible species of the type (TMM)Fe($PF_3)_x$($CO)_{3-x}$.[34]

The high-resolution infrared spectrum as well as

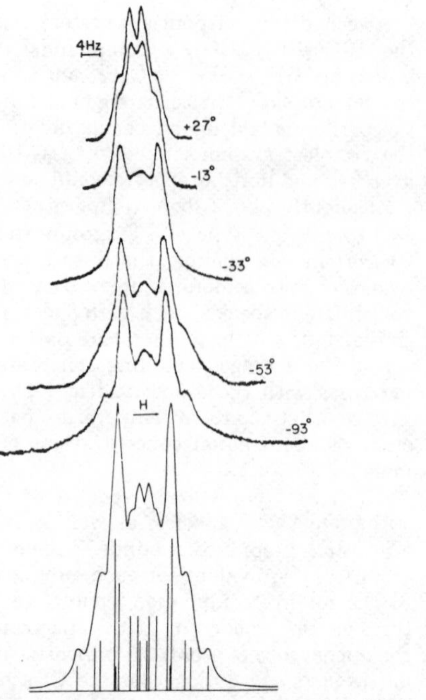

Figure 3. Variable-temperature ^{19}F spectra of TMMFe($PF_3)_3$. Bottom spectrum: computer-simulated low-temperature limit at slightly different scale.

proton nmr suggests that all three carbonyls in (TMM)Fe($CO)_3$ are equivalent. The fluorine spectra of the mono, bis, and tris phosphine complexes also show one, two, and three equivalent PF_3 groups, respectively. These patterns do not change with temperature.

The nonrigid character of the substituted (TMM)Fe($CO)_3$ system is revealed by temperature-dependent proton nmr. At 100°, the spectrum of the monophosphine shows six equivalent protons coupled to a single phosphorus atom. At low temperatures, the spectrum broadens, then sharpens to a more complex pattern due to three types of protons in a 1:1:1 ratio.[34]

The process most probably responsible for this behavior can be seen by viewing the molecule down the iron–methane carbon axis (Figure 2). It is apparent that the methylene protons have different positions relative to the PF_3 group, and hence are no longer chemically equivalent. If there is hindered motion of the PF_3 group relative to the TMM ligand, the resulting transformations through all three identical structures will allow each methylene proton to spend equal time in each of the three possible positions relative to PF_3. If the rate of interconversion is fast enough, all protons will appear equivalent on the nmr time scale. The complex coupling pattern at low temperatures can be interpreted[34] in terms of these structures.

The bis and tris phosphines show comparable evidence of hindered rotation, and at high temperatures the proton spectra show a triplet and quartet respectively indicating equivalent protons coupled to two and three phosphorus atoms. As the temperature is

(32) J. D. Warren, M. A. Busch, and R. J. Clark, *Inorg. Chem.*, **11**, 452 (1972).
(33) G. F. Emerson, K. Ehrlich, W. P. Giering, and P. C. Lauterbur, *J. Amer. Chem. Soc.*, **88**, 3172 (1966).
(34) R. J. Clark, M. R. Abraham, and M. A. Busch, *J. Organometal. Chem.*, **35**, C33 (1972).

lowered, the spectrum of the tris phosphine undergoes the changes shown in Figure 3. In this case, as the rate of interconversion is slowed, the methylene protons remain chemically equivalent, but become magnetically nonequivalent due to different spatial relationships with each of the three PF_3 groups. A computer calculation of the low-temperature spectrum was found to agree quite well with the pattern observed at −93°. The proton spectrum of the bis phosphine also shows broadening at lower temperatures, and the limiting pattern is even more complex.

The authors gratefully acknowledge past financial support from the Atomic Energy Commission and current support from the National Science Foundation (Grant-20276). R. J. C. also particularly thanks Dr. E. O. Brimm who interested him in this area of research, initially against R. J. C.'s will. R. J. C. also wishes to thank his students whose dissertations are cited in the references.

Addendum (May 1976)

Since the publication of this Account, there have been several studies of stereochemical non-rigidity using the PF_3 group as the probe into the molecule. As an example of recently published work, we cite the investigations of M. A. Bennett and co-workers.[35] They have investigated a series of rhodium compounds such as $Rh_2(PF_3)_6$(acetylene) and found them to be fluxional in nature.

H. Mahnke has studied the $Fe(PF_3)_x(CO)_{5-x}$ compounds further.[36] These, along with the $Fe(CO)_5$ parent, undergo intramolecular rearrangement so rapidly that they cannot be studied by NMR, even at quite low temperatures. However, by a combined variable temperature IR and NMR study (the latter yielding only time-averaged data), an indication of the rate of rearrangement in these molecules was obtained. This technique depended upon a shift in isomer population in the various $Fe(PF_3)_x(CO)_{5-x}$ compositions as the temperature varied. A frequency of exchange between axial and equatorial sites in $Fe(CO)_5$ at −20° of 1.1×10^{10} sec^{-1} was calculated. A surprisingly high ^{13}C chemical shift of 17.7 ppm between axial and equatorial sites was indicated by this analysis.

(35) M. A. Bennett, R. N. Johnson, and T. W. Turney, *Inorg. Chem.*, **15**, 90 (1976) and subsequent papers.
(36) H. Mahnke, R. J. Clark, R. Rosanske, and R. K. Sheline, *J. Chem. Phys.*, **60**, 2997 (1974).

Transition Metal Atom Inorganic Synthesis in Matrices

Geoffrey A. Ozin* and Anthony Vander Voet

Lash Miller Chemistry Laboratory and Erindale College, University of Toronto, Toronto, Ontario, Canada

Received September 12, 1972

Reprinted from ACCOUNTS OF CHEMICAL RESEARCH, *6, 313 (1973)*

The general techniques of matrix isolation have been well documented since the pioneering work of Whittle, Dows, and Pimentel.[1] Briefly, the matrix isolation technique may be described as a method for studying molecules or molecular fragments (radicals or ions) which, although possibly unstable at room temperature, are stabilized at cryogenic (4.2–20 K) temperatures in solid matrices. These matrices may be inert or reactive with respect to the molecule or fragment in question.

The species under study may be produced prior to condensation by high-temperature vaporization, by microwave or electrical discharge techniques, or *in situ* by photolysis studies. Conventional forms of optical spectroscopy as well as electron spin resonance and Mössbauer methods are then employed to study the trapped or "matrix-isolated" species. The reader is referred to recent reviews[2-9] and a book[10] for detailed descriptions of the matrix-isolation technique.

Although most of the elements of the periodic table can be produced in a monatomic form and have been at some time the subject of study by vapor equilibrium, uv-visible, and mass spectrometric techniques, only very recently has this wealth of physical data been utilized in a practical sense by the synthetic chemist.

Gaseous metal atoms characteristically have small activation energies for reactions with each other in interatomic collisions, the effects of which can be minimized by using high-vacuum conditions. The cocondensation of metal atoms with other compounds in an inert matrix, onto a surface cooled to cryogenic temperatures, can result in reactions of the type

$$M + X \rightarrow \text{ionic species}$$
$$M + X \rightarrow \text{molecular species}$$
$$M + X \rightarrow \text{free radical species}$$

Transition metal atoms tend to undergo reactions of the second type—*i.e.*, formation of binary species—and it is these with which this Account will be primarily concerned.

Cocondensation reactions of this type are currently being examined by matrix infrared and Raman spectroscopy.[11-14] Although matrix uv-visible,[15,16] electron spin resonance, and Mössbauer[17] spectroscopy are potentially useful, they are playing only a minor role in these studies. Whenever possible, the experiments are supplemented by variable concentration studies, isotopic substitutions, observations of spectral changes when diffusion is allowed to occur at higher temperatures, and matrix Raman depolarization measurements.[18]

With these methods, it is proving feasible to synthesize entirely new compounds which would be difficult, if not impossible, to prepare and stabilize by other methods, to control and monitor the courses of chemical reactions, and to study their reactive intermediates. The results of many of these studies involving transition metal atoms have important implications in the fields of chemisorption, homogeneous and heterogeneous catalysis, nitrogen fixation, and synthetic and naturally occurring oxygen carriers.

Binary Transition Metal Complexes, M(XY)

Using high-dilution experimental conditions, a number of carbonyl ($M(CO)$), dinitrogen ($M(N_2)$), and dioxygen ($M(O_2)$) complexes have recently been isolated and identified in rare gas matrices.

Geoffrey Ozin received his B.Sc. degree from Kings College, London, and in 1967 his D.Phil. degree from Oriel College, Oxford University, under Professor Ian Beattie. Following this, he spent 2 years as an ICI postdoctoral fellow in Professor Beattie's laboratories at Southampton University, where he worked in the fields of single-crystal and high-temperature gas-phase laser Raman spectroscopy and main group inorganic chemistry. In 1969 he joined the faculty of Erindale College, University of Toronto. He now holds the rank of Associate Professor and his present research involves the investigation of the products of matrix condensation reactions between transition metal atoms and a variety of small gaseous molecules at cryogenic temperatures, using matrix isolation infrared, laser Raman, and uv–visible spectroscopic methods.

Anthony Vander Voet received his B.Sc. degree and M.Sc. degree at the University of Alberta. Following this, he spent 1967–1969 as Professor of Inorganic Chemistry and Head of the Chemistry Department at the Universidad Javeriana, Bogota, Colombia. He then returned to Canada, where he is presently completing his Ph.D. degree at the University of Toronto, under the supervision of Dr. Ozin.

(1) E. Whittle, D. A. Dows, and G. C. Pimentel, *J. Chem. Phys.*, **22**, 1943 (1954).
(2) W. Weltner, Jr., *Advan. High Temp. Chem.*, **2**, 85 (1969).
(3) H. E. Hallam, *Annu. Rep. Progr. Chem.*, *A*, **67**, 117 (1970).
(4) A. J. Barnes and H. E. Hallam, *Quart. Rev., Chem. Soc.*, **23**, 392 (1969).
(5) J. W. Hastie, R. H. Hague and J. L. Margrave, *Spectrosc. Inorg. Chem.*, **1**, 58 (1970).
(6) D. E. Milligan and M. E. Jacox, *Advan. High Temp. Chem.*, **4**, 1 (1971).
(7) G. A. Ozin, *Spex Speaker*, **16** (4) (1971).
(8) J. S. Ogden and J. J. Turner, *Chem. Brit.*, **7**, 186 (1971).
(9) L. Andrews, *Annu. Rev. Phys. Chem.*, **22**, 109 (1972).
(10) B. Meyer, "Low Temperature Spectroscopy," Elsevier, New York, N. Y., 1971.
(11) E. P. Kündig, M. Moskovits, and G. A. Ozin, *J. Mol. Struct.*, **14**, 137 (1972).
(12) G. A. Ozin and A. Vander Voet, *J. Chem. Phys.*, **56**, 4768 (1972).
(13) See, for example: (a) L. Andrews, *J. Chem. Phys.*, **50**, 4288 (1969); (b) R. R. Smardzewski and L. Andrews, *ibid.*, **57**, 1327 (1972); (c) R. C. Spiker, L. Andrews, and C. Trindle, *J. Amer. Chem. Soc.*, **94**, 2401 (1972).
(14) See, for example: (a) J. S. Anderson, A. Bos, and J. S. Ogden, *J. Chem. Soc. D*, 1381 (1971); (b) A. J. Hinchcliffe, J. S. Ogden, and D. D. Oswald, *J. Chem. Soc., Chem. Commun.*, 338 (1972).
(15) R. L. DeKock, *Inorg. Chem.*, **10**, 1205 (1971).
(16) L. Andrews and G. C. Pimentel, *J. Chem. Phys.*, **47**, 2905 (1967).
(17) A. Bos, A. T. Howe, D. W. Dale, and L. W. Becker, *J. Chem. Soc., Chem. Commun.*, 730 (1972).
(18) (a) J. W. Nibler and D. A. Coe, *J. Chem. Phys.*, **55**, 5133 (1971); (b) H. Huber, G. A. Ozin, and A. Vander Voet, *Nature (London), Phys. Sci.*, **232**, 166 (1971).

Monodinitrogen Complexes, M(N₂). With the increasing number of reports of stable and matrix-isolated transition metal compounds containing coordinated dinitrogen molecules, there has also arisen considerable interest in the mode of bonding and nature of the interaction between the metal and the ligands. Although X-ray crystallography serves to determine unambiguously the type of bonding (end-on or sideways) in stable species, in such cases where diffraction studies are not possible (*i.e.*, matrix-isolation systems), vibrational spectroscopic investigations using mixed isotopically substituted ligands, $^{14}N^{15}N$, are able to distinguish between the bonding types (see later).

X-Ray crystallography has shown that the dinitrogen ligand in transition metal–dinitrogen complexes studied to date is always bonded in an "end-on" linear or near-linear M—N≡N skeleton.

Direct evidence for the existence of a compound containing the dinitrogen molecule bonded in a "sideways fashion" has not previously been reported in the literature. It was, however, postulated[19] as a transition state in the intramolecular $^{14}N^{15}N$ exchange of the cations $[Ru(NH_3)_5(^{14}N^{15}N)]^{2+}$–$[Ru(NH_3)_5(^{15}N^{14}N)]^{2+}$ as evidenced in the time-dependent intensity changes in the Ru–N₂ stretching region. It is noteworthy that Brintzinger's dinitrogen-bridged binuclear titanocene complex $(C_5H_5)_2$-$TiN_2Ti(C_5H_5)_2$ shows no absorption in the infrared attributable to the NN stretching mode. Of the two suggested dinitrogen-bridged centrosymmetric structures, *i.e.*

Ti—N≡N—Ti

end-on

$$\text{Ti} \diagdown \overset{N}{\underset{N}{\parallel\parallel}} \diagup \text{Ti}$$

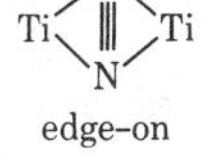

edge-on

the latter was favored on the basis of chemical intuition.[20]

Using statistical mixtures of nitrogen isotopes $^{14}N_2$–$^{14}N^{15}N$–$^{15}N_2$ = 1:2:1 diluted in Ar, the expected patterns (see Chart I) of NN absorptions are ideally: (a) "side-on" bonded dinitrogen—*three* equally spaced lines with absorption intensities in the ratio 1:2:1; (b) "end-on" bonded dinitrogen—*four* lines, with a closely spaced central *doublet*, with the absorption intensities of the four lines are 1:1:1:1. The reason for the four-line pattern with a central doublet for "end-on" bonded N₂ as opposed to a 1:2:1 triplet for the "side-on" bonded is related to the non-equivalence of the mixed isotopic species $M^{14}N≡^{15}N$ and $M^{15}N≡^{14}N$ and the absence of such isotopic effects in the case of

$$M—\overset{^{14}N}{\underset{^{15}N}{\parallel\parallel}}$$

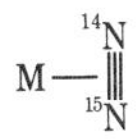

In the case of "end-on" bonded N₂, it is worth noting that the magnitude of the splitting of the central doublet (3.8 cm^{-1} in $Ni^{14}N≡^{15}N/Ni^{15}N≡^{14}N$,[21,22] 1.7 cm^{-1} in $Rh^{14}N≡^{15}N/Rh^{15}N≡^{14}N$,[23] and 4.0 cm^{-1} in $[Ru(NH_3)_5(^{14}N^{15}N)]^{2+}/[Ru(NH_3)_5(^{15}N^{14}N)]^{2+}$[24]) can be used to estimate the bond stretching force constant of the metal–nitrogen bond. This has been done for NiN_2 and has considerable relevance to chemisorption studies of gaseous N₂ on nickel (see, for example, the work of Eischens[25]).

The data collected so far (as shown in Table I)

Table I

Fe—N≡N, 2020 cm^{-1} [23c]	$Co—\overset{N}{\underset{N}{\parallel\parallel}}$, 2100 cm^{-1} [26]	Ni—N≡N, 2090 cm^{-1} [21,22]
	Rh—N≡N, 2155–2158 cm^{-1} [23c]	Pd—N≡N, 2211–2215 cm^{-1} [21]
		Pt—N≡N, 2166–2170 cm^{-1} [27]

yield the surprising result that both modes of bonding occur. These data for cobalt provide the only "direct" observation indicating that dinitrogen can be bonded to a transition metal in a "sideways" fashion. It is important to note that all of these monodinitrogen compounds were synthesized under virtually identical matrix conditions and that the observed structural variations from end-on to side-on bonded N₂ must reflect properties characteristic of the metal atoms.

At this early stage in the experiments it is not possible to offer a rationale for this remarkable set of data, although it is worth commenting on the extended Hückel MO calculations that have been performed on both end-on and side-on FeN_2 and that predict that the end-on configuration will be favored from an energy point of view.[28] The n_σ and n_π orbi-

Chart I

SIDE-ON BONDED DINITROGEN

END-ON BONDED DINITROGEN

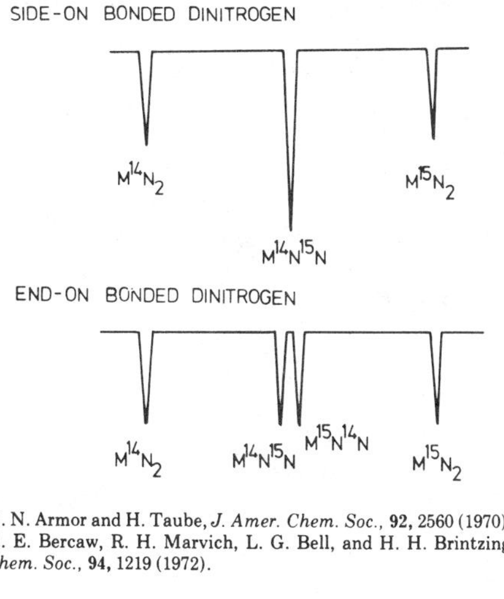

(19) J. N. Armor and H. Taube, *J. Amer. Chem. Soc.*, **92**, 2560 (1970).

(20) J. E. Bercaw, R. H. Marvich, L. G. Bell, and H. H. Brintzinger, *J. Amer. Chem. Soc.*, **94**, 1219 (1972).

(21) H. Huber, E. P. Kündig, M. Moskovits, and G. A. Ozin, *J. Amer. Chem. Soc.*, **95**, 332 (1973).

(22) M. Moskovits and G. A. Ozin, *J. Chem. Phys.*, **55**, 1251 (1973).

(23) Unpublished work done at the University of Toronto (1972): (a) H. Huber, E. P. Kündig, M. Moskovits, and G. A. Ozin; (b) E. P. Kündig and G. A. Ozin; (c) G. A. Ozin and A. Vander Voet.

(24) Yu. J. Borodko, A. K. Shilova, and A. E. Shilov, *Russ. J. Phys. Chem.*, **44**, 349 (1970).

(25) R. P. Eischens, *Accounts Chem. Res.*, **5**, 75 (1972).

(26) G. A. Ozin and A. Vander Voet, *Can. J. Chem.*, **51**, 637 (1973).

(27) G. A. Ozin, M. Moskovits and E. P. Kündig, *Can. J. Chem.*, in press.

(28) K. B. Yatsimirskii and Yu. A. Kraglyak, *Dok. Akad. Nauk SSSR*, **186**, 885 (1969).

tal populations for N–N and N–Fe bonds in FeN_2 suggest that the NN stretching frequency of side-on N_2 should lie at a considerably lower frequency than end-on N_2. In actual fact the NN stretching frequency of CoN_2 lies about 80 cm^{-1} *above* FeN_2 and 10 cm^{-1} above NiN_2.

Monodioxygen Complexes, $M(O_2)$. Using techniques similar to those described for $M(N_2)$, Ni, Pd, and Pt atoms have been cocondensed with dilute $^{16}O_2$, $^{16}O_2$–$^{18}O_2$, and $^{16}O_2$–$^{16}O^{18}O$–$^{18}O_2$ in Ar.[29] In all cases matrix infrared spectra were obtained characteristic of complexes containing a single molecule of dioxygen having oxygen atoms in equivalent environments; that is

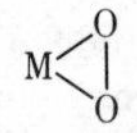

(*i.e.*, in each case using $^{16}O_2$–$^{16}O^{18}O$–$^{18}O_2$–Ar = 1:2:1:800, a 1:2:1 *triplet* of O–O stretching modes was observed). As found for $M(O_2)$ and $(O_2)M(O_2)$ (see later) and many synthetic dioxygen carriers, the "side-on" mode of bonding appears to be the more common form, which is not unexpected for transition metals in relatively low oxidation states.[30] The O–O stretching frequencies corresponding to the three isotopic molecules $M(^{16}O_2)$, $M(^{16}O^{18}O)$, and $M(^{18}O_2)$ are shown in Table II.

Table II

	Frequencies, cm^{-1}		
Molecule	Ni	Pd	Pt
$M(^{16}O_2)$	966.2	1024.0	926.6
$M(^{16}O^{18}O)$	940.1	995.5	901.4
$M(^{18}O_2)$	913.6	967.0	875.0

The bonding of O_2 to Ni, Pd, and Pt in these cocondensation reactions can be discussed on the basis of the π bonding scheme of Chatt and Dewar.[31,32] A delocalized three-center molecular orbital scheme involving the symmetry-allowed metal orbitals and an in-plane π^* of O_2 describes the situation more satisfactorily than a localized model in terms of O_2^{2-} or O_2^-. The degree of back-bonding which occurs will depend upon the relative energies and the amount of overlap between the orbitals.

We now try briefly to relate Ibers'[33] rationale for the factors affecting reversible oxygen uptake in synthetic dioxygen carriers to our data for binary dioxygen complexes. Ibers proposes that increased electron density at the metal, upon changing the metal in a complex, for example Ir > Rh > Co in the series in Table III, will assure increasing O_2 uptake properties. This would appear to be borne out by the above crystallographic data and is in fact also reflected in the "anomalous" order of the O–O stretching frequencies for $Ni(O_2)$, $Pd(O_2)$, and $Pt(O_2)$ which

Table III

	O–O, Å	M–O, Å	Chemical properties
$[Co(O_2)(2{=}phos)_2]BF_4$	1.42	1.902 1.871	Irreversible[34]
$[Rh(O_2)(diphos)_2]PF_6$	1.418	2.026 2.025	Reversible[35]
$[Ir(O_2)(diphos)_2]PF_6$	1.625	1.961 1.990	Irreversible[35]

parallel (in an inverse fashion) the order of O–O bond lengths in the Co, Rh, and Ir complexes given in Table III, that is

r(O–O), Å	Rh < Co < Ir
ν(O–O), cm^{-1}	Pd > Ni > Pt

1:2 Binary Transition Metal Complexes, $M(XY)_2$

The cocondensation of transition metal atoms with more concentrated matrices than those previously described often leads to higher stoichiometry complexes. In addition, diffusion-controlled warm-up experiments of MXY in the presence of excess XY may lead to stepwise reactions of the type

$$MXY + XY \rightarrow M(XY)_2$$
$$M(XY)_2 + XY \rightarrow M(XY)_3 \text{ etc.}$$

In the case of Ni, Pd, and Pt, dicarbonyl ($M(CO)_2$[15,36]), bisdinitrogen ($M(N_2)_2$[21]), and bisdioxygen ($M(O_2)_2$[29]) complexes have been synthesized by one or either of the above routes and have been characterized in some detail using isotopic substitution. For example, the bisdioxygen complexes of Ni, Pd, and Pt are formed when the metal atoms are cocondensed with either pure O_2 at 4.2–10 K or with dilute O_2–Ar matrices which have been annealed to 30 K.[29] All three molecules contain sideways-bonded O_2 in which the two dioxygen molecules as well as the oxygen atoms are equivalent. For example, in $^{16}O_2$–Ar matrices the single line at 1115.5 cm^{-2} assigned to $(^{16}O_2)Pd(^{16}O_2)$ splits into a "six"-line spectrum in $^{16}O_2$–$^{16}O^{18}O$–$^{18}O_2$–Ar matrices (Figure 1) which corresponds to all the possible combinations of the ^{16}O–^{18}O isotopes (Table IV) with an intensity distribution tending to favor a D_{2d} tetrahedral "spiro"

Table IV

Molecule	ν_{obsd},[27] cm^{-1}	ν_{calcd},[27] cm^{-1}	Assignment
$(^{16}O_2)Pd(^{16}O_2)$	1115.5	1110.9	ν_{OO}
$(^{16}O_2)Pd(^{16}O^{18}O)$	1092.2	1094.2	ν_{OO}
$(^{16}O^{18}O)Pd(^{16}O^{18}O)$	1080.5	1079.9	ν_{OO}
$(^{16}O_2)Pd(^{18}O_2)$	1067.2	1074.4	ν_{OO}
$(^{16}O^{18}O)Pd(^{18}O_2)$	1060.5	1062.5	ν_{OO}
$(^{18}O_2)Pd(^{18}O_2)$	1048.5	1047.7	ν_{OO}
$(^{16}O_2)Pd(^{16}O_2)$	504.0	504.8	ν_{PdO}

(29) H. Huber and G. A. Ozin, *Can. J. Chem.*, **50**, 3746 (1972).
(30) R. Mason, *Nature (London)*, 217, 543 (1968).
(31) J. Chatt, *J. Chem. Soc.*, 2939 (1953).
(32) M. J. S. Dewar, *Bull. Soc. Chim. Fr.*, **18**, C71 (1951).
(33) J. A. McGinnity, R. J. Doedens, and J. A. Ibers, *Inorg. Chem.*, **6**, 2247 (1967).
(34) N. W. Terry, E. L. Amma, and L. Vaska, *J. Amer. Chem. Soc.*, **94**, 653 (1972).
(35) J. A. McGinnity, N. C. Payne, and J. A. Ibers, *J. Amer. Chem. Soc.*, **91**, 6301 (1969).
(36) E. P. Kündig, M. Moskovits, and G. A. Ozin, *Can. J. Chem.*, **50**, 3587 (1972).

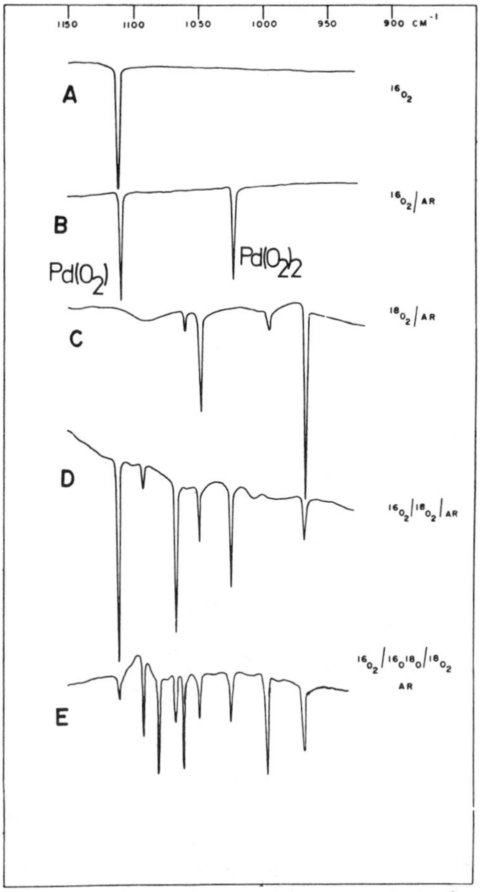

Figure 1. Observed infrared spectra of products of cocondensation of Pd atoms with (A) $^{16}O_2$; (B) $^{16}O_2$-Ar (1:200); (C) $^{18}O_2$-Ar (1:200); (D) $^{16}O_2$-$^{18}O_2$-Ar (1:1:400); (E) $^{16}O_2$-$^{16}O^{18}O$-$^{18}O_2$-Ar (1:2:1:800), showing the presence of Pd(O_2) and (O_2)Pd(O_2). (Note that the $^{18}O_2$ contains 7% of $^{16}O^{18}O$ and is most apparent in spectra C and D.)

Chart II

Bonding Scheme for (O_2)M(O_2)

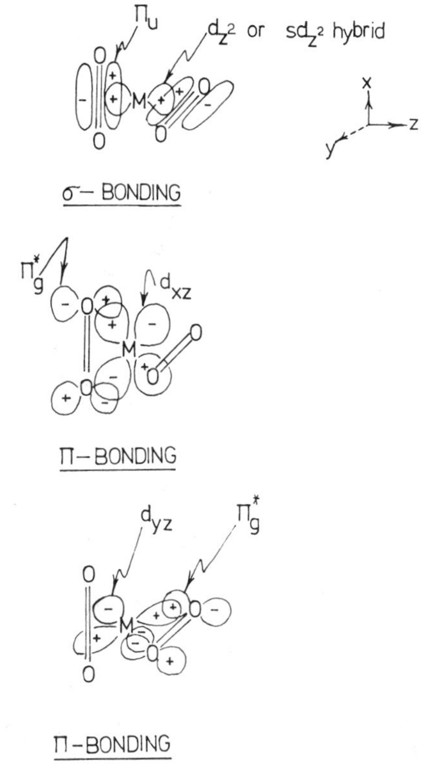

structure. The nonobservation of a splitting for cis-trans possibilities of the ($^{16}O^{18}O$)M($^{16}O^{18}O$) molecule also supports the D_{2d} assignment (see Chart II). This model would also be favored on a Dewar-Chatt π-bonded scheme[32,33] from maximum π-bonding overlap considerations illustrated schematically above.

1:3 Binary Transition Metal Complexes, M(XY)$_3$

The 1:3 complexes M(XY)$_3$ have been characterized only in the case of CO and N_2.[15,21,27,36] In the case of the tricarbonyl species the structure conforms to that expected on the basis of Gillespie's electron-pair repulsion model[37] and Kettle's overlap arguments,[38] that is, D_{3h} triangular planar. A similar situation appears to exist for Ni(N_2)$_3$, Pd(N_2)$_3$, and Pt(N_2)$_3$.

Binary Tetracarbonyl and Tetradinitrogen Complexes

Experience is proving that the cocondensation of metal atoms with *pure* reactive matrix gases such as CO, N_2, and O_2 produces the complex with the highest stoichiometry. In the case of pure CO, it proved possible to synthesize the previously unknown tetracarbonyls of Pd and Pt.

Both matrix infrared[15,39,40] (including mixed $^{12}C^{16}O$-$^{13}C^{16}O$ isotopic substitution) and Raman[11] (including matrix Raman depolarization measurements) experiments have been reported for the Ni-, Pd-, and Pt-CO cocondensation reactions and define the complexes to be M(CO)$_4$ with regular T_d symmetry.

When Co atoms were cocondensed with pure CO (Figure 2), the matrix infrared and Raman spectra[23b] indicated the major product to be the free radical Co(CO)$_4$. The matrix spectra are consistent with a D_{2d} tetrahedral structure (*i.e.*, a squashed tetrahedron). This distortion is not unexpected for a d^9 Jahn-Teller system.

The only tetradinitrogen species unambiguously identified to date is Ni(N_2)$_4$ formed in the cocondensation reaction of Ni atoms with pure N_2 at 4.2–10 K or N_2-Ar = 1:10 deposited at 30 K.[21] The infrared and Raman spectral data in argon and the excellent agreement between the calculated and observed NN stretching frequencies and infrared absorption intensities for all isotopic molecules Ni($^{14}N_2$)$_n$($^{15}N_2$)$_{4-n}$

(37) R. J. Gillespie and R. S. Nyholm, *Quart. Rev., Chem. Soc.*, **7**, 339 (1957).
(38) S. F. A. Kettle, *J. Chem. Soc. A*, 420 (1966).
(39) J. H. Darling and J. S. Ogden, *Inorg. Chem.*, **11**, 666 (1972).
(40) H. Huber, E. P. Kündig, M. Moskovits, and G. A. Ozin, *Nature (London), Phys. Sci.*, **235**, 98 (1972).

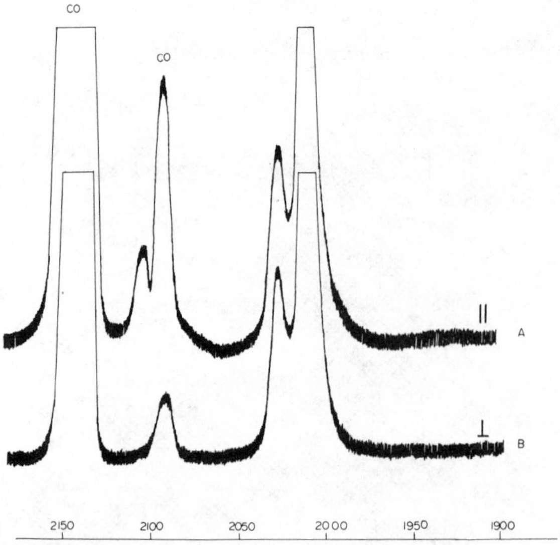

Figure 2. The matrix Raman spectrum of the products of the cocondensation reaction of cobalt atoms and CO at 4.2 K: (A) parallel and (B) crossed polarizations showing the presence of $Co(CO)_4$.

provided unambiguous evidence for the existence of $Ni(N_2)_4$ having regular tetrahedral symmetry and hence "end-on" bonded dinitrogen.

Binary Penta- and Hexacarbonyl Complexes

Using metal atom–CO cocondensation reactions DeKock[15] examined the Ta–CO system, Weltner[41] the U–CO system, and Turner[42] the Cr–CO system. The matrix infrared spectra together with the diffusion-controlled warm-up data, in all three cases, indicated a process in which a monocarbonyl species is initially formed which successively takes on CO ligands to form all of the intermediate binary carbonyls $M(CO)_n$ up to $n = 6$. Not unexpectedly the spectra were complex, and as a result vibrational assignments were tentative and structural conclusions regarding the intermediates could not be made with any degree of certainty. In this context it is worth noting that a more controlled route to binary pentacarbonyls,[42] for example, $M(CO)_5$ where M = Cr, Mo, and W, is by matrix uv photodetachment of a CO ligand from the parent hexacarbonyl $M(CO)_6$.

Binary Mixed Carbonyl–Dinitrogen Complexes of Nickel, $Ni(CO)_n(N_2)_{4-n}$, $n = 1$–3

A classic example which demonstrates the versatility of metal atom synthesis involves the reaction of nickel atoms with mixed CO–N_2 matrices. Matrix infrared and Raman spectra have been recorded for the products of the above reaction together with $^{12}C^{18}O$ and $^{15}N_2$ isotopic substitution and provide evidence for the formation of all mixed species, $Ni(CO)_n(N_2)_{4-n}$, where $n = 1$–3.[27] This direct synthesis can be compared with the uv photolysis of $Ni(CO)_4$

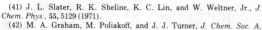

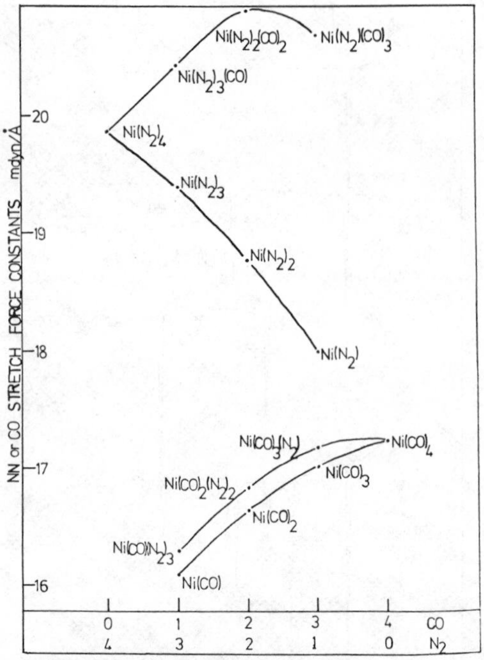

Figure 3. Graphical representation of the variation of the stretching force constants k_{CO} and k_{N_2} for the series of compounds $Ni(CO)_n(N_2)_{4-n}$ (n = 0–4) indicating the effect of replacement of dinitrogen or carbonyl ligands on tetracoordinated nickel in the matrix.

in N_2–Ar matrices[43] which exclusively yielded $Ni(CO)_3(N_2)$.

The vibrational assignments for the ligand stretching modes in $Ni(CO)_n(N_2)_{4-n}$ to the individual molecules are relatively clear-cut. A Cotton–Kraihanzel analysis[44] which can be applied with fair success to pure carbonyl and dinitrogen species has been performed on each molecule, including their $^{12}C^{18}O$ and $^{15}N_2$ isotopically substituted counterparts, in an attempt to examine trends within the various k_{NN} and k_{CO} force constants. The results of these calculations are illustrated graphically in Figure 3.

The order of k_{NN} bond stretching force constants is found to be: $Ni(CO)_3(N_2) < Ni(CO)_2(N_2)_2 > Ni(CO)(N_2)_3 > Ni(N_2)_4$. This is the expected trend when the weaker π acceptor N_2 is successively replaced by the stronger π acceptor CO. A similar but inverse effect is observed for the k_{CO} bond stretching force constants: $Ni(CO)_4 > Ni(CO)_3(N_2) > Ni(CO)_2(N_2)_2 > Ni(CO)(N_2)_3$ (see ref 27 for a detailed explanation of these force constant trends).

Conclusion

The cocondensation of metal atom vapors in their monatomic form with small gaseous molecules can be effectively studied using matrix isolation spectroscopic techniques. From the magnitudes of the shifts of the stretching frequencies of the coordinated mol-

(41) J. L. Slater, R. K. Sheline, K. C. Lin, and W. Weltner, Jr., *J. Chem. Phys.*, **55**, 5129 (1971).
(42) M. A. Graham, M. Poliakoff, and J. J. Turner, *J. Chem. Soc. A*, 2939 (1971).
(43) A. J. Rest, *J. Organometal. Chem.*, **40**, C76 (1972).
(44) F. A. Cotton and C. S. Kraihanzel, *J. Amer. Chem. Soc.*, **84**, 4432 (1962).

ecule XY as compared to those of the free molecule, we know that the products are "real" molecules of definable stoichiometries, with conventional chemical bonding and structures, and not, for example, just weakly interacting van der Waals species. As mentioned in the introduction, the only differences between the "matrix" synthetic method and "normal" syntheses are the choice of a lower temperature and of a rigid inert "solvent."

The advantages of the technique include control of the stoichiometry of the products, the cryogenic stabilization of normally unstable chemical species, the study of chemical reactions under diffusion-controlled conditions, and the synthesis of entirely new compounds. The applications of these studies can be found in any system where the interaction between a metal atom and a gaseous molecule is of importance, *i.e.*, catalysis, chemisorption, and biological-fixation processes. The future of "matrix" synthesis holds many possibilities as there are yet many interactions to be studied. Perhaps the most exciting chemical possibilities include the chemistry of many of the reactive species formed: hydrogenation (*e.g.*, $M\text{-}N_2 + H_2$, M-olefin + H_2), simultaneous cocondensation of two high-temperature species (SiO (monomer) + M (atom)), reactions of metal atoms with free radicals (M (atom) + CH_2), etc. The possibilities are real since, although experimentally unusual, the technique has now been made quite routine.

We wish to thank Professor A. G. Brook and Professor E. A. Robinson for financial support and continued interest in this work; Dr. Martin Moskovits, Mr. E. P. Kündig, and Mr. H. Huber for helpful discussion and technical assistance; and the National Research Council of Canada and the Research Corporation for financial assistance.

Addendum (May 1976)

Mononuclear transition metal complexes recently synthesized by cocondensation reactions include $Ti(CO)_6$;[45] $V(CO)_n$, n being 1–6;[46] $Mn(CO)_5$;[47] $Re(CO)_5$;[48] $Rh(CO)_4$;[49] $Ir(CO)_4$;[49] $Co(CO)_n$,[50,51] $Pd(CO)_n$,[36,52,53] and $Pt(CO)_n$,[53] n being 1–4; $Cu(CO)_n$[54] and $Ag(CO)_n$,[55] n being 1–3; Au(CO), $Au(CO)_2$, and Au(CO)(OC);[56] $Ti(N_2)_6$;[45] $V(N_2)_6$;[46] $Rh(N_2)_n$, n being 1–4;[57] $Pt(N_2)_n$, n being 1–3;[58,59] $Ag(O_2)$;[60] $Au(O_2)$;[61] $Cr(O_2)_2$;[62] $Cu(O_2)_2$;[62] $Ag(O_4)$;[60] $Ni(N_2)_n(O_2)$, $Pd(N_2)_n(O_2)$, and $Pt(N_2)_n(O_2)$, n being 1–2;[63] $Ag(C_2H_4)(O_2)$;[64] $Ag(CO)(O_2)$;[65] $Cu(C_2H_4)$;[66] $Ag(C_2H_4)$;[64] $Au(C_2H_4)$;[67] $Ni(C_2H_4)_n$[68] and $Cu(C_2H_4)_n$,[66] n being 1–3; $Ni(C_2H_4)_3$;[69] $Pd(C_2H_4)_3$;[69] $Co(C_2H_4)_3$;[69] $Fe(cod)_2$,[69] where "cod" stands for cycloocta-1,5-diene; $Co(cod)_2$;[69] $Cr(C_6H_6)_2$;[70] $Fe(C_6H_6)_2$;[69] $Ti(C_6H_6)_2$;[71] $Pd(norbornene)_3$;[69] $Ni(CS)_n$, n being 1–4;[72] $Ni(PN)_4$;[73] and $Ag(PN)_n$, n being 1–2.[73]

Binuclear transition metal complexes recently synthesized by metal atom cocondensation reactions include $V_2(CO)_{12}$;[46] $Cr_2(CO)_{10}$;[74] $Re_2(CO)_{10}$;[74] $Mn_2(CO)_{10}$;[47] $Co_2(CO)_8$;[50] $Ir_2(CO)_8$;[49] $Rh_2(CO)_8$;[49] $Ni_2(CO)_7$;[75] $Cu_2(CO)_6$;[54] $Ag_2(CO)_6$;[55] $Mn_2(CO_t)$,[47] where "CO_t" is terminally-bonded CO; $Ni_2(CO_t)$;[75] $Cu_2(CO_t)$;[75] $Mn_2(CO_b)$,[47] where "CO_b" is bridge-bonded CO; $Ni_2(CO_b)$;[75] $Cu_2(CO_b)$;[75] $Mn_2(CO_b)_2$;[47] $V_2(N_2)_{12}$;[46] and $Rh_2(N_2)_8$.[57]

First row transition metal diatomic molecules recently synthesized and characterized by metal atom matrix diffusion and concentration techniques include Sc_2,[76] Ti_2,[76] V_2,[77] Cr_2,[78] Mn_2,[79] Fe_2,[79,80] Co_2,[80] Ni_2,[79,80] and Cu_2.[54,81]

(45) R. Busbey, W. Klotzbücher, and G. A. Ozin, *Inorg. Chem.* (1976) submitted.

(46) T. A. Ford, H. Huber, E. P. Kündig, M. Moskovits, and G. A. Ozin, *Inorg. Chem.*, **15**, 1666 (1976) and *J. Am. Chem. Soc.*, **98**, 3176 (1976).

(47) H. Huber, E. P. Kündig, G. A. Ozin, and A. J. Poë, *J. Am. Chem. Soc.*, **97**, 308 (1974).

(48) E. P. Kündig and G. A. Ozin, *J. Am. Chem. Soc.*, **96**, 5585 (1974).

(49) L. Hanlan and G. A. Ozin, *J. Am. Chem. Soc.*, **96**, 6324 (1974).

(50) L. Hanlan, H. Huber, E. P. Kündig, B. McGarvey, and G. A. Ozin, *J. Am. Chem. Soc.*, **97**, 7054 (1975).

(51) O. Crichton, M. Poliakoff, A. J. Rest, and J. J. Turner, *J. Chem. Soc., Dalton Trans.*, 1321 (1973).

(52) J. H. Darling and J. S. Ogden, *J. Chem. Soc., Dalton Trans.*, 1079 (1973).

(53) E. P. Kündig, M. McIntosh, M. Moskovits, and G. A. Ozin, *J. Am. Chem. Soc.*, **95**, 7324 (1973).

(54) H. Huber, E. P. Kündig, M. Moskovits, and G. A. Ozin, *J. Am. Chem. Soc.*, **97**, 2097 (1975).

(55) D. McIntosh and G. A. Ozin, *J. Am. Chem. Soc.*, **98**, 3167 (1976) and *Inorg. Chem.*, **15**, 1669 (1976).

(56) D. McIntosh and G. A. Ozin, *Inorg. Chem.* (1976) in press.

(57) G. A. Ozin and A. Vander Voet, *Can. J. Chem.*, **51**, 637 (1973).

(58) E. P. Kündig, M. Moskovits, and G. A. Ozin, *Can. J. Chem.*, **51**, 2710 (1973).

(59) D. W. Green, J. Thomas, and D. M. Gruen, *J. Chem. Phys.*, **58**, 5453 (1973).

(60) M. McIntosh and G. A. Ozin, *Inorg. Chem.* (1976) in press.

(61) M. McIntosh and G. A. Ozin, *Inorg. Chem.* (1976) in press.

(62) J. H. Darling, M. B. Garton-Sprenger, and J. S. Ogden, *J. Chem. Soc. Faraday Symp.*, **8**, 75 (1973).

(63) W. Klotzbücher and G. A. Ozin, *J. Am. Chem. Soc.*, **95**, 3790 (1974); *idem.* **97**, 3965 (1975).

(64) D. McIntosh and G. A. Ozin, *J. Organomet. Chem.* (1976) in press.

(65) H. Huber and G. A. Ozin, *Inorg. Chem.* (1976) in press.

(66) H. Huber, D. McIntosh, and G. A. Ozin, *J. Organomet. Chem.*, C50, 112 (1976).

(67) D. McIntosh and G. A. Ozin, *J. Organomet. Chem.* (1976) in press.

(68) W. Power and G. A. Ozin, *J. Am. Chem. Soc.* (1976) in press.

(69) P. L. Timms, paper delivered at the ACS Meeting "Atomic Species in Chemical Synthesis", Chicago, 1975.

(70) D. M. Gruen, J. W. Boyd, and J. M. Lavoie, *J. Chem. Phys.*, **60**, 4088 (1974).

(71) M. T. Anthony, M. L. H. Green, and D. Young, *J. Chem. Soc., Dalton Trans.*, 1419 (1975) and references therein, *see* A. J. Downs et al.

(72) H. Huber, M. Moskovits, and G. A. Ozin, *Prog. Inorg. Chem.*, **19**, 105 (1975).

(73) P. L. Timms, *Angew. Chem. Int. Ed.*, **14**, 275 (1975).

(74) E. P. Kündig, M. Moskovits, and G. A. Ozin, *Angew. Chem. Int. Ed. Engl.*, **14**, 292 (1975) and M. Moskovits and G. A. Ozin in "Cryochemistry", John-Wiley, New York, 1976.

(75) M. Moskovits and J. Hulse, *Surf. Sci.*, **57**, 125 (1976).

(76) R. Busbey, W. Klotzbücher, and G. A. Ozin, *J. Am. Chem. Soc.*, **98**, 4013 (1976).

(77) T. A. Ford, H. Huber, W. Klotzbücher, E. P. Kündig, M. Moskovits, and G. A. Ozin, *J. Chem. Phys.* (1976) in press.

(78) E. P. Kündig, M. Moskovits, and G. A. Ozin, *Nature*, **254**, 503 (1975).

(79) T. C. De Vore, A. Ewing, H. F. Franzen, and V. Calder, *Chem. Phys. Lett.*, **35**, 78 (1975).

(80) C. McClaren, W. Klotzbücher, and G. A. Ozin, unpublished data.

(81) G. A. Ozin, *Appl. Spectrosc.* (Oct. 1976).

On the Interpretation of the Optical Spectra of Hexahalogen Complexes of 4d and 5d Transition Metal Ions

Peter C. Jordan

Department of Chemistry, Brandeis University, Waltham, Massachusetts 02154

Received August 6, 1973

Reprinted from ACCOUNTS OF CHEMICAL RESEARCH, *7, 202 (1974)*

The electronic spectra of octahedral hexahalide complexes of 4d and 5d transition metal ions have been of interest for many years.[1] The interpretation of these spectra (and those of many similar compounds) has provided much of the impetus for the development of ligand-field theory and for the application of molecular orbital theory to extremely complex systems. These theories provide the framework for understanding bonding and assigning transitions in these complexes. Recent high-resolution spectra,[2-9] in which some line widths are less than 1 cm^{-1}, allow exceptionally stringent tests of the assignment schemes which have been proposed; this analysis will be the focus of this Account.

A large number of these complexes are known; their composition is MX^{m-}, where *m* may be 0, 1, 2, or 3. While not all such compounds have been observed, there appears to be only one restriction to species that can be synthesized. If we represent the electronic structure as $M^{6-m}(X^-)_6$, the outer electronic shell of the metal ion is composed of d electrons. No isolatable species exist with more than six d electrons in this shell. Neutral species other than hexafluorides are uncommon (WCl_6 is the only example), and few uninegative complexes are known.

The neutral species have been studied in gas phase and in solution at room temperature. The complex ions have been studied in numerous solvents at room temperature, in glasses, in various crystalline hosts at temperatures as low as 1.3 K, and as pure single crystals. The spectroscopic techniques that have been used include electron paramagnetic resonance, optical, infrared, Raman, and reflectance spectroscopy, magnetic circular dichroism measurements, and Zeeman studies. With such an arsenal of experimental tools it would appear that assignment of the electronic transitions should be relatively straightforward. Due to the experimental and theoretical complexity of this problem, such is not the case.

The electronic spectra of these species are superficially quite similar.[1b] At low energy there is a series of sharp, weak transitions having molar extinction coefficients, ϵ, of 100 or less (region I) in gas phase or HX solution at room temperature. At intermediate energies numerous broad intense transitions are found for which ϵ may be as large as 9000 (region II). In the far-uv region extremely intense, very broad structureless transitions are observed with $\epsilon \gtrsim 20{,}000$ (region III). For example, in $IrCl_6^{2-}$ region I extends to $\sim$17,000 cm^{-1}, region II lies between $\sim$17,000 and $\sim$35,000 cm^{-1}, and region III is beyond $\sim$35,000 cm^{-1}. The pattern is general, with details depending upon the individual complex. If the central metal ion has six d electrons, region I is absent and region II spectra are much less intense.

Interpretation of these spectra is based upon the ideas of ligand-field theory.[10] In its simplest form, this theory treats the complex as a central metal ion surrounded by six univalent halide ions. The outer d orbitals of the metal ion are perturbed by the ligand field of octahedral symmetry created by the halide ions. The degeneracy is partially removed, the orbitals being split into two groups which can be classified according to their transformation properties under the operations of the group O_h: the d_{xz}, d_{yz}, and d_{xy} orbitals form a set of t_{2g} orbitals and the d_{z^2} and $d_{x^2-y^2}$ orbitals form a set of e_g orbitals. In general the t_{2g} level is lower in energy.

The systems we consider correspond to strong ligand-field coupling, *i.e.*, the $t_{2g} \rightarrow e_g$ excitation energy is large. As the transition metal ions of interest have *n* ($\leq$6) d electrons, only the t_{2g} subshell is occupied in the ground state, which is therefore $t_{2g}{}^n$. Electrostatic interaction and spin–orbit coupling split the states in this configuration. In the strong-field coupling limit these states are in 1:1 correspondence to those arising from the familiar p^{6-n} configuration.[11] Figure 1 gives a schematic correlation diagram for the $t_{2g}{}^4$ (p^2) system as a function of the spin–orbit coupling constant, ζ. When ζ is small, Russell–Saunders coupling is applicable; when it is large, *j–j* coupling is appropriate. We use Griffith's notation[12] in the two limiting cases and the Bethe notation[13] (Γ_i) in the intermediate region. The accidental degeneracy of Γ_3 and Γ_5 states, corresponding to atomic states with $J = 2$ ($^1D_2$ and $^3P_2$), arises be-

Peter Jordan received his B.S. from Caltech in 1957 and his Ph.D. from Yale University in 1960. He then spent a year at Cambridge University and 3 years at the University of California at La Jolla. In 1964 he joined the Chemistry Department at Brandeis University, where he is now Associate Professor. His main interests are in statistical mechanics, but he also maintains a continuing interest in molecular spectroscopy and quantum chemistry.

(1) (a) C. K. Jorgensen, *Progr. Inorg. Chem.*, **12**, 106 (1970); (b) D. S. Martin, Jr., *Advan. Chem. Ser.*, **No. 98**, 74 (1971). These articles provide recent detailed review of the field.

(2) P. B. Dorain and R. G. Wheeler, *J. Chem. Phys.*, **45**, 1172 (1966). Two lower energy, relatively weak transitions not discussed in this paper have also been observed.

(3) P. B. Dorain, H. H. Patterson, and P. C. Jordan, *J. Chem. Phys.*, **49**, 3845 (1968).

(4) H. H. Patterson and P. B. Dorain, *J. Chem. Phys.*, **52**, 849 (1970).

(5) I. N. Douglas, *J. Chem. Phys.*, **51**, 3066 (1969).

(6) J. R. Dickinson, *et al.*, *J. Chem. Phys.*, **56**, 2668 (1972).

(7) S. B. Peipho, *et al.*, *J. Chem. Phys.*, **57**, 982 (1972).

(8) S. B. Peipho, *et al.*, *Mol. Phys.*, **24**, 609 (1972).

(9) W. H. Inskeep, R. W. Schwartz, and P. N. Schatz, *Mol. Phys.*, **25**, 805 (1973).

(10) J. S. Griffith, "The Theory of Transition Metal Ions," Cambridge University Press, New York, N. Y., 1964.

(11) See ref 10, p 238 ff.

(12) See ref 10, p 226 ff; p 242; p 396 ff.

(13) G. F. Coster, *et al.*, "Properties of the Thirty-Two Point Groups," MIT Press, Cambridge, Mass., 1963.

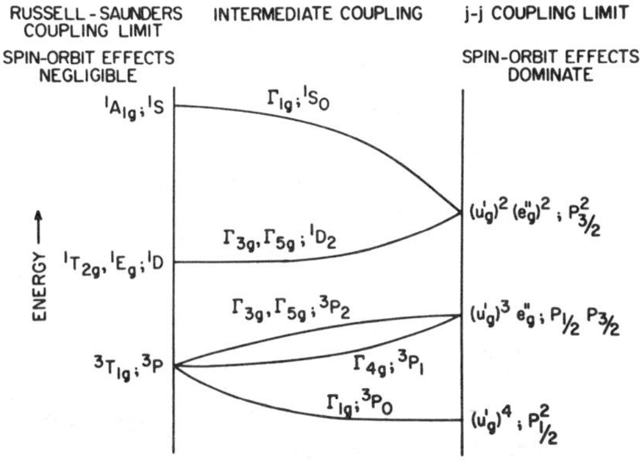

Figure 1. Correlation diagram for states arising from the electronic configuration t_{2g}^4 as a function of spin-orbit coupling. The isomorphism with states arising from the p^2 configuration is indicated in each coupling region. Notation indicates the symmetry of the states.

cause we limit consideration to the t_{2g}^4 configuration; it is removed in a more complete treatment. Using correlation diagrams such as this one it is possible to account for the region I spectra, as was first done by Van Vleck[14] and Tanabe and Sugano[15] for iron group complexes. The low intensity of these spectra is due to the fact that pure electronic transitions within these configurations are parity forbidden. In molecular systems they may be activated by coupling with an ungerade vibrational mode and are thus expected to be weak.

If one only considers the possibility of d–d transitions, there are serious difficulties in interpreting region II spectra. If we follow Moffitt's[16] suggestion and assume that these are due to $t_{2g} \rightarrow e_g$ excitations, they would be parity forbidden and could only be activated by vibronic coupling. Such a mechanism does not account for the large extinction coefficients observed in some of the transitions. Transitions of the type 5d → 6p (or 4d → 5p) are parity allowed but may be ruled out on energetic grounds; they are expected to occur in the vacuum ultraviolet.[17] A totally different viewpoint is needed; this was introduced by Jorgensen[18] who suggested that region II spectra arose *via* electron transfer from a ligand orbital to a metal t_{2g} orbital.

Of the outer orbitals of the halide ligands, only the p orbitals appear to be important in understanding the spectra.[17] They may be grouped into two sets: six σ orbitals directed along the metal–ligand axes and 12 π orbitals perpendicular to these axes. The octahedral field of the ligand ions interacts with these orbitals and splits their degeneracy just as it does that of the metal d orbitals. Figure 2 presents a molecular orbital energy level diagram (including both ligand p and metal d orbitals) based upon semiempirical calculations.[17] There is some disagreement as to the location of the σe_g orbital[19] and the πt_{2g} orbital,[6-9] but the ordering in Figure 2 remains

(14) R. Finkelstein and J. H. Van Vleck, *J. Chem. Phys.*, **8**, 790 (1940).
(15) Y. Tanabe and S. Sugano, *J. Phys. Soc. Jap.*, **9**, 753, 766 (1954).
(16) W. Moffitt, *et al.*, *Mol. Phys.*, **2**, 109 (1959).
(17) F. A. Cotton and C. B. Harris, *Inorg. Chem.*, **6**, 376 (1967); Document No. 9156, ADI Auxiliary Publication Project, Library of Congress, Washington, D. C..
(18) C. K. Jorgensen, *Mol. Phys.*, **2**, 309 (1959).
(19) T. P. Sleight and C. R. Hare, *J. Phys. Chem.*, **72**, 2207 (1968).

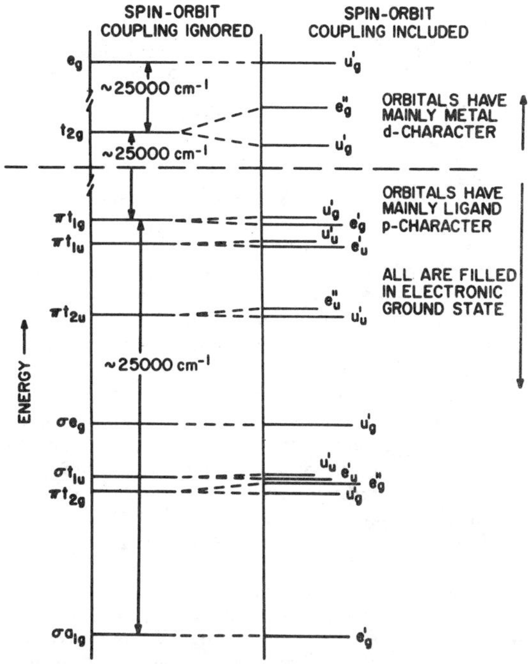

Figure 2. Relative energy of spectroscopically significant molecular orbitals for MX_6^{2-} ions. The energies of ligand MO's are taken from ref 17 for $OsCl_6^{2-}$. The notation on the left denotes the symmetry of spatial orbitals; that on the right is for spin orbitals.

the most likely. Orbitals of the same symmetry interact so that the πt_{1u} orbital incorporates some σt_{1u} character, etc.

In addition to perturbations induced by the octahedral ligand field, the molecular orbitals are further perturbed by spin–orbit coupling. Just as an atomic p orbital is split into doublet ($j = 1/2$) and quartet ($j = 3/2$) states, a t_1 or t_2 orbital splits into doublet (e' or e'') and quartet (u') states. The magnitude of the splitting depends upon both atomic number and ionic charge, increasing with increasing atomic number and with increasing positive charge. A qualitative splitting pattern is also incorporated in Figure 2. It should be noted that spin–orbit coupling need not affect all the ligand molecular orbitals equally; the amount of the splitting depends greatly on the extent of mixing of orbitals of the same symmetry.

On the basis of the MO picture given in Figure 2, it is possible to account for the large extinctions seen in region II since many parity-allowed transitions are possible. These involve excitation of an electron from an ungerade ligand MO to the metal t_{2g} orbital, *i.e.*, an electron-transfer transition. However, with the quantity of high-resolution data now available, there are far more sensitive tests than extinction coefficients alone from which to infer reasonable assignments. Just as the $t_{2g} \rightarrow e_g$ hypothesis cannot account for all region II features, we shall see that the electron-transfer hypothesis is equally fallible. Arguments based mainly upon analogy or plausibility have a nasty habit of breaking down when subjected to critical tests. The central problem is determining selection rules to limit the possible assignments. In $OsCl_6^{2-}$ there are only six major bands in region II while 34 different $t_{2g} \rightarrow e_g$ transitions and 90 different electron-transfer transitions are possible.

Region III is attributed to electron transfer from a ligand orbital to the metal e_g orbital which is consistent with the enormous intensities of these bands. The high-resolution data required to make more precise assignments are presently unavailable.

Before proceeding, it is necessary to have a general feeling for the nature of the experimental data that are available. A great number of studies have been made at room temperature, but these do not provide the sort of information we need. Under such conditions only broad band absorptions are observed; details of the individual vibronic transitions which comprise the band are completely obscured. Only at low temperature is the vibronic structure resolved. The importance of this vibronic data cannot be overemphasized. Spectroscopic assignments can never be proved; instead, just as with the deduction of kinetic mechanisms, arguments are based upon negative inference. The process of assignment is rather like a trial—determination should be beyond reasonable doubt. The more evidence that is available, the fewer assignments that are tenable. While we shall not ignore other data, we shall mainly consider low-temperature spectra of various M(IV) ions in single cubic crystals for the following reasons: more studies have been carried out on these systems than on any others; at low temperature individual vibronic lines can be seen; the cubic environment ensures that stringent selection rules are in force.

The available data have been obtained by two different procedures, each with its own limitations. Photographic recording of spectra provides unsurpassed resolution.[2-4,20] There is a wealth of detail; the energy and line width of a transition can be determined with exceptional accuracy. Due to the peculiarities of photographic plates it is, however, extremely difficult to determine relative intensity in any but the most qualitative terms. Spectrophotometric recording on a double-slit instrument generally allows accurate measurement of extinction coefficients.[6-9] However, it is not possible to obtain anything like the resolution available photographically.

Both techniques have been used to study the magnetic field dependence of the transitions. The photographic approach is based upon Zeeman effect studies.[20] Not only does this yield information about the shift of the energy levels in a field but, by making absorbance measurements using polarized light, the possible assignments are greatly limited since there are generally extremely severe restrictions on selection rules. The experimental difficulty is that the energy shifts must be at least as large as the line widths or no effect can be observed.

The spectrophotometric method uses the phenomenon of magnetic circular dichroism (MCD),[6-9] that is, in a magnetic field all molecules become optically active. The magnitude of this induced optical activity depends upon three factors:[21] the magnetic moment of the ground state; the magnetic moment of the excited state; and the amount of magnetically induced mixing of ground and excited states with states with nonzero magnetic moments. It is this last contribution which ensures that *all* molecules are optically active in a magnetic field.

It is possible to distinguish these factors experimentally: The first leads to a temperature dependence (type C features in MCD), the difference between the first two leads to a dispersion line shape (type A), and the third leads to an absorption line shape (type B). The advantage of MCD is that the phenomenon is observable even for broad band transitions. The difficulty is that assignments based upon MCD are less constrained than ones based upon Zeeman studies. The problem is one of resolution: individual Zeeman features on a photographic plate coalesce to a single MCD feature if the instrument's slit width exceeds the splitting of the lines.

Magnetic data are commonly given in terms of g values. The shift in an energy level due to a magnetic field is defined as

$$\Delta E = |\beta_0| g H m_J \quad (1)$$

where β_0 is the Bohr magneton, H is the field strength, and m_J is the azimuthal quantum number. For singlets m_J is 0; for doublets it is $\pm\frac{1}{2}$; for triplets it is 0, ± 1; and for quartets it is $\pm\frac{1}{2}$, $\pm\frac{3}{2}$.

In addition to the instrumental problems common to any spectroscopic analysis, there are difficulties particular to the study of doped crystals. As these are solid solutions the choice of host crystal is crucial. Only with the host crystals Cs_2ZrCl_6 and Cs_2ZrBr_6 is extensive vibronic structure found; presumably these show the least distortion from cubic symmetry.[22] The species for which good vibronic data are available are few: $RuCl_6^{2-}$,[4] $ReCl_6^{2-}$,[2] $OsCl_6^{2-}$,[3,7] $IrCl_6^{2-}$,[5,7,20] $OsBr_6^{2-}$,[9] and $IrBr_6^{2-}$.[6] Our discussion is limited to these systems, and we emphasize the spectra of Os(IV) in Cs_2ZrCl_6 for which the most extensive and best resolved data are available. Even with the "perfect host crystal" there remain problems in solid-state spectroscopy. Determination of molar extinction coefficients is only approximate, as one cannot run a doped crystal and a blank crystal of the same thickness, surface properties, and strain characteristics simultaneously.[23] Finally the vibrational spectrum is no longer that of an isolated MX_6^{2-} octahedron; the crystal lattice modifies both vibrational frequencies and vibronic selection rules.[24]

Analysis

A catalog of observed region II transitions and their most probable assignments is given in Table I. A band is described as very broad if it has no vibrational structure, as broad if the only vibrational feature appears to be a progression in the symmetric stretching frequency, $\nu_1(a_{1g})$, as sharp when individual vibronic features are seen, and as very sharp when there is an embarrassment of vibronic detail.

In this section we shall mainly discuss the spectra

(20) R. Massuda and P. B. Dorain, *J. Chem. Phys.*, **59**, 5652 (1973).

(21) A. D. Buckingham and P. J. Stephens, *Annu. Rev. Phys. Chem.*, **17**, 399 (1966). This review provides an excellent description of magnetic circular dichroism and detailed discussion of the effects which contribute to an MCD pattern.

(22) S. B. Peipho, *et al.*, *Mol. Phys.*, **19**, 781 (1970). Comparison of the spectra in this paper with those of ref 6 dramatically points out the effect of the host crystal.

(23) P. N. Schatz *et al.*, *Symp. Faraday Soc.*, **No. 3**, 14 (1969).

(24) G. O'Leary and R. B. Wheeler, *Phys. Rev. B*, **1**, 4409 (1970). This paper provides a clear discussion of the lattice dynamics of cubic crystals such as Cs_2ZrCl_6 which form anti-fluorite lattices.

Table I
Catalog of Observed Region II Spectra (Data Are of Low Temperature Measurements on Doped Crystals of Cs_2ZrCl_6 or Cs_2ZrBr_6)

Species	Band	Energy, 10^3 cm^{-1}	Intensity[h]	Structure[h,i]	Probable assignment[j]	Bases for attribution[k]
$ReCl_6^{2-}$[a]	A[g]	28.2	vw	br	$\pi t_{1g} \rightarrow u_g'$, V	E, I, S, A
	B	29.3	w	sh	$^4T_{2g}$, V (?)	E, S
	5	30.0	m	br	$\pi t_{1u} \rightarrow u_g'$	I, S, A
	6	31.5	m	sh	$^4T_{1g}$, V (?)	E, S
	8	33.7	vs	br	$\pi t_{1u} \rightarrow e_g''$; $\pi t_{2u} \rightarrow u_g'$	E, I, S, A
	9	35.7	vs	br	$\pi t_{2u} \rightarrow e_g''$	E, I, S, A
$OsCl_6^{2-}$[b]	3	24.1	w	sh	$\Gamma_{4g} + \nu_4$	E, RI, S, MCD
					$\pi t_{1g} \rightarrow e_g''$, V	E, S, A
	4	25.5	s	br	u_u' $(\pi t_{1u}) \rightarrow e_g''$	E, I, MCD
	6	27.2	m	vsh	$\Gamma_{5g} + \nu_4(2)$	E, RI, MCD, *g*
	8	28.7	s	sh	$e_u''(\pi t_{2u}) \rightarrow e_g''$; $u_u'(\pi t_{2u}) \rightarrow e_g''$; $\pi t_{2u} \rightarrow e_g''$, V	E, I, MCD
	9	32.1	m	sh	$\Gamma_{5g} + \nu_4$	E, S, RI, MCD
					$(\pi t_{2u}, u_g') \rightarrow (e_g'')^2$ (?)	E
	10	35.7	m	br	$\sigma e_g \rightarrow e_g''$, V (?)	E
$RuCl_6^{2-}$[c]	3	18.3	w	sh	$\pi t_{1g} \rightarrow e_g''$, V	E, S, A
	4	19.1	s	br	$u_u'(\pi t_{1u}) \rightarrow e_g''$	E, I, S, A
	5	20.4	m	br	$(u_u'(\pi t_{1u}), u_g') \rightarrow (e_g'')^2$	E, I, S
	6	21.9	m	br	$(u_u'(\pi t_{1u}), u_g') \rightarrow (e_g'')^2$ (?)	E, S
	7	22.8	s	sh	$\pi t_{2u} \rightarrow e_g''$ then $(\pi t_{2u}, u_g') \rightarrow (e_g'')^2$	E, I, S, A
$IrCl_6^{2-}$[d]	D	17.9	w	vbr	$\pi t_{1g} \rightarrow e_g''$, V	E, S, MCD, A
	C	19.3	s	br	$\pi t_{1u} \rightarrow e_g''$	E, S, I, MCD
	B	20.8	m	vsh	d–d, V	S, MCD, *g*, Z, A; NI
	A	22.9	vs	vsh	$\pi t_{2u} \rightarrow e_g''$	E, I, MCD, A
					d–d, V	P, Z, S; NI
$OsBr_6$[e]	1 + 2	17.5	w	vbr	$u_g'(\pi t_{1g}) \rightarrow e_g''$	E, S, A
	3	18.5	s	vbr	$u_u'(\pi t_{1u}) \rightarrow e_g''$	E, S, I, MCD
	4	20.0	m	vsh	d–d, V	S, MCD; NI
	5	22.2	s	sh, then	$e_u''(\pi t_{2u}) \rightarrow e_g''$	E, S, I, MCD
				br	$e_u'(\pi t_{1u}) \rightarrow e_g''$, V	E, MCD
	6	24.0	s	sh	$u_u'(\pi t_{2u}) \rightarrow e_g''$	E, S, I, MCD
$IrBr_6^{2-}$[f]	1 + 2	11.8	w	vbr	$u_g'(\pi t_{1g}) \rightarrow e_g''$, V	E, S, MCD, A
	3 + 4	13.0	s	br	$u_u'(\pi t_{1u}) \rightarrow e_g''$	E, I, S, MCD
	5	14.5	w	sh	d–d, V (?)	S, MCD; NI
	6	17.1	s	vsh	$e_u''(\pi t_{2u}) \rightarrow e_g''$	E, I, S, MCD
	7	~17.6	s	br	$e_u'(\pi t_{1u}) \rightarrow e_g''$, V	E, MCD
	8	18.7	s	vsh	$u_u'(\pi t_{2u}) \rightarrow e_g''$	E, I, S, MCD

[a–f] Band nomenclature corresponds with the notation of ref 2–5, 9, and 6, respectively. [g] Bands A and B were not reported in ref 2. [h] m, medium; s, strong; w, weak; br, broad; sh, sharp; v, very. [i] See text for discussion of terminology. [j] Charge-transfer assignments are identified by subshells (see Figure 2). d–d transitions are noted by their most intense vibronic state (if such precision is warranted). Other vibronic transitions are indicated by V. Assignments marked (?) are tentative. [k] The bases are energy (E); intensity (I); relative intensity of vibronic states (RI); vibronic structure (S); MCD; *g* value (*g*); polarization (P); Zeeman studies (Z); and analogy with other systems (A). Where positive identification has not been possible, assignment is based on negative inference (NI).

of $OsCl_6^{2-}$ to indicate the way in which the data—frequency, intensity, relative intensity, MCD pattern, *g* value, polarization, and systematic trends—can be used to limit assignment possibilities. We choose this system because its spectra are the richest and best resolved and its ground state is singlet (see Figure 1), conditions which limit assignment possibilities. Our fundamental criterion for assessing a proposed assignment is consistency; the absence of an allowed transition is *prima facie* evidence against an interpretive scheme. Correlation of spectroscopic data with theoretical predictions can only be used to corroborate an assignment.

The most intense features in the $OsCl_6^{2-}$ spectra are bands 4 and 8 (see Table I). Because of their strength these transitions must be electric dipole allowed charge-transfer (CT) processes; the lowest lying of these transitions should then be $\pi_{1u} \rightarrow t_{2g}$ and $\pi t_{2u} \rightarrow t_{2g}$ CT excitations (see Figure 2). Since the ground state of $OsCl_6^{2-}$ is Γ_{1g}, the only electric dipole allowed transitions occur to Γ_{4u} states. In Figure 3 we present the energy level diagram for the possible Γ_{4u} states arising from the two low-lying CT excitations. The orbitals are ordered according to MO calculations,[17] with orbital excitation energies chosen to fit the positions of bands 4 and 8. We assume that spin-orbit coupling is the dominant interaction and ignore effects such as metal-ligand electrostatic interactions.[25] The spin-orbit parameters used are $\zeta \sim 600$ cm^{-1} for chloride[6] and $\zeta \sim 2400$ cm^{-1} for Os(IV).[3] Since ζ_{Os} is large, *j-j* coupling is a good approximation, and the ground state would be $(u_g')^4$ (see Figure 1). The Γ_{4u} levels can then be classified according to the subshells involved in a particular transition, *e.g.*, $u_u' \rightarrow e_g''$, $(e_u', u_g') \rightarrow (e_g'')^2$, etc. Note that two Γ_{4u} states arise from each $(u_u', u_g') \rightarrow (e_g'')^2$ transition. The relative intensity of transitions in each CT manifold and the *g* values of the excited states have been included in the diagram. Since the ground state of $OsCl_6^{2-}$ contains about 10% of the $(u_g')^2(e_g'')^2$ configuration,[3] the *j-j* coupling description is only approximate. The "two-electron" transitions are no longer forbidden but may have significant intensity.

(25) This is a serious limitation whose validity can only be assessed when more data are available.

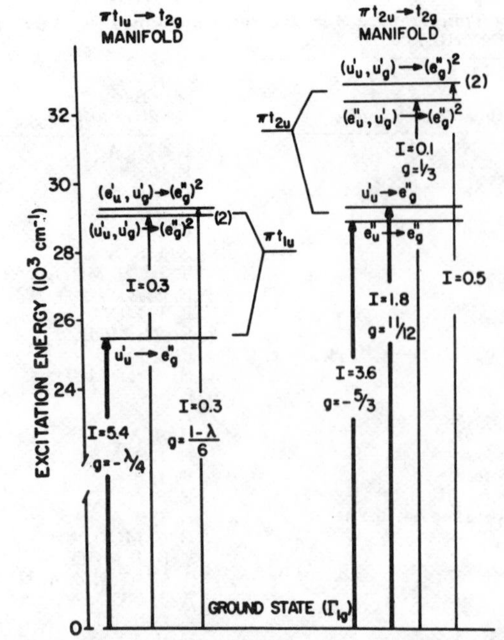

Figure 3. Energy level diagram for Γ_{4u} states arising from $\pi t_{1u} \rightarrow t_{2g}$ and $\pi t_{2u} \rightarrow t_{2g}$ charge-transfer excitations in $OsCl_6^{2-}$. Classification of levels is discussed in text. Intense transitions are noted by heavy arrows. Relative intensities and g values (where meaningful) are included. Total intensity in each manifold is arbitrarily set to 6. For reference, energies before and after including metal spin–orbit coupling are shown in the center. An orbital reduction factor of unity has been assumed.

Figure 3 has been calculated assuming a simple CT model. Neglect of electrostatic effects which would shift the energy levels and split the remaining degeneracies probably does not change any major qualitative features. However. the g values of the individual $(u_u',u_g') \rightarrow (e_g'')^2$ states depend on the size of such effects; for this reason no g values are given for these states. Finally, g values for states in the $\pi t_{1u} \rightarrow t_{2g}$ manifold are given in terms of a parameter λ which is a measure of σ–π mixing of the two t_{1u} orbitals; it varies between $-1 \leq \lambda \leq 2$.[26]

The CT model predicts three very intense transitions. The low-energy band involves a single electronic state for which the probable g value suggests a $-A$ feature in MCD; this accords with the observed properties of band 4.[8] The upper band should be composed of two electronic transitions split by ~450 cm^{-1}; the lower of these transitions should be about twice as intense; the MCD should show a $-A$ feature followed by a $+A$ feature. These predictions are in good accord with the properties of band 8.[8] The actual splitting seems closer to ~1200 cm^{-1}; part of this difference may be accounted for when ligand–metal electrostatic interactions are considered.[27]

There is extensive structure associated with band 8 probably due in part to vibronically activated transitions involving other electronic states in this region of the $\pi t_{2u} \rightarrow t_{2g}$ manifold.[8] From Figure 3 we see that the upper Γ_{4u} states in the $\pi t_{1u} \rightarrow t_{2g}$ manifold should occur in the vicinity of band 8. Since these are similar to the state which gives rise to band 4, they should be rather broad and quite weak in comparison to the intense $\pi t_{2u} \rightarrow t_{2g}$ features. While they are likely to be masked, they may account for some of the band 8 structure.

The only low-lying Γ_{4u} state in the $\pi t_{1u} \rightarrow t_{2g}$ manifold arises from the subshell excitation $u_u' \rightarrow e_g''$. However, a vibronically activated progression based upon the subshell excitation $e_u' \rightarrow e_g''$ is expected to commence between 25,250 and 26,600 cm^{-1}, depending upon the value of λ. As such features are found in most other systems[6,7,9] studied, it seems that this transition is masked by the excitation to the Γ_{4u} state; this indicates that λ is small, between 0 and 0.3 for this system.

Band 6 contains an unbelievable amount of fine structure; there are at least 14 separate vibronic progressions.[3] We consider only the most intense based upon lines at 27,080 (6-D) and 27,157 cm^{-1} (6-G).[3] Higher members of both progressions 6-D and 6-G are split,[3,28] 6-D into four lines. As individual electronic states in d^4 systems such as $OsCl_6^{2-}$ cannot be quartets, the feature 6-D must be a vibronic transition based upon degenerate electronic and vibrational states. Analysis of the splitting pattern of 6-D indicates that the activating vibration is $\nu_4(t_{1u})$.[3] Analysis of the d–d model showed that for $t_{2g} \rightarrow e_g$ transitions in d^4 systems there are certain approximate selection rules: vibronic transitions to Γ_{1g}, Γ_{2g}, and Γ_{3g} states are forbidden and those to Γ_{4g} states are much weaker than those to Γ_{5g} states.[29] On this basis the lines 6-D and 6-G were assigned as transitions to $\Gamma_{5g} + \nu_4(t_{1u})$ states with g values of +0.12 and −0.31, respectively. This agreed with the relative intensity of the transitions[29] and with the MCD pattern, a $-A$ feature followed by a $+A$ one.[8] The MCD may also be accounted for if band 6 arises from a $\pi t_{2g} \rightarrow t_{2g}$ CT excitation.[8] Then lines 6-D and 6-G would be transitions to $\Gamma_{4g} + \nu_4(t_{1u})$ vibronic states with $g = -1.67$ and $+0.92$, respectively. However recent measurements in fields as high as 70 kG show no magnetically induced splitting.[30] Using line widths as a guide, $|g| \lesssim 0.3$ for both states, which rules out the CT hypothesis. The surprising feature of this $t_{2g} \rightarrow e_g$ identification is that band 6 is a moderately intense, vibronically activated band which must borrow its intensity not from the nearby bands 4 or 8 but rather from the region III transitions some 15,000 cm^{-1} to the blue.[29]

Band 3 shows structure superposed upon a broad background. The intense resolved features are a progression of sharp lines beginning at 24,087 cm^{-1} (3-A) and a pair of broader lines beginning at 24,777 cm^{-1} (3-G).[3] As we expect the $\pi t_{1g} \rightarrow t_{2g}$ CT band to occur in this region (see Figure 2) we may compute the properties of the possible vibronic transition on the assumption that they are activated by vibronic mixing with the nearby Γ_{4u} state, band 4. The re-

(26) Previous authors introduced a bonding criterion to suggest that $\lambda > 1$ (see ref 22). Overlap was used as a basis for fixing the relative phases in the bonding and antibonding t_{1u} orbitals. This assumes that interactions with the core (one-electron effects) dominate, which need not be so. Electrostatic (two-electron) interactions are surely important in determining the extent of σ–π mixing; if these dominate, λ may even become negative, though this is unlikely.

(27) P. C. Jordan, in preparation.

(28) P. B. Dorain, private communication.

(29) P. C. Jordan, H. H. Patterson, and P. B. Dorain, *J. Chem. Phys.*, **49**, 3858 (1968). When the parameters of the theory are reduced to their simplest forms, reasonable estimates of the reduced matrix elements which appear lead to W/V ratios between 20 and 500, this can qualitatively account for the low intensity of vibronic transitions to Γ_{4g} states.

(30) P. B. Dorain, private communication.

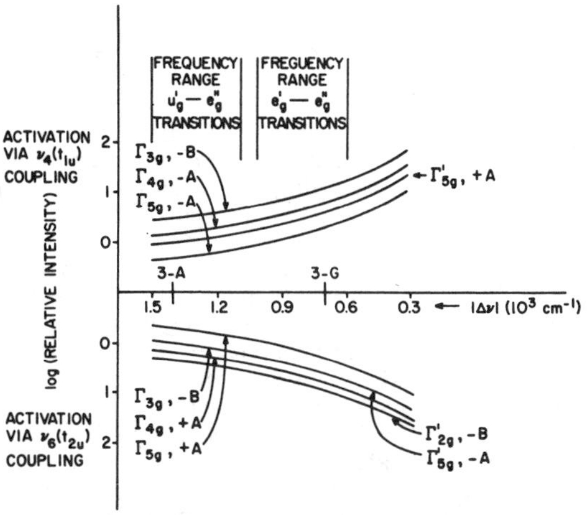

Figure 4. Relative intensities of transitions in the $\pi t_{1g} \rightarrow t_{2g}$ manifold assuming activation by vibronic mixing with the nearby $\pi t_{1u} \rightarrow t_{2g}$ Γ_{4u} state: $\Delta\nu$ is the separation of the vibronic state from the Γ_{4u} state. The nature of the MCD features predicted (±A, ±B) is included. The frequency ranges shown refer to $OsCl_6^{2-}$, assuming electrostatic effects are small. For reference, the positions of lines 3-A and 3-G in $OsCl_6^{2-}$ are also indicated.

sults are given in Figure 4 for both $\nu_4(t_{1u})$ and $\nu_6(t_{2u})$ activation. The observed MCD shows +A features for both 3-A and 3-G.[8] The most obvious assignments based upon splitting and MCD, 3-A as $\Gamma_{4g} + \nu_6$ and 3-G as $\Gamma_{5g}' + \nu_4$, are inconsistent with the relative intensity pattern as there should then be intense $\Gamma_{5g}' + \nu_6$ and $\Gamma_{2g}' + \nu_6$ lines with −A and −B MCD features; none are found. Other assignment schemes lead to similar contradictions. The d-d hypothesis suggests only one state with a magnetic moment in this region,[3,29] the most intense of the $\Gamma_{4g} + \nu_4$ vibronic states with $g = +0.58$ which is consistent with the MCD of either state. We assign it to 3-A which is very sharp, reminiscent of the more intense $t_{2g} \rightarrow e_g$ transition of band 6. The broad background and some structure presumably does arise from the $\pi t_{1g} \rightarrow t_{2g}$ CT excitations. The broader resolved feature 3-G could be the $\Gamma_{5g}' + \nu_4(t_{1u})$ vibronic state as has been proposed.[8] The absence of resolved features corresponding to the $u_g' \rightarrow e_g''$ states is then puzzling but may be accounted for in two ways: if these are strongly overlapped, a broad absorption is likely; electrostatic mixing acts to increase the intensity of the $\Gamma_{5g}' + \nu_4$ state. The possibility of other CT excitations, $\pi t_{2g} \rightarrow t_{2g}$ and $\sigma e_g \rightarrow t_{2g}$, has also been considered. Such attributions lead to inconsistencies between MCD and relative intensity, run counter to trends in other MX_6^{2-} systems, and contradict the predictions of MO theory.[17]

Band 9 contains three major progressions commencing at 32,097 (9-A), 32,156 (9-B), and 32,208 cm^{-1} (9-C). The first is intense and shows a +B feature in MCD, the second is slightly more intense with a +A MCD, and the third is weak with a +A MCD. Higher members in the progression 9-B split into four components, indicative of vibronic excitation.[31] The data may be correlated on the basis of a $\sigma e_g \rightarrow t_{2g}$ CT assignment but there are three serious drawbacks.[8] Gross features of CT spectra for various

(31) P. B. Dorain, private communication.

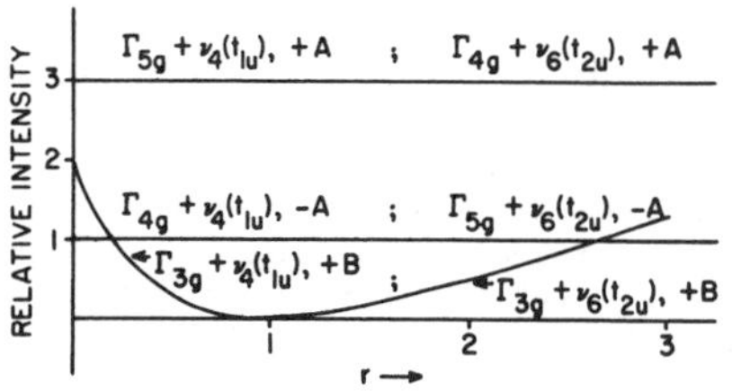

Figure 5. Relative intensities of transitions in the $\sigma e_g \rightarrow t_{2g}$ manifold assuming activation by vibronic mixing with the $\pi t_{2u} \rightarrow t_{2g}$ Γ_{4u} states; the parameter r is discussed in the text. The nature of the MCD features predicted is included. Intensities for ν_4 and ν_6 activated transitions are normalized separately.

MCl_6^{2-} compounds in the 5d transition metal series should be similar.[8,32] An analogous transition should be seen in $IrCl_6^{2-}$. None is found,[5,7,32] which suggests that this band is due to an effect specific to a d^4 system. MO calculations place the transition at higher energy (Figure 2).[17] Assuming the transition is activated by vibronic mixing with the Γ_{4u} states in band 8, we can characterize the possible vibronic transitions. The results are shown in Figure 5 in terms of a parameter, r, $r = (E - E_e)/(E - E_u)$. E is the energy of the vibronic line, and E_e and E_u are the frequencies of the no-phonon lines of the $e_u'' \rightarrow e_g''$ and $u_u' \rightarrow e_g''$ transitions in the $\pi t_{2u} \rightarrow t_{2g}$ manifold. Reasonable values for r are ~1.5. CT assignments consistent with the intensity and MCD of lines 9-B and 9-C require 9-A to be a weak transition and the occurrence of a fourth line of intermediate intensity with a −A feature in MCD. The d-d hypothesis predicts a moderately intense $\Gamma_{5g} + \nu_4$ state with $g = -0.82$;[3,29] this can account for the vibronic structure and MCD of line 9-B. If this is correct, the other two lines may correspond to two Γ_{4u} states of the upper levels of the $\pi t_{2u} \rightarrow t_{2g}$ manifold which are expected in this region (Figure 3). Under some circumstances electrostatic effects could account for the observed pattern of two lines, one with $g = 0$.

Little is known of band 10. It is a rather structureless transition of moderate intensity which is possibly a vibronically activated CT transition. As there is reason to believe the $\sigma t_{1u} \rightarrow t_{2g}$ transition occurs at 38,000 cm^{-1},[33] attribution of this band to a $\sigma e_g \rightarrow t_{2g}$ transition is suggested by MO calculations (Figure 2).[17]

The analysis we have given has correlated *all* features predicted to have moderate or large intensity with observed spectroscopic lines. Absent features (such as two of the $\Gamma_{5g} + \nu_4(t_{1u})$ d-d states) occur at energies where they are masked by very intense transitions.[3,29] The only tentative assignments are in bands 9 and 10. Transitions in other complexes have been assigned using similar arguments but with less assurance as the data do not limit attributions as strictly as in $OsCl_6^{2-}$. A few of these assignments warrant closer attention.

It has been proposed that bands 1 and 2 and band 4 in $OsBr_6^{2-}$ are $u_g' \rightarrow e_g''$ and $e_g' \rightarrow e_g''$ transitions, respectively, in the $\pi t_{1g} \rightarrow t_{2g}$ manifold.[9] While energetically reasonable, the vibronic structure of the bands is vastly different, which is unlikely for states arising from similar electronic configura-

(32) B. D. Bird, P. Day, and E. A. Grant, *J. Chem. Soc. A*, 100 (1970).
(33) G. C. Allen, *et al.*, *Inorg. Chem.*, 11, 787 (1972).

tions. An analysis, similar to that made for band 3 in $OsCl_6^{2-}$, indicates that this CT assignment of band 4 cannot simultaneously account for the MCD and the relative intensities. The possibility of a d-d assignment should be considered even though this raises other questions. Why is there just one when four or five are expected? Can the large shift in $t_{2g} \rightarrow e_g$ excitation energy from that in $OsCl_6^{2-}$ be rationalized?

In $IrBr_6^{2-}$ bands 6 and 8 have been assigned to the two components of the $\pi t_{2u} \rightarrow t_{2g}$ transition.[6] The very different vibronic structure of these two bands can be understood if each is activated by coupling with the no-phonon line of that band. In the lower transition ($e_u'' \rightarrow e_g''$) only a_{1g} and t_{1g} lattice modes are active. This accounts for both the MCD and the vibronic spacing if $\nu_9(t_{1g})$ is ~ 40 cm^{-1}, which is reasonable.[9] In the upper transition ($u_u' \rightarrow e_g''$), all vibrations are active and a far different pattern is possible. The suggestion that band 5 is the upper spin-orbit component of the $\pi t_{1g} \rightarrow t_{2g}$ CT transition[6] is unlikely for reasons similar to those given in the discussion of $OsBr_6^{2-}$.

In $IrCl_6^{2-}$ the very intense band A is extraordinary. Its intensity and energy relative to band C suggest, by analogy with all other MX_6^{2-} systems considered, that it arises from a $\pi t_{2u} \rightarrow t_{2g}$ excitation.[7,22] Therefore one expects two well-separated spin-orbit components, the lower twice as intense, with a $-$A feature followed by a +A feature in MCD. Nothing like this is observed; instead there is a doublet, split by only 5 cm^{-1}, with the lower frequency component far more intense which exhibits the expected MCD pattern.[7] If an extremely strong Ham effect[34] were operative this might be rationalized with the proposed CT assignment.[7] However, high-resolution studies of the ν_1 progressions based upon this doublet show that the intense line contains at least three components and cannot be correlated with a no-phonon state.[20] Furthermore, Zeeman studies indicate that neither feature obeys no-phonon selection rules.[20] If the $\pi t_{2u} \rightarrow t_{2g}$ CT states and some vibronic states arising from a $t_{2g} \rightarrow e_g$ transition were nearly degenerate, these anomalies might be accounted for. Band B is also interesting. Hot band studies suggest a vibronic transition. Tentative assignment[7] to a $\pi t_{2g} \rightarrow t_{2g}$ CT transition is unlikely: the MCD would require inversion of the spin-orbit levels of the πt_{2g} orbital;[7] a similar suggestion for $OsCl_6^{2-}$ was shown to be untenable; Zeeman studies indicate some magnetic moments far too small for components of a $\pi t_{2g} \rightarrow t_{2g}$ CT state.[20] Douglas's suggestion[5] of a d-d transition seems more reasonable, although the low magnetic moments cannot be accounted for by his assignment to a $^4T_{1g}$ strong coupling state.

(34) F. Ham, *Phys. Rev.*, 138, A1727 (1965).

In summation, critical analysis of the structure of the region II spectra in MX_6^{2-} systems indicates that few assignments of intermediate intensity bands can be made with confidence. There are even unanswered questions about the gross features of some intense bands. While many assignments can be ruled out, positive identifications are often difficult to make.

Further MCD studies on $RuCl_6^{2-}$ and $ReCl_6^{2-}$ would provide tests of our proposed assignments. Zeeman studies on the sharp bands in $IrBr_6^{2-}$ and $OsBr_6^{2-}$ would help determine whether these are CT or d-d transitions. Advances in crystal growing methods could provide better host lattices and therefore even higher resolution spectra.

A detailed theoretical study of $IrCl_6^{2-}$ is surely needed. Analysis of metal-ligand electrostatic interactions would be most useful to corroborate assignments in d^4 systems. Finally, a more complete treatment of the vibronic coupling problem including the effects of the crystal lattice should be undertaken.

I wish to thank Professor Paul Dorain for many helpful discussions and suggestions.

Cobalt-59 Nuclear Quadrupole Resonance Spectroscopy

Theodore L. Brown

School of Chemical Sciences and Materials Research Laboratory, University of Illinois, Urbana, Illinois 61801

Received April 4, 1974

Reprinted from Accounts of Chemical Research, *7, 408 (1974)*

Since the original observation in 1950 of a pure nuclear quadrupole resonance (nqr) transition in a solid,[1] a substantial literature dealing with nqr spectroscopy has developed.[2-4] Nevertheless, it is fair to say that it remains a rather specialized tool. In part this derives from the inherent limitations of the technique. Pure nuclear quadrupole resonance spectroscopy must be carried out on crystalline solids. The nqr spectrum is simpler, and therefore less liable to provide interesting and instructive detail than is often seen in nuclear magnetic resonance spectra of solutions. In addition, the instrumental requirements are considerably more varied than is the case for nmr. Finally, the acquisition of nqr data has proven quite difficult for several nuclear quadrupole systems which would be of great chemical interest, *e.g.*, ^{14}N, ^{2}H.

Theodore L. Brown is professor of chemistry in The School of Chemical Sciences, University of Illinois—Urbana. He was born in Green Bay, Wis., in 1928, received a B.S. degree in Chemistry from The Illinois Institute of Technology, and a Ph.D. from Michigan State University. Since then he has been a member of the faculty at Illinois. His research interests lie in the areas of inorganic and organometallic chemistry, with emphasis on nuclear quadrupole resonance spectroscopy, and kinetics and mechanistic studies of transition-metal organometallic compounds.

(1) H. G. Dehmelt and H. Krüger, *Naturwissenschaften*, 37, 111 (1950).
(2) T. P. Das and E. L. Hahn, "Nuclear Quadrupole Resonance Spectroscopy," Academic Press, New York, N. Y., 1958.
(3) E. Schempp and P. J. Bray, in "Physical Chemistry, An Advanced Treatise," Vol. IV, D. Henderson, Ed., Academic Press, New York, N. Y., 1970, Chapter II.
(4) E. A. C. Lucken, "Nuclear Quadrupole Coupling Constants," Academic Press, New York, N. Y., 1969.

Table I
Selected Quadrupolar Nuclides of Transition Elements[a]

Isotope	% abundance	Spin I	Electric quadrupole moment, eQ[b]
^{51}V	99.8	$\frac{7}{2}$	−0.04
^{53}Cr	9.54	$\frac{3}{2}$	+0.04[c]
^{55}Mn	100	$\frac{5}{2}$	0.40[d]
^{59}Co	100	$\frac{7}{2}$	0.40
^{63}Cu	69.1	$\frac{3}{2}$	−0.16
^{65}Cu	30.9	$\frac{3}{2}$	−0.15
^{95}Mo	15.72	$\frac{5}{2}$	0.12
^{97}Mo	9.46	$\frac{5}{2}$	1.1
^{101}Ru	16.98	$\frac{5}{2}$	?
^{105}Pd	22.2	$\frac{5}{2}$	0.73[e]
^{185}Re	37.1	$\frac{5}{2}$	2.8
^{187}Re	62.9	$\frac{5}{2}$	2.6
^{191}Ir	38.5	$\frac{3}{2}$	1.5
^{193}Ir	61.5	$\frac{3}{2}$	1.5

[a] Data from "Handbook of Chemistry and Physics," 54th ed, R. C. Weast, Ed., Chemical Rubber Company, Cleveland, Ohio, 1973, unless otherwise indicated. [b] Units are 1×10^{-24} cm^2. [c] J. A. Thompson, R. P. Scharenberg, W. R. Lutz, and R. D. Larsen, *Phys. Rev. C*, **7**, 1413 (1973). [d] E. Handrick, A. Steudel, and H. Walther, *Phys. Lett. A*, **29**, 486 (1969). [e] H. Leelavathi, S. Ismail, and K. H. Channappa, *Ind. J. Pure Appl. Phys.*, **10**, 820 (1972).

Despite these limitations, nqr spectroscopy is an important technique because it provides almost unique information about charge distribution. Furthermore, the continuing development of pulse techniques as the basis for commonly available laboratory instrumentation has provided the possibilities for much more sensitive and reliable methods for detection of nqr transitions in solids.

Many of the transition elements possess in significant abundances isotopes with $I > \frac{1}{2}$, a condition necessary for the presence of a nonzero nuclear quadrupole moment (Table I). Nuclear quadrupole resonance spectroscopy utilizing these nuclides is a potentially valuable, but largely untapped, source of bonding and structural information. The nqr spectroscopy of cobalt-59 is perhaps best developed of that for any transition element. This account of the present state of the art regarding cobalt-59 nqr spectroscopy provides an indication of the scope of the technique and its potential for application to many other elements.

Theory

For a nucleus with spin $I > \frac{1}{2}$, an energy term arises due to the interaction of the nuclear quadrupole moment, eQ, with the components V_{ij} of the electric field gradient (efg) at the nucleus.[5] The Hamiltonian is written as

$$\mathcal{H}_Q = \tfrac{1}{6}\sum_{i,j} V_{ij}Q_{ij} \qquad (1)$$

In the principal axis system, the off-diagonal elements of the field gradient tensor V_{ij} are zero. The divergence theorem leads to the LaPlace equation, $V_{xx} + V_{yy} + V_{zz} = 0$, *i.e.*, the tensor is traceless. Thus, it possesses two independent components. The largest efg component is defined as V_{zz}; it is frequently referred to as eq_{zz} or just q. The departure of the efg tensor from axial symmetry is given by the asymmetry parameter η (eq 2). The axes are chosen

$$\eta = \frac{V_{xx} - V_{yy}}{V_{zz}} \qquad (2)$$

so that $|V_{xx}| < |V_{yy}| < |V_{zz}|$; this means that $0 < \eta < 1$. Using these conventions, the most common form of the Hamiltonian is given by

$$\mathcal{H}_Q = \frac{e^2q_{zz}Q}{4I(2I-1)}[3(I_z^2 - I^2) + \eta(I_x^2 - I_y^2)] \qquad (3)$$

The product $e^2q_{zz}Q$ or $e^2q_{zz}Q/h$ is termed the quadrupole coupling constant. The quadrupole coupling constant and the asymmetry parameter, η, reflect the two independent components of the field gradient tensor.

The energy level diagram for spin $I = \frac{7}{2}$ is shown in Figure 1. It follows from the expression for $\mathcal{H}_Q$, eq 3, that the energy depends only on the absolute value of m, the nuclear magnetic spin quantum number. The selection rules for quadrupole transitions in the limiting case $\eta = 0$ require that $\Delta m = \pm 1$. When η departs significantly from 0, a mixing of spin states occurs, and other transitions become allowed. Transitions between the allowed nuclear quadrupole energy states can be produced by interaction of the magnetic component of a radiofrequency field of appropriate frequency with the nuclear magnetic moment, which is collinear with the nuclear quadrupole moment. When $\eta = 0$ the three allowed transitions for $I = \frac{7}{2}$ are in the frequency ratio 3:2:1 (Figure 1). As the asymmetry parameter departs from 0, however, the transition energies change, as shown in Figure 1.

In the presence of a magnetic field, the degeneracy of m states of equal absolute value is removed. We will not be concerned, however, in this account with experiments involving significant Zeeman contributions to the Hamiltonian.

Origins of Field Gradients

The observed quadrupole coupling constant represents the product of the nuclear quadrupole moment eQ and the major component V_{zz} ($=eq_{zz}$) of the electric field gradient tensor. The quantity q_{zz} represents the expectation value for the one-electron field gradient operator, taken over the entire charge distribution external to the nucleus

$$q_{zz} = \left\langle \psi_{el} \left| \frac{3\cos^2\theta - 1}{r^3} \right| \psi_{el} \right\rangle + \sum_i Z_i \left(\frac{3\cos^2\theta_i - 1}{r_i^3} \right) \qquad (4)$$

The summation is taken over all of the nuclei and electrons in the space surrounding the nucleus in question. In practice, in dealing with neutral molecules or polyatomic ions in lattices, it is frequently necessary to consider only those atoms and electrons in the molecule or ion containing the quadrupolar nucleus. The effect of external nuclear charges is largely canceled by the spherical electron distributions centered on those nuclei. Further, because of the r^{-3} dependence of the field gradient operator, only charges in the close vicinity of the quadrupolar nucleus affect the field gradient significantly. Ac-

(5) See ref 4, Chapters 3–7, and ref 2 and 3 for more detailed exposition of theory.

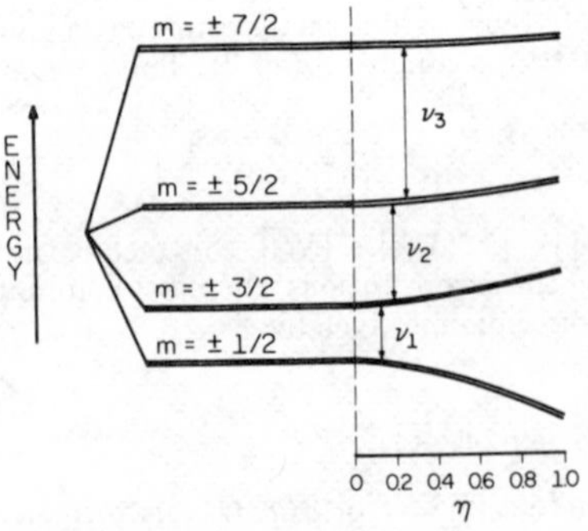

Figure 1. Nuclear quadrupole energy level diagram for $I = 7/2$.

cordingly, it is generally possible to consider the field gradient as arising from electrons in valence orbitals centered on the quadrupolar nucleus, and—in some cases—a set of effective charges on immediately adjacent atoms.

Both valence electrons and charges external to the atom (ion) containing the quadrupolar nucleus can contribute to polarization of the inner core electrons to produce Sternheimer shielding or antishielding contributions to the field gradients.[6]

Occupancy of the 4s, 4p, and 3d orbitals of the first-row transition element may be expected to lead to significant nonzero contributions to q_{zz}. Table II shows the calculated field gradient due to a single electron in a metal $3d_{z^2}$ orbital (q_{320}) for three isoelectronic metals with a $4s^2 3d^6$ configuration (^5D state), based upon Clementi's Hartree–Fock SCF atomic functions.[7,8] The value for q_{320} increases markedly with increasing nuclear charge. The table also shows the value of q_{320} for Fe(0) in the $4s^1 3d^6 4p^1$ configuration (^7P state).[8] The change in configuration does not produce a large change in q_{320}. The field gradient due to the $4p_z$ electron, q_{410}, for the ^7P state of Fe is calculated to be -4.97×10^{-15} esu cm^{-3}.[8] This is not a great deal smaller than q_{320}. Thus, although in general the 4p orbitals have much lower occupancies than the 3d, it may not always be justifiable to dismiss the 4p electron contribution to q_{zz}.

$$q_{zz} = q_{320}[N_{d_{z^2}} + \tfrac{1}{2}(N_{d_{xz}} + N_{d_{yz}}) - (N_{d_{xy}} + N_{d_{x^2-y^2}})] + q_{410}[N_{p_z} - \tfrac{1}{2}(N_{p_x} + N_{p_y})] \quad (5)$$

The major component of the electric field gradient may be expressed in terms of valence orbital populations as in eq 5.[9]

Accounting for the quadrupole coupling constant in terms of the valence orbital populations in this manner amounts to the assumption that external charge distributions are of negligible consequence. This is likely to be a safe assumption in many organometallic compounds of cobalt, in which the metal has a near-zero charge, and the charges on atoms immediately adjacent to the metal are also small. On the other hand, in coordination compounds of Co(II) and

Table II
Comparative Calculated Field Gradients for a 3d Electron

Atom (ion)	Configuration	State	q_{320}, 10^{15} esu cm^{-3}
Mn^-	$4s^2 3d^6$	^5D	−6.70
Fe	$4s^2 3d^6$	^5D	−9.23
Co^+	$4s^2 3d^6$	^5D	−12.16
Fe	$4s^1 3d^6 4p^1$	^7P	−8.79

Co(III), the external charge distribution around the metal may make a major contribution to the field gradient. It is essentially impossible to separate contributions to the field gradient due to point charges located external to the valence orbitals from contributions due to donation of electrons into valence orbitals of the metal. The "donated-charge" model[10] treats the elements of the field gradient tensor as arising from a summation of effective contributions from individual ligands, as in eq 6–8, where the angles θ_i and ϕ_i are measured from the Z and X axes to each of the i ligands. In these equations the parameter $[L_i]$ associated with each ligand may be viewed as the sum of two contributing terms

$$V_{zz} = \sum_i [L_i](3\cos^2\theta_i - 1) \quad (6)$$

$$V_{xx} = \sum_i [L_i](3\sin^2\theta_i \cos^2\phi - 1) \quad (7)$$

$$V_{yy} = \sum_i [L_i](3\sin^2\theta_i \sin^2\phi_i - 1) \quad (8)$$

$$[L_i] = e^2Q\left[\frac{q_i(1 - \gamma_{r_i})}{r_i^{3}} + \frac{q'_i(1 - R)}{\langle r'_i{}^3\rangle}\right] \quad (9)$$

The symbol γ_{r_i} represents the Sternheimer correction factor for charges q_i resident on the ligand atom i at distance r_i.[6] Since these charges are external to the valence orbitals of the central atom, γ_{r_i} will in general be negative, *i.e.*, an antishielding contribution. The Sternheimer shielding correction for charge donated by the ligands to the valence orbitals is represented by R. The "lattice" or external charge contribution to [L] may be viewed as an effective point charge q_i located at the donor atom, at distance r_i from the quadrupolar nucleus. The covalency contribution is more subtle; $q'_i/\langle r'_i{}^3\rangle$ represents the expectation value for the radial component of the field gradient due to charge donated from L into valence orbitals of the central atom.

Even when the bond to L is quite polar, the second, covalency, term is the more important, because $\langle r'_i{}^{-3}\rangle$ is much larger than $r_i{}^{-3}$. The magnitudes of the Sternheimer terms are not well known. R is in general on the order of 0 to 0.2. γ_{r_i}, as computed for free atoms and ions, is large at large distances from the central nucleus.[6] (For example, γ_∞ has been estimated to be about 8 for Co metal.[11]) At typical metal–ligand distances γ_{r_i} may be far from attaining the maximum value. Furthermore, the effect of the ligand field on the central atom may be to quench antishielding contributions.[12,13]

(6) Reference 4, Chapter 3.
(7) G. Malli and S. Fraga, *Theor. Chim Acta*, **6**, 54 (1966).
(8) C. D. Pribula, T. L. Brown, and E. Münck, *J. Amer. Chem. Soc.*, **96**, 4149 (1974).
(9) T. L. Brown, P. A. Edwards, C. B. Harris, and J. L. Kirsch, *Inorg. Chem.*, **8**, 763 (1969).
(10) See R. V. Parish, *Prog. Inorg. Chem.*, **15**, 124 (1972), for discussion.
(11) T. P. Das and M. Pomerantz, *Phys. Rev.*, **123**, 2070 (1961).
(12) H. W. deWijn, *J. Chem. Phys.*, **44**, 810 (1966).

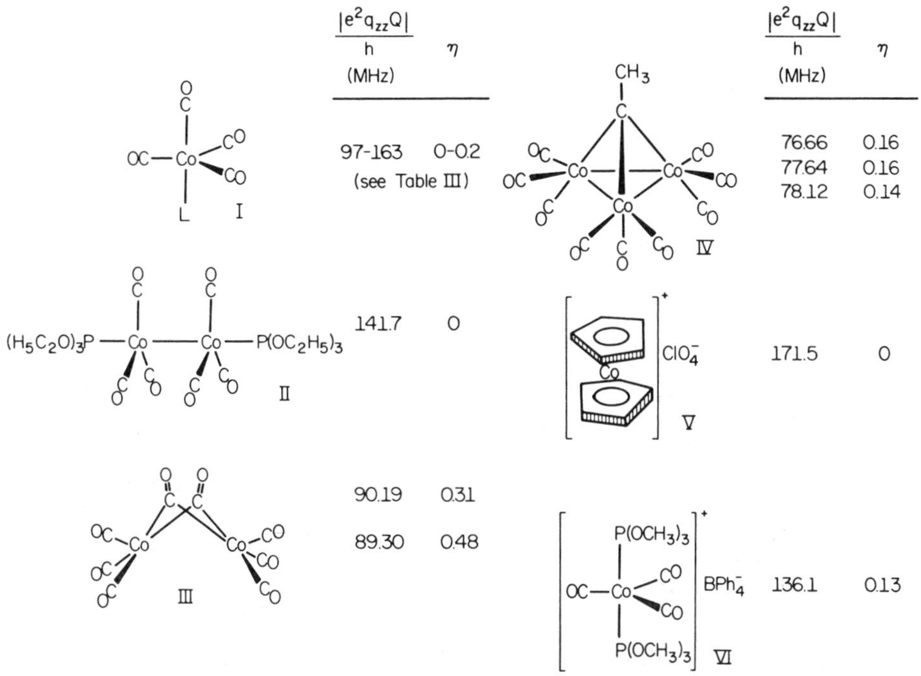

Figure 2. Geometrical structures and field-gradient tensor parameters for several organometallic cobalt compounds.

This empirical approach, which treats the contribution from each ligand as a parameter characteristic of that ligand, gives rise to the additive partial field gradient (pfg) model, which has been employed extensively in interpretation of ^{57}Fe and ^{119}Sn Mössbauer quadrupole splittings.[10,14,15] We shall remark upon applications of this model to ^{59}Co nqr spectra in a later section.

Five-Coordinate Co(I) Compounds

During the past few years an extensive body of ^{59}Co nqr data for 5-coordinate cobalt carbonyl compounds has accumulated. It is useful to think of these compounds as derived from a parent $Co(CO)_5^+$ species isostructural with $Fe(CO)_5$ and $Mn(CO)_5^-$, which are trigonal bipyramidal. From Mössbauer spectra[16,17] and ^{55}Mn nqr spectra,[8] respectively, the field gradients eq_{zz} for the latter two species are +4.4 and 2.2, in units of 10^{15} esu cm^{-3}. Stable $Co(CO)_5^+$ salts are not known to date, but phosphite- and phosphine-substituted derivatives are. Comparisons of eq_{zz} for these and the analogous iron compounds suggest that eq_{zz} for $Co(CO)_5^+$ should be on the order of 5.4×10^{15} esu cm^{-3},[8] corresponding to an e^2Qq_{zz}/h of about 156 MHz.

The sign of q_{zz} in $Fe(CO)_5$ is known from the Mössbauer work to be positive. Since q_{320} and q_{410} (eq 5) are both negative, this requires that the negative terms within the brackets dominate. (It might be noted parenthetically that the 3d orbital populations make the major contribution to q_{zz}.) The evidence strongly suggests that q_{zz} has the same sign for the cobalt species. A larger value of eq_{zz} for the cobalt as compared with the iron species is consistent with the calculated effect of increasing nuclear charge on the values for q_{320}, as indicated in Table II.

Replacement of all five CO groups by $P(OCH_3)_3$ groups results in only a very small change in field gradient at cobalt; e^2Qq_{zz}/h for $[Co(P(OCH_3)_3)_5]BPh_4$ is 153.7 MHz.

Replacement of one of the axial CO groups in $Co(CO)_5^+$ by a ligand of formal negative charge produces a series of trigonal 5-coordinate cobalt species of the geometry shown as I in Figure 2. Table III lists values of quadrupole coupling constants for several $Co(CO)_4X$ compounds.[9,18–21] These results show that groups which are poor σ donors and/or strong π acceptors cause large quadrupole coupling constants at the metal. This is consistent with the assumption that the d_{xy} and $d_{x^2-y^2}$ orbitals are more highly populated than the d_{z^2} and d_{xz}, d_{yz}. Strongly σ-donor groups should increase the populations of the d_{z^2} orbital, thus decreasing the deficiency of electron density in the region of positive contribution to q_{zz}. On the other hand, π-acceptor ligands, by removing electron density principally from d_{xz} and d_{yz}, should increase this difference, and thus increase q_{zz}. The σ-donor and π-acceptor characteristics of the ligands thus work in opposite directions in the sense that σ donation toward cobalt from axial ligands tends to decrease q_{zz}, whereas π-acceptor action tends to increase it.

(13) G. Burns and E. G. Wikner, *Phys. Rev.*, **121**, 155 (1961).
(14) G. M. Bancroft and R. H. Platt, *Advan. Inorg. Chem. Radiochem.*, **15**, 59 (1972).
(15) G. M. Bancroft, *Coord. Chem. Rev.*, **11**, 247 (1973).
(16) R. L. Collins and R. Pettit, *J. Amer. Chem. Soc.*, **85**, 2332 (1963).
(17) P. Kienle, *Phys. Verd.*, **3**, 33 (1963).
(18) D. D. Spencer, J. L. Kirsch, and T. L. Brown, *Inorg. Chem.*, **9**, 235 (1970).
(19) A. N. Nesmeyanov, G. K. Semin, E. V. Bryuchova, K. N. Anisimov, N. E. Kolobova, and V. N. Khandozhko, *Izv. Akad. Nauk SSSR, Ser. Khim.*, 1936 (1969).
(20) H. W. Spiess and R. K. Sheline, *J. Chem. Phys.*, **53**, 3036 (1970).
(21) T. E. Boyd and T. L. Brown, *Inorg. Chem.*, **13**, 422 (1974).

Table III
Cobalt-59 Nqr Data for $Co(CO)_4X$ Compounds

X^a	e^2Qq_{zz}, MHz^b	η	Ref
$-SnCl_3$	161.45	0.0	9
$-SnBr_3$	159.88	0.072	9, 20
$-SnI_3$	153.04	0.06	9
$-GeCl_3$^{c}	161.48	0.030	9, 20
	159.37	0.039	
$-SiCl_3$	130.67	0.13	9
$-HgCo(CO)_4$^{c}	112.0	0.05	18
	110.8	0.0	
$-SiPh_3$	101.09	0.0	9
$-GePh_3$	109.63	0.05	21
$-SnPh_3$	104.00	0.285	9, 20
$-PbPh_3$	110.82	0.05	9
$-Sn(CH_3)_3$	96.8	0.03	18

^{a} Ph = C_6H_5. ^{b} At 298 K. Data at 77 K are reported in ref 19. ^{c} Separate resonances due to crystallographically distinct species.

For compounds of the form $X_nSn[Co(CO)_4]_{4-n}$ (X = Cl, Br, CH_3; n = 1, 2, or 3), the variation in quadrupole coupling constant at cobalt[18,19] correlates smoothly with the highest frequency CO stretching mode for the compounds measured in hexane solution. In the case of neither observable, however, is it readily possible to separate the σ and π effects.

Other Organometallic Cobalt Compounds

Figure 2 shows nqr data for several organometallic compounds in which the chemical environment and symmetry about cobalt change. The nqr data listed beside each structure clearly reflect the symmetry of the environment about the metal. In I and II the cobalt is in an essentially axial environment. In $Co_2(CO)_8$, III, on the other hand, the environment is decidedly nonaxial. Mooberry, Pupp, Slater, and Sheline[22] have carried out an elegant single-crystal ^{59}Co broad-line nmr study of $Co_2(CO)_8$.

The three crystallographically nonequivalent cobalt atoms in the solid-state structure of IV[23] possess closely similar field gradient tensor parameters.[24] The essentially axial symmetry about cobalt in structures V and VI is reflected in the near-zero values for η. Several other cationic cobalt carbonyl species have been reported.[8]

Cobalt (III) Complexes

Among the earliest ^{59}Co nqr measurements of chemical interest were those on diamagnetic, low-spin d^6 complexes of cobalt(III).[26-28] In these compounds the metal is surrounded by six coordinated ligands in a roughly octahedral array. In some attempts to interpret the nqr data explicit account has been taken of three contributions to the efg tensor at cobalt: (a) the valence electrons of the central metal ion, (b) the ionic charges on the ligand atoms bonded to the central atom, and (c) the ionic charges external to the complex ion (lattice charges). In the compound $[Co(NH_3)_5Cl]Cl_2$, for example,[29] the chlorine nqr resonances provide, *via* the Townes–Dailey model, an estimate of the ionicity of the Co–Cl bond. Contribution b was evaluated by assuming that net charges reside at the sites of the ligated atoms and that the operative value for $1 - \gamma$ is 8. Contribution c is easily evaluated from a knowledge of the crystal structure, assuming that $1 - \gamma_\infty = 8$; it proves to be negligibly small. Thus, contributions a and b must sum to give the observed field gradient at cobalt. If it is assumed that $q > 0$, as seems much the more reasonable choice,[30] the charges on the atoms involved are: Co, −0.48; N, +0.63; Cl, −0.67. The last of these comes from the assumption that the Townes–Dailey model is applicable to the nqr data for the chlorine bound to cobalt. The fact that the charges on nitrogen and cobalt turn out unreasonable is probably the result of the assumptions made in evaluating contribution b. Unfortunately, there is at the present no compelling theoretical model on the basis of which one might choose an appropriate value for $1 - \gamma_r$ in dealing with ligands, and $1 - \gamma_\infty$ in dealing with external lattice charges. For this reason it is useful to formulate a model for the efg tensor at the metal in terms of partial field gradient contributions from the ligands attached to the metal, as expressed in eq 6–8. The concept of partial field gradient parameters leads at once to a few simple predictions.[30,31] If it is assumed that the pfg parameter for a ligand is relatively insensitive to the other ligands bound to the metal and that 90° ligand–metal–ligand bond angles obtain around the metal, then the quadrupole coupling constant for CoA_5X should be about half that for *trans*-CoA_4X_2 and equal to that for *cis*-CoA_4X_2.

Table IV
^{59}Co Nqr Data for Cobalt (III) Complexes

Compound	$e^2q_{zz}Q/h$	η	Ref
$[Co(NH_3)_5Cl]Cl_2$	31.74	0.215	29
trans-$[Co(NH_3)_4Cl_2]Cl$	59.23	0.136	28
cis-$[Co(en)_2Cl_2]NO_3$	33.71	0.173	32
trans-$[Co(en)_2Cl_2]NO_3$	62.78	0.132	28
trans-$[Co(en)_2Cl_2]Cl$	60.63	0.272	28
trans-$[Co(en)_2Cl_2]ClO_4$	60.22,	0.149	33
	59.92	0.149	
trans-$[Co(en)_2Cl_2]Cl \cdot H_2O_5{}^+Cl^-$	71.73	0.222	26

Table IV shows data for several cobalt(III) complexes.[26,28,29,32,33] The first and second pairs of compounds illustrate the "rules" stated above. The last four compounds in the table show the degree to which the ions external to the coordination sphere itself affect the efg at the metal. Except for the last compound listed, in which strong hydrogen bonding to the axial chlorides seems possible,[34] the field gra-

(22) E. S. Mooberry, M. Pupp, J. L. Slater, and R. K. Sheline, *J. Chem. Phys.*, **55**, 3655 (1971).
(23) P. W. Sutton and L. F. Dahl, *J. Amer. Chem. Soc.*, **89**, 261 (1967).
(24) T. L. Brown and J. P. Yesinowski, unpublished observations. ^{59}Co field gradient tensor parameters in $C_6H_5CCo_3(CO)_9$ and $ClCCo_3(CO)_9$ are very similar to those for $CH_3CCo_3(CO)_9$.
(25) (a) J. Voitländer and R. Longino, *Naturwissenschaften*, **46**, 664 (1959); (b) J. Voitländer, H. Klocke, R. Longino and H. Thiene, *ibid.*, **49**, 491 (1962).
(26) H. Hartmann, M. Fleissner, and H. Sillescu, *Theor. Chim. Acta*, **2**, 63 (1964).
(27) H. Hartmann and H. Sillescu, *Theor. Chim. Acta* **2**, 371 (1964).
(28) I. Watanabe and Y. Yamagata, *J. Chem. Phys.*, **46**, 407 (1967).
(29) I. Watanabe, H. Tanaka, and T. Shimizu, *J. Chem. Phys.*, **52**, 4031 (1970).
(30) G. M. Bancroft, *Chem. Phys. Lett.*, **10**, 449 (1971).
(31) R. R. Berrett and B. W. Fitzimmons, *J. Chem. Soc. A*, 525 (1967).
(32) I. Watanabe, *J. Chem. Phys.*, **57**, 3014 (197).
(33) T. B. Brill and Z. Z. Hugus, Jr., *J. Phys. Chem.*, **74**, 3022 (1970).
(34) Brill and Hugus account for the larger value for *trans*-$[Co(en)_2Cl_2]Cl \cdot H_2O_5{}^+Cl^-$ in terms of a larger calculated lattice contribution. However, they employ the estimated free atom value of 8[11] for ($1 - \gamma_\infty$).

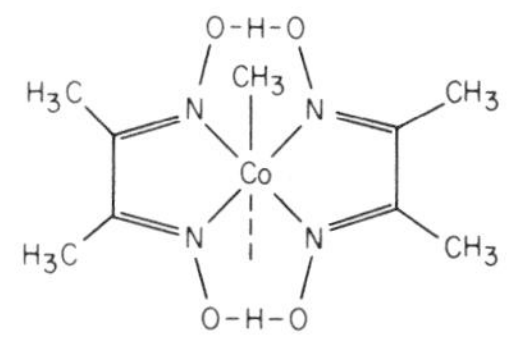

Figure 3. Structure of bis(dimethylglyoximato)cobalt(III) complexes (cobaloximes).

dient is not markedly altered by a change in anion. The asymmetry parameter is, however, noticeably affected by lattice forces, as evidenced by the nonzero values for $Co(NH_3)_5Cl^{2+}$ and *trans*-$Co(NH_3)_4Cl_2^+$. There is reason to believe, on the basis of calculations,[29,33] that the external lattice charge contribution to the field gradient is ordinarily too small to cause much effect on either $e^2q_{zz}Q/h$ or η. Nonzero values for η may be due to small but significant displacements of ligands away from fourfold axial symmetry. In the absence of specific interactions between ligands and external charges, and where the geometry about the metal can be well defined, it should thus be possible to interpret both $e^2q_{zz}Q/h$ and η in terms of the electronic interactions between metal and ligands.

The series of cobalt(III) complexes known as cobaloximes[35] afford an interesting series for study of ^{59}Co nqr because their geometrical structures are well defined. It is thus possible to employ the nqr data to learn something about the interaction of the ligands with the central metal.[36] The general structure of the cobaloximes is shown as in Figure 3. The cobalt is coordinated by a planar array of nitrogens which are part of two dimethylglyoxime (dh) monoanions. The axial ligands may be varied over a wide range in terms of donor capabilities.

Because the planar nitrogen atoms make two distinct angles with the metal (about 80 and 100°),[37–41] the cobaloximes do not possess fourfold axial symmetry. They should thus not be expected in general to possess small asymmetry parameters. Use of the additive ligand pfg model makes it possible to define the field gradient tensor at Co in terms of the pfg parameters for the planar nitrogens, [N], and the average pfg parameter of the two axial ligands, [X] = ½([A] + [B]).

$$eq_{kk} = 4[X] - 4[N] \quad (10)$$

$$eq_{jj} = -2[X] + 4[N][3\cos^2(40°) - 1] \quad (11)$$

$$eq_{ii} = -2[X] + 4[N][3\sin^2(40°) - 1] \quad (12)$$

The simplest initial assumption is that [N] is independent of the pfg parameters for the axial ligands. It develops, however, that for a series of complexes CoN_4AB, in which the axial ligands A and B are varied widely, the experimental data can be fitted to pfg parameters [N] and [X] only if it is assumed that [N] decreases as the ratio [N]/[X] grows smaller, *i.e.*, as the axial ligands become more strongly donating. This result is in contrast to the conclusions reached on the basis of Mössbauer studies of many low-spin Fe(II) complexes, in which the pfg parameters of ligands are generally assumed independent of the pfg parameters of other ligands on the same metal.

Table V
Ratio of pfg Parameters of the Planar Nitrogen Ligands to Axial Ligands in [Co(dh)₂AB] Complexes[36]

A, B	[N] / ½([A] + [B])
Br, Br	4.3
Cl, Cl	3.6
Cl, NC_5H_5	3.1
Cl, PPh_3	2.8
Cl, $P(n\text{-}C_4H_9)_3$	1.9
CH_3, CH_3OH	1.8
$CHCl_2$, $S(CH_3)_2$	1.6
CH_3, $N(CH_3)_3$	1.4
CH_3, IMID[a]	1.35
CH_3, NC_5H_5	1.10
CH_3, PPh_3	0.73
CH_3, $S(CH_3)_2$	0.68
CH_3, $AsPh_3$	0.67
CH_3, $P(OCH_2)_3CC_2H_5$	0.51
CH_3, $P(OCH_3)_3$	0.56
CH_3, $P(n\text{-}C_4H_9)_3$	0.54

[a] IMID = 1-(2-trifluoromethylphenyl)imidazole.

The complex possessing the most weakly donating axial ligand system observed is represented by $Co(dh)_2Cl_2^-$ (dh = dimethylglyoximato monoanion), and the strongest by $CH_3Co(dh)_2P(n\text{-}C_4H_9)_3$. If it is assumed that the pfg parameter for [N] varies smoothly between these two extreme cases, it is possible to construct a graph of eq_{zz} and η as a function of the ratio [N]/[X]. Such a graph,[36] corresponding to 80 and 100° N–Co–N bond angles, is shown in Figure 4. From the observed values for $e^2q_{zz}Q/h$ and η it is possible in principle to locate positions on the curves which give agreement between calculated and observed quantities.

If the field gradient parameters could be rigorously assumed to depend only on the effects of the immediately coordinated atoms, accurate values for $e^2q_{zz}Q/h$ and η would make it possible to assign quite accurately values to [N]/[X].

Despite certain limitations, it is possible to obtain an extensive and fairly reliable list of [N]/[X] values for a series of $Co(dh)_2AB$ complexes (Table V).

If we assume that for a series of axial ligands the variations in effective distances are small with respect to the other variations, and that covalency dominates in determining the pfg value, then the variation in axial ligands which occasions variation in [N]/[X] may be associated with changes in the strength of covalent binding of the axial ligand to the metal.

One expects that the paramagnetic contributions should dominate in determining the ^{59}Co nmr chemical shifts in cobaloximes, and that these contributions should decrease as the ligand field produced by

(35) G. N. Schrauzer, *Accounts Chem. Res.*, **1**, 97 (1968).
(36) R. A. LaRossa and T. L. Brown, *J. Amer. Chem. Soc.*, **96**, 2072 (1974).
(37) (a) J. S. Swanson, Ph.D. Thesis, University of Illinois, Urbana, Ill., 1971; (b) W. W. Adams and P. G. Lenhert, *Acta Crystallogr., Sect. B*, **29**, 2412 (1973).
(38) P. G. Lenhert, *J. Chem. Soc., Chem. Commun.* 980 (1967).
(39) K. S. Viswanathan and N. R. Kuncher, *Acta Crystallogr.*, **14**, 675 (1961).
(40) D. L. McFadden and A. T. McPhail, *J. Chem. Soc., Dalton Trans.*, 363 (1974).
(41) S. Brücker and L. Rondaccio, *J. Chem. Soc., Dalton Trans.*, 1017 (1974).

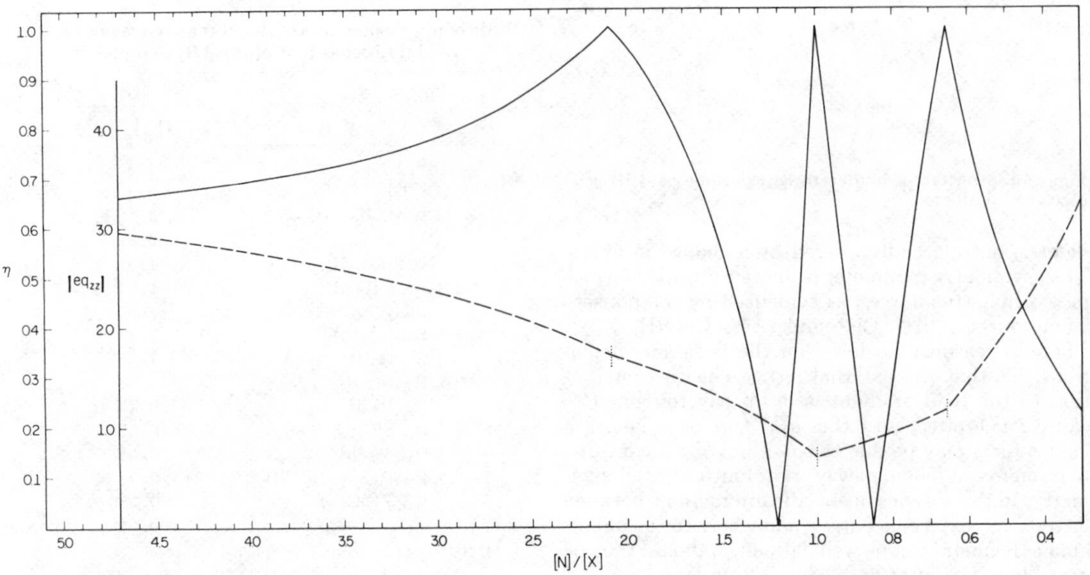

Figure 4. Scaled field-gradient and asymmetry parameter *vs.* ratio of ligand pfg parameters for $[CoN_4L_2]$ complexes (∠N–Co–N = 80 and 100°).

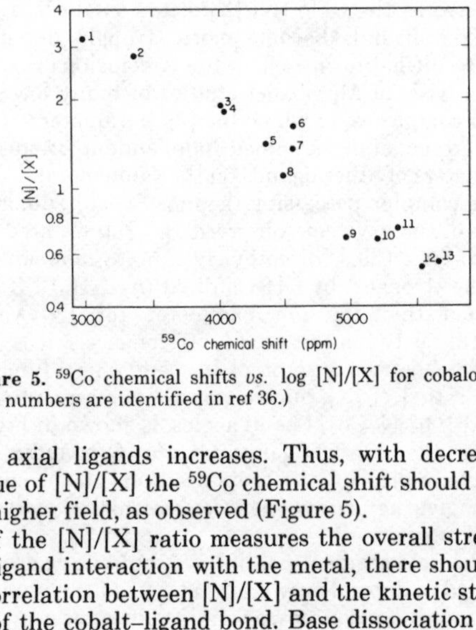

Figure 5. ^{59}Co chemical shifts *vs.* log [N]/[X] for cobaloximes. (The numbers are identified in ref 36.)

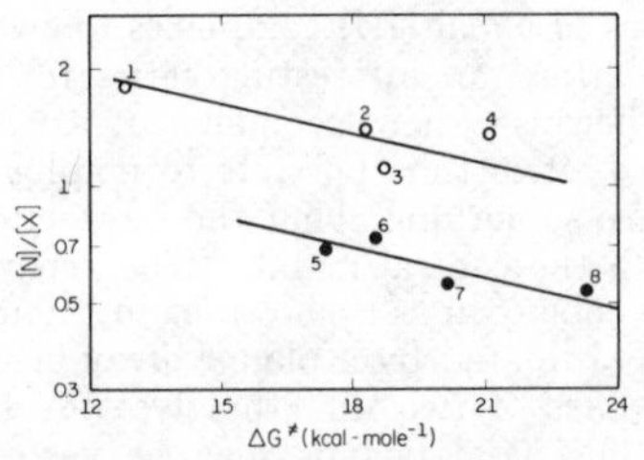

Figure 6. Log [N]/[X] *vs.* ΔG^* for dissociation of axial ligand from $CH_3Co(dh)_2L$ compounds. L groups as follows: 1, CH_3OH; 2, $N(CH_3)_3$; 3, NC_5H_5; 4, IMID; 5, $S(CH_3)_2$; 6, $P(C_6H_5)_3$; 7, $P(OCH_3)_3$; 8, $P(n\text{-}C_4H_9)_3$. (The values of ΔG^* for the CH_3OH and NC_5H_5 complexes were estimated from the values obtained for the $[CH_3Co(tmed)CH_3OH]BPh_4$ and $CH_3Co(dh)_2$(3-fluoropyridine) complexes, respectively.[42c] Only small corrections are involved.)

the axial ligands increases. Thus, with decreasing value of [N]/[X] the ^{59}Co chemical shift should move to higher field, as observed (Figure 5).

If the [N]/[X] ratio measures the overall strength of ligand interaction with the metal, there should be a correlation between [N]/[X] and the kinetic stability of the cobalt–ligand bond. Base dissociation from $CH_3Co(dh)_2L$ complexes has been found[42] to be purely dissociative; rupture of the cobalt–ligand bond is the major contributor to the enthalpy of activation.[43] In Figure 6 is shown a graph of log [N]/[X] *vs.* ΔG^* for base dissociation. It is clear that all the compounds do not obey a single monotonic relationship. On the other hand, it is possible to discern a regular relationship in the expected direction for the third-row bases. Similarly, the second-row bases seem to follow a relationship of similar slope, but displaced upward in terms of [N]/[X].

The different behavior of the second-row bases from those in the third row may arise from different relative importances of electrostatic and covalent components of the metal–ligand bond in affecting [L]. One would expect the covalent component, which gives rise to the larger field gradient contribution, to be larger for the heavier bases. Thus [N]/[X] should be smaller. The changed relative importances of the two contributions to the bonding need not, however, lead to a very great difference in metal–ligand bond strength, as reflected in ΔG^*.

The different behavior of the second and third row bases might also arise in part from π bonding to the third row axial ligands. The fact that cobalt in the methylatocobaloximes is capable of π bonding has been suggested on the basis of several lines of evidence. The presence of an additional energy term ascribable to π bonding has been noted in the cobalt–$P(OCH_3)_3$ interaction.[44]

Future Prospects

The examples described above demonstrate the potential usefulness of nqr as a tool in determining certain characteristics of the environment around a

(42) (a) T. L. Brown, L. M. Ludwick, and R. S. Stewart, *J. Amer. Chem. Soc.*, **94**, 384 (1972); (b) R. J. Guschl and T. L. Brown, *Inorg. Chem.*, **12**, 2815 (1973); (c) R. J. Guschl and T. L. Brown, *ibid.*, **13**, 959 (1974).

(43) W. Trogler, R. C. Stewart, and L. G. Marzilli, *J. Amer. Chem. Soc.*, **96**, 3697 (1974).

(44) R. L. Courtwright, R. S. Drago, J. A. Nusz, and M. S. Nozari, *Inorg. Chem.*, **12**, 2809 (1973).

transition metal. It is easy to imagine that the technique could have great value in probing the environment about cobalt in many other cobalt(III) compounds, including vitamin B_{12}, methylcobalamin, coenzyme B_{12}, and others. Similarly, compounds of several other transition elements might be studied. Unfortunately, the techniques usually employed by chemists in the past have lacked the requisite sensitivity to permit observation of nqr spectra for many nuclei of interest.

The most promising approaches for the future seem to lie in the use of double resonance techniques,[45,46] capable of providing enormously increased sensitivity in observing quadrupolar transitions, and permitting observation of transitions of very low frequency, such as in ^{2}H nqr spectra, where the transitions are observed in the 100–160-kHz region.[47] A most spectacular demonstration of the high sensitivity inherent in one of the techniques is the observation of ^{17}O nqr in natural abundance (0.037%) in several quinones, by means of an adiabatic demagnetization in the lab frame, level-crossing, double-resonance experiment.[48] Extension of these techniques to observation of the quadrupole resonance spectra of transition elements in diamagnetic compounds seems entirely feasible.

Our research efforts in nqr have been supported by The Advanced Research Project Agency through contract SD-131 with the Materials Research Laboratory, and by the National Science Foundation through Contract GH 33634 and through Grants GP 6396X and GP 30256X. I am indebted to many of my former students for their collaborative efforts, but most especially to Cheryl Pribula and Robert LaRossa.

(45) S. R. Hartmann and E. L. Hahn, *Phys. Rev.*, **128,** 2042 (1962).
(46) R. E. Slusher and E. L. Hahn, *Phys. Rev.*, **166,** 332 (1968).
(47) J. L. Ragle and K. L. Sherk, *J. Chem. Phys.*, **50,** 3553 (1969).
(48) Y.-N. Hsieh, J. C. Koo, and E. L. Hahn, *Chem. Phys. Lett.*, **13,** 563 (1972).
(49) L. S. Chia, W. R. Cullen, M. C. L. Gerry, and E. C. Lerner, *Inorg. Chem.*, **14,** 2975 (1975).

Addendum (May 1976)

A study of ^{59}Co nqr spectra of derivatives of dicobalt octacarbonyl has recently appeared.[49] The most exciting new development since the Account was published has been the emergence of double resonance techniques. The spin echo double resonance (SEDOR) technique has been used to observe the ^{2}D quadrupole coupling constant in $DMn(CO)_5$.[50] The experiment yielded the unexpected benefit of an accurate value for the D—Mn distance. The SEDOR technique has wide potential application to organometallic systems.

a second double resonance technique that has produced highly interesting nqr data is adiabatic demagnetization in the lab frame. For detection of the quadrupole transitions of an abundant but difficult nuclear system such as ^{14}N, adiabatic demagnetization is combined with level crossing. Edmonds and co-workers have produced several very nice ^{14}N and ^{2}D nqr studies using this technique.[51] We have used it to observe the ^{14}N nqr spectra of a variety of coordinated nitrogen bases, including glyoximato,[52] pyridine,[53] phenanthroline, nitrato,[53] and thiocyanato. Quadrupole transitions as low in frequency as 100 kHz are readily detected in diamagnetic systems. In an interesting double resonance experiment the ^{14}N and ^{25}Mg quadrupolar transitions in chlorophyll-a and magnesium phthalocyanine were observed.[54]

Double resonance techniques may be applied to detection of quadrupole transitions of relatively rare spin systems such as ^{25}Mg or ^{67}Zn. They thus offer the potential to greatly extend the range of applicability of the nqr technique to many additional areas of organometallic and bioinorganic chemistry.

(50) P. S. Ireland and T. L. Brown, *J. Magn. Reson.*, **20,** 300 (1975).
(51) (a) D. T. Edmonds, *Pure Appl. Chem.*, **40,** 193 (1974). (b) M. J. Hunt, A. L. MacKay, and D. T. Edmonds, *Chem. Phys. Lett.*, **34,** 473 (1975).
(52) Y. N. Ysieh, P. S. Ireland, and T. L. Brown, *J. Magn. Reson.*, **21,** 445 (1976).
(53) Y. N. Hsieh, G. V. Rubenacker, C. P. Cheng, and T. L. Brown, *J. Am. Chem. Soc.*, in press.
(54) O. Lumpkin, *J. Chem. Phys.*, **62,** 3281 (1975).

The Coordination Chemistry of Nitric Oxide

Richard Eisenberg* and Carol D. Meyer

Department of Chemistry, University of Rochester, Rochester, New York 14627

Received July 22, 1974

Reprinted from ACCOUNTS OF CHEMICAL RESEARCH, *8, 26 (1975)*

Complexes of nitric oxide have been known for centuries, yet never has interest in transition metal nitrosyls been as keen as at present. In part a natural adjunct to the growth of interest in organometallic chemistry, especially metal carbonyls and complexes of π acids, the marked surge of research in this area has been stimulated by several important developments during the past decade.

Prior to 1960 the interactions of nitric oxide with transition metal ions were studied principally as a way of synthesizing nitrosyl complexes, and the bonding of NO to transition metal ions was envisioned in one of several possible ways, Lewis structures I–IV, of which the first three correspond to ter-

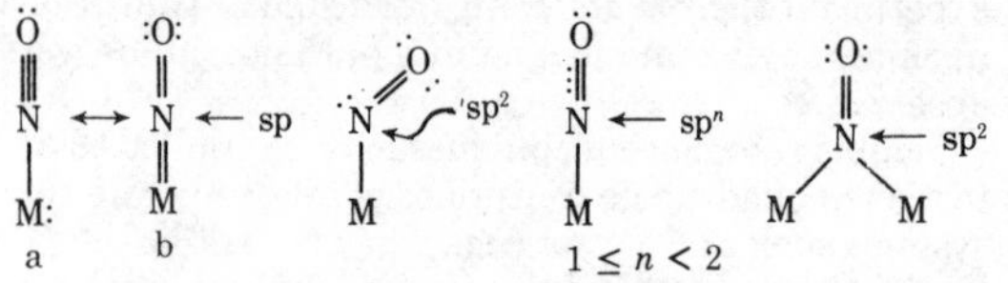

linear, I　　bent, II　　III　　bridging, IV

minal nitrosyls.[1–3] Structures I and II are commonly called the *linear* and *bent,* or NO^+ and NO^-, modes of nitric oxide coordination, respectively. The latter designation arises from the formalism used in coordination chemistry in which the electron pair in the metal–ligand σ bond is associated entirely with the ligand for the purpose of assigning ligand charge and metal oxidation state.[4]

In the assignment of formal oxidation state the existence of metal–ligand π bonding is ignored, but since π bonding is vital to the linear mode of NO coordination (structure Ib), the NO^+ designation has been extensively criticized as not representative of physical reality regarding charge distributions, bond strengths, and the like. It is worth remembering, however, that formalisms in general are merely useful constructs. In nitrosyl and organometallic chemistry, the available formalisms (the oxidation state formalism and the electroneutral approach[1,5]) are not intended as substitutes for detailed bonding descriptions.

Coordination by NO^+ is directly analogous to metal–carbonyl bonding (CO and NO^+ are isoelectronic) with its synergistic coupling of σ and π bonding. Coordination by NO^-, on the other hand, was proposed by Sidgwick[1] as analogous to that by halide ion and structurally similar to organic nitroso compounds. While several possibilities for NO coordination had thus been recognized, the prevailing view prior to 1960 was that in all but a few systems the linear mode of coordination was followed, as in the historically important nitroprusside ion $Fe(NO)(CN)_5^{2-}$ [1,2] and the nitrosyl carbonyls $Co(NO)(CO)_3$ and $Fe(NO)_2(CO)_2$.[6] In none of the other systems, such as $Co(NO)(NH_3)_5^{2+}$,[7] was *definitive* evidence available establishing any other structural arrangement for metal–NO bonding.

A significant turning point in nitrosyl chemistry was achieved in 1967 with the report by Ibers and Hodgson[8] of the first structurally documented bent nitrosyl complex, V. This structure unequivocally es-

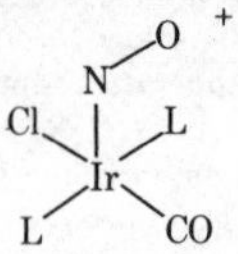

V

tablished the structural duality of coordinated NO, and in doing so confirmed the bent mode of metal–nitrosyl bonding (II) suggested years earlier.[1] Moreover, since V had been prepared from the reaction of NO^+ with $IrCl(CO)(PPh_3)_2$ in which the metal complex served as an electron-pair donor, it was seen that NO^+ possessed amphoteric character and could function as either an electron-pair acceptor[8] or an electron-pair donor with concomitant π back-bonding (I).

The interconvertibility of the linear and bent modes of nitric oxide coordination was postulated in 1969 by Collman and coworkers[9] in the context of the catalytic activity of certain metal nitrosyls. Since the linear and bent modes of bonding (I and II) differ by

Richard Eisenberg is Associate Professor at the University of Rochester. He was born in New York City and received his A.B. and Ph.D. degrees from Columbia University, the latter in 1967. That year he joined the faculty of Brown University, and 6 years later he moved to Rochester. He is an Alfred P. Sloan Foundation Fellow. His research interests are in organotransition metal chemistry, structure, and synthesis.

Carol D. Meyer, a native of New Jersey, received a B.Sc. degree from Bucknell University in 1971 and is currently finishing Ph.D. studies at Brown University. Besides studies of the catalyzed reactions of NO, her PH.D. research has involved investigation of the substituted 2-oxa-5-norbornyl cation.

(1) N. V. Sidgwick and R. W. Bailey, *Proc. Roy. Soc., Ser. A,* **144,** 521 (1934); N. V. Sidgwick, "Chemical Elements and Their Compounds," Vol. I, Clarendon Press, Oxford, 1950, p 685.

(2) T. Moeller, *J. Chem. Educ.,* **23,** 542 (1946); T. Moeller, "Inoganic Chemistry," Wiley, New York, N.Y., 1952, p 598.

(3) (a) J. Lewis, R. J. Irving, and G. Wilkinson, *J. Inorg. Nucl. Chem.,* **7,** 32 (1958); (b) W. P. Griffith, J. Lewis, and G. Wilkinson, *ibid.,* **7,** 38 (1958).

(4) J. E. Huheey, "Inorganic Chemistry: Principles of Structure and Reactivity," Harper and Row, New York, N.Y., 1972, Chapter 8.

(5) F. A. Cotton and G. Wilkinson, "Advanced Inorganic Chemistry," 3rd ed, Wiley-Interscience, New York, N.Y., 1972, Chapter 22.

(6) R. L. Mond and A. E. Wallis, *J. Chem. Soc.,* **121,** 22 (1922); A. A. Blanchard, *Chem. Rev.,* **26,** 409 (1940); J. S. Anderson and W. Hieber, *Z. Anorg. Allg. Chem.,* **208,** 238 (1932).

(7) (a) J. Sand and O. Genssler, *Ber. Deutsch. Chem. Ges.,* **36,** 2083 (1903); (b) A. Werner and P. Karrer, *Helv. Chim. Acta,* **1,** 54 (1918).

(8) D. J. Hodgson and J. A. Ibers, *Inorg. Chem.,* **7,** 2345 (1968); D. J. Hodson, N. C. Payne, J. A. McGinnety, R. G. Pearson, and J. A. Ibers, *J. Amer. Chem. Soc.,* **90,** 4486 (1968).

(9) J. P. Collman, N. W. Hoffman, and D. E. Morris, *J. Amer. Chem. Soc.,* **91,** 5659 (1969).

an electron pair on the nitrosyl ligand, interconversion would alternately withdraw and donate an electron pair to the metal center. On bending of the nitrosyl, an electron pair would be withdrawn from the metal, creating a vacant coordination site and thus fulfilling a key criterion of catalytic activity.[10] This proposal gave great impetus to further studies on transition metal nitrosyls, especially in relation to homogeneous catalysis.[10,11]

The advances in the chemistry of coordinated nitric oxide were accompanied by mounting concern over the role of NO as a major air pollutant. Each year 10^6 tons of nitrogen oxides (NO and NO_2) are produced in fossil fuel combustion processes, mainly as NO, according to eq 1.[12] Environmental control of NO_x has pursued two lines of investigation involving, first, the modification of combustion conditions to minimize the formation of nitrogen oxides and, second, the development of catalysts for the facile conversion of NO into less harmful chemical entities.

$$\tfrac{1}{2}N_2 + \tfrac{1}{2}O_2 \longrightarrow NO \quad (1)$$

$$\Delta H_f{}^\circ = 21.6 \text{ kcal/mol}; \Delta G_f{}^\circ = 20.72$$

Despite its thermodynamic instability, NO is kinetically inert with respect to decomposition and reduction, and requires the presence of a catalyst for many of its reactions. These catalysts are invariably metals or metal oxides such as CuO, Rh_2O_3, Co_3O_4, Pt, and $La_{1-x}Pb_xMnO_3$,[13-15] which may or may not be supported. The need to know how NO interacts with the metal centers of a catalyst surface has also served as a stimulus to studies during the past decade on the reactions of NO with metal complexes in solution.

The activity in the chemistry of metal–nitric oxide complexes has been accompanied by a number of excellent reviews.[16-22] Several of these are comprehensive, while others are specialized in the areas of nitrosyl structures,[21] NO bonding,[19] organometallic nitrosyls,[17] synthetic methods,[22] and the reactions of coordinated NO^+.[20] The purpose of this Account is to present a perspective on the coordination chemistry of nitric oxide with emphasis on the relationship of structure to chemical reactivity and on the reactions of NO promoted or catalyzed by metal complexes.

(10) J. P. Collman, *Accounts Chem. Res.*, **1**, 136 (1968); J. P. Collman and W. R. Roper, *Advan. Organometal Chem.*, **7**, 53 (1968).
(11) For a number of excellent reviews, see *Advan. Chem. Ser.*, **70** (1968). Also see: (a) J. A. Osborn, F. H. Jardine, J. F. Young, and G. Wilkinson, *J. Chem. Soc. A*, 1711 (1966), and subsequent papers by Wilkinson and coworkers; (b) J. Halpern, *Accounts Chem. Res.*, **3**, 386 (1970).
(12) W. Bartok, A. R. Crawford, and A. Skopp, *Chem. Eng. Progr.*, **67**, 64 (1971).
(13) (a) E. R. S. Winter, *J. Catal.*, **22**, 158 (1971); (b) C. S. Howard and F. Daniels, *J. Phys. Chem.*, **62**, 215 (1958).
(14) (a) M. Shelef, K. Otto, and H. Gandhi, *Atmos. Environ.*, **3**, 107 (1969); (b) K. Otto and M. Shelef, *J. Phys. Chem.* **76**, 37 (1972), and references therein; (c) M. Shelef and K. Otto, *J. Catal.*, **10**, 408 (1968); (d) M. Shelef and H. S. Gandhi, *Ind. Eng. Chem., Prod. Res. Develop.*, **11**, 2 (1972).
(15) (a) R. L. Klimisch and G. J. Barnes, *Environ. Sci. Technol.*, **6**, 543 (1972); (b) R. J. H. Voorhoeve, J. P. Remeika, and D. W. Johnson, Jr., *Science*, **180**, 62 (1973).
(16) B. F. G. Johnson and J. A. McCleverty, *Progr. Inorg. Chem.*, **7**, 277 (1966).
(17) W. P. Griffith, *Advan. Organometal. Chem.*, **7**, 211 (1968).
(18) N. G. Connelly, *Inorg. Chim. Acta Rev.*, **6**, 48 (1972).
(19) J. H. Enemark and R. D. Feltham, *Coord. Chem. Rev.*, **13**, 339 (1974).
(20) J. Masek, *Inorg. Chim. Acta Rev.*, **3**, 99 (1969).
(21) B. A. Frenz and J. A. Ibers, *MTP (Med. Tech. Publ. Co.) Int. Rev. Sci., Phys. Chem., Ser. One*, **11**, 33 (1972).
(22) K. G. Caulton, *Coord. Chem. Rev.*, in press.

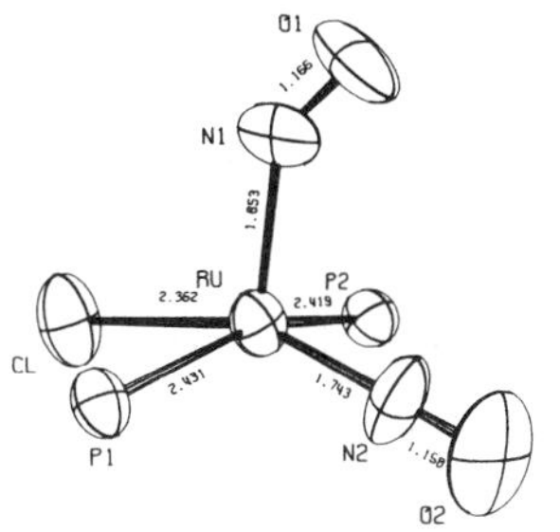

Figure 1. The inner coordination geometry of $Ru(NO)_2Cl(PPh_3)_2^+$.

The Structures of Nitrosyls

In order to understand the reactivity of coordinated nitric oxide, one must first consider the molecular and electronic structures of transition metal nitrosyls and the interrelationship of nitrosyl bonding mode and coordination geometry. We focus on five-coordinate nitrosyls because they provide a most fertile ground for inquiry. The two principal geometries in five-coordination are the square pyramid (SP) and the trigonal bipyramid (TBP), and for sterically unencumbered systems the energetic difference between these geometries is not thought to be great.[23] Examination of the representative structures presented in Table I reveals that, for the platinum group systems, if the complex has an SP geometry with an apical nitrosyl then the NO ligand bonds in a bent manner, whereas if the complex has a TBP geometry the nitrosyl coordinates linearly. In all of these cases the complexes are isoelectronic, as described below.

Representative of the first of these structural types is the complex $Ru(NO)_2Cl(PPh_3)_2^+$, VI,[24] which was

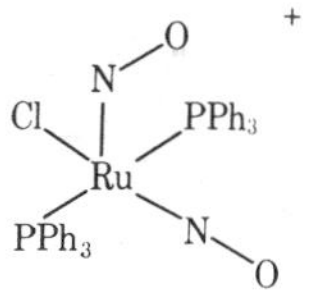

VI

synthesized by reaction of ethanolic solutions of NO^+-PF_6^- with the Stiddard and Townsend complex $Ru(NO)Cl(PPh_3)_2$.[25] This reaction was carried out in order to prepare a dinitrosyl system possessing both linear and bent modes of nitric oxide coordination. The structure of VI is shown in Figure 1. As hoped, the complex is SP with an angularly coordinated NO^- in the apical position and a linearly bonded NO^+ in the basal plane.[24] A comparison of the bonding parameters for the two metal–nitrosyl units shows that the M–N distance for a linear nitrosyl is shorter than that for a bent NO (see Figure 1) because of greater metal nitrosyl π bonding in the former unit. The nitrosyl stretching frequencies ν_{NO} of 1845 and 1687 cm^{-1} for the linear and bent nitrosyls,

(23) E. L. Muetterties and R. A. Schunn, *Quart. Rev., Chem. Soc.*, **20**, 245 (1966).
(24) (a) C. G. Pierpont, D. G. VanDeveer, W. Durland, and R. Eisenberg, *J. Amer. Chem. Soc.*, **92**, 4760 (1970); (b) C. G. Pierpont and R. Eisenberg, *Inorg. Chem.*, **11**, 1088 (1972).
(25) M. H. B. Stiddard and R. E. Townsend, *Chem. Commun.*, 1372 (1969).

Table I
Representative Five-Coordinate Group VIII Nitrosyl Complexes

Complex[a]	No. of elec-trons	Coordination geometry[b]	M–N–O angle, deg	M–N distance, Å	ν_{NO}	Ref
$Fe(NO)(mnt)_2^-$	20	Distorted SP–ap NO[c]	180 (0)	1.61 (1)	1867	*d*
$Fe(NO)(Me_2dtc)_2$	21	SP–apNO	170.4 (6)	1.720 (5)	1690	*e*
$Fe(NO)(mnt)_2^{2-}$	21	SP–apNO	165 (2)	1.58 (2)	1645	*d*
$Ru(NO)_2ClL_2^+$	22	SP–apNO	138 (2)	1.85 (2)	1687	24
		baNO	178 (2)	1.74 (2)	1845	
$RuH(NO)L_3$	22	TBP–axNO	176 (1)	1.79 (1)	1645	27
$Ru(NO)(diphos)_2^+$	22	TBP–eqNO	174 (1)	1.74 (1)	1673	26
$Os(NO)_2(OH)L_2^+$	22	SP–apNO	127 (2)	1.98 (5)	1632	*f*
		baNO	*ca.* 180	1.71 (4)	1842	
$Os(NO)(CO)_2L_2^+$	22	TBP–eqNO	177 (1)	1.89 (1)	1750	*g*
$Co(NO)(Me_2dtc)_2$	22	SP–axNO	135 (1)	1.75 (1)	1630	*h*
$Co(NO)(diars)_2^{2+}$	22	TBP–eqNO	179	1.68	1852	41
$Co(NO)Cl_2L'_2$	22	Distorted TBP–eqNO[i]	165 (1)	1.70 (1)	1735, 1630	44
$Rh(NO)Cl_2L_2$	22	SP–apNO	125 (1)	1.91 (1)	1620	*j*
$Ir(NO)Cl_2L_2$	22	SP–apNO	123 (2)	1.94 (2)	1560	*k*
$Ir(NO)Cl(CO)L_2^+$	22	SP–apNO	124.1 (9)	1.97 (1)	1680	8
$IrH(NO)L_3^+$	22	TBP–axNO	175 (3)	1.68 (3)	1715	*l*

[a] L = PPh_3; mnt = maleonitriledithiolate; Me_2dtc = *N,N*-dimethyldithiocarbamate; L′ = $PMePh_2$. [b] ap = apical; ax = axial; ba = basal; eq = equatorial; SP = square pyramidal; TBP = trigonal bipyramidal. [c] Trans S–Fe–S bond angles are 162.4 (1) and 147.9 (1)°. [d] D. G. Van-Derveer, *et al*, American Crystallography Association Meeting Abstracts, Vol. 1, P4 (1973). [e] (At −80°) G. R. Davies, *et al., J. Chem. Soc.*, 1275 (1970). [f] J. M. Waters and K. R. Whittle, *Chem. Commun.*, 518 (1971). [g] G. R. Clark, K. R. Grundy, W. R. Roper, J. M. Waters, and K. R. Whittle, *J. Chem. Soc., Chem. Commun.*, 119 (1972). [h] J. H. Enemark and R. D. Feltham, *J. Chem. Soc., Dalton Trans.*, 718 (1972). [i] Trans angles are P–Co–P = 168 (1) and Cl–Co–Cl = 108.4 (3)°. [j] S. Z. Goldberg, C. Kubiak, C. D. Meyer, and R. Eisenberg, *Inorg. Chem.*, submitted for publication. [k] D. M. P. Mingos and J. A. Ibers, *ibid.*, **10**, 1035 (1971). [l] D. M. P. Mingos and J. A. Ibers, *ibid.*, **10**, 1479 (1971).

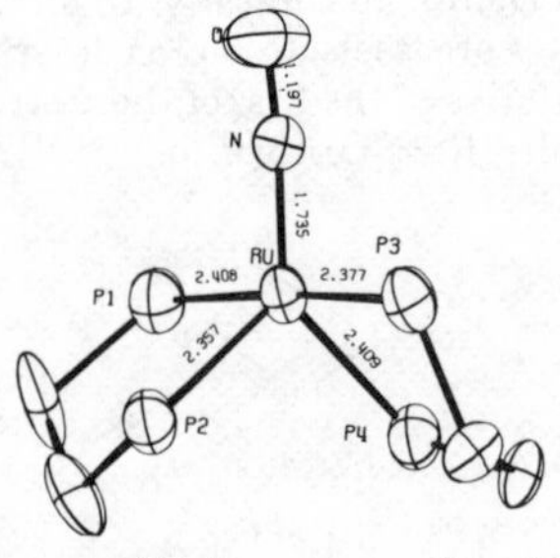

Figure 2. The inner coordination geometry of $Ru(NO)(diphos)_2^+$.

respectively, are also consistent with the notions of metal–nitrosyl bonding embodied in structures I and II.

Representative examples of the TBP linear NO structures are $Ru(NO)(diphos)_2^+$ (VII)[26] and RuH-$(NO)(PPh_3)_3$ (VIII).[27] Complex VII was originally

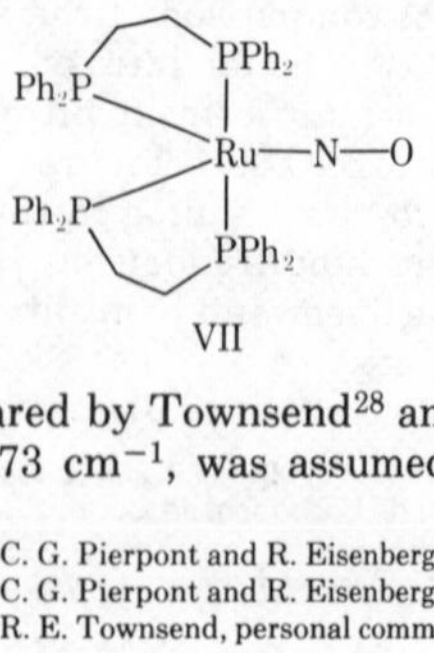

VII

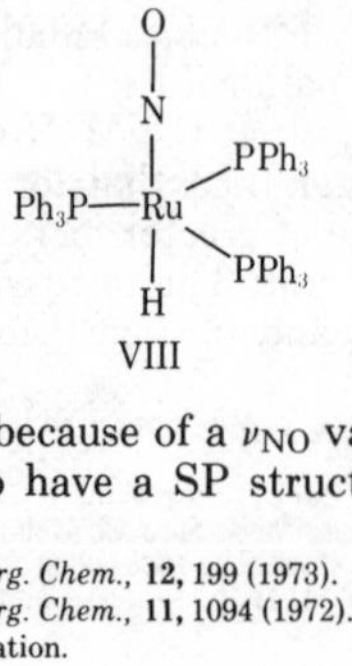

VIII

prepared by Townsend[28] and, because of a ν_{NO} value of 1673 cm^{-1}, was assumed to have a SP structure similar to V and VI. However, the structure as shown in Figure 2 turns out to be TBP with a linearly coordinated nitrosyl in the equatorial position and bidentate diphos ligands spanning equatorial and axial positions.

Complex VIII is the triphenylphosphine member of the series $RuH(NO)L_3$, L = tertiary phosphine, synthesized by Wilson and Osborn.[29] All of these systems have ν_{NO} in the range 1620–1645 cm^{-1}, and all except the methyldiphenylphosphine member of the series are effective as hydrogenation and isomerization catalysts of terminal olefins. At least two of the systems also exhibit fluxional behavior.[29] As predicted from solution studies, the structure of VIII is TBP with an axial nitrosyl linearly coordinated.[27] Because of the low steric requirements of the hydride, the complex shows an expected distortion toward pseudotetrahedral geometry.

Complexes VI–VIII and the related Os and Ir complexes of Table I thus exhibit only three different structural arrangements and support the empirical conclusion that in five-coordinate nitrosyls the coordination geometry and the mode of nitric oxide coordination are intimately related.

The Bonding and Bending of Nitrosyls

To develop notions of bonding and bending in nitrosyls requisite for understanding this relationship, one must go beyond the simple picture conveyed by structures I–III to molecular orbital theory. Our approach to this problem was inspired mainly by Walsh's study[30] of triatomic molecules such as linear

(26) C. G. Pierpont and R. Eisenberg, *Inorg. Chem.*, **12**, 199 (1973).
(27) C. G. Pierpont and R. Eisenberg, *Inorg. Chem.*, **11**, 1094 (1972).
(28) R. E. Townsend, personal communication.
(29) S. T. Wilson and J. A. Osborn, *J. Amer. Chem. Soc.*, **93**, 3068 (1971).

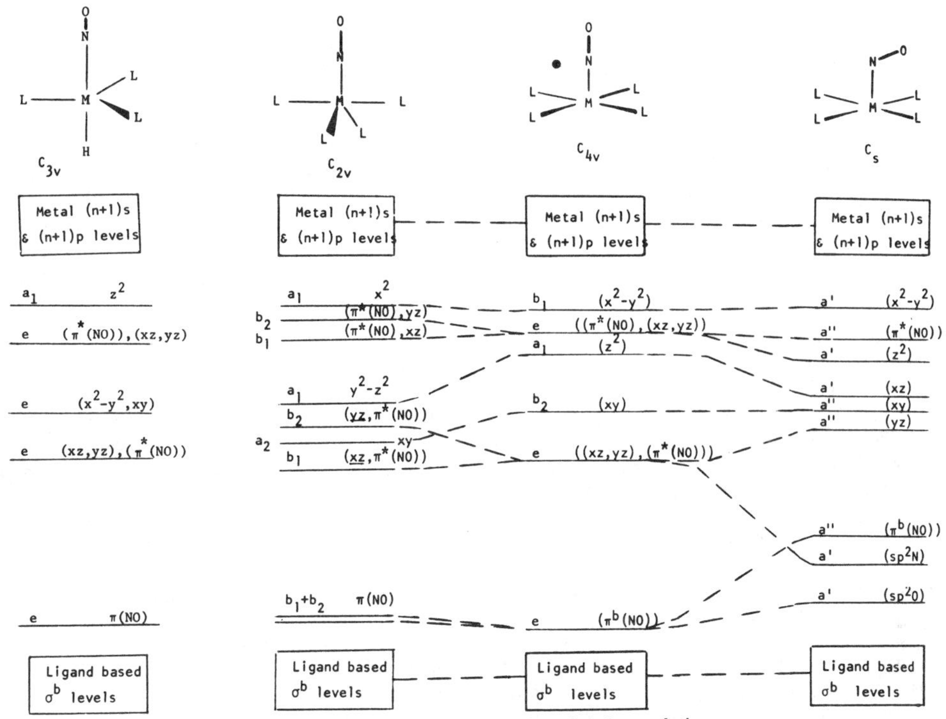

Figure 3. Molecular orbital energy level orderings for five-coordinate nitrosyls and their correlations.

NO_2^+ and bent NO_2^- in which energy levels between the limiting geometries were correlated. While Walsh's analysis dealt solely with the B–A–B bond angle in AB_2 systems,[30] a similar treatment of metal nitrosyls has to contend with more degrees of freedom than the M–N–O angle, the most important being the coordination geometry about the metal.[31]

To simplify the correlations, only one structural parameter was allowed to vary at a time with all others held constant. The results, which are shown in Figure 3, are based on qualitative MO energy level orderings for the three observed structure types, SP bent apical NO (C_s symmetry), TBP linear equatorial NO (C_{2v}) and TBP linear axial NO (C_{3v}), plus that for a square-pyramidal linear apical NO structure of C_{4v} symmetry. The energy levels of the C_{4v} structure are correlated with those of the SP bent NO complex as the M–N–O unit bends[31] and with those of the TBP-equatorial NO complex as the ligand-field geometry varies.[26]

In considering the two SP structures, one sees that the π bonding e level ((xz,yz), π^*(NO)) splits in C_s symmetry with the a′ member connecting to a stable nitrosyl localized level not present in C_{4v} and the a″ member correlating with the metal yz level. Consequently, the d_{z^2} level in C_{4v}, which is strongly σ^* with respect to the metal–nitrosyl bond, correlates with the largely nonbonding xz level in C_s. The first correlation results in a change in orbital character from a metal-based π function to one localized on the nitrosyl which helps explain the oxidative addition of NO^+ to d^8 complexes to yield bent nitrosyl or NO^- systems.[32]

The key correlation between the energy levels of the C_{4v} and C_{2v} structures is that between the σ^* d_{z^2} orbital in C_{4v} and the less strongly antibonding a_1 level in C_{2v} designated as $y^2 - z^2$. (The unconventional label $y^2 - z^2$ arises because of the change in principal symmetry axis in going from the ideal TBP to the C_{2v} TBP-equatorial NO structure.) Correlations involving the C_{3v} structure are not shown because of the need to designate a specific rearrangement mechanism and the difficulties attendant in deriving orbital energies for the intermediate structures.

The major conclusions of the correlation diagrams of Figure 3 become evident when electrons are placed into the level orderings. For five-coordinate nitrosyl complexes of the platinum metals, each system contains 22 electrons counted *assuming* NO^+ in the following way: ten electrons in the ligand σ^b functions, four electrons in the π^b(NO) set, and a d^8 metal configuration. In C_{4v} symmetry this number of electrons would require occupation of the σ^* d_{z^2} level, and the structure thus becomes unstable relative to the alternative geometries. By bending the nitrosyl (C_s structure), the system avoids placing electrons into a strongly antibonding orbital but at the cost of metal–nitrosyl π bonding. On the other hand, variation of the ligand field to TBP allows the maintenance of a strong metal–nitrosyl π interaction but reduces the

(30) A. D. Walsh, *J. Chem. Soc.*, 2266 (1953).

(31) C. G. Pierpont and R. Eisenberg, *J. Amer. Chem. Soc.*, **93**, 4905 (1971).

(32) In this regard, we note that the delocalized π bonding e level in C_{4v} is traditionally assigned to the metal (xz,yz) set, leading to the NO^+ formulation of the nitrosyl regardless of the per cent nitrosyl π^* character calculated for the molecular orbitals of this level.

effectiveness of σ bonding in the entire structure.[33] Hence, five-coordinate 22-electron nitrosyl complexes have several stable structural alternatives. The factors which lead to a particular arrangement, while not fully delineated, relate to the steric and electronic effects of other ligands in the system and the nature of the central metal. A second and related conclusion is that, unless sterically constrained, the structures of C_s, C_{2v}, and C_{3v} symmetry may not represent deep energetic minima and that interconversion between them may be facile.

For complexes constrained by steric and electronic effects to be SP, an additional conclusion can be drawn. Whereas 22-electron systems are required to have a bent nitrosyl structure, 20-electron systems should have linear M–N–O units in order to maximize both σ and π interactions, and 21-electron systems should be slightly bent. While the available evidence presented in Table I tends to support this conclusion, there is a paucity of structural data on such systems.

Correlation diagrams similar to Figure 3 have been presented by other investigators, and this approach has been used in dealing with four-coordinate nitrosyl complexes and dinitrosyl systems.[19,34,35] The major differences between the published correlations for the five-coordinate NO complexes involve the position of the nitrosyl localized level in the C_s structure and its consequent correlations. Current efforts in this area are being devoted to developing a simple, usable theory which will have predictive capabilities.[19,35]

Interrelating Reactivity and Structure

The chemical reactivity of coordinated NO depends on the mode of bonding. Thus, linearly coordinated nitrosyls with ν_{NO} greater than ca. 1850 cm^{-1} undergo nucleophilic attack at the nitrosyl N atom as shown by reactions 2–4,[36–38] while bent nitrosyls undergo electrophilic attack as in reactions 5 and 6.[39,40]

$$Ru(NO) (bipy)_2Cl^{2+} + 2OH^- \rightleftharpoons$$
$$(\nu_{NO}\ 1931\ cm^{-1})$$
$$Ru(NO_2) (bipy)_2Cl + H_2O \quad (2)$$

$$IrCl_3(NO)L_2{}^+ + ROH \rightleftharpoons IrCl_3(RONO)L_2 + H^+ \quad (3)$$
$$(\nu_{NO}\ 1945\ cm^{-1};\ L = PPh_3)$$

$$Ru(NO) (diars)_2Cl^{2+} + N_2H_4 \longrightarrow$$
$$(\nu_{NO}\ 1883\ cm^{-1})$$
$$Ru(N_3) (diars)_2Cl + H_2O + 2H^+ \quad (4)$$

$$OsCl(NO) (CO)L_2 + HCl \rightleftharpoons OsCl_2(HNO) (CO)L_2 \quad (5)$$

$$Co(NO)(salen) + B + \tfrac{1}{2}O_2 \longrightarrow Co(NO_2)(salen)B \quad (6)$$
$$B = pyr,\ 2\text{-Me-pyr},\ Et_3N,\ PrNH_2$$

Linearly coordinated nitrosyls having lower ν_{NO} values, indicative of significant contributions from resonance structure Ib and decreased electrophilicity of the nitrosyl N atom, are inert to nucleophilic attack.

The conversion of NO$^+$ to NO$^-$ has been demonstrated by adding coordinating anion to the coordinatively saturated TBP-equatorial NO complex Co(NO)(diars)$_2{}^{2+}$ (diars = o-phenylenebis(dimethylarsine)), thereby necessitating the transfer of an electron pair from the metal to the nitrosyl, giving the bent nitrosyl product Co^{III}(NO)(diars)$_2$X$^+$ (X = Cl$^-$, Br$^-$, NCS$^-$).[41] In the platinum group systems, a less easily rationalized conversion involves the reaction of CO with IrCl(NO)L$_2{}^+$ to give V.[42]

The reversible interconversion of linear and bent nitrosyls in stereochemically nonrigid molecules has attracted considerable attention, but to date the best evidence for such interconversion remains indirect. For example, when Ru(NO)$_2$ClL$_2{}^+$ (VI) is synthesized by the reaction of Ru(^{15}NO)ClL$_2$ with ^{14}NO$^+$BF$_4{}^-$,[43] the label is completely scrambled, implying rapid equilibration through the proposed mechanism 7.[24b] For the fluxional nitrosyl hydride

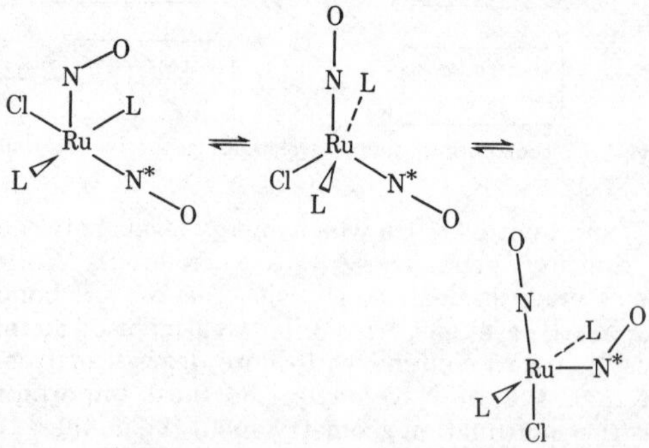

$$(7)$$

systems RuH(NO)L$_3$,[29] rapid interconversion between the TBP-axial NO structure of VIII and one in which two of the phosphines remain equivalent suggests the possible intermediacy of a SP-apical NO structure which would necessitate nitrosyl bending. Finally, Collman and coworkers have described a "paradox" in explaining two different nitrosyl stretching frequencies observed in complexes Co(NO)Cl$_2$L$_2$, where L is tertiary phosphine.[44] These complexes appear to undergo intramolecular redox reactions involving nitrosyl bending in solution and on solid surfaces, although a structure determination of a typical member of the series, Co(NO)Cl$_2$(P-MePh$_2$)$_2$, reveals only one isomer in the bulk crystal.[44]

The significance of the interconversion of linear and bent NO's lies in the transfer of the electron pair between the metal and the nitrosyl, allowing the

(33) If only σ bonding is assumed for the M–L interactions in 22-electron M(NO)L$_4$ systems, then the trigonal-bipyramidal linear NO structure will have two orbitals occupied which are weakly σ^* with respect to bonds in the equatorial plane, whereas the square-planar bent NO structure will have no M–L σ^* orbitals occupied.

(34) D. M. P Mingos, Inorg. Chem., 12, 1209 (1973).

(35) R. Hoffmann, M. M. L. Chen, M. Elian, A. R. Rossi, and D. M. P. Mingos, Inorg. Chem., 13, 2666 (1974).

(36) J. B. Godwin and T. J. Meyer, Inorg. Chem., 10, 471 (1971); 10, 2150 (1971).

(37) C. A. Reed and W. R. Roper, J. Chem. Soc., Dalton Trans., 1243 (1972).

(38) P. G. Douglas, R. D. Feltham, and H. G. Metzger, J. Amer. Chem. Soc., 93, 84 (1971).

(39) K. R. Grundy, C. A. Reed. and W. R. Roper, Chem. Commun., 1501 (1970).

(40) S. G. Clarkson and F. Basolo, Inorg. Chem., 12, 1528 (1973); J. Chem. Soc., Chem. Commun., 670 (1972).

(41) J. H. Enemark and R. D. Feltham, Proc. Nat. Acad. Sci. U.S., 69, 3534 (1972).

(42) C. A. Reed and W. R. Roper, Chem. Commun., 1459 (1969).

(43) J. P. Collman, P. Farnham, and G. Dolcetti, J. Amer. Chem. Soc., 93, 1788 (1971).

(44) C. P. Brock, J. P. Collman, G. Dolcetti, P. H. Farnham, J. A. Ibers, J. E. Lester, and C. A. Reed, Inorg. Chem., 12, 1304 (1973).

metal center to achieve coordinative unsaturation in a unique way.[9] Since this is a prime criterion in catalytic activity,[10] it has been suggested that NO can activate a metal center more so than the corresponding carbonyl which exhibits no tendency to bend.[9] Catalytically active group VIII nitrosyls include $MH(NO)L_3$ (M = Ru, Os),[29] $Rh(NO)(PPh_3)_3$,[9] and $Fe(NO)_2(CO)_2$[45] in reactions such as hydrogenation, isomerization, and diolefin dimerization, but as yet no definitive link between catalytic activity of nitrosyls and NO bending has been established. Further study in this vein is clearly warranted.

Reactions of Nitric Oxide Promoted by Metal Complexes

The simplest interaction of NO with transition metal ions is that leading to the formation of nitrosyls in which the NO ligand coordinates in either the linear or bent manner, as illustrated by reactions 8 and 9.[6,7] In analyzing the more complex reactions

$$Co_2(CO)_8 + 2NO \longrightarrow 2Co^{-I}(NO)(CO)_3 + 2CO \quad (8)$$

$$CoCl_2 + NH_3 + NO \longrightarrow [Co^{III}(NO^-)(NH_3)_5]Cl_2 \quad (9)$$

below, the initial formation of a nitrosyl species will be assumed.

Examples of nitric oxide disproportionation promoted by metal complexes in solution are now well established and are exemplified by reactions 10–13

$$CoCl_2 + 2en + 3NO \longrightarrow Co(en)_2(NO_2)Cl_2 + N_2O \quad (10)$$

$$IrCl(N_2)L_2 + 4NO \longrightarrow IrCl(NO)(NO_2)L_2 + N_2 + N_2O \quad (11)$$

$$RhClL_2L' + 4NO \longrightarrow RhCl(NO)(NO_2)L_2 + N_2O + L' \quad (12)$$

$$L' = CO,\ PPh_3$$

$$IrBr(NO)_2L_2 + 2NO \longrightarrow IrBr(NO)(NO_2)L_2 + N_2O \quad (13)$$

($L = PPh_3$),[46–48] in which the product NO_2 remains coordinated to the metal center. Reaction 10[46] is the most easily comprehensible since it is unencumbered with additional steps or complicating side reactions. In (10), the initial formation of the bent nitrosyl species, $[Co(NO)(en)_2Cl]Cl$,[49] is followed first by electrophilic attack of free NO on the coordinated NO^- and subsequently by reaction with a second NO molecule leading to products.[46] It is uncertain whether initial NO attack is through the nitrogen or oxygen end of the molecule (the dipole moment of NO is only 0.16 D), but the fact that the product nitro complex and the reactant $Co(III)NO^-$ species in (10) have the same stereochemistry is proposed as supporting the notion that the Co–N bond remains intact throughout the reaction sequence and that the free NO molecule attacks the bent nitrosyl with its oxygen end.[46]

Reactions 11–13 may also proceed by electrophilic attack on a bent nitrosyl, but initial NO substitution is needed in (11)[47] and (12)[48] to yield the common reactive system $M(NO)_2XL_2$ (IX) which is suggested by the occurrence of reaction 13.[47] Since IX has two more valence-shell electrons than $Ru(NO)_2Cl(PPh_3)_2^+$ (VI), it must have at least one bent

Scheme I

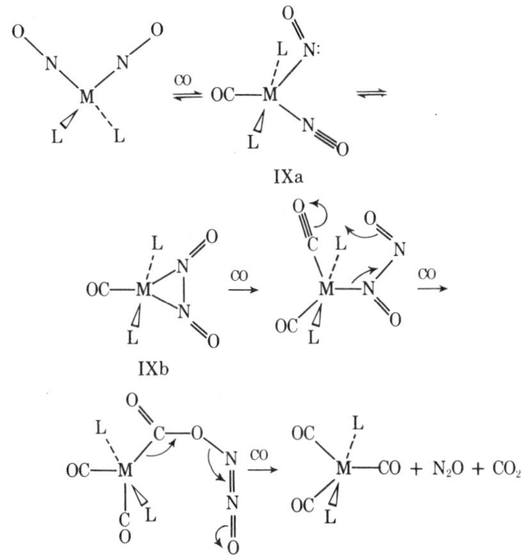

nitrosyl which can serve as the reactive site in the NO disproportionation reaction. Alternatively, Ibers and Haymore[47] have proposed that, in systems such as IX, the two nitrosyls function as a cis dinitrogen dioxide ligand. An equilibrium between $M(N_2O_2)$ and $M(NO^+)(NO^-)$ can then be envisioned to connect these two formulations (*i.e.*, XIa $\rightleftharpoons$ XIb) and explain the observed reaction chemistry.

In 1973 Johnson and Bhaduri[50] showed that the dinitrosyl complex $Ir(NO)_2L_2^+$ reacts with CO reducing the nitrosyl ligands to N_2O and forming CO_2 *via* reaction 14. This reduction of coordinated NO has re-

$$Ir(NO)_2L_2^+ + 4CO \longrightarrow Ir(CO)_3L_2^+ + N_2O + CO_2 \quad (14)$$

cently been extended to other dinitrosyl systems.[47] In rationalizing reaction 14, Johnson and Bhaduri[50] proposed a mechanism involving a nitrene intermediate, but since the initial addition of CO to $Ir(NO)_2L_2^+$ results in a complex isoelectronic with IX, an alternative mechanism based on $M(NO^+)(NO^-)$ or $M(N_2O_2)$ becomes plausible, as in Scheme I.[50a] By reacting the product complex with NO, the reactant dinitrosyl can be regenerated and a continuous catalytic cycle becomes possible.[50]

Reactions of Nitric Oxide Catalyzed by Metal Complexes

The use of homogeneous catalysts to effect conversion of NO into less noxious products is a recent and environmentally significant development in the coordination chemistry of nitric oxide. Inspired by the report of reaction 12 in which a Rh(I) complex is apparently oxidized to Rh(III) by NO[48] (the nitrosyl ligand in the product appears to coordinate as NO^-)

(45) J. P. Candlin and W. H. Hanes, *J. Chem. Soc. C*, 1856 (1968).
(46) D. Gwost and K. G. Caulton, *Inorg. Chem.*, **13**, 414 (1974).
(47) B. L. Haymore and J. A. Ibers, *J. Amer. Chem. Soc.*, **96**, 3325 (1974).
(48) W. B. Hughes, *Chem. Commun.*, 1126 (1969).
(49) D. A. Snyder and D. L. Weaver, *Inorg. Chem.*, **9**, 2760 (1970).

(50) B. F. G. Johnson and S. Bhaduri, *J. Chem. Soc., Chem. Commun.*, 650 (1973).
(50a) Johnson, Bhaduri, and coworkers have recently revised their mechanistic proposal [*J. Chem. Soc., Chem. Commun.*, 809 (1974)] to coincide with that of Ibers and Haymore[47] involving a M–(N_2O_2) *intermediate* based on certain isotope labeling experiments. However, careful analysis of their data renders proof of this mechanism inconclusive at this time.

Table II
Reduction of NO by CO Catalyzed by Solutions of $[Rh(CO)_2Cl_2]^-$

Run	Catalyst solution[a]	Time, hr	ΔP_{NO}[b]	ΔP_{CO}	P_{CO_2}	P_{N_2O}
I	$RhCl_3 \cdot xH_2O$	20.5	110	66	52	39
II	$Ph_4As[Rh(CO)_2(Cl)_2]$/4ml of HCl	23	314	149	138	128
III	$[Rh(CO)_2(\mu\text{-}Cl)]_2$/4 ml of HCl	23	316	149	138	120
IV	Recharge of run II	23	304	143	145	125
V	$Ph_4As[Rh(CO)_2Cl_2]$	23	47	33	9	14
VI	$Ph_4As[Rh(CO)_2Cl_2]$/~40 mmol of HCl gas	21.5	46	16	9	5
VII	$Ph_4As[Rh(CO)_2Cl_2]$/3 ml of H_2O	23	243	100	98	91
VIII	$Ph_4As[Rh(CO)_2Cl_2]$/1.8 ml of H_2O and 1.4 g of LiCl	23	167	79	80	71
IX	$Ph_4As[Rh(CO)_2Cl_2]$/6.3 ml of 70% $HClO_4$	24	262	129	139	117

[a] All solutions contained 1 m*M* Rh in 100 ml of ethanol. Amounts of other components are specified. Initial ratio of NO:CO was 4:3 unless otherwise stated. CO was added first. HCl refers to a 37% aqueous solution. [b] Partial pressures are in mmHg and were calculated from calibration plots using vpc with estimated errors of ±2% of reported values. ΔP_{NO} and ΔP_{CO} are the changes in partial pressure of reactant gases from $t = 0$ to the time indicated for each run; P_{CO_2} and P_{N_2O} are the partial pressures of the product gases at the end of the time period.

and the well-known reduction of $RhCl_3$ by CO according to eq 15,[51] we decided to investigate catalysis

$$RhCl_3 + 3CO + H_2O \longrightarrow [RhCl_2(CO)_2]^- + CO_2 + 2H^+ + Cl^- \quad (15)$$

of the reduction of NO by CO using ethanolic solutions of $RhCl_3 \cdot xH_2O$. In this mixed-gas study, the reaction vessel was charged to an initial pressure of *ca.* 620 mm in a 4:3 NO:CO ratio and the reaction progress was monitored by gas chromatographic analysis of the gas mixture and by system pressure decrease. As shown in run I of Table II, catalysis of reaction 16 was indeed observed.[52] However, the

$$2NO + CO \longrightarrow N_2O + CO_2 \quad (16)$$

reaction was characterized by an 8–10-hr induction period during which the solution changed in color from dark red to olive green and only slow product evolution occurred.

With the finding that this induction period could be dramatically shortened either by reaction with CO prior to exposure to the mixed gases or by addition of aqueous HCl to the solution, it became apparent that initial reduction of $RhCl_3$ *in situ* according to reaction 15 was prerequisite to catalytic activity. Investigation of the catalytic properties of $[RhCl_2(CO)_2]^-$ confirmed this idea and established unequivocally that this species is the true catalyst in the $RhCl_3$ system.[53]

The results of our studies on the homogeneous $[RhCl_2(CO)_2]^-$ catalyst system are summarized in Table II. In contrast to run I, formation of the green catalytic intermediate from this anion (runs II and III) is complete within 2 hr. During the ensuing 6 hr conversion continues at a nearly uniform rate to consume 50% of the reactant gases, corresponding to the production of *ca.* 15 mmol of products per mmol of Rh, within the first 8 hr of exposure to the mixed gases. A marked decrease in the rate of product evolution is then observed with regeneration of the $[RhCl_2(CO)_2]^-$ anion which can be recovered in virtually quantitative yield as its $AsPh_4^+$ salt. Alternatively, the system can be recharged with NO–CO (run IV), and the results of runs II and III repeated. The existence of an equilibrium between $[RhCl_2(CO)_2]^-$ and a presumed green catalytic intermediate was established from the isosbestic point at 360 nm obtained monitoring the uv spectrum of a dilute solution of the carbonyl anion charged under a CO–NO atmosphere as shown in Figure 4.

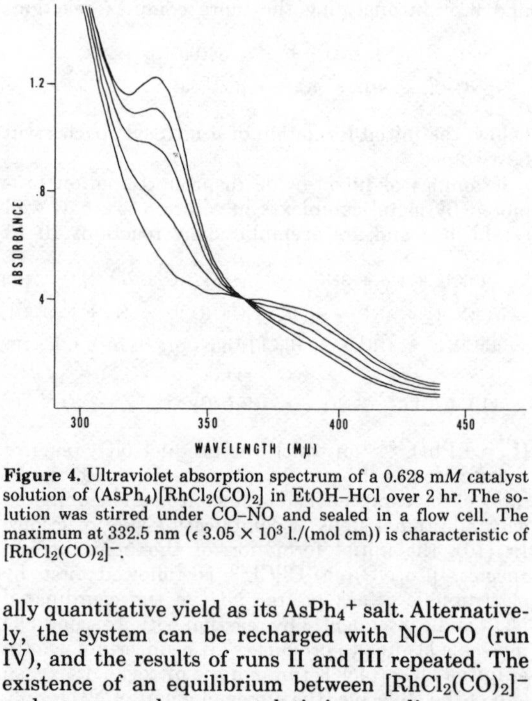

Figure 4. Ultraviolet absorption spectrum of a 0.628 m*M* catalyst solution of $(AsPh_4)[RhCl_2(CO)_2]$ in EtOH–HCl over 2 hr. The solution was stirred under CO–NO and sealed in a flow cell. The maximum at 332.5 nm (ϵ 3.05 × 10^3 l./(mol cm)) is characteristic of $[RhCl_2(CO)_2]^-$.

Since initial reaction of $[RhCl_2(CO)_2]^-$ under the mixed gases obviously involves NO substitution or adduct formation, we have investigated the reaction of $[RhCl_2(CO)_2]^-$ with NO alone and find that rapid stoichiometric evolution of CO_2 and N_2O occurs with the formation of a dark red solution which catalyzes the continued slow reduction of NO. This latter observation is consistent with the report by Stanko and coworkers[54] discussed below on the $RhCl_3$ catalyzed disproportionation of NO. Re-formation of $[RhCl_2(CO)_2]^-$ under CO is accompanied by N_2O and

(51) B. R. James, G. L. Rempel, and F. T. T. Ng, *J. Chem. Soc. A*, 2454 (1969), and references therein; J. A. Stanko, G. Petrov, and C. K. Thomas, *Chem. Commun.*, 1100 (1969).
(52) J. Reed, Jr., and R. Eisenberg, *Science*, **184**, 568 (1974).
(53) C. D. Meyer and R. Eisenberg, *J. Am. Chem. Soc.*, **98**, 1364 (1976).
(54) J. A. Stanko, C. A. Tollinche, and T. H. Risby, *Can. J. Chem.*, in press; J. A. Stanko, personal communication.

Scheme II

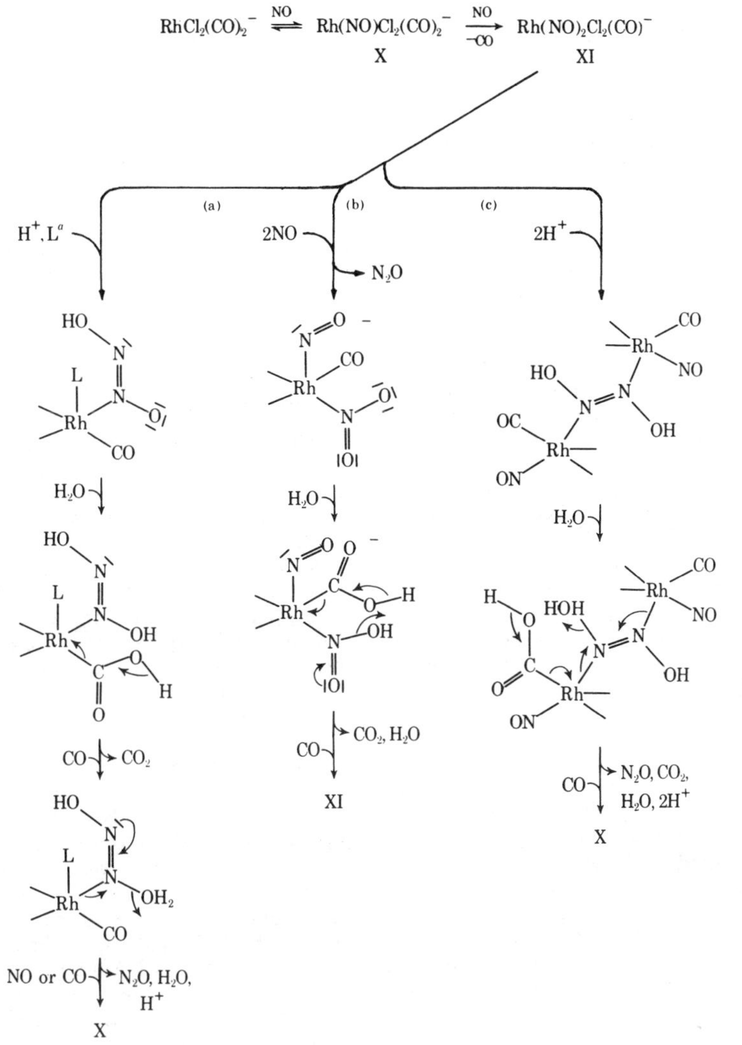

a L = CO, NO, or ROH.

CO_2 evolution, and the reversibility of this reaction sequence, as suggested by Johnson and Bhaduri,[50] is fundamental to the observation of a continuous redox cycle in which both the carbonyl and nitrosyl species are precursors to the same catalytic intermediate formed in the presence of mixed gases.

The importance of water and acid in the catalytic process has been investigated (runs V–IX), and both are required for maximum activity. The requirement for water suggests intimate involvement in the catalytic cycle. In view of the established requirement for water in the formation of CO_2 *via* reaction 15[51] and the formation of water in the catalyzed disproportionation of NO described below,[54] it seems likely that water is both consumed in the oxidation of CO and produced in the reduction of NO.

Several mechanistic possibilities consistent with these observations are outlined in Scheme II, in which rapid formation of a nitrosyl adduct is followed by a slow substitution reaction to form the reactive dinitrosyl carbonyl intermediate, XI. Possible mechanistic pathways which lead to product evolution are shown as a–c in which oxidation of Rh concomitant with reduction of coordinated NO leads to the formation of species containing either coordinated hyponitrite, nitrite ion, or the corresponding acids. Incorporation of water followed by reductive elimination of CO_2 and acid-catalyzed decomposition of the oxy acids completes the cycle.

In order to ascertain the importance of binuclear intermediates in the catalytic cycle, we investigated the variation in reaction rate as a function of initial metal ion concentration. Using dilute solutions (less than 6×10^{-4} *M*) the product evolution curves were linear over at least a 20-hr period and the rate dependence on Rh was found to be first order.[53] While the transitory formation of dimeric species as depicted in path c cannot be excluded, it is apparent that the rate-determining step involves only monomeric species.

The likely involvement of dinitrosyls in the above catalytic cycle receives support from the report by Ibers and Haymore[47] in which catalysis of reaction 16 by $Ir(NO)_2(PPh_3)^+$ and other dinitrosyl complexes is observed. However, no mechanistic studies have been made as yet on these systems.[50a] With a view to understanding the interrelationship of structure and reactivity, we have investigated the catalytic properties of various phosphine– and diolefin–rhodium complexes in Me_2SO[55] and find promotion of the reduction of NO by CO to be a rather general phenomenon. Included in this study were $[RhL_2Cl]_2$, $RhBrL_3$, $RhClL_3$, $RhCl(CO)L_2$, $[Rh(NBD)Cl]_2$, $[Rh(COD)Cl]_2$, and $Rh(NO)L_3$ (L = PPh_3; NBD = norbornadiene; COD = 1,5-cyclooctadiene). While none of the systems is as active as $[RhCl_2(CO)_2]^-$–(EtOH)–HCl, they all exhibit rapid product formation within the first 20 hr and approach the same slow limiting rate thereafter. Initial rapid product formation could *not* be reproduced on recharging these systems.

Recently Stanko and coworkers[54] have observed the catalyzed disproportionation of NO using ethanolic solutions of $RhCl_3$ to give N_2O and ethyl nitrite. A mechanism proposed by Stanko[54] involves the formation of a Rh^{III}–$(NO^+)(NO^-)$ intermediate followed by NO^- assisted attack of ethanol on NO^+ to give ethyl nitrite and the nitrosyl hydride ligand, HNO, which upon displacement and dimerization decomposes to give N_2O and H_2O. Alternative possibilities can be formulated, and the catalyzed disproportionation and reduction reactions may actually have steps in common.

Finally, Nunes and Powell[56] reported in 1970 the Cu(I)-catalyzed reduction of NO by $SnCl_2$. The reduction products were N_2O and hydroxylamine in ratios strongly influenced by Cu(I) and Sn(II) concentrations. The kinetic results were interpreted in terms of the formation of the catalytic intermediates Cl_3Sn–$Cu(NO)Cl_2^{2-}$, Cl_3Sn–$Cu(N_2O_2)Cl_2^{2-}$, and $[Cu_2(SnCl_3)_2Cl_4(N_2O_2)]^{4-}$, with the immediate precursors of both products being two-nitrogen species.

Concluding Comments

Studies in the coordination chemistry of nitric oxide over the last few years have been numerous, and significant progress has been achieved in establishing the structural systematics of metal nitrosyls, in developing a comprehensive bonding description of them, and in elucidating and exploring the reaction chemistry of coordinated NO.

A most intriguing aspect of nitrosyl chemistry which will command attention in the future will be the more complicated reactions of nitric oxide with metal complexes which go beyond the synthesis of M–NO bonds. That complexes in solution have been used as promoting agents and true catalysts for NO disproportionation and reduction seems only a beginning. Understanding how these reactions occur and applying that knowledge to the design of new, more efficient catalyst systems will be the goals of more exhaustive studies. Other avenues of inquiry will include extension to catalytically active complexes of the first transition series, the homogeneously catalyzed reduction of NO by other gaseous species, and perhaps the simultaneous reduction of NO and SO_2 by CO.

We thank the National Science Foundation for support of this research; R.E. gratefully acknowledges the Alfred P. Sloan Foundation for a Fellowship. The efforts of previous collaborators whose names appear in the references are greatly appreciated, in particular, Dr. C. G. Pierpont who was an original motivating force to our work in nitrosyls and Dr. J. Reed who initiated our catalytic studies.

Addendum (May 1976)

The green catalytic intermediate in the carbon monoxide reduction of NO catalyzed by $[RhCl_2(CO)_2]^-$ is confirmed to be the dinitrosyl carbonyl complex XI by its solution infrared spectrum.[57] Bands are observed at 2095(s), 1715(m), and 1680(s) cm^{-1} assignable to carbonyl and nitrosyl stretching frequencies consistent with this formulation. Efforts to obtain an esr spectrum of any paramagnetic intermediates have yielded at best a weak signal at $\langle g \rangle = 2.04$, corresponding to a species present to less than 0.5% of the total Rh concentration.[57] This is taken as further support for the formulation of XI as a diamagnetic dinitrosyl.

An isotope labeling study has been carried out using $H_2{}^{18}O$ to determine unequivocally if water serves as the oxygen transfer agent in the $[RhCl_2(CO)_2]^-$ catalysis of reaction (16).[58] This study used a flow reactor apparatus in which the NO/CO mixture was bubbled through the catalyst solution, thus sweeping the product CO_2 and N_2O gases from the solution before isotopic equilibration could occur. The results show that one of the CO_2 oxygen atoms originates from the water in the reaction medium. The formation of a Rh(III) hydroxycarbonyl species is thus established. This presumably arises from nucleophilic attack of H_2O on a Rh(III)–CO intermediate as in (17). The very small amounts of doubly labeled CO_2 further indicate that XII exhibits no tendency for back reaction and exchange of the carbonyl oxygen relative to decarboxylation and product formation. This study thus confirms water's postulated role in the catalysis of (16) and further underscores the mechanistic complexity of what is a stoichiometrically simple reaction.

$$Rh^{III}\text{—}CO + H_2{}^{*}O \xrightarrow[-H^+]{} Rh^{III}\text{—}C(=O)(\overset{*}{O}H) \longrightarrow Rh^{I} + CO^{*}O + H^{+} \quad (17)$$

XII

(55) C. D. Meyer, J. Reed, and R. Eisenberg, in "Organotransition–Metal Chemistry," Y. Ishii and M. Tsutsui, Ed., Plenum Press, New York, N.Y., 1975, p. 369.

(56) T. L. Nunes and R. E. Powell, *Inorg. Chem.*, **9**, 1912 (1970).

(57) D. E. Hendriksen and R. Eisenberg, *Inorg. Chem.*, submitted for publication.

(58) D. E. Hendriksen and R. Eisenberg, *J. Am. Chem. Soc.*, **98**, 4662 (1976).

Inorganic Chemistry with Ylides

Hubert Schmidbaur

Anorganisch-chemisches Laboratorium der Technischen Universität München, D 8000 Munich, Germany

Received July 8, 1974

Reprinted from ACCOUNTS OF CHEMICAL RESEARCH, *8*, 62 (*1975*)

Modern nomenclature uses the term "ylide" for a class of compounds which was first considered by Staudinger[1] in the 1920's and was later developed by Wittig[2] and his collaborators after World War II. The most important representatives of this class are the phosphorus ylides, bonding in which is depicted:[2]

$$\underset{\text{ylide}}{R_3\overset{+}{P}-\overset{-}{C}H_2} \longleftrightarrow \underset{\text{ylene}}{R_3P{=}CH_2}$$

According to this representation, two canonical formulas are of principal significance for proper description of the structure and bonding. One of these, the *ylide formula* in the narrow sense, emphasizes the dipolar zwitterionic nature involving an *onium center* at elements like phosphorus, sulfur, nitrogen, or arsenic, next to a *carbanionic function,* which may be at least partially delocalized into suitable substituents. In the *ylene formula,* on the other hand, a true double bond is postulated between the onium center and the ylidic carbon, thus reducing or even eliminating the formal charges at these atoms.[3,4]

The application of modern physical techniques[5] and the results of sophisticated theoretical calculations[6-8] have made it increasingly clear that the ylide formula predominates in the ground states of these and related molecules. Most of the early investigators successfully used this description for most of their problems of structure and reactivity and for the rationalization of reaction mechanisms.[2-4] Therefore it is with justification that the term "ylide" is used nowadays almost exclusively in the literature.

When working with ylides we are, therefore, dealing with stable compounds containing a special type of carbanion! These carbanions are not associated with metallic cations, as in many organometallic reagents, but with an onium group fixed in a certain position of the system and giving rise to new specific features.

It is only 20 years since ylides, predominantly those of phosphorus and sulfur, were introduced as reagents into organic synthesis. Today they rank among the most important tools of the preparative organic chemist both in research and in industry, as, *e.g.*, in the Wittig olefin synthesis.

Until recently, however, inorganic and organometallic chemists seem to have neglected the great potential of ylides. To judge from the literature, there was no more than preliminary study of the role of ylides in reactions with inorganic components before 1965. Various groups have since been attracted more and more by the many fascinating aspects of this new branch of organometallic chemistry.

Starting from simple silicon and other main-group derivatives of ylides, this chemistry has now expanded to transition metals as possible coordination centers for ylidic carbanions. Ylides are considered as versatile ligands to metals in their various oxidation states. Ylide chemistry with silicon was first applied to organic and organometallic synthesis, and the latter aspect has been found promising in the field of transition-metal catalysis.

Silicon in Ylide Chemistry

In view of the many successful applications of ylides in organic chemistry, it must have appeared logical to extend this chemistry to silicon, the homolog of carbon. Experienced organometallic chemists made this first important step and discovered the high affinity of ylidic carbanions toward silicon,[9-13] germanium,[13] and tin.[9,12,13] It soon appeared, however, that silicon exerted a pronounced stabilizing effect on these carbanions[13,14] which was contrary to all expectations based on electronegativity arguments and the relative inductive effects of alkyl and silyl groups.

This phenomenon is particularly obvious from a comparison of the structures of some corresponding alkylated and silylated ylides[15] (Scheme I). The deprotonation of phosphonium and sulfoxonium[16,17] cations by organometallic bases occurs selectively at the least alkylated α-carbon atom, (**1** → **2**, **3** → **4**). On the other hand, silylated onium cations are converted into ylides having the silicon substituent di-

Hubert Schmidbaur was born at Landsberg am Lech, Bavaria, in 1934. He studied at the University of Munich, where he obtained the Dr. rer. nat. degree under Max Schmidt. He served as assistant and lecturer at Munich and then at Marburg, as Professor of Inorganic Chemistry at the University of Würzburg (1967–1973), and is now Professor of Inorganic and Analytical Chemistry at the Technical University of Munich. He was recipient of the Dozentenpreis des Verbands der Chemischen Industrie in 1966 and of the 1974 Frederic Stanley Kipping Award in Organosilicon Chemistry sponsored by Dow Corning Corporation.

(1) H. Staudinger and J. Meyer, *Helv. Chim. Acta,* **2,** 635 (1919).
(2) G. Wittig and G. Geissler, *Justus Liebigs Ann. Chem.,* **580,** 44 (1953); G. Wittig and U. Schöllkopf, *Chem. Ber.,* **87,** 1318 (1954).
(3) A. W. Johnson, "Ylid Chemistry," Academic Press, New York, N.Y., 1966.
(4) G. M. Kosolapoff and L. Maier, Ed., "Organophosphorus Chemistry," Wiley, New York, N.Y., 1972.
(5) See the section salt-free ylides below.
(6) R. Hoffmann, D. B. Boyd, and S. Z. Goldberg, *J. Amer. Chem. Soc.,* **92,** 3929 (1970).
(7) J. Absar and J. R. Van Wazer, *J. Amer. Chem. Soc.,* **94,** 2382 (1972).
(8) D. B. Boyd and R. Hoffmann, *J. Amer. Chem. Soc.,* **93,** 1063 (1971).
(9) D. Seyferth and S. O. Grim, *Chem. Ind.* (*London*), 849 (1959); *J. Amer. Chem. Soc.,* **83,** 1610 (1961); **83,** 1613 (1961).
(10) H. Gilman and R. A. Tomasi, *J. Org. Chem.,* **27,** 3647 (1962).
(11) D. Seyferth and G. Singh, *J. Amer. Chem. Soc.,* **87,** 4156 (1965).
(12) D. Seyferth, G. Singh, and R. Suzuki, *Pure Appl. Chem.,* **13,** 1596 (1966).
(13) N. E. Miller, *J. Amer. Chem. Soc.,* **87,** 390 (1965); *Inorg. Chem.,* **4,** 1458 (1965).
(14) H. Schmidbaur and W. Tronich, *Chem. Ber.,* **100,** 1032 (1968).
(15) H. Schmidbaur and W. Malisch, *Chem. Ber.,* **103,** 3007 (1970).
(16) H. Schmidbaur and G. Kammel, *Chem. Ber.,* **104,** 3252 (1971).
(17) H. Schmidbaur and W. Kapp, *Chem. Ber.,* **105,** 1203 (1972).

Scheme I

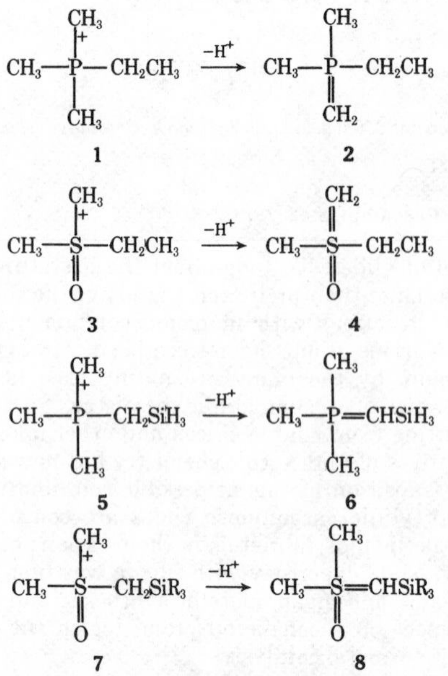

rectly attached to the carbanion generated in the process (5 → 6, 7 → 8), even if alternatives are available. This phenomenon does not have its origin in a steric effect, since simple SiH_3 substituents[15] behave the same as bulky SiR_3 groups.[13,14]

Essentially the same behavior is encountered with arsenic ylides[13,18] and with various organosilicon, -germanium,[14] -tin,[14] -phosphino,[19,20] -arsino,[19] -stibino,[19] and -phosphonium or -arsonium[19,21] substituents at the ylidic carbon.

Among the most striking consequences of the "silicon effect" in ylides is the course of some transylidation[15,22] reactions. Transylidation[23] occurs between ylides and onium salts when the competing ylides differ significantly in basicity or when there are large differences in lattice energies of the onium salts. Equation 1 gives an example which demonstrates the reduced basicity of an ylide bearing a silicon substituent as compared to the alkylated analog. (The lattice energies of the salts in this particular case should be very similar.)

$$(C_2H_5)_3P{=}CHCH_3 + [(C_2H_5)_3\overset{+}{P}CH_2SiH_3]Cl^- \xrightarrow{100\%} (C_2H_5)_3P{=}CHSiH_3 + [(C_2H_5)_4\overset{+}{P}]Cl^- \quad (1)$$

The stabilizing effect of main-group elements like silicon in ylides and the related reduction of ylide basicity have been interpreted in terms of a d-orbital participation in the Si–carbanion bonding[13,14] and on simple electrostatic grounds.[15] The latter concept emphasizes the increase in space around the carbanionic center and the corresponding reduction of repulsive forces between the high electron density of the ylidic carbon and the bonding electrons of the σ framework upon introduction of the larger third-row elements.

Though neither of these pictures is very satisfactory, it is gratifying that such recent physical data as ^{13}C resonance studies[24] and photoelectron work[25] fully support the more qualitative findings from synthesis and chemical reactions.

Most of the questions about the bonding of silicon in ylides concern the general problem of interactions between this element and atoms like fluorine, oxygen, and nitrogen, which are known to form exceedingly strong bonds to silicon.[26] All of these have at least one lone pair of electrons.

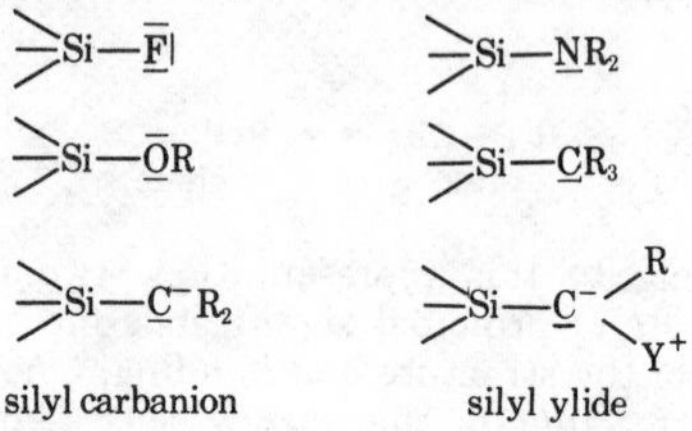

silyl carbanion   silyl ylide

The Role of Silyl Ylides in Synthesis

Shortly after their first preparation, silylated ylides were found to be very useful in synthesis. In fact it appeared that some very important basic members of the ylide series were initially accessible only *via* their silyl derivatives.[18,27]

Salt-free trialkylphosphonium alkylides were obtained for the first time in a pure state through a *desilylation* process using alcohols or silanols[27] (eq 2). (Since then, the more traditional methods of synthesis have been developed further and, with some modifications, are now also applicable[28].) The same was true for the initial synthesis of trimethylarsenic methylide, $(CH_3)_3AsCH_2$, which was obtained through desilylation of a silylated precursor.[18] Dimethylsulfoxonium methylide, $(CH_3)_2S(O)CH_2$, is accessible *via* various routes, one of which is again a desilylation process.[29]

$$R_3P{=}CHSiMe_3 + MeOH \longrightarrow R_3P{=}CH_2 + Me_3SiOMe \quad (2)$$

$$R = CH_3, C_2H_5, i\text{-}C_3H_7, n\text{-}C_3H_7, n\text{-}C_4H_9, C_6H_5$$
$$Me = CH_3$$

The silylated precursors, representing the starting materials for eq 2, are now readily available with a large variety of substituents.[14,15,22,26,27,30,31] Besides trialkylsilyl groups, a number of di-, tri-, and even tetrafunctional silyl components have been used as ligands to the ylidic carbanions attached to phosphorus and sulfur. These lead to new types of

(18) H. Schmidbaur and W. Tronich, *Inorg. Chem.*, **7**, 168 (1968).
(19) H. Schmidbaur and W. Tronich, *Chem. Ber.*, **101**, 3545 (1968).
(20) H. Schmidbaur and W. Malisch, *Chem. Ber.*, **104**, 150 (1971). *Angew. Chem.*, **82**, 84 (1970); *Angew. Chem., Int. Ed. Engl.*, **9**, 77 (1970).
(21) G. H. Birrum and C. N. Matthews, *J. Amer. Chem. Soc.* **88**, 4198 (1966).
(22) H. Schmidbaur, H. Stühler, and W. Vornberger, *Chem. Ber.*, **105**, 1084 (1972).
(23) H. J. Bestmann, "Neuere Methoden präp. org. Chem.," Vol. V, Verlag Chemie, Weinheim/Berstr., 1967.
(24) H. Schmidbaur and F. H. Kohler, unpublished results.
(25) K. A. Ostoja-Starzewski, H.tom Dieck, and H. Bock, *J. Organomet. Chem.*, **65**, 311 (1974).
(26) H. Schmidbaur, *Advan. Organometal. Chem.*, **9**, 260 (1970).
(27) H. Schmidbaur and W. Tronich, *Angew. Chem.*, **79**, 412 (1967); *Angew. Chem., Int. Ed. Engl.*, **6**, 448 (1967); *Chem. Ber.*, **101**, 595 (1967).
(28) R. Köster, D. Simić, and M. A. Grassberger, *Justus Liebigs Ann. Chem.*, **739**, 211 (1970).
(29) H. Schmidbaur und W. Tronich, *Tetrahedron Lett.*, 5335 (1968).
(30) H. Schmidbaur and W. Malisch, *Chem. Ber.* **102**, 83 (1969).
(31) H. Schmidbaur and W. Malisch, *Chem. Ber.*, **103**, 3448 (1970).

difunctional open-chain and cyclic ylides, as indicated by examples **9–16**. Among these compounds the

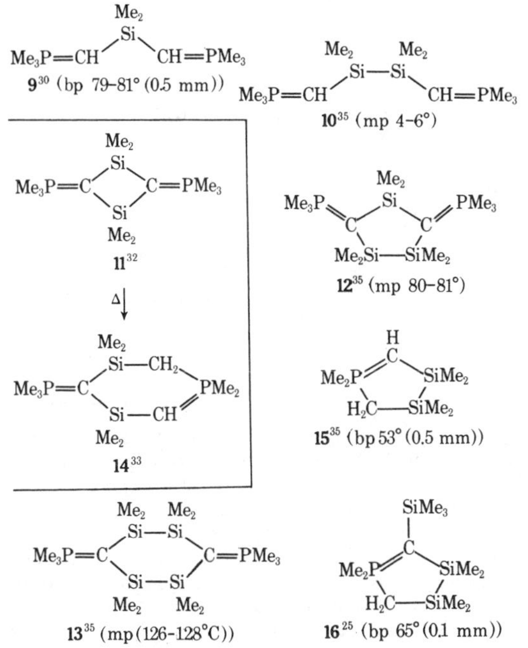

9[30] (bp 79–81° (0.5 mm)); **10**[35] (mp 4–6°); **11**[32]; **12**[35] (mp 80–81°); **14**[33]; **15**[35] (bp 53° (0.5 mm)); **13**[35] (mp(126–128°C)); **16**[25] (bp 65° (0.1 mm))

bis(phosphoranylidene)disilacyclobutanes of type **11** are of special interest.[32] Here two carbanions are members of a small strained ring system, but obviously the presence of the two silicon atoms provides stabilization to the system.[25] The carbon analogs have never been observed.

At elevated temperature and in the presence of excess ylide as a catalyst, the four-membered ring undergoes an isomerization through ring expansion[33] (**11** → **14**). The mechanism of this catalysis involves the excess ylide simply as a deprotonating and reprotonating agent (eq 3). The consequence of this new

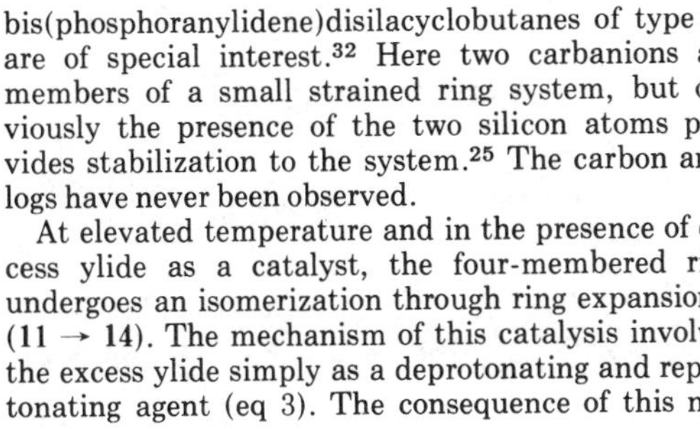

(3)

B = base (ylide)

type of carbanionic rearrangement involving silicon[34] is to provide an even better separation of the two centers of negative charge.

The disilanyl species **10, 12, 13, 15,** and **16**[35,36] were investigated to obtain information on possible delocalization of the carbanionic charge into the Si–Si moieties, a potential (p → d → d) system, but no direct evidence was found.[36,25]

A broad variety of compounds was obtained through the new process of *transsilylation*.[20] This method provides a means of introducing more complicated substituents, starting from the simple trimethylsilyl ylides:

$$R_3P{=}C(SiMe_3)_2 + 2Me_2SiCl_2 \xrightarrow{-2Me_3SiCl} R_3P{=}C(SiMe_2Cl)_2$$
$$\xrightarrow{2MeSiCl_3} R_3P{=}C[SiMeCl_2]_2$$
$$\xrightarrow{2SiCl_4} R_3P{=}C[SiCl_3]_2$$
$$\xrightarrow{2MeSi(CH_2Cl)Cl_2} R_3P{=}C[SiMe(CH_2Cl)Cl]_2$$
$$\xrightarrow{2PCl_3} R_3P{=}C[PCl_2]_2$$

The last example[20] shows that other elements can, like phosphorus, be easily attached to the ylides by this technique. All of these are high-yield reactions, and the sole by-product is the volatile trimethylchlorosilane. Other phosphorus-, arsenic- and antimony-substituted ylides available[14,21,37] are of the following types.

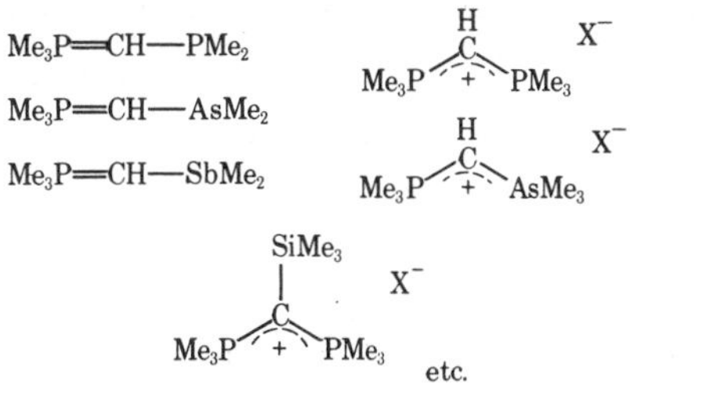

In this connection it should be pointed out that the *heavy* group IV and V elements are found to occur almost exclusively as disubstituted ylides. Thus, whereas a species of formula **18** is easily prepared, the corresponding compound **17** could not be isolated[14] (eq 4). It is probably not simply a coinci-

$$\underset{\mathbf{17}}{2Me_3P{=}CHSnMe_3} \longrightarrow \underset{\mathbf{18}}{Me_3P{=}C(SnMe_3)_2} + Me_3P{=}CH_2 \qquad (4)$$

dence that similar findings have been reported for stannyldiazoalkanes:[38] $|N{\equiv}N^+{-}\underline{C}^-[Sn(CH_3)_3]_2$. Further work on this problem is desirable.

Desilylation-Generated Salt-Free Ylides

The isolation of salt-free high-purity ylides through desilylation (eq 2) provided an excellent basis for a new systematic investigation of the properties of these fascinating compounds. For example, $(CH_3)_3P{=}CH_2$, shortly after its preparation as a salt-free substance, was subjected to detailed ^1H, ^{13}C, and ^{31}P nmr,[39-41] infrared and Raman,[42] dipole

(32) H. Schmidbaur and W. Malisch, *Chem. Ber.*, **103**, 97 (1970).
(33) W. Malisch and H. Schmidbaur, *Angew. Chem.*, **86**, 554 (1974).
(34) R. West, R. Lowe, H. F. Stewart, and A. Wright, *J. Amer. Chem. Soc.*, **93**, 282 (1971), and earlier papers.
(35) H. Schmidbaur and W. Vornberger, *Angew. Chem.*, **82**, 773 (1970); *Angew. Chem., Int. Ed. Engl.*, **9**, 737 (1970).
(36) H. Schmidbaur and W. Vornberger, *Chem. Ber.*, **105**, 3173 (1972); 105, 3187 (1972).
(37) K. Issleib and R. Lindner, *Justus Liebigs Ann. Chem.*, **699**, 40 (1966).
(38) M. F. Lappert, J. Lorberth, and J. S. Poland, *J. Chem. Soc. A*, 2954 (1970).
(39) H. Schmidbaur and W. Tronich, *Chem. Ber.*, **101**, 604 (1968).
(40) H. Schmidbaur, W. Buchner, and D. Scheutzow, *Chem. Ber.*, **106**, 1251 (1973).
(41) K. Hildenbrand and H. Dreeskamp, *Z. Naturforsch. B*, **28**, 226 (1973).
(42) W. Sawodny, *Z. Anorg. Allg. Chem.*, **368**, 284 (1969).

moment,[43] and photoelectron studies.[25,44] From this careful work a better insight into the nature of bonding,[25] and also into the features of the rapid inter- and (possibly) intramolecular proton exchange, was obtained. Although much information has been accumulated to date, no specific mechanism for this prototropy has been established, nor has a solution been found by theoretical calculations.[6,8]

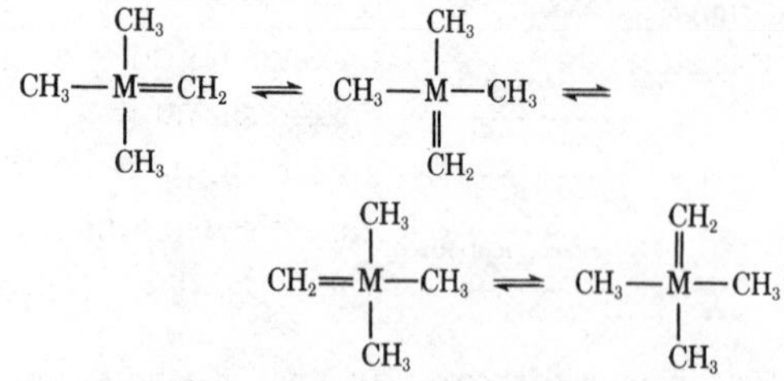

$$CH_3-\underset{\underset{CH_3}{|}}{\overset{\overset{CH_3}{|}}{M}}=CH_2 \rightleftharpoons CH_3-\underset{\underset{CH_2}{\|}}{\overset{\overset{CH_3}{|}}{M}}-CH_3 \rightleftharpoons$$

$$CH_2=\underset{\underset{CH_3}{|}}{\overset{\overset{CH_3}{|}}{M}}-CH_3 \rightleftharpoons CH_3-\underset{\underset{CH_3}{|}}{\overset{\overset{CH_2}{\|}}{M}}-CH_3 \quad (5)$$

$$M = P, As$$

Tetraalkylalkoxyphosphoranes and -fluorophosphoranes

The desilylation-generated salt-free ylides gave access, for the first time, to tetraalkylalkoxyphosphoranes and related compounds, through carefully controlled addition of alcohols, hydrogen fluoride,

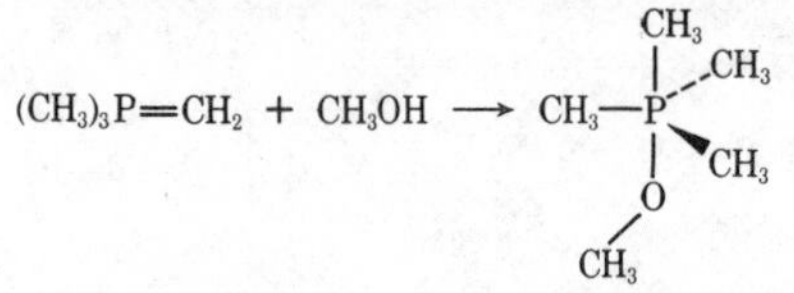

$$(CH_3)_3P=CH_2 + CH_3OH \longrightarrow CH_3-\underset{\underset{CH_3}{\overset{|}{O}}}{\overset{\overset{CH_3}{|}}{P}}\overset{\diagup CH_3}{\diagdown CH_3} \quad (6a)$$

$$(C_2H_5)_3P=CHCH_3 \xrightarrow{HF} (C_2H_5)_4PF \quad (6b)$$

etc.[45-48] Some of these molecules were easily characterized as containing pentacoordinate phosphorus atoms.[47] At low temperatures they are rigid on the nmr time scale and axial and equatorial alkyl groups can be clearly distinguished by ^1H and ^{13}C techniques.[47,49] This is the first case in which a direct comparison of axial and equatorial P–CH_3 bonding was possible. Excess alcohol converts R_4POR' compounds into phosphonium salts of hydrogen-bonded hydropolyalkoxide anions of types $R_4P^+H\text{-}(OR')_2$.[47,50]

Silylated ylides react similarly *with organic carbonyl compounds* rather than paralleling the classic Wittig olefination process, because of a successful competition of silicon with phosphorus for the carbonyl oxygen atoms. Some typical examples have been described and can be explained on this basis.[10,51] The underlying principle is therefore the

(43) H. Schmidbaur, unpublished experiments.
(44) Esr studies have been carried out on arylated ylide radicals.
(45) H. Schmidbaur and H. Stühler, *Angew. Chem.*, **84**, 166 (1972); *Angew. Chem., Int. Ed. Engl.*, **11**, 145 (1972).
(46) H. Schmidbaur, K. H. Mitschke, and J. Weidlein, *Angew. Chem.*, **84**, 165 (1972); *Angew. Chem., Int. Ed. Engl.*, **11**, 144 (1972).
(47) H. Schmidbaur, H. Stühler, and W. Buchner, *Chem. Ber.*, **106**, 1238 (1973).
(48) H. Schmidbaur, K. H. Mitschke, W. Buchner, H. Stühler, and J. Weidlein, *Chem. Ber.*, **106**, 1226 (1973).
(49) H. Schmidbaur, W. Buchner, and F. H. Köhler, *J. Amer. Chem. Soc.*, **96**, 6208 (1974).
(50) H. Schmidbaur and H. Stühler, *Chem. Ber.*, **107**, 1420 (1974).
(51) H. Schmidbaur and H. Stühler, *Angew. Chem.*, **85**, 344 (1973); *Angew. Chem., Int. Ed. Engl.*, **12**, 320 (1970).

same as in the desilylation reactions with alcohols (see examples 1 and 2).

Example 1

$$3(C_6H_5)_2C=O + 2(CH_3)_3P=CHSi(CH_3)_3 \longrightarrow$$
$$(CH_3)_3P=O + (CH_3)_3SiOSi(CH_3)_3 +$$
$$2(C_6H_5)_2C=CH_2 + (CH_3)_2P-\underset{\underset{O}{\|}}{CH}=C(C_6H_5)_2$$

Example 2

$$3(CH_3)_2C=O + 2(CH_3)_3P=CHSi(CH_3)_3 \longrightarrow$$
$$2(CH_3)_3PO + (CH_3)_3SiOSi(CH_3)_3 +$$
$$(CH_3)_2C=CH_2 + CH_3-\underset{\underset{CH_3}{|}}{C}=CH-\underset{\underset{CH_3}{|}}{C}=CH_2$$

Ylide Derivatives of Other Main-Group Elements

Apart from the homologs of silicon and phosphorus, some of the *group III and group II metals* appear to be suitable partners for reactions with ylides. Whereas with boron compounds only straightforward addition reactions had been observed in previous work,[52-54] there is now evidence for a much more complicated system with the compounds of aluminum, gallium, indium, and thallium. In addition to simple 1:1 adducts of the classical type,[27] a new variation of the transylidation process is observed[55] (eq 8) in secondary reactions. In this case, unlike the

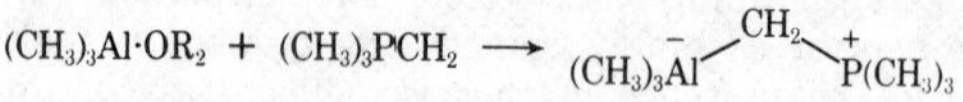

$$(CH_3)_3Al\cdot OR_2 + (CH_3)_3PCH_2 \longrightarrow (CH_3)_3\overset{-}{Al}\overset{\overset{CH_2}{\diagup}}{\diagdown}\overset{+}{P}(CH_3)_3 \quad (7)$$

$$(CH_3)_2GaCl + (CH_3)_3PCH_2 \longrightarrow (CH_3)_2\overset{\overset{CH_2}{\diagup}}{\underset{\underset{Cl}{\diagdown}}{Ga}}\overset{+}{P}(CH_3)_3 \quad (8)$$

$$+(CH_3)_3PCH_2 \downarrow -[(CH_3)_4P]Cl$$

$$\left[(CH_3)_2\overset{\overset{CH_2}{\diagup}}{\diagdown}\overset{+}{\underset{\underset{CH_2^-}{|}}{P}}(CH_3)_2 \right]$$

$$2x \downarrow$$

heterocyclic structure diagram:
$$Me_2Ga \cdots GaMe_2$$

sublimes 90° (0.1 mm)

behavior of the silylated phosphonium salts (eq 1), the deprotonation occurs at one of the alkyl groups attached to phosphorus, followed by ring closure through intercomplexation. Thus, the ylidic function is shifted away from the initial metal substituent.

The resulting heterocyclic species have alternating onium and metalate centers bridged by alkylene groups. In this structure, which is also widespread in transition-metal chemistry (see below), all the carbanionic centers can be accommodated at the acceptor sites of the metals, a feature masking the inherent ylidic character of the species and converting them into a special type of organometallic com-

(52) M. F. Hawthorne, *J. Amer. Chem. Soc.*, **80**, 3480 (1958); **83**, 367 (1961).
(53) D. Seyferth and S. O. Grim, *J. Amer. Chem. Soc.*, **83**, 1613 (1961).
(54) D. Seyferth, S. O. Grim, and T. O. Read, *J. Amer. Chem. Soc.*, **82**, 1510 (1960).
(55) H. Schmidbaur and H. J. Fuller, *Chem. Ber.*, **107**, 3674 (1974).

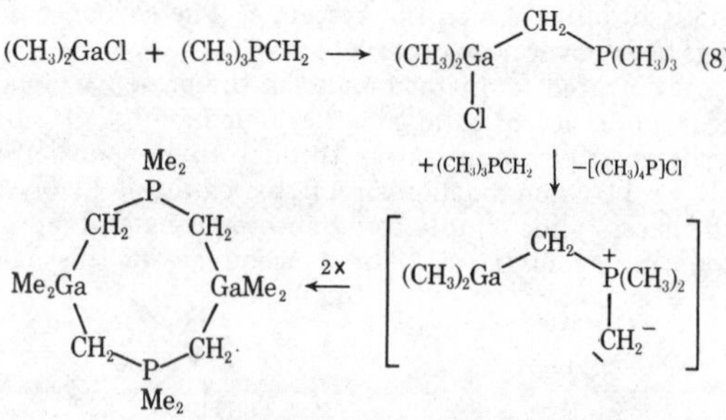

pound. The lack of a suitable acceptor site at the group IV elements obviously makes a similar reaction pathway unfavorable, and generation of a free ylidic function is the sole alternative.

Magnesium alkyls have also recently been found to undergo reactions of this type with phosphorus ylides.[56] With these reagents the even higher acceptor capacity of the metal leads to the formation of a novel organomagnesium polymer:

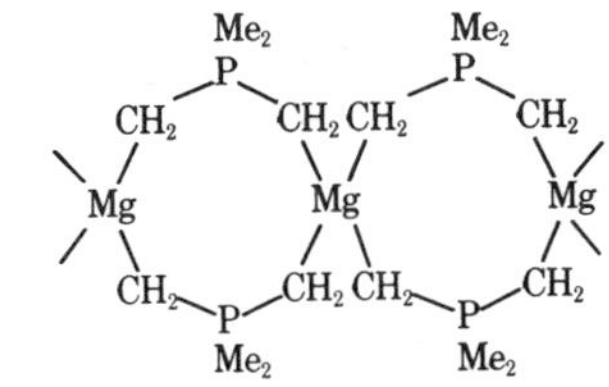

The products of this reaction are themselves extremely reactive, and their chemistry is currently under more extensive investigation. Reactions with transition metal halides lead to ylide derivatives of these elements, as illustrated in eq. 9. The products of this type are discussed below.

$$[Mg\{(CH_2)_2P(CH_3)_2\}_2]_n + 2nCuCl \xrightarrow{-nMgCl_2}$$
$$n[Cu(CH_2)_2P(CH_3)_2]_2 \quad (9)$$

The alkali metals have, for a long time, been known to form addition compounds with ylides;[1,3,4] indeed the stability of some of these adducts is a major problem in the synthesis of "salt-free" ylides.[3,4] Lithium salts in particular are strongly bonded to the ylidic carbanions, but little has been reported on the nature of these adducts[57] (eq 10). An nmr study of lithiated ylides[58] has, however, shown that metalation again occurs in the side chains and that there is equivalence of the two carbanionic centers in solution (see eq. 11). From studies of the ^{1}H–

$$(CH_3)_3PCH_2 \xrightarrow[L_n]{Li^+X^-} (CH_3)_3PCH_2Li^+L_nX^- \quad (10)$$

$(L_n$ = solvation shell)

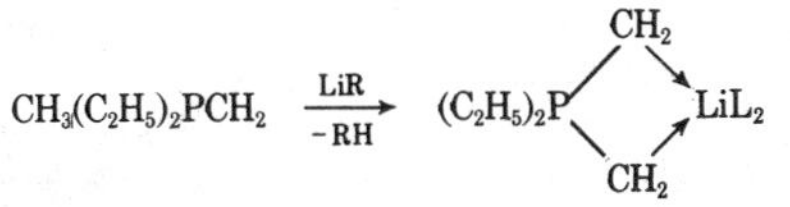

$$CH_3(C_2H_5)_2PCH_2 \xrightarrow[-RH]{LiR} (C_2H_5)_2P\langle{}_{CH_2}^{CH_2}\rangle LiL_2 \quad (11)$$

^{13}C and ^{1}H–^{31}P couplings a direct covalent interaction between the metal and the carbon atoms[58] was deduced. The state of oligomerization is unknown.

Ylide complexes of the salts of the heavier alkali metals are of much lower thermal stability, and the metalation products with these metals are very salt-like in character. A cesium derivative probably contains true $(CH_3)_2P(CH_2)_2^-$ anions in a lattice with Cs^+ cations.[59]

Transition Metal Chemistry of Ylides

Though a confusing variety of different types of products was initially obtained from reactions of yl-

ides with transition-metal derivatives, a general scheme which allows their classification is now slowly appearing. Again, the ylides may occur as simple terminal ligands (as in A), as bridging groups (as in B), or as chelating moieties (as in C). Both A and B had been previously found with main-group elements, but C is a new configuration.

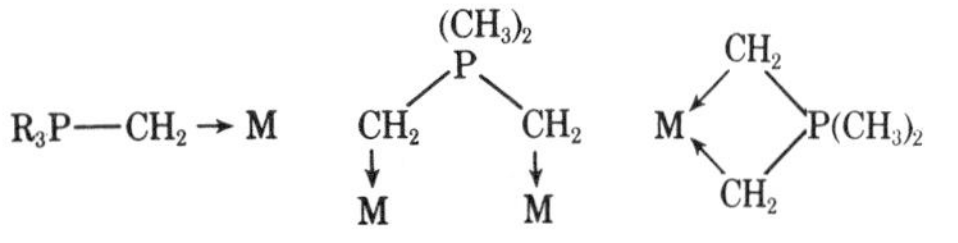

The most remarkable feature of transition-metal ylide chemistry is, however, the unusual thermal stability of the underlying organometallic bonds, M–C–P$^+$, –S$^+$, etc. This is not particularly striking with the new ylide derivatives of the d^{10} metal ions of zinc, cadmium,[56,60] and mercury[9,61-63] because the organometallic compounds of these elements are generally very stable anyway. Indeed, the high tendency toward formation and the stability of the Zn and Cd ylides merely reflects these characteristics. The general scheme for their formation parallels that of the magnesium species, and it is assumed that the principal products involve the B-type structure.

$$n(CH_3)_2Zn + 2n(CH_3)_3PCH_2 \xrightarrow{-2CH_4}$$

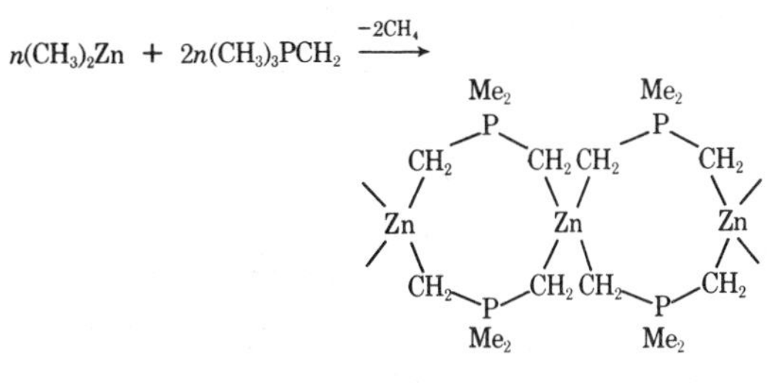

With mercury, simple addition to give A-type products is observed in most cases.[61,62]

The inherent instability generally associated with *coinage metal* organometallics is, however, not reflected by the properties of the ylide complexes of copper, silver, and gold.[63-65] Copper(I) halides and their phosphine complexes were shown[64] to form 1:2 coordination compounds containing two Cu–C bonds, which are thermally stable well above room temperature, *e.g.*

$$[(CH_3)_3P-CH_2CuCH_2-P(CH_3)_3]^+Cl^-$$

colorless crystals, dec 80°

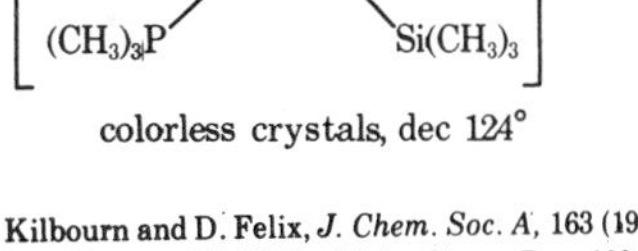

colorless crystals, dec 124°

(56) H. Schmidbaur and J. Eberlein, *Chem. Ber.*, in press.
(57) A. Piskala, M. Zimmermann, G. Fouquet, and M. Schlosser, *Collect. Czech. Chem. Commun.*, 36, 1482 (1971).
(58) H. Schmidbaur and W. Tronich, *Chem. Ber.*, 101, 3556 (1968).
(59) H. Schmidbaur, unpublished observations.

(60) B. T. Kilbourn and D. Felix, *J. Chem. Soc. A*, 163 (1969).
(61) H. Schmidbaur and K. H. Räthlein, *Chem. Ber.*, 106, 102 (1973).
(62) N. A. Nesmeyanov, V. M. Novikov, and D. A. Reutov, *J. Organometal. Chem.*, 4, 202 (1965), and references contained therein.
(63) D. R. Mathiason and N. E. Miller, *Inorg. Chem.*, 7, 709 (1968).
(64) H. Schmidbaur, J. Adlkofer, and W. Buchner, *Angew. Chem.*, 85, 448 (1973); *Angew. Chem., Int. Ed. Engl.*, 12, 415 (1973).
(65) H. Schmidbaur and R. Franke, *Angew. Chem.*, 85, 449 (1973); *Angew. Chem., Int. Ed. Engl.*, 12, 416 (1973).

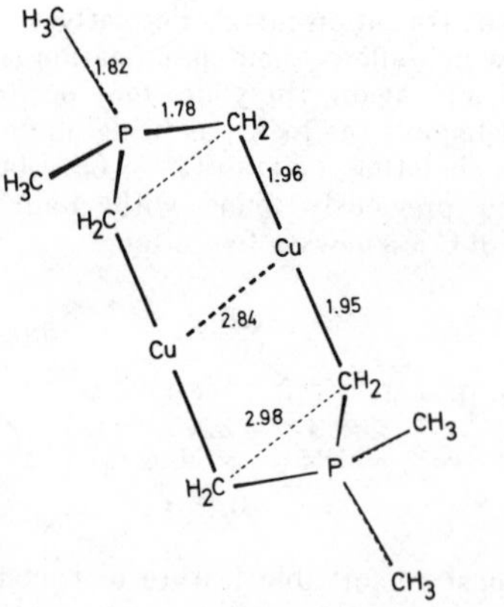

Figure 1.

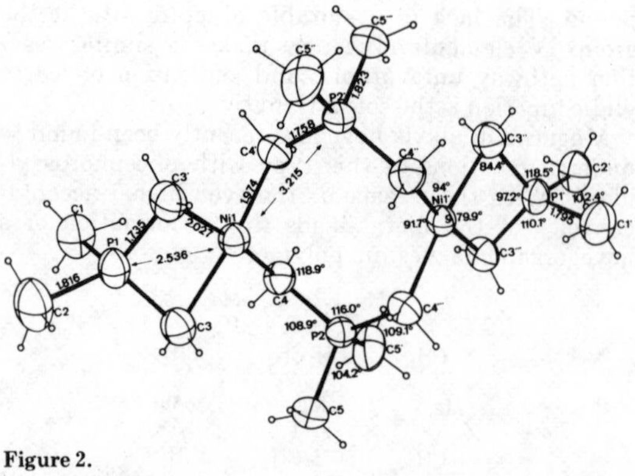

Figure 2.

Even molecular species, such as the ylide complex of trimethylsilylmethylcopper(I), are easily prepared and may be kept at 20° either as a solid or in solution: $(CH_3)_3SiCH_2CuCH_2P(CH_3)_3$, dec 30°. The structures of these compounds are illuminated by analytical and spectroscopic data, and a linear arrangement of the two Cu–C bonds was proposed. This proposal has now been confirmed in an X-ray analysis of the secondary product of the reaction of CuCl with excess $(CH_3)_3PCH_2$. Transylidation is again observed,[64] leading to a cyclic dimer:

$$2CuCl \ + \ 4(CH_3)_3PCH_2 \xrightarrow{-2[((CH_3)_4P]Cl}$$

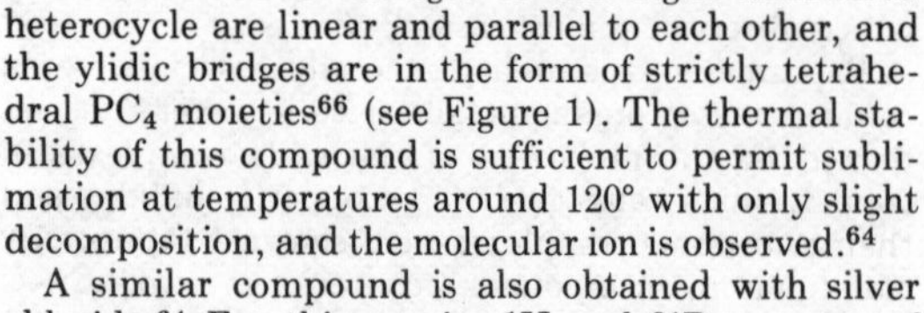

mp 132–136°; subl 100° (0.1 mm)

The two C–Cu–C linkages of this eight membered heterocycle are linear and parallel to each other, and the ylidic bridges are in the form of strictly tetrahedral PC_4 moieties[66] (see Figure 1). The thermal stability of this compound is sufficient to permit sublimation at temperatures around 120° with only slight decomposition, and the molecular ion is observed.[64]

A similar compound is also obtained with silver chloride.[64] For this species ^{1}H and ^{31}P nmr are of special interest, because the ^{1}H–C–^{107,109}Ag and ^{31}P–C–^{107,109}Ag couplings can be observed directly, providing evidence for the presence of covalent Ag–C

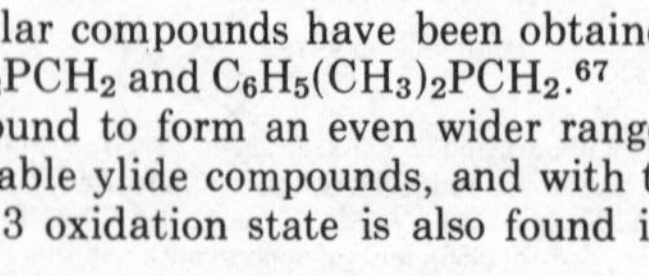

mp 153–155°; subl 150° (0.1 mm)

bonds. Similar compounds have been obtained from $CH_3(C_6H_5)_2PCH_2$ and $C_6H_5(CH_3)_2PCH_2$.[67]

Gold is found to form an even wider range of exceedingly stable ylide compounds, and with this element the +3 oxidation state is also found in novel

metal ylides of square-planar d^8 configurations. In Scheme II appear several examples and the simple relations between some of them.[65,68]

From divalent nickel a whole series of stable organometallic compounds has been detected through the reaction of various nickel alkyl and nickel halide complexes with ylides.[69] These range from simple dimethylnickel monomers[70] to bridged and cage-type molecules, in which the ylides sometimes occur as the only ligands to the metal[71] (Scheme III).

The products of the reaction of $[(CH_3)_3P]_3NiCl_2$ with $(CH_3)_3PCH_2$ are particularly interesting because, depending on the conditions of crystallization or sublimation, several isomers may be obtained.[71] The gas phase contains monomers or dimers, whereas two different dimers were observed (and isolated) from solution. One of the isomers (20) was subjected to a crystal structure determination and its constitution fully established[72] (see Figure 2). This isomer contains the ylidic ligands $(CH_3)_2P(CH_2)_2^-$ both in the bridging position (B) and in the chelating position (C). It is, therefore, a very important example for illustrating the variety of ways in which ylides can interact with metals.

Zerovalent nickel had been previously shown[73,74] to accommodate ylide molecules in its coordination sphere, but to date only the terminal (monodentate) mode of interaction (A) has been confirmed. A crystal structure[75] has been reported for a compound in the carbonyl series, and the spectra of these complexes have been carefully studied.[74]

$$(C_6H_{11})_3P\text{—}CHCH_3$$
$$\downarrow$$
$$Ni(CO)_3$$

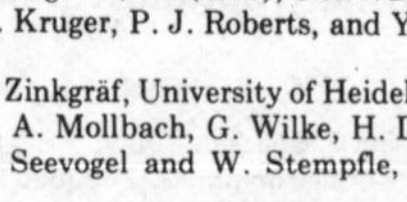

The reactions of palladium and platinum complexes with $(CH_3)_3PCH_2$ have yielded only salt-like

(66) G. Nardin, L. Randaccio, and E. Zangrando, *J. Organometal. Chem.* in press.

(67) H. Schmidbaur, J. Adlkofer, and M. Heimann, *Chem. Ber.*, 107, 3697 (1974).

(68) Dissertation, R. Franke, University of Würzburg, 1974.

(69) Dissertation, H. H. Karsch, University of Würzburg, 1974.

(70) H. H. Karsch, H. F. Klein, and H. Schmidbaur, *Chem Ber.*, 106, 93 (1973).

(71) H. H. Karsch and H. Schmidbaur, *Angew. Chem.*, 85, 910 (1973); *Angew. Chem., Int. Ed. Engl.* 12, 853 (1973); *Chem. Ber.*, 107, 3684 (1974).

(72) D. J. Brauer, C. Kruger, P. J. Roberts, and Y. H. Tsay, *Chem. Ber.*, 107, 3706 (1974).

(73) Dissertation, K. Zinkgräf, University of Heidelberg, 1968.

(74) F. Heydenreich, A. Mollbach, G. Wilke, H. Dreeskamp, E. G. Hoffmann, G. Schroth, K. Seevogel and W. Stempfle, *Isr. J. Chem.*, 10, 293 (1972).

(75) C. Krüger, *Angew. Chem.*, 84, 412 (1972); *Angew. Chem., Int. Ed. Engl.*, 11, 387 (1972).

Scheme II[a]

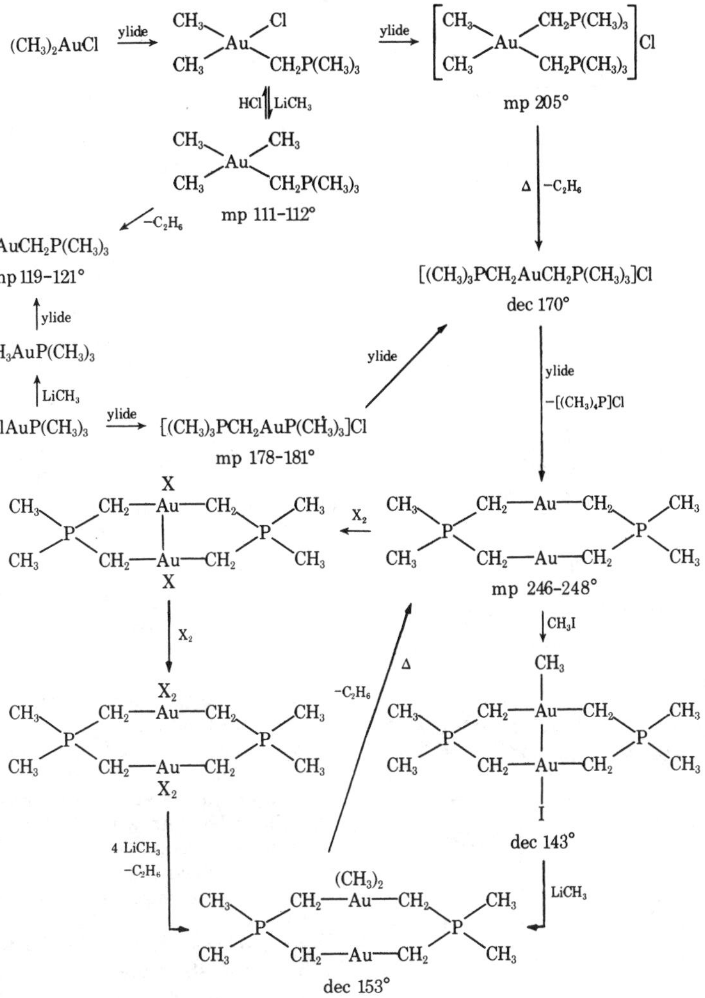

[a] Ylide = $(CH_3)_3PCH_2$, X = halogen.

products,[76] in which two different modes of interaction of ylidic carbanions with the metals are encountered. Though here the ylides fully occupy the first coordination sphere of the metals, there is only partial transylidation by excess ylide. Two of the ligands are left as monodentate groups, and one is converted into a bidentate moiety. The products are stable to air and moisture and have comparatively high decomposition temperatures (eq 12). In all of these compounds the metal atoms are again found to form no less than four stable σ bonds to carbon.

Recent work has also demonstrated that palladium can function as a coordination center for weakly basic[77] or unstable ylides[78-80] of sulfur and nitrogen.

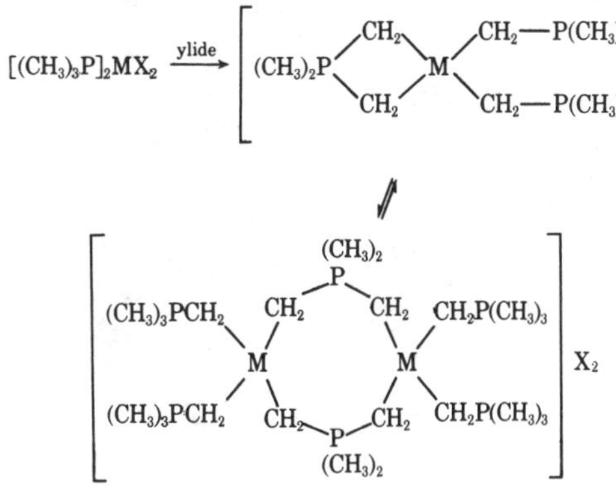

(12)

M = Pd, Pt; X = Cl, Br

The interaction in these cases has been clearly established by careful X-ray diffraction studies.

Of the elements in the cobalt group, only cobalt itself has been studied with respect to its behavior in

(76) J. Adlkofer, Dissertation, University of Würzburg, 1974.
(77) P. Bravo, G. Fronza, G. Gaudiano, and C. Ticozzi, *Gazz. Chim. Ital.*, **103**, 623 (1973); *J. Organometal. Chem.*, **74**, 143 (1974).
(78) M. Keeton, R. Mason, and D. R. Russell, *J. Organometal. Chem.*, **33**, 259 (1971).
(79) N. A. Bailey, R. D. Gillard, M. Keeton, R. Mason, and D. R. Russell, *Chem. Commun.*, 396 (1966).
(80) R. D. Gillard, M. Keeton, R. Mason, M. P. Pillrow, and D. R. Russell, *J. Organometal. Chem.*, **33**, 247 (1971).

Scheme III

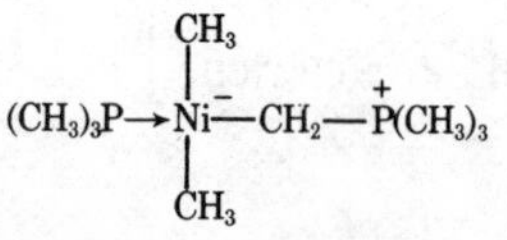

orange crystals dec 70°

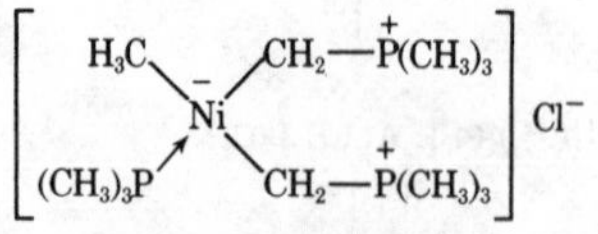

yellow powder dec 95°

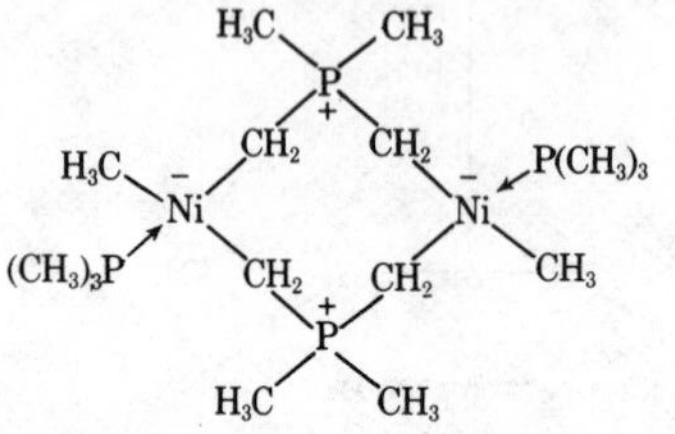

orange liquid mp −5°

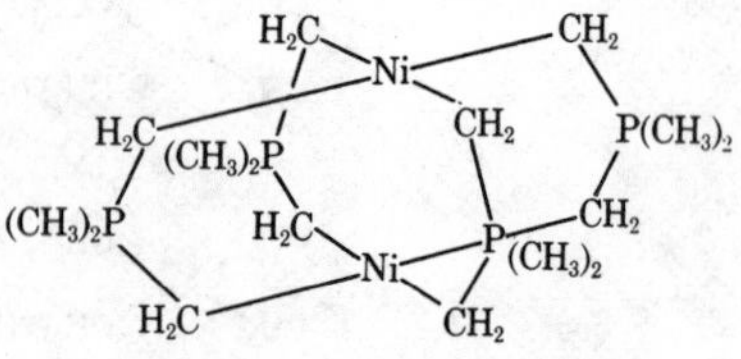

21, yellow crystals dec 87°

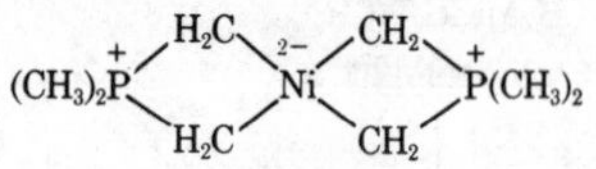

19 (gas phase)

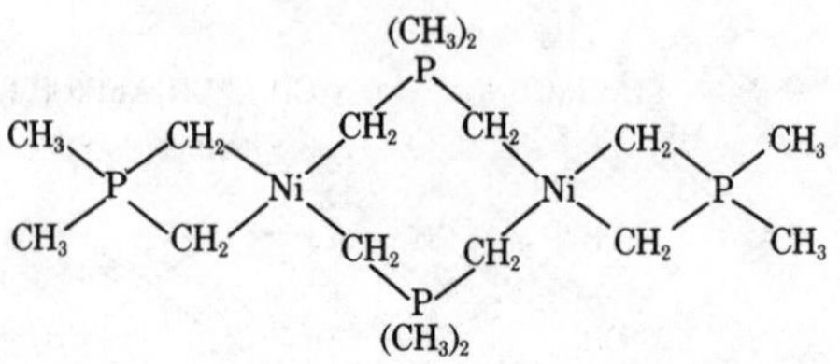

20, yellow crystals dec 123°

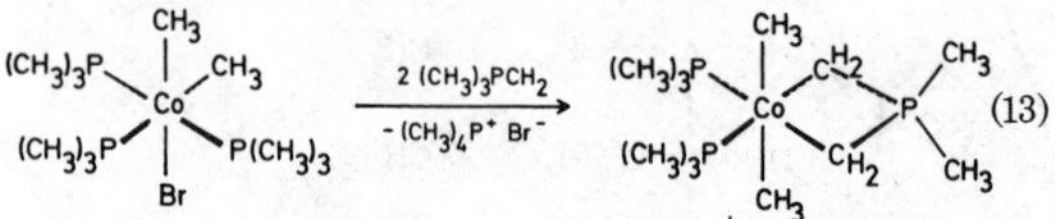

22

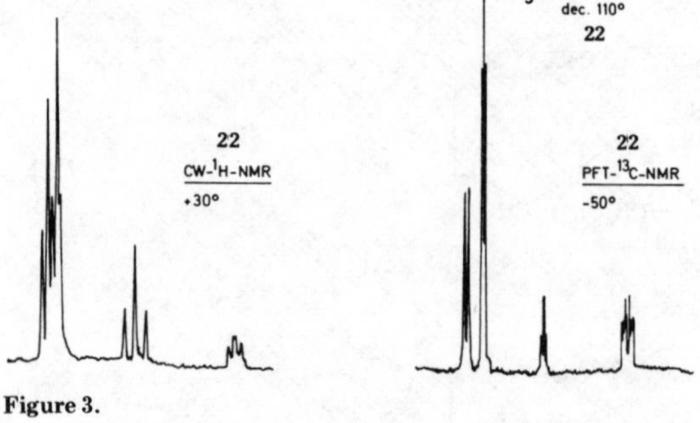

Figure 3.

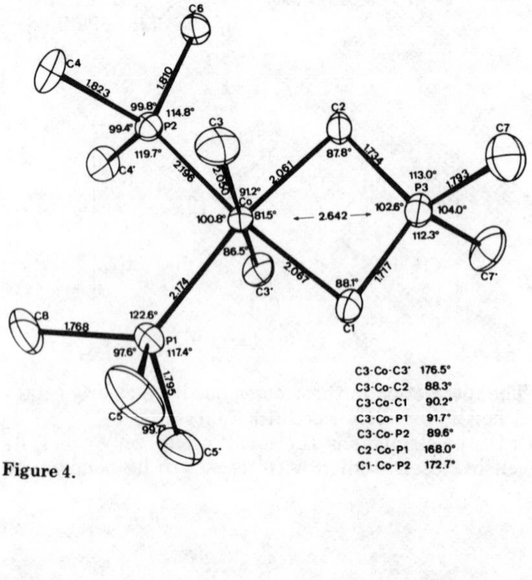

Figure 4.

reactions with ylides.[69] An important example of ylide coordination to an octahedral d^6 metal center was detected as part of a broader series of experiments, and its structure has been investigated in considerable detail by spectroscopic and diffraction methods.[72,81] As with some of the complexes of the chromium group, the ylide in **22** was found to function as a chelating ligand occupying one octahedral edge but with significant stereochemical differences in the configuration of the four-membered ring compared to that in the nickel compound **20**. Whereas the tilted structure in **20** suggests a pseudo-phosphaallyl type of interaction, the planarity of the

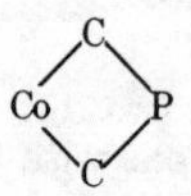

ring for the cobalt species suggests a more regular σ-type bonding (Figures 3 and 4).

Iron, manganese, and vanadium and their homologs have attracted relatively little attention and only the reactions of their carbonyl compounds with ylides have been thoroughly studied.[82-84] Much more information is rapidly becoming available, however, on the ylide complexes of chromium, molybdenum, and tungsten. Among syntheses of a wide range of compounds, a recent report[85] on that of

(81) H. H. Karsch, H. F. Klein, C. G. Kreiter, and H. Schmidbaur, *Chem. Ber.*, **107**, 3602 (1974).

(82) H. Alper and R. A. Partis, *J. Organometal. Chem.*, **44**, 371 (1972).

(83) D. K. Mitchell, W. D. Korte, and W. C. Kaska, *Chem. Commun.*, 1384 (1970); *J. Amer. Chem. Soc.*, **96**, 2847 (1974).

(84) S. Z. Goldberg, E. N. Duesler, and K. N. Raymond, *Chem. Commun.*, 826 (1971).

(85) E. Kurras, U. Rosenthal, H. Mennenga, and G. Ohme, *Angew. Chem.*, **85**, 913 (1973); *Angew. Chem., Int. Ed. Engl.*, **12**, 854 (1973).

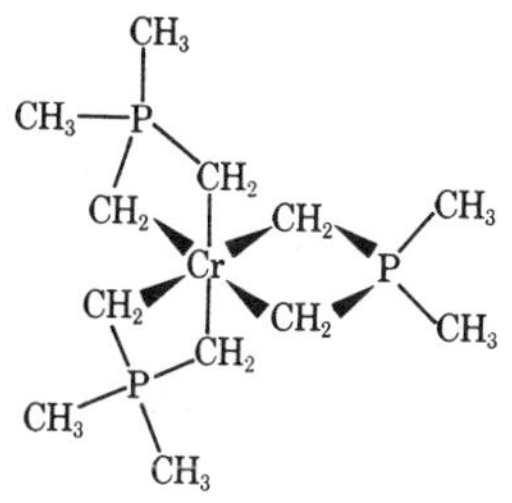

chromium(III) tris(dimethylphosphonium bismethylide) seems to be of special significance. In this octahedral d^3 complex, $(CH_3)_2P(CH_2)_2^-$ again appears as the sole ligand comprising all six Cr–C σ bonds. The stability of this molecule shows the great potential of metal ylide chemistry, from which many more exciting results are expected.

The author gratefully acknowledges the cooperation of a series of coworkers, who have contributed to the work from his own laboratory reported in this Account.

Addendum (May 1976)

Recently the field of inorganic chemistry with ylides has grown considerably. The structures of two of the compounds of the earlier work have been confirmed by x-ray diffraction:

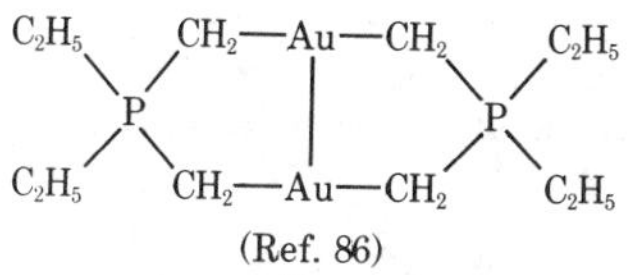

(Ref. 86)

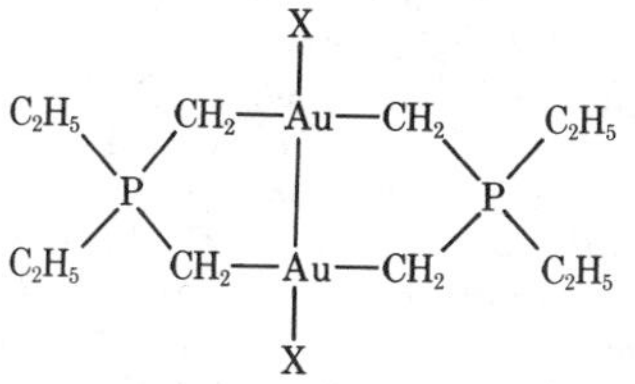

(Ref. 87)

This result was final proof for the existence of the previously unknown Au_2^{4+} unit in these complexes. In another attempt to provide evidence for the unusual Au(II) oxidation state, Mössbauer and ESCA spectroscopy were used,[88] and the earlier conclusions were confirmed.

Arsenic analogs of the dimeric phosphorus ylide complexes of the three coinage metals have been obtained from similar reactions:

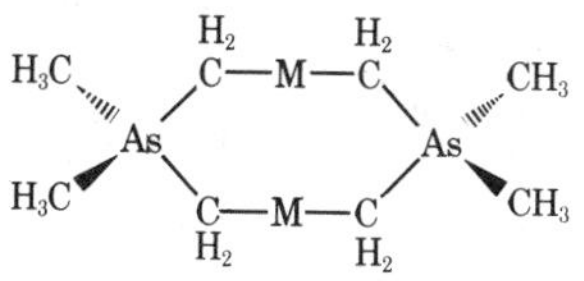

M = Cu, Ag, Au

This result indicates tat many of the phosphorus derivatives should have very similar arsenic counterparts.[89]

Thus, e.g., triphenylphosphonium and -arsonium methylide have been shown to form both stable 1:1 and 1:2 coordination compounds with copper and silver halides of the following type:[91,92]

$(C_6H_5)_3Y—CH_2—M—X$

$[(C_6H_5)_3Y—CH_2—M—CH_2—Y(C_6H_5)_3]X$

M = Cu, Ag
X = Cl, Br
Y = P, As

The successful synthesis of the new arsenic ylides $(C_2H_5)_3As{=}CH_2$ and $(C_6H_5)_3As{=}CH_2$ has been reported and their properties described.[93,94] A careful study of their ^1H- and ^{13}C-NMR spectra led to the proposal of a pyramidal structure of the ylidic carbanion,[95] which was supported by the results of a photoelectron investigation.[96] Mixed methyl/phenyl-substituted phosphorus ylides $(CH_3)_n(C_6H_5)_{3-n}P{=}CH_2$[97] and carbodiphosphoranes $(CH_3)_n(C_6H_5)_{3-n}P{=}C{=}P(C_6H_5)_{3-n}(CH_3)_n$ have been prepared and their reactions studied.[98,99] Among these the ambident nature of $(CH_3)_3P{=}C{=}P(CH_3)_3$ is of special interest. The following equations[100] are illustrative:

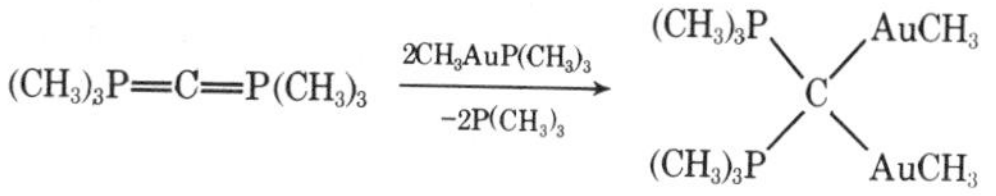

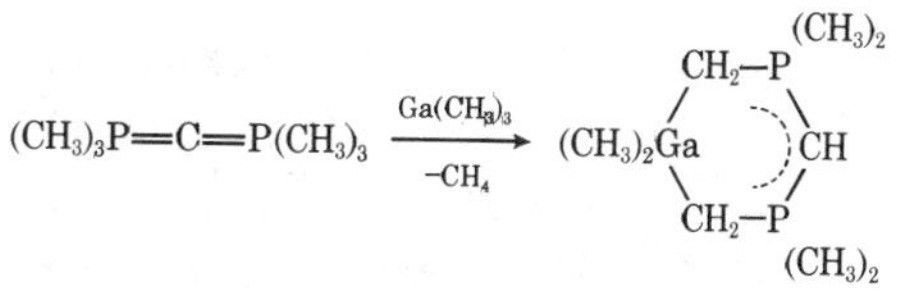

Similar products were also found with a new nitrogen-containing double ylide,[100] e.g.:

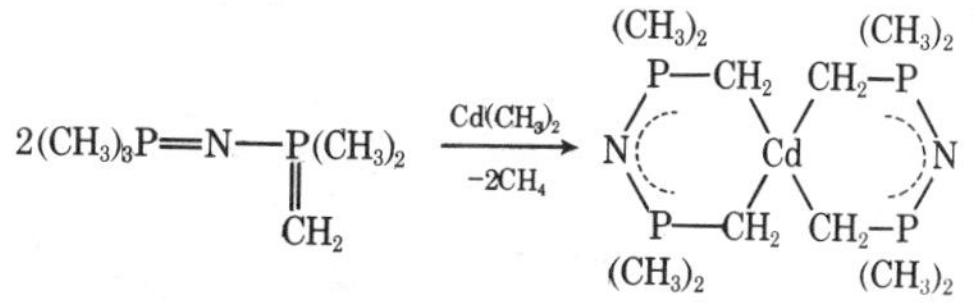

(86) H. Schmidbaur, J. R. Mandl, V. Bejenke, and G. Huttner, *Chem. Ber.*, in preparation.
(87) H. Schmidbaur, J. R. Mandl, A. Frank, and G. Huttner, *Chem. Ber.*, **109**, 466 (1976).
(88) H. Schmidbaur, J. R. Mandl, F. E. Wagner, D. F. van der Vondel, and G. P. van der Kelen, *Chem. Commun.*, **1976**, 170.
(89) H. Schmidbaur and W. Richter, *Chem. Ber.*, **108**, 2656 (1975).
(90) W. Richter and H. Schmidbaur, *Chem. Ber.*, in preparation.
(91) Y. Yamamoto and H. Schmidbaur, *J. Organomet. Chem.*, **96**, 133 (1975).
(92) Y. Yamamoto and H. Schmidbaur, *J. Organomet. Chem.*, **97**, 479 (1975).
(93) Y. Yamamoto and H. Schmidbaur, *Chem. Commun.*, **1975**, 668.
(94) W. Richter and H. Schmidbaur, unpublished results.
(95) H. Schmidbaur, W. Richter, W. Wolf, and F. H. Köhler, *Chem. Ber.*, **108**, 2649 (1975).
(96) K. A. Ostoja-Starzewski, W. Richter, and H. Schmidbaur, *Chem. Ber.*, **109**, 473 (1976).
(97) H. Schmidbaur and M. Heimann, *Z. Naturforsch.*, **29b**, 485 (1974).
(98) H. Schmidbaur and O. Gasser, *J. Am. Chem. Soc.*, **97**, 6281 (1975).
(99) M. S. Hussain and H. Schmidbaur, *Z. Naturforsch.*, **31b**, 149 (1976).
(100) H. Schmidbaur and O. Gasser, *Angew. Chem.*, in preparation.
(101) H. Schmidbaur and H. J. Füller, *Angew. Chem.*, in preparation.

The important work by Kaska et al.,[102] Ishii, Itoh et al.,[103] Tanaka, Matsubayashi et al.,[104] and others can only be briefly mentioned here. A forthcoming review will present a more complete coverage of the pertinent literature.

(102) W. C. Kaska, D. K. Mitchell, R. F. Reichelderfer, and W. D. Korte, *J. Am. Chem. Soc.*, **96**, 2847 (1974).
(103) (a) H. Nishiyama, K. Itoh, and Y. Ishii, *J. Organomet. Chem.*, **87**, 29 (1975); (b) K. Itoh, H. Nishiyama, T. Ohnishi, and Y. Ishii, *J. Organomet. Chem.*, **76**, 401 (1974).
(104) H. Koczuka, G. Matsubayashi, and T. Tanaka, *Inorg. Chem.*, **14**, 253 (1975); **15**, 417 (1976); **13**, 443 (1974).

Solid-State Pressure Effects on Stereochemically Nonrigid Structures

John R. Ferraro*

Chemistry Division, Argonne National Laboratory, Argonne, Illinois 60439

Gary J. Long

Department of Chemistry, University of Missouri—Rolla, Rolla, Missouri 65401

Received July 25, 1974

Reprinted from ACCOUNTS OF CHEMICAL RESEARCH, *8, 171 (1975)*

The geometric structure of a molecule depends upon the magnitude of the energy barrier which prevents conversion into any of its geometric isomers.

In an extensive and impressive group of papers,[1,2] Pearson has presented a series of symmetry rules which may be used to predict the most stable structure of a molecule,[1] its structural rigidity, and its mode of reaction.[2] The rules, which are an extension of the work of Bader,[3] are based upon the use of perturbation theory and group theory to evaluate the effect of a vibrational distortion on the ground-state geometric configuration of a molecule. The theory supporting these symmetry rules has been extensively developed during the past few years.[1-4]

However, in general, experimental verification of the rules has been based upon molecular interconversions or reactions which occur either in solution or in the isolated gaseous state. The reliability of these orbital symmetry rules in predicting a stereochemically rigid or flexible geometric structure in the solid state has not as yet been experimentally confirmed.

This Account presents evidence which indicates that these symmetry rules are useful for predicting solid-state structural interconversions between geometric isomers at high pressure. Due to space limitation, an extension of the ideas on this topic will be presented elsewhere.[5] Furthermore, this Account provides a new approach to the classification of the various effects which are observed in coordination compounds under high pressure.

In the approach developed by Pearson,[1] it is first assumed that all first-order structural distortions of the molecular geometry have occurred. These distortions include any first-order Jahn–Teller distortions required to produce a nondegenerate electronic ground state, and any vibrational distortions which can occur along the totally symmetric normal vibrational modes. On the basis of this assumption, the energy of an initial molecular configuration in the presence of a distortion may be expressed as

$$E = E_0 + \frac{1}{2}Q^2 \int \psi_0 \left| \frac{\partial^2 v}{\partial Q^2} \right| \psi_0 \mathrm{d}\tau + \sum_k \frac{\left[Q \int \psi_0 \left| \frac{\partial v}{\partial Q} \right| \psi_k \mathrm{d}\tau \right]^2}{E_0 - E_k} \tag{1}$$

$$E = E_0 + f_{00}Q^2 + f_{0k}Q^2 \tag{2}$$

where Q is a measure of the magnitude of the displacement of the initial molecular configuration along a normal coordinate which is then designated

John R. Ferraro received his B.S. and Ph.D. at Illinois Institute of Technology, and his M.S. at Northwestern University. He has been at Argonne National Laboratory since 1948, and is presently a senior chemist. His research interests are in the area of molecular spectroscopy of inorganic and coordination complexes.

Gary J. Long, who is an Associate Professor of Chemistry at the University of Missouri—Rolla, was born in Binghamton, N.Y., in 1941. He received his B.S. degree from Carnegie-Mellon University and his Ph.D. with Professor W. A. Baker, Jr., from Syracuse University where he was an NIH Doctoral Research Fellow. His research has dealt mainly with investigation of the magnetic, electronic, and structural properties of transition-metal complexes. During the 1974–1975 academic year Dr. Long is on sabbatical leave at the Inorganic Chemistry Laboratory at Oxford University in England.

(1) R. G. Pearson, *J. Am. Chem. Soc.*, **91**, 4947 (1969); *J. Chem. Phys.*, **53**, 2986 (1970); *Pure Appl. Chem.*, **27**, 145 (1971).
(2) R. G. Pearson, *J. Am. Chem. Soc.*, **91**, 1252 (1969); *ibid.*, **94**, 8287 (1972); *Acc. Chem. Res.*, **4**, 152 (1971).
(3) R. F. W. Bader, *Can. J. Chem.*, 40, 1164 (1962); *Mol. Phys.*, **3**, 137 (1960).
(4) H. C. Longuet-Higgins, *Proc. Roy. Soc. London, Ser. A*, **235**, 537 (1956).
(5) G. J. Long and J. R. Ferraro, *Inorg. Chem.*, to be submitted for publication.

as the Q coordinate, and V is the nuclear–nuclear and nuclear–electronic potential energy. The remaining symbols have their usual meanings and are discussed below. The reader should consult ref 1 and 2 for further details on the derivation and use of this expression.

In eq 1 and 2, the first term, E_0, is the minimized energy of the initial undistorted molecular configuration. The term $f_{00}Q^2$ represents the distortion-induced change in the energy of the initial electronic configuration and is always positive because the initial energy is minimized for the undistorted initial molecular configuration. This term provides a restoring force which would tend to remove the distortion and return the molecule to its initial configuration. The term $f_{0k}Q^2$ is always negative because it in effect changes the initial wave function, ψ_0, to fit the nuclear coordinates of the distorted molecular configuration. The sum of the second and third terms is the experimental force constant for the normal coordinate Q.

Three cases are possible, depending upon the relative magnitudes of the values of f_{00} and f_{0k}. If $f_{00} \gg f_{0k}$, the initially chosen molecular configuration is stable and no distortion is expected. If $f_{00} \cong f_{0k}$, the initial molecular configuration is potentially unstable and *may* spontaneously distort along the normal coordinate Q to a new molecular configuration. Finally, if $f_{00} \ll f_{0k}$, the initial molecular configuration is unstable and will spontaneously change *via* normal coordinate Q to a relatively more stable configuration. From these results, we can see that a knowledge of the magnitude of f_{0k} can provide us with some information about the relative stability of the initial molecular configuration.

At this point, as discussed in detail by Pearson,[2] some rather extensive approximations must be made in order to evaluate the probable magnitude of f_{0k}. The first of these approximations involves limiting the summation in the f_{0k} term to only the one or two lowest excited electronic states (represented by ψ_k) available to the molecular configuration. This approximation will only be good if the energy of the remaining terms is much greater than E_0. The second approximation involves replacing the total ground-state electronic wave function, ψ_0, with the highest occupied molecular orbital, ψ_{HOMO}, for the configuration. Similarly, the total first excited state wave function, ψ_k, is replaced by the lowest unoccupied molecular orbital, ψ_{LUMO}, for the configuration, etc.

We may now determine whether the integrals in the f_{0k} terms are zero or nonzero by making use of the symmetry of the wave functions and the symmetry of the operator, $\partial v/\partial Q$, which will be the same as that of the normal coordinate, Q.[1] The integral will be nonzero only if the direct symmetry product

$$\Gamma_{\psi_{\mathrm{HOMO}}} \times \Gamma_Q \times \Gamma_{\psi_{\mathrm{LUMO}}}$$

contains the totally symmetric irreducible representation. In Pearson's papers, this requirement is stated in terms of the transition density, ρ_{0k}, which is the product, $\psi_0\psi_k$, of the two wave functions and represents the amount of electronic charge transferred within the molecule as a result of nuclear motion. In terms of the transition density the integral will be nonzero only if the symmetry of the transition density is the same as that of Q, i.e.,

$$\Gamma_{\rho_{0k}} = \Gamma_{\psi_0} \times \Gamma_{\psi_k} = \Gamma_{\psi_{\mathrm{HOMO}}} \times \Gamma_{\psi_{\mathrm{LUMO}}} = \Gamma_Q$$

Apart from the symmetry requirement, the value of f_{0k} will depend upon the energy difference, $E_0 - E_k$. Pearson[1] has suggested that an energy gap of the order of 4 eV between the HOMO and the LUMO is sufficiently small enough to indicate the possibility of a structural instability. Hence, with the proper symmetry, and with a small enough energy gap, a change in molecular configuration may occur. However, there is an additional requirement for the occurrence of a molecular interconversion, namely, that distortion along the normal coordinate, Q, must, if continued, lead to an alternative structure.

In summary, there are three basic requirements which, if satisfied, would lead to a possible structural interconversion. First, the symmetry product must be correct, i.e., the symmetry of the transition density, ρ_{0k}, must be the same as that of a normal coordinate, Q. Second, the energy gap between the HOMO and one or more of the excited-state molecular orbitals must be of the order of 4 eV or less. Finally, the normal coordinate, Q, must lead to a viable alternate structure.

We believe that these ideas, as developed by Pearson, may be useful in predicting whether or not a structural interconversion will occur in a solid which is subjected to high pressures.[5] It should be noted that, in general, bending force constants are much smaller than stretching force constants, and as a result, for normal modes involving vibrations, f_{0k} is more likely to be larger than f_{00}, which in this case should be small. Thus, molecular rearrangements involving bending modes and hence changes in bond angles are more likely to occur than those involving stretching modes.

On the basis of the above arguments, a solid complex is more likely to show a structural interconversion along a normal coordinate at high pressure if the symmetry of that coordinate is the same as that of the transition density, ρ_{0k}, as determined for the HOMO and LUMO of the complex. The lack of such agreement may not prevent such an interconversion, but it might make an alternative structural change more likely. One must also, at this point, consider the changes in the energy gap between the HOMO and the excited-state molecular orbitals at high pressure. In general, these energies are expected to shift with the application of pressure.

Structural Interconversion in the Solid State

As emphasized above, most structural interconversions have been studied in solution. Nuclear magnetic resonance spectroscopy has been an ideal tool for such studies because often the equilibrium established between labile structures can be shifted in favor of one structure by a change in temperature. Until recently, few high-pressure studies of solid-state structural interconversions of complexes have used vibrational and electronic spectroscopy. The Mössbauer effect has, however, been used extensively by Drickamer and his coworkers.[6]

(6) H. G. Drickamer and C. W. Frank in "Electronic Transitions and the High Pressure Chemistry and Physics of Solids," Chapman and Hall, London, 1973.

Table I
Solid-State Pressure-Induced Structural Transformations

Compound[a]	Coordination No.	Approximate local symmetry, ambient pressure	Structural transformation	Approximate transformation pressure, kbar	Spectroscopic probe	Remarks
$Ni(BzPh_2P)_2Cl_2$	4	T_d	No change		Electronic Far-ir	
$Ni(BzPh_2P)_2Br_2$	4	1/3 square planar 2/3 T_d	Square planar	20	Electronic Far-ir	Reversible
$Ni(Qnqn)Cl_2$	4	T_d	Dimeric	2	Electronic Far-ir	Irreversible Dimerization
$CuCl_4^{2-}$ [b]	4	Flattened T_d	Square planar	20	Far-ir	Reversible
$Ni(CN)_5^{3-}$ [c]	5	SQP + TBP	SQP	7	Ir(2100-cm^{-1} region)	Reversible
$[NiLX]^+$, $NiLX_2$, $[NiL_2X]^+$, NiL_3X_2	5	SQP + TBP	TBP	Onset of P	Electronic	Reversible

[a] Abbreviations: Bz, benzyl; Ph, phenyl; Qnqn, *trans*-2-(2′-quinolyl)methylene-3-quinuclidinone; L, organic ligand; X, halide or pseudohalide; SQP, square pyramidal; TBP, trigonal bipyramidal. [b] With Cs^+ and $(CH_3)_2CHNH_3^+$. [c] In $[Cr(en)_3][Ni(CN)_5]\cdot1.5H_2O$.

It is reasonable to assume that solid-state interconversions involve considerably larger energy effects than those observed in solution. For solid complexes, in addition to the symmetry effects discussed above, molecular packing, lattice forces, ligand flexibilities, metal–ligand bond distances, d–d electronic transition energies, orbital overlap and orientation effects, and hydrogen bonding among other factors must also be considered. High pressure is known to effect many of these factors[6-8] and will favor the structure with a smaller packing volume. High-pressure effects are observed to shorten the metal–ligand bond distance and to increase the average ligand field strength, $10Dq$.[7-9] In the cases involving high-spin complexes, this increase in $10Dq$ may be sufficient to overcome the electron spin pairing energy and produce a low-spin complex.

Of particular interest is the effect of high pressure on the infrared absorption bands of a solid complex. A reduction in the metal–ligand bond distance shifts the vibrational bands to higher energy. For bending modes, which might possibly transform one structure into another, the effects of pressure may be smaller and conceivably the associated band may shift to a lower energy. It is also possible that, at high pressure, normally forbidden modes may become allowed (in a lower site symmetry), and if this mode yields a structural interconversion, the conversion may then become allowed. Thus, it is of interest to examine the solid-state rigidity of various molecules with differing stereochemical configurations at high pressure.

High-Pressure Studies of Several Solids in Different Symmetries

Solid-state structural transformations obtained for several representative solids at high pressure are presented in Table I. From these results, it may be concluded that structural interconversions are possible for transition-metal complexes in the solid state. The interconversions are all reversible, with the exception of that for the $Ni(Qnqn)Cl_2$ complex.

We now propose a new scheme for the classification of the types of behavior observed in transition-metal compounds at high pressure (see Table II).[6,9-22] The four behavior classes are based primarily upon the presence or absence of a structural and/or electronic change in the complex between ambient and high pressure. Class 1 compounds exhibit neither large structural nor electronic changes, but they would include compounds which show small effects, such as slight unit cell contractions, minor crystallographic changes in space group, small changes in crystal-field parameters, and small shifts in charge-transfer bands. Class 2 compounds exhibit significant structural changes with, at most, minor electronic changes, whereas the reverse situation holds for class 3 compounds. Classes 2 and 3 may be further subdivided as shown in Table II depending upon the absence or presence of a coordination number change, etc. Class 4 includes compounds with *both* electronic structural changes at high pressure and, of course, could have many subdivisions if necessary based upon the presence or absence of each electronic and structural factor.

The behavior of various selected transition-metal complexes at high pressure will now be discussed in

(7) J. R. Ferraro, D. W. Meek, E. C. Siwiec, and A. Quattrochi, *J. Am. Chem. Soc.*, **93**, 3862 (1971).
(8) H. G. Drickamer, *Solid State Phys.*, **17**, 1 (1965).
(9) G. J. Long and J. R. Ferraro, *Inorg. Nucl. Chem. Lett.*, **10**, 393 (1974).
(10) J. R. Ferraro, K. Nakamoto, J. T. Wang, and L. Lauer, *J. Chem. Soc., Chem. Commun.*, 266 (1973).
(11) G. J. Long and J. R. Ferraro, *J. Chem. Soc., Chem. Commun.*, 719 (1973); G. J. Long and D. L. Coffen, *Inorg. Chem.*, **13**, 270 (1974).
(12) R. W. Vaughan and H. G. Drickamer, *J. Chem. Phys.*, **47**, 468 (1967).
(13) R. D. Willett, J. R. Ferraro, and M. Choca, *Inorg. Chem.*, **13**, 2919 (1974).
(14) P. J. Wang and H. G. Drickamer, *J. Chem. Phys.*, **59**, 559 (1973).
(15) L. J. Basile, J. R. Ferraro, M. Choca, and K. Nakamoto, *Inorg. Chem.*, **13**, 496 (1974).
(16) C. B. Bargeron, M. Avinor, and H. G. Drickamer, *Inorg. Chem.*, **10**, 1338 (1971).
(17) D. C. Fisher and H. G. Drickamer. *J. Chem. Phys.*, **54**, 4825 (1971).
(18) J. R. Ferraro and J. Takemoto, *Appl. Spectrosc.*, **28**, 66 (1974).
(19) C. B. Bargeron and H. G. Drickamer, *J. Chem. Phys.*, **55**, 3471 (1971).
(20) C. W. Frank and H. G. Drickamer, *J. Chem. Phys.*, **56**, 3551 (1972).
(21) P. J. Wang and H. G. Drickamer, *J. Chem. Phys.*, **59**, 713 (1973).
(22) L. J. Basile, J. H. Enemark, R. D. Feltham, J. R. Ferraro, and T. E. Nappier, unpublished data.

Table II
Behavior Classes for Pressure-Induced Solid-State Changes

	Structural change		Electronic change			
Behavior class	Geometric change	CN change	Spin-state change	Oxidation-state change	Examples[a]	Ref
1	No	No	No	No	Green $Ni(BzPh_2P)_2Cl_2$	10
					$[Ni(Qnqn)Cl_2]_2$	11
					$Co(Qnqn)Cl_2$	9
					FeS_2	12
2A	Yes	No	No	No	Green $Ni(BzPh_2P)_2Br_2$	10
					Several $CuCl_4^{2-}$	13–14
					$Ni(CN)_5^{3-}$	15
2B	Yes	Yes	No	No	$Ni(Qnqn)Cl_2$	11
					$Co(py)_2Cl_2$	68
3A	No	No	Yes	No	$Mn(Fe)S_2$	16
					$Fe(phen)_2(N_3)_2$	17
					$Fe(phen)_2(NCS)_2$	17–19
3B	No	No	No	Yes	$Fe(acac)_3$	20
					$Cu(OXin)_2$	21
					Hemin	6
4	Yes		Yes		$Co(NO)(Ph_2CH_3P)_2Cl_2$	22

[a] Abbreviations: see Table I; also phen, phenanthroline; acac, acetylacetonate; OXin, 8-hydroxyquinoline.

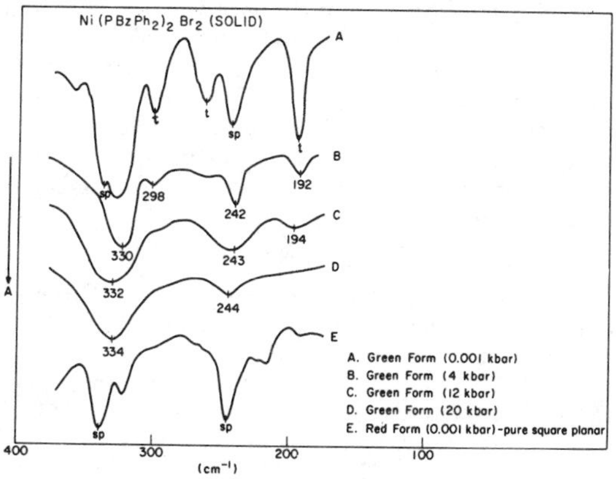

Figure 1. Skeletal vibrations in the green and red isomers of $Ni(BzPh_2P)_2Br_2$ at ambient and high pressure; t = tetrahedral; sp = square planar. Green isomer: A, ambient pressure; B, 4 kbar; C, 12 kbar; D, 20 kbar. Red isomer: E, ambient pressure (pure sp isomer).

terms of their coordination number and behavior type. Later, we will publish a more extensive evaluation, classification, and review of all such high-pressure studies.

Four-Coordinate Complexes

The two complexes, dichloro- and dibromobis(benzyldiphenylphosphine)nickel(II), $Ni(BzPh_2P)_2X_2$, may each be prepared as both red and green isomers.[23] Both of the red complexes are the diamagnetic square-planar forms of the complex. However, there are substantial differences between the two green isomers. The green bromide isomer (with a reduced magnetic moment of 2.70 μ_B at room temperature) has been shown by single-crystal X-ray analysis[24] to contain one square planar and two tetrahedral nickel atoms per unit cell. The magnetic moment of the green chloride isomer (3.23 μ_B at room temperature) and its spectroscopic properties reveal that it is fully tetrahedral in coordination geometry.[23]

Both the electronic and infrared absorption spectra of the two paramagnetic green isomers were studied as a function of pressure.[10] The results for the ν_{Ni-N} and ν_{Ni-X} vibrational bands are presented in Figure 1. The green $Ni(BzPh_2P)_2Cl_2$ isomer retains its tetrahedral coordination geometry at all pressures and shows no indication of any conversion to a square-planar geometry at high pressure. However, the green $Ni(BzPh_2P)_2Br_2$ isomer is transformed from the above-mentioned mixture of tetrahedral and square-planar coordination geometries at ambient pressure, to the purely square-planar red isomer at high pressure.[10] This reversible pressure-induced structural transformation is essentially complete at ca. 20 kbar and represents class 2A behavior. In this instance, the change in the spin state of the nickel ion occurs as a result of the geometric structural change and not directly as a consequence of the high pressure.

In another high-pressure study,[11] it was possible to irreversibly convert the paramagnetic violet pseudotetrahedral nickel complex, $Ni(Qnqn)Cl_2$, into its yellow paramagnetic binuclear $[Ni(Qnqn)Cl_2]_2$ isomer. In these complexes, the Qnqn ligand is

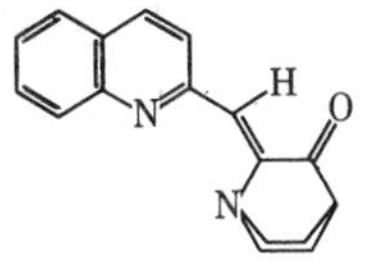

Qnqn

trans-2-(2′-quinolyl)methylene-3-quinuclidinone. Both the yellow and violet isomers have been pre-

(23) M. C. Browning, J. R. Mellor, D. J. Morgan, S. A. J. Pratt, L. E. Sutton, and L. M. Venanzi, *J. Chem. Soc.*, 693 (1962).

(24) B. T. Kilbourn, H. M. Powell, and J. A. C. Darbyshire, *Proc. Chem. Soc., London*, 207 (1963); B. T. Kilbourn and H. M. Powell, *J. Chem. Soc. A*, 1688 (1970).

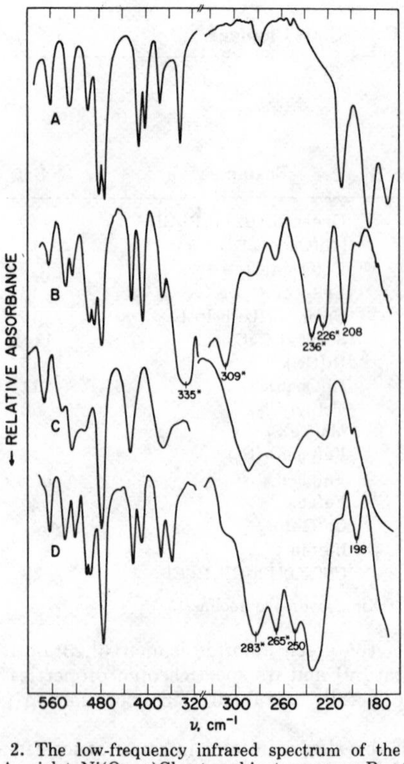

Figure 2. The low-frequency infrared spectrum of the ligand, Qnqn, A; violet $Ni(Qnqn)Cl_2$ at ambient pressure, B; at ca. 20 kbar, C; and at ambient pressure after the release of pressure, D. * indicates an isotope-sensitive band.

pared directly,[11] and the X-ray structure[25] of the yellow binuclear isomer has revealed two bridging and two terminal chlorine ligands and bidentate coordination for Qnqn. The application of pressure to the violet monomeric complex causes the two nickel-chloride nonbonded distances to decrease to a point where the two additional bridging chlorine bonds are formed, and the yellow binuclear complex results. The low-frequency infrared spectra of the ligand and the violet complex are presented in Figure 2. The spectrum of the complex clearly reveals the irreversible changes in both the ν_{Ni-Cl} and ν_{Ni-N} vibrational bands as a function of pressure. The electronic absorption spectrum of the violet isomer also reveals the expected changes in the d–d bands at high pressure.

This is the first example of such an irreversible pressure-induced structural transformation known to us. The irreversibility of this transformation may result from the bond energy of the two additional chlorine bridging bonds, which would make the reverse transformation thermodynamically unfavorable. This transformation involves both a change in coordination number and a change in coordination geometry and represents class 2B behavior. The yellow dimeric $[Ni(Qnqn)Cl_2]_2$ exhibits only minor changes at high pressure and is in class 1.[11]

The room-temperature preparation of $[(CH_3)_2CHNH_3]_2CuCl_4$ has been found[26] from X-ray studies to contain one copper ion in a square-planar

(25) G. J. Long and E. O. Schlemper, *Inorg. Chem.*, 13, 279 (1974).
(26) D. N. Anderson and R. G. Willett, *Inorg. Chim. Acta*, 8, 167 (1974).

Table III
Structural Inferences from Pressure Effects on Several Five-Coordinate Complexes

Complex[a]	Type of ligand	$d\nu/dp$, cm^{-1}/kbar	Structure
[NiLX]Y (24)[b]	Tetradentate	33–70	TBP
[PdLX]Y (2)	Tetradentate	33–81	TBP
[PtLX]Y (1)	Tetradentate	27	TBP
[CoLX]Y (1)	Tetradentate	7	SQP
$[NiLX_2]$ (3)	Tridentate	9–32	Intermediate
$[NiL_2X]Y$ (5)	Bidentate	9–32	Intermediate
$[CoL_2X]$ (2)	Bidentate	Very slight shift	SQP
$[NiL_3X_2]$ (6)	Monodentate	8–29	Intermediate
$[CoL_3X_2]$ (2)	Monodentate	8–23	Intermediate

[a] Abbreviations: see Table I; also, Y, polyatomic anion. [b] Number of compounds studied.

configuration and two copper ions in tetrahedrally distorted square-planar configurations. The crystal is held together by hydrogen bonding from the isopropylammonium ions. At high pressures the coordination geometry of the two tetrahedrally distorted copper ions is reversibly converted to a square-planar geometry.[13] The conversion is observed as a change in the ν_{Cu-Cl} and δ_{ClCuCl} vibrational bands. Confirmation for the conversion was also found in the change occurring in the electronic region.[13,14] A similar structural conversion is also found[14] in Cs_2CuCl_4 and Cs_2CuBr_4. These compounds exhibit a geometric structural change with no change in coordination number or spin-state and belong to class 2A.

Five-Coordinate Complexes

An X-ray diffraction study[27] of the $[Cr(en)_3][Ni(CN)_5]\cdot 1.5H_2O$ complex has shown that its unit cell contains two crystallographically independent $[Ni(CN)_5]^{3-}$ ions, one with a regular square-pyramidal geometry, and one with a distorted trigonal-bipyramidal geometry. Dehydration of the complex converts all of the $[Ni(CN)_5]^{3-}$ ions to the square-pyramidal geometry.[27] When this compound was subjected to pressures of ca. 7 kbar at 78 K, the coordination geometry of the trigonal-bipyramidal $[Ni(CN)_5]^{3-}$ ion was converted reversibly to the square-pyramidal geometry.[15] The infrared spectrum of this compound at ambient and high pressure is presented in Figure 3. In order to prevent the dehydration of the complex at high pressure—presumably a result of localized heating produced by the 6X beam condenser used with the pressure cell—these studies were made at 78 K. In this complex, the reversible transformation represents behavior class 2A in a five-coordinate complex.

An extensive high-pressure study of many five-coordinate nickel(II) complexes with ligands ranging from monodentate to tetradentate has revealed several nonrigid structures in the solid state.[7] The results for several metal ions are presented in Table III and reveal that "tripod-like" tetretradentate ligands prefer the trigonal-bipyramidal structure. The importance of the larger number of chelate rings, and the increased entropy and free energy of

(27) K. N. Raymond, P. W. R. Corfield, and J. A. Ibers, *Inorg. Chem.*, 7, 1362 (1968).

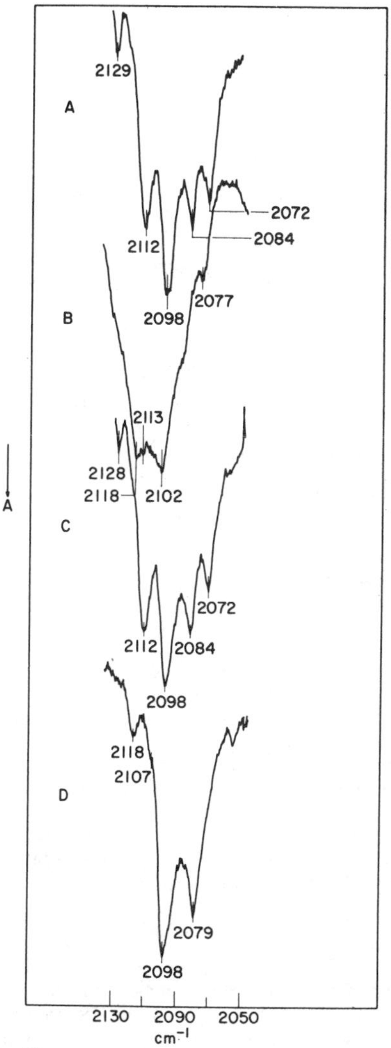

Figure 3. The cyanide stretching vibrational bands in $[Cr(en)_3][Ni(CN)_5] \cdot 1.5H_2O$ at 78 K and ambient pressure, A; at 78 K and *ca.* 7-kbar, B; and at 78 K and ambient pressure after release of high pressure, C. Spectrum D is that of $[Cr(en)_3][Ni(CN)_5]$ at ambient temperature and pressure.

formation for tetradentate ligand complexes of the type [NiLX]Y, are indicated by the more numerous trigonal-bipyramidal structures. As the number of chelate rings is reduced, stability decreases, and the tendency to form intermediate five-coordinate complexes results.[28] NiL_3X_2 complexes with no chelate rings are unstable and dissociate in solution, whereas, the application of high pressure tends to distort these solids toward the distorted intermediate five-coordinate geometry. In these five-coordinate complexes, a gradual change from class 1 behavior (with small values of $d\nu/dp$) to class 2A behavior is observed.

The five-coordinate square-pyramidal complex Fe(NO)(salen) (salen = *N,N'*-ethylenebis(salicylidenimine)) has been shown to contain iron in an intermediate spin state ($S = 3/2$) and to exhibit spin equilibrium at low temperature.[29] Mössbauer spectral results indicate that the complex most likely contains Fe(III) and NO^-, although this formulation is still open to question. A recent study of the NO vibrational absorption band as a function of pressure has revealed a shift to lower frequency at high pressure.[30] These results appear to be consistent with a change in spin state for the iron ion which may be accompanied by a change in NO oxidation state. This compound would appear to fit into class 3, but additional studies will be required to confirm and refine this classification because structural changes may also be significant.

Two isomers of $Co(NO)(Ph_2CH_3P)_2Cl_2$ are known.[31] One of these isomers is trigonal bipyramidal and contains Co(I) and NO^+ ions—most likely with a linear Co–N–O bond. The second isomer is square pyramidal and contains Co(III) and NO^- with a bent Co–N–O bond. The NO vibrational absorption band occurs at ca. 1750 cm^{-1} in the first isomer and at ca. 1650 cm^{-1} in the second. A preliminary study indicates the structural conversion of the trigonal-bipyramidal isomer to the square-pyramidal isomer at high pressure.[22] This represents class 4 behavior with significant structural and electronic changes.

Six-Coordinate Complexes

To date we have not been successful in changing the coordination number or geometry of an octahedral (or close to octahedral) complex at high pressure. However, several octahedral high-spin complexes have been reversibly converted—at least in part—to the analogous low-spin octahedral complexes at high pressure.[17,18,32] Table IV summarizes some of these results. The effect of pressure on the skelatal vibrations in $Fe(phen)_2(NCS)_2$ is shown in Figure 4. The initial conversion from the high-spin to the low-spin state has been explained[6,17] by the increase in ligand-field potential with pressure until it exceeds the electron pairing energy. This initial effect is accompanied by the back-donation of the metal t_{2g} electrons into the π^* orbitals of the ligand. With a further increase in pressure this back donation is reduced by the accessibility of π electrons from the ligand.[6]

In an extensive series of papers,[12,16,17,20,33-36] Drickamer and his coworkers have studied the change in spin state with pressure of a wide variety of iron compounds by utilizing Mössbauer effect spectroscopy. In one of these studies,[12] the Mössbauer effect spectrum of FeS_2 as a function of pressure gave changes in the chemical isomer shift and quadrupole splitting which indicated small and continuous changes with increasing pressure. This be-

(28) J. R. Ferraro and K. Nakamoto, *Inorg. Chem.*, **11**, 2290 (1972).

(29) A. Earnshaw, E. A. King, and L. F. Larkworthy, *J. Chem. Soc. A*, 2459 (1969).

(30) R. D. Feltham, J. H. Enemark, P. L. Johnson, L. J. Basile, J. R. Ferraro, and H. H. Wickman, unpublished data.

(31) C. P. Brock, J. P. Collman, G. Dolcetti, P. H. Farnham, J. A. Ibers, J. E. Lester, and C. A. Reed, *Inorg. Chem.*, **12**, 1304 (1973).

(32) J. S. Wood, *Prog. Inorg. Chem.*, **16**, 227 (1972).

(33) S. C. Fung and H. G. Drickamer, *J. Chem. Phys.*, **51**, 4350, 4360 (1969).

(34) D. C. Grenoble and H. G. Drickamer, *J. Chem. Phys.*, **55**, 1624 (1971).

(35) D. C. Grenoble, C. W. Frank, C. B. Bargeron, and H. G. Drickamer, *J. Chem. Phys.*, **55**, 1633 (1971).

(36) A. R. Champion and H. G. Drickamer, *J. Chem. Phys.*, **47**, 2591 (1967).

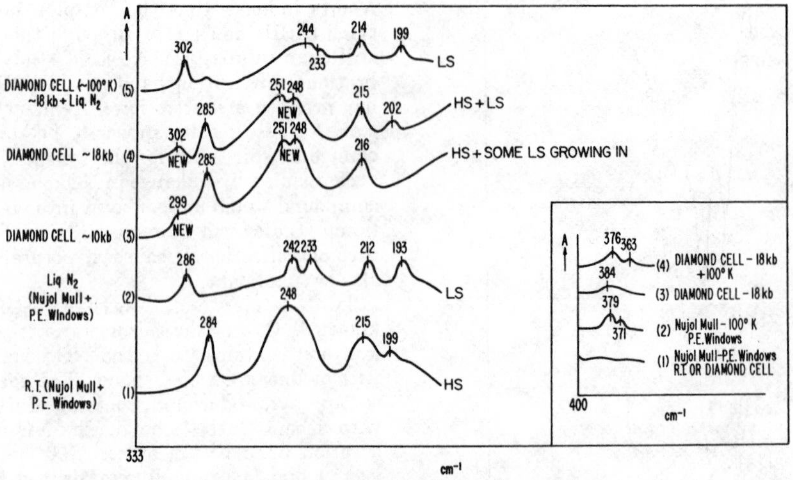

Figure 4. The skeletal vibrations in $Fe(phen)_2(NCS)_2$ as a function of pressure and temperature.

Table IV
High-Pressure Spin-State Interconversions

Compound[a]	Central atom coordination no.	No. of unpaired electrons	Pressure effect	Conversion pressure, kbar	Experimental probe
Fe(NO)(salen) (d^5)	5	3	High spin ⟶ low spin	21	NO stretching region
$Fe(phen)_2(NCS)_2$ (d^6)	6	2	High spin ⟶ low spin	18	Skeletal far-ir and Mössbauer effect.
$Fe(phen)_2(NCSe)_2$ (d^6)	6	2	High spin ⟶ low spin	8–10	Skeletal far-ir
$Fe(bipy)_2(NCS)_2$ (d^6)	6	2	High spin ⟶ low spin	15	Skeletal far-ir
$Co(nnp)(NCS)_2$ (d^7)[b]	5	3	High spin ⟶ low spin	4	Skeletal far-ir

[a] Abbreviations: see Tables I and II; also bipy, bipyridyl; salen, *N,N′*-ethylenebis(salicylidenimine); nnp is $(C_2H_5)_2N(CH_2)_2$-$NH(CH_2)_2P(Ph)_2$. [b] L. Sacconi and J. R. Ferraro, *Inorg. Chim. Acta,* **9**, 49 (1974).

havior would indicate class 1 behavior for FeS_2. However, a related Mössbauer effect study[16] of ^{57}Fe as a substitutional impurity in MnS_2, which has the same cubic structre as FeS_2 with a larger lattice parameter, revealed a distinct change in the electronic spin state of the Fe(II) ion as a function of pressure. The Mössbauer spectrum of this material is shown in Figure 5. The substitutional Fe(II) ion, which exists predominately in the high-spin state at ambient pressure, is gradually converted to low-spin Fe(II) as the pressure is increased. The change is rapid from ca. 50 to 120 kbar, and above 120 kbar the iron(II) is essentially all in the low-spin state. The change in spin state at high pressure has been attributed[16] to an increase in the ligand-field strength at high pressure with the resulting spin pairing of the six d electrons in the t_{2g} orbitals of the iron(II) ion. This provides additional evidence for the pressure dependence of the ligand field.[7-9] This compound exhibits class 3A behavior with a change in spin state from high spin to low spin with increasing pressure.

In a related study of the high-pressure Mössbauer spectra of a series of metal ferrocyanides, Fung and Drickamer[33] observed evidence for the conversion of low-spin iron(II) to its high-spin state at very high pressure and high temperature. This surprising observation was explained on the basis of a reduced amount of metal-to-ligand back-bonding at the higher pressures. Associated with this decrease in back-bonding would be a weakening of the π bonding between the filled t_{2g} orbitals of the iron(II) and the orbitals of the cyanide ligands. This would result in an increase in the energy of the low-spin state relative to the high-spin state, and an increased population—especially at high temperature—of the high-spin state.

The preceding argument indicating a decreased amount of back-bonding at high pressure is further supported by the work of Fisher and Drickamer[17] on iron(II)—phenanthroline complexes, and by Bargeron and Drickamer[19] on similar substituted phenanthroline complexes. For these ligands, all of the tris complexes are low spin at ambient pressure, but a small amount of conversion to the high-spin complex was observed at high pressure. The bis(phenanthroline)iron(II) halide and pseudohalide complexes are high spin at ambient pressure and tend to convert to the low-spin state with increasing pressure. However, at very high pressures, the rate of conversion either slows or reverses with increasing pressure, depending upon the back-bonding ability of the halide or pseudohalide ligand. Specifically, for $Fe(phenanthroline)_2(N_3)_2$, the amount of the low-spin complex is at a maximum at ca. 40 kbar[17] and decreases at higher pressures, presumably because of the decreasing amount of back-bonding. All of these complexes exhibit class 3A behavior.

A study of the high-pressure Mössbauer spectra of several related ferricyanide complexes[33,36] revealed the reduction of the Fe(III) ion to the expected low-spin iron(II) state at intermediate pressures. This was confirmed in Prussian Blue which had been prepared with selectively labeled ^{57}Fe sites. At higher pressures, once again the low-spin iron(II) ion was converted to the high-spin iron(II) ion. These materials exhibit class 3B behavior at lower pressures, and class 3A behavior at higher pressures. In a related study[20] of 12 ferric β-diketone complexes, Frank and Drickamer have observed a reversible reduction of iron(III) to iron(II) at high pressure. The degree of reduction is related to the change in the Mössbauer isomer shift with pressure and the nature of the ligands. An additional agreement between the intensity of the charge-transfer absorption band and the extent of reduction was also observed. These compounds exhibit class 3B behavior at all pressures studied.

To date, as would be predicted by the Bader and Pearson model,[1-3] no octahedral complexes have been converted at high pressure to complexes with a different coordination number or coordination geometry. However, because such successful structural interconversions have occurred with distorted four-coordinate complexes, it is planned to study a series of highly distorted octahedral complexes at high pressure.

There has been, as of this writing, no detailed study of any seven- or eight-coordinate complexes as a function of pressure.

Discusion of the Nonrigidity of Solids at High Pressure

It may be concluded that solid-state high-pressure structural transformations are possible in transition-metal complexes. All of the transformations examined thus far have been reversible, with the exception of that in $Ni(Qnqn)Cl_2$, in which a dimer is formed at high pressure. In this complex, two additional bonds are formed on dimerization, and they contribute to the stability of the high-pressure phase.

The probability of producing structural interconversions with pure or nearly pure tetrahedral and octahedral complexes is predicted, on the basis of the theoretical considerations discussed above, to be small. For undistorted tetrahedral complexes, this prediction is borne out by experiment. Attempts[37] to convert complexes of nearly tetrahedral symmetry have been unsuccessful. The results of our work, however, indicate that the pressure-induced conversions of tetragonally distorted tetrahedral complexes are possible, in particular where an asymmetric ligand field is observed by the central atom (e.g., a complex involving several types of ligands).

It is possible that distorted six-coordinate complexes may also behave similarly.[38-40] The results

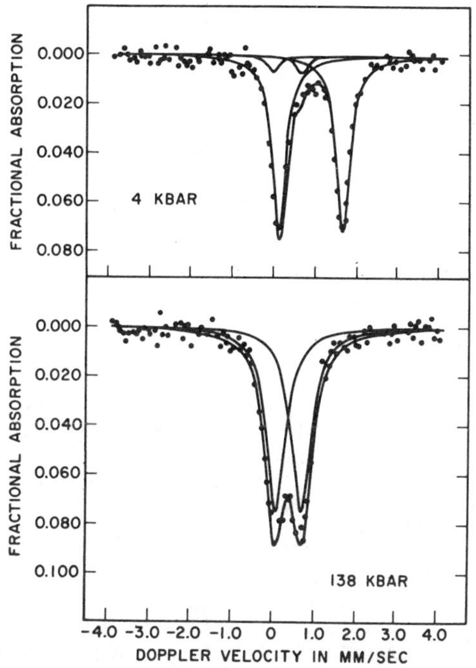

Figure 5. The Mössbauer effect spectrum of $Mn(^{57}Fe)S_2$ at 4 and 138 kbar.

observed to date for four- and six-coordinate complexes are not surprising because the energy barrier to rearrangement for true tetrahedral and octahedral structures is certainly high. For the $Ni(BzPh_2P)_2Br_2$ complex, the unpaired electrons may contribute to the lowering of the energy difference between the distorted tetrahedral and the square-planar configurations. This effect would be superimposed upon the beneficial effect of a starting structure which is distorted toward the square-planar geometry.

For five-coordinate complexes the energy barrier for structural interconversion is small, and many examples have been reported in which the trigonal-bipyramidal and square-pyramidal isomers both exist.[32,41-45] This is apparently also true in the solid state, because our high-pressure studies indicate that interconversion is readily obtained. In systems containing a tripod-like tetradentate ligand, the ligand flexibility favors the trigonal-bipyramidal structure. For the $Ni(CN)_5^{3-}$ ion, a structure which is distorted in the direction of the pressure-stable square-pyramidal phase, the monodentate cyanide ion permits the rearrangement to occur.

Nonrigid configurations for seven- and eight-coordinate complexes have been demonstrated in solution studies.[46-57] However, no solid-state high-pres-

(37) J. R. Ferraro, unpublished data.
(38) R. Eisenberg and J. A. Ibers, *J. Am. Chem. Soc.*, 87, 3776 (1965).
(39) A. E. Smith, G. N. Schrauzer, V. P. Hayweg, and W. Heinrich, *J. Am. Chem. Soc.*, 87, 5798 (1965).
(40) R. Eisenberg and H. B. Gray, *Inorg. Chem.*, 6, 1844 (1967).
(41) E. L. Muetterties, *Acc. Chem. Res.*, 3, 266 (1970).
(42) R. R. Holmes, *Acc. Chem. Res.*, 5, 296 (1972).
(43) E. L. Muetterties, *Rec. Chem. Prog.* 31, 51 (1970).
(44) E. L. Muetterties and R. A. Schunn, *Q. Rev., Chem. Soc.*, 20, 245 (1966).
(45) E. L. Muetterties and C. M. Wright, *Q. Rev., Chem. Soc.*, 21, 109 (1967).
(46) L. Malatesta, M. Fermi, and V. Valenti, *Gazz. Chim. Ital.*, 94, 1278 (1964); E. B. Fleischer, A. E. Gebala, D. R. Swift, and P. A. Tasker, *Inorg. Chem.*, 11, 2775 (1972).
(47) J. L. Hoard and J. V. Silverton, *Inorg. Chem.*, 2, 235 (1963).
(48) R. V. Parish, *Coord. Chem. Rev.*, 1, 439 (1966).
(49) F. Klanberg, D. R. Eaton, L. J. Guggenberger, and E. L. Muetterties, *Inorg. Chem.*, 6, 1271 (1967).

sure studies have been reported; several of these investigations are presently under way in our laboratories.

The authors acknowledge with thanks the research contributions of Dr. L. J. Basile of Argonne National Laboratory during the course of this work. G. J. Long thanks Argonne National Laboratory for a Faculty Research Participation Award during the summer of 1973 and the Inorganic Chemistry Laboratory of Oxford University for assistance. The authors thank Ms. A. A. Long for typing the manuscript. This research was performed under the auspices of the U.S. Atomic Energy Commission.

Addendum (May 1976)

The discussion on the importance of structural distortions is further illustrated by comparing the tetragonally distorted tetrahedral $[(CH_3)_2CHNH_3]_2CuCl_4$ complex with the square planar $[(C_2H_5)_2NH_2]_2CuCl_4$ complex. The second complex does not change at high pressure (Class 1 behavior). However, the first complex is converted at high pressure to the square planar configuration (Class 2A behavior).[13] The structural differences in the two complexes may in part reflect the stronger hydrogen bonding in the square planar compound. Willett et al.[58] have indicated that as the hydrogen bonding increases between a substituted ammonium cation and the $CuCl_4^{2-}$ anion, the "trans" Cl-Cu-Cl angles in $CuCl_4^{2-}$ increase, and the anion approximates more closely the square planar configuration. As expected, the unit cell volume of the square planar structure is less than that of the tetrahedral or pseudotetrahedral structure and is reflected in an increase in density.

Data for a series of $CuCl_4^{2-}$ compounds with anions varying from Cs^+, where no hydrogen bonding is possible, to $Pt(NH_3)_4^{2+}$, where extensive hydrogen bonding exists, are presented in Table V. It is apparent that, as the hydrogen bonding increases, the "trans" Cl-Cu-Cl angle increases, the density increases, and the structure approaches the square planar configuration. Higher temperatures would tend to decrease the hydrogen bonding, increase the unit cell volume, and stabilize the more tetrahedral configuration. In this regard, it is noted[13] that both of the above mentioned complexes have high temperature D_{2d} phases. In the same manner, high pressures will tend to promote hydrogen bonding, decrease the unit cell volume, and stabilize the square planar configuration, as is observed for the two copper complexes. For other systems pertaining to this discussion, see references 66 and 67.

Table V
Geometric Parameters for Several $CuCl_4^{2-}$ Compounds

Compound	Average Trans Cl-Cu-Cl Angle	Density	Structure[a]
Cs_2CuCl_4[58,60,61]	124	—	$T\text{–}T_d$
$[(CH_3)_4N]_2CuCl_4$[58,60]	128	1.40	$T\text{–}T_d$
$[(\phi CH_2)N(CH_3)_3]_2CuCl_4$[62]	132	1.36	$T\text{–}T_d$
$[(C_2H_5)_3NH]_2CuCl_4$[63]	137	1.33	$T\text{–}T_d$
$(C_{13}H_{19}N_2OS)_2CuCl_4$[64]	143	1.43	$T\text{–}T_d$
$[(CH_3)_2CHNH_3]_2Cu\text{-}Cl_4$[13,58,59]	135	1.50	⅓ SQPL + ⅔ D_{2d}
$[(C_2H_5)_2NH_2]_2Cu\text{-}Cl_4$[13,58,59]	162	1.70	SQPL
$[Pt(NH_3)_4]CuCl_4$[65]	180	—	SQPL

[a] Abbreviations: $T\text{–}T_d$, tetragonally distorted tetrahedral; SQPL, square planar.

(50) S. J. Lippard, *Prog. Inorg. Chem.*, **8**, 109 (1966); R. V. Parish and R. G. Perkins, *J. Chem. Soc.*, 345 (1967).
(51) E. L. Muetterties, *Inorg. Chem.*, **12**, 1963 (1973).
(52) H. H. Claassen, E. L. Gasner, and H. Selig, *J. Chem. Phys.*, **49**, 1803 (1968).
(53) G. R. Rossman, F. D. Tsay, and H. B. Gray, *Inorg. Chem.*, **12**, 824 (1973).
(54) J. L. Hoard, T. A. Hamor, and M. D. Glick, *J. Am. Chem. Soc.*, **90**, 3172 (1968).
(55) K. O. Hartman and F. A. Miller, *Spectrochim. Acta*, **24**, 669 (1968).
(56) B. R. McGarvey, *Inorg. Chem.*, **5**, 476 (1966).
(57) R. G. Hayes, *J. Chem. Phys.*, **44**, 2210 (1966).
(58) R. D. Willett, J. A. Haugen, J. Lebsack, and J. Morrey, *Inorg. Chem.*, **13**, 2510 (1974).
(59) D. N. Anderson and R. D. Willett, *Inorg. Chim. Acta*, **8**, 167 (1974).
(60) B. Morosin and E. C. Lingafelter, *J. Phys. Chem.*, **65**, 50 (1961).
(61) L. Helmholtz and R. F. Kruh, *J. Am. Chem. Soc.*, **74**, 1176 (1952).
(62) M. Bonamico, G. Dessy, and A. Vaciago, *Theor. Chim. Acta*, **7**, 367 (1967).
(63) J. Lamatte-Brasseur, O. Dideberg, and L. DuPont, *Cryst. Struct. Commun.*, **1**, 313 (1972).
(64) A. C. Bonamartini, M. Nardelli, C. Palmieri, and C. Pellizzi, *Acta Crystallogr. Sec. B*, **27**, 1775 (1971).
(65) W. E. Hatfield and T. S. Piper, *Inorg. Chem.*, **3**, 841 (1964).
(66) J. P. Steadman and R. D. Willett, *Inorg. Chim. Acta*, **4**, 367 (1970).
(67) R. L. Harlow, W. J. Wells, G. W. Watt, and S. H. Simonsen, *Inorg. Chem.*, **13**, 2106 (1974).
(68) C. Postmus, J. R. Ferraro, A. Quattrochi, K. Shobatake, and K. Nakamoto, *Inorg. Chem.*, **81**, 1851 (1969).

Spectroscopic Investigations of Excited States of Transition-Metal Complexes

Glenn A. Crosby

Chemical Physics Program, Department of Chemistry, Washington State University, Pullman, Washington 99163

Received January 9, 1975

Reprinted from ACCOUNTS OF CHEMICAL RESEARCH, *8, 231 (1975)*

Photoluminescence spectroscopy has been a standard method of inquiry for the investigation of excited states of organic materials for over 30 years. Complemented by theoretical developments, analyses of fluorescence and phosphorescence phenomena have led to orbital and spin classifications of excited states, symmetry assignments, and an enormous amount of detailed information about rate constants, energy levels, vibrational structures, and relaxation phenomena.[1,2] Moreover, the information obtained from luminescence measurements on organic materials has laid a firm foundation for the magnificent developments in organic photochemistry[3,4] and the use of organic materials as probes for studying energy transfer in gases, solutions, and solids. Luminescence is a primary tool for unraveling the details of relaxation processes occurring in large molecular systems.

The use of emission spectroscopy for probing the excited states of inorganic complexes containing organic ligands is of relatively recent origin. First extensively applied to rare earth chelates[5] and chromium compounds,[6] the power of emission techniques has only recently been demonstrated in the quest to define the full range of types and properties of the excited states of transition-metal complexes. Reasons for this late development are not difficult to find. Unlike organic molecules, notably the aromatics, that have electronic structures virtually ensuring the incidence of easily observable photoluminescence, many of the common transition-metal complexes possess electronic structures that promote rapid radiationless loss of absorbed energy and thus produce no emission. This statement holds for the bulk of complexes of the first transition series, and these are precisely the elements where the chemistry is best defined. Luminescent complexes of the 3d elements are known, but the number is relatively small. Moreover, the observed luminescence from a series often owes its origin to a state or states arising from a particular electronic configuration of the metal atom that is common to the entire group of active materials.[7]

A second important factor retarding the study of transition-metal complexes by emission techniques is the requirement of specialized optical and cryogenic equipment. Although some transition-metal complexes emit in the visible region, the bulk of them luminesce in the red or near-infrared region of the spectrum where sensitivity is low and, for many materials, the quantum yield also is low. Frequently, complexes achieve good luminescence efficiencies only at sub-77 K temperatures, ranges not yet routinely available in many laboratories.

Structural Prerequisites for Photoluminescence

If one accepts a nominal minimum energy difference of 10 kK between ground and emitting states as a practical condition for the observation of luminescence, then probable candidates for the emission experiment can be selected by electronic considerations. One is also guided by the rule: "In the absence of photochemistry from upper excited states, emission from a transition-metal complex with an unfilled d shell will occur from the lowest electronic excited state in the molecule or from those states that can achieve a significant Boltzmann population relative to the lowest excited state". This generalization, originally inferred from sparse evidence on d^3 and d^6 complexes, appears to be withstanding the test of new experimental data as long as conventional light sources are employed and one's purview is restricted to chemical complexes and does not include ions imbedded in hard lattices.[8]

The Utility of the nd^6 Configuration

The requirement of a large energy gap between ground and excited state is rarely satisfied by complexes of the first transition row. If one moves to the second and third transition series, however, and confines attention to complexes with metal ions having nd^6 configurations, the basic requirement is generally met. Furthermore, by a judicious choice of ligands and central metal ions, one has the capability of designing series of complexes in which the nature of the lowest excited state(s) can be stipulated. According to the rule enunciated above the emission will originate from the lowest excited level(s), and thus the properties of the excited state(s) can then be revealed through a detailed investigation of the observed luminescence.

Glenn A. Crosby was born in Pennsylvania and studied at Waynesburg College for his B.S. degree. Following receipt of the Ph.D. from University of Washington, Seattle, in 1954, he spent 2 postdoctoral years with Michael Kasha at Florida State University. From 1957 to 1967, he was on the faculty at the University of New Mexico, and then moved to Washington State University, where he is Professor of Chemistry and Chemical Physics. Professor Crosby's research interests concern classification of excited states of complexes, interactions of excited states with magnetic and electric fields, and the design and investigation of materials displaying extended interactions in the solid state.

(1) M. Kasha, *Radiat. Res., Suppl.*, **2**, 243 (1960).
(2) M. A. El Sayed, *Acc. Chem. Res.*, **1**, 8 (1968).
(3) N. J. Turro, "Molecular Photochemistry", W. A. Benjamin, Inc., New York, N.Y., 1965.
(4) R. O. Kan, "Organic Photochemistry", McGraw-Hill, New York, N.Y., 1966.
(5) G. A. Crosby, *Mol. Cryst.*, **1**, 37 (1966).
(6) L. S. Forster, *Transition Met. Chem.*, **5**, 1 (1969).
(7) P. D. Fleischauer and P. Fleischauer, *Chem. Rev.*, **70**, 199 (1970).
(8) J. N. Demas and G. A. Crosby, *J. Am. Chem. Soc.*, **92**, 7262 (1970).

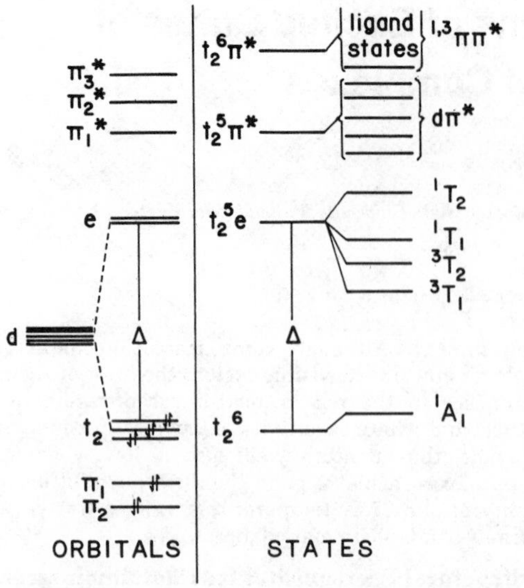

Figure 1. Schematic orbital and state diagram for a d^6 electronic system perturbed by an octahedral environment. A strong-field description is assumed. Shown also are bonding and antibonding π orbitals located on the ligands. Splitting due to spin–orbit coupling has been ignored.

A schematic representation of the electronic situation for nd^6 complexes is depicted in Figure 1 where, for simplicity, octahedral microsymmetry has been assumed. One visualizes a central metal ion surrounded by six negative groups supplied by the ligands. It is important to realize that most of the systems investigated are not octahedral and the symmetries range from C_s to O_h. Many different elements in various oxidation states can occupy the central coordination site, and the spin–orbit coupling constants can range from ~0.5 kK (Co^{3+}) to ~4 kK (Ir^{3+}).[9] Chemically, the luminescent species range from anions through molecular species to highly positively charged cations. Common to all the species considered here, however, are a metal ion with a d^6 spin-paired electronic configuration and a ligand system containing at least one moiety with a π structure. The two electronic systems are depicted in Figure 1. In the ground state all electrons are generally paired as indicated.

Formal descriptions of the low-lying excited states of nd^6 complexes can be derived from successive one-electron promotions. There are basically four orbital types expected. For clarity we use the scheme based on octahedral microsymmetry.

dd States. These can be visualized as arising from promotions of an electron from t_2 to e, an excitation confined (essentially) to the metal ion. The promoted electron either has its spin parallel or antiparallel to the spin of its partner remaining in the t_2 core. Configurational energy is determined by Δ, the octahedral ligand-field parameter. Δ is a function of the ligand strengths, the position of the metal ion in the periodic table, and the oxidation state of the ion.[10] *Thus, to a degree, Δ is at our command,* and the average energy of the cluster of ligand-field states can be chemically controlled. The splitting into the four T states is a measurement of interelectronic repulsion and is usually given in terms of Racah parameters.

dπ* States. These arise from the excitation of a metal (t_2) electron to a π^* antibonding orbital located on the ligand system. One can view this process as a transfer of electronic charge from the central ion to the ligands, i.e., as an incipient oxidation of the metal ion. From this point of view one expects such states to be related to the redox potential of the complex—the more easily oxidized, the lower should lie the $d\pi^*$ transitions. Thus, a change of metal ion, while preserving the d^6 configuration, will drastically affect the position of the $d\pi^*$ states.

$\pi\pi^*$ States. States of this description are associated with the promotion of an electron in a π bonding orbital of the ligand system to a π^* antibonding orbital also on the ligand system. These, as depicted in Figure 1, usually lie at relatively high energies and are substantially of ligand character. The metal ion plays a minor role in the definition of their energies but can affect other sensitive properties such as decay times and quantum yields.

πd States. Formally, such states are expected to arise from a promotion of electronic charge from the ligand π system to the metal (e) orbital. For the complexes investigated in the author's laboratory no firm evidence for them has been acquired,[11] and they have not been included in the level system depicted in the figure.

Thus, simple model considerations delineate the orbital types of excited states expected in metal complexes. Complemented by spin-coupling rules, they also define the possible spin labels for the electronic terms. For nd^6 complexes both singlet and triplet states are expected to arise from each excited configuration, and appropriate multiplicity symbols have been attached to the dd and $\pi\pi^*$ states in the figure. For the $d\pi^*$ states no spin label has been included, since there is strong evidence that S is not even approximately a "good" quantum number for most complexes.[12] For the heavier elements S loses meaning as a label for dd states as well. We will return to this point later. A complete labeling of an excited state would include the final symmetry label of the irreducible representation of the symmetry group to which the complex belongs.

The energy-level sequence in Figure 1 is schematic only. The relative ordering can be altered by switching metal ions, exchanging ligands, modifying the ligands, or varying the geometry, i.e., by employing chemistry. There are also distinct possibilities of perturbing level structures by solvent effects, pressure variations, magnetic and electric interactions, etc. Thus, one has available a wide range of chemical and physical means to modify the electronic properties in predictable ways. The final descriptions of the states of a particular complex rest, as always, with experiment. We reiterate that the important feature to notice is that the *lowest excited states,* i.e., those that are responsible for the luminescence, can be specified

(9) J. S. Griffith, "The Theory of Transition-Metal Ions", University Press, Cambridge, England, 1964.

(10) C. J. Ballhausen, "Introduction to Ligand Field Theory", McGraw-Hill, New York, N.Y., 1962.

(11) M. Gouterman, L. K. Hanson, G.-E. Khalil, R. Leenstra, and J. W. Buchler, *J. Chem. Phys.*, in press.

(12) G. A. Crosby, K. W. Hipps, and W. H. Elfring, Jr., *J. Am. Chem. Soc.*, **96**, 629 (1974).

chemically. Then, the powerful tools of luminescence spectroscopy become available for defining their properties.

Goals of Current Research

In the author's laboratory the research effort has encompassed chemical synthesis, physical measurements, and the construction of theoretical models. The guiding philosophy has been to focus on the general problem of defining the nature of the excited states of transition-metal complexes as revealed by spectroscopic measurements, primarily photoluminescence. The immediate goals have been (a) to develop experimental methods commensurate with the task of determining energy-level splittings, symmetries, radiative and radiationless rate constants, geometrical parameters, spin labels, magnetic characteristics, etc. of the excited states of transition-metal complexes, (b) to formulate criteria that will allow the lowest excited states of transition-metal complexes to be characterized empirically by orbital type and spin label from a study of optical properties, (c) to generate quantitative models of the excited states that will not only correlate data deduced from diverse experimental methods but will have genuine predictive capability, and (d) to arrive at a degree of sophistication of the descriptions of excited states such that their roles in photochemical, electrochemical, and chemical transformations can be understood.

The ultimate goal of the research has been to arrive at a degree of understanding such that metal complexes can be engineered to possess desirable optical, electrooptical, photochemical, electrochemical, and chemical properties.

Principal Experimental Methods

Absorption Spectroscopy. This is especially valuable if the spectra are measured at 77 K where better resolution of band structure is frequently noted. The measurement of absorption spectra as a function of solvent polarity is also an important diagnostic tool for recognizing the natures of excited states. Standard equipment is used, and cell designs for low-temperature measurements are available.[13]

Emission Spectroscopy. The basic apparatus requires (1) source of uv-visible light for excitation and either a filter train or a small monochromator for limiting the bandpass, (2) a detection system consisting of a scanning monochromator and a photomultiplier. For most materials a red-sensitive sensor and phase-locked electronics are highly desirable. The emission intensity vs. wavelength is usually displayed on a chart recorder and is frequently corrected for instrument response and reported linear in frequency. The sample is usually a solution of the complex in a solvent that forms a rigid glass at 77 K. For certain types of emission the solid material can be suspended in liquid nitrogen and excited directly. Emitted light is sometimes observed at room temperature, but the luminescence efficiency almost invariably improves as the temperature is lowered to 77 K. For some substances considerable improvement in resolution and large spectral changes occur at still lower temperatures. Sub-4.2 K data may be necessary to provide detailed descriptions of some systems.

Quantum Yield Measurements. In principle the quantum yield is a simple luminescence efficiency measurement. One counts the number of photons emitted by the sample and compares this with the number of photons absorbed from the source. Basically the apparatus employed for this determination is the same as that used for standard emission measurements. Considerably more calibration, data reduction, stabilization, and standardization are required.[14,15]

Decay Time Measurements. In this experiment a short burst of radiation is used to excite the sample, and the transient emitted light is monitored electronically. The exponential signal is frequently displayed on an oscilloscope, photographed, and plotted on a semilogarithmic scale. The slope yields the mean decay time for the ensemble. For transition-metal complexes the measured decay times are frequently in the microsecond range, and the nitrogen laser has provided a convenient source of pulsed (fwhh, ~20 nsec) uv (337 nm) radiation.

Temperature-Dependent Measurements. Since the spectra obtained from transition-metal complexes are usually quite diffuse, direct measurement of term splittings is often impossible. Valuable information on electronic excited states can be obtained from monitoring emission spectra, quantum yields, and decay times as a function of temperature and employing computer techniques to extract molecular parameters. Indeed, for closely spaced excited states (~10 cm^{-1}) it is often necessary to achieve quite low temperatures (~1.5 K) to obtain energy-level splittings and rate constants for individual levels. A variable temperature helium dewar with the sample in contact with the escaping gas must be employed if data accurate enough for computer analysis are to be acquired.[16]

Other Methods. In principle those experimental techniques commonly employed for the investigation of the excited states of organic materials could be used on inorganic complexes. These include excited-state EPR,[17] phosphorescence microwave double resonance,[18] excited-state electronic absorption spectroscopy,[19] polarization measurements,[20] and infrared optical double resonance.[21] Few complexes have been investigated by these methods, and the field is ripe for exploitation.

Often the use of several techniques is required to unravel the nature of a given set of energy levels producing luminescence in a transition-metal complex or a series of complexes. The point of view adopted in the rest of this short review will be to proceed from a coarse classification of excited states, as demanded by the simpler measurements, to a final brief description of the types of information available with the use of more complicated techniques. There will be a con-

(13) G. A. Crosby and B. Pankuch, *Chem. Instrum.*, **2**, 329 (1970).

(14) J. N. Demas and G. A. Crosby, *J. Phys. Chem.*, **75**, 991 (1971).

(15) G. A. Crosby, J. N. Demas, and J. B. Callis, *J. Res. Natl. Bur. Stand., Sect. A*, **76**, 561 (1972).

(16) R. W. Harrigan and G. A. Crosby, *Spectrochim. Acta, Part A*, **26**, 2225 (1970).

(17) S. W. Mao and N. Hirota, *Mol. Phys.*, **27**, 309 (1974).

(18) J. van Egmond and J. H. van der Waals, *Mol. Phys.*, **26**, 1147 (1973).

(19) G. Porter and M. A. West, *Tech. Chem., Part 2*, **6**, 367 (1974).

(20) I. Fujita and H. Kobayashi, *Inorg. Chem.*, **12**, 2758 (1973).

(21) D. C. Baker, W. H. Elfring, and G. A. Crosby, Abstracts, 29th Annual Northwest Regional Meeting of the American Chemical Society, Cheney, Wash., June 1974.

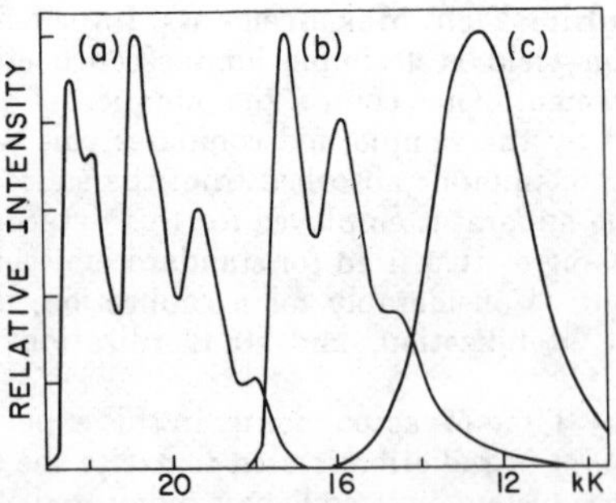

Figure 2. Luminescence from transition-metal complexes at 77 K exemplifying $^3\pi\pi^*$, $d\pi^*$, and ^3dd origin, respectively. (a) $[Rh(phen)_3](ClO_4)_3$ in water–methanol glass, (b) $[Ru(bpy)_3]Cl_2$ in ethanol–methanol glass, (c) solid $[RhCl_2(phen)_2]Cl$.

comitant rise in sophistication of the models employed to rationalize the observations.

Experimental Classification of Excited States

Since the orbital and spin characteristics of the lowest lying excited states in transition-metal complexes not only govern the emission spectroscopy but are most probably involved in controlling the photochemical and some chemical properties as well,[22,23] a key problem is the formulation of experimental criteria for assigning appropriate labels to observed excited states. The approach has been to synthesize complexes whose lowest excited configurations are fairly certain on general structural grounds and to use the spectroscopic properties to formulate rules for recognizing analogous configurations and states in more complex situations. These investigations have been carried out systematically on d^6 ions, and the criteria have been formulated for experiments carried out at the convenient temperature of 77 K.[24]

^{3}dd Excited States. If the average ligand-field parameter is sufficiently small and the oxidation potential of the metal ion is high, one has a disposition of lowest states pictured in Figure 1. The lowest excited configuration is dd, and one expects luminescence to arise from the lowest term, formally $^3T_1$. These conditions are satisfied by $[RhCl_2(phen)_2]^+$ (phen = 1,10-phenanthroline) whose spectrum is displayed in Figure 2. The spectroscopic behavior of this ion is typical of a broad class of complexes whose lowest (emitting) excited state has been classified as ^{3}dd.[25,26]

Emission is broad and structureless. It is relatively matrix independent but is rarely observed at room temperature. After flash excitation, the mean decay time ranges from 10 to 500 μsec and is temperature dependent. The emission is characterized by a low quantum yield that can be enhanced considerably by eliminating high-frequency vibrations from the coordination sphere.[27] The energy of the luminescence is a function of the average ligand-field strength about the central ion.[25]

dπ* Excited States. By switching to an ion of the 4d or 5d transition series and choosing one that is easily oxidized, one modifies the energy-level sequence to produce a $d\pi^*$ configuration lowest. The prototype systems are ruthenium(II) complexes containing 2,2′-bipyridine (bpy) or 1,10-phenanthroline (phen) as coordinated ligands.[28] The well-studied spectrum of $[Ru(bpy)_3]^{2+}$ is shown in Figure 2. Extensive studies of emission spectra from states arising from a $d\pi^*$ configuration have identified the following characteristic features at 77 K.

The emission is intense and highly structured. Often a single vibrational progression is the dominant feature of the band. Frequently the luminescence is observable at room temperature in a rigid matrix and sometimes in fluid solution.[29] The decay time is ~1–10 μsec at 77 K. Usually the emission lies in the tail of a strong absorption region and suffers strong reabsorption when the pure solid material is excited. The quantum yields often exceed 1,[30] and thus a short radiative life (<100 μsec) is indicated. If the complex possesses a dipole moment, the emission can be shifted substantially by varying the hydroxylic content of the rigid glass.[31] Because of the dominant role played by spin–orbit coupling, no spin label is meaningful.[12]

$^3\pi\pi^*$ Excited States. These states are easily observed from complexes containing metal ions with closed shells.[32] They are also the emitting levels in rare-earth complexes of trivalent lutetium, lanthanum, and gadolinium.[33,34] To engineer a complex possessing an open d shell that has $^3\pi\pi^*$ states lowest, one must choose a system with a large value of Δ and a metal ion that is difficult to oxidize. Rhodium(III) complexes of bipyridine and phenanthroline satisfy these requirements. Both the $d\pi^*$ and dd states are raised in energy sufficiently such that the lowest (emitting) states in the complexes become $^3\pi\pi^*$.[25,26]

Emission spectra originating from $^3\pi\pi^*$ states in complexes often resemble those emanating from the uncoordinated ligands both with regard to energy and to band structure. In the complex the band is generally red shifted (≤1 kK) from its position observed from the corresponding ligand.[25,35] The decay time at 77 K is usually in the millisecond range and is considerably shorter (~10^2) than the decay of the uncoordinated ligand phosphorescence. The energy of the emission is not highly solvent dependent. [Since coordination drastically affects the nonbonding electrons of the ligands, a ligand exhibiting $^3n\pi^*$ phosphorescence would be expected to produce a complex exhibiting $^3\pi\pi^*$ emission. The similarities noted

(22) V. Balzani, private communication.
(23) W. H. Elfring, Jr. and G. A. Crosby, Abstracts, 29th Annual Northwest Regional Meeting of the American Chemical Society, Cheney, Wash., June 1974.
(24) G. A. Crosby, R. J. Watts, and D. H. W. Carstens, *Science*, **170**, 1195 (1970).
(25) D. H. W. Carstens and G. A. Crosby, *J. Mol. Spectrosc.*, **34**, 113 (1970).
(26) J. E. Hillis and M. K. DeArmond, *J. Lumin.*, **4**, 273 (1971).
(27) T. R. Thomas, R. J. Watts, and G. A. Crosby, *J. Chem. Phys.*, **59**, 2123 (1973).
(28) D. M. Klassen and G. A. Crosby, *J. Chem. Phys.*, **48**, 1853 (1968).
(29) F. E. Lytle and D. M. Hercules, *J. Am. Chem. Soc.*, **91**, 253 (1969).
(30) J. N. Demas and G. A. Crosby, *J. Am. Chem. Soc.*, **93**, 2841 (1971).
(31) J. N. Demas, T. F. Turner, and G. A. Crosby, *Inorg. Chem.*, **8**, 674 (1969).
(32) D. M. Hercules, Ed., "Fluorescence and Phosphorescence Analysis: Principles and Applications", Wiley, New York, N.Y., 1966, Chapter 4.
(33) G. A. Crosby, R. E. Whan, and R. M. Alire, *J. Chem. Phys.*, **34**, 743 (1961).
(34) G. A. Crosby, R. J. Watts, and S. J. Westlake, *J. Chem. Phys.*, **55**, 4663 (1971).
(35) R. J. Watts, G. A. Crosby, and J. L. Sansregret, *Inorg. Chem.*, **11**, 1474 (1972).

above could well disappear.]

A Case History. Use of the listed criteria to assign the lowest excited states of a related set of complexes is exemplified by the case of the three related $5d^6$ iridium(III) complexes.[24] The luminescence from *cis*-$[IrCl_2(5,6\text{-Mephen})_2]^+$ has a measured decay of 190 ± 5 μsec in an alcoholic glass, possesses pronounced structure, and is slightly red shifted from the emission from the 5,6-dimethyl-1,10-phenanthroline molecule itself. It is assigned to a $^3\pi\pi^*$ origin. The characteristic structure of the emission from *cis*-$[IrCl_2(phen)_2]^+$, the short decay time (~7 μsec), and the high quantum yield (0.50) label the luminescence as $d\pi^*$ in origin. By the criteria stated above, the broad structureless emission emanating from $[IrCl_4(phen)]^-$ in the solid state is clearly ^{3}dd in origin.

The assignments of the lowest excited states of the three iridium(III) complexes show the efficacy of using complementary experimental methods for arriving at a unique orbital labeling of excited states for well-defined configurations. When significant configurational mixing occurs, the state assignment problem becomes considerably more complicated, and more interesting.

Manifestations of Configuration Interaction

In the previous section it was implicitly assumed that an excited state could be assigned a unique orbital label. In the dd and $\pi\pi^*$ cases a spin labeling was also considered to be meaningful for the lowest excited configuration. This situation obtains when the emitting states are relatively isolated from electronic states arising from other configurations. Complexes have been synthesized, however, whose spectroscopic properties indicate that significant $d\pi^*$–$\pi\pi^*$ interaction is occurring.[35,36] Its incidence is signalled by a near-coincidence of uncoordinated ligand $^3\pi\pi^*$ emission and complex emission and a strong dependence of measured decay time (77 K) on the polarity of the rigid glass medium. When a $^3\pi\pi^*$ level lies lowest but is strongly perturbed by $d\pi^*$ states lying nearby, configuration interaction is most easily recognized and can be treated in a semiquantitative way. For $[IrCl_2(5,6\text{-Mephen})_2]^+$ a factor of 7 in measured decay time has been introduced simply by switching solvents. This has led to a semiquantitative model of $d\pi^*$–$\pi\pi^*$ interaction and established a criterion for deciding upon the appropriateness of attaching a single configuration label to the emitting levels. Cases where this is manifestly impossible have been described.

Clearly one expects dd–$d\pi^*$ and dd–$\pi\pi^*$ configuration interaction to occur in complexes and indeed to be of major importance in those systems where a near-coincidence of two zero-order configurational energies occurs. The observable spectroscopic manifestations should be a "mixing" of band types, subtle changes in emission characteristics with changes in solvent, and values for excited-state parameters that lie intermediate between those characteristic of "pure" states. Evidence for dd–$d\pi^*$ configuration interaction exists,[37] but no systematic spectroscopic study has yet been reported. A clear case of dd–$\pi\pi^*$ interaction has not yet been seen by the author. The recent photochemical results of Balzani et al. strongly point toward a near-coincidence of several types of states.[22] It is important to emphasize that configuration interaction may be revealed only in subtle changes of experimentally measured quantities, and detailed experimentation and data reduction may be required to achieve a quantitative measure of it.

(36) R. J. Watts and G. A. Crosby, *Chem. Phys. Lett.*, **13**, 619 (1972).
(37) W. H. Elfring, Jr., unpublished work, this laboratory.

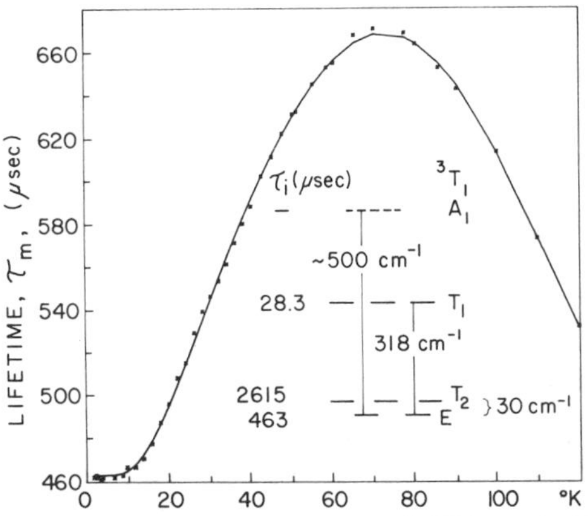

Figure 3. Measured decay times of solid $K_3Co(CN)_6$ as a function of temperature. ×, data points; —, computer fit. Decay constants and energy gaps obtained by a least-squares fit of the equation $\tau_m = [2 + 3e^{-\epsilon_{T2}/kT} + 3e^{-\epsilon_{T1}/kT}]/[(2/463) + (3/\tau_{T_2})e^{-\epsilon_{T2}/kT} + (3/\tau_{T_1})e^{-\epsilon_{T1}/kT}]$. (See ref 38.)

In our discussion of state labeling and classifying we have formulated criteria explicitly based on data obtained at 77 K, since most information is available for that temperature. As shown later, the "states" are really clusters of levels with populations controlled by Boltzmann statistics. Nonetheless, the criteria still remain valid for the reason that splittings for many complexes turn out to be smaller than kT at 77 K, and thus an average property is displayed. This statement is justified a fortiori in the later sections where the detailed splittings and properties of individual electronic levels are described.

Quantification of Excited State Properties

Level Splittings, Rate Constants, and Symmetry Assignments. Once the configurational identity (and sometimes the spin label) for an emitting term has been verified, considerable additional detailed information about the emitting states can be acquired by a study of the temperature dependences of the band structure, the decay time, and the quantum yield of the luminescence. The matching of these experimental results through computer analyses to theoretical models has begun to generate quantitative descriptions of the excited states.

Ligand-Field States. In Figure 3 is a plot of the measured decay time of the luminescence of $K_3Co(CN)_6$ as a function of temperature.[38] Not only is the mean life of the luminescence highly temperature dependent, but the observed decays are exponential at all temperatures reached. This behavior strongly points toward a manifold of excited states producing the emission whose occupation numbers are controlled by equilibrium statistics at all temperatures. Computer analysis of the emitted light under

(38) K. W. Hipps and G. A. Crosby, *Inorg. Chem.*, **13**, 1543 (1974).

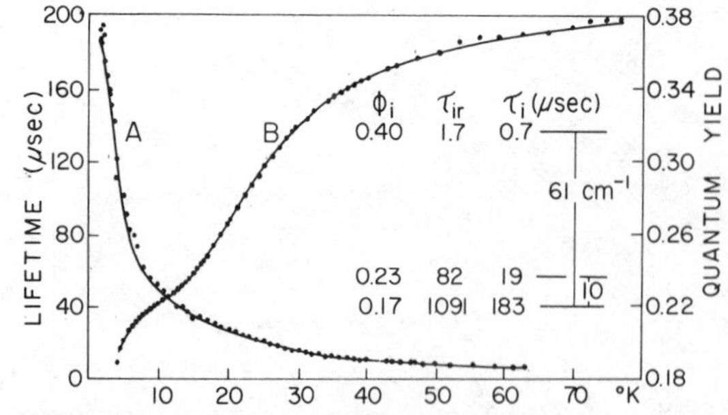

Figure 4. Temperature dependence of the calculated (—) and observed (···) lifetime and quantum yield of $[Ru(bpy)_3]^{2+}$ in poly(methyl methacrylate). Lifetimes, curve A, are read from the left ordinate; quantum yields, curve B, from the right. Energies, quantum yields, and decay constants obtained by computer fit of the experimental data. (See ref 40.)

these assumptions has produced energies and decay parameters for the individual levels. Comparison with simple crystal-field calculations supplied the final group theoretical labels for the states that are included in the figure. Thus the $^3T_1$ term pictured in Figure 1 is interpreted to be a cluster of levels split by spin–orbit coupling, each possessing unique decay parameters. A value for the spin–orbit coupling constant for diamagnetic cobalt(III) has also been obtained. When temperature-dependent quantum-yield data are available for a system (see below), highly important additional information on both radiative and radiationless processes can be extracted.

Computer analysis of the temperature dependence of the emission decay time of ruthenocene, which also displays dd luminescence, has also been carried out in the 4.2–77 K temperature range.[39] The derived level pattern has been rationalized on the basis of a $^3E_1$ (dd) term (D_5), split by ~0.5 kK by spin–orbit coupling into $A_2 + A_1 + E_1 + E_2$ components in order of increasing energy. Information on the degree of covalency in the molecule in the excited configuration has also been furnished by the analysis.

Charge-Transfer States. The temperature dependences of both the quantum yield and the decay time of the intense luminescence of $[Ru(bpy)_3]^{2+}$ are displayed in Figure 4. One sees that the decay time increases monotonically with decreasing temperature below 77 K, a behavior exactly opposite to that for $K_3Co(CN)_6$. The emission from the former ion, now firmly established as arising from a $d\pi^*$ configuration, has been subjected to detailed investigations.[40–46] Simultaneous computer analyses of both the quantum yield and decay curves displayed in Figure 4 have produced level splittings and both radiative and radiationless decay parameters for the individual levels. They are included in the figure.

Geometrical Distortions of Excited States. Although vibronic analyses of band structures have been carried out for numerous diatomic molecules and organic substances, subjecting emission bands of transition-metal complexes to detailed analyses has just begun. The lack of effort is a result of the paucity of emission spectra and the general diffuseness of most observed emission bands. Because of the huge geometrical changes that may occur upon excitation, especially t → e excitation, and the probable role of excited states in controlling photochemistry (and chemistry), it is of considerable interest to measure excited-state geometries quantitatively.

The emission spectrum of $K_3Co(CN)_6$, which is devoid of detail at 77 K, displays considerable vibrational structure at 4.2 K, where the emission originates from the *lowest* (E) spin–orbit component of the $^3T_1$ term.[38] An analysis of this band has been carried out by the use of a simple spectral fitting technique. Relative intensities of the vibronic transitions were assigned by the use of Franck–Condon factors obtained from harmonic oscillator models of the ground (A_1) and excited (E) states. The analysis has yielded the approximate geometry of the excited complex and a quantitative measure of its expanded geometry. It has also yielded a measure of the weakening of the (M–CN) harmonic force constant in the excited state and has located the probable frequency of the nonvertical $A_1 \rightarrow E(^3T_1)$ transition.

A similar analysis of the band contour of the luminescence observed from ruthenocene has also been completed.[39] Detailed information on the lowest excited level has been obtained. Other molecules are under investigation. Because of the huge Stoke's shifts exhibited by these and many other complexes, such studies should play a key role in establishing a connection between spectroscopic properties and photochemical reactivities.

Excited-State Charge Distributions. Systematic investigations of the emitted light from series of related complexes of ruthenium(II) with bipyridine and phenanthroline have not only permitted group theoretical labels for the emitting levels to be inferred but also have led to the development of a coupling model for $d\pi^*$ excited states. The model appears to be applicable to a broad class of complexes and is capable of correlating information derived from diverse experimental methods.[40,41]

The ion (parent) coupling model views the $d\pi^*$ excited states of a d^6 complex as the result of coupling the low-lying electronic states of the corresponding d^5 complex (obtained from a one-electron oxidation) with an optical electron residing in an antibonding orbital on the ligand π system. The coupling scheme has been quantified mathematically.[44] The observed splittings of the lowest group of $d\pi^*$ emitting states are determined by exchange integrals and a mixing parameter that is dependent on the spin–orbit coupling parameter of the central metal ion in the complex. Comparison of the experimentally derived splittings with the theoretical parameters has allowed a quantitative measurement to be made of the relative importance of spin–orbit vs. electrostatic interactions in defining the final electronic states. Moreover, since

(39) G. A. Crosby, G. D. Hager, K. W. Hipps, and M. L. Stone, *Chem. Phys. Lett.*, **28**, 497 (1974).
(40) R. W. Harrigan, G. D. Hager, and G. A. Crosby, *Chem. Phys. Lett.*, **21**, 487 (1973).
(41) R. W. Harrigan and G. A. Crosby, *J. Chem. Phys.*, **59**, 3468 (1973).
(42) G. D. Hager and G. A. Crosby, *J. Am. Chem Soc.*, **97**, 7031 (1975).
(43) G. D. Hager, R. J. Watts, and G. A. Crosby, *J. Am. Chem. Soc.*, **97**, 7037 (1975).
(44) K. W. Hipps and G. A. Crosby, *J. Am. Chem. Soc.*, **97**, 7042 (1975).
(45) D. C. Baker and G. A. Crosby, *Chem. Phys.*, **4**, 428 (1974).
(46) D. C. Baker, K. W. Hipps, and G. A. Crosby, manuscript in preparation.

the spin–orbit contributions are, in principle, obtainable from EPR measurements on the oxidized complex, then the formalism provides a mechanism for relating the excited-state properties of a given complex with the ground-state properties of the species generated by a one-electron oxidation. If one then exploits known trends in spin–orbit coupling and crystal-field parameters with oxidation state and position in the periodic table, the way is open to draw generalizations about the optical and magnetic properties of excited states applicable to large numbers of transition-metal complexes.

Excited-State Interactions

Vibronic Coupling. The lowest state of the $d\pi^*$ emitting manifolds of trigonal ruthenium(II) complexes has been assigned A_1 symmetry in the group D_3.[41] Transitions to the ground state from an A_1 state are electronically forbidden, and the coupling mechanism leading to the finite radiative life can be inferred from the behavior of the luminescence at very low temperatures. Figure 5 shows the luminescence from $[Ru(bpy)_3]^{2+}$ doped in $[Zn(bpy)_3]SO_4$ lattice at 4.2 K and at 1.65 K. The intense, short-lived, Franck–Condon allowed band observed at 4.2 K (assumed to come from E) is replaced by a less intense, long-lived band at 1.65 K whose first peak is displaced 415 cm^{-1} toward lower frequencies from the 4.2 K spectrum. We have interpreted this behavior in terms of a vibronic coupling mechanism lending intensity to the formally forbidden A_1 state, which is essentially the only state populated at the lowest temperature.[45] The proposed explanation is indicated in insert a in Figure 5. Although the experiments have not been carried out, it is obvious that linear polarization measurements at very low temperatures should yield valuable information on vibronic coupling in molecules containing heavy metal ions.

Magnetic Field Interactions. As indicated in Figure 5 an external field applied along the trigonal axis of a complex should induce a first-order splitting of the E level and cause a second-order mixing of the A_1 and A_2 levels. These effects should produce a qualitative dependence of both the observed spectrum *and* the measured decay times on magnetic field strength. The spectral changes are shown in Figure 5 where the spectrum observed at 1.65 K can be essentially switched back to that observed at 4.2 K by the application of a 75-kG field *without* changing the temperature.[45]

The decay time of the photoluminescence is also magnetic field dependent. A monotonic decrease in the measured life of the emission at a fixed temperature occurs as the external field is increased. These results have been treated quantitatively and provide detailed information on the eigenkets of the excited states.[46]

Magnetic effects of these magnitudes have threefold importance: (a) they show that the optical properties can be modulated with external fields and thus provide impetus for additional investigations of complexes for possible uses as electrooptical materials, (b) they provide a means of supplying quantitative experimental tests of the validity of proposed theoretical models for excited states in complexes containing heavy metal ions, and (c) they provide a means for obtaining *g* values for excited-state parameters that are not easily measured by standard resonance methods.

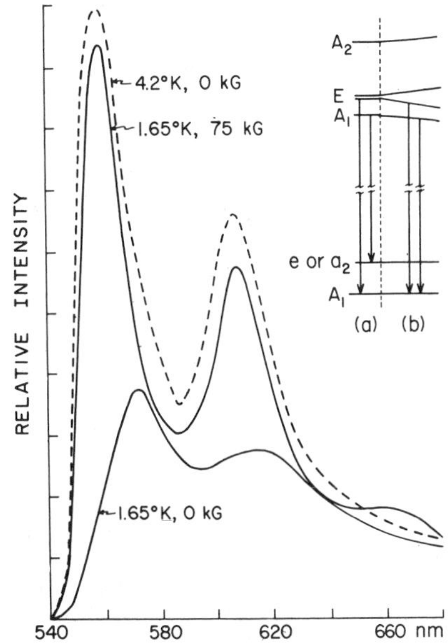

Figure 5. Emission spectra of $[Ru(bpy)_3]^{2+}$ in $[Zn(bpy)_3]SO_4 \cdot 7H_2O$ crystal lattice in zero magnetic field at 1.65 and 4.2 K and in 75-kG field at 1.65 K. (a) Energy-level diagram of $[Ru(bpy)_3]^{2+}$; (b) magnetic field dependence of the emitting levels with magnetic field parallel to principal axis. ↓, allowed transitions. (See ref 45).

Research Trends

From a perusal of current research publications one can recognize several distinct lines of investigation and also discern areas where a modest effort could yield substantial progress.

Investigations of Mechanisms and Models. The validity of current models for describing the excited states of transition-metal complexes should be subjected to much further experimental verification. In particular, the role of spin–orbit coupling in controlling excited-state properties has not received adequate attention. The spin-based language developed for organics[47] may not even be appropriate at all for certain types of states possessing highly interesting and unusual spectroscopic, photochemical, and chemical properties. Mechanistic studies of relaxation phenomena in complexes are receiving attention.[26,48] Attempts to correlate radiative and radiationless properties with theoretical models have met limited success, but the real problem is lack of systematic data derived from numerous systems. The interpretation of relaxation rates in complexes in terms of current theories of radiationless processes is complicated by the still uncertain role of spin–orbit coupling and configuration interaction in dictating the gross features of the optically probed states. Subtle changes, such as deuterium substitution,[27] indicate that much can be learned from the use of luminescence as a tool, but few detailed investigations have been carried out.

(47) M. Kasha, *Discuss. Faraday Soc.*, **No. 9**, 14 (1950).
(48) M. K. DeArmond, *Acc. Chem. Res.*, **7**, 309 (1974).

Extension to Other Ions and Configurations. The bulk of the detailed information on excited states of transition-metal complexes has been derived from d^6 and d^3 configurations. The data from complexes with ions of other electronic configurations are relatively sparse. Nonetheless, many d^5, d^8, and d^{10} complex ions, for example, are highly luminescent, and use of a variety of experimental techniques holds high promise for elucidating the natures of their excited states. Even within the class of complexes with d^3 and d^6 configurations, much could be learned from studies of the luminescence from mixed-ligand materials, a topic that has received little attention but holds great promise for ascertaining the nature and extent of ligand–ligand interactions in complexes.[49]

Application of New Techniques. Although the methods outlined above for probing excited states have produced much valuable information, recent developments in instrumentation point to a host of feasible new measurements. The application of lasers, especially the tunable dye varieties, offers great promise for investigating such important phenomena as wavelength-dependent quantum yields, slow thermal relaxation steps, the incidence of magnetic field dependent relaxation rates, and many others. For nonluminescent materials lasers provide the opportunity to measure excited-state properties by viewing transient absorption spectra and monitoring depletion of the ground state.[50] For luminescent complexes such techniques as circularly polarized luminescence, phosphorescence microwave double resonance, infrared optical double resonance, infrared absorption spectroscopy of excited states, and emission polarization measurements have great potential for delineating excited-state properties.

Spectroscopy and Photochemistry. Perhaps the most exciting and explosive development involving excited states is the current effort to correlate photophysical properties with photochemical behavior.[51] The use of materials that act as energy donors or reducing agents when excited and *exhibit luminescence in solution* has opened up a host of possibilities for investigating photochemical mechanisms.[52] Coupled with the use of high power lasers and single-photon counting equipment such substances will certainly play key roles in unravelling inorganic photochemical pathways. Yet to be exploited, however, is the use of photochemical results to clarify the nature of spectroscopically nonobserved excited states.

(49) W. Halper and M. K. DeArmond, *J. Lumin.*, **5**, 225 (1972).
(50) D. Magde and M. W. Windsor, *Chem. Phys. Lett.*, **27**, 31 (1974).
(51) G. Malouf and P. C. Ford, *J. Am. Chem. Soc.*, **96**, 601 (1974).
(52) J. N. Demas and A. W. Adamson, *J. Am. Chem. Soc.*, **93**, 1800 (1971).

Spectroscopy and Electrochemistry. The observation of electrochemical luminescence (ecl) from inorganic complexes and its correlation with photoluminescence have joined the fields of inorganic emission spectroscopy and electrochemistry.[53] This bridge will undoubtedly facilitate the flow of information between the two fields, to the benefit of both. By a systematic correlation of the polarographic behavior with the spectroscopic properties, one has the real possibility of relating electronic structures to redox potentials, a necessary step toward tying spectroscopy to chemical thermodynamics.

Spectroscopy and Chemistry. Still to be found is a firm connection between the nature of excited states, as revealed by spectroscopic studies, and the chemical behaviors of inorganic complexes. Although many thermal reactions are not expected to go through excited-state intermediates, some may, and the spectroscopic probe should be a valuable tool for studying mechanisms. In the author's laboratory we have attempted to establish such connections for very limited classes of reactions, but the waters are still muddy. The quest continues.

Material Design

A long sought goal of this investigator has been to design complexes, especially solids, that will exhibit unusual electronic properties. The knowledge of excited states of complexes that has already accumulated can now be used to engineer potentially valuable materials. A combined effort, integrating synthesis and physical measurements guided by semiquantitative theoretical models, appears to be the kind of program required. Finding interesting substances to characterize spectroscopically is no problem whatsoever; identifying those that will elucidate fundamental principles or lead to valuable new materials is a far more difficult task, but the potential rewards certainly make the effort worthwhile.

The investigations from the author's laboratory described herein were entirely funded by the U.S. Air Force Office of Scientific Research (present grant, AFOSR-72-2207). This support is gratefully acknowledged. I wish also to express my indebtedness to my current and former students and postdoctoral colleagues for their extensive collaboration and their dedication to high standards.

(53) N. E. Tokel-Takvoryan, R. E. Hemingway, and A. J. Bard, *J. Am. Chem. Soc.*, **95**, 6582 (1973).

Organometallic Electrochemistry

RAYMOND E. DESSY* AND LEO A. BARES

Virginia Polytechnic Institute and State University, Blacksburg, Virginia 24061

Received November 1, 1971

Reprinted from ACCOUNTS OF CHEMICAL RESEARCH, *5, 415 (1972)*

Although the areas of inorganic and organic electrochemistry have been explored rather intensively, only during the last 5 years has organometallic electrochemistry been probed in breadth and depth. Electrochemical techniques have been employed in organometallic syntheses,[1,2] as an ancillary tool in electronic structural studies,[2–11] and in a miscellany of kinetic/mechanistic studies.[12–19] This Account, which is not intended as a review of the field, will be directed primarily to work previously performed or in progress in this laboratory. It is aimed at a broad survey of organometallic electrochemistry. Special attention will be given to (a) the available pathways in organometallic electrochemical reductions, and (b) the physical consequences of the addition (or abstraction) of an electron from an organometallic molecule.

Pathways for Reduction

A systematic study[20] indicates that the types of electrochemical behavior shown in Scheme I are to be found in organometallic species; R is a σ- or π-bonded organic residue, M is a metal, and Q is another ligand such as R, MR, or halogen.

The exploration has involved electrochemical examination of the parent under three-electrode conditions in dimethoxyethane with tetrabutylammonium perchlorate as supporting electrolyte and with all potentials referenced to 10^{-3} M $Ag^+|Ag$. Electrochemical reversibility has been studied by triangular voltammetry at a hanging Hg drop. The n values (number of electrons/step) have been determined by exhaustive controlled-potential electrolyses, and chemical reversibility has been evaluated by reoxidation, followed by ir and uv spectroscopic examination of the solutions. (A very readable discussion of these techniques, used as tools by the experimental inorganic chemist, may be found in ref 2.)

Organometallic derivatives of main-group elements tend to extrude carbanions or radicals, or to abstract hydrogen, as the following typical equations indicate.[21]

$$RHgCl \xrightarrow{e} RHg\cdot + Cl^- \xrightarrow{e} R:^- + Hg \quad \text{Anion extrusion}$$

$$CCl_3HgCl \xrightarrow{e} Cl^- + CCl_3Hg\cdot \longrightarrow CCl_3\cdot + Hg$$

$$\qquad\qquad\qquad\qquad \searrow CCl_3H \quad \text{Radical extrusion}$$

$$Ph_3SnSiPh_3 \xrightarrow{e} Ph_3Sn:^- + Ph_3Si \xrightarrow[\text{H abstraction}]{\text{solvent}} Ph_3SiH$$

$$Ph_2SiCl_2 \xrightarrow{2e} Ph_2SiH_2$$

Homodimetallic Species (R_mMMR_m) and Their Anions ($R_mM:^-$)

Derivatives of groups IV and V give rise to some interesting metalloid anion species which are synthetically useful; *e.g.*

$$R_3MX \longrightarrow R_3M:^- + X^-$$
$$(M = \text{Ge, Sn, Pb; } X = \text{halogen, acetate})$$
$$R_2MX \longrightarrow R_2M:^- + X^- \ (M = \text{As, Sb, P})^{20,22}$$

These species result from the reactions

$$2R_mMX \xrightarrow{2e} 2X^- + 2R_mM\cdot \longrightarrow R_mMMR_m \xrightarrow{2e} 2R_mM:^-$$

and parallel classical metal and metal/amalgam reductions. All of the systems show a one-electron reduction near -1 V, typical of $R_mMX \rightarrow R_mM\cdot + X^-$, with a second wave at more cathodic potentials that can be identified with R_mMMR_m.

Raymond E. Dessy has been at Virginia Polytechnic Institute and State University since 1966, as Professor of Chemistry. He came there from the University of Cincinnati and before that he was a postdoctoral fellow with M. S. Newman and instructor at The Ohio State University. He received his B.S. in pharmacy in 1953 and a Ph.D. in chemistry in 1956 from the University of Pittsburgh. He was an Alfred P. Sloan Fellow, 1962–1964. His research interests revolve around study of fast organometallic reaction mechanisms.

Leo A. Bares is presently completing Ph.D. requirements under R. E. Dessy at Virginia Polytechnic Institute and State University.

(1) B. L. Laube and C. D. Schmulbach, *Progr. Inorg. Chem.*, **14,** 65 (1971), and references therein.
(2) J. B. Headridge, "Electrochemical Techniques for Inorganic Chemists," Academic Press, New York, N. Y., 1969, and references therein.
(3) C. Elschenbroich and M. Cais, *J. Organometal. Chem.*, **18,** 135 (1959).
(4) L. F. Warren and M. F. Hawthorne, "The Chemistry of Bis-[π-(3)-1,2-dicarbollyl]metallates of Nickel and Palladium," *U. S. Clearinghouse Fed. Sci. Tech. Inform.*, AD-698136, AD 1969, and references therein.
(5) J. A. McCleverty, *Progr. Inorg. Chem.*, **10,** 49, (1968), and references therein.
(6) (a) S. P. Gubin, *Pure Applied Chem.*, **23,** 463 (1970), and references therein; (b) S. P. Gubin, S. A. Smirnova, and R. I. Denisovich, *J. Organometal. Chem.*, **30,** 257 (1971).
(7) S. P. Gubin, S. A. Smirnova, and R. I. Denisovich, *ibid.*, **20,** 229 (1969).
(8) A. N. Nesmeyanov, *et al.*, *ibid.*, **20,** 169 (1969).
(9) T. J. Meyer and J. A. Ferguson, *J. Chem. Soc. D*, 623 (1971).
(10) R. Pribil, Jr., J. Masek, and A. A. Vicek, *Inorg. Chim. Acta*, **5,** 57 (1971).
(11) J. Masek, *Inorg. Chim. Acta Rev.*, 99 (1969).
(12) T. J. Meyer and J. A. Ferguson, *Inorg. Chem.*, **10,** 1025 (1971).
(13) R. L. Middaugh and F. Farha, Jr., *J. Amer. Chem. Soc.*, **88,** 4147 (1966).
(14) R. J. Wiersema and R. L. Middaugh, *ibid.*, **89,** 5078 (1967); **91,** 2074 (1969).
(15) J. A. McCleverty, *et al.*, *J. Organometal. Chem.*, **80,** C75 (1971).
(16) T. Matsumoto, M. Sato, and A. Schmiru, *Bull. Chem. Soc. Jap.*, **44,** 1720 (1971).
(17) G. Paliani, S. M. Murgia, and G. Cardaci, *J. Organometal. Chem.*, **30,** 221 (1971).
(18) G. Piazza and G. Paliani, *Z. Phys. Chem.* (*Frankfurt am Main*), **71,** 91 (1970).
(19) R. F. Broman and R. W. Murray, *Anal. Chem.*, **37,** 1408 (1965).
(20) R. E. Dessy, R. B. King, and M. Waldrop, *J. Amer. Chem. Soc.*, **88,** 5112 (1966).
(21) R. E. Dessy, W. Kitching, and T. Chivers, *ibid.*, **88,** 453 (1966).
(22) R. E. Dessy, W. Kitching, and T. Chivers, *ibid.*, **88,** 467 (1966).

Scheme I

RMQ $\xrightarrow{\text{e}}$ RMQ$\left(\frac{\cdot}{\cdot}\right)$ $\xrightarrow{\text{e}}$ RMQ$\left(=\right)$ Stable radical anion or
singlet-state dianion

$\downarrow$ −e   RM· + Q⁻

RMQ·⁺

$\xrightarrow{\text{e}}$ RM:⁻ $\longrightarrow$ R:⁻ + M   Metalloid anion production or
carbanion extrusion

RM·

$\xrightarrow{\text{HA}}$ RMMR   Coupling

$\longrightarrow$ RMH   Hydrogen abstraction

R· + M   Radical extrusion

Hg

$\longrightarrow$ RHgR   Electrode incorporation

RM·

$\longrightarrow$ R₂M + M   Disproportionation

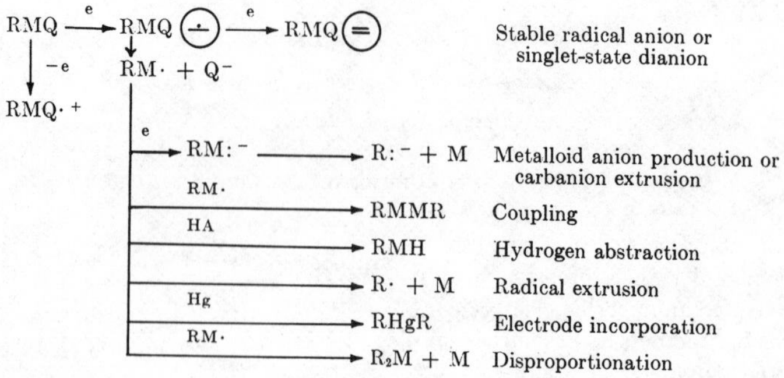

Table I

Catenated bond[a]	$E_{1/2}$	D_{m-m}, kcal/mole
si–si	No red	42
ge–ge	-3.5^{b}	37
sn–sn	-2.9^{b}	34
pb–pb	-2.0^{b}	
sb–sb	-2.5^{b}	30
bi–bi	-2.3^{b}	25

[a] Lower case element symbols refer to the atom and one valence.
[b] Ph derivatives.

These metalloid anions are related to the series of metallic anions derived from the transition series by electrochemical reduction of the corresponding metal–metal bonded dimers,[23-25] e.g.

$$[\text{CpM(CO)}_3]_2 \xrightarrow{2e} \text{CpM(CO)}_3:^- \quad (\text{M} = \text{Cr, Mo, W})$$

$$[\text{CpM(CO)}_2]_2 \xrightarrow{2e} \text{CpM(CO)}_2:^- \quad (\text{M} = \text{Fe, Ru})$$

$$[\text{CpNi(CO)}]_2 \xrightarrow{2e} \text{CpNi(CO)}:^-$$

$$[(\text{OC})_5\text{Mn}]_2 \xrightarrow{2e} (\text{OC})_5\text{Mn}:^-$$

Each M–M bond system has a characteristic half-wave potential and there appears to be a primitive correlation between half-wave potential and bond strength; see Table I for examples.

Metal and Metalloid Anion Nucleophilicities. Figure 1 shows the good correlation between the rate of nucleophilic displacement of halide ion from an organic halide and the oxidation potential of the anion at a platinum electrode. This linear free energy relationship is of the type of Edwards' four-parameter equation

$$\log k/k_0 = aE_N + bH_N$$

In this expression, a and b are empirically determined coefficients, E_N is the standard electrode potential for the nucleophile involved, H_N is defined by the equation $H_N = pK_a + 1.74$, and k_0 refers to the corresponding process in water. Normally, E_N dominates the right-hand side of the equation. Oxidation is a fair model for nucleophilicity, since the latter does involve formal loss of electrons by the attacking nucleophile. This corre-

(23) R. E. Dessy, P. M. Weissman, and R. L. Pohl, J. Amer. Chem. Soc., **88**, 5117 (1966).
(24) R. E. Dessy and P. M. Weissman, ibid., **88**, 5124, 5129 (1966).
(25) R. E. Dessy, R. L. Pohl, and R. B. King, ibid., **88**, 5121 (1966).

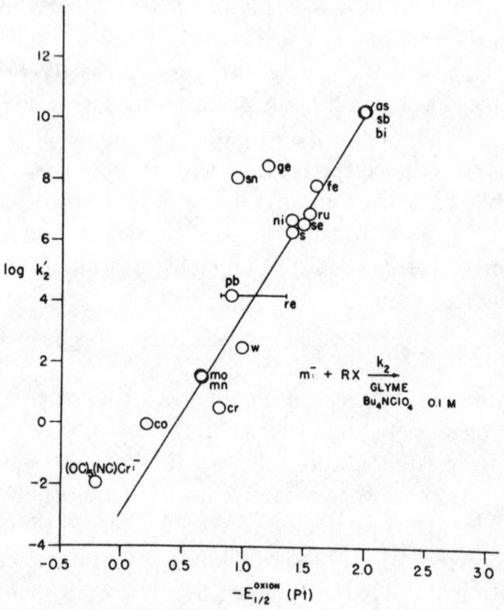

Figure 1. Rates of nucleophilic substitution processes for organometallic anions vs. oxidation potential.

lation should have considerable synthetic use, and its extension to new anions is easily implemented. It should be noted that correlations involving Hg microelectrodes fail because of the process

$$\text{RM:}^- \xrightarrow[\text{Hg}]{-e} (\text{RM})_2\text{Hg}$$

particularly when M is a transition metal.

Heterodimetallic Species ($R_mMM'R_m$). The electrochemical characterization of heterodimetallic compounds resulting from the reaction

$$\text{R–M–X} + \text{R'M:}^- \longrightarrow \text{R–M–M'–R'} + \text{X}^-$$

is very easy. Considering the half-wave potentials of the processes

$$\text{RM–MR} \xrightarrow{2e} 2\text{RM:}^-$$

$$\text{R'M'–M'R'} \xrightarrow{2e} 2\text{R'M':}^-$$

when the reduction potential for the electrochemical scission of RM–M'R' lies between the two homodimetallic parents or is more anodic, the reduction will proceed as

$$\text{RM–M'R'} \xrightarrow{1e} \text{RM:}^- + \text{R'M'·}$$

where R′M′· is associated with the parent reducing most cathodically. In a few cases (all containing Ph_3M moieties) $n = 2$, and the reduction proceeds to give both RM:$^-$ and R′M:$^-$.

It has been possible, utilizing the reduction potentials for homo- and heterodimetallic species as identifying criteria, to examine whether the reactions

$$\text{R–M:}^- + \text{R'–M'–X} \longrightarrow$$

and

$$\text{R'–M':}^- + \text{R–M–X} \longrightarrow$$

commute; *i.e.*, do two apparently equal pathways to the same product actually yield a unique compound? Figure 2 indicates that commutation is not always observed. (In the figure, lower case element symbols will be used to refer to the atom involved and one valence.) Further studies indicate that this is because of the following processes.

$$\text{RM:}^- + \text{R'–M'–Q} \longrightarrow \text{R'M'}\cdot + \text{R–M}\cdot + \text{Q}^- \qquad \text{(1e transfer)}$$

$$\text{RM:}^- + \text{R'–M'–Q} \longrightarrow \text{R–M–Q} + \text{R'–M':}^- \qquad \text{(Q exchange)}$$

$$\text{RM:}^- + \text{R–M–M'–R'} \longrightarrow \text{R–M–M–R} + \text{R'M':}^- \qquad \text{(metalloid anion displacement)}$$

A more detailed study[26] indicates that the nucleophilicity of the metallic or metalloidal anions is a key to predicting whether the displacement reaction shown above will occur.

These facts should be useful in predicting which reactions would be most successful in an attempt to create species having a polymetallic backbone, for example

$$[(\pi\text{-}C_5H_5)Mo(CO)_3]_2SnCl_2 \xrightarrow{(\pi\text{-}C_5H_5)Fe(CO)_2:^-} [(\pi\text{-}C_5H_5)Fe(CO)_2]_2Sn[Mo(CO)_3(\pi\text{-}C_5H_5)]_2 \xleftarrow{(\pi\text{-}C_5H_5)Mo(CO)_3:^-} [(\pi\text{-}C_5H_5)Fe(CO)_2]_2SnCl_2$$

As might be expected from available data, the bottom reaction has been reported to yield the desired compound. The top reaction fails because $CpFe(CO)_2$:$^-$ displaces $CpMo(CO)_3$:$^-$ anion from any intermediate fe–mo compound.

Structural Reorganization. Many cases studied, particularly those involving acetylene–iron carbonyl complexes, the bipyridyliron dinitrosyls, and corresponding isoelectronic cobalt carbonyl nitrosyls mentioned later, show electrochemical reversibility *via* cyclic voltammetry; however, exhaustive controlled-potential reduction yields a solution whose oxidation potential is anodic with respect to the previously determined reduction potential. Yet, controlled-potential oxidation regenerates starting material. This appears to be akin to some of the reorganization or structural changes reported by Busch in his study of the electrochemistry of macrocyclic complexes[27] and to the geometric changes occurring on change in charge state

	$CpFe(CO)_2$:$^-$	Ph_3Sn:$^-$	Ph_3Pb:$^-$	$(OC)_5Mn$:$^-$	$CpMo(CO)_3$:$^-$
$CpFe(CO)_2I$	1 fe–fe	7 sn–fe	9 pb–fe	11 N.R.	13 N.R.
Ph_3SnCl	6 fe–sn	2 sn–sn	15 ?	17 mn–mn sn–sn mn–sn	19 mo–sn
Ph_3PbOAc	8 fe–pb	14 sn–sn pb· → Ph_4Pb	3 pb· → Ph_4Pb	21 mn–pb	23 not mo–pb
$(OC)_5MnCl$	10 ?	16 ?	20 not mn–pb	4 N.R.	25 N.R.
$CpMo(CO)_3Cl$	12 N.R.	18 sn–sn mo:$^-$ mo–Cl	22 pb–pb mo:$^-$ mo–Cl	24 mn–mn mo:$^-$ mo–Cl	5 N.R.

Figure 2. Organometallic anion–organometallic halide reaction products.

reported for classical coordination compounds by Gray and Dahl.[5] The organic counterparts are the change in structure of cyclooctatetraene upon reduction (tub to planar geometry) and the bond angle–length changes reported for the tetracyanoquinonedimethide radical anion–neutral system. It is known in coordination chemistry that extensive structural changes lead to slow electron-exchange rates. This Franck–Condon-like limitation of electron-transfer rate is also seen in the cyclooctatetraene (COT) area. COT and COT·$^-$ exchange slowly (different geometry), but COT^{2-} and COT·$^-$ exchange rapidly (both planar).

Figure 3 indicates the extent of one-electron transfer between a large number of organometallic species. Complete transfer (CT) and no transfer (NT) are found in accord with Latimer redox concepts, where homoexchange ($A + A\cdot^- \rightleftharpoons A\cdot^- + A$) is involved. It is obvious that some exchange rates are slow compared to the diffusion limit, and that these slow rates occur where the above-mentioned electrochemical reversibility anomaly is seen. Equilibrium constants determined on opposite sides of the tieline are also not consistent for these species. Unfortunately, single-crystal X-ray work in the organometallic area has been hampered by crystallization problems and air sensitivity. At present the nature of the distortions involved is not known. However, the question of charge and spin delocalization has been carefully examined in those compounds formed from the group VI metals (Cr, Mo, and W) and from Fe and Ru. Both groups form a well-defined series of mononuclear and dinuclear species.

Physical Consequences of Organometallic Redox Processes

A meticulous study has been made of the following series of charged species involving mononuclear compounds by employing the indicated techniques to probe the nature of (1) the location of spin density in the radical species (esr);[28,29] (2) the locus of charge distribution (nmr);[30,31] (3) the nature of the electric field gradient at

(26) R. E. Dessy and R. L. Pohl, *J. Amer. Chem. Soc.*, **90**, 2005 (1968).
(27) D. H. Busch, *et al.*, *ibid.*, **92**, 400 (1970).
(28) R. E. Dessy and L. Wieczorek, *ibid.*, **91**, 4963 (1969).
(29) R. E. Dessy, J. C. Charkoudian, T. P. Abeles, and A. L. Rheingold, *ibid.*, **92**, 3947 (1970).

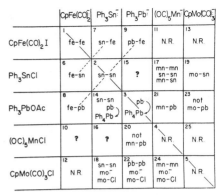

$A\cdot^- + B \underset{K}{\rightleftharpoons} B\cdot^- + A; \quad A\cdot^- + A \xrightarrow[k]{} A + A\cdot^-$

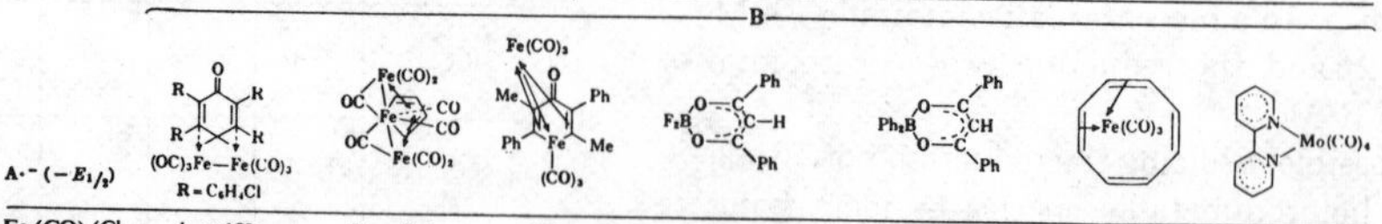

$A\cdot^-$ ($-E_{1/2}$)							
$Fe_2(CO)_7$(Cl-$C_6H_4C)_4$ (1.2)	$k < 10^7$	$K = 0.3$	NT	NT	NT	NT	NT
$Fe_2(CO)_6$-$(C_6H_5C)_4$ (1.3)	$K = 30$	$k \leq 10^7$	$K = 0.1$	NT	NT	NT	NT
$Fe_2(CO)_7$-$(PhC_2Me)_2$ (1.5)	CT	$K = 1$	$k < 10^7$	SOME T	NT	NT	NT
F_2B-$(PhCO)_2CH$ (1.6)	CT	CT	CT	$k \approx 10^7$	...	NT	NT
Ph_2B-$(PhCO)_2CH$ (1.7)	CT	CT	CT	...	$k \approx 10^7$	NT	NT
COT-Fe-$(CO)_3$ (2.0)	CT	CT	CT	CT	CT	...	NT
(bipy)Mo-$(CO)_4$ (2.2)	CT	CT	CT	CT	CT	...	$k = 1.8 \times 10^9$

Figure 3. Electron-transfer processes in organometallic systems.

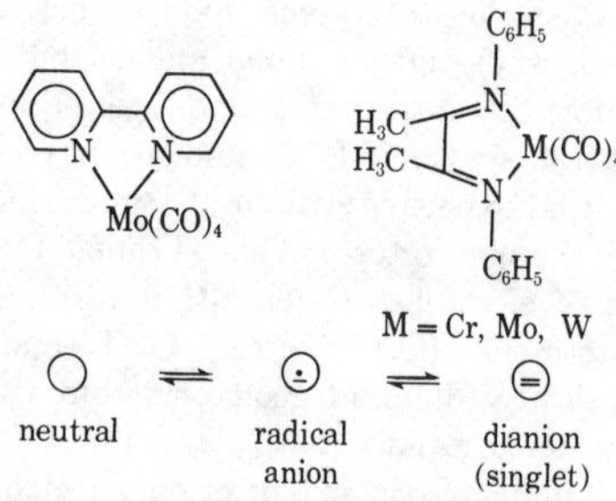

the core metal atom (Mössbauer);[29] (4) the method of transmission of charge density (ir).[28,29]

Octahedral Cases. The typical radical anion formed from a neutral precursor complex shows an esr hyperfine splitting pattern which greatly resembles that of the radical anion of the ligand itself. Hyperfine coupling constants derived from spectral simulation studies indicate that relatively small changes in hyperfine coupling constants are observed. This is exemplified in the case of 2,2′-bipyridyl and 2,2′-bipyridylmolybdenum tetracarbonyl radical anions shown. The

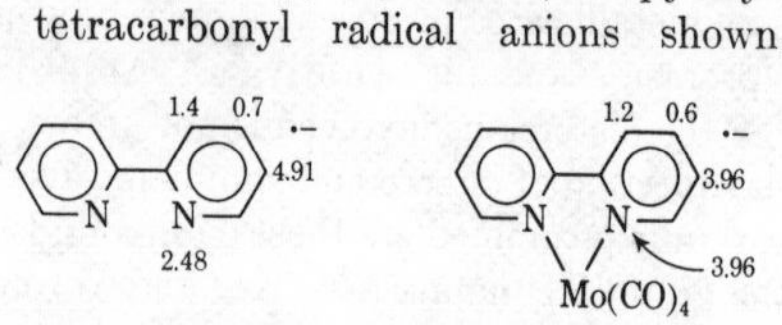

hyperfine coupling, (gauss)

conclusion is that spin density in the coordinated ligand radical ion is not appreciably different from the uncoordinated ligand radical ion ($a = \rho Q$, where a = hyperfine coupling in gauss, ρ is the spin density at an aromatic carbon center, and Q is an empirical parameter, ~25 G for aromatic protons). Nmr studies on the dianion species,[31] however, indicate that only a small amount of charge density resides in the organic ligand. Relative to the neutral precursors, incremental upfield shifts of approximately 0.2 ppm for aromatic protons are observed in the dianion species of the diacetylanil metal complexes ($\Delta\delta = qK$,[32] where $\Delta\delta$ is the change in chemical shift, q is the excess charge at an aromatic carbon center, and K is a constant, approximately 10 ppm for aromatic protons). Roughly 10% of the charge density for each of the two electrons added in dianion formation is retained in the organic ligand moiety of the complex. Although this is in marked contrast to the findings in aromatic organic radical ion–neutral systems, where a good correlation exists between spin and charge density, it is not, perhaps, very surprising when one views the pronounced ability of metal atom systems to serve as charge sinks and distributors, in comparison to organic systems.

Electronic absorption spectral studies tend to support the concept of charge and spin separation in these radical systems.[33] Aromatic diimine intraligand bands and metal to ligand (here, ligand refers only to the diimine) and metal to carbonyl charge-transfer bands have been identified. Shifts in these bands upon electrochemical reduction of diimine group VIb tetracarbonyls suggest that the electron reversibly added in forming the radical anion enters a π^* orbital predominantly ligand in character, but that much of the resultant charge is transmitted to the carbonyls.

The pertinent charge transfer bands for 1,10-phenanthrolinechromium tetracarbonyl are

	$M \rightarrow \pi_{CO}^*$	$M \rightarrow \pi_L^*$
○	30,700 cm^{-1}	20,400 cm^{-1}
⊖	~35,800 cm^{-1}	~23,800 cm^{-1}

The band assignments for the neutral complex were tentatively reported previously[34] and further sub-

(30) R. E. Dessy, A. L. Rheingold, and G. D. Howard, *J. Amer. Chem. Soc.*, **94**, 746 (1972).
(31) R. E. Dessy, J. C. Charkoudian, and A. L. Rheingold, *ibid.*, **94**, 738 (1972).
(32) G. Fraenkel, R. E. Carter, A. McLachlan, and J. H. Richards, *ibid.*, **82**, 5846 (1960).
(33) An extensive study of the electronic absorption spectra and electronic structures on neutral compounds, radical anions, radical cations, and singlet-state dianions of diimine derivatives of group VIb hexacarbonyls, $Fe(NO)_2(CO)_2$, and $Co(CO)_3(NO)$ and on (olefin)-iron tetracarbonyls is near completion.
(34) H. Saito, J. Fujita, and K. Saito, *Bull. Chem. Soc. Jap.*, **41**, 359 (1968).

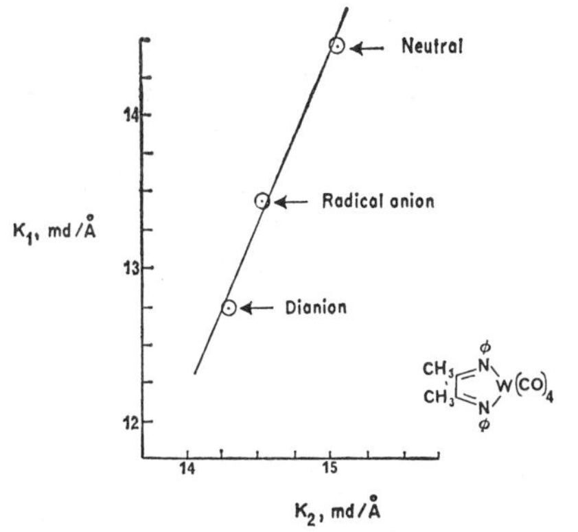

Figure 4. Carbonyl stretch force constants *vs.* charge.

stantiated in this laboratory by extensive ligand substitutional perturbations. Band assignments for the radical anion have been made by comparing the spectrum of the complex radical anion to that of the ligand radical anion. The same types of band structures, attributable to ligand $\pi \rightarrow \pi^*$ transitions, are clearly discernible in both spectra, although they do not occur at precisely the same energies. By subtracting out the $\pi \rightarrow \pi^*$ bands, it is possible to locate, with confidence, the charge-transfer bands. The 20,400-cm^{-1} band, assigned as M $\rightarrow \pi_L^*$, is the lowest energy band in the spectrum of the neutral complex. Consequently, the electron added during reduction enters a lowest unoccupied molecular orbital predominantly ligand in character. If the charge of this electron is localized on the ligand of the complex radical anion, one would predict, on the basis of gross electrostatic effects, that the M $\rightarrow \pi_L^*$ band would experience a more dramatic hypsochromic shift than would the M $\rightarrow \pi_{CO}^*$ band. This is not observed to be the case. The M $\rightarrow \pi_{CO}^*$ band shifts to higher energy by roughly 5100 cm^{-1} compared to a blue shift of about 3400 cm^{-1} for the M $\rightarrow \pi_L^*$ band. This observation suggests that much of the charge due to the reduction electron is transmitted *via* the metal to the carbonyls.

Charge distribution can be further studied by observing changes in the CO infrared stretching vibrations. Infrared studies, using the Cotton–Kraihanzel force field approximation method,[28,29,31] indicate that as electrons are added to these molecules the CO stretching force constant decreases *for all* of the attached CO molecules. One finds experimentally that k_1 is affected more than k_2 (Figure 4) for all group VI derivatives, in *contrast* to the prediction based on a π-only mechanism of charge transmission to and through the metal atom.[35]

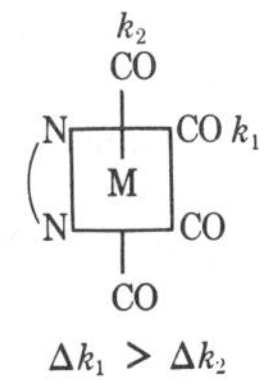

This has led to the suggestion that an anisotropic σ mechanism of transfer is involved. Unfortunately, the

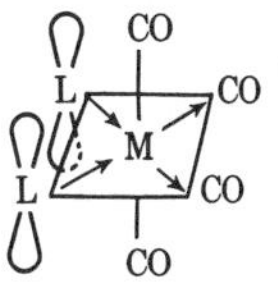

study of the electric field gradient and charge distribution at the central core metal atom has not been possible in this series due to the lack of a nucleus appropriate for Mössbauer studies.

Trigonal-Bipyramidal Cases. However, in the (olefin)iron tetracarbonyl molecules (exemplified by the

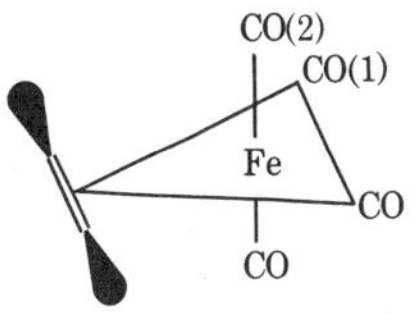

structure) it has been possible to employ Mössbauer spectroscopy to study the transfer mechanism.[29] Esr studies indicate that spin density in the radical anion is largely confined to the olefin ligand (see Table II). Ir

Table II

Ligand	Olefin[a] radical anion a_H, G	‖Fe(CO)$_4$ radical anion a_H, G
Methyl fumarate	6.7	5.1
Methyl cinnamate	5.5	5.0

[a] M. Baizar, Monsanto Central Research, unpublished observations.

studies show shifts of the CO stretch force constant, upon reduction of these complexes, to lower values as expected. The corresponding Mössbauer data reveal clearly the mode of charge transmission. The isomer shift is positive upon reduction, indicating a lower s-electron density at the core metal atom resulting from increased s-electron shielding due to ligand $\rightarrow$ metal π donation (Table III).

Table III

Mössbauer Parameters[a]

	IS,[a,b] mm/sec		QS,[b] mm/sec	
Compound	Neutral molecule	Radical anion	Neutral molecule	Radical anion
$Fe(CO)_5$[c]	0.201		2.60	
Maleic anhydride-$Fe(CO)_4$	0.266	0.28	1.359	0.87
N-Phenylmaleimide-$Fe(CO)_4$	0.248	0.37	1.540	0.73
N,N-Dimethylacrylamide-$Fe(CO)_4$	0.260	0.30	1.654	1.26
Dimethyl fumarate-$Fe(CO)_4$	0.263	0.39	1.563	0.79

[a] All measurements made at −196°, relative to sodium nitroprusside. [b] Neutral compounds ±0.005; radical anions ±0.02. [c] R. L. Collins and R. Petit, *J. Amer. Chem. Soc.*, **85,** 2332 (1963).

(35) F. A. Cotton, *Inorg. Chem.*, **3**, 702 (1964), and references therein.

Tetrahedral Cases. The tetrahedral species $L_2Fe(NO)_2^{\cdot}$ and $L_2Co(CO)NO$ are of interest since a good "spin label" is present in the form of the nitrosyl *nitrogens* that can test spin density in this area.[31] The systems form the charge states

$$L_2Co(CO)(NO) \xrightarrow{e} \dot{\ominus}$$

$$\dot{\oplus} \xleftarrow{-e} L_2Fe(NO)_2 \xrightarrow{e} \dot{\ominus} \xrightarrow{e} \circleddash\!\!\!=$$

Hyperfine coupling constants are indicated in Table IV.

Table IV

	a_M, G	a_N, G (no. of nuclei)	a_H, G (no. of nuclei)
Bipyridylcobalt carbonyl nitrosyl radical anion	6.9	3.4 (2)	6.9 (2)
(Di-2-pyridyl ketone)iron dinitrosyl radical anion		2.4 (2)	4.8 (2) 0.6 (2)

No *NO* hyperfine coupling is observable. The similarity between hyperfine coupling in the ligand radical anion and coordinated radical anion again suggests spin localization in the ligand π system. Mössbauer data indicate that charge transmission again is σ in nature, negative isomer shifts indicating increased electron density around iron (Table V).

Table V
Mössbauer Parameters[a]

(Di-2-pyridyl ketone)-iron dinitrosyl	IS, mm/sec	QS, mm/sec
Radical cation	0.64	0.66
Neutral	0.52	0.72
Radical anion	0.43	0.97

[a] All measurements made at −196°, relative to sodium nitroprusside.

Studies on the organic carbonyl stretching frequency in the (di-2-pyridyl ketone)iron and -cobalt complexes[31] indicate that the radical-cation, neutral, and radical-anion species all have the same organic carbonyl group absorbing at 1661 cm^{-1}, although the NO frequencies shift by 40 cm^{-1} per unit charge. Since this carbonyl should be extremely sensitive to charge density, one has further proof that in L–M–CO systems the coordinated ligand is a good spin sink, but a poor charge sink.

Effect of Charge on Fluxional Behavior and Geometry. Some geometric and energetic information is available for charge states of compounds from the group VI metals and Fe and Ru which involve *dimetallic* bridged species involving M–M bonds. The accepted structures for these compounds are shown in Figure 5.[30]

In all cases, we are dealing with systems that show electro-reversible two-electron reduction to singlet-state dianions, where neutral species and dianion are in equilibrium with radical anion.

$$\bigcirc + \circleddash\!\!\!= \rightleftharpoons 2\dot{\ominus}$$

Compounds of structure I constitute $X_6AA'X'_6$ spin systems where $J_{XX'} = 0$, and compounds of structure II constitute $X_3X'_3AA'X'_3X'''_3$ systems factorable into two $X_3AA'X_3$ systems if ${}^4J_{HH'}$ is taken to be immeasureably small. Two parameters are important in analyzing these systems: $L = |J_{XA} - J_{XA'}|$; $N = |J_{XA} + J_{XA'}|$. Usually ${}^4J_{HP} < {}^2J_{HP}$, and it is assumed that $L = N$.

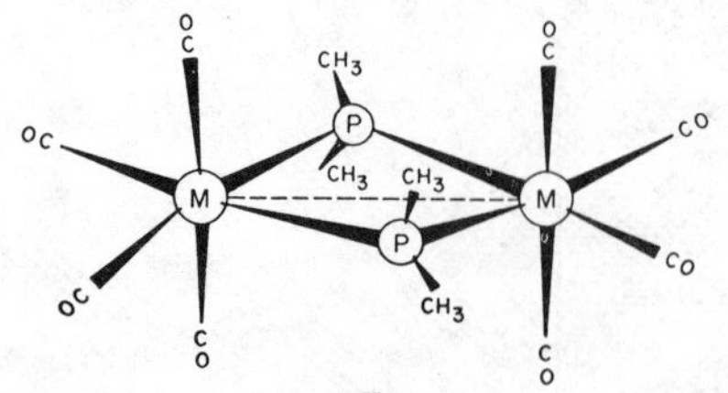

I
M = Cr, Mo, W

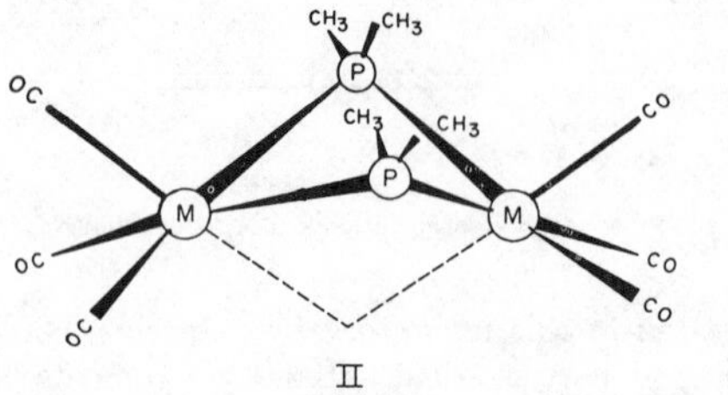

II
M = Fe, Ru

Figure 5.

The X part of the $X_nAA'X'_n$ system has a deceptively simple appearance when L and N are of similar magnitude. When $J_{AA'} \ll L$ the spectrum is a doublet. When $J_{AA'} \gg L$ it is a triplet.

Compounds of type I show doublet spectra, characteristic of a low $J_{PP'}$. Compounds of type II show a triplet nmr spectrum, characteristic of a high $J_{PP'}$, and division of the methyl hydrogens into two sets; one is exo to the dihedral angle, the other endo.

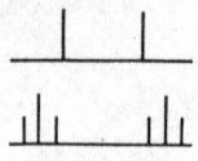

compound I
compound II

Upon electrochemical reduction to the radical anion, compounds of type I exhibit an esr spectrum which is characteristic of a high phosphorus coupling (1:2:1 triplet). Compounds of type II show an esr spectrum in which the exo hydrogens are still distinguished from the endo hydrogens, indicating that during the esr time scale of observation the dihedral angle is maintained.

Further reduction to the dianion species gives rise to solutions of I and II which exhibit strong P–P coupling. In the case of compounds of structure II, however, the nmr spectra are strongly temperature dependent (Figure 6). In retrospect, it seems obvious that the molecules are undergoing inversion and belong to a new class of fluxional molecules. The activation parameters for the inversion process are given in Table VI. Cotton has recently reported another member of this series, $(R_2GeCo(CO)_3)_2$.[36]

The mechanism of the fluxional motion is still under

(36) R. D. Adams and F. A. Cotton, *J. Amer. Chem. Soc.*, **92**, 5003 (1970).

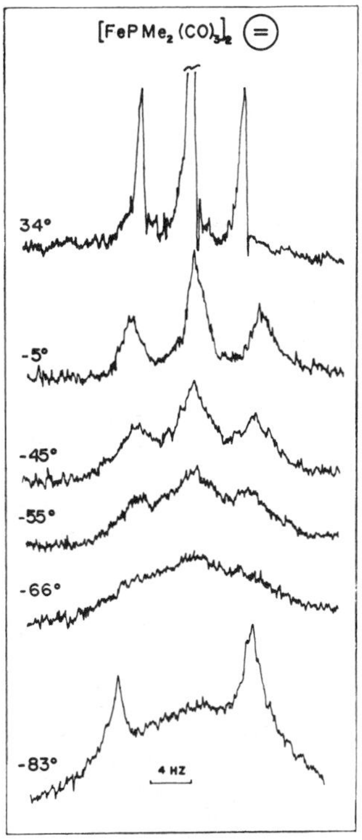

Figure 6. Nmr spectra of $[FePMe_2(CO)_3]_2^{2-}$ as a function of temperature.

Table VI
Kinetic and Thermodynamic Parameters[a]

Compd II	T_c[a]	$E^{\ddagger}$, kcal/mole	$\Delta S^{\ddagger}$, eu	k_1, 0°, l./sec	$J_{pp'}$ cps
Fe ○	338 ± 2	2.1 ± 0.2	−45 ± 1	15.3	85
Fe ⊜	207 ± 2	2.2 ± 0.3	−39 ± 2	352	>500

[a] Coalescence temperature, °K.

study. The observations suggest the following explanation for the large change in kinetic parameters for inversion upon addition of the two electrons: occupation of an MO largely σ^* in character involving metal orbitals leading to an increased M–M distance and an increased charge on the metal. The nmr chemical shift data and the Mössbauer spectra are in agreement with occupation of an orbital with high σ character. The methyl protons show a strong shift to higher field upon reduction, and the isomer shift of the iron is in the direction of increasing s-electron density at the nucleus. The consequence of σ^* occupation and addition of charge to the core metal atom would be a repulsion between the atoms and a lengthening of the Fe–Fe distance. The only way this can be achieved is to shorten the P–P distance and increase phosphorus coupling, as observed.

Consequently, one reaches the conclusion that the addition of electrons to a previously unoccupied molecular orbital results in a change in molecular geometry.

The Future

Thus far, the chemical reactivity and catalytic behavior of electrochemically generated organometallic species have not been fully exploited. Some of the avenues that need exploration hinge on the following observations.

Several of the species can induce polymerization of olefins. Bispyridinemolybdenum tetracarbonyl reduces in a multielectron process to give the radical anion of 2,2′-bipyridylmolybdenum tetracarbonyl. Since chemical reduction of pyridine yields the 4,4′-bipyridyl, a stereospecificity has been achieved by the template action of the core metal.

Although the precoupling lifetime of the radical anions derived from negatively substituted olefins

$$C{=}C{-}Q \xrightarrow{e} C{=}C{-}Q\cdot^{-} \longrightarrow (-C{-}C{-}Q)_2$$

is of the order of milliseconds, the lifetime of the species coordinated to $Fe(CO)_4$ is measured in months. The source of this stability must lie in the spin–charge distribution examined above.

It would thus seem that organometallic electrochemistry is a multifaceted area. The electrochemical results can be used to predict the pathways to new compounds. The large number of charge states existing for many species permits the study of spin and charge delocalization in a molecular species with minimal perturbation of the coordination sphere. Electrochemical template syntheses and tailoring of catalytic activity remain to be explored.

This work has been supported by the Air Force Office of Scientific Research and the National Science Foundation.

Addendum (May 1976)

Recently an extensive study of the electronic absorption spectra of the neutral complexes, radical anions, radical cations, and singlet-state dianions of diimine derivatives of group VIB hexacarbonyls and $Fe(NO)_2(CO)_2$ was completed.[37] Pseudo-octahedral complexes of the type $LM(CO)_4$ (M = Cr, Mo, W; L = 1,10-phenanthroline, 2,2′-bipyridyl, diacetyldianil, methyl- and phenyl-substituted 1,10-phenanthrolines, and 2,2′-bipyridyls) and pseudo-tetrahedral complexes of the type $LFe(NO)_2$ (L represents the same ligands as in $LM(CO)_4$, excluding diacetyldianil) were included in the study. It was found that linear correlations exist between the first electrochemical reductive half-wave potentials and the lowest energy electronic transitions, assigned as metal-to-diimine-charge-transfer-transitions, in the neutral spectra for all the phenanthroline and bipyridyl complexes. Analysis of the esr spectra of the reduced bipyridyl complexes and the corresponding reduced diimines reveals that the unpaired spin densities in the same positions of the free diimine radical anion and the complex radical anion are very similar.[28,29] Comparisons of the complete electronic spectra of the diimine radical anions with the corresponding bipyridyl and phenanthroline complex radical anion spectra show striking similarities, particularly in the low-energy regions, indicating that in the reduction of $LM(CO)_4$ and $LFe(NO)_2$, it is the diimine of the complex and not the complex as a whole that is actually being reduced. Thus, electrochemical data, coupled with the electonic spectra of the neutral complexes, esr data, and electronic spectra of the complex radical anions all provide evidence indicating that reductive electrons enter pure ligand orbitals in all the phenanthroline and bipyridyl complexes and a molecular orbital predominantly ligand in character in the diacetyldianil complexes.

Furthermore, the electronic spectra of both neutral and reduced 1,10-phenanthroline and methyl-substituted 1,10-phenanthrolines have been calculated by open- and closed-shell Self-Consistent Field molecular orbital methods,[38] with the theoretical and experimental spectra comparing very favorably. A linear correlation has also been found between the first reductive half-wave potentials of the phenanthroline complexes and the calculated energies of the first virtual orbitals of the free phenanthrolines. Consequently, electrochemical half-wave potential data and electronic spectra of both neutral and reduced complexes and corresponding ligands, in conjunction with open- and closed-shell S.C.F.-M.O. calculations, indicate that the ligand molecular orbital involved in the reduction of the bipyridyl and phenanthroline complexes is the first virtual orbital of the diimine, relatively unaltered by complexation. This, then, means that the phenanthrolines and bipyridyls do not engage in any pi-bonding to the metal in the corresponding $LM(CO)_4$ and $LFe(NO)_2$ complexes. These conclusions are quite in contrast to cases in which phenanthrolines and bipyridyls apparently do engage in pi-bonding to a metal[39,40] or are generally purported to do so.[41,42] It is concluded from these results that when the phenanthroline and bipyridyl complexes are reduced, any charge transmitted to the carbonyls or nitrosyls must be transmitted to the metal from the diimine by a sigma mechanism, further substantiating the conclusions reached on the basis of observed changes in CO and NO force constants. Publications in which pi-bonding is analyzed or invoked without in-depth experimental substantiation should be viewed with caution.

Organometallic electrochemistry is providing new routes and insights in the areas of bonding, catalysis, and synthesis, and even a brief survey of the literature will provide a number of key entry points to the reader interested in pursuing the exciting efforts of others in the area.[43–45]

(37) Presented in part at 8th Great Lakes Regional Meeting of the American Chemical Society, June 1974, in press.

(38) R. L. Flurry, "Molecular Orbital Theories of Bonding in Organic Molecules", Dekker, New York, 1968.

(39) P. Day and N. Sanders, *J. Chem. Soc. (A)*, 1530 (1967).

(40) E. Konig, *Coord. Chem. Rev.*, **3**, 471 (1968).

(41) B. Hutchinson and K. Nakamoto, *Inorg. Chim. Acta*, **3**, 591 (1969).

(42) F. A. Cotton and G. Wilkinson, "Advanced Inorganic Chemistry", 3rd Ed., Wiley, New York, 1972, p. 723.

(43) R. W. Callahan, G. M. Brown, and T. J. Meyer, *Inorg. Chem.*, **14**, 1443 (1975).

(44) G. R. Langford, M. Akhtar, P. D. Ellis, A. G. MacDiarmid, and J. D. Odom, *Inorg. Chem.*, **14**, 2937 (1975).

(45) B. K. Teo, M. B. Hall, R. F. Fenske, and L. F. Dahl, *Inorg. Chem.*, **14**, 3103 (1975).

Electron-Transfer Mechanisms for Organometallic Intermediates in Catalytic Reactions

Jay K. Kochi

Department of Chemistry, Indiana University, Bloomington, Indiana 47401

Received May 13, 1974

Reprinted from Accounts of Chemical Research, *7, 351 (1974)*

Organometallic intermediates play a crucial role in a variety of organic and biochemical reactions, particularly those catalyzed by transition-metal complexes.[1] Although a number of qualitative mechanisms have been proposed, there is almost no fundamental insight into how bonds between carbon and metals are made and broken in organometallic species.

Organic reaction mechanisms are generally characterized by two-electron changes (*i.e.*, electron-pair processes), whether they proceed by ionic or free-radical pathways. In inorganic chemistry, on the other hand, one-electron changes (*i.e.*, electron-transfer processes) are well established in a variety of oxidation–reduction reactions. To bridge this dichotomy, we have felt that the concept of electron transfer should play an even wider role in the reactions of organometallic intermediates than heretofore suspected.[2]

We have emphasized the importance of one-electron processes in catalytic reactions by examining the reactions of organometallic model compounds. In this Account we wish to describe mainly our studies aimed at probing some of the problems presented in the study of alkyl transfers in organometallic intermediates.

Reactions involving discrete one-electron changes represent the most direct method for mechanistic studies of complex organometallic processes. Our initial approach to studying these reactions is to deal directly with various oxidation–reduction processes of the free-radical intermediates, which are necessarily constrained to one-equivalent changes as we described in a recent review.[3] Unambiguous methods for producing and reacting transient alkyl radicals are the basis for these studies. Thus, the reaction of an alkyl radical (R) with a metal complex (MX_n) may lead to the formation of an alkyl–metal bond, ligand transfer, or electron transfer, as described in reactions 1–3, respectively.[4]

$$R\cdot + M^nX_n \longrightarrow R{-}M^{n+1}X_n \quad (1)$$

$$R\cdot + M^nX_n \longrightarrow R{-}X + M^{n-1}X_{n-1} \quad (2)$$

$$R\cdot + M^nX_n \longrightarrow R^{\pm} + M^{n\mp 1}X_n \quad (3)$$

Jay K. Kochi was born in Los Angeles and educated at Cornell University and UCLA. He obtained his Ph.D. degree at Iowa State University with George Hammond in 1952. After an instructorship at Harvard University and NIH Special Fellowship at Cambridge University, he was at Shell Development Co. in Emeryville, Calif., until 1962. He reentered academic work at Case Institute of Technology (later Case Western Reserve University) and moved to Indiana University in 1969 where he is the Earl Blough Professor of Chemistry. His research has included mechanistic studies of organic reactions catalyzed by metal complexes, photochemistry of organometallic compounds, and application of esr spectroscopy to transient organic and organometallic free radicals and to the mechanism of homolytic reactions.

In each case, the metal nucleus undergoes a formal change in oxidation state of one. A homolytic substitution reaction shown in eq 4 is also possible

$$R\cdot + M^nX_n \longrightarrow R{-}M^nX_{n-1} + X\cdot \quad (4)$$

in which the oxidation state of the metal is unchanged.

Alkylation of metal complexes by alkyl radicals presented in eq 1 has been shown to occur in the reaction between copper(II) complexes and alkyl radicals, in which alkylcopper(III) species are important intermediates in the oxidation of alkyl radicals to alkenes given in eq 5.[3] Similarly, an alkylchromium(III) species is the key intermediate formed during the reduction of alkyl radicals by chromium(II) reagent to alkanes.[5] Finally, four-coordinate phosphoranyl(IV) intermediates arise during the reactions of phosphorous(III) compounds and free radicals, and such an addition–elimination sequence forms the basis of a number of ligand substitution reactions.[6] All of these examples represent facile reactions, with a second-order rate constant greater than 10^4 M^{-1} sec^{-1}, since they must compete with other radical reactions which generally limit the lifetimes of alkyl radicals in solution to less than a millisecond.[7]

$$R\cdot + Cu^{II}(OAc)_2 \longrightarrow [R{-}Cu^{III}(OAc)_2] \longrightarrow R(-H) + Cu^{I}OAc + HOAc \quad (5a)$$

The rise and fall of the oxidation state of the metal resulting from the attachment of a free radical to the metal nucleus followed by the subsequent loss of a ligand from the metastable organometallic intermediate constitute the basis for a number of catalytic

(1) M. M. Taqui Khan and A. E. Martell, "Homogeneous Catalysis by Metal Complexes," Vol. I and II, Academic Press, New York, N. Y., 1974; G. N. Schrauzer, Ed. "Transition Metals in Homogeneous Catalysis," Marcel Dekker, New York, N. Y., 1971.

(2) This notion merely represents a resuscitation of the Michaelis postulate which has been largely neglected by organic chemists after its universality was refuted (R. Stewart, "Oxidation Mechanisms," W. A. Benjamin, New York, N. Y., 1964, p 6 ff).

(3) J. K. Kochi, "Free Radicals," Wiley-Interscience, New York, N. Y., 1973, Chapter 11; J. K. Kochi, A. Bemis, and E. L. Jenkins, *J. Amer. Chem. Soc.*, **90**, 4616 (1968).

(4) Coordination around the metal center hereinafter will be largely unspecified unless required for the discussion (L and X generally denote neutral and anionic ligands, respectively). Oxidation numbers are included only as a bookkeeping device, and are not necessarily intended to denote actual changes in oxidation states (*cf.* (a) C. K. Jorgensen, "Oxidation Numbers and Oxidation States," Springer-Verlag, New York, N. Y., 1969; (b) J. P. Collman, *Accounts Chem. Res.*, **1**, 136 (1968); (c) J. Halpern, *ibid.*, **3**, 386 (1970).

(5) J. K. Kochi and D. D. Davis, *J. Amer. Chem. Soc.*, **86**, 5264 (1964); J. K. Kochi and J. W. Powers, *ibid.*, **92**, 137 (1970).

(6) J. K. Kochi and P. J. Krusic, *J. Amer. Chem. Soc.*, **91**, 3944 (1969); K. U. Ingold and B. P. Roberts, "Free Radical Substitution Reactions," Wiley-Interscience, New York, N. Y., 1971.

(7) K. U. Ingold, ref 3, Chapter 2.

processes. However, it is not necessary for all oxidation–reduction reactions to occur by inner-sphere processes at the metal site. Thus, the ligand transfer reaction[8] in eq 2 may proceed by a one-step process involving direct displacement on the ligand center by a free radical shown in eq 6. Such an inner-sphere

$$R\cdot + Cu^{II}Br_2 \longrightarrow [R\cdots Br\cdots CuBr]^* \longrightarrow R{-}Br + Cu^{I}Br, \text{ etc.} \quad (6)$$

mechanism proceeding by the bridged activated complex has been described for a number of wholly inorganic systems.[9] The microscopic reverse process, *viz.*, the 1-equiv oxidation of a metal complex, is also a fairly common mode of reaction of organic halides, *e.g.*[5,10,11]

$$Cr^{II} + R{-}Br \longrightarrow Cr^{III}Br + R\cdot \quad (7)$$

In a similar manner, the substitution reaction represented in eq 4 may proceed by two distinct pathways. A mechanism involving addition to form an organometallic intermediate $R{-}M^{n+1}X_n$ followed by elimination has been described for three-coordinate phosphorus and arsenic compounds by the observation of transient paramagnetic phosphorus(IV) and arsenic(IV) species by electron spin resonance (esr) spectroscopy.[12] Alternatively, alkyl radicals may react with a metal complex in a one-step process involving direct displacement of a ligand. Thus, no paramagnetic ^{11}B or ^{27}Al intermediates are observed by esr spectroscopy during the facile metathesis of *tert*-butoxy radicals and trialkylboranes and -alanes even when the reactions are carried out at temperatures as low as −100°.[13] The clear mechanistic dis-

$$(CH_3)_3B + t\text{-BuO}\cdot \longrightarrow t\text{-BuOB}(CH_3)_2 + CH_3\cdot$$

tinction between a substitution process proceeding by addition–elimination or direct displacement rests on the detection of a metal-centered intermediate by the use of esr techniques which can effectively probe for paramagnetic species present in concentrations less than 10^{-6} *M* and at low temperatures.[14] Substitution reactions such as reaction 4 form the basis of the radical-chain process for the efficient addition of alkylboron and -aluminum compounds to unsaturated compounds.[15]

Finally, the outer-sphere electron-transfer process represented in eq 3 may be involved in the oxidation of radicals capable of producing more or less stable ionic entities such as trityl, tropyl, or pyryl.[3] More commonly, the formation of carbocations follows the heterolysis of an alkylmetal intermediate such as[16]

$$PhCH_2CH_2\cdot + Cu^{II} \longrightarrow PhCH_2CH_2Cu^{III} \longrightarrow$$

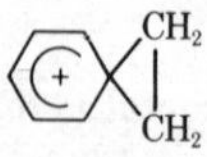

$+ Cu^{I}$ (5c)

Similarly, except in unusual cases,[17] outer-sphere electron-transfer processes for the reduction of radicals by metal complexes directly to carbanions is not favored relative to the formation of an alkylmetal intermediate followed by reaction at the ligand site[18]

$$ClCH_2CH_2\cdot + Cr^{II} \longrightarrow$$

$$ClCH_2CH_2Cr^{III} \begin{cases} \xrightarrow{H_3O^+} ClCH_2CH_3 + Cr^{III} \\ \longrightarrow CH_2{=}CH_2 + Cr^{III}Cl \end{cases}$$

With this background established for one-electron transfer reactions of free radicals and metal complexes, we wish now to examine the mechanisms of more complex organometallic reactions and catalytic processes in which 2-equiv changes may appear to be involved in the overall transformation.

Mechanistic Studies of Alkyl Transfers from Organometallics

There are a number of basic transformations which are of primary importance in the reactions of organometallic intermediates, including: (1) alkyl transfer and aromatic metallations, (2) oxidative addition, (3) reductive elimination, and (4) β elimination and its reverse, addition.[19-21] Most of these processes are commonly considered to proceed by two-electron pathways. In this section we wish to present alternative one-electron mechanisms for these processes, with the principal emphasis on establishing some criteria for distinguishing among various mechanisms.

A. Electrophilic and Electron-Transfer Mechanisms in Alkyl Transfer and Aromatic Substitution. The dichotomy between two-electron and one-electron mechanisms in organometallic reactions is well-illustrated in this Account by two processes: (1) alkyl transfer from σ-alkylmetals and (2) aromatic substitution.

1. Alkyl Transfer from Organometals. Electrophilic processes represent by far the most common path-

(8) J. K. Kochi and D. M. Mog, *J. Amer. Chem. Soc.*, **87**, 522 (1965).

(9) (a) H. Taube, *Advan. Inorg. Chem., Radiochem.*, **1**, 1 (1959). (b) The mechanisms drawn for the two inner-sphere processes, one occurring at the metal center in eq 5 and the other on the ligand site in eq 6, are not readily differentiated, however, since there are examples in which the alkylmetal intermediate undergoes reductive elimination as shown.[10]

$$CH_3 + Cu^{II}(OAc)_2 \longrightarrow CH_3Cu^{III}(OAc)_2 \longrightarrow CH_3OAc + Cu^{I}OAc \quad (5b)$$

(10) C. L. Jenkins and J. K. Kochi, *J. Amer. Chem. Soc.*, **94**, 843, 856 (1972).

(11) P. B. Chock and J. Halpern, *J. Amer. Chem. Soc.*, **91**, 582 (1969).

(12) P. J. Krusic, W. Mahler, and J. K. Kochi, *J. Amer. Chem. Soc.*, **94**, 6033 (1972); A. G. Davies, D. Griller, and B. P. Roberts, *J. Organometal. Chem.*, **38**, C8 (1972).

(13) P. J. Krusic and J. K. Kochi, *J. Amer. Chem. Soc.*, **91**, 3942 (1969); K. U. Ingold, *J. Chem. Soc., Perkin Trans. 2*, 420 (1973); A. G. Davies and B. P. Roberts, ref 3, Chapter 10.

(14) (a) J. K. Kochi and P. J. Krusic, *Chem. Soc., Spec. Publ.*, **24**, 147 (1970). (b) A less direct indication of the formation of the formation of a metastable adduct lies in the observation of β-cleavage shown below for phosphoranyl(IV) species.[12]

$$(CH_3)_3CO\cdot + P^{III}R_3 \longrightarrow (CH_3)_3COP^{IV}R_3 \longrightarrow (CH_3)_3C\cdot + OP^{V}R_3$$

(15) H. C. Brown, M. M. Midland, and G. W. Kabalka, *J. Amer. Chem. Soc.*, **93**, 1024 (1971); G. W. Kabalka and R. F. Daley, *ibid*, **95**, 4428 (1973).

(16) J. K. Kochi, *Pure Appl. Chem.*, **4**, 377 (1971).

(17) K. Okamoto, K. Komatsu, O. Murai, and O. Sakaguchi, *Tetrahedron Lett.*, 4989 (1972).

(18) J. K. Kochi, D. M. Singleton, and L. J. Andrews, *Tetrahedron*, **24**, 3503 (1968); *J. Amer. Chem. Soc.*, **89**, 6547 (1967); **90**, 1582 (1968).

(19) See ref 4b, c.

(20) P. S. Braterman and R. J. Cross, *Chem. Soc. Rev.*, **2**, 271 (1973); M. C. Baird, *J. Organometal. Chem.*, **64**, 289 (1974); W. Mowat, A. Shortland, G. Yagupsky, N. J. Hill, M. Yagupsky, and G. Wilkinson, *J. Chem. Soc., Dalton Trans.*, 533 (1972).

(21) Various types of insertion reactions and rearrangements should also be included.

Table I
Rates of Acetolysis of CH_3-Pb and CH_3CH_2-Pb Bonds at 20°

Bond	Compound	Rate constant, 10^6 sec^{-1} k	k'[a]	Relative rate[b] Found	Calcd[c]
CH_3–Pb	$(CH_3)_4Pb$	7.3	1.8	1.00	1.00
	$(CH_3)_3PbCH_2CH_3$	10	3.3	1.8	1.8
	$(CH_3)_2Pb(CH_2CH_3)_2$	13	6.5	3.6	3.4
	$CH_3Pb(CH_2CH_3)_3$	11	11	6.1	6.2
CH_3CH_2–Pb	$(CH_3CH_2)_4Pb$	4.9	1.2	1.00	1.00
	$(CH_3CH_2)_3PbCH_3$	2.1	0.70	0.56	0.57
	$(CH_3CH_2)_2Pb(CH_3)_2$	0.88	0.44	0.35	0.33
	$CH_3CH_2Pb(CH_3)_3$	0.21	0.21	0.17	0.18

[a] Normalized for each alkyl group. [b] Per alkyl group based on R_4Pb = 1.00. [c] Constant multiplicative factor of 1.84 for CH_3-Pb and 1.75 for CH_3CH_2-Pb.

way by which carbon–metal bonds are cleaved.[22] We have chosen organolead compounds for the study of alkyl-transfer reactions, because they incorporate the best features of organometallic systems, *i.e.*, they are reactive but sufficiently substitution stable and well behaved in solution to allow for quantitative study.

(a) Tetraethyllead in acetic acid solutions undergoes protonolysis according to eq 8 with a pseudo-

$$Et_4Pb + HOAc \longrightarrow Et_3PbOAc + EtH \quad (8)$$

first-order rate constant of 4.9×10^{-6} sec^{-1} at 20° and shows a deuterium kinetic isotope effect of 9.[23] The trends in Table I for a series of methyl and ethyllead compounds indicate that the rates of acetolysis of a particular alkyl–lead bond increase as the methyl groups are successively replaced with ethyl groups in the trialkyllead leaving group.[23] Thus, the reactivity of an ethyl–lead bond is the greatest in tetraethyllead and the least in ethyltrimethyllead. Similarly, the CH_3-Pb bond is the least reactive in tetramethyllead and most reactive in methyltriethyllead. In fact, a constant multiplicative factor S separates the rates of acetolysis between any two contiguous members in the series $(CH_3)_{4-n}Pb(CH_2CH_3)_n$ where n = 0, 1, 2, 3, or 4, as listed in the fifth column of Table I. This correlation represents a linear free energy relationship in which the factor S is the sensitivity of the particular alkyl–lead cleavage to methyl or ethyl substitution in the departing trialkylplumbonium ion, *i.e.*

$$k(R) = k_0(R)S^m = k_0(R)\exp(m\Delta\Delta F^*/RT) \quad (9)$$

in which $k_0(R)$ represents the cleavage rate of R_4Pb (where R = CH_3 or CH_3CH_2), m is the number of ethyl groups, and $\Delta\Delta F^*$ is the change in the free energy of activation as a result of an ethyl substitution. In particular, the replacement of each methyl group by an ethyl group leads to an approximate doubling of the rate of acetolysis or lowering the free energy of the transition state by about 0.4 kcal mol^{-1}.

When the relative reactivities of methyl and ethyl groups are compared under conditions of constant leaving group (R_3Pb^+), we find that the ratios of rates of methyl and ethyl cleavage are more or less invariant among all the methylethyllead compounds. This rate ratio (R_0) measures the intrinsic difference between the rates of CH_3-Pb and CH_3CH_2-Pb cleavage. The values of R_0 can also be obtained under conditions in which the absolute rate constants $k(R)$ are unknown and unobtainable. Thus, the following four relative rates can be obtained from studies involving three intramolecular comparisons together with one intermolecular competition (M = methyl, E = ethyl), and it can be shown that

$$[R(M_4/E_4)][R(ME_3)][R(M_2E_2)][R(M_3E)] = R_0^4 \quad (10)$$

The value of R_0 for the acetolysis of methylethyllead compounds computed in this manner is in good agreement with the value obtained from the absolute rate measurements. The procedure outlined in eq 10 can be used to obtain R_0 for the protonolysis of CH_3-Pb and CH_3CH_2-Pb bonds by triflic acid which is too fast to measure accurately by absolute methods. The same general method can also be applied to the cleavage of tetraalkyllead by other electrophiles such as copper(I) and copper(II).[23,24] In every case, CH_3-Pb cleavage is significantly faster than CH_3CH_2-Pb, as shown by the values of R_0 in Table II, and this factor represents a mechanistic criterion for electrophilic cleavage of an alkyl–lead bond. The change of electrophile from CH_3CO_2H to $CH_3CO_2H_2^+$ (triflic acid) is accompanied by a large increase in rate (Table II), but there is also a parallel increase in selectivity, R_0. This trend is opposite to that expected on the basis of a direct proton transfer from the acid to the tetraalkyllead moiety *via* an open transition state. According to the Hammond postulate, the stronger acid in such a formulation should lead to a decrease in selectivity relative to the weaker acid, since bond making (*i.e.*, R···H) will not have progressed as far in the transition state. From such a basis, we conclude that the conjugate base X must also be involved directly in the proton-transfer process, as represented in a closed four-centered transition state,[25] and it would be supported by a

(22) F. R. Jensen and B. Rickborn, "Electrophilic Substitution in Organomercurials," McGraw-Hill, New York, N. Y., 1968; M. H. Abraham, *Compr. Chem. Kinet.*, **12**, 1 (1972).
(23) N. A. Clinton, and J. K. Kochi, *J. Organometal. Chem.*, **42**, 229 (1972); N. A. Clinton, H. C. Gardner, and J. K. Kochi, *ibid.*, **56**, 227 (1973).
(24) N. A. Clinton and J. K. Kochi, *J. Organometal. Chem.*, **42**, 241 (1972); *ibid.*, **56**, 243 (1973).

Table II
Intrinsic Reactivities of CH$_3$–Pb and CH$_3$CH$_2$–Pb Bonds in Electrophilic Cleavages in Acetic Acid Solutions

Electrophile	Rate constant,[a] l. mol^{-1} sec^{-1}	R_0	S
HOAc	2.8×10^{-5}	8.7	1.80
HOTf(H$_2$OAc$^+$)	$>2 \times 10^{-1}$	48	1.77
Cu^{I}OAc	3.3×10^{-2}	~56	1.85
Cu^{II}Cl$_2$	10^{-1}	26	1.8
(Mass Spectral Cracking)		0.3	0.7

[a] For reaction of tetraethyllead.

Table III
Reactivity Parameters for PbMe$_n$Et$_{4-n}$ and IrCl$_6{}^{2-}$

PbMe$_n$Et$_{4-n}$	k,[a] l. mol^{-1} sec^{-1}	Alkyl chloride (mmol)			IP,[b] eV	E,[c] V
		EtCl	MeCl	EtCl/MeCl		
PbEt$_4$	25	0.101			8.13	1.67
PbEt$_3$Me	11	0.104	0.0015	24	8.26	1.75
PbEt$_2$Me$_2$	3.3	0.101	0.0040	25	8.45	1.83
PbEtMe$_3$	0.57	0.092	0.012	24	8.65	2.00
PbMe$_4$	0.02		0.105		8.90	2.13

[a] Second-order rate constant. [b] Ionization potential. [c] Electrochemical oxidation potential.

$$R_4Pb + HX \longrightarrow \left[R_3Pb \overset{R}{\underset{X}{\diagdown}} H \right] \longrightarrow R_3PbX + HR$$

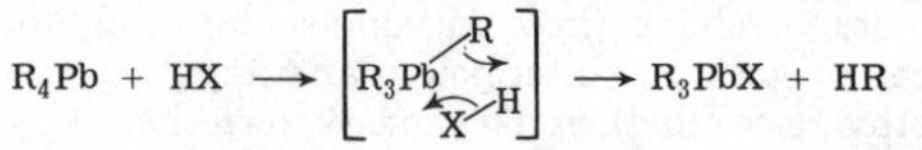

stereochemical study. The catalytic cleavage of alkyllead by copper(I) complexes represents a similar electrophilic process.[23]

(b) Electrophilic processes, however, are not the only routes available for alkyl transfers. Thus, facile alkyl transfers also occur from the same series of organoleads to hexachloroiridate(IV), which is known to participate in outer-sphere oxidations involving one-electron changes.[26] For example, the oxidative cleavage of tetraalkyllead by 2Ir^{IV}Cl$_6{}^{2-}$ occurs rapidly to afford alkyl chloride and 2Ir^{III}. The kinetics

$$R_4Pb + 2Ir^{IV}Cl_6{}^{2-} \xrightarrow{\text{HOAc}} R_3PbOAc + RCl + 2Ir^{III}$$

are first order in each reactant. Two important criteria can be used to distinguish this alkyl transfer from the more conventional electrophilic processes described above. First, the rate of reaction of PbMe$_n$Et$_{4-n}$ with IrCl$_6{}^{2-}$ increases successively as methyl is replaced by ethyl groups (see $n = 4$ to 0 in Table III, column 2). Second, a given ethyl group is cleaved approximately 25 times faster than a methyl group (column 5). Both of these reactivity trends are diametrically opposed to those established for an electrophilic cleavage as discussed above in Tables I and II. These results suggest that the rate-limiting step with IrCl$_6{}^{2-}$ occurs prior to alkyl transfer; we propose the mechanism given in Scheme I in which the slow step in eq 11 involves charge transfer. Indeed, there is a linear correlation of the rates (log k)

(25) M. H. Abraham and J. A. Hill, *J. Organometal. Chem.*, **7**, 11 (1967).
(26) H. C. Gardner and J. K. Kochi, *J. Amer. Chem. Soc.*, **96**, 1982 (1974).

Scheme I

$$R_4Pb + Ir^{IV}Cl_6{}^{2-} \longrightarrow R_4Pb\cdot{}^+ + Ir^{III}Cl_6{}^{3-} \quad (11)$$

$$R_4Pb\cdot{}^+ \xrightarrow{\text{fast}} R\cdot + R_3Pb^+ \quad (12)$$

$$R\cdot + Ir^{IV}Cl_6{}^{2-} \xrightarrow{\text{fast}} RCl + Ir^{III}Cl_5{}^{2-}, \text{ etc.} \quad (13)$$

of PbMe$_n$Et$_{4-n}$ with the one-electron oxidation potentials, which are related to the electron-detachment process, $R_4Pb \rightarrow R_4Pb\cdot{}^+ + \epsilon$. There is also a striking relationship with the vertical ionization potentials of R$_4$Pb determined by He(I) photoelectron spectroscopy shown in Figure 1.[26] Selectivity in the transfer of alkyl groups according to Scheme I occurs during fragmentation of the cation radical in a fast subsequent step which is consistent with a mass

$$\left[R_2Pb \overset{\cdot{}^+}{\underset{CH_2CH_3}{\overset{CH_3}{\diagdown}}} \right] \overset{k'(CH_3)}{\underset{k'(CH_3CH_2)}{\rightleftarrows}} \begin{matrix} R_2PbCH_2CH_3{}^+ + CH_3\cdot (14a) \\ R_2PbCH_3{}^+ + CH_3CH_2\cdot (14b) \end{matrix}$$

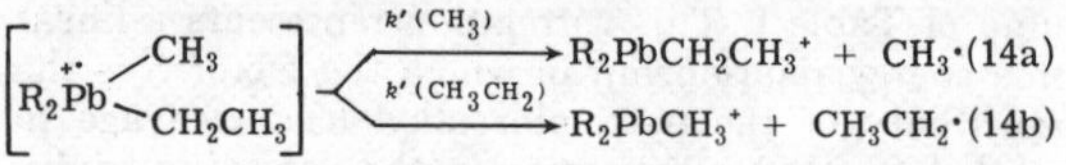

spectral study (Table II) and in accord with expectations based on differences in bond strengths of ethyl and methyl compounds.

Examination of the electron spin resonance spectrum during the reaction with IrCl$_6{}^{2-}$ did not reveal the presence of the cation radical PbEt$_4\cdot{}^+$, which must be a highly unstable intermediate even at temperatures as low as $-20°$. Nonetheless, the formation of ethyl radicals in high yields is evident from spin-trapping experiments with nitrosoisobutane. The use of IrCl$_6{}^{2-}$ as efficient scavenger in the ligand-transfer oxidation of alkyl radicals is implied in Scheme I (eq 13) by the isolation of alkyl chlorides in high yields (even in the presence of a hundredfold excess of bromide ion). Strikingly, separate experiments show that ethyl radicals generated unambiguously from the thermolysis of propionyl peroxide are quantitatively converted by IrCl$_6{}^{2-}$ to ethyl chloride.

(c) The subtle distinction between electrophilic and charge-transfer processes in alkyl transfers from organometallics is also shown in the very rapid reductions of copper(II) acetate, triflate, and chloride by tetraethyllead which proceed at widely different rates in acetic acid.[24] In all of these cases the ethyl group suffers an overall 2-equiv oxidation such that two Cu(II) are required, *i.e.*

$$Et_4Pb + 2Cu^{II}X_2 \longrightarrow Et_3PbX + 2Cu^{I}X + Et_{ox}$$

where Et_{ox} = ethylene, ethyl acetate, and ethyl chloride. It is noteworthy that the products (Et_{ox}) are identical with those obtained from the unique reductions of the same Cu(II) complexes by ethyl radicals generated by independent methods developed in earlier studies.[3] We conclude that the ethyl radical is the prime intermediate formed in the reaction of tetraethyllead with Cu(II) as it is with hexachloroiridate above. Moreover, the stoichiometry of the two reactions is the same in that the oxidation of each tetraethyllead requires 2 equiv of copper(II) and iridium(IV).

$$Et_4Pb + Cu^{II}X_2 \longrightarrow Et_3PbX + Et\cdot \quad (15)$$

$$Et\cdot + Cu^{II}X_2 \xrightarrow{\text{fast}} Et_{ox} + Cu^{I}X \quad (16)$$

There are essentially two mechanisms by which ethyl radicals can be formed during the reduction of Cu(II) by tetraethyllead (eq 15). An outer-sphere electron transfer in Scheme II is equivalent to the

Scheme II

$$Et_4Pb + Cu^{II}X_2 \longrightarrow Et_4Pb\cdot^{+} + Cu^{I}X_2^{-} \quad (17)$$

$$Et_4Pb\cdot^{+} \xrightarrow{\text{fast}} Et_3Pb^{+} + Et\cdot\text{, etc.} \quad (12)$$

mechanism previously shown for the stoichiometrically analogous reduction of hexachloroiridate(IV) in Scheme I. Alkyl transfer from tetraethyllead to Cu(II) by an electrophilic process represented in eq 19 is an alternative pathway for reduction. The alkyl radical is formed in this mechanism by a rapid homolysis of an alkylcopper(II) intermediate. In order

Scheme III

$$Et_4Pb + Cu^{II}X_2 \longrightarrow Et_3PbX + EtCu^{II}X \quad (18)$$

$$EtCu^{II}X \xrightarrow{\text{fast}} Et\cdot + Cu^{I}X\text{, etc.} \quad (19)$$

to distinguish between Schemes II and III for the reduction of copper(II), selectivity studies were carried out with various methylethyllead compounds (Table II). An electrophilic pattern consistent only with Scheme III is clearly established in the facile reaction of tetraalkyllead compounds with copper(II) chloride. Therefore, the formation of alkyl radicals during the reduction of copper(II) is associated with the ready homolysis of a metastable alkylcopper(II) intermediate in a subsequent step (eq 19). These conclusions are enlightening in view of prejudices favoring Scheme II by its expected relationship to the mechanism of hexachloroiridate reduction.

2. Aromatic Substitution. Metalation of alkenes and arenes represents a general method of electrophilic catalysis in substitution reactions at trigonal carbon centers. A particularly relevant example is the oxidative substitution of arenes by Pb(IV), in which the detection and isolation of aryllead(IV) intermediates support a mechanism such as[27]

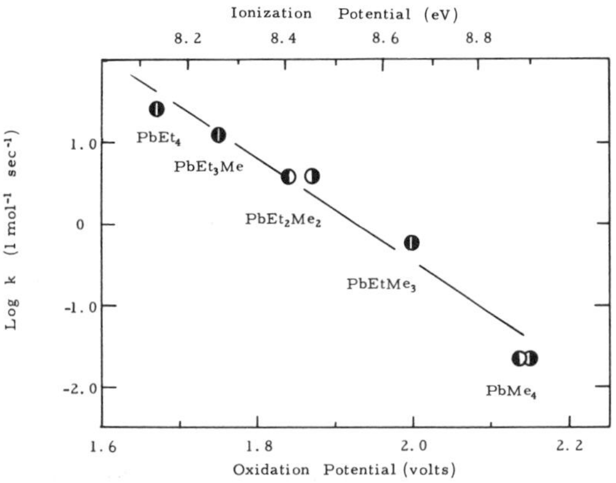

Figure 1. Correlation of the rates of oxidation of $PbMe_nEt_{4-n}$ by $IrCl_6^{2-}$ in acetonitrile with the electrochemical oxidation potential, ◐ (lower scale), and the vertical ionization potential, ◑ (upper scale).

Scheme IV

$$ArH + Pb^{IV}(O_2CCF_3)_4 \longrightarrow ArPb^{IV}(O_2CCF_3)_3 + CF_3CO_2H \quad (20)$$

$$ArPb^{IV}(O_2CCF_3)_3 \longrightarrow ArO_2CCF_3 + Pb^{II}(O_2CCF_3)_2 \quad (21)$$

Oxidative substitution by this scheme requires the metal oxidant to initially effect electrophilic addition. Indeed, arenes can be metalated by a number of metal carboxylates, such as those derived from mercury(II), thallium(III), lead(IV), and palladium(II). The high yields of aryl esters obtained from the oxidation of arenes including toluene with lead(IV) trifluoroacetate are traced directly to the reductive elimination from the aryllead(IV) intermediate in eq 21.

Benzene is also readily oxidized by cobalt(III) in high yields to phenyl trifluoroacetate in trifluoroacetic acid solutions at room temperature according to eq 22.[28] The kinetics of the oxidative substitu-

$$C_6H_6 + 2Co^{III}(O_2CCF_3)_3 \longrightarrow C_6H_5O_2CCF_3 + 2Co^{II}(O_2CCF_3)_2 \quad (22)$$

tion, however, indicate that only one Co(III) is involved in, or prior to, the rate-limiting transition state. The kinetic isotope effect, retardation by Co(II), and esr studies support a mechanism involving two successive one-electron transfers such as that presented in Scheme V.

The electron-transfer Scheme V cannot be distinguished *a priori* from the electrophilic Scheme IV. Both mechanisms depend on π-electron availability. For example, the formation of arene cation radicals in Scheme V is determined by the ionization poten-

(27) J. R. Campbell, J. R. Kallman, J. T. Pinhey, and S. Sternhell, *Tetrahedron Lett.*, 1763 (1972); 5369 (1973).

(28) J. K. Kochi, R. T. Tang, and T. Bernath, *J. Amer. Chem. Soc.*, **95**, 7114 (1973).

Scheme V

$$\text{arene} + Co^{III} \xrightleftharpoons{slow} \text{arene}^{+\cdot} + Co^{II} \quad (23a)$$

$$\text{arene}^{+\cdot} + CF_3CO_2H \longrightarrow \underset{H \quad O_2CCF_3}{\text{cyclohexadienyl}} + H^+ \quad (23b)$$

$$\underset{H \quad O_2CCF_3}{\text{cyclohexadienyl}} + Co^{III} \longrightarrow C_6H_5O_2CCCF_3 + H^+ + Co^{II} \quad (23c)$$

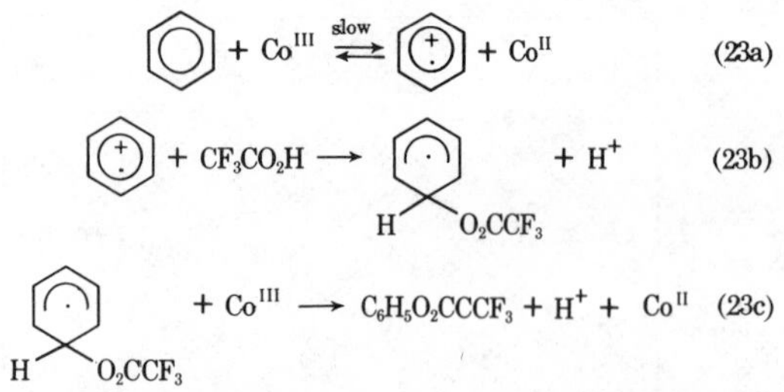

Figure 2. Esr spectrum obtained during the reaction of lead(IV) trifluoroacetate with 2,5-dimethyl-2,4-hexadiene in trifluoroacetic acid at 25°. Proton nmr field markers are in kHz.

tials of the respective arenes, which are directly related to the easily measurable absorption frequencies of the charge-transfer complexes. In turn, the rates of electrophilic substitution of arenes in Scheme IV can also be related to charge-transfer frequencies by a Hammett relationship (using σ^+), provided the substituent causes only a small perturbation on the benzene ring. Thus, ionization potentials and Hammett parameters are indirectly related in these benzenoid systems insofar as the orbital from which the electron is removed by charge transfer has the same symmetry as the orbital which participates in electrophilic attack. In other words, mechanistic distinctions between rate-limiting electron transfer (eq 23a) and electrophilic addition (eq 20) cannot easily be made on the basis of substituent effects. Electron-releasing substituents facilitate and electron-attracting substituents hinder both oxidation processes. Although the differences between the two mechanistic schemes appear to be slight, they can be distinguished in individual cases such as toluene, in which the toluene radical cation reacts differently than the tolylmetal intermediate in Schemes V and IV, respectively.[28]

Judging by this dichotomy, it is not surprising, therefore, to find examples of oxidative substitutions of arenes which proceed competitively *via* both mechanisms. Indeed, reaction of arenes with Tl(III), which is a well-established 2-equiv oxidant like Pb(IV), is generally considered to be an electrophilic process since the one-electron route involves the highly metastable Tl(II) species.[29] Nonetheless, a variety of arene cation radicals detectable by esr are generated under thallation conditions and may be implicated as reaction intermediates in an apparently electrophilic metalation process.[30] The metalation of olefins may involve similar 1-equiv processes, since the esr spectrum of the alkene cation radical can be derived by mixing them with thallium(III) or lead(IV) trifluoroacetates as shown in Figure 2. Finally, the concept of electron transfer in organometallic reactions should not be restricted to electrophilic processes. The generalization to nucleophilic processes with the roles of donors and acceptors reversed has been discussed recently.[30b]

B. Alkylation of Metal Complexes by Oxidative Addition. Alkylation of metal centers may be

achieved by oxidative addition of alkyl derivatives in which the increase in the coordination number of the metal by addition of an alkyl group is accompanied by an increase in its oxidation state.[19] Oxidative addition by free radicals leading to 1-equiv changes was described in the introduction to this Account. Those reactions of alkyl halides involving overall 2-equiv changes (*i.e.*, the alkyl group is electrophilic) have been postulated to proceed *via* an S_N2 process, a concerted three-center addition, or a free-radical chain process to conform to the observation of inversion, retention or racemization at the alkyl center.[31] Inversion is a usual consequence of the S_N2 process and requires the metal to act as a nucleophile, although displacement of a radical is also possible.[32,33] A three-center process leading to retention of configuration, however, is formally akin to an attack by a metal acting as an electrophilic species,[34] and it is unexpected in a process involving oxidation of the metal. Electron transfer contributions, however, may obscure such a limitation.[33] Indeed, the facility with which successive 1-equiv chain reactions take place may have been underestimated due to their high sensitivity to adventitious impurities, particularly molecular oxygen, which may cause inhibition, retardation, or catalysis.[31] The recent reports of a radical chain process in oxidative addition of alkyl bromides to Ir(I) and Pt(0) complexes are particularly relevant examples, in which the propagation sequence involves a ligand transfer reaction (eq 24b).[35] We are

$$R\cdot + Ir^{I} \longrightarrow RIr^{II} \quad (24a)$$

$$RIr^{II} + RBr \longrightarrow RIr^{III}Br + R\cdot, \text{ etc.} \quad (24b)$$

seeking evidence for transient alkyliridium(II) and platinum(I,III) species by esr studies.

C. Reductive Elimination from Alkylmetals:

(29) A. McKillop and E. C. Taylor, *Advan. Organometal. Chem.*, **11**, 147 (1973).

(30) (a) I. H. Elson and J. K. Kochi, *J. Amer. Chem. Soc.*, **95**, 5061 (1973). (b) The limitation of space does not permit a discussion of electron-transfer processes in nucleophilic reactions of organometals. For a recent report, see W. A. Nugent, F. Bertini, and J. K. Kochi, *J. Amer. Chem. Soc.*, **96**, 4945 (1974).

(31) J. A. Labinger, R. J. Braus, D. Dolphin, and J. A. Osborn, *Chem. Commun.*, 612 (1970); R. C. Pearson and W. R. Muir, *J. Amer. Chem. Soc.*, **92**, 5519 (1970); J. A. Labinger, A. V. Kramer, and J. A. Osborn, *ibid.*, **95**, 7908 (1973).

(32) R. E. Dessy and L. A. Bares, *Accounts Chem. Res.*, **5**, 415 (1972); F. R. Jensen, V. Madan, and D. H. Buchanan, *J. Amer. Chem. Soc.*, **92**, 1414 (1970).

(33) *Cf.* M. D. Johnson, *et al.*, *J. Chem. Soc., Chem. Commun.*, 685 (1972); 163 (1973); J. H. Espenson and T. D. Sellers, *J. Amer. Chem. Soc.*, **96**, 94 (1974); K. M. Nicholas and M. Rosenblum, *ibid.*, **95**, 4449 (1973).

(34) D. S. Matteson, "Organometallic Reaction Mechanisms," Academic Press, New York, N. Y., 1974.

(35) J. S. Bradley, D. E. Connor, D. Dolphin. J. A. Labinger, and J. A. Osborn, *J. Amer. Chem. Soc.*, **94**, 4043 (1972); J. A. Osborn in "Prospects in Organotransition Metal Chemistry," Plenum Press, New York, N. Y., in press. *Cf.* also M. Lappert and P. W. Lednor, *J. Chem. Soc., Chem. Commun.*, 948 (1973). We have recently shown that typical diamagnetic d^{10} complexes such as those of nickel(0) and platinum(0) are capable of undergoing facile one-electron transfers (I. H. Elson, D. G. Morrell, and J. K. Kochi, to be published).

Coupling and Disproportionation. Reductive elimination as the reverse of oxidative addition involves both the reduction of the oxidation number as well as the coordination number of the organometallic intermediate. Conceptually, homolysis of an alkyl–metal bond represents the simplest example of a 1-equiv process in reductive elimination. Thermodynamic data on the strengths of transition-metal–alkyl bonds are scarce, but indirect evidence suggests that the bond dissociation energies are relatively high, being more generally like those of the main group metals.[36] The facile homolysis of alkylcopper(II) described earlier in eq 19 appears to be an exception rather than the rule.

Reductive elimination of alkylmetals involving an overall 2-equiv transformation can proceed in a concerted manner without the involvement of alkyl free radicals.[20,37] Thus, in a series of trialkyl(triphenylphosphine)gold(III) complexes, a pair of cis alkyl groups are reductively coupled to alkane by a first-order elimination process involving a prior dissociation of the phosphine ligand.[38] The absence of alkene

$$CH_3CH_2Au^{III}(CH_3)_2L \longrightarrow CH_3CH_2CH_3 + CH_3Au^{I}L \quad (25)$$

and products derived from attack on solvent indicate that alkyl radicals are not intermediates in the reductive coupling. Analogous reductive eliminations have been observed in platinum(IV) and nickel(II) complexes.[39]

Reductive elimination of alkylmetals may also proceed by disproportionation of the alkyl groups. Thus, the thermolysis of di-*n*-butylbis(triphenylphosphine)platinum(II) occurs by a first-order intramolecular process and affords an equimolar mixture of butane and butene-1, but no octane obtains as a result of reductive coupling.[40] A three-coordinate

$$(n\text{-}C_4H_9)_2Pt^{II}L_2 \longrightarrow n\text{-}C_4H_{10} + 1\text{-}C_4H_8 + (Pt^0L_2)$$

platinum(II) intermediate similar to that in the isoelectronic gold(III) complex above has been postulated.

Reductive elimination by a 2-equiv coupling or disproportionation of the alkyl derivative of Ni(II), Pt(IV,II), Au(III), and similar metal complexes by various intramolecular processes is not mechanistically unreasonable,[20] since the reduced metal species is available as a fairly stable entity, *i.e.*, Ni(0), Pt(II,0), Au(I), etc.[41] However, there are a number of alkylmetals such as those of Cu(I), Ag(I), Au(I), Ir(I), etc., in which the metal cannot readily sustain a 2-equiv reduction. The overall reduction requires two metal centers as will be shown for alkylcopper(I), -silver(I), and -gold(I) complexes.

Alkyl(triphenylphosphine)gold(I) complexes are amenable to kinetic study since they are relatively stable and exist in solution as monomeric species.[44] *n*-Alkylgold(I) complexes undergo reductive coupling similar to the silver(I) analogs in very high yields according to eq 26. The rates of the decomposition

$$2CH_3CH_2AuL \longrightarrow CH_3CH_2CH_2CH_3 + 2Au^0 + 2L \quad (26)$$

of $CH_3(PPh_3)Au$ in decalin solution follow first-order kinetics for approximately two half-lives. The first-order rate constant is independent of the concentration of $CH_3(PPh_3)Au$ in the range of 0.01–0.04 *M* and the presence of oxygen, but the rate is retarded by the addition of triphenylphosphine. The formation of ethane in near-quantitative yields as the sole hydrocarbon product even in the presence of hydrogen donor solvents and molecular oxygen strongly supports a molecular process for the methyl coupling. The monomeric nature of $CH_3(PPh_3)Au$ in solution, the first-order thermolysis, and the retardation suggest a rate-limiting loss of ligand. Rapid reaction of the coordinatively unsaturated methylgold(I) species with another CH_3AuPPh_3 in a subsequent step would account for all the known facts,

$$CH_3AuL \xrightarrow{k_1} CH_3Au + L$$

$$CH_3Au + CH_3AuL \longrightarrow CH_3CH_3 + 2Au^0 + L$$

but a further description of the coupling process is undefined. One or several transients such as the binuclear $CH_3AuAuCH_3(L)$ and $Au(CH_3)_2AuL$ are likely intermediates.[42] Alkylsilver(I) complexes similarly afford coupled products on reductive elimination.[43,44]

D. β Elimination from Alkylmetals. The thermal decomposition of alkylcopper(I) differs from that of the silver(I) and gold(I) analogs in that no alkyl coupling is observed, and only products of disproportionation result, as represented in eq 27.[45] The ab-

$$2CH_3CH_2Cu^I \longrightarrow CH_3CH_3 + CH_2{=}CH_2 + 2Cu^0 \quad (27)$$

sence of coupled dimer in the decomposition of ethylcopper(I) species in eq 27 indicates alkyl radicals *per se* are not prime intermediates, since the bimolecular disproportionation and combination of ethyl radicals are relatively invariant with the medium and favor coupling, *i.e.*, $k_d/k_c = 0.18$.[46] Free-radical routes for decomposition are available, however, in those alkylcopper(I) analogs such as the more thermally stable neophyl derivatives which have no available β hydrogens.[47] Labeling studies with β,β-dideuteriobutylcopper(I) in ether show a specific intermolecular transfer of a β deuterium, which has been accommodated by a two-step mechanism in-

(36) N. J. Friswell and B. G. Gowenlock, *Advan. Free Radical Chem.*, **1**, 39 (1965); **2**, 1 (1967). *Cf.* also D. M. P. Mingos, *J. Chem. Soc., Chem. Commun.*, 165 (1972); M. C. Baird, *J. Organometal. Chem.*, **64**, 289 (1974); P. S. Braterman and R. J. Cross, *J. Chem. Soc., Dalton Trans.*, 657 (1972); P. J. Davidson, M. F. Lappert, and R. Pearce, *Accounts Chem. Res.*, **7**, 209 (1974).

(37) However, for a radical chain mechanism in the 2-equiv reductive elimination of alkylmercury(II), see G. M. Whitesides, *et al.*, *J. Amer. Chem. Soc.*, **92**, 6611 (1970); **96**, 870 (1974).

(38) A. Tamaki, S. A. Magennis, and J. K. Kochi, *J. Amer. Chem. Soc.*, **95**, 6487 (1973).

(39) M. P. Brown, R. J. Puddephatt, and C. E. E. Upton, *J. Organometal. Chem.*, **49**, C61 (1973); T. Yamamoto, A. Yamamoto, and S. Ikeda, *J. Amer. Chem. Soc.*, **93**, 3350 (1971).

(40) G. M. Whitesides, J. G. Gaasch, and E. R. Stedronsky, *J. Amer. Chem. Soc.*, **94**, 5258 (1972). See also *ibid.*, **95**, 4451 (1973).

(41) The alkene-induced homolysis of dialkylplatinum(II) is noteworthy [N. G. Hargeaves, R. J. Puddephatt, L. H. Sutcliffe, and P. J. Thomson, *J. Chem. Soc., Chem. Commun.*, 861 (1973)]. *Cf.* also ref 39b.

(42) A. Tamaki and J. K. Kochi, *J. Organometal. Chem.*, **61**, 441 (1973).

(43) M. Tamura and J. K. Kochi, *J. Amer. Chem. Soc.*, **93**, 1483 (1971).

(44) (a) G. M. Whitesides, D. Bergbreiter, and P. E. Kendall, *J. Amer. Chem. Soc.*, **96**, 2806 (1974); (b) G. M. Whitesides, C. P. Casey, and J. K. Krieger, *ibid.*, **93**, 1379 (1971).

(45) M. Tamura and J. K. Kochi, *J. Organometal. Chem.*, **42**, 205 (1972).

(46) M. J. Gibian and R. C. Corley, *Chem. Rev.*, **73**, 441 (1973).

(47) G. M. Whitesides, E. J. Panek, and E. R. Stedronsky, *J. Amer. Chem. Soc.*, **94**, 232 (1972).

volving prior β elimination of copper hydride.[48] The

$$CH_3CH_2CD_2CH_2Cu^IL \longrightarrow CH_3CH_2CD{=}CH_2 + DCuL \quad (28a)$$

$$CH_3CH_2CD_2CH_2|CuL + DCuL \longrightarrow CH_3CH_2CD_2CH_2D + 2Cu^0 + 2L \quad (28b)$$

reactivity of alkylcopper to reductive protonolysis in eq 28b is, however, relatively insensitive to structure and increases from *n*-butyl to *sec*-butyl to *tert*-butyl in the order 1:1.3:3.3.[47] Aggregation of alkylcopper species may contribute to this apparent lack of selectivity, and indeed reductive elimination may even proceed directly within the cluster itself.[49] Alkylcopper(I) species are more stable in THF solutions compared to ether, but otherwise afford the same products on decomposition. The rates in THF show an unusual and marked inhibition and autocatalysis. The esr spectrum obtained during the thermolysis has been attributed to a transient Cu^ICu^0 mixed-valence intermediate which we incorporated as a catalytic species in the Whitesides formulation.[50] However, there are several attractive mechanistic implications in an alternative free-radical *chain* process, in which autocatalysis is due to labilization of an alkylcopper(I) bond by Cu(0) in eq 29a, followed by abstraction of a β hydrogen in eq 29b.

$$RCu^I + Cu^0 \rightleftharpoons RCu^ICu^0 \longrightarrow R\cdot + 2Cu^0 \quad (29a)$$

$$R\cdot + CH_3CH_2Cu^I \longrightarrow RH + \cdot CH_2CH_2Cu^I \longrightarrow CH_2{=}CH_2 + Cu^0 \quad (29b)$$

The driving force for the removal of a reactive hydrogen located in a β position relative to a metal, as shown in eq 29b, has been previously explored in esr studies of the main-group derivatives such as Si, Ge, and Sn.[51] The resulting β-metal-substituted alkyl radicals have rather unique conformational and structural properties which are related to hyperconjugative and homoconjugative interactions between the metal substituent and the radical center on carbon shown in **1**. The distinction between these struc-

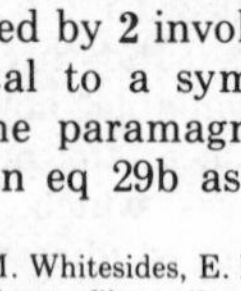

1, M = Si, Ge, Sn

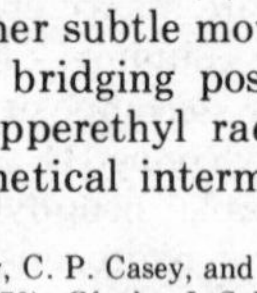

2, M = Cu(I), Ni(0), etc.

tures (**1**) and the common diamagnetic π-olefin complexes of transition metals[52] such as Ni(0) and Cu(I) illustrated by **2** involves a rather subtle movement of the metal to a symmetrical bridging position. Indeed, the paramagnetic β-copperethyl radical presented in eq 29b as a hypothetical intermediate in the decomposition of ethylcopper(I) may represent the missing link between structures such as **1′** and **2′**.[52b] Although these structures remain speculative

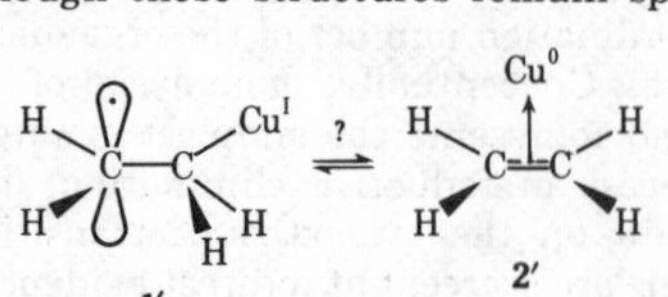

1′ 2′

at this juncture, we think that they introduce the provocative notion that paramagnetic metal species may be viable intermediates in other organometallic processes such as the addition of metal hydrides to olefins, *i.e.*

$$M{-}C{-}C\cdot + MH \longrightarrow M{-}C{-}C{-}H + M\cdot \quad (30a)$$

$$M\cdot + C{=}C \rightleftharpoons M{-}C{-}C\cdot, \text{ etc.} \quad (30b)$$

and the reverse process, β elimination.[53] Esr spectroscopy represents the ideal technique to probe for these metastable structures, and we are engaged in the pursuit of such intermediates in catalytic reactions.

Alkyl Transfers from Organometallic Intermediates in Catalytic Processes

The oxidation–reduction reactions of organometallic intermediates presented in the foregoing description can be applied, in combination, to a variety of catalytic processes. We apply them in this Account to the metal-catalyzed alkyl-transfer reactions of Grignard reagents originally investigated by Kharasch and coworkers.[54] The catalytic reactions between labile organometals [RM] and alkyl halides [RX] can be generally classified into two categories, coupling in eq 31a and disproportionation in eq 31b,

$$R{-}m + R{-}X \xrightarrow{\text{cat.}} \begin{cases} R{-}R + mX & (31a) \\ RH + R({-}H) + mX & (31b) \end{cases}$$

depending on the catalyst. For example, silver(I) and copper(I) are effective catalysts in the coupling of alkyl groups, whereas iron effects only disproportionation except when aryl and vinylic halides are employed. Each catalyst shows unique features which are best described within the following mechanistic context.

A. Homo Coupling with Silver(I). Silver is an effective catalyst for the coupling of Grignard reagents and alkyl halides, and it is especially useful when both alkyl groups are the same.[55] When different alkyl groups are employed, a mixture of three coupled products is obtained. Disproportionation be-

$$RMgX + RX \xrightarrow{Ag^I} R{-}R + MgX_2$$

comes increasingly important with secondary and tertiary groups, independently of whether they are derived from the Grignard reagent or the alkyl halide. The rate of production of butane from ethyl-

(48) G. M. Whitesides, E. R. Stedronsky, C. P. Casey, and J. San Fillippo, Jr., *J. Amer. Chem. Soc.*, **92**, 1426 (1970). *Cf.* also J. Schwartz and J. B. Cannon, *ibid.*, **96**, 2276 (1974), for alkyliridium(I).

(49) M. Tamura and J. K. Kochi, *J. Amer. Chem. Soc.*, **93**, 1485 (1971).

(50) K. Wada, M. Tamura, and J. K. Kochi, *J. Amer. Chem. Soc.*, **92**, 6656 (1970).

(51) P. J. Krusic and J. K. Kochi, *J. Amer. Chem. Soc.*, **91**, 6161 (1969); **93**, 846 (1971); T. Kawamura, P. Meakin, and J. K. Kochi, *ibid.*, **94**, 8065 (1972); T. Kawamura and J. K. Kochi, *ibid.*, **94**, 648 (1972); *J. Organometal. Chem.*, **47**, 79 (1973).

(52) (a) E. O. Fischer, *et al.*, "Metal π-Complexes," Vol. I and II, Elsevier, Amsterdam, 1966, 1972; R. G. Salomon and J. K. Kochi, *J. Amer. Chem. Soc.*, **95**, 1889 (1973). (b) Hg(I) may be another example for σ and π interactions [F. R. Jensen and H. E. Guard, *J. Amer. Chem. Soc.*, **90**, 3250 (1968); P. A. W. Dean, *et al.*, *J. Chem. Soc., Chem. Commun.*, 626 (1973)]. See ref 34, p 298.

(53) Trapping or observation of radical intermediates in chain reactions will depend on the efficiency of the propagation sequence such as eq 29 or 30. These metal hydride and alkylmetal species may be excellent chain-transfer agents, as shown for group IVb hydrides [*cf.* H. G. Kuivila, *Accounts Chem. Res.*, **1**, 299 (1968)].

(54) M. S. Kharasch and O. Reinmuth, "Grignard Reagents of Nonmetallic Substances," Prentice-Hall, New York, N. Y., 1954.

(55) M. Tamura and J. K. Kochi, *Synthesis*, 303 (1971).

magnesium bromide and ethyl bromide is roughly first order in silver and ethyl bromide, but zero order in Grignard reagent.[43a] The reactivity of alkyl halides follows the order: *tert*-butyl > isopropyl > *n*-propyl bromide in the ratio 20:3:1. Structural variations in the Grignard reagent show no apparent systematic trend.

The results can be accommodated by Scheme VI, in which the coupling arises from alkylsilver(I) intermediates generated *via* two largely independent pathways.

Scheme VI

$$R'MgX + Ag^I \longrightarrow R'Ag^I + MgX_2 \quad (32)$$

$$RAg^I, R'Ag^I \longrightarrow [R{-}R, R'{-}R, R'{-}R'] + 2Ag^0 \quad (33)$$

$$Ag^0 + R{-}X \longrightarrow R\cdot + Ag^IX \quad (34)$$

$$R\cdot + Ag^0 \longrightarrow RAg^I \text{ etc.} \quad (35)$$

The rate-limiting step in this mechanism is given by eq 34 in which the alkyl halide is responsible for the reoxidation of silver(0) produced in eq 33. The reactivity of alkyl halides in eq 34 follows the order of stability of alkyl radicals, *i.e.*, tertiary > secondary > primary. This slow step is closely akin to the production of alkyl radicals by the ligand-transfer reduction of alkyl halides with other reducing metal complexes described earlier in eq 7.[5,10]

Previous studies have shown that alkyl radicals are not involved in the reductive dimerization of alkylsilver(I) in eq 33.[42,43] More direct evidence for the selective formation of alkyl radicals from the alkyl halide is shown by trapping experiments as well as stereochemical studies. Thus, the catalytic reaction of *cis*-propenylmagnesium bromide with methyl bromide yielded *cis*-butene-2, in accord with the retention of stereochemistry during the reductive coupling of vinylsilver(I) complexes.[43] On the other hand the

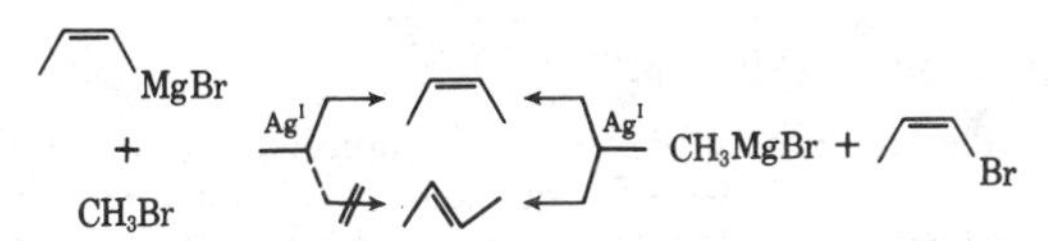

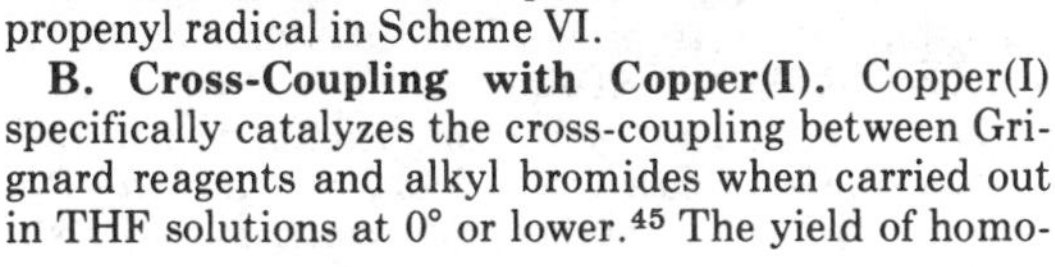

reverse combination, *cis*-propenyl bromide and methylmagnesium bromide, is catalytically converted to a mixture of *cis*- and *trans*-butene-2, consistent with the formation and rapid isomerization of the 1-propenyl radical in Scheme VI.

B. Cross-Coupling with Copper(I). Copper(I) specifically catalyzes the cross-coupling between Grignard reagents and alkyl bromides when carried out in THF solutions at 0° or lower.[45] The yield of homo-

$$RMgX + R'X \xrightarrow{Cu^I} R{-}R' + MgX_2$$

dimers, R–R and R′–R′, under these conditions is negligibly small. This cross-coupling reaction is most facile with primary alkyl halides, but, unlike silver(I) catalysis, the secondary and tertiary alkyl halides are generally inert and give poor yields of coupled products and mainly disproportionation. The structure of the Grignard reagent is not as important, in analogy with the cross-coupling observed with lithium dialkylcuprates.[56]

(56) *Cf.* A. E. Jukes, *Advan. Organometal. Chem.*, **12**, 215 (1974), and C. R. Johnson and G. A. Dutra, *J. Amer. Chem. Soc.*, **95**, 7783 (1973).

The coupling of ethylmagnesium bromide and ethyl bromide to *n*-butane follows overall third-order kinetics, being first order in each component and the copper(I) catalyst.[45] There is no evidence for alkyl radicals in the copper(I)-catalyzed coupling process, and we propose the following two-step mechanism

Scheme VII

$$RMgBr + Cu^IBr \longrightarrow RCu^I + MgBr_2 \quad (32')$$

$$RCu^I + R'Br \longrightarrow R{-}R' + Cu^IBr \quad (36)$$

The rate-limiting step 36 can be shown independently by examining the stoichiometric reaction of alkylcopper(I) directly with organic halides. However, the extent to which *decomposition* of the alkylcopper(I) intermediate (described earlier in eq 27) competes with the catalytic coupling reaction introduces disproportionation products. The latter involves a copper(0)-catalyzed sequence[49] similar to that observed with iron (*vide infra*), and it is especially important with secondary and tertiary alkyl systems. The effects of structural variation are consistent with a rate-limiting step involving nucleophilic displacement of halide in eq 36.[56] The involvement of a nucleophilic copper(I) center, *i.e.*, oxidative addition, followed by reductive elimination has direct analogy to the mechanism which has been established with the analogous gold(I) catalyst.[57] Organo-

$$RCu^I + R'X \longrightarrow R(R')Cu^{III}X \quad (37a)$$

$$R(R')Cu^{III}X \longrightarrow R{-}R' + Cu^IX \quad (37b)$$

copper(III) intermediates presented in eq 37 are formally related to the species discussed earlier (eq 5) in the association of alkyl radicals with copper(II) complexes, with both showing a marked propensity for reductive elimination. Although the direct observation of these highly metastable intermediates is unlikely, the analogous organogold intermediates are more stable and can be isolated or observed directly by nmr.[59]

Catalysis by Iron. Alkyl disproportionation is the sole reaction observed during the iron-catalyzed reaction of ethylmagnesium bromide and ethyl bromide.[60] The catalyst is a reduced iron species formed

$$CH_3CH_2MgBr + CH_3CH_2Br \xrightarrow{Fe^I} CH_3CH_3 + CH_2{=}CH_2 + MgBr_2$$

(57) Oxidative addition of alkyl halides to alkyl(triphenylphosphine)gold(I) follows the expected pattern: $CH_3I > EtI > i\text{-}PrI$, *e.g.*[58]

$$CH_3Au^I + CH_3CH_2I \xrightarrow{slow} CH_3(CH_3CH_2)Au^{III}I$$

$$CH_3(CH_3CH_2)Au^{III}I + CH_3Au^I \longrightarrow (CH_3)_2AuCH_2CH_3 + IAu^I$$

The subsequent reductive elimination of trialkylgold(III) complexes to coupled dimer was described earlier (*cf.* eq 25.)

(58) A. Tamaki and J. K. Kochi, *J. Chem. Soc., Dalton Trans.*, 2620 (1973); *J. Organometal. Chem.*, **64**, 411 (1974).

(59) The parallel between copper(I) and gold(I) is further shown in the behavior of the corresponding cuprate(I) and aurate(I) complexes. Thus, alkylgold(I) reacts with an equimolar amount of alkyllithium to afford an isolable lithium dialkylaurate(I). The anionic dimethylaurate(I) species formed in this manner is at least 10^6 times more reactive to oxidative addition of methyl iodide than the neutral methyl(triphenylphosphine)gold(I).[58] The same pattern is qualitatively established with organocopper(I) species in comparing the coupling reaction in eq 36 with that reported for lithium dialkylcuprates.[56]

(60) M. Tamura and J. K. Kochi, *J. Organometal. Chem.*, **31**, 289 (1971); *Bull Chem. Soc. Jap.*, **44**, 3063 (1971).

in situ by the reaction of iron(II,III) with Grignard reagent, and effective in concentrations as low as 10^{-5} *M*. Although the reaction has limited synthetic utility, it merits study since it can provide insight into some of the complications involved with organometallic intermediates.

The rate of reaction shows first-order dependence on the concentration of iron and ethyl bromide, but is independent of the concentration of ethylmagnesium bromide. The rate, however, varies with the structure of the Grignard reagent, and disproportionation usually results except when the alkyl group is methyl, neopentyl, or benzyl, none of which possesses β hydrogens. The reactivities of the alkyl bromides (*tert*-butyl> isopropyl > *n*-propyl) as well as the kinetics are the same as the silver-catalyzed coupling described above and suggest a similar mechanism.[60,61]

Scheme VIII

$$Fe^I + RBr \longrightarrow Fe^{II}Br + R\cdot \qquad (38a)$$

$$R\cdot + Fe^I \longrightarrow RFe^{II} \qquad (38b)$$

$$R'MgBr + Fe^{II}Br \longrightarrow R'Fe^{II} + MgBr_2 \qquad (38c)$$

$$RFe^{II}, R'Fe^{II} \longrightarrow$$
$$[RH, R'H, R(-H), R'(-H)] + 2Fe^I, etc. \qquad (38d)$$

According to this postulate, the difference between coupling with silver and disproportionation with iron rests on the decomposition of the alkylmetal intermediate. Indeed, it has been shown separately that the decomposition of alkylsilver(I) proceeds by reductive coupling. Unfortunately, the highly unstable alkyliron intermediate in Scheme VIII is not yet accessible to independent study, but the somewhat analogous dialkylmanganese(II) species undergoes similar reductive disproportionation by a mechanism[62] reminiscent of alkylcopper(I) described in eq 27. Selective trapping of alkyl radicals from the alkyl halide component during the course of the catalytic disproportionation is the same as the previous observation with silver,[42] and it indicates that the prime source of radicals in the Kharasch reaction lies in the oxidative addition of alkyl halide to reduced iron in eq 38a. Separate pathways for reaction of isopropyl groups derived from the organic halide and the Grignard reagent are also supported by deuterium labeling studies which show that they are not completely equilibrated.[63] Furthermore, the observation of CIDNP (AE multiplet effect) in the labeled propane and propene derived only from the alkyl halide component can be attributed to a bimolecular disproportionation of isopropyl radicals arising from diffusive displacements. However, the latter can only be a minor fate of the alkyl radicals derived from the alkyl halide, since the coupled dimer is not formed in amounts required by the bimolecular reaction of alkyl radicals.[62,64]

Cross-coupling of Grignard reagents with 1-alkenyl halides, in marked contrast to alkyl halides, occurs readily with the reduced iron catalyst, especially that derived from tris(dibenzoylmethido)iron(III), as well as a recently reported nickel catalyst.[65,66] *n*-Propyl- and *n*-hexylmagnesium bromides react with vinyl bromide to afford pentene-1 and octene-1, respectively. Similarly, cyclohexylmagnesium bromide

$$RMgBr + \underset{}{>}C{=}C\underset{}{<}^{Br} \xrightarrow{Fe^I} \underset{}{>}C{=}C\underset{}{<}^{R} + MgBr_2$$

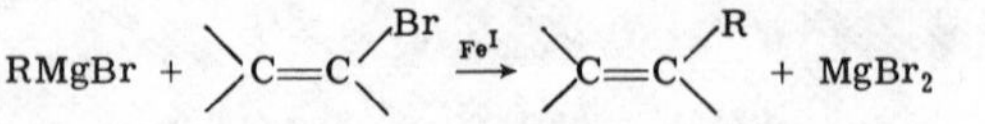

produces propenylcyclohexane in high yields from propenyl bromide. The reaction is stereospecific, since *cis*- and *trans*-propenyl bromides afford *cis*- and *trans*-butene-2, respectively. Secondary and even tertiary alkyl Grignard reagents can be coupled in excellent yields with other 1-alkenyl bromides including β-bromostyrene.[65] The iron-catalyzed reaction of Grignard reagents with 1-alkenyl halides can be differentiated from the reaction with alkyl halides. Thus, a mixture of propenyl bromide and ethyl bromide on reaction with methylmagnesium bromide afforded butene-2, but no crossover products such as pentene-2 or propylene. The latter certainly would have resulted if a propenyliron species *per se* were involved in the catalytic process. Cross-coupling under these circumstances clearly merits further study.

I wish to thank my coworkers, especially Nye Clinton, Ian Elson, Hugh Gardner, Akihiro Tamaki, and Masuhiko Tamura, for their notable contributions to the studies reported in this Account. I am also grateful to the National Science Foundation and the Petroleum Research Fund, administered by the American Chemical Society, for financial support of this work.

Addendum (May 1976)

The dichotomy between electrophilic and electron transfer mechanisms in the cleavage of organometals has been elaborated with tetraalkyllead compounds.[67] Examination of the sterically less-hindered dialkylmercury allows a wider range of alkyl groups to be studied, encompassing the gamut of methyl, ethyl, isopropyl, and tert-butyl groups.[68] A quantitative relationship for electrophilic cleavage is developed for acetolysis [R-HgR' + HOAc → RH + AcO-HgR'] by a single equation: $\log k/k_0 = L + C$, where k_0 and k are the rate constants for cleavage of MeHgMe and R-HgR', respectively, L is the leaving group constant for HgR', and C is the cleaved group constant for R.

(61) The oxidation number in the reduced iron species has not been established. We tentatively favor iron(I) based on some recent studies. Reductive elimination may then proceed from mono- and dialkyliron(II) in eq 38d.

(62) M. Tamura and J. K. Kochi, *J. Organometal. Chem.*, **29**, 111 (1971).

(63) R. B. Allen, R. G. Lawler, and H. R. Ward, *J. Amer. Chem. Soc.*, **95**, 1692 (1973).

(64) The large enhancement possible in CIDNP may not reflect its chemical importance until they are *quantitatively* related. A small amount of radical combination leading to CIDNP may have been overlooked in the chemical studies.

(65) M. Tamura and J. K. Kochi, *J. Amer. Chem. Soc.*, **93**, 1487 (1971); S. Neumann, unpublished results.

(66) K. Tamao, K. Sumitani, and M. Kumada, *J. Amer. Chem. Soc.*, **94**, 4374 (1972). See also J. Klein and R. Levene, *ibid.*, **94**, 2520 (1972); M. F. Semmelhack and L. Ryono, *Tetrahedron Lett.*, 2967 (1973).

(67) H. C. Gardner and J. K. Kochi, *J. Am. Chem. Soc.*, **97**, 1855 (1975).

(68) (a) W. A. Nugent and J. K. Kochi, *J. Am. Chem. Soc.*, **98**, 273 (1976); (b) *Idem.*, **98**, in press.

The applicability of such a linear free energy relationship indicates that steric interactions caused by the leaving group are unimportant in acetolysis. **L** reflects only an electronic effect and shows a striking relationship with the vertical ionization potentials of RHgR′ determined by He(I) photoelectron spectroscopy. Incremental changes in **L** on proceeding from R′ = Me < Et < *i*-Pr < *t*-Bu become "saturated," in contrast to the additive relationship characteristic of Taft σ^* parameters for the same alkyl substituents. The origin of the saturation pattern for alkyl substituents is traced to a polarizability factor, examined by comparing the ionization potentials of various alkyl compounds RX with those of organometals (Rm).

The cleaved group parameter **C** varies in a nonsystematic order R = Me < Et > *i*-Pr > *t*-Bu. **C** includes contributions from electronic effects which are opposed to steric effects at the reaction site. The latter is negligible in electrophilic cleavages proceeding via a prior electron transfer in which the outer-sphere process is rate limiting. Under these circumstances electronic effects are dominant, and the relative rates of cleavage of Me, Et, *i*-Pr, and *t*-Bu groups from organometals follow the "saturation" pattern. These results together with the known stereochemical retention of configuration support the notion of a three-center (triangular) transition state for protonolysis in which sizeable charge is developed on both the cleaved group and the leaving group.

Acetolysis studies on dialkylmercury have demonstrated not only the importance of the cleaved group but also the leaving group in electrophilic substitution. Alkyl groups are excellent probes for measuring these electronic effects quantitatively, and the correlations of the rates with the ionization potentials show that a positive charge is developed on both the leaving group and the cleaved group. These facets of the reactivity of dialkylmercurials toward protonic electrophiles may be extended more generally, since it has been long recognized that nucleophilic reactivity is influenced by the polarizability of the nucleophile. Thus, the Edwards oxybase equation contains a term related to the oxidation potential of the nucleophile as well as a term related to its basicity. The molecular orbital analog of the Edwards equation has been developed by Klopman, in which electrostatic and covalent terms are the counterparts to basicity and polarizability, respectively.

Organometallic nucleophiles are σ-donors and have negligible basicity in the Edwards sense. Thus, we anticipate that the nucleophilic reactivity of organometals using either the Edwards or Klopman model, should reduce to an equation in which electron release by alkyl groups is the important consideration. The latter, in essence, represents a "virtual" ionization of the carbon–metal bond by the electrophile since it can be directly related to the energetics of electron detachment. The three-center (triangular) transition state is an adequate model at this juncture.

Furthermore, *hard* electrophiles such as Brönsted acids as well as *soft* electrophiles such as tetracyanoethylene,[69] hexachloroiridate(IV),[70] and peroxides[30b] evoke the same electron demand from the organometal in the absence of steric factors. This similarity underscores the strong caveat that the oft-cited correlations between activation parameters (e.g., log *k*) and ionization or oxidation potentials by themselves represent insufficient proof that a reaction proceeds by an electron transfer mechanism.

The mechanisms of metal-catalyzed coupling of alkyl groups have been examined with iron and nickel complexes.[70,71] The importance of electron transfer processes in the labilization of alkyl–nickel intermediates has been described.

Electron spin resonance spectra of paramagnetic olefin–metal complexes of aluminum, silver(0), and copper(0) in matrices as well as niobium(IV) and tantalum(IV) in solution have been reported recently.[72,73]

(69) (a) H. C. Gardner and J. K. Kochi, *J. Am. Chem. Soc.*, **97**, 5026 (1975); (b) *Idem.*, **98**, 2460 (1976); (c) J. Y. Chen, H. C. Gardner, and J. K. Kochi, *J. Am. Chem. Soc.*, **98**, in press.

(70) (a) S. M. Neumann and J. K. Kochi, *J. Org. Chem.*, **40**, 599 (1975); (b) R. S. Smith and J. K. Kochi, *J. Org. Chem.*, **41**, 502 (1976).

(71) D. G. Morrell and J. K. Kochi, *J. Am. Chem. Soc.*, **97**, 7262 (1975).

(72) I. H. Elson and J. K. Kochi, *J. Am. Chem. Soc.*, **97**, 1262 (1975).

(73) P. H. Kasai and D. McLeod, Jr., *J. Am. Chem. Soc.*, **97**, 5611, 6602 (1975).

Role of the Bridging Ligand in Inner-Sphere Electron-Transfer Reactions

Albert Haim

Department of Chemistry, State University of New York, Stony Brook, New York 11794

Received February 13, 1975

Reprinted from ACCOUNTS OF CHEMICAL RESEARCH, *8, 264 (1975)*

Some redox reactions between metal complexes in solution are known as *inner-sphere* reactions. They are characterized by the formation of a binuclear complex (an intermediate and/or a transition state) along the redox pathway between reactants and products. A central feature of the binuclear complexes is a ligand, known as the bridging ligand,[1] that forms part of the coordination spheres of both the oxidizing and the reducing metal ions.

A contrasting category is that of *outer-sphere* reactions in which an electron appears to be transferred directly from one complex to the other, without interpenetration of the coordination shells of the metal ions.

The direct demonstration of the inner-sphere mechanism is based, in major part, on ligand-transfer studies,[2–4] but the detection of binuclear intermediates in which the metals exhibit the initial[5–7] or the final oxidation states[8–11] has also provided important information. The elementary steps in the pathway connecting reactants to products are usually represented as in eq 1–3.[12,13]

$$M^{III}L_5X + N^{II}L'_6 \rightleftharpoons L_5M^{III}XN^{II}L'_5 + L' \quad (1)$$

$$L_5M^{III}XN^{II}L'_5 \rightleftharpoons L_5M^{II}XN^{III}L'_5 \quad (2)$$

$$S + L_5M^{II}XN^{III}L'_5 \rightleftharpoons M^{II}L_5S + N^{III}L'_5X \quad (3)$$

Equation 1 is a substitutional step and results in the formation of the precursor binuclear complex where the two metal ions are bridged by the ligand X. Activation of the precursor complex results in a configuration appropriate for electron transfer. Transfer takes place under Franck–Condon restrictions, and is followed by deactivation with formation of the successor[12] or postcursor[14] complex. Finally, the successor binuclear complex dissociates into mononuclear products, and the overall redox reaction is consummated.

When reaction 1 or 3 is rate determining, the overall process is substitution controlled. When the rate-determining step is reaction 2, it is electron transfer controlled. The faster vanadium(II) reductions of carboxylatoamminecobalt(III) complexes are examples of substitution-controlled redox reactions with eq 1 rate determining.[15–17] The chromium(II) reductions of some chlororuthenium(III) complexes feature eq 3 as the rate-determining step.[9,10]

In the present Account, we are concerned with the role of the bridging ligand in electron-transfer reactions, and therefore focus attention on those systems for which eq 2 controls the reaction rate.

Effect of Hydrogen Ion Concentration on Reaction Rates

Unless there is strong association between the reactants,[5,6,18] most one-electron redox reactions between metal complexes follow mixed second-order kinetics. External anions, e.g., those not included in the coordination spheres of the metal ions, often affect reaction rates, but such effects are not considered here, and the reader is referred to the most recent review on this subject.[19] The only external reagent affecting reaction rates that is of importance in the present discussion is the hydrogen ion.

Can Water Serve as a Bridging Ligand? When the dominant form of one of the reactants is present as a protonated form, parallel acid-independent and inverse acid paths are often encountered. Thus, the $Co(NH_3)_5OH_2^{3+}$–Cr^+ reaction in a sodium perchlorate medium obeys the rate law[20]

$$(k_0 + k_{-1}/[H^+])[Co(NH_3)_5OH_2^{3+}][Cr^{2+}] \quad (4)$$

which is interpreted as two parallel pathways with activated complexes of composition $[CrCo(NH_3)_5OH_2^{5+}]^‡$ and $[CrCo(NH_3)_5OH^{4+}]^‡$. However, from a reexamination[21] of this system using lithium perchlorate, it was concluded that the k_0 term is, most probably, the manifestation of a medium effect.

Albert Haim was born in France in 1931. He received the Industrial Chemistry degree in 1954 from the University of Uruguay, and the Ph.D. degree in 1960 from the University of Southern California where his research was directed by Wayne K. Wilmarth. After a postdoctoral year with Henry Taube at Stanford University, he joined The Pennsylvania State University and, in 1966, moved to the State University of New York at Stony Brook, where he is now Professor of Chemistry.

(1) H. Taube, H. Myers, and R. L. Rich, *J. Am. Chem. Soc.*, **75**, 4118 (1953).
(2) H. Taube, *Adv. Inorg. Chem. Radiochem.*, **1**, 1 (1959).
(3) J. P. Candlin, J. Halpern, and S. Nakamura, *J. Am. Chem. Soc.*, **85**, 2517 (1963).
(4) A. Haim and N. Sutin, *J. Am. Chem. Soc.*, **88**, 5343 (1966).
(5) R. D. Cannon and H. Gardiner, *J. Am. Chem. Soc.*, **92**, 3800 (1970).
(6) R. D. Cannon and A. Gardiner, *Inorg. Chem.*, **13**, 390 (1974).
(7) E. S. Gould, *J. Am. Chem. Soc.*, **94**, 4360 (1972).
(8) A. Haim and W. K. Wilmarth, *J. Am. Chem. Soc.*, **83**, 509 (1961).
(9) W. G. Movius and R. G. Linck, *J. Am. Chem. Soc.*, **91**, 5394 (1969).
(10) D. Seewald, N. Sutin, and K. O. Watkins, *J. Am. Chem. Soc.*, **91**, 7307 (1969).
(11) B. Grossman and A. Haim, *J. Am. Chem. Soc.*, **92**, 4835 (1970).
(12) N. Sutin, *Acc. Chem. Res.*, **1**, 225 (1968).
(13) It is assumed that N^{II} accepts substitution more readily than M^{III}, whereas the reverse situation obtains for N^{III} and M^{II}.
(14) R. G. Linck, *MPT Int. Rev. Sci., Inorg. Chem., Ser. One*, **9**, 303 (1971).
(15) H. J. Price and H. Taube, *Inorg. Chem.*, **7**, 1 (1968).
(16) C. Hwang and A. Haim, *Inorg Chem.*, **9**, 500 (1970).
(17) T. J. Przystas and A. Haim, *Inorg. Chem.*, **11**, 1016 (1972).
(18) D. Gaswick and A. Haim, *J. Am. Chem. Soc.*, **93**, 7347 (1971).
(19) J. E. Earley, *Prog. Inorg. Chem.*, **13**, 243 (1970).
(20) A. Zwickel and H. Taube, *J. Am. Chem. Soc.*, **81**, 1288 (1959).
(21) D. T. Toppen and R. G. Linck, *Inorg. Chem.*, **10**, 2635 (1971).

A similar conclusion had been previously reached in a study of the Fe^{3+}–Cr^{2+} reaction[22] in a lithium perchlorate medium.[23] Moreover, a reexamination[24] of the Cr^{2+}–Cr^{3+} exchange reaction confirmed the earlier[25] upper limit of $2 \times 10^{-5}\ M^{-1}\ sec^{-1}$ for the acid-independent term.

On the basis of this information, it was suggested[21] that water does not act as a bridging ligand in reductions by Cr^{2+}. In view of the extremely low basicity of coordinated water, it seems unlikely that it can serve as a bridging ligand in other redox reactions. Therefore, we suggest that, if water is the only potential bridging ligand, an acid-independent term in the rate law corresponds to an outer-sphere pathway. Support for this hypothesis is obtained from isotope fractionation studies and the effect of external anions on the $Co(NH_3)_5OH_2^{3+}$–V^{2+} reaction,[26] and from comparisons of the $Co(NH_3)_5OH_2^{3+}$–Eu^{2+} reaction with known outer-sphere reactions.[27]

What Are the Consequences of Protonation of the Bridging Ligand? Complexes with ligands containing basic sites can exhibit either rate acceleration or retardation with increasing hydrogen ion concentration. The rate retardation found in the chromium(II)[28] and vanadium(II)[17] reductions of *cis*-$Co(en)_2(HCO_2)_2^+$ and in the $Co(NH_3)_5OCOCH_3^{2+}$–Cr^{2+} reaction[29] is interpreted as an equilibrium effect, hydrogen ion and the reducing agent competing for the same site (presumably the carbonyl oxygen) in the bridging ligand. Rate acceleration is found in the

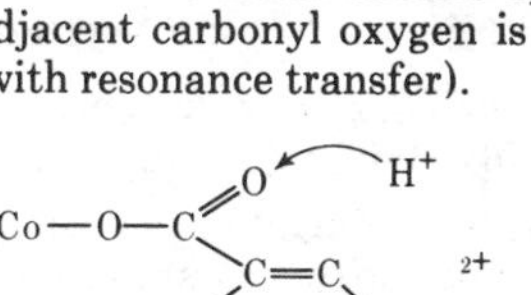

chromium(II) reductions of fumaratopentaamminecobalt(III)[30] and *p*-formylbenzoatopentaamminecobalt(III).[31] In the former case, improved conjugation between the two metal centers upon protonation of the adjacent carbonyl oxygen is invoked (remote attack with resonance transfer).

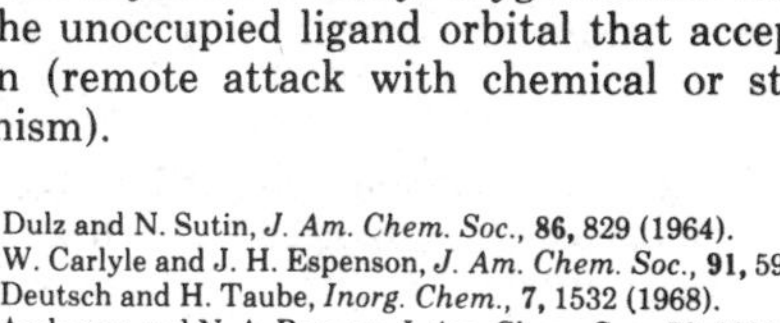

In the second case, chromium(II) is known to attack the formyl oxygen, and it is postulated that protonation of the adjacent carbonyl oxygen lowers the energy of the unoccupied ligand orbital that accepts the electron (remote attack with chemical or stepwise mechanism).

(22) G. Dulz and N. Sutin, *J. Am. Chem. Soc.*, **86**, 829 (1964).
(23) D. W. Carlyle and J. H. Espenson, *J. Am. Chem. Soc.*, **91**, 599 (1969).
(24) E. Deutsch and H. Taube, *Inorg. Chem.*, **7**, 1532 (1968).
(25) A. Anderson and N. A. Bonner, *J. Am. Chem. Soc.*, **76**, 3826 (1954).
(26) H. Diebler, P. Dodel, and H. Taube, *Inorg. Chem.*, **5**, 1688 (1966); P. Dodel and H. Taube, *Z. Phys. Chem. (Frankfurt am Main)*, **44**, 92 (1965).
(27) F. F. Fan and E. S. Gould, *Inorg. Chem.*, **13**, 2647 (1974).
(28) J. R. Ward and A. Haim, *J. Am. Chem. Soc.*, **92**, 475 (1970).
(29) M. B. Barrett, J. H. Swinehart, and H. Taube, *Inorg. Chem.*, **10**, 1983 (1971).
(30) H. Diaz and H. Taube, *Inorg. Chem.*, **9**, 1304 (1970), and references therein.
(31) A. Zanella and H. Taube, *J. Am. Chem. Soc.*, **94**, 6403 (1972).

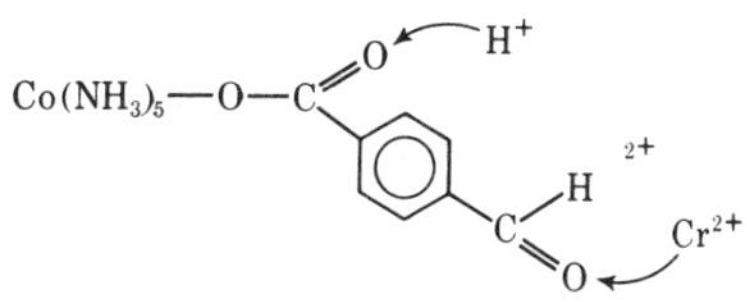

Finally, medium effects can manifest themselves as changing rates with changing hydrogen ion concentration, giving rise to apparent acid-dependent pathways; therefore, utmost care must be exercised in choosing ionic media to control ionic strength.[32,33]

Mechanistic Ambiguities for Inverse Acid Pathways. The chromium(II) reductions of aquopentaamminecobalt(III) and salicylatopentaamminecobalt(III) feature an inverse acid term in the rate laws. The mechanistic implication is that removal of a proton provides a favorable pathway for formation of the precursor binuclear complex. However, there

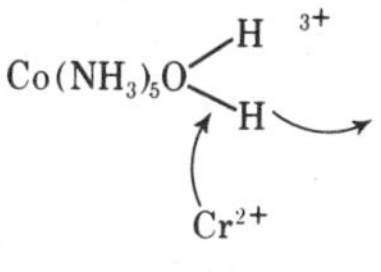

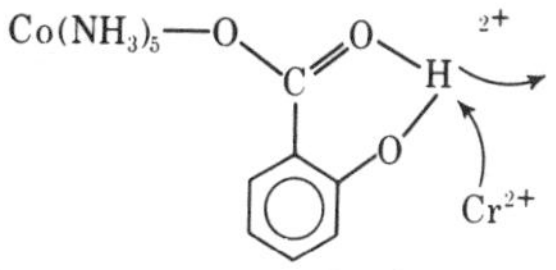

are mechanistic ambiguities associated with these pathways, and the two systems (which have been incorrectly discussed in the literature) will be used to illustrate the problems involved.[14,34]

Two mechanisms,[35] A and B, can be formulated for the $1/[H^+]$ pathway of the $Co(NM_3)_5OH_2^{3+}$–Cr^{2+} reaction. The value of k_{-1} (see eq 4) is interpreted as

Mechanism A

$$Co(NH_3)_5OH_2^{3+} + Cr^{2+} \rightleftharpoons (NH_3)_5CoOHCr^{4+} + H^+ \quad \text{rapid equilibrium, } Q_p \quad (5)$$

$$(NH_3)_5CoOHCr^{4+} \longrightarrow \text{products} \quad \text{slow, } k_{et} \quad (6)$$

Mechanism B

$$Co(NH_3)_5OH_2^{3+} \rightleftharpoons Co(NH_3)_5OH^{2+} + H^+ \quad \text{rapid equilibrium, } Q_a \quad (7)$$

$$Co(NH_3)_5OH^{2+} + Cr^{2+} \rightleftharpoons (NH_3)_5CoOHCr^{4+} \quad \text{rapid equilibrium, } Q'_p \quad (8)$$

$$(NH_3)_5CoOHCr^{4+} \longrightarrow \text{products} \quad \text{slow, } k_{et} \quad (9)$$

Q_pk_{et} in mechanism A and as $Q_aQ'_pk_{et}$ in mechanism B. The ambiguity arises because, in both mechanisms, the steps preceding the rate-determining electron transfer are rapid equilibria. Under these circumstances, the rate law specifies the compositon of the activated complex, but not the order of aggregation and/or dissociation of the species that produce it.[36-38] The two mechanisms are depicted diagrammatically in Figure 1. It has been argued[14] that, if dissociation of the proton does not obtain and mechanism A is operative, then dividing k_{-1} by Q_a leads

(32) T. W. Newton and B. B. Baker, *J. Phys. Chem.*, **67**, 1425 (1963).
(33) C. Lavallee and E. Deutsch, *Inorg. Chem.*, **11**, 3133 (1972).
(34) A. Liang and E. S. Gould, *J. Am. Chem. Soc.*, **92**, 6791 (1970).
(35) A third mechanism[14] involving preequilibrium formation of a water-bridged binuclear complex is neglected on the basis of our discussion (vide supra) of the unlikely role of water as a bridging ligand.
(36) E. L. King in "Catalysis", P. H. Emmett, Ed., Vol. II, Reinhold, New York, N.Y., 1955, p 337.
(37) T. W. Newton and F. B. Baker, *Adv Chem Ser.*, **No. 71**, 268 (1967).
(38) A. Haim, *Inorg. Chem.*, **5**, 2081 (1966).

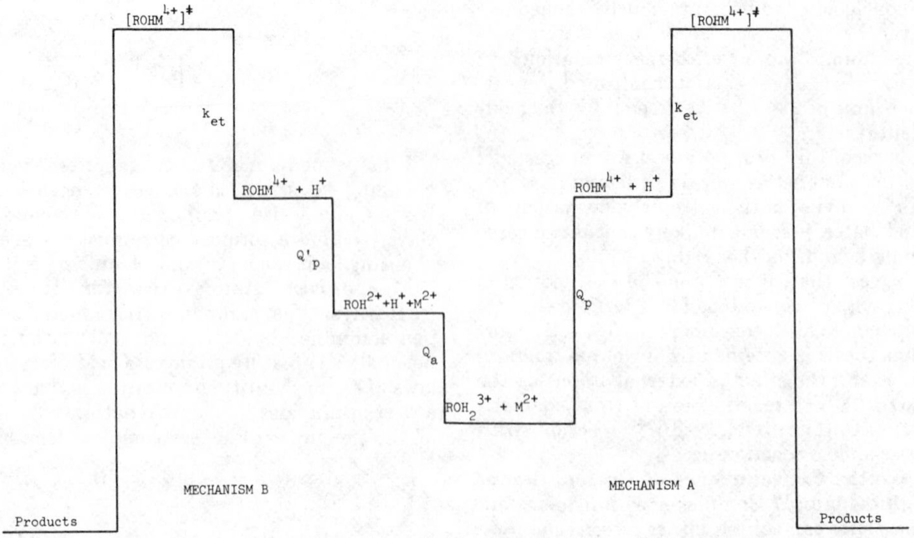

Figure 1. Free-energy profile for Mechanisms A and B for the reaction $Co(NH_3)_5OH_2^{3+} + Cr^{2+}$ (R = $Co(NH_3)_5$, M = Cr).

to a fictitious value of a second-order rate constant (Q_pk_{et}/Q_a) which would, in turn, lead to incorrect comparisons with genuine second-order rate constants (say for the $Co(NH_3)_5Cl^{2+}$–Cr^{2+} reaction). This argument is correct provided that proton dissociation according to eq 7 is precluded.

However, the proton-transfer reactions of aquo complexes are rapid and reversible and *all* the species involved in reactions prior to the electron transfer are in equilibrium. Consequently, whether the precursor complex $(NH_3)_5CoOHCr^{4+}$ is formed predominantly by reaction 5, by reaction 8, or by comparable contributions from the two reactions, it is perfectly meaningful to inquire (and make comparisons) about the free-energy difference between $Co(NH_3)_5OH^{2+}$ + Cr^{2+} and the transition state $[(NH_3)_5CoOHCr^{4+}]^{\ddagger}$.

In the case of the salicylatopentaamminecobalt(III)–chromium(II) reaction, the precursor complex P_B can be formed either by reaction between the deprotonated complex II and chromium(II) or by pro-

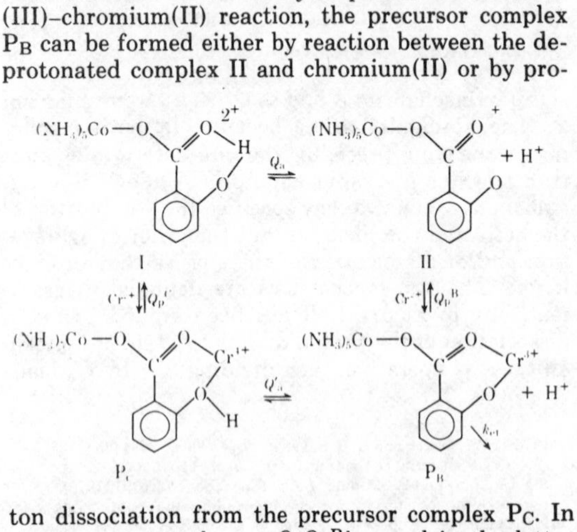

ton dissociation from the precursor complex P_C. In the former case, $k_{-1} = Q_aQ_p{}^Bk_{et}$, and in the latter case, $k_{-1} = Q_p{}^CQ'_ak_{et}$. Using $k_{-1} = 0.03$ sec^{-1} and pQ_a = 10.2,[34] then $Q_p{}^Bk_{et} = 4.80 \times 10^8$ M^{-1} sec^{-1}. Since this value is higher, by several orders of magnitude, than second-order rate constants for chromium(II) reductions of other carboxylato complexes and approaches the value for a diffusion-controlled reaction, it was concluded[34] that P_B was formed by proton loss from P_C rather than by association between II and Cr^{2+}. But since I, II, and P_C are connected by rapid equilibria, it is apparent that, in order to preclude a rapid equilibrium between II, Cr^{2+}, and P_B, a barrier for substitution into Cr^{2+} by II must exist. However, there appears to be no electronic or geometric feature in II to prevent rapid substitution into the coordination sphere of Cr^{2+}, especially in view of the postulated rapid substitution by I. Therefore, provided that the acid–base and chromium(II) substitution equilibria are rapid[39] compared to electron transfer, this system is completely analogous to the $Co(NH_3)_5OH_2^{3+}$–Cr^{2+} system.

Geometric Considerations. Adjacent and Remote Attacks, Doubly Bridging, Chelation

Early in the development of the field, there was interest in determining the sites used by the bridging ligand to bind the two metal ions. For example, in the reduction of cyanoamminecobalt(III) complexes by chromium(II),[40] the only available lone pair is on the nitrogen atom of CN^-, and consequently, bridging occurs as shown. On the other hand, for an oxidant

$$(NH_3)_5Co{-}C{\equiv}N{:}^{2+} \leftarrow Cr^{2+}$$

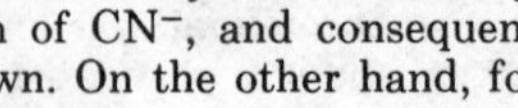

remote attack

such as fumaratopentaamminecobalt(III), the reductant can bind either the oxygen atom near the cobalt(III) center or one of the oxygen atoms of the carboxylate group far from the cobalt.[41] Attack by the

(39) It may be, however, that the equilibria involving I are not rapidly established. A. C. Dash and R. K. Nanda, *Inorg. Chem.*, 12 2024 (1973), report that the reaction of I with Al^{3+} to produce $(NH_3)_5CoOCOC_6H_4OAl^{4+}$ reaches completion in approximately 10 min at 28°.

(40) J. P. Birk and J. H. Espenson, *J. Am. Chem. Soc.*, **90,** 1153 (1968).

(41) Attack at the oxygen atom bound to the cobalt has been shown to be inoperative: K. L. Scott and A. G. Sykes, *J. Chem. Soc., Dalton Trans.*, 1832 (1972).

reducing agent at the lead-in atom of the bridging ligand or, in the case of carboxylate complexes, at the carbonyl oxygen nearest the oxidizing center is referred to as *adjacent*.[42] The term *remote* attack denotes binding by the reductant at any other site.

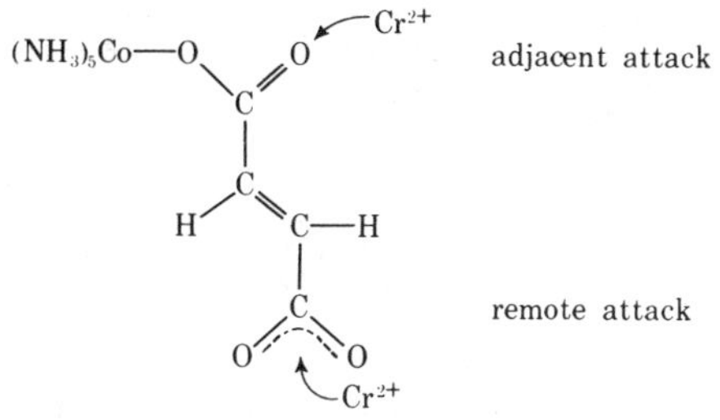

Evidently, monoatomic ligands can only display adjacent attack. For polyatomic ligands, remote attack provides a fairly common reaction pathway. Thus, chromium(II) reacts with isothiocyanatopentaamminecobalt(III)[43] and isonicotinamidepentaamminecobalt(III)[44] by remote attack (eq 10 and 11).

$$Co(NH_3)_5NCS^{2+} + Cr^{2+} \rightleftharpoons (NH_3)_5CoNCSCr^{4+} \quad (10)$$

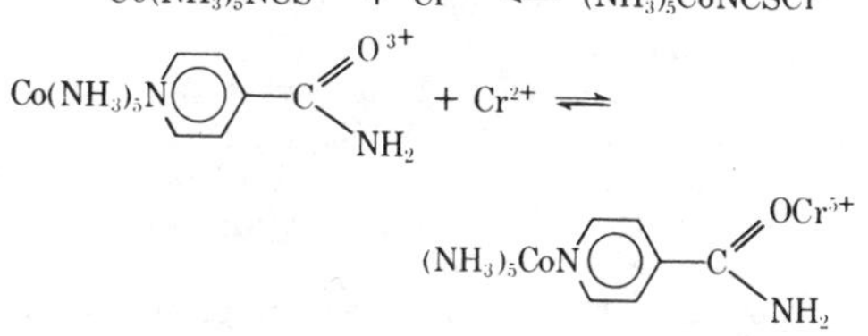

(11)

Only two firmly established examples of adjacent attack at the lead-in atom have been reported. The reduction of $Co(NH_3)_5SCN^{2+}$ by $Co(CN)_5{}^{3-}$ proceeds exclusively *via* adjacent attack (eq 12)[45] and reduc-

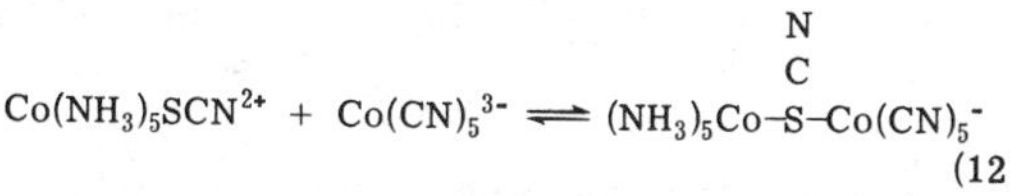

(12)

tion by Cr^{2+} features parallel adjacent and remote pathways (eq 13).[43]

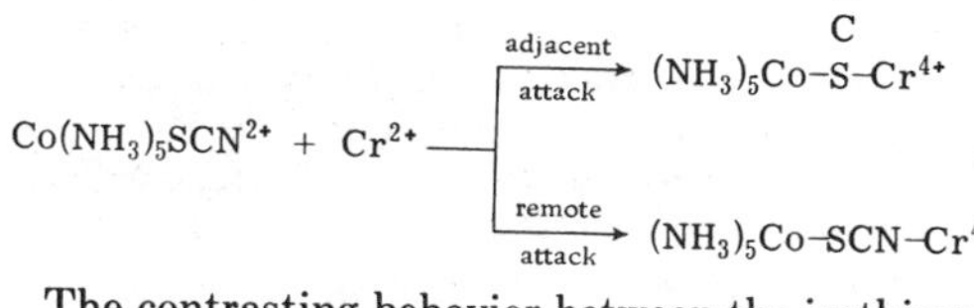

(13)

The contrasting behavior between the isothiocyanate (only remote attack) and thiocyanate (at least some adjacent attack) complexes can be rationalized on the basis of the electronic structures. The Co–N–C bond in $[Co(NH_3)_5NCS]Cl_2$ is linear.[46] Therefore, a pair of electrons is not available for reaction via adjacent attack, and reaction proceeds via remote attack at sulfur. For the thiocyanato complex, both the sul-

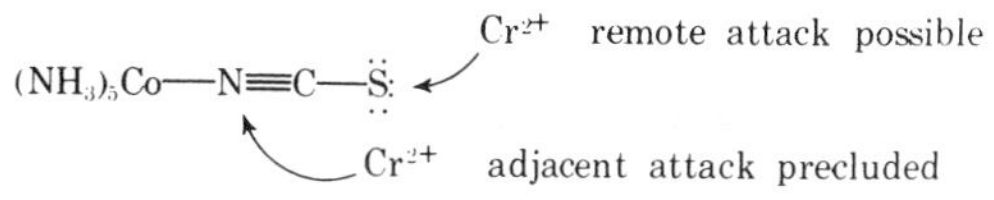

fur and nitrogen atoms have a lone pair available, and it is not unexpected that parallel adjacent and remote attacks by chromium(II) obtain. The aston-

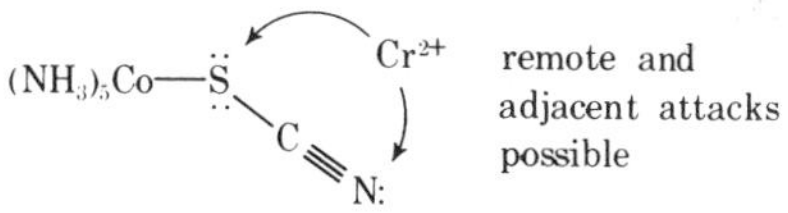

ishing feature, however, is the efficiency of the reaction for attack at sulfur (k = 8.0 × 10^4 M^{-1} sec^{-1} compared to 1.9 × 10^5 M^{-1} sec^{-1} for attack at nitrogen). On the basis of thermodynamic and steric factors,[43] a value of 10^3 for the ratio of remote to adjacent attacks by Cr^{2+} on $Co(NH_3)_5SCN^{2+}$ is a reasonable estimate. The observed ratio 2.4 is substantially smaller, and, therefore, since the rate of remote attack seems to be in line with rates of reactions of similar compounds (k for $Co(NH_3)_5N_3{}^{2+} + Cr^{2+}$ is 3 × 10^5 M^{-1} sec^{-1}), it appears that an unusually high reactivity is associated with adjacent attack by Cr^{2+} on $Co(NH_3)_5SCN^{2+}$. The high electron-mediating ability of sulfur bound to cobalt(III) has been observed in other Co(III)–Cr(II) systems,[47] but the factor or factors responsible for the high rates are not apparent.

In the reactions of *cis*-$Cr(N_3)_2{}^+$, *cis*-$Co(NH_3)_4(N_3)_2{}^8$, and *cis*-$Co(en)_2(HCO_2)_2{}^+$ with Cr^{2+},[28,48,49] two ligands are transferred from oxidant to reductant, and it is inferred that doubly bridged transition states are involved. In the reduction of

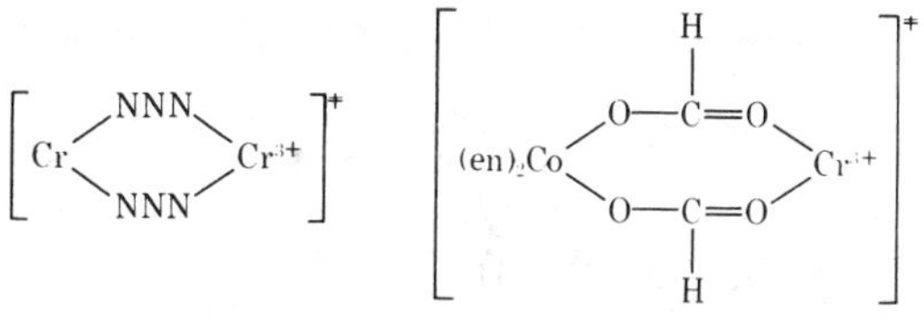

malonatopentaamminecobalt(III) by Cr^{2+},[33,50] the kinetically controlled product is the chelated malonatochromium(III) complex,[51] and therefore chelation obtains in the transition state. These effects are ac-

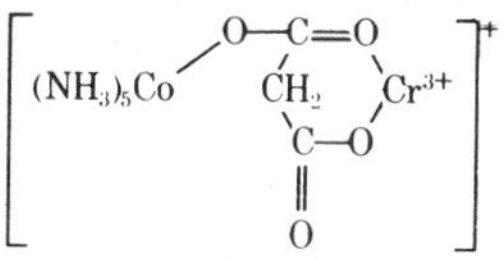

companied by considerable increases in rates,[28,50] presumably because of the increased stability of the precursor complexes. Additional geometric details about precursor complexes or transition states are lacking. Thus, it is not known whether the Co–Cl–Cr bond is linear or not in the transition state for the most famous inner-sphere reaction ($Co(NH_3)_5Cl^{2+}$ + Cr^{2+}), although both linear[52] and angular[53] halide

(42) The designation adjacent attack for reaction at the carbonyl oxygen near the oxidant is inconsistent with the designation used for ligands other than carboxylate. This admittedly arbitrary definition is widely used because of historical reasons.[2]
(43) C. Shea and A. Haim, *J. Am. Chem. Soc.*, **93**, 3055 (1971).
(44) F. Nordmeyer and H. Taube, *J. Am. Chem. Soc.*, **90**, 1162 (1968).
(45) C. Shea and A. Haim, *Inorg. Chem.*, **12**, 3013 (1973).
(46) M. R. Snow and R. F. Boomsma, *Acta Crystallogr., Sect. B*, **28**, 1908 (1972).
(47) R. H. Lane and L. E. Bennett, *J. Am. Chem. Soc.*, **92**, 1089 (1970).
(48) R. Snellgrove and E. L. King, *J. Am. Chem. Soc.*, **84**, 4609 (1962).
(49) A. Haim, *J. Am. Chem. Soc.*, **88**, 2324 (1966).
(50) G. Svatos and H. Taube, *J. Am. Chem. Soc.*, **83**, 4172 (1961).
(51) D. Huchital and H. Taube, *Inorg. Chem.*, **4**, 1660 (1965); M. V. Olson and C. E. Behnke, *ibid.*, **13**, 1329 (1974).

Table I
Rate Constants for Selected Reactions of Cr(II) with Co(III) and Cr(III) Complexes (25°, $\mu = 1.0\,M$)

Oxidant	k, M^{-1} sec^{-1}	$k_{Co(NH_3)_5L}/k_{Cr(H_2O)_5L}$
$(NH_3)_5CoNCS^{2+}$	19	1.4×10^5
$(NH_3)_5CoF^{2+}$	9×10^5	3.4×10^7
$(NH_3)_5CoOH^{2+}$	1.6×10^6	2.3×10^6
(NH₃)₅Co—N⟨⟩C(O)NH₂ ³⁺	17.4	10
(NH₃)₅Co—O—C(O)—C(H)=C(CO₂H)(H) ²⁺	1.32	0.4
(NH₃)₅Co—O—C(O)—C(H)=C(H)(CO₂H) ²⁺	2×10^2	50

bridges have been substantiated by X-ray diffraction studies of stable binuclear complexes.

Chemical Mechanism. Direct Participation of Bridging Ligand

Following the activation of the precursor binuclear complex, electron transfer takes place. Two extreme roles can be envisaged for the bridging ligand.[54] The electron (or hole) can be transferred *to* the ligand, and, in a subsequent step, from the ligand radical intermediate to the oxidizing center (or reducing center). Alternately, at no time is the electron (or hole) in a bound state of the ligand, and the bridge acts simply as a mediator. The mechanistic designations for the two cases are chemical, radical or stepwise mechanism and resonance or exchange mechanism, respectively.

Generation of Bound Radical by Pulse Radiolysis. Evidence for the direct participation of ligands comes from pulse radiolytic studies of *p*-nitrobenzoatopentaamminecobalt(III). Hydrated electrons react rapidly with the complex and a metastable intermediate, which decays by a first-order process to cobalt(II), is produced.[55] The intermediate is assumed to be the cobalt(III) complex of the radical ion derived by reduction of the ligand, and the first-order decay corresponds to intramolecular electron transfer from the radical to the cobalt(III) center.

$$Co(NH_3)_5O-\overset{O}{\underset{\parallel}{C}}-\langle\rangle-NO_2^{2+} \xrightarrow[8 \times 10^{10}\ M^{-1}\ sec^{-1}]{e^-_{aq}}$$

$$Co(NH_3)_5O-\overset{O}{\underset{\parallel}{C}}-\langle\rangle-NO_2^+ \xrightarrow[2.6 \times 10^3\ sec^{-1}]{H^+}$$

$$Co^{2+} + NH_4^+ + O_2C-\langle\rangle-NO_2^-$$

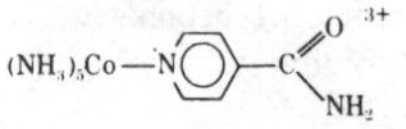

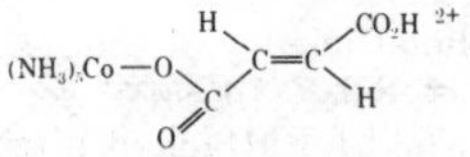

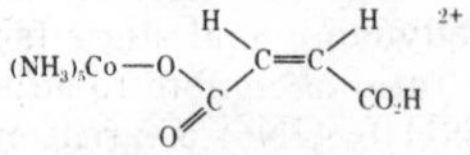

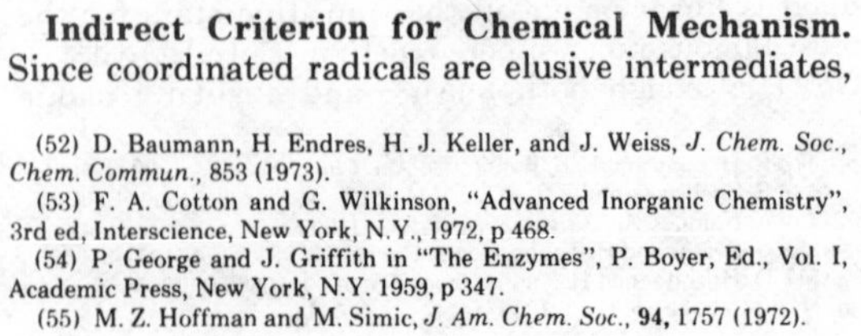

Indirect Criterion for Chemical Mechanism. Since coordinated radicals are elusive intermediates,

an indirect criterion has been developed to recognize the operation of a chemical mechanism.[42] This involves comparisons of reductions of cobalt(III) and chromium(III) complexes with the same reductant and via the same bridging ligand. In the chemical mechanism, electron transfer takes place *to* the ligand, and, insofar as the ligand orbitals are not much affected by coordination, the rate of electron transfer to the ligand will be rather insensitive to the nature of the oxidant. Conversely, for the resonance mechanism, important changes in the inner coordination shell of the oxidant are part of the activation process, and rates will depend on the identity of the oxidant.

Some results are presented in Table I. In selected cases (F⁻, OH⁻, NCS⁻), where thermodynamic arguments[56] rule out the chemical mechanism, rate ratios k_{CoA_5L}/k_{CrW_5L} are in the range 10^5–10^7. This is in accord with the expectation of strong discrimination in rate with respect to oxidizing center for the resonance mechanism. In contrast, the rate ratio for isonicotinamide as the bridging ligand is small (10), and it is inferred that the chemical mechanism obtains. Fumarate and maleate also give low k_{CoA_5L}/k_{CrW_5L} ratios (0.4 and 50), and it was concluded[30,57] that the radical mechanism is operative.

Although some reservations have been expressed in the use of rate ratios as a diagnostic of resonance vs. chemical mechanism,[58] and it is important to consider carefully the nature of the metal orbitals accepting the electron in making rate comparisons,[59] there is no question that the chemical mechanism provides an accessible and favorable pathway for electron transfer. In particular, when the electron must be transferred through long distances and there is mismatch in the symmetries of the donor, acceptor, and ligand orbitals, the distortions about the two metal centers needed to meet Franck–Condon restrictions may be uncoupled,[60] and the stepwise mechanism becomes preferred.

Resonance Mechanism. Thermodynamic and Kinetic Contributions to the Role of the Bridging Ligand

In the resonance exchange mechanism, the bridging ligand brings together the two metal centers and mediates the electron transfer. In trying to assess the role of the bridging ligand, one must decide what quantitative measurement will be used. Since we are concerned with reactivity, it seems appropriate to turn to rate constants. For simple inner-sphere reactions where the kinetics are mixed second order, the electron-transfer step is rate determining, and the resonance transfer mechanism obtains, the measured rate coefficient is the product of the equilibrium constant Q_p for precursor complex formation and the rate constant k_{et} for electron transfer within the binuclear unit. The role of the bridging ligand is there-

oxidant + reductant ⇌

        precursor complex   rapid, Q_p

precursor complex ⟶ successor complex   slow, k_{et}

(52) D. Baumann, H. Endres, H. J. Keller, and J. Weiss, *J. Chem. Soc., Chem. Commun.*, 853 (1973).
(53) F. A. Cotton and G. Wilkinson, "Advanced Inorganic Chemistry", 3rd ed, Interscience, New York, N.Y., 1972, p 468.
(54) P. George and J. Griffith in "The Enzymes", P. Boyer, Ed., Vol. I, Academic Press, New York, N.Y. 1959, p 347.
(55) M. Z. Hoffman and M. Simic, *J. Am. Chem. Soc.*, **94**, 1757 (1972).

(56) H. Taube and E. S. Gould, *Acc. Chem. Res.*, **2**, 321 (1969).
(57) M. V. Olson and H. Taube, *Inorg. Chem.*, **9**, 2072 (1970).
(58) R. Davies and R. B. Jordan, *Inorg. Chem.*, **10**, 2432 (1971).
(59) R. G. Gaunder and H. Taube, *Inorg. Chem.*, **9**, 2627 (1970).
(60) H. Taube, *Pure Appl. Chem.*, **24**, 289 (1970).

Table II
Rate Constants, Equilibrium Constants and Intrinsic Barriers for Reactions of Azido and Thiocyanato Complexes that Proceed via Remote Attack (25°)

Reaction	k_1, M^{-1} sec^{-1}	K	k_{intr}, M^{-1} sec^{-1}
$Co(NH_3)_5NCS^{2+} + Cr^{2+} \rightleftharpoons Co(NH_3)_5^{2+} + CrSCN^{2+}$	19	5.2×10^4	8.3×10^{-2}
$Co(NH_3)_5N_3^{2+} + Cr^{2+} \rightleftharpoons Co(NH_3)_5^{2+} + CrN_3^{2+}$	3×10^5	3.7×10^{11}	0.49
$Co(NH_3)_5SCN^{2+} + Cr^{2+} \rightleftharpoons Co(NH_3)_5^{2+} + CrNCS^{2+}$	1.9×10^5	4.7×10^{15}	2.8×10^{-3}
$Co(NH_3)_5NCS^{2+} + Fe^{2+} \rightleftharpoons Co(NH_3)_5^{2+} + FeSCN^{2+}$	$<3 \times 10^{-6}$	6.4×10^{-16}	$<1.2 \times 10^2$
$Co(NH_3)_5N_3^{2+} + Fe^{2+} \rightleftharpoons Co(NH_3)_5^{2+} + FeN_3^{2+}$	8.8×10^{-3}	3.6×10^{-8}	46
$Co(NH_3)_5SCN^{2+} + Fe^{2+} \rightleftharpoons Co(NH_3)_5^{2+} + FeNCS^{2+}$	0.12	5.8×10^{-5}	15.8
$Cr(OH_2)_5NCS^{2+} + Cr^{2+} \rightleftharpoons Cr^{2+} + Cr(OH_2)_5SCN^{2+}$	1.4×10^{-4}	3.3×10^{-6}	7.5×10^{-2}
$Cr(OH_2)_5N_3^{2+} + Cr^{2+} \rightleftharpoons Cr^{2+} + Cr(OH_2)_5N_3^{2+}$	6	1	6
$Cr(OH_2)_5SCN^{2+} + Cr^{2+} \rightleftharpoons Cr^{2+} + Cr(OH_2)_5NCS^{2+}$	40	3.0×10^5	7.5×10^{-2}

fore dual. It brings the metal ions together (thermodynamic contribution) and mediates the transfer of the electron (kinetic contribution). The thermodynamic effect can be understood on the basis of the usual considerations about stability constants of complexes. The kinetic effect manifests itself in the reorganization energy, the interaction energy, and the symmetry properties of donor, carrier, and acceptor orbital.[60,61]

A somewhat different approach is based on the linear-free-energy relationship between rate constants and equilibrium constants. The efficiency of bridging ligands is measured by the intrinsic barrier to electron transfer obtained by correcting observed rate constants for the overall free energy of reaction.[12]

A third approach makes use of stability constants and rate constants to obtain relative stabilities of transition states.[62] Effectively, this approach corrects the rate constants for differences in stabilities between the reactants.

Symmetrical and Unsymmetrical Bridging Ligands. Azide and Thiocyanate. Importance of Precursor Complex Stability. The reductions of $Co(NH_3)_5N_3^{2+}$ by Cr^{2+} [63] or Fe^{2+} [64] are considerably faster than the corresponding reductions of $Co(NH_3)_5NCS^{2+}$.[63,65] Similarly, the exchange of chromium between CrN_3^{2+} and Cr^{2+} is considerably faster than the corresponding exchange reaction of $CrNCS^{2+}$.[66,67] In these systems, both for reactants and for products, the nitrogen-bonded isothiocyanato complexes are the more stable linkage isomers. The early interpretation of the kinetic results (see Table II) was that the reactions proceed by inner-sphere remote attack and produce the thermodynamically unstable sulfur-bonded thiocyanato product and the stable nitrogen-bonded (by necessity) azido complex. The difference in the thermodynamic stability of the products was then invoked to account for the slow reactions of the isothiocyanato vs. the azido complexes.

This argument can be placed on a quantitative basis and shown[68] to be insufficient to account for the trends. Rate constants for reactions with different overall standard free-energy changes can be corrected to obtain the intrinsic barrier[12] $k_{intr} = k/K^{1/2}$, where k is the observed rate constant and K the equilibrium constant for the reaction. The values obtained, included in Table II, show that in every case the corrected azide-mediated reactions are faster than those mediated by thiocyanate. Since equal intrinsic barriers would have been predicted from the simple thermodynamic argument, additional factors must be operative, and it was suggested that the differences in the stabilities of the precursor complexes can account for the trends.[68] Since $k = Q_p k_{et}$ and $k_{intr} = k/K^{1/2}$, it can be shown[68] that $k_{intr}^{N_3}/k_{intr}^{SCN} = Q_p^{N_3}/(Q_p^{NCS}Q_p^{SCN})^{1/2}$ and $k_{intr}^{N_3}/k_{intr}^{NCS} = Q_p^{N_3}/(Q_p^{NCS}Q_p^{SCN})^{1/2}$, where Q_p^L represents the equilibrium constant for formation of the precursor complex bridged by L. Since Cr^{2+} is a hard acid, it displays a strong discrimination in favor of binding nitrogen over sulfur. Therefore, $Q_p^{N_3}/(Q_p^{NCS}Q_p^{SCN})^{1/2} > 1$, and the intrinsic rate constants for azide acting as a bridge are larger than the corresponding constants for thiocyanate or isothiocyante.

Reactions of Halogen Complexes. "Normal" and "Inverted" Orders. Relative Stabilities of Transition States. Because of historical reasons, the reactivity pattern $I^- > Br^- > Cl^- > F^-$ is referred to as "normal", whereas the opposite trend is called "inverted" or "reverse". For the reductions of $Co(NH_3)_5X^{2+}$, $Cr(NH_3)_5X^{2+}$, and $Cr(H_2O)_5X^{2+}$ complexes (X is a halogen) by Cr^{2+}, the normal order is observed.[62] For the reductions of $Co(NH_3)_5X^{2+}$ by Fe^{2+} and Eu^{2+} and of $Ru(NH_3)_5X^{2+}$ by Cr^{2+}, the reverse order obtains.[69] These trends are based on comparisons of second-order rate constants. A different approach for comparing a series of reactions uses relative stabilities of transition states.[62] Thus, a useful quantity in comparing the $CrCl^{2+}$–Cr^{2+} and CrF^{2+}–Cr^{2+} reactions is the "equilibrium quotient" $Q^{\ddagger}_{F,Cl}$ for the reaction

$$[CrFCr^{4+}]^{\ddagger} + Cl^- \overset{Q^{\ddagger}_{F,Cl}}{\rightleftharpoons} [CrClCr^{4+}]^{\ddagger} + F^- \quad (14)$$

The value of $Q^{\ddagger}_{F,Cl}$ can be computed as $k_{Cl}Q_{Cl}/$

(61) N. Sutin in "Inorganic Biochemistry", G. L. Eichhorn, Ed., Vol. 2, Elsevier, Amsterdam, 1973, p 611.
(62) A. Haim, *Inorg. Chem.*, **7**, 1475 (1968).
(63) J. P. Candlin, J. Halpern, and D. L. Trimm, *J. Am. Chem. Soc.*, **86**, 1019 (1964).
(64) A. Haim, *J. Am. Chem. Soc.*, **85**, 1016 (1963).
(65) J. Espenson, *Inorg. Chem.*, **4**, 121 (1965).
(66) D. L. Ball and E. L. King, *J. Am. Chem. Soc.*, **80**, 1091 (1958).
(67) R. Snellgrove and E. L. King, *Inorg. Chem.*, **3**, 288 (1964).
(68) D. P. Fay and N. Sutin, *Inorg. Chem.*, **9**, 1291 (1970). Note that in this article both k and k_{et} are taken to be proportional to $K^{1/2}$.
(69) H. Taube, "Electron Transfer Reactions of Complex Ions in Solution", Academic Press, New York, N.Y., 1970, p 51.

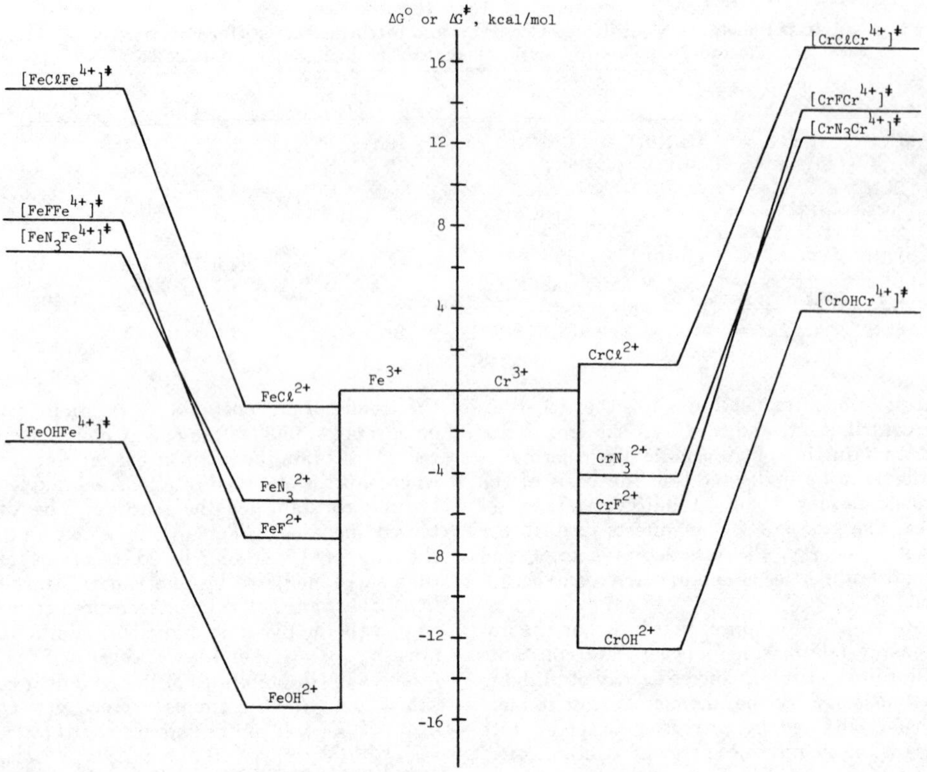

Figure 2. Relative stabilities of chromium and iron complexes and transition states.

$k_F Q_F$ from

$$CrF^{2+} + Cr^{2+} \overset{k_F}{\rightleftharpoons} [CrFCr^{4+}]^{\ddagger} \quad (15)$$

$$CrCl^{2+} + Cr^{2+} \overset{k_{Cl}}{\rightleftharpoons} [CrClCr^{4+}]^{\ddagger} \quad (16)$$

$$Cr^{3+} + Cl^- \overset{Q_{Cl}}{\rightleftharpoons} CrCl^{2+} \quad (17)$$

$$Cr^{3+} + F^- \overset{Q_F}{\rightleftharpoons} CrF^{2+} \quad (18)$$

Effectively, values of $Q^{\ddagger}_{X,Y}$ provide a correction of the rate constant ratio k_X/k_Y for the difference in free energies of the ground states, and thus are a measure of the relative affinities of transition states for the bridging ligands X and Y.[62,69] For example, the equilibrium quotients for

$$[CrFCr^{4+}]^{\ddagger} + Cl^- \rightleftharpoons [CrClCr^{4+}]^{\ddagger} + F^-$$

$$CrF^{2+} + Cl^- \rightleftharpoons CrCl^{2+} + F^-$$

are 6×10^{-3} and 5×10^{-6}, respectively. The smaller discrimination of the transition states indicates either a substantial degree of bond breaking in the activation process or a higher permeability to electron transfer of Cl^- compared to F^-. In the case of F^- and OH^- the opposing trend is observed, the transition states displaying a higher discrimination than the ground states.

$$[CrFCr^{4+}]^{\ddagger} + OH^- \rightleftharpoons [CrOHCr^{4+}]^{\ddagger} + F^-$$

$$Q_{F,OH} = 2 \times 10^7$$

$$CrF^{2+} + OH^- \rightleftharpoons CrOH^{2+} + F^-$$

$$Q_{F,OH} = 2 \times 10^5$$

This comparison suggests either a substantial degree of bond making in the transition state or a higher electron permeability of OH^- vs. F^-. Finally, a comparison between N_3^- and F^- reveals a reversal in affinities in going from ground states to transition states. The constants for

$$[CrFCr^{4+}]^{\ddagger} + N_3^- \rightleftharpoons [CrN_3Cr^{4+}]^{\ddagger} + F^-$$

$$CrF^{2+} + N_3^- \rightleftharpoons CrN_3^{2+} + F^-$$

are 10 and 0.04, respectively, and more efficient electron transfer by N_3^- than by F^- is indicated. The trends for the Fe^{2+}–FeX^{2+} exchange reactions are entirely analogous, and are presented, together with those for the Cr^{2+}–CrX^{2+} reactions, in Figure 2.

Orbital Symmetry Considerations

The relative electron-mediating ability of bridging ligands depends on the identity of the two metal centers involved in the redox reaction, and attempts have been made to correlate reactivity trends and electronic structure.[54,70,71] The importance of the symmetry of donor and acceptor orbitals in determining the changes in bond lengths attending activation was noted,[54] as was the importance of matching the symmetry of metal and ligand orbitals in facilitating electron transfer.[70] The pertinent data are not very extensive, but some generalizations have been made.[14] For electron transfer between e_g orbitals, Cl > N_3^- ≫ $CH_3CO_2^-$. For transfer from an e_g to a t_{2g}

(70) J. Halpern and L. E. Orgel, *Disc. Faraday Soc.*, **29**, 32 (1960).
(71) H. Taube, *Proc. Robert A. Welch Conf. Chem. Res.*, **5** (1962).

orbital, $N_3^- > Cl^- \sim CH_3CO_2^-$. For transfer from a t_{2g} to an e_g orbital, $N_3^- \gg Cl$.

A possible explanation of the observed trends is based on orbital symmetry arguments.[14,72] When the symmetries of the orbitals of the metal ions that donate and accept the electron are the same, bridging ligands with orbitals of matching symmetry may provide a lower energy pathway for electron transfer. Thus, if the reductant has the electron to be donated in an e_g-type orbital and the lowest acceptor orbital in the oxidant is also of e_g symmetry, then chloride, presumably being a σ carrier, is a "better" bridging ligand than azide or acetate. Conversely, when the electronic configurations of oxidant and reductant are such that the donor and/or acceptor orbital is of t_{2g} symmetry, the azide and acetate become effective bridging ligands, presumably because of favorable overlap of the t_{2g} orbital with the π system of azide or acetate.

The faster rate of the pentaammineisonicotinamideruthenium(III)-Cr^{2+} reaction as compared to the corresponding cobalt(III) reaction has also been rationalized on the basis of symmetry considerations.[72] In the former system, oxidant and bridging ligand have matching π symmetry (Ru(III) with a $t_{2g}{}^5e_g{}^0$ electronic configuration accepts the incoming electron in a π-type orbital and the lowest unoccupied molecular orbital of the ligand has π symmetry), and once the electron is given up by the reductant to the bridging ligand, the overall electron transfer reaction is consummated. On the other hand, for the Co(III)–Cr(II) system, where oxidant and reductant have acceptor and donor orbitals of σ symmetry, but the carrier orbital has π symmetry, the resonance mechanism may become prohibitive, perhaps because of the difficulty in meeting simultaneously Franck–Condon and symmetry restrictions. Under these circumstances, a stepwise or chemical mechanism for electron transfer provides the lower energy pathway.

What Ligands Can Act as Bridges?

A basic, unshared pair of electrons in the bridging ligand is a necessary condition for the formation of a precursor binuclear complex and, therefore, for an inner-sphere electron transfer. This is, apparently, also a sufficient condition for reaction via adjacent attack, perhaps because direct exchange is possible between metal ions bridged by a single atom. For example, halides, OH^-, and SCN^- bound to an oxidant have a basic lone pair and act as bridging ligands. Bound water and oxygen-bonded urea have a lone pair in the lead-in atom, but the basicity of the oxygen is very low; CN^- and NCS^- (assuming that rehybridization does not take place) have no lone pair on C or N, respectively. Therefore, oxidants bound to the last four ligands do not feature-inner sphere reductions *via* adjacent attack.

The availability of σ lone pair for binding the incoming metal ion is not a sufficient condition, however, for reaction via remote attack. Thus, in the Cr^{2+} reductions of the linkage isomers of formamidopentaamminecobalt(III),[73] the reaction of the O-bonded isomer proceeds by an outer-sphere mechanism, whereas the reaction of the conjugate base of the N-bonded isomer features inner-sphere, remote attack.

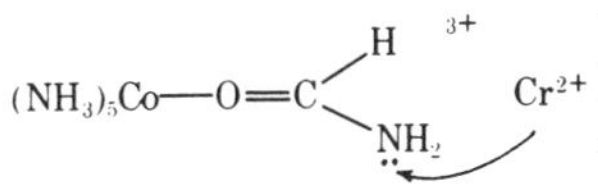

Remote attack precluded. N and O are not part of a conjugated system when Cr^{2+} binds at N

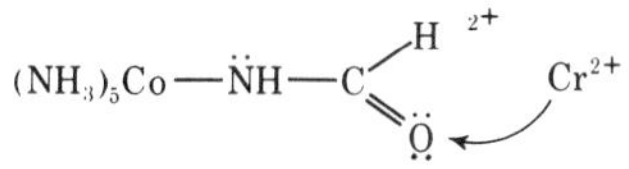

Remote attack possible. O and N are part of a conjugated system

The suggestion was made[73,74] that necessary conditions for inner-sphere electron transfer *via* remote attack are a donor atom in the bridging ligand that has a lone pair of electrons available for σ bonding to the incoming metal, and that the two donor atoms of the bridging ligand must be part of a conjugated system extending between the two metal centers.

It must be noted that, although the conditions outlined above may be necessary to ensure remote attack, they are not sufficient. Thus, $Co(NH_3)_5O_2CC_6H_4$-*p*-CO_2^+ and $Co(NH_3)_5O_2CC_6H_4$-*p*-CHO^{2+} feature a basic, remote oxygen in conjugation with the oxygen bound to the cobalt(III), but only in the reaction of the latter complex with Cr^{2+} is remote attack operative.[60] The difference in the two systems has been ascribed to the ease of reduction of the *p*-formylbenzoate ligand. Whether facile reduction of the bridging ligand is a prerequisite for electron transfer through extended conjugated systems has not been established, but it is noteworthy that all cases of bridged electron transfer through more than three atoms involve easily reduced ligands. However, since the comparisons are restricted to Co(III)–Cr(II) reactions, it may be that, because the symmetry of donor and acceptor orbitals (σ) does not match the symmetry of the carrier orbital (π), reaction can only take place via the chemical mechanism, and, consequently, an easily accessible lowest unoccupied molecular orbital of the ligand becomes necessary.

The Real Thing. Intramolecular Electron Transfer

It has been widely recognized that measurements of rates of intramolecular electron transfer within binuclear complexes could provide considerable insight into the details of the mechanism of electron transport between metal ions across ligands. Although a great deal of effort appears to have been expended in various laboratories to reach this goal, only in the last year has unequivocal evidence for intramolecular electron transfer been obtained. The difficulties can be traced to low equilibrium constants for the formation of precursor binuclear complexes from mononuclear reactants, to competitive outer-sphere pathways in systems where precursor complexes would have been expected to be sufficiently stable to produce deviations from mixed second-order kinetics,[75] or in general to the difficulty in finding appropriate pairs of metal ions with the appropriate coordination

(72) H. Taube, *Ber. Bunsenges. Phys. Chem.*, **76**, 964 (1972).

(73) R. J. Balahura and R. B. Jordan, *J. Am. Chem. Soc.*, **92**, 1533 (1970).

(74) R. J. Balahura and R. B. Jordan, *J. Am. Chem. Soc.*, **93**, 625 (1971).

(75) The deviations from second-order kinetics observed in labile systems[5,6,18] where there is appreciable association between the reactants have been interpreted on the basis of intramolecular electron transfer. However, the alternate mechanism whereby dissociation of the ion pair is followed by bimolecular electron transfer cannot be ruled out.

Table III
Rate Constants for Intramolecular Electron Transfer
in $(NH_3)_5Co^{III}LRu^{II}(NH_3)_4OH_2^{4+}$

Complex	k, sec$^{-1\ a}$
Co^{III}O$_2$C—⟨N⟩—N Ru^{II}	1×10^2 ^{b}
Co^{III}O$_2$CCH$_2$—⟨N⟩—N Ru^{II}	1.6×10^{-2}
Co^{III}O$_2$C ⟨N⟩ N Ru^{II}	1.6×10^{-3}
Co^{III}O$_2$CCH$_2$ ⟨N⟩ N Ru^{II}	5.5×10^{-3}

^{a} At 25°, 1 M toluenesulfonic acid. ^{b} In 0.1 M CF$_3$CO$_2$H; may have SO$_4^{2-}$ instead of H$_2$O in the coordination sphere of Ru(II).

spheres and oxidation potentials to yield binuclear complexes that undergo internal electron transfer competitively with dissociation.

In order to circumvent these difficulties, a series of binuclear Ru(III)–Co(III) complexes was synthesized and treated with a stoichiometric deficiency of a rapid one-electron reducing agent.[76] This produces a binuclear precursor complex in situ and, provided that electron transfer within the binuclear complex is slow compared to its rate of generation, the intramolecular electron transfer can be measured. This approach proved to be successful and to have a broad scope in the case of a series of Co(III)–Ru(II) binuclear complexes bridged by pyridinecarboxylate ligands.[76] The results, summarized in Table III, show the importance of conjugation (decrease by a factor of 10^5 in going from 4-pyridinecarboxylate to 3-pyridinecarboxylate), the insulating effect of a CH$_2$ group (decrease by a factor of 10^4 in going from 4-pyridinecarboxylate to 4-pyridineacetate), and even the possibility of the actual electron-transfer path by-passing the bridging ligand and proceeding by an outer-sphere mechanism (3-pyridineacetate being somewhat faster than 3-pyridinecarboxylate, and subject to rate acceleration with increasing [H$^+$]). Very recently, another successful approach to the measurement of intramolecular electron-transfer rates made use of the high affinity[77] of the Fe(CN)$_5^{3-}$ moiety for nitrogen heterocycles and of the newly[78] devised synthesis of

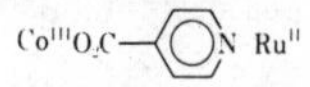

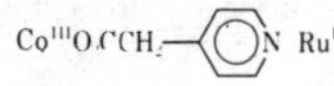

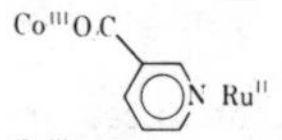

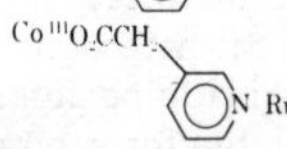

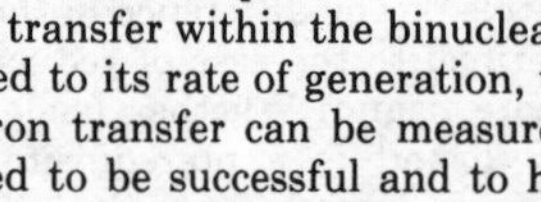

$$Co(NH_3)_5N\text{⟨}⟩\text{⟨}⟩N^{3+}$$

The rate constants for the formation and dissociation of the binuclear precursor and for intramolecular electron transfer have been obtained.[79] This ap-

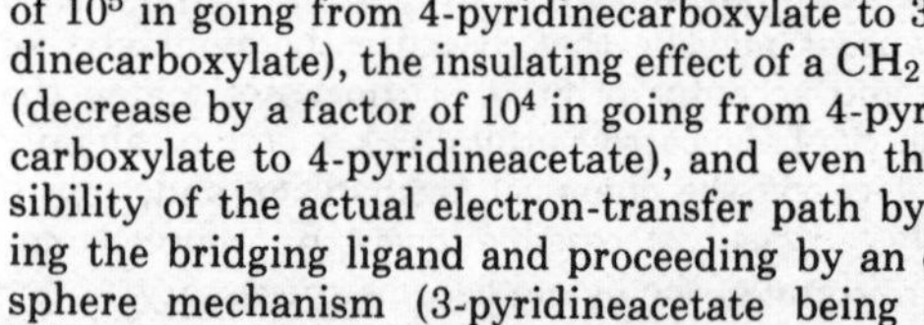

$$Co(NH_3)_5(4,4'\text{-bpy})^{3+} + Fe(CN)_5OH_2^{3-} \underset{k_d}{\overset{k_f}{\rightleftharpoons}}$$

$$(NH_3)_5Co^{III}N\text{⟨}⟩\text{⟨}⟩NFe^{II}(CN)_5 \overset{k_{et}}{\longrightarrow}$$

$$(NH_3)_5Co^{II}N\text{⟨}⟩\text{⟨}⟩NFe^{III}(CN)_5$$

$$k_f = 5.5 \times 10^3 \text{ sec}^{-1}; \quad k_d = 4.5 \times 10^{-3} \text{ sec}^{-1};$$
$$k_{et} = 2.6 \times 10^{-3} \text{ sec}^{-1}$$

proach is being extended to other cobalt(III) complexes, and the importance of electronic and steric considerations is already apparent.[80] For the pentaamminepyridinecarboxylatocobalt(III) + Fe(CN)$_5$OH$_2^{3-}$ systems, with N in the 4 position, $k_{et} = 1.7 \times 10^{-4}$ sec^{-1}; with N in the 3 position, $k_{et} < 3 \times 10^{-5}$ sec^{-1}; and with N in the 2 position, the binuclear complex is not formed.

The author is grateful to the National Science Foundation for continuing support of his researches into the mechanism of electron-transfer reactions. The dedication, enthusiasm, and productivity of students and postdoctoral associates have made that continuing support possible.

Addendum (May 1976)

There is a great deal of activity in studies of intramolecular electron-transfer within binuclear complexes. Thermal studies center on Co^{III}-Fe^{II} [81,82] and Co^{III}-Ru^{II} [83] complexes bridged by derivatives of 4,4'-bipyridine. Photochemical studies focus on Co^{III}-Cu^I [84] and Ru^{II}-Cu^{II} [85] complexes bridged by alkenoic acids and pyrazine, respectively.

The work in progress cited in reference 80 has been published,[81] and new measurements of intramolecular electron transfer rates have been carried out for **1**, where N- - - -N is bifunctional ligand **2**, **3**, or **4**.

$$(NH_3)_5Co^{III}N\text{- - -}NFe^{II}(CN)_5$$
1

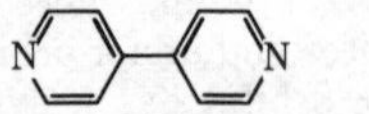

N⟨⟩—⟨⟩N
2

N⟨⟩—CH=CH—⟨⟩N
3

N⟨⟩—CH$_2$CH$_2$—⟨⟩N
4

The rate constants for electron transfer are compared with the shift in the metal to ligand charge transfer band of the pentacyanoferrate(II)-heterocycle caused by coordination of the remote nitrogen atom to the Co(NH$_3$)$_5^{3+}$ moiety. Kinetic and spectroscopic information indicate the extent of "communication" between the two metal centers.[82]

Similarly, the rates of electron transfer in **5**, where N- - - - -N is **2**, **3**, **4**, or **6**, are correlated with the ex-

$$(NH_3)_5Co^{III}N\text{- - -}NRu^{II}(NH_3)_5^{5+}$$
5

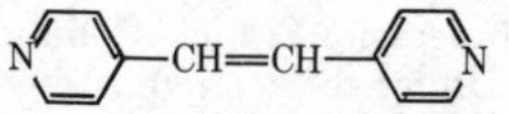

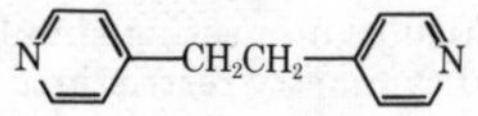

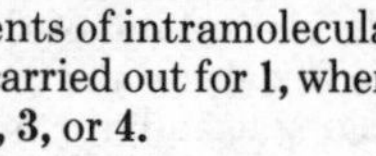

N⟨⟩—S—⟨⟩N
6

tinction coefficients of the corresponding mixed valence species $(NH_3)_5Ru^{III}N\text{- - - -}NRu^{II}(NH_3)_5^{5+}$. The in-

(76) S. S. Isied and H. Taube, *J. Am. Chem. Soc.*, **95**, 8198 (1973).

(77) H. E. Toma and J. M. Malin, *Inorg. Chem.*, **12**, 1039 (1973); *ibid.*, **12**, 1080 (1973).

(78) A. Miralles, Ph.D. Thesis, State University of New York, Stony Brook, N.Y., July 1974.

(79) D. Gaswick and A. Haim, *J. Am. Chem. Soc.*, **96**, 7845 (1974).

(80) J. J. Jwo and A. Haim, work in progress.

(81) J. J. Jwo and A. Haim, *J. Am. Chem. Soc.*, **98**, 1172 (1976).

(82) J. J. Jwo and A. Haim, Abstracts, Centennial Meeting of the American Chemical Society, New York, April 1976, Paper INORG 15.

tensities of the intervalence transitions are a measure of the electron coupling between the exchanging centers.[83]

The photochemical excitation of the $Cu^I(d) \rightarrow L(\pi^*)$ charge transfer band in 7, where L = $^-O_2CCH=CH_2$, $^-O_2CCH=CHCO_2H$, $^-O_2CH=CHC_6H_5$, $^-O_2CH_2$-$CH_2CH=CH_2$ results in the efficient production of Co^{II} and Cu^{II} [4]. The results are rationalized in terms of electron transfer mediated by π-delocalized orbitals for conjugated ligands or by direct overlap of olefin π-antibonding orbitals with metal orbitals for non-conjugated ligands.[84]

There is no net photodecomposition of 8, but flash photolysis results in the transient bleaching of the metal to ligand charge transfer band of 8. It is believed[85] that a photostimulated electron-transfer produces 9 which

$$(NH_3)_5Co^{III}LCu^I(OH_2)_5{}^{3+}$$
7

$$(NH_3)_5Ru^{II}N\bigcirc NCu^{II}(OH_2)_5{}^{4+}$$
8

$$(NH_3)_5Ru^{III}N\bigcirc NCu^I(OH_2)_5$$
9

undergoes internal electron transfer with regeneration of 8.

(83) H. Taube, Abstracts, Centennial Meeting of the American Chemical Society, New York, April 1976, Paper INORG 90.

(84) J. K. Farr, L. G. Hulett, R. H. Lane, and J. K. Hurst, *J. Am. Chem. Soc.*, 97, 2654 (1975).

(85) V. A. Durante and P. C. Ford, *J. Am. Chem. Soc.*, 97, 6898 (1975).

The Olefin Metathesis Reaction

Nissim Calderon

Research Division,† The Goodyear Tire and Rubber Company, Akron, Ohio 44316

Received March 24, 1971

Reprinted from Accounts of Chemical Research, 5, 127 (1972)

Olefin metathesis is a catalytically induced reaction wherein olefins undergo bond reorganization, resulting in a redistribution of alkylidene moieties (eq 1).[1]

All metathesis catalysts are derived from transition metal compounds. It is convenient to classify these catalysts into two main groups: (a) heterogeneous catalysts—transition metal oxides or carbonyls deposited on high-surface-area supports,[3] (b) homoge-

After completing his undergraduate education and earning the M.Sc. degree at the Hebrew University of Jerusalem, Israel, Nissim Calderon came to the United States and enrolled in the then newly created Ph.D. program in Polymer Science at the University of Akron. His research in Jerusalem and Akron involved synthesis of organometallics and their application as polymerization catalysts. After earning his Ph.D. in 1932, he joined the Synthetic Rubbers department of the Goodyear Tire & Rubber Comyany, where his main activity has been research in the area of transition metal-olefin chemistry. In 1967 he was appointed Section Head in the Basic Polymer Research Department in charge of new elastomers.

† Contribution No. 469.

(1) "Olefin disproportionation" is the name first selected by the authors of ref 2 to describe the *overall* process of the [metathesis + isomerization] of acyclic olefins using heterogeneous metal oxide catalysts. At the outset of the discovery of homogeneous catalysts capable of inducing a "clean" metathesis reaction, it became evident that the term "olefin disproportionation" was inadequate to properly describe the nature and scope of the reaction, and was even misleading in certain cases. Since the basic process on hand is an alkylidene interchange, it was decided to adopt the name "olefin metathesis," as this name properly conveys the nature and scope of the reaction. In its specific applications (as shown throughout the present Account) it can be utilized to: (a) disproportionate olefins; (b) polymerize cycloolefins; (c) prepare catenanes and other macrocyclics; (d) synthesize dienes and trienes. Hence, olefin disproportionation is to be considered a special case of the more general olefin metathesis reaction.

(2) R. L. Banks and G. C. Bailey, *Ind. Eng. Chem., Prod. Res. Develop.*, 3, 170 (1964).

(3) For a detailed review see G. C. Bailey, *Catal. Rev.*, 3, 37 (1969).

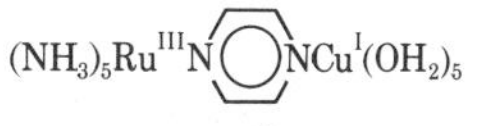

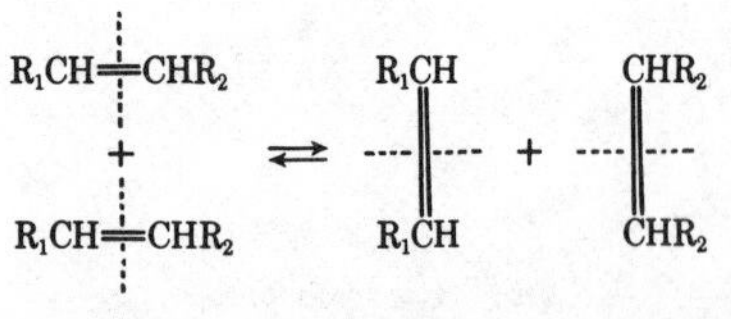

$$R_1CH \overset{=}{=} CHR_2 \qquad R_1CH \qquad CHR_2$$
$$+ \qquad \rightleftharpoons \qquad \cdots | \cdots + \cdots | \cdots \qquad (1)$$
$$R_1CH \overset{=}{=} CHR_2 \qquad R_1CH \qquad CHR_2$$

neous catalysts—transition metal salts or coordination compounds in combinations with selected organometallic derivatives or Lewis acids.[4,5]

The first to report processes involving olefin metathesis were Eleuterio, who described the ring-opening polymerization of cycloolefins[6] by a MoO_3-Al_2O_3 catalyst, and Banks and Bailey,[2] who employed various heterogeneous catalysts for the disproportionation of olefins at high temperatures. In 1963 and subsequently, Natta, et al.,[7-9] reported the use of homogeneous catalysts derived from $MoCl_5$ or WCl_6 and organoaluminum derivatives for the ring-opening polymerization of cycloolefins.

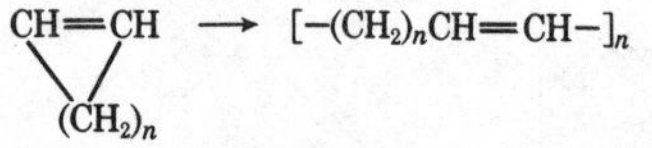

$$\underset{(CH_2)_n}{CH=CH} \longrightarrow [-(CH_2)_nCH=CH-]_n \qquad (2)$$

These early literature citations do not indicate that the respective workers recognized the fact that the basic chemistry involved in the cycloolefin polymerization by ring opening, and in the seemingly unrelated olefin disproportionation reaction, is actually the same.

In a U. S. patent application filed in 1966, Calderon and Chen[10] reported that W/Al homogeneous catalyst combinations are highly effective in promoting the olefin metathesis reaction. The products from metathesis of a mixture of 2-butene and 2-butene-d_8 showed that the redistribution process proceeds *via* an interchange of alkylidene moieties as depicted in eq 1. A detailed discussion of this aspect of the reaction was later presented in formal publications.[4,5] Shortly thereafter, Bradshaw and coworkers,[11] who studied the metathesis reaction over heterogeneous catalyst systems, concluded that their results supported a "quasicyclobutane" intermediate (eq 3). Critical experi-

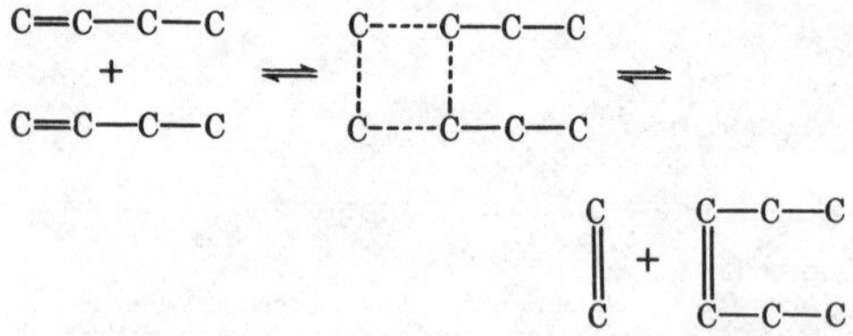

$$\underset{C=C-C-C}{\overset{C=C-C-C}{}} + \rightleftharpoons \quad \boxed{\begin{matrix} C\cdots C-C-C \\ \vdots \quad \vdots \\ C\cdots C-C-C \end{matrix}} \rightleftharpoons$$

$$\underset{C}{\overset{C}{\|}} + \underset{C-C-C}{\overset{C-C-C}{\diagdown\diagup}} \qquad (3)$$

ments to substantiate the quasi-cyclobutane structure were not offered.

The present Account is intended to describe: (a) the salient features of the metathesis reaction by homogeneous catalysts; (b) suggested mechanistic schemes; and (c) useful applications.

Salient Features

Redistribution by Transalkylidenation. Mass spectral studies[5] show that olefin metathesis does not proceed by transalkylation as depicted in eq 4. Rather

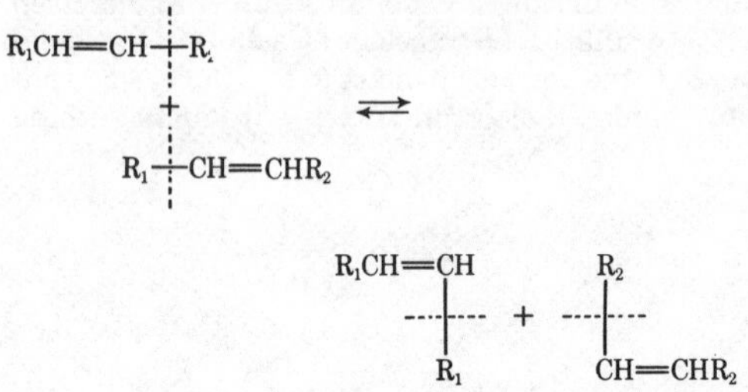

$$R_1CH=CH \!\!-\!\! R_2$$
$$+ \qquad \rightleftharpoons$$
$$R_1 \!\!-\!\! CH=CHR_2$$

$$\underset{R_1}{\overset{R_1CH=CH}{\cdots|\cdots}} + \underset{CH=CHR_2}{\overset{R_2}{\cdots|\cdots}} \qquad (4)$$

the data are consistent with transalkylidenation according to eq 1. A further confirmation of the transalkylidenation mechanism was provided by Mol, et al.,[12] who carried out metathesis experiments involving [2-^{14}C]propylene over supported metal oxide catalysts. Analysis of the metathesis products obtained from [1-^{14}C]propylene and [3-^{14}C]propylene excluded formation of any π-allyl intermediates during the redistribution process.[13]

Random Distribution of Alkylidene Groups. Olefin metathesis is a process wherein bond energy values do not differ substantially between the various equilibrating components in the system; hence, entropy considerations should dominate this essentially thermoneutral reaction[14] and thus afford a statistical distribution of products when carried to equilibrium. A series of metathesis experiments on various mixtures of 2-pentene and 6-dodecene, in which the relative concentrations of the [$CH_3CH=$], [$C_2H_5CH=$], and [$C_5H_{11}CH=$] were varied, demonstrated that the concentrations of the anticipated reaction components are in excellent agreement with those predicted for a random scrambling of constituents (see Figure 1).

Macrocyclization of Cycloolefins. Metathesis of cycloolefins should result in the formation of high molecular weight polymeric rings; in other words, the ring-opening polymerization of cycloolefins by tungsten-based catalysts can be considered a special case of the olefin metathesis reaction.[15] Earlier proposals that cycloolefin polymerizations, catalyzed by tungsten, proceed by cleavage of carbon–carbon single bonds α to the double bond[8] would not appear to be valid.

(4) N. Calderon, H. Y. Chen, and K. W. Scott, *Tetrahedron Lett.*, 3327 (1967).

(5) N. Calderon, E. A. Ofstead, J. P. Ward, W. A. Judy, and K. W. Scott, *J. Amer. Chem. Soc.*, **90**, 4133 (1968).

(6) H. S. Eleuterio, U. S. Patent 3,074,918 (1963).

(7) G. Natta, G. Dall'Asta, G. Mazzanti, and G. Mortoni, *Makromol. Chem.*, **69**, 163 (1963).

(8) G. Natta, G. Dall'Asta, and G. Mazzanti, *Angew. Chem.*, **76**, 765 (1964).

(9) G. Natta, G. Dall'Asta, I. W. Bassi, and G. Carella, *Makromol. Chem.*, **91**, 87 (1966).

(10) N. Calderon and H. Y. Chen, U. S. Patent 3,535,401 (1970).

(11) C. P. C. Bradshaw, E. J. Howman, and L. Turner, *J. Catal.*, **7**, 269 (1967).

(12) J. C. Mol, J. A. Moulijn, and C. Boelhouwer, *Chem. Commun.*, 633 (1968).

(13) A. Clark and C. Cook, *J. Catal.*, **15**, 420 (1969).

(14) G. Calingaert and H. A. Beatty, *J. Amer. Chem. Soc.*, **61**, 2748 (1939).

(15) K. W. Scott, N. Calderon, E. A. Ofstead, W. A. Judy, and J. P. Ward, Abstracts, 155th National Meeting of the American Chemical Society, San Francisco, Calif., April 1968; see also *Advan. Chem. Ser.*, **No. 91**, 399 (1969).

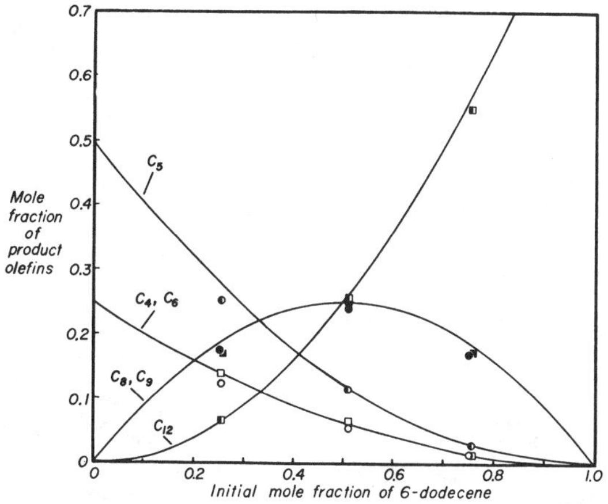

Figure 1. Metathesis of 2-pentene with 6-dodecene. solid lines represent theory for ideal random composition: (○) C_4, 2-butene; (◐) C_5, 2-pentene; (□) C_6, 3-hexene; (●) C_8, 2-octene; (■) C_9, 3-nonene; (◧) C_{12}, 6-dodecene.

The following important implications, which bear directly on the nature of cycloolefin polymerization, follow from the features discussed in the preceding paragraphs. (a) Ring-opening polymerization should possess the basic features of equilibrium polymerization. (b) Macrocyclic species should be present in the polymerization mixture at equilibrium, according to eq 5. (c) In the absence of side reactions the applica-

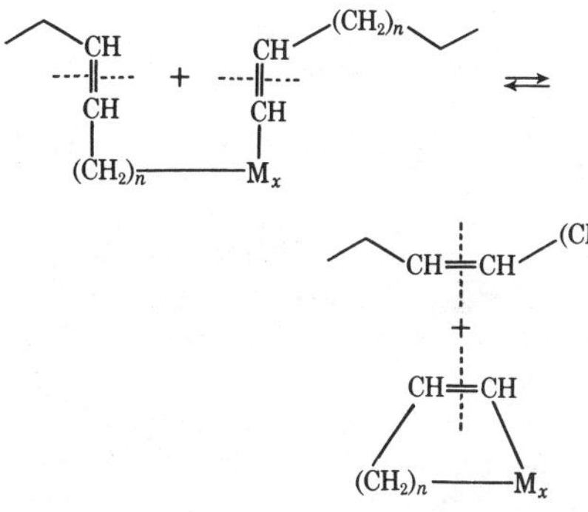

(5)

M_x represents x repeated units

tion of the olefin metathesis reaction to cycloolefins should yield only macrocyclic species. (d) Acyclic vinylenic compounds should lead to ring scission, resulting in open-chain polymer molecules. (e) Ring–chain equilibrium occurs when a, b, and d are operative.

Cis–Trans Equilibria. The intimate relationship between the olefin metathesis reaction and cis–trans interconversions is apparent from the early metathesis work involving both homogeneous and heterogeneous catalysts. Metathesis experiments on pure *cis-* and *trans-*2-pentene with the catalyst systems $C_2H_5AlCl_2$–WCl_6–C_2H_5OH and $C_2H_5AlCl_2$–$py_2Mo(NO)_2Cl_2$ demonstrated[5,16] that, at equilibrium, a thermodynamically favored cis/trans composition for the respective 2-butenes, 2-pentenes, and 3-hexenes is obtained.

(16) W. B. Hughes, *Chem. Commun.*, 431 (1969).

Mechanistic Schemes

A scheme proposed for the olefin metathesis reaction consists of three main processes.[5] The first is bis-olefin–metal complex formation as depicted in eq 6.

$$WCl_6 + C_2H_5OH + C_2H_5AlCl_2 + 2RCH{=}CHR' \longrightarrow$$

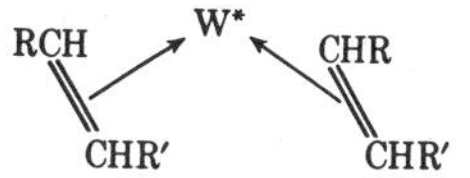

(6)

Experimentally, WCl_6 and C_2H_5OH are combined prior to addition of the $C_2H_5AlCl_2$. During this step, which presumably affords $WCl_5OC_2H_5$, a distinct color change from dark blue to red burgundy is observed. It is accompanied by the evolution of a stoichiometric amount of HCl.[17] The moderately stable $WCl_5OC_2H_5$ slowly decomposes into C_2H_5Cl[18] and an orange crystalline precipitate, presumably $WOCl_4$.

When the organoaluminum compound comes into contact with the tungsten component, an active catalyst is formed instantaneously. The activity of this catalyst prepared in the absence of olefin decays at a moderate rate.[19] By analogy to classical Ziegler–Natta catalysts,[20] one postulates the following sequence of reactions.

$$WCl_6 + C_2H_5AlCl_2 \longrightarrow C_2H_5WCl_5 + AlCl_3 \quad (7)$$

$$C_2H_5WCl_5 \longrightarrow WCl_5 + C_2H_5\cdot \quad (8)$$

$$2C_2H_5\cdot \longrightarrow C_2H_6 + C_2H_4 \quad (9)$$

$$2C_2H_5\cdot \longrightarrow C_4H_{10} \quad (10)$$

The formation of ethane and ethylene (in addition to ethyl chloride) during the reaction of $WCl_5OC_2H_5$ and $C_2H_5AlCl_2$ has been confirmed experimentally.[21] By a sequence similar to eq 7–10, WCl_5 may undergo further reduction to WCl_4. Moreover, pentavalent tungsten may undergo disproportionation, as depicted in eq 11. The presence of $AlCl_3$, formed during the re-

$$2WCl_5 \rightleftharpoons WCl_6 + WCl_4 \quad (11)$$

duction of W(VI), may give rise to associations *via* μ-chloride bonding (eq 12) and acid–base type equilibria (eq 13).

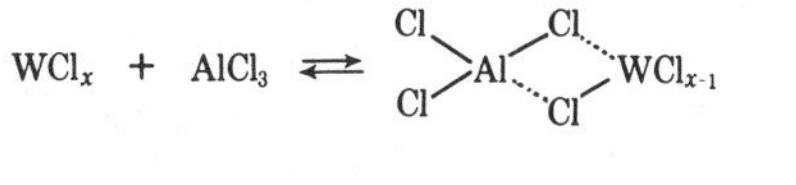

(12)

$$WCl_x + AlCl_3 \rightleftharpoons WCl_{x-1}^{\ +} + AlCl_4^{\ -} \quad (13)$$

It is suspected that $AlCl_3$, whenever present, does play an active role in the metathesis catalyst. Indirect evidence points to this contention. (a) The activity of the binary C_4H_9Li–WCl_6 catalyst system[22] can be increased by at least 100-fold if an equimolar

(17) K. F. Castner, private communication.
(18) E. A. Ofstead, private communication.
(19) N. Calderon and D. D. Bates, unpublished results.
(20) See a review chapter by D. O. Jordan in "The Stereochemistry of Macromolecules," Vol. 1, A. D. Ketely, Ed., Marcel Dekker, New York, N. Y., 1967, Chapter 1, pp 1–45.
(21) W. A. Judy, unpublished results.
(22) J. L. Wang and H. R. Menapace, *J. Org. Chem.*, **33**, 3794 (1968).

amount of $AlCl_3$ is added to the reaction[21] (Li:W:Al molar ratio of 2:1:1). (b) WCl_4 obtained by reduction of WCl_6 with H_2 at high temperatures is an inactive metathesis catalyst; however, in combination with $AlCl_3$ (Al:W molar ratio of 2–4:1), a highly active catalyst, free of any organometallic component, is obtained.[21] (c) An active metathesis catalyst is also obtainable from $AlCl_3$–WCl_6 or $AlBr_3$–WCl_6 combinations[21,23] (Al:W molar ratio of 2–8:1). Apparently, the aluminum component does not function as a reducing agent.

The fragmentary results available suggest that the formation of active sites involves removal of chloride ligands from the coordination sphere of the tungsten, thus providing coordination sites for the incoming olefinic ligands. This can be accomplished by either a reduction sequence (eq 7–10), or by an acid–base equilibrium (eq 13), or both. In either case the aluminum component appears to play an important role. It may be further speculated that, by association *via* μ-chloride bonding (eq 12), the aluminum component retards the oligomerization tendencies of reduced tungsten chlorides.

The series of reactions encompassed by eq 6 would terminate with formation of the complex bearing two olefinic ligands in a cis configuration about the tungsten (W*). The complexation process is thought to be a stepwise process. Hughes,[24] who conducted a kinetic study of the metathesis of 2-pentene, interpreted his results in terms of five basic equilibria. The kinetic

$$\text{Mo*} + \text{RCH{=}CHR}' \underset{k_{-1}}{\overset{k_1}{\rightleftharpoons}} \text{Mo*(RCH{=}CHR}') \quad (14)$$

$$\text{Mo*(RCH{=}CHR}') + \text{RCH{=}CHR}' \underset{k_{-2}}{\overset{k_2}{\rightleftharpoons}} \text{Mo*(RCH{=}CHR}')_2 \quad (15)$$

$$\text{Mo*(RCH{=}CHR}')_2 \underset{k_{-3}}{\overset{k_3}{\rightleftharpoons}} \text{Mo*(RCH{=}CHR)(R}'\text{CH{=}CHR}') \quad (16)$$

$$\text{Mo*(RCH{=}CHR)(R}'\text{CH{=}CHR}') \underset{k_{-4}}{\overset{k_4}{\rightleftharpoons}} \text{Mo*(RCH{=}CHR)} + \text{R}'\text{CH{=}CHR}' \quad (17)$$

$$\text{Mo*(RCH{=}CHR)} \underset{k_{-5}}{\overset{k_5}{\rightleftharpoons}} \text{Mo*} + \text{RCH{=}CHR} \quad (18)$$

results are in agreement with the suggestion that the formation of the monoolefinic complex (eq 14) is more facile than the formation of the bisolefin complex (eq 15).

The Transalkylidenation Step. The mode of bond scission which is observed in the metathesis reaction is suggestive of a concerted reaction pathway proceeding *via* a four-centered transition state. Equation 19 illustrates the transalkylidenation step using layman's symbols of dotted lines and arrows. The only real contention intended to be conveyed by eq 19 is that, during the concerted process, there is a transition state

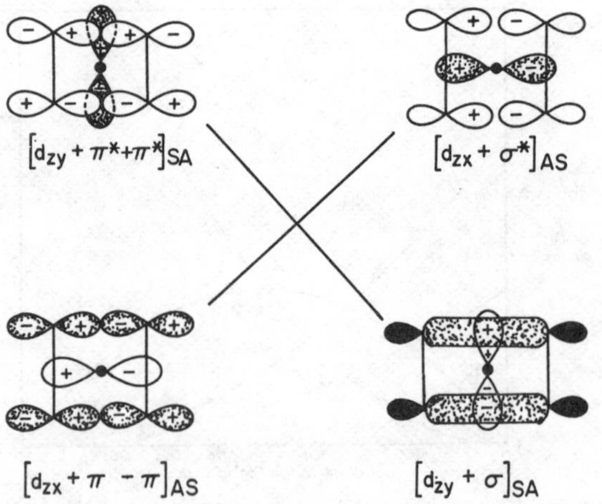

Figure 2. The removal of symmetry restrictions through a relocalization of ligand–metal AS and SA electron density (F. Mango, *et al., J. Amer. Chem. Soc.*, **93**, 1123 (1971)).

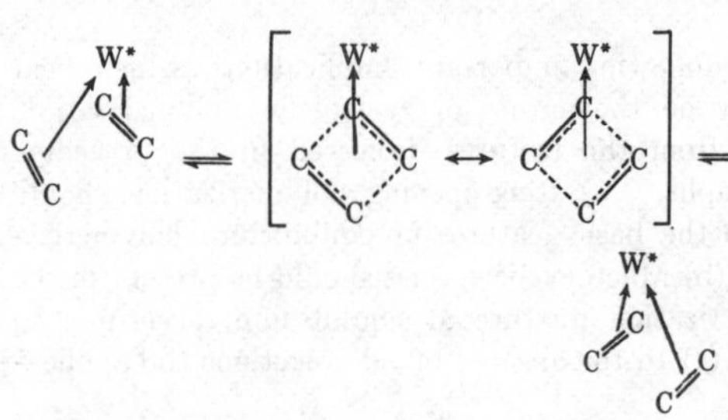

(19)

characterized by having all four carbons equally related to the metal.

If the process described in eq 19 is accepted as being a true concerted one, the formation of the quasi-cyclobutane transition state and its transformation into a bisolefin–metal complex should be viewed as cycloaddition reactions; hence, the principles of orbital symmetry conservation[25] of Woodward and Hoffmann must be considered here.

According to Mango[26,27] a transition metal complex, having atomic d orbitals of the proper symmetries and an available electron pair, can conceivably switch a symmetry-forbidden $[2_s + 2_s]$ cycloaddition to a symmetry-allowed transformation; if so, a concerted reaction pathway does exist for the conversion of two olefinic ligands coordinated to a transition metal into a four-membered ring. Figure 2 illustrates Mango's "forbidden-to-allowed" transformation. A simultaneous injection of a pair of electrons from the d_{zy} metal orbital into the $\pi^* + \pi^*$ antibonding combination and withdrawal of a pair of electrons from the bonding π–π combination by the d_{zx} metal orbital can be executed with conservation of orbital symmetry. The net result is a filled SA σ and a vacant AS σ^* orbital.

(23) P. R. Marshall and B. J. Ridgewell, *Eur. Polym. J.*, **5**, 29 (1969).
(24) W. B. Hughes, *J. Amer. Chem. Soc.*, **92**, 532 (1970).
(25) For a complete presentation see R. B. Woodward and R. Hoffmann, "The Conservation of Orbital Symmetry," Verlag Chemie, Weinheim, Germany, 1970.
(26) F. D. Mango and J. H. Schachtschneider, *J. Amer. Chem. Soc.*, **89**, 2484 (1967).
(27) F. D. Mango, *Advan. Catal.*, **20**, 291 (1969).

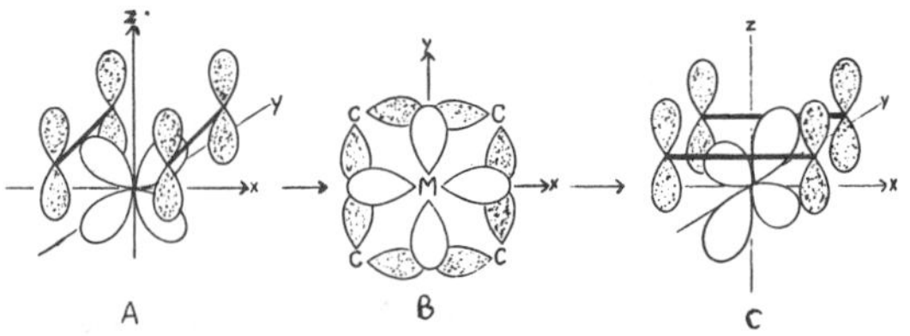

Figure 3. Transalkylidenation *via* tetramethylene–metal transition state.

Should the quasi-cyclobutane transition state (eq 19) be considered a ground-state cyclobutane? Not necessarily. Lewandos and Pettit,[28] in a recent publication, suggested that the transalkylidenation process proceeds *via* a transition state in which: "the bonding is most conveniently described as resulting from the interaction of a basic set of metal atomic orbitals and four methylenic units." Figure 3 illustrates the transformation *via* the "tetramethylene–metal" transition state as suggested by Lewandos and Pettit.

Arguments concerning the conservation of orbital symmetry, involved with rendering an "allowed" status to the process depicted in Figure 3, are not presented here for the sake of brevity. The important feature of the proposed mechanism is that the transformation of the bisolefin complex into the tetramethylene–metal transition state is accomplished by forward donation of four electrons from filled ligand orbitals to empty metal orbitals and back donation of four electrons from filled metal orbitals to empty ligand orbitals. Consequently, the carbon–carbon σ bonds of the initial olefins are ruptured concurrently with the π bonds, so that no genuine cyclobutane molecule is ever realized along the reaction coordinate.

The Olefin Exchange Step. In order to account for the high rates of reaction which are observed at very low catalyst levels, a rapid olefin exchange (eq 20) that alternates with the transalkylidenation step (eq 19) was proposed.[5] This scheme is somewhat different from the

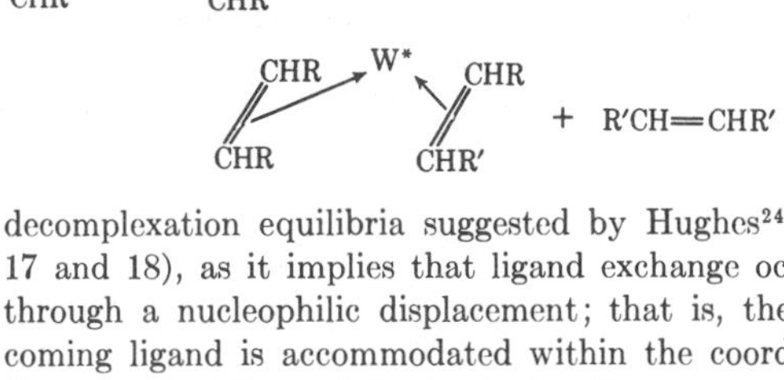

(20)

decomplexation equilibria suggested by Hughes[24] (eq 17 and 18), as it implies that ligand exchange occurs through a nucleophilic displacement; that is, the incoming ligand is accommodated within the coordination sphere of the metal prior to the disengagement of the leaving ligand.

A careful analysis of products obtained during the early stages of cycloolefin polymerizations suggests that the olefin exchange step is slower than the transalkylidenation step. The propagation process proposed for the cycloolefin polymerization comprises two steps[15] which represent the application of eq 19 and 20

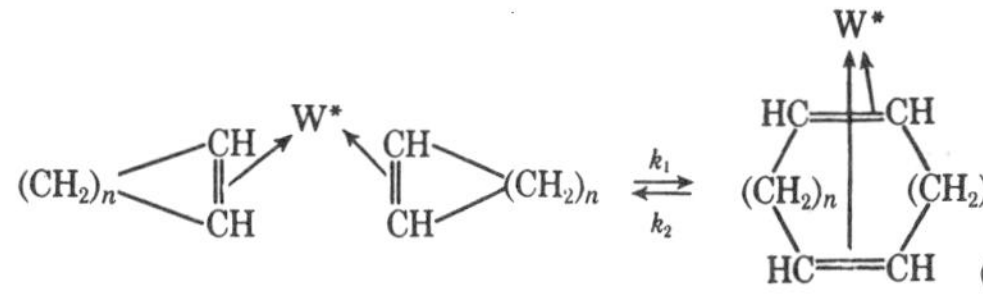

(21)

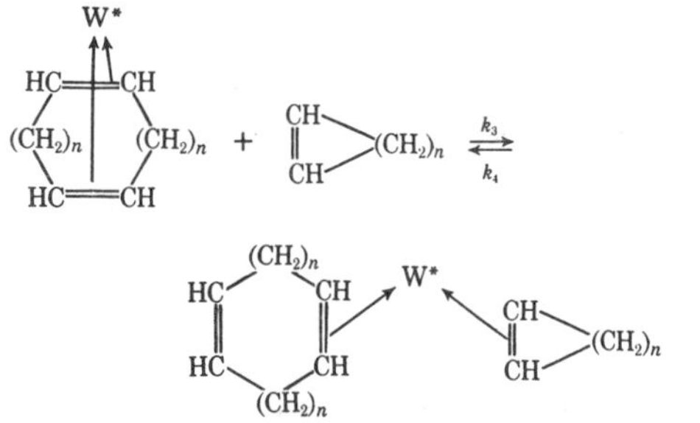

(22)

to cycloolefins (eq 21 and 22). Reiteration of these two steps would lead to the formation of macrocyclic polymers. One may also visualize a transfer process involving an olefin-exchange step (eq 23).

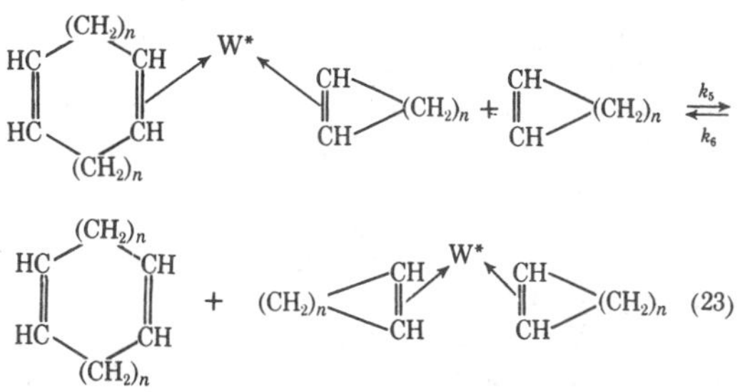

(23)

If the exchange step ($k_3 \ldots k_6$) were faster than the transalkylidenation step (k_1, k_2), the transfer-to-monomer process would predominate in the early stages of the reaction. This should result in a preponderance of low molecular weight oligomers, with little or no high molecular weight product formed initially. Under such circumstances, one would expect the number-average molecular weight of the product to remain very low until the final stage of the polymerization.

However, experimental results do not support this scheme. High molecular weight polymer was observed during early stages of the ring-opening polymerization of cyclooctene ($C_2H_5AlCl_2$–WCl_6–C_2H_5OH catalyst[29]). Similar results have been observed in the polymerization of cyclopentene by tungsten-based catalysts.[18,30] These results can be accommodated by assuming that the transalkylidenation step is much faster than the olefin-exchange step—an assumption which is contrary to that of Hughes,[24] who concluded that the rate-determining step is in fact the transalkylidenation step (eq 16). However, this difference may be reconciled by the fact that different catalyst systems were employed in these studies (W *vs.* Mo).

(28) G. S. Lewandos and R. Pettit, *Tetrahedron Lett.*, 789 (1971).

(29) N. Calderon, E. A. Ofstead, and W. A. Judy, paper presented at the Central Regional Meeting of the American Chemical Society, Akron, Ohio, May 1968.

(30) G. Pampus, J. Witte, and M. Hoffmann, *Rev. Gen. Caout. Plast.*, **47**, 1343 (1970).

Synthetic Applications

Olefin Synthesis. The applicability of the olefin metathesis reaction in areas such as enhancing the market value of petrochemical streams is obvious. A process for converting surplus propylene to ethylene and butene has been commercialized.

The effect of substitution on ease of participation in the metathesis reaction is: $[CH_2=] > [RCH_2CH=] > [(R)_2CHCH=] > [(R)_2C=]$. In addition, chlorine substitution at vinylic sites deactivates the double bond toward the metathesis reaction.[31]

Polyalkenamers. A homologous series of linear unsaturated polyalkenamers is provided by the application of the olefin metathesis reaction to cycloolefins of the general formula $[(CH_2)_nCH=CH]$, where $n = 2$, 3, 5, 6 ... and higher, with the exception of cyclohexene which does not undergo ring-opening polymerization. Unsaturated alicyclic monomers possessing more than one double bond undergo polymerization, provided that the double bonds are not conjugated.[32] Depending on the structure of the repeat units and the configuration of the double bonds, polyalkenamers may possess properties ranging from amorphous elastomeric to crystalline plastics.[33]

Careful analysis of experimental results[34] suggests that the high molecular weight polyalkenamers are essentially linear and not macrocyclic. To accommodate these results the participation of traces of acyclic olefins in metathesis steps during polymerization is suspected.

Macrocyclics and Catenanes. By proper selection of reaction conditions (high dilutions) it was shown that the metathesis of cycloolefins affords relatively high yields of macrocyclic compounds.[35] Of special interest is the cyclic dimer of cyclooctene, 1,9-cyclohexadecadiene, which is convertible to the respective musk-like ketone.[36]

Wolovsky[37] and Ben-Efraim, Batich, and Wasserman[38] carried out concurrent mass spectrometric studies on the macrocyclic species obtained from the metathesis of cyclododecene. Both concluded that interlocking ring systems are present in the mixture. The formation of these exquisite cyclic structures was accounted for[37,38] by assuming an intramolecular

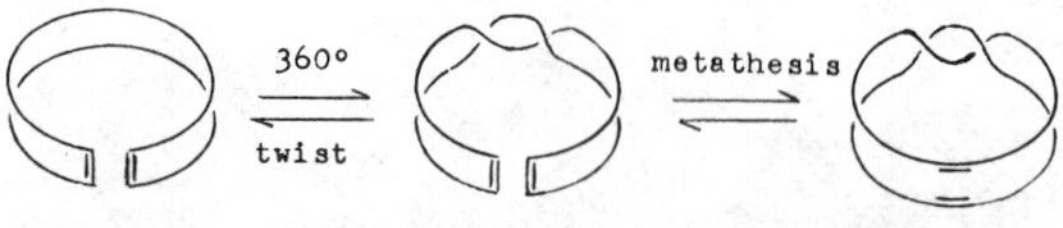

Figure 4. Formation of catenanes.

metathesis of a "strip" twisted by 360°, as illustrated in Figure 4.

Perfectly Alternating Copolymers. Substituted cycloolefins can be polymerized to high molecular weight polymers,[32] thus providing a convenient route for the synthesis of certain perfectly alternating copolymers. For example

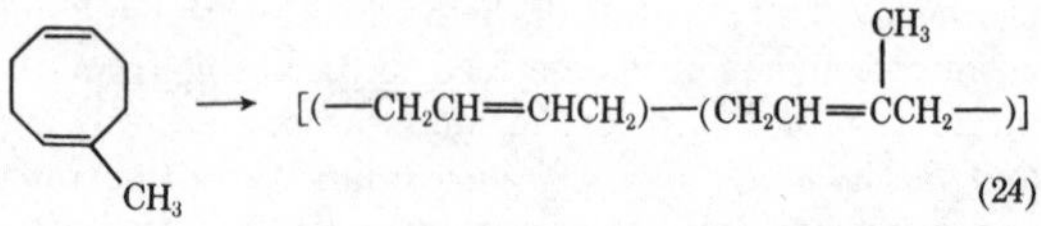

$$\longrightarrow \quad [(-CH_2CH=CHCH_2)-(CH_2CH=\overset{CH_3}{\underset{|}{C}}CH_2-)] \tag{24}$$

The preparation of several diene alternating copolymers by this method has been reported by Ofstead.[31]

Liquid Polymers, Dienes, and Triene Syntheses. The cross-metathesis of cycloolefins with acyclic olefins may be utilized as a method for synthesis of liquid polymers, dienes, and trienes.[39,40] Michajlov and Harwood[41] have utilized the process for the characterization of polymer structures. For example, they were able to determine the monomer sequence distribution in a variety of butadiene–styrene copolymers. By employing high 2-butene:copolymer ratios they were able to exhaustively cleave copolymer chains into low molecular weight fragments which were then subjected to analysis by gas chromatography. The various cleavage products were characteristic of specific sequences of the monomers in the parent copolymer.

Conclusion

As with other newly discovered catalytic reactions, the surge in development of useful applications for the olefin metathesis reaction outpaced the elucidation of the reaction mechanism. Undoubtedly, as more researchers focus their attention on this relatively new process, an expansion of the scope of its applications will occur. Concurrently, one hopes that critical experimentation will be conducted that will clarify at least some of the mechanistic aspects proposed to date.

I wish to thank my colleagues at the Research Division of The Goodyear Tire and Rubber Company, in particular Drs. E. A. Ofstead and K. W. Scott, who contributed many useful ideas related to the olefin metathesis reaction.

(31) E. A. Ofstead, Paper presented at the 4th International Synthetic Rubber Symposium, London, Sept 30, 1969; SRS4, No. 2, p 42.

(32) N. Calderon, E. A. Ofstead, and W. A. Judy, *J. Polym. Sci.*, Part A-2, **5**, 2209 (1967).

(33) N. Calderon and M. C. Morris, *ibid.*, Part A-2, **5**, 1283 (1967).

(34) K. W. Scott, N. Calderon, E. A. Ofstead, W. A. Judy, and J. P. Ward, Princeton University Conference, Advances in Polymer Science and Materials, Nov 21, 1968; *Rubber Chem. Technol.*, **44**, No. 5 (1971); see also K. W. Scott, N. Calderon, and T. A. Ofstead, 11th Proceedings of the International Institute of Synthetic Rubber Producers, Inc., May 1970.

(35) N. Calderon, U. S. Patents 3,439,056 and 3,439,057 (1969).

(36) L. G. Wideman, *J. Org. Chem.*, **33**, 4541 (1968).

(37) R. Wolovsky, *J. Amer. Chem. Soc.*, **92**, 2132 (1970).

(38) D. A. Ben-Efraim, C. Batich, and E. Wasserman, *ibid.*, **92**, 2133 (1970).

(39) W. J. Kelly, unpublished results.

(40) E. A. Zuech, W. B. Hughes, D. H. Kubick, and E. T. Kittleman, *J. Amer. Chem. Soc.*, **92**, 528 (1970).

(41) L. Michajlov and H. J. Harwood, Abstracts, 160th National Meeting of the American Chemical Society, Chicago, Ill., Sept 1970; *Polym. Prepr., Amer. Chem. Soc., Div. Polym. Chem.*, **11**, 1198 (1970).

Addendum (May 1976)

Carbene-to-Metallocycle Mechanistic Scheme. Lately, several reports disclosed results that are incompatible with mechanistic pathways that assume a pairwise interchange of alkylidene moieties. Thus, the aforementioned quasicyclobutane and tetramethylene–metal transition states, as well as the metallocyclopentane intermediate proposed by Grubbs,[42] have been challenged by a non-pairwise carbene-to-metallocycle scheme (eq. 25):

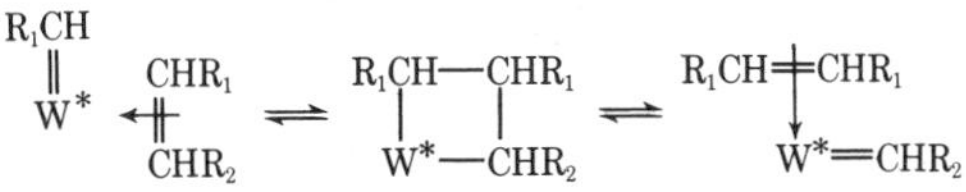
(25)

The first to suggest a transalkylidenation via a non-pairwise carbene–metal intermediate were Chauvin and co-workers.[43] Critical experiments designed to distinguish between pairwise and non-pairwise schemes have been attempted by various researchers.[44–48] The cross metathesis study of 1,7-octadiene with 1,7-octadiene-d_4 reported by Grubbs[48] stands alone as the one that cannot be readily accounted for by pairwise schemes. This subject is extensively discussed in a recent review.[49]

Evidence for carbene involvement in olefin metathesis was offered by Cardin,[50] who isolated a compound having an electron-rich alkylidene group coordinated to Rh^{I} as a side product in the disproportionation of unsaturated amines with rhodium phosphine. Dolgoplosk[44] was able to prepare active metathesis catalyst by coreacting WCl_6 or WCl_4 with $C_6H_5CHN_2$. Recently, Casey[51] prepared and isolated $(C_6H_5)_2C{:}W(CO)_5$. Upon heating this stable carbene–metal complex with an olefin, Casey demonstrated a net transfer of alkylidene group from the olefin to the diphenyl carbene in addition to cyclopropane derivatives (eq. 26):

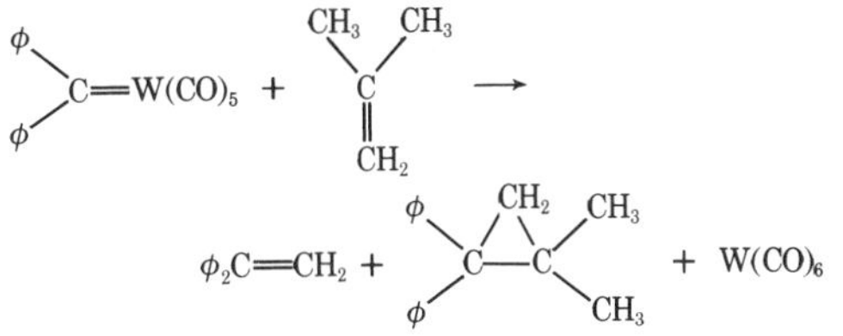
(26)

Selectivity in Cross Metathesis Reactions. It was disclosed[47,52,53] that certain catalysts which display a low apparent metathesis activity with terminal olefins, when used on mixtures of terminal and internal olefins, lead to the selective formation of cross metathesis products. Thus, 1-pentene inhibits the self-metathesis of 2-pentene to yield a preponderance of the cross reaction products. This observation was reconfirmed by Muetterties.[54] To rationalize this phenomenon it has been suggested that terminal olefins are, in fact, more reactive than internal olefins, but they do prefer to undergo regenerative metathesis via metallocycle intermediates (a) or (b) (eq. 27):

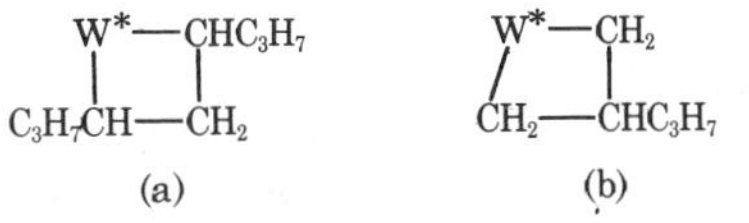
(27)

In an attempt to resolve whether regenerative metathesis of terminal olefins proceed via (a) or (b), the cross metathesis reaction of 1-pentene and cyclopentene was examined. The results strongly suggest that intermediate (a) is favored. This is consistent with the assertion that RCH=W* is energetically more stable than CH_2=W*.

(42) R. H. Grubbs and T. K. Brunck, *J. Am. Chem. Soc.*, **94**, 2538 (1972).
(43) J. L. Herisson and Y. Chauvin, *Makromol. Chem.*, **141**, 161 (1970).
(44) B. A. Dolgoplosk, K. L. Makovetsky, T. G. Golenko, Y. V. Korshak, and E. I. Timyakova, *Eur. Polym. J.*, **10**, 901 (1974).
(45) K. W. Scott, N. Calderon, E. A. Ofstead, W. A. Judy, and J. P. Ward, *Rubber Chem. Technol.*, **44**, 1341 (1971).
(46) T. J. Katz and J. McGinnis, *J. Am. Chem. Soc.*, **97**, 1592 (1975).
(47) W. J. Kelly and N. Calderon, *J. Macromol. Sci. Chem.*, **A9**, 911 (1975).
(48) R. H. Grubbs, P. L. Burk, and D. C. Carr, *J. Am. Chem. Soc.*, **97**, 3265 (1975).
(49) N. Calderon, E. A. Ofstead, and W. A. Judy, *Angew. Chem.*, **88**, 433 (1976).
(50) D. J. Cardin, M. J. Doyle, and M. F. Lappert, *Chem. Commun.*, 927 (1972).
(51) C. P. Casey and T. J. Burkhardt, *J. Am. Chem. Soc.*, **95**, 5833 (1973); *ibid.*, **96**, 7808 (1974).
(52) N. Calderon, Paper presented at the Japan-U.S. seminar on: "Unsolved Problems in Ionic Polymerization"—October 1974, Hakone, Japan.
(53) W. J. Kelly, *Am. Chem. Soc.*, Div. Pet. Chem. Prepr., **17**, H32 (1972).
(54) M. T. Mocella, M. A. Busch, and E. L. Muetterties, *J. Am. Chem. Soc.*, **98**, 1283 (1976).

Addition Reactions of Butadiene Catalyzed by Palladium Complexes

Jiro Tsuji

Basic Research Laboratories, Toray Industries Inc., Kamakura, Japan

Received June 1, 1972

Reprinted from ACCOUNTS OF CHEMICAL RESEARCH, ***6***, *8 (1973)*

Remarkable advances have been made in the last 20 years in organic synthesis by using transition-metal complexes. Many novel synthetic transformations have been discovered which seemed impossible by conventional methods.

Among the several transition metals commonly used in organic synthesis, palladium occupies a unique position. Palladium compounds are versatile reagents or catalysts for organic synthesis, as has been made apparent by the many new synthetic reactions discovered since the invention of the famous Wacker process in the late 1950's.

In the last 10 years, extensive synthetic studies have been carried out involving palladium compounds and mainly olefinic and aromatic substrates. A particular feature of many of these reactions is oxidation with divalent palladium, and nucleophilic substitution and addition reactions of olefins are typical examples. A previous Account by the author treated mainly this type of reaction.[1]

On the other hand, recently a different type of reaction catalyzed by palladium complexes has been discovered which involves oligomerization, mainly dimerization, and telomerization of butadiene. Oligomerization reactions of butadiene catalyzed by transition-metal complexes, particularly nickel complexes, which give various cyclic and linear oligomers, have been studied extensively and quite thoroughly, especially by Wilke.[2,3] However, studies on the oligomerization and telomerization of butadiene using palladium catalysts were initiated only recently. They show that palladium-catalyzed reactions of butadiene are both unique and useful.

Palladium complexes catalyze cocyclization of butadiene with heteropolar double bonds and telomerization with nucleophiles. These reactions, which are impossible or difficult to achieve by other transition-metal complexes, constitute the principal substance of this Account.

Cocyclization Reactions of Butadiene with Heteropolar Double Bonds

A novel reaction of butadiene catalyzed by palladium complexes is the cocyclization of 2 mol of butadiene with a heteropolar double bond to form a six-membered heterocyclic compound, which is expressed by the following general scheme.

Jiro Tsuji was born in Shigaken, Japan, in 1927. He received his B.S. at Kyoto University and then came to this country as a Fulbright Exchange student, where he obtained his Ph.D. at Columbia, with Gilbert Stork. His main research interests have been in organic synthesis, especially of natural products, and in organotransition metal chemistry. Dr. Tsuji is presently a Senior Research Associate at Basic Research Laboratories, Toray Industries, Inc.

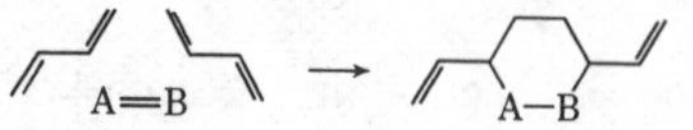

Cocylizations of butadiene catalyzed by nickel complexes are known, but they involve only olefinic and acetylenic bonds.[4,5] The palladium-catalyzed reaction is a unique example of metal-catalyzed cocyclization involving unsaturated bonds to hetero atoms.

The first example of the cocyclization involved 2 mol of butadiene and 1 mol of aldehyde to give 2-substituted 3,6-divinyltetrahydropyrans. In addition to the pyrans, 1-substituted 2-vinyl-4,6-heptadien-1-ols are also formed. Four research groups independently found this reaction.[6-9]

Palladium is an active catalyst for decarbonylation of aldehydes.[10] It is assumed that the first step in decarbonylation is an oxidative addition of the aldehyde to the palladium catalyst to form the acyl–palladium complex (eq 1). Generally, acyl–metal com-

$$\text{R-CHO} + \text{Pd} \longrightarrow \left[\text{RCO-}\overset{|}{\underset{|}{\text{Pd}}}\text{-H} \longrightarrow \text{R-}\overset{\text{CO}}{\underset{|}{\overset{|}{\text{Pd}}}}\text{-H} \right] \longrightarrow$$
$$\text{RH} + \text{Pd} + \text{CO} \quad (1)$$

plexes undergo olefin insertion.[11] In our attempt to synthesize ketones by the insertion of olefins into the acyl–palladium bonds formed from aldehydes, reactions of butadiene with aldehydes in the presence of the palladium complex of $(C_6H_5)_3P$ were carried out. The products isolated were not the expected ketones, but rather 2-substituted 3,6-divinyltetrahydropyrans

$CH_2{=}CHCH{=}CH_2$ + RCHO $\longrightarrow$

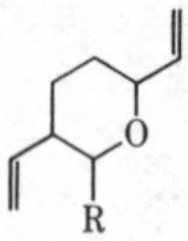

1

+ $CH_2{=}CHCHCH_2CH{=}CHCH{=}CH_2$ (with RCHOH on C-3)

2

(1) J. Tsuji, *Accounts Chem. Res.*, **2**, 144 (1969).
(2) G. Wilke, *Angew. Chem.*, **75**, 10 (1963).
(3) G. Wilke, B. Bogdanovic, P. Hardt, P. Heimbach, W. Keim, M. Kroner, W. Oberkirch, K. Tanaka, E. Steinrucke, D. Walter, and H. Zimmermann, *Angew. Chem.*, **78**, 157 (1966).
(4) P. Heimbach and W. Brenner, *Angew. Chem.*, **78**, 983 (1966).
(5) W. Brenner, P. Heimbach, K. J. Ploner, and F. Thomel, *Angew. Chem.*, **81**, 744 (1969).
(6) K. Ohno, T. Mitsuyasu, and J. Tsuji, *Tetrahedron Lett.*, 67 (1971).
(7) R. M. Manyik, W. E. Walker, K. E. Atkins, and E. S. Hammack, *Tetrahedron Lett.*, 3813 (1970).
(8) P. Haynes, *Tetrahedron Lett.*, 3687 (1970).
(9) H. A. Jung, *British Patent*, application Feb 25, 1970.
(10) J. Tsuji and K. Ohno, *J. Amer. Chem. Soc.*, **90**, 94 (1968).
(11) R. F. Heck, *Accounts Chem. Res.*, **2**, 10 (1969).

(**1**) and 1-substituted 2-vinyl-4,6-heptadien-1-ols (**2**), in high yields.

The reaction of benzaldehyde with butadiene at 80° for 10 hr in the presence of a catalytic amount of palladium acetate ($Pd(OCOCH_3)_2$) and $(C_6H_5)_3P$ gave 1-phenyl-2-vinyl-4,6-heptadien-1-ol and 2-phenyl-3,6-divinyltetrahydropyran (*ca.* 90% yield), which were separated by distillation. Gas chromatographic assay of the pyran showed the presence of four stereoisomers, and three of them were isolated by column chromatography and characterized by nmr spectra.[12] They were 2-phenyl-*trans*-3,*cis*-6-divinyltetrahydropyran (**3**), 2-phenyl-*cis*-3,*cis*-6-divinyltetrahydropyran (**4**), and 2-phenyl-*cis*-3,*trans*-6-divinyltetrahydropyran (**5**). The nmr studies revealed that the large 2-phenyl group takes an equatorial conformation in the preferred structure, and the rate of inversion of the pyran ring is very slow. For example, in the nmr spectrum of **4**, when H_5 was irradiated, the H_2 proton became a quartet. If rapid inversion took place, the quartet would collapse to a triplet.[13] On the other hand, rapid inversion was observed with 2-methyl-3,6-divinyltetrahydropyran.

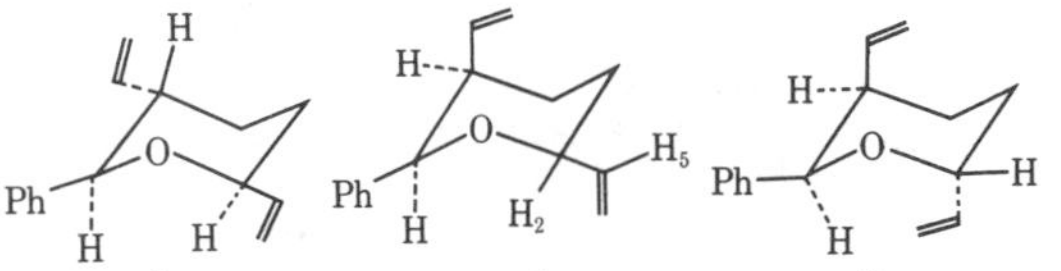

3 **4** **5**

Various palladium-phosphine complexes can be used as effective catalysts. Most simply, $Pd(OCOCH_3)_2$ or palladium acetylacetonate can be used with $(C_6H_5)_3P$. Also π-allylpalladium chloride and $[(C_6H_5)_3P]_2PdCl_2$ are active catalysts in the presence of an excess of bases which remove the chloride ions from the coordination sphere of palladium. Zerovalent complexes such as $[(C_6H_5)_3P]_4Pd$ and bis(triphenylphosphine)(maleic anhydride)palladium are also active. As the ligand, $(C_6H_5)_3P$ is most conveniently used. Trialkylphosphines and triarylarsines are less active. The reaction can be carried out even at room temperature in solvents such as tetrahydrofuran, benzene, and dimethylformamide.

It was found that the ratio of the two products (**1**:**2**) can be controlled by changing the molar ratio of triphenylphosphine to palladium in the system. The unsaturated alcohol was obtained as a main product when the ratio was near unity. By increasing the amount of $(C_6H_5)_3P$, the relative amount of the pyran increased and the pyran was formed nearly selectively when the ratio was above 2. Also it was observed that triarylarsines tend to increase the relative amount of the unsaturated alcohol.[14]

Aromatic aldehydes, furfural, and aliphatic aldehydes, including formaldehyde, take part in the reaction. On the other hand, attempted reactions with common ketones such as acetone were not successful, and only 1,3,7-octatriene was formed. Some specific ketones, however, were found to take part in the reaction.[12] Perfluoroacetone gave 2,2-bis(trifluoromethyl)-3,6-divinyltetrahydropyran (**6**). In addition, other products (**7**, **8**, **9**) were obtained. Dihydropyrans **8** and **9** were formed by a Diels–Alder-type reaction. 1,1,1-Trifluoro-2-trifluoro-

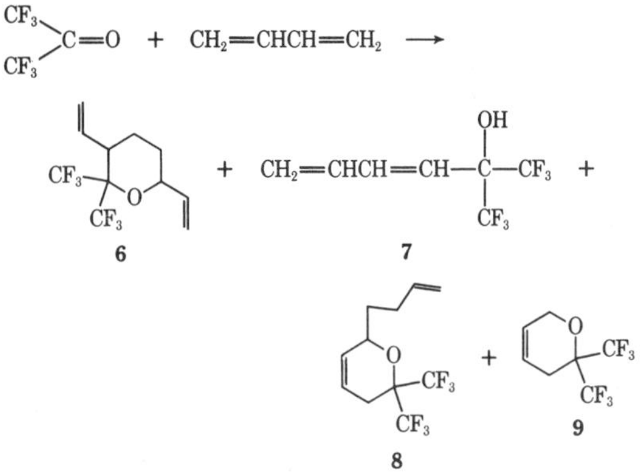

6 **7** **8** **9**

methyl-*trans*- 3,5-hexadien-2-ol (**7**) was formed by the 1:1, rather than 2:1, addition of butadiene to the ketone. A similar 1:1 product was obtained by the

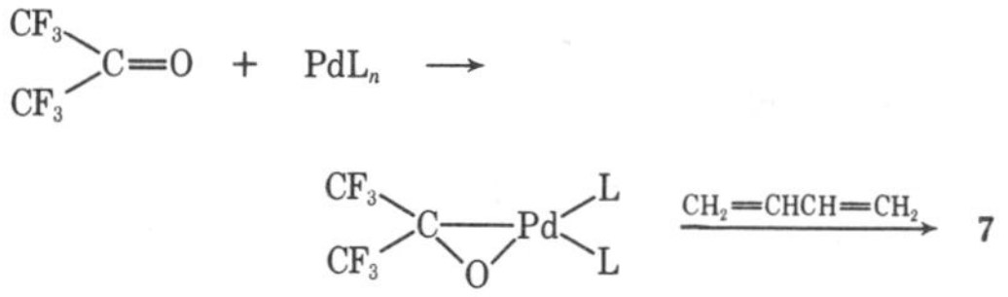

10

reaction of butadiene and acetone in the presence of nickel complex, which was consumed stoichiometrically.[15] Other active ketones like α-diketones, such as biacetyl and benzil, gave the corresponding pyrans (**11**).

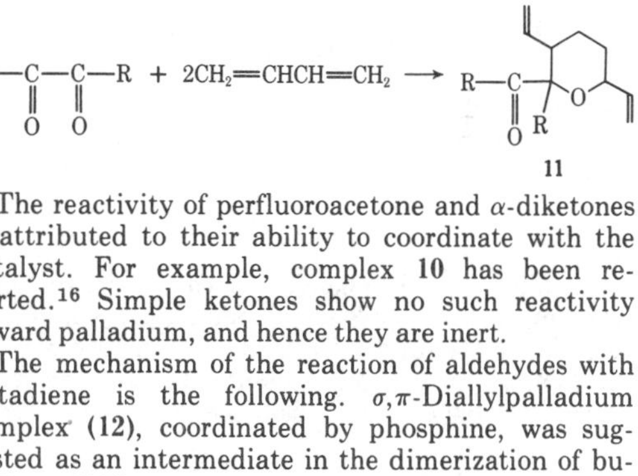

11

The reactivity of perfluoroacetone and α-diketones is attributed to their ability to coordinate with the catalyst. For example, complex **10** has been reported.[16] Simple ketones show no such reactivity toward palladium, and hence they are inert.

The mechanism of the reaction of aldehydes with butadiene is the following. σ,π-Diallylpalladium complex (**12**), coordinated by phosphine, was suggested as an intermediate in the dimerization of butadiene.[17] This intermediate seems to play an important role in the reaction of aldehydes. In the presence of aldehyde, insertion of the carbonyl group into the palladium–σ-allyl bond takes place giving an alkoxide-type complex (**13**). This intermediate has an equilibrium between π-allyl (**13**) and σ-allylpalladium (**14**) structures depending on the concentration of the phosphine. The former complex is expected to decompose through hydrogen transfer from the 4 po-

(12) K. Ohno, T. Mitsuyasu, and J. Tsuji, *Tetrahedron*, **28**, 3705 (1972).

(13) J. A. Pople, W. G. Schneider, and H. J. Bernstein, "High Resolution Nuclear Magnetic Resonance," McGraw-Hill, New York, N. Y., 1959, p 218.

(14) K. Ohno and J. Tsuji, unpublished work.

(15) P. Heimbach, P. W. Jolly, and G. Wilke, *Advan. Organometal. Chem.*, **8**, 39 (1970).

(16) B. Clark, M. Green, R. B. L. Osborn, and F. G. A. Stone, *J. Chem. Soc. A*, 168 (1968).

(17) S. Takahashi, H. Yamazaki, and N. Hagihara, *Mem. Inst. Sci. Ind. Res. Osaka Univ.*, **25**, 125 (1968).

Scheme I

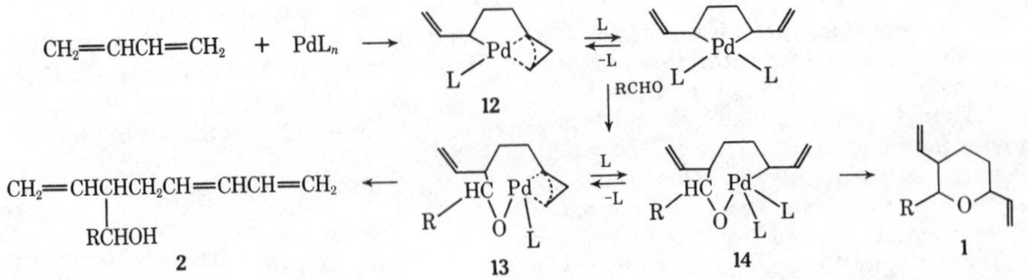

sition of the π-allyl moiety to the oxygen to give the unsaturated alcohol 2. The pyran 1 is formed from the latter complex through ligand coupling (Scheme I). Existence of the postulated equilibrium between σ-allyl- and π-allylpalladiums is supported by the fact that the product distribution varies markedly with the molar ratio of $(C_6H_5)_3P$ to Pd. The relative extent of reaction by each pathway depends mainly on the relative ratio of the two catalytic components.

Isocyanates also take part in the cocyclization with butadiene.[18] The isocyanate C=N bond, rather than C=O bond, is involved in the reaction to give divinylpiperidones in high yield and selectivity. Use of a catalytic amount of the palladium complex suffices. For example, a benzene solution of phenyl isocyanate and isoprene in the presence of bis(triphenylphosphine)(maleic anhydride)palladium was shaken at 100° for 20 hr, and crystalline *trans*-3,6-diisopropenyl-1-phenyl-2-piperidone (**15**) as well as oily *cis*-3,6-diisopropenyl-1-phenyl-2-piperidone (**16**) were obtained. In this reaction, tail-to-tail coupling of isoprene takes place selectively.

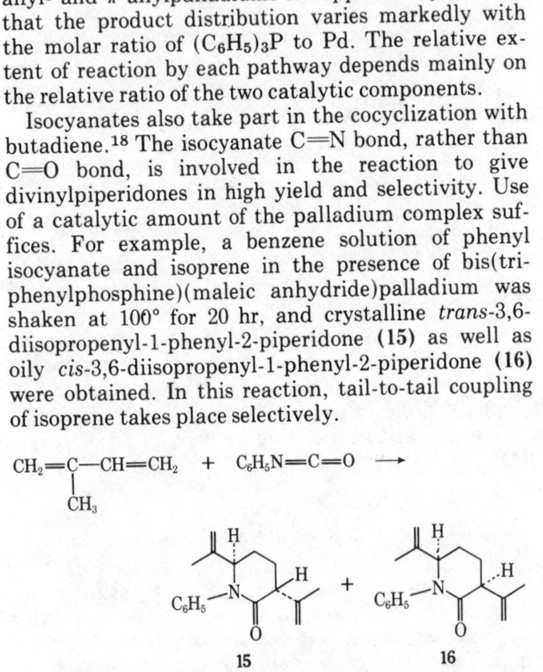

Butadiene and phenyl isocyanate gave equal amounts of (*Z*)-3-ethylidene-1-phenyl-6-vinyl-2-piperidone (**17**) and its *E* isomer **18** in 75% yield. In the reaction of butadiene, double-bond migration to the conjugated position took place.

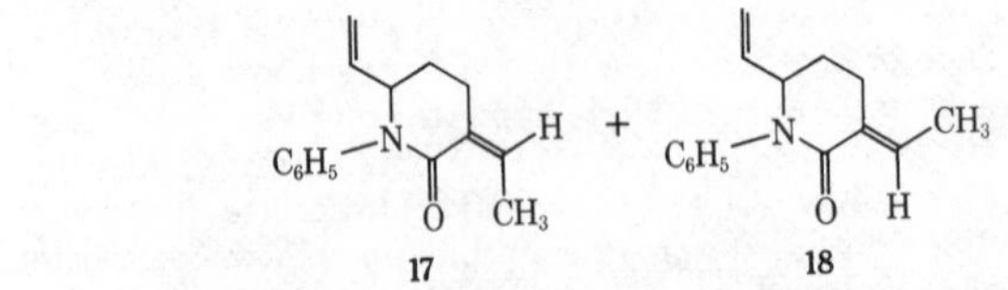

For this reaction, $Pd(OCOCH_3)_2$ and palladium acetylacetonate are most conveniently used with $(C_6H_5)_3P$ as the catalyst. Platinum complexes of phosphine have the same activity for the piperidone synthesis, although they are less active in the pyran formation reaction.

(18) K. Ohno and J. Tsuji, *Chem. Commun.*, 247 (1971).

Dimerization and Telomerization of Butadiene

Formation of *n*-dodecatetraene from butadiene in 30% yield by using bis(π-allyl)palladium is the first report of the palladium-catalyzed oligomerization of butadiene.[3] Later studies revealed that palladium-catalyzed reactions of butadiene proceed quite differently from reactions catalyzed by other transition-metal complexes. Linear dimerization is the main reaction with palladium catalysts, and no cyclodimerization or cyclotrimerization is possible.

The most interesting and important feature observed in the palladium-catalyzed dimerization of butadiene is the formation of telomers with various nucleophiles. As is well known, simple olefins coordinated to divalent palladium compounds react with nucleophiles to give vinylic compounds.[1] These reactions are oxidations with divalent palladium. The nucleophiles so far known to react with olefins are water,[19] alcohols,[19] carboxylic acids,[19] amines,[20-23] ammonia,[24] enamines,[25] active methylene compounds with two negative groups,[26] azides,[24] and cyanide.[27] Also CO reacts with olefin complexes of palladium.[28] It has been found that butadiene dimerizes with efficient incorporation of the same nucleophiles mainly at a terminal position to give 1-substituted 2,7-octadienes.

$$2CH_2{=}CHCH{=}CH_2 + YH \rightarrow CH_2{=}CHCH_2CH_2CH_2CH{=}CHCH_2Y$$

In this sense, the dimerizations catalyzed by palladium complexes are somewhat different from oligomerization reactions of butadiene catalyzed by other transition-metal complexes. With nickel, cobalt, iron, and other catalysts, it is generally not possible to introduce functional groups efficiently into oligomers. Recently, it was reported that alcohols,[29] amines,[30,31] and active methylene compounds[32] can

(19) J. Tsuji, *Advan. Org. Chem.*, **6**, 109 (1969).
(20) G. Paiaro, A. De Renzi, and R. Palumbo, *Chem. Commun.*, 1150 (1967).
(21) R. Palumbo, A. De Renzi, A. Panunzi, and G. Paoaro, *J. Amer. Chem. Soc.*, **91**, 3874 (1969).
(22) E. Stern and M. L. Spector, *Proc. Chem. Soc. London*, 370 (1961).
(23) H. Hirai, H. Sawai, and S. Makishima, *Bull. Chem. Soc. Jap.*, **43**, 1148 (1970).
(24) M. Tada, Y. Kuroda, and T. Sato, *Tetrahedron Lett.*, 2871 (1969).
(25) J. Tsuji, H. Takahashi, and M. Morikawa, *Tetrahedron Lett.*, 4387 (1965).
(26) H. Takahashi and J. Tsuji, *J. Amer. Chem. Soc.*, **90**, 2387 (1968).
(27) Y. Odaira, T. Oishi, T. Yukawa, and S. Tsutsumi, *J. Amer. Chem. Soc.*, **88**, 4105 (1966).
(28) J. Tsuji, M. Morikawa, and J. Kiji, *J. Amer. Chem. Soc.*, **86**, 4851 (1964).
(29) T. C. Shields and W. E. Walker, *Chem. Commun.*, 193 (1971).
(30) P. Heimbach, *Angew. Chem., Int. Ed. Engl.*, **7**, 882 (1968).
(31) R. Baker, D. E. Halliday, and T. N. Smith, *Chem. Commun.*, 1583 (1971).
(32) R. Baker, D. E. Halliday, and T. N. Smith, *J. Organometal. Chem.*, **36**, C61 (1972).

be introduced into butadiene and its dimers by using some specific nickel complexes to give butenyl and octadienyl ethers and corresponding compounds. However, the selectivity of the reactions is not high, and a mixture of telomers is obtained.

Furthermore, it should be pointed out that this reaction of butadiene is not an oxidation, which consumes divalent palladium, but catalytic. In contrast to the reactions of ethylene and other monoolefins, which reduce divalent palladium stoichiometrically and are catalytic only when used with a Cu(II)-O_2 cooxidation system, this type of reaction of butadiene represents true catalysis by palladium. Thus no metallic palladium is deposited during the reaction carried out under appropriate conditions.

Another noteworthy feature is that $PdCl_2$, widely used to catalyze reactions of simple olefins, is not an effective catalyst of oligomerization; $Pd(OCOCH_3)_2$ and palladium acetylacetonate can be used conveniently. That is, palladium complexes without coordinated halogen can be used as satisfactory catalysts. When $PdCl_2$ is used, bases must also be added in order to eliminate the coordinated chloride ion from the coordination sphere of palladium. Zerovalent palladium complexes such as $[(C_6H_5)_3P]_4Pd$ or bis(triphenylphosphine)(maleic anhydride)palladium are also used.

Dimerization of butadiene to form 1,3,7-octatriene (**19**) proceeds in aprotic solvents in a high yield using, for example, the above zerovalent palladium complexes as catalysts.[33,34] The reaction is faster in isopropyl alcohol.

$$2CH_2{=}CHCH{=}CH_2 \rightarrow \underset{\mathbf{19}}{CH_2{=}CHCH_2CH_2CH{=}CHCH{=}CH_2}$$

Telomerization of butadiene with carboxylic acids in the presence of palladium salts and alkali phenoxide was first claimed in patents.[35] Also the reaction of phenol and butadiene, catalyzed by $PdCl_2$ under basic conditions, gave a small amount of phenoxybutene. When pyridine was added, 1-phenoxyoctadiene was obtained in a low yield.[36,37] These studies showed that palladium is a catalyst of telomerization of butadiene, but it seems likely that the activity of the palladium catalyst is low in the absence of proper ligands. A major breakthrough was the discovery that ligands are very important in the palladium-catalyzed telomerization. Especially, $(C_6H_5)_3P$ has a profound effect on the reaction, increasing selectivity and activity of the palladium catalyst. In the following, telomerizations with various nucleophiles are surveyed.

Reactions of Alcohols, Phenols, and Water. Butadiene reacts with alcohols to form 1-alkoxy-2,7-octadiene (**20**) as a main product, accompanied by a small amount of 3-alkoxy-1,7-octadiene (**21**).[38-40] Methanol is the most reactive alcohol. Steric effects seem to have a large influence on the reaction, and secondary alcohols react to only a small extent. Almost no ether formation was observed with tertiary alcohols. Very facile reaction takes place with phenols to give phenyl octadienyl ether in high yield.[37,41]

$$CH_2{=}CHCH{=}CH_2 + ROH \rightarrow \underset{\mathbf{20}}{CH_2{=}CHCH_2CH_2CH_2CH{=}CHCH_2OR} + \underset{\mathbf{21}}{CH_2{=}CHCH_2CH_2CH_2\underset{OR}{CH}CH{=}CH_2}$$

From the standpoint of industrial application, the most promising process which might be developed by the telomerization of butadiene is synthesis of *n*-octanol. For this purpose, telomerization with water is most desirable. It is well known that simple olefins react with water very easily in the presence of $PdCl_2$ to give carbonyl compounds. Actually the reaction of water with olefins is responsible for the rapid emergence of the modern organic chemistry of palladium. The oxidative reaction of butadiene with water using $PdCl_2$ and $CuCl_2$ gives crotonaldehyde and 3-ketobutyraldehyde[42] (eq 2). The telomerization of butadiene with water catalyzed by palladium complexes, however, is rather tricky, and it was only recently that water was shown to be incorporated into the dimer.

$$CH_2{=}CHCH{=}CH_2 + H_2O \rightarrow CH_3CH{=}CHCHO + CH_3C(O)CH_2CHO \quad (2)$$

Under the usual conditions, reaction of water with butadiene takes place only to a small extent. Efficient telomerization with water was found to occur in the presence of a considerable amount of CO_2 in solvents like butyl alcohol, acetone, and acetonitrile, as follows.[43]

$$CH_2{=}CHCH{=}CH_2 + H_2O \longrightarrow \begin{cases} \underset{\mathbf{22}}{CH_2{=}CHCH_2CH_2CH_2CH{=}CHCH_2OH} + \mathbf{19} \\ \underset{\mathbf{23}}{CH_2{=}CHCH_2CH_2CH_2CHOHCH{=}CH_2} + \underset{\mathbf{24}}{\text{octadienyl ethers}} \end{cases}$$

The effect of CO_2 is remarkable. For example, reaction of butadiene (1 mol), water (2 mol), and CO_2 (0.5 mol) in acetone (1.8 mol) for 2 hr gave the following yields of the above compounds: **22** (69%), **23** (7%), **19** (13%), and **24** (10%). The role of CO_2 in facilitating the telomerization with water is not clear. Also CO_2 showed some effect on the simple dimerization of butadiene catalyzed by palladium or platinum catalysts.[44]

In an attempt to synthesize 1-octanol, the reaction of trimethylsilanol with butadiene was carried out and 1-trimethylsiloxy-2,7-ocatdiene was obtained, which was hydrolyzed very easily to give 2,7-octadienol (**22**).[45]

(33) S. Takahashi, T. Shibano, and N. Hagihara, *Tetrahedron Lett.*, 2451 (1967).
(34) S. Takahashi, T. Shibano, and N. Hagihara, *Bull. Chem. Soc. Jap.*, **41**, 454 (1968).
(35) E. J. Smutny, U. S. Patent, 3,350,451 (1967), 3,407,224 (1968).
(36) E. J. Smutny, *J. Amer. Chem. Soc.*, **89**, 6793 (1967).
(37) E. J. Smutny, H. Chung, K. C. DeWhirst, W. Keim, T. M. Shryne, and H. E. Thyret, *Amer. Chem. Soc. Div. Petrol. Chem., Prepr.*, **14**, B100 (1969).
(38) S. Takahashi, H. Yamazaki, and N. Hagihara, *Bull. Chem. Soc. Jap.*, **41**, 254 (1968).
(39) Mitsubishi Chemical Industries, Netherlands Patent, 68,16008.
(40) British Petroleum Co., Netherlands Patent, 69,15738, 69,15582.
(41) E. J. Smutny, U. S. Patent, 3,518,318 (1970), 3,518,315 (1970).
(42) J. Smidt, W. Hafner, R. Jira, J. Sedlmeier, R. Sieber, R. Rüttinger, and H. Kojer, *Angew. Chem.*, **71**, 176 (1959).
(43) K. E. Atkins, W. E. Walker, and R. M. Manyik, *Chem. Commun.*, 330 (1971).
(44) J. F. Kohle, L. H. Slaugh, and K. L. Nakamaye, *J. Amer. Chem. Soc.*, **91**, 5905 (1969).
(45) S. Takahashi, T. Shibano, and N. Hagihara, *Kogyo Kagaku Zasshi*, **72**, 1798 (1969).

$2CH_2=CHCH=CH_2 + (CH_3)_3SiOH \rightarrow$

$\qquad CH_2=CHCH_2CH_2CH_2CH=CHCH_2OSi(CH_3)_3 \xrightarrow{H_2O} 22$

Reaction of Carboxylic Acids. The reaction of acetate anion with simple olefins in the presence of $PdCl_2$ or $Pd(OCOCH_3)_2$ has been studied extensively; vinylic acetates are the main products. On the other hand, the reaction of acetate with butadiene has received less attention. Butadiene reacts with acetate to give butadienyl monoacetate.[22] Also, formation of crotyl acetate and 3-acetoxy-1-butene in nearly equal amounts by using $PdCl_2$ and $CuCl_2$ was reported.[46] Further studies revealed that octadienyl acetate is formed by using $PdCl_2$ and a base as catalyst.[35]

$CH_2=CHCH=CH_2 + CH_3CO_2H \rightarrow$

$\qquad CH_3CH=CHCH_2OCOCH_3 + CH_3CH(OCOCH_3)CH=CH_2$

Later it was found that addition of acetate to butadiene to form 1-acetoxy-2,7-octaiene **(25)** as a main product and 3-acetoxy-1,7-octadiene **(26)** as a minor product proceeds more efficiently when a palladium catalyst is used combined with $(C_6H_5)_3P$ as a ligand.[33,47]

$CH_2=CHCH=CH_2 + CH_3CO_2H \rightarrow$

$\qquad CH_2=CHCH_2CH_2CH_2CH=CHCH_2OCOCH_3 +$

$\qquad\qquad\qquad\qquad 25$

$\qquad CH_2=CHCH_2CH_2CH_2CH(OCOCH_3)CH=CH_2$

$\qquad\qquad\qquad\qquad 26$

The reaction carried out without proper solvents is too slow to be satisfactory from a practical standpoint, but the reaction is faster under selected conditions. Selection of solvents and their relative amounts seems to be important.[48] A more remarkable effect was observed by addition of basic compounds. Sodium or potassium acetate accelerates the reaction especially at high concentrations.[49] A marked effect of addition of a molar quantity of tertiary amines such as 2-(N,N-diethylamino)ethanol has been reported.[50] For example, the reaction of butadiene, acetic acid, and 2-(N,N-diethylamino)ethanol(4.0 mol each) in the presence of palladium acetylacetonate and $(C_6H_5)_3P$ (3.0 mmol each) gave essentially complete conversion of butadiene after 2 hr at 90° to yield **25** (71%), **26** (21%), and **19** (8%).

$(C_6H_5)_3P$ is a good ligand for the reaction, and sometimes it is used in a large excess over palladium to secure prolonged life of the catalyst. Some phosphites such as trimethylol propanephosphite **(27)** have favorable effects on the rate and gave the high selectivity value of 25.[50] Reaction times of 2 hr at 50° with this ligand gave an 81% yield of 25 and 9% of **26** without forming octatriene. However, this ligand seems to be decomposed during the reaction.

$$C_2H_5-C\overset{\displaystyle CH_2O}{\underset{\displaystyle CH_2O}{-CH_2O-}}P$$

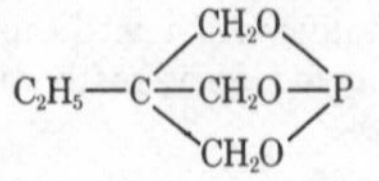

$$27$$

(46) T. Inagaki, Y. Takahashi, S. Sakai, and Y. Ishii, *Bull. Jap. Petrol. Inst.*, **13**, 73 (1971).

(47) D. R. Bryant and J. E. McKeon, U. S. Patent, 3,534,088 (1970).

(48) T. Arakawa and H. Miyake, *Kogyo Kagaku Zasshi*, **74**, 1143 (1971).

(49) T. Mitsuyasu and J. Tsuji, German Offen., 2040,708 (1972).

(50) W. E. Walker, R. M. Manyik, K. E. Atkins, and M. L. Farmer, *Tetrahedron Lett.* 3817 (1970).

Another possible application of the telomerization is synthesis of terpenes. Telomerization of isoprene with acetate or water, if carried out stereospecifically, would be an attractive synthetic method for terpene alcohols. However, so far it has not been possible to synthesize terpenes such as geranyl acetate with high selectivity from isoprene and acetic acid. A mixture of products is always obtained and separation of useful compounds is not easy.[51] Moreover, the reactivity of isoprene is much lower than that of butadiene.

Carboxylic acids other than acetic acid react similarly, but with somewhat different reactivity. An exception is formic acid, and formate ester is formed in only trace amount.[52] A reaction of formic acid with butadiene in the presence of triethylamine, catalyzed by $Pd(OCOCH_3)_2$, leads to the formation of 1,6-octadiene. When $(C_6H_5)_3P$ was added, a mixture of 1,7-octadiene and 1,6-octadiene (1:2) was obtained. CO_2 is released during the reaction and the stoichiometry is as in eq 3.

$2CH_2=CHCH=CH_2 + HCO_2H \rightarrow$

$\qquad CH_2=CHCH_2CH_2CH_2CH=CHCH_3 + CO_2 \quad (3)$

Reaction of Ammonia and Amines. Direct addition of ammonia to olefinic bonds would be an attractive method of amine synthesis if it could be carried out smoothly, but so far the facile addition of ammonia to simple olefins had not been reported. Like water, ammonia does not react with butadiene smoothly under usual conditions, and selection of proper conditions is important. A reaction of aqueous ammonia (28%, 5 g) with butadiene (32 g) in acetonitrile (60 ml) in the presence of $Pd(OCOCH_3)_2$ (63 mg) and $(C_6H_5)_3P$ (261 mg) at 80° for 10 hr gave tri-2,7-octatrienylamine **(28)** (29 g) as a main product, accompanied by a small amount of di-2,7-octadienylamine **(29)** (1.2 g). Also isomeric triamine **30** was formed as a by-product.[53]

$CH_2=CHCH=CH_2 + NH_3 \longrightarrow$

$\qquad (CH_2=CHCH_2CH_2CH_2CH=CHCH_2)_3N +$

$\qquad\qquad\qquad\qquad 28$

$\qquad (CH_2=CHCH_2CH_2CH_2CH=CHCH_2)_2NH +$

$\qquad\qquad\qquad\qquad 29$

$\qquad (CH_2=CHCH_2CH_2CH_2CH=CHCH_2)_2-N$

$\qquad\qquad\qquad\qquad\qquad | $

$\qquad\qquad\qquad CH_2=CHCH_2CH_2CH_2CHCH=CH_2$

$\qquad\qquad\qquad\qquad\qquad 30$

Selection of proper solvents and their amounts are important. Acetonitrile, dimethylformamide and *tert*-butyl alcohol are most suitable. Water seems to have a favorable effect on the reaction, and aqueous ammonia is superior to pure ammonia.

Primary and secondary amines react smoothly with butadiene, and their hydrogens are replaced by 2,7-octadienyl groups to give secondary and tertiary amines.[33,34]

Reaction of Enamines. Enamines have dipolar double bonds and their cyclization to give four- and

(51) K. Suga, S. Watanabe, and K. Hijikata, *Aust. J. Chem.*, **24**, 197 (1971).

(52) S. Gardner and D. Wright, *Tetrahedron Lett.*, 163 (1972).

(53) T. Mitsuyasu, M. Hara, and J. Tsuji, *Chem. Commun.*, 345 (1971).

six-membered rings is known. Also enamines are reactive nucleophiles. In view of these dual reactivities of enamines, the reaction of butadiene with enamines using palladium catalyst was investigated.

Reactions of the pyrrolidine enamines of cyclohexanone and cyclopentanone with butadiene proceed smoothly in acetonitrile in the presence of $Pd(OCOCH_3)_2$ and $(C_6H_5)_3P$.[54] The products isolated after hydrolysis were 2-(2,7-octadienyl)cyclohexanone accompanied by 2,6-di(2,7-octadienyl)cyclohexanone and the corresponding cyclopentanone derivatives.

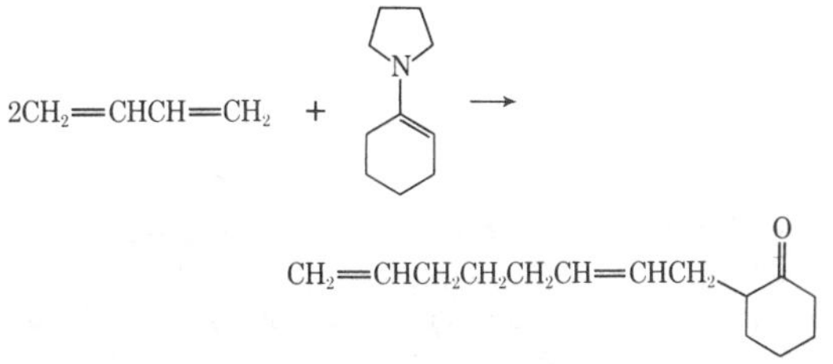

Reaction of Active Methylene Compounds. Compounds with a methylene or methyne group to which are attached two electronegative groups, such as carbonyl, alkoxycarbonyl, formyl, cyano, nitro, and sulfonyl groups, react with butadiene smoothly and their acidic hydrogens are replaced with 2,7-octadienyl groups to give mono- and disubstituted compounds (**31** and **32**).[39,55] Also branched telomers (**33**) are formed as by-products.

$$CH_2{=}CHCH{=}CH_2 + CH_2Y_2 \longrightarrow$$

$$CH_2{=}CHCH_2CH_2CH_2CH{=}CHCH_2CHY_2 \; (\mathbf{31}) +$$

$$(CH_2{=}CHCH_2CH_2CH_2CH{=}CHCH_2)_2CY_2 \; (\mathbf{32}) +$$

$$CH_2{=}CHCH_2CH_2CH_2CH(CH{=}CH_2)CHY_2 \; (\mathbf{33})$$

The reactions with β-diketones, α-formyl ketones, malonates, α-formyl, α-cyano, α-keto, and α-nitro esters, and related compounds have been carried out using mainly $[(C_6H_5)_3P]_2PdCl_2$ and sodium phenoxide as catalysts. In these reactions, when a bidentate ligand such as ethylenebis(diphenylphosphine) was used, 1 mol of butadiene reacted with 1 mol of active methylene compound to give butenyl derivatives.[56]

$$CH_2{=}CHCH{=}CH_2 + CH_2Y_2 \rightarrow CH_3CH{=}CHCH_2CHY_2 + CH_2{=}CHCH(CH_3)CHY_2$$

In general, methylene or methyne groups activated by one electronegative group are not active for the telomerization; simple ketones and esters, for example, are inert. However, nitroalkanes react with butadiene smoothly, and their α hydrogens are replaced with 2,7-octadienyl groups.[53] For example, a reaction of nitromethane with butadiene in the presence of $[(C_6H_5)_3P]_2PdCl_2$ and sodium hydroxide in isopropyl alcohol at room temperature gave 9-nitro-1,6-nonadiene (**34**), 9-nitro-1,6,11,16-heptadecatetraene (**35**), and 9-nitro-9-(2,7-octadienyl)-1,6,11,16-heptadecatetraene (**36**), accompanied by a small amount of branched products. The relative amounts of these

$$CH_2{=}CHCH{=}CH_2 + CH_3NO_2 \rightarrow$$

$$CH_2{=}CHCH_2CH_2CH_2{=}CH_2CH_2NO_2 \; (\mathbf{34}) +$$

$$(CH_2{=}CHCH_2CH_2CH{=}CHCH_2)_2CHNO_2 \; (\mathbf{35}) +$$

$$(CH_2{=}CHCH_2CH_2CH_2CH{=}CHCH_2)_3CNO_2 \; (\mathbf{36})$$

products can be controlled by adjusting the reaction time and the ratio of reactants. The reaction is also applicable to other nitroalkanes. Hydrogenation of these unsaturated nitroalkanes gives corresponding long-chain amines, which are novel in that they have a primary amino group at the middle of the molecule, rather than at the terminal position as in common amines. Following is one example.

$$(CH_3CH_2CH_2CH_2CH_2CH_2CH_2CH_2)_2CHNH_2$$

Transfer of Allylic Groups

Exchange of allylic groups was observed in the reaction of allyl acetate and sodium propionate in the presence of $PdCl_2$ to give allyl propionate.[57] During the course of studies on the telomerization of butadiene, it was found that the 2,7-octadienyl group can be transferred to other active hydrogen compounds. For example, 1-acetoxy-2,7-octadiene reacts with diethylamine under palladium catalysis to yield *N,N*-diethyl-2,7-octadienylamine by transfer of the allylic group.[50,58] The reaction can be extended to allylic functional moieties in general, including alcohols, ethers, esters, and amines. For example, allyl alcohol reacts with diethylamine in the presence of palladium acetylacetonate and $(C_6H_5)_3P$ at 50° for 30 min to yield allyldiethylamine[59] (eq 4). Allylic

$$CH_2{=}CHCH_2OH + (C_2H_5)_2NH \rightarrow CH_2{=}CHCH_2N(C_2H_5)_2 \quad (4)$$

ethers and esters react with acetylacetone to give carbon allylated products.[60] Allylation of enamines with allyl acetate is another example[54] (eq 5). Even

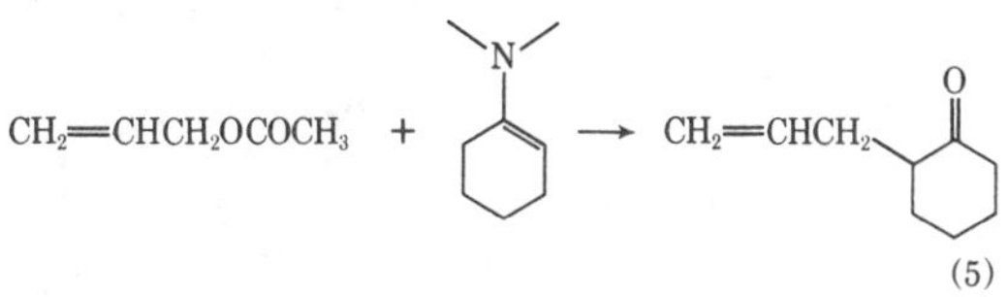

(5)

allyldiethylamine reacts with acetylacetone to give 3-allylacetylacetone and 3,3-diallylacetylacetone by C–N bond cleavage.[59]

In these allyl transfer reactions, C–O and C–N bonds are cleaved. Thus simple elimination of allylic

(54) J. Tsuji, special lecture presented at 23rd International Congress of Pure and Applied Chemistry, July 1971.

(55) G. Hata, K. Takahashi, and A. Miyake, *Chem. Ind.*, 1836 (1969); *J. Org. Chem.*, **36**, 2116 (1971).

(56) K. Takahashi, G. Hata, and A. Miyake, *Chem. Ind.*, 488 (1971); *Bull. Chem. Soc. Jap.*, **45**, 1183 (1972).

(57) A. Sabel, J. Smidt, R. Jira, and H. Prigge, *Chem. Ber.*, **102**, 2939 (1969).

(58) T. S. Shryne, E. J. Smutny, and D. P. Stevenson, U. S. Patent, 3,493,617 (1970).

(59) K. E. Atkins, W. E. Walker, and R. M. Manyik, *Tetrahedron Lett.*, 3821 (1970).

(60) G. Hata, K. Takahashi, and A. Miyake, *Chem. Commun.*, 1392 (1970); *Bull. Chem. Soc. Jap.*, **45**, 230 (1972).

groups is also possible in the absence of nucleophiles by using palladium catalyst. 1-Phenoxy-2,7-octadiene is converted into 1,3,7-octatriene by heating with palladium complexes of phosphine.[36,37] Similarly, acetic acid is eliminated from 1-acetoxy-2,7-octadiene by the catalysis of palladium.

$$CH_2{=}CHCH_2CH_2CH_2CH{=}CHCH_2OX \rightarrow CH_2{=}CHCH_2CH_2CH{=}CHCH{=}CH_2 + XOH$$

$$X = CH_3CO,\ C_6H_5,\ CH_3$$

Reaction of Hydrosilanes

Telomerization with hydrosilanes, namely hydrosilation of butadiene, proceeds in the presence of palladium catalysts. This reaction is somewhat different from other palladium-catalyzed telomerizations discussed earlier. In the hydrosilation of butadiene, 1 or 2 mol of butadiene react depending on the kind of silane.[61-63] Trimethylsilane reacts with 2 mol of butadiene, catalyzed by bis(triphenylphosphine)(maleic anhydride)palladium, to give 1-trimethylsilyl-2,6-octadiene in 98% yield (eq 6). In this reaction, the 2,6-

$$2CH_2{=}CHCH{=}CH_2 + HSi(CH_3)_3 \rightarrow CH_3CH{=}CHCH_2CH_2CH{=}CHCH_2Si(CH_3)_3 \quad (6)$$

octadienyl rather than the 2,7-octadienyl derivative was formed. On the other hand, 1 mol of butadiene reacts with trichlorosilane at 100° for 6 hr to give 1-trichlorosilyl-2-butene in 93.5% yield.

$$CH_2{=}CHCH{=}CH_2 + HSiCl_3 \rightarrow CH_3CH{=}CHCH_2SiCl_3$$

Unlike the telomerizations mentioned previously, the hydrosilation of butadiene is possible by using various palladium compounds including $PdCl_2$ without base, and even metallic palladium when combined with various aryl- and alkylphosphines.

Carbonylation of Butadiene

Palladium is an active catalyst for carbonylation of simple olefins.[1] In the carbonylation of butadiene, 1 mol of butadiene reacts with CO by using $PdCl_2$ as the catalyst to give 3-pentenoate.[64,65] Further studies revealed that the carbonylation of butadiene can be made to give 3,8-nonadienoate by using different palladium catalysts.[54,66,67]

$$CH_3CH{=}CHCH_2CO_2R \leftarrow CH_2{=}CHCH{=}CH_2 + CO + ROH \rightarrow CH_2{=}CHCH_2CH_2CH_2CH{=}CHCH_2CO_2R$$

The essential factor which differentiates the monomeric and dimeric carbonylations is the presence or absence of a halide ion coordinated to palladium. With halide-free complexes, such as $Pd(OCOCH_3)_2$ and $(C_6H_5)_3P$, 3,8-nonadienoate is obtained almost selectively. In a typical run, a reaction of butadiene (20 g), isopropyl alcohol (30 ml), and CO (50 atm) at 110° for 16 hr gave isopropyl 3,8-nonadienoate (32.5 g). The reaction proceeds smoothly even in *tert*-butyl alcohol to give the *tert*-butyl ester in a high yield.

In the carbonylation reaction, 2 mol of butadiene form the diallylic palladium complex and then CO insertion takes place to give 3,8-nonadienoate. When chloride ion coordinates with the palladium, formation of the diallylic complex is not possible, and only monomeric complex is formed, CO insertion to which gives 3-pentenoate.

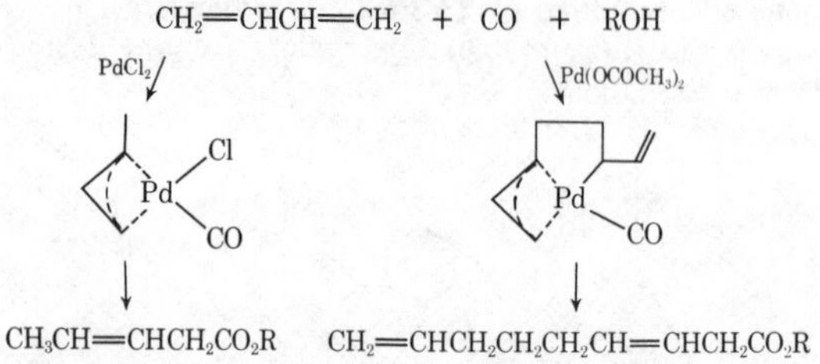

CO insertion into π-allylpalladium chloride is a well-established reaction.[68] Carbonylation of allyl chloride catalyzed by $PdCl_2$ gives 3-butenoyl chloride *via* the π-allylpalladium complex as an intermediate.[69,70] Carbonylation of allyl chloride in the presence of an excess of butadiene, catalyzed by π-allylpalladium chloride in benzene, produced a mixture of 3,7-octadienoyl chloride and 3-butenoyl chloride.[71] The former was formed by insertion of butadiene into π-allylpalladium bond, followed by CO insertion.

$$CH_2{=}CHCH_2Cl + CH_2{=}CHCH{=}CH_2 + CO \rightarrow CH_2{=}CHCH_2CH_2CH{=}CHCH_2COCl + CH_2{=}CHCH_2COCl$$

Mechanism of the Telomerization Reactions

Unlike nickel, the size of a palladium atom precludes cyclization, and a rapid hydrogen transfer occurs to give linear oligomers and telomers. The σ,π-diallylpalladium complex **12** coordinated by phosphine is postulated as the intermediate in the dimerization. 1-Substituted 2,7-octadienes are formed from this intermediate by 1,6 addition of nucleophiles (YH) to the allylic ligand. In other words, this intermediate is protonated at C_6 *via* σ-allylpalladium bond fission by protic medium (YH) and attacked by the nucleophiles at C-1 of the π-allylpalladium bond to yield the corresponding telomers. 3,6 addition gives rise to branched telomers. The fact that 1-methoxy-6-deuterio-2,7-octadiene was formed by the reaction of butadiene with CH_3OD can be explained by this mechanism.[38]

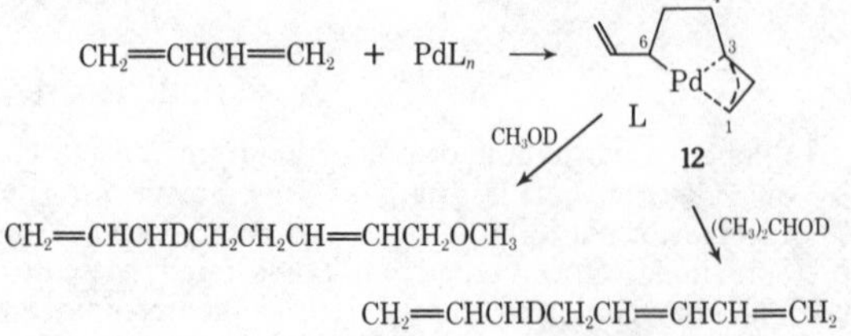

(61) S. Takahashi, T. Shibano, and N. Hagihara, *Chem. Commun.*, 161 (1969).
(62) S. Takahashi, T. Shibano, H. Kojima, and N. Hagihara, *Organometal. Chem. Syn.*, 1, 193 (1971).
(63) M. Hara, K. Ohno, and J. Tsuji, *Chem. Commun.*, 247 (1971).
(64) J. Tsuji, J. Kiji, and S. Hosaka, *Tetrahedron Lett.*, 605 (1964).
(65) S. Hosaka and J. Tsuji, *Tetrahedron*, **27**, 3821 (1971).
(66) W. E. Billups, W. E. Walker, and T. C. Shields, *Chem. Commun.*, 1067 (1971).
(67) J. Tsuji, M. Hara, and Y. Mori, *Tetrahedron*, **28**, 3721 (1972).
(68) J. Tsuji, J. Kiji, and M. Morikawa, *Tetrahedron Lett.*, 1811 (1963).
(69) J. Tsuji, J. Kiji, S. Imamura, and M. Morikawa, *J. Amer. Chem. Soc.*, **86**, 4350 (1964).
(70) W. T. Dent, R. Long, and G. H. Whitefield, *J. Chem. Soc.*, 1588 (1964).
(71) D. Medema, R. van Helden, and C. F. Kohll, *Inorg. Chim. Acta*, **3**, 255 (1969).

Formation of 1,3,7-octatriene is possible from the same intermediate complex by hydrogen migration from C-4 to C-6. When the reaction was carried out in $(CH_3)_2CHOD$, the octatriene obtained was thought to be monodeuterated at C-6.[34] From this result, it is reasonable to assume that the addition of a proton of isopropyl alcohol to the C-6 atom and the elimination of the proton from C-4 result in an increasing rate of hydrogen migration from C-4 to C-6, and hence enhancement of the rate of octatriene formation is possible.[17]

Stoichiometric reactions of π-allylpalladium chloride with nucleophiles are known. For example, the complex reacts with diethyl malonate to give diethyl allylmalonate. Also reactions of the complex with enamines of cyclohexanone produced 2-allylcyclohexanone.[25,72]

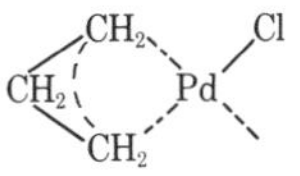

$+ \ CH_2(CO_2R)_2 \longrightarrow$

$CH_2{=}CHCH_2CH(CO_2R)_2 \ + \ Pd \ + \ Cl^-$

In the catalytic reactions of butadiene, the original π-allylic ligand comes off at the end of the cycle, and subsequent cycles involve formally a reversible oxidation–reduction between the palladium(0) complex and the π-allylic palladium(II) complex derived from butadiene. This is possible in the absence of the strongly coordinated chloride ion.

Related Reactions

Reaction of butadiene in benzene using π-allylpalladium acetate as the catalyst yielded the trimer, *n*-1,3,6,10-dodecatetraene, with selectivity of 70% and conversion of 30% in 22 hr.[73] A minor product was 1,3,7-octatriene. When the reaction was carried out in methanol, a mixture of octadienyl, dodecatrienyl, hexadecatetraenyl, and higher methyl ethers was obtained. π-Allylpalladium acetylacetonate is not an active catalyst for the trimerization. An acetate-bridged palladium complex (**37**) was isolated, which was found to be the active catalyst of the trimerization. Also stable chelate complexes of palladium (**38**) are catalysts for the trimerization of butadiene in dimethylformamide or Me_2SO (60% yield, 30% conversion) at 70°.[74] Salts of palladium with noncomplexing anions such as ClO_4^- and BF_4^- convert butadiene into a mixture of divinylcyclobutanes.[75]

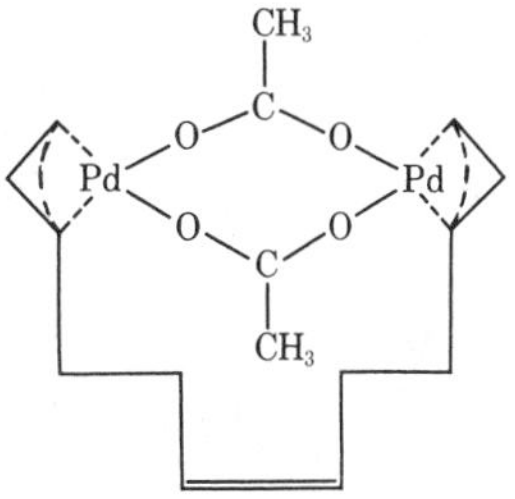

37

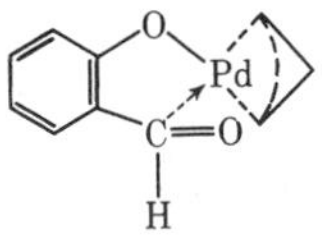

38

Stereoselective synthesis of 1,4-hexadiene from butadiene and ethylene has been reported.[76] The catalyst was prepared by the reaction of $[(C_6H_5)_3P]_2PdCl_2$ with diisobutylaluminum chloride.

$$CH_2{=}CHCH{=}CH_2 + CH_2{=}CH_2 \rightarrow CH_2{=}CHCH_2CH{=}CHCH_3$$

I deeply appreciate the diligent efforts and careful work of my coworkers, mentioned in the references.

(72) J. Tsuji, H. Takahashi, and M. Morikawa, *Kogyo Kagaku Zasshi*, **69**, 920 (1966).

(73) D. Medema and R. van Helden, *Recl. Trav. Chim. Pays-Bas*, **90**, 324 (1971).

(74) A. S. Astakhova, A. S. Berenblyum, L. G. Korableva, I. P. Rogachev, B. G. Rogachev, and M. L. Khidekel, *Izv. Akad. Nauk SSSR, Ser. Khim.*, 1362 (1971); *Chem. Abstr.*, **75**, 88750b (1971).

(75) E. G. Chepaikin and M. L. Khidekel, *Izv. Akad. Nauk SSSR, Ser. Khim.*, 1129 (1971); *Chem. Abstr.*, **75**, 87760t (1971).

(76) W. Schneider, *Amer. Chem. Soc., Div. Petrol. Chem. Prepr.* **14**, B89 (1969).

Palladium(II)-Catalyzed Exchange and Isomerization Reactions

Patrick M. Henry

Department of Chemistry, University of Guelph, Guelph, Ontario, Canada

Received June 12, 1972

Reprinted from Accounts of Chemical Research, **6**, 16 (1973)

In the last dozen years considerable advances have been made in homogeneous catalysis in general and in Pd(II) catalysis in particular. No doubt much of the impetus in Pd(II) catalysis research was provided by disclosure of the Wacker process for manufacture of acetaldehyde by Smidt and coworkers in 1959[1] (eq 1). This process is now the preferred method of man-

$$C_2H_4 + \tfrac{1}{2}O_2 \xrightarrow[\substack{CuCl_2 \\ H_2O}]{PdCl_2} CH_3CHO \qquad (1)$$

ufacturing acetaldehyde. Also, ketones can be produced by oxidation of higher olefins (eq 2).

$$RCH{=}CHR' + \tfrac{1}{2}O_2 \longrightarrow RCOCH_2R' \qquad (2)$$

This concentrated effort on Pd(II) catalysis has resulted in discovery of a number of new reactions. One which has commercial possibilities is the vinyl ester synthesis[2] (eq 3). Closely related is the olefin arylation reaction[3] (eq 4).

$$C_2H_4 + 2OAc^- \xrightarrow[O_2]{Pd(II)} CH_2{=}CHOAc + HOAc \qquad (3)$$

R—⟨◯⟩—HgCl + PdCl_2 + RCH=CH_2 ⟶

RCH=CH—⟨◯⟩—R + Pd^0 + HgCl_2 + HCl (4)

Carbon monoxide is involved in many of the new reactions such as the aromatic acid synthesis[4] (eq 5)

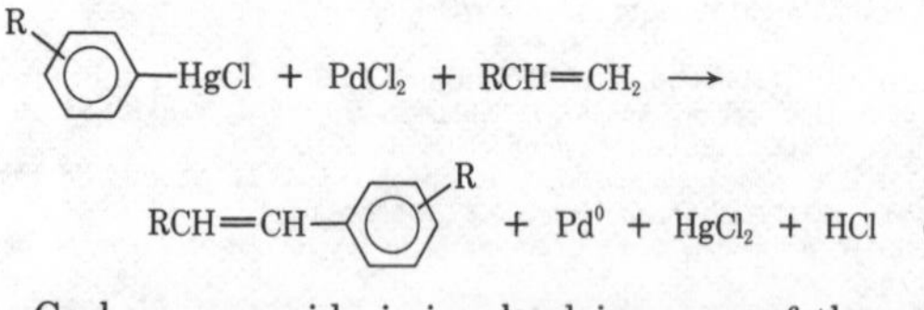

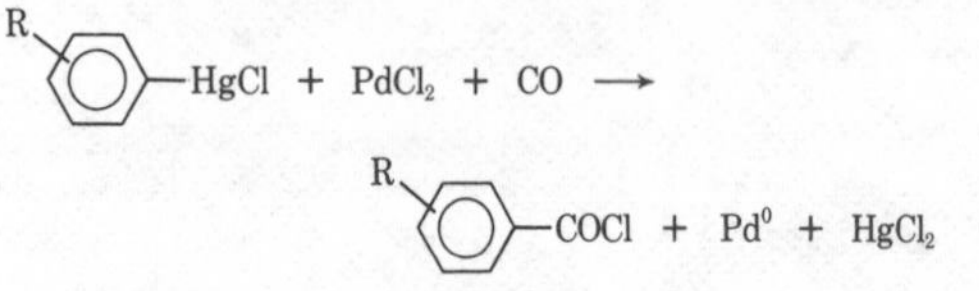

R—⟨◯⟩—HgCl + PdCl_2 + CO ⟶

R—⟨◯⟩—COCl + Pd^0 + HgCl_2 (5)

and the isocyanate synthesis[5] (eq 6). In some cases

$$RNH_2 + PdCl_2 + CO \rightarrow RNCO + 2HCl + Pd^0 \qquad (6)$$

both olefin and CO are involved, as in the carbonylation of olefins[6] (eq 7). A reaction which involves nei-

$$RCH{=}CH_2 + CO + PdCl_2 \dashrightarrow RCHClCH_2COCl + Pd^0 \qquad (7)$$

ther olefin nor CO is the aromatic coupling reaction[7] (eq 8).

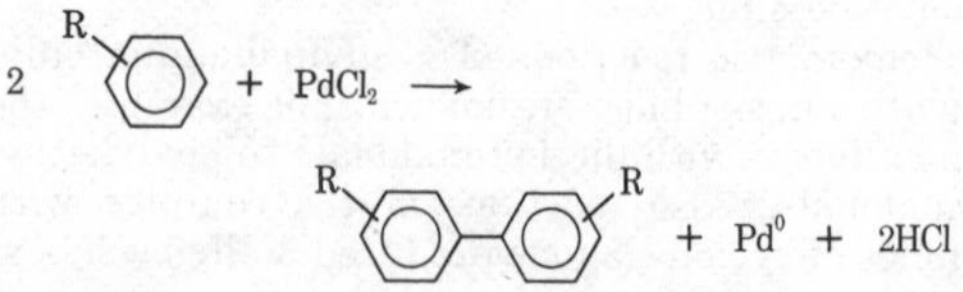

The examples given above are only a few of the many new Pd(II)-catalyzed reactions.[8]

Patrick Henry received his B.S. and M.S. degrees from De Paul University, Chicago, Ill., and his Ph.D. degree from Northwestern University in 1956. He then joined Hercules, Inc., where he worked mainly on metal ion catalyzed reactions. In October 1971 he joined the chemistry faculty at the University of Guelph. His major research interests are in the areas of organic oxidation by metal ions, coordination chemistry, and mechanisms of homogeneous catalysis and reaction of metal ions with oxygen.

2 R—⟨◯⟩ + PdCl_2 ⟶

R—⟨◯⟩—⟨◯⟩—R + Pd^0 + 2HCl (8)

In spite of these synthetic advances in Pd(II) chemistry, relatively little effort has been expended on the mechanisms of Pd(II) catalysis. In fact, the only reaction which has been extensively studied is the basic reaction of the Wacker process (eq 9). The

$$PdCl_4^{2-} + C_2H_4 + H_2O \rightarrow Pd^0 + CH_3CHO + 2HCl + 2Cl^- \qquad (9)$$

results of several kinetic studies[9-11] of this reaction are summarized in eq 10. It is generally agreed that

$$-d[C_2H_4]/dt = \\ k[PdCl_4^{2-}][C_2H_4]/([H^+][Cl^-]^2) \qquad (10)$$

the mechanism for this reaction is that given by eq 11-14. The important feature of this mechanism is

$$PdCl_4^{2-} + C_2H_4 \underset{}{\overset{K_1}{\rightleftharpoons}} PdCl_3(C_2H_4)^- + Cl^- \qquad (11)$$

$$PdCl_3(C_2H_4)^- + H_2O \overset{K_2}{\rightleftharpoons} PdCl_2(OH)(C_2H_4)^- + HCl \qquad (12)$$

$$Cl_2(H_2O)Pd{-}CH_2CH_2OH \xrightarrow{fast}$$

$$Pd^0 + CH_3CHO + HCl + Cl^- + H_2O \qquad (14)$$

the cis insertion of the elements of Pd(II) and coordinated OH across the olefinic double bond (eq 13) to give the β-hydroxyethylpalladium(II) alkyl (or

(1) J. Smidt, R. Jira, J. Sedlmeier, R. Sieber, R. Rüttinger, and H. Kojer, *Angew. Chem., Int. Ed. Engl.*, **1**, 80 (1962).

(2) I. I. Moiseev, M. N. Vargaftik, and Ya. K. Sirkin, *Dokl. Akad. Nauk SSSR*, **133**, 377 (1960).

(3) R. F. Heck, *J. Amer. Chem. Soc.*, **90**, 5518 (1968), and following papers.

(4) P. M. Henry, *Tetrahedron Lett.*, 2285 (1968).

(5) E. W. Stern and M. L. Spector, *J. Org. Chem.*, **31**, 596 (1966).

(6) J. Tsuji, M. Morikawa, and J. Kiji, *J. Amer. Chem. Soc.*, **86**, 8451 (1964).

(7) R. van Helden and C. Verberg, *Recl. Trav. Chim. Pays-Bas*, **84**, 1263 (1965).

(8) For recent reviews see: (a) E. W. Stern, *Catal. Rev.*, **1**, 74 (1968); (b) J. Tsuji, *Accounts Chem. Res.*, **2**, 144 (1969); (c) F. R. Hartley, *Chem. Rev.*, **69**, 799 (1969); (d) P. M. Henry, *Trans. N. Y. Acad. Sci.*, **33**, 41 (1971); (e) P. M. Maitlis, "Organic Chemistry of Palladium," Vol. II, Academic Press, New York, N. Y., 1971.

(9) I. I. Moiseev, M. N. Vargaftik, and Ya. K. Sirkin, *Dokl. Akad. Nauk SSSR*, **153**, 140 (1963).

(10) R. Jira, J. Sedlmeier, and J. Smidt, *Justus Liebigs Ann. Chem.*, **693**, 99 (1966).

(11) P. M. Henry, *J. Amer. Chem. Soc.*, **86**, 3246 (1964); **88**, 1595 (1966).

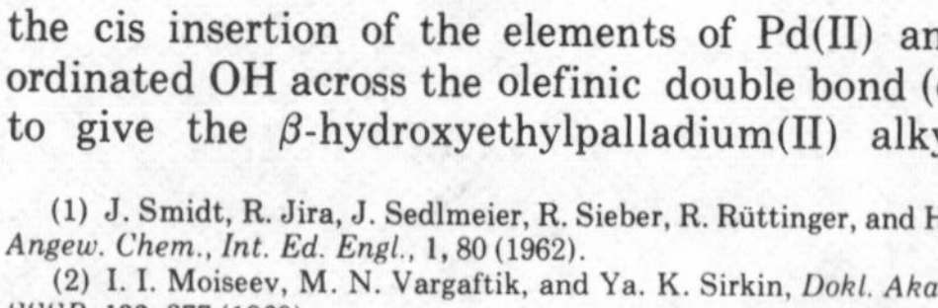

hydroxypalladation adduct, in analogy to the well-known hydroxymercuration adducts[12]). The kinetic expression (eq 10) is consistent with the need for both coordinated ethylene and OH since it contains the $[H^+]$ and $[Cl^-]^2$ terms in the numerator of the rate expression required by eq 11 and 12.[13]

The only other system which has been studied to any extent is the oxidation of olefins to enol and allylic acetates in acetic acid (AcOH). The initial product distributions for several straight-chain olefins indicated that the reaction proceeds *via* an acetoxypalladation route[14] analogous to that proposed for the Wacker reaction. With ethylene the reaction scheme would be as given by eq 15. Because of the

$$C_2H_4 + Pd(OAc)_2 \rightarrow AcOPdCH_2CH_2OAc \rightarrow HPdOAc + CH_2{=}CHOAc \quad (15)$$

complexities of the system, an interesting kinetic study[15] was unable definitely to determine if the acetoxypalladation proceeds *via* coordinated acetate addition to coordinated olefin in a fashion similiar to that postulated for hydroxypalladation (eq 13).

No doubt experimental difficulties are one reason Pd(II) catalysis has not been studied more thoroughly. The rates are often inconveniently slow and, since most are oxidative in nature, they are complicated by the effects of precipitation of Pd metal.

One type of Pd(II)-catalyzed reactions free of these complications comprises the vinylic and allylic exchange reactions represented by eq 16 (X or Y =

$$CH_2{=}CHX \text{ or } CH_2{=}CHCH_2X + Y^- \xrightleftharpoons{Pd(II)} CH_2{=}CHY \text{ or } CH_2{=}CHCH_2Y + X^- \quad (16)$$

OOCR, Cl, OR, NR_2 etc.). Since this reaction is nonoxidative in nature, these reactions do not precipitate Pd metal, and the rates are in a convenient range for measurement at 25°. Moreover, stereochemical evidence for mechanism can readily be obtained in these systems. Therefore the author chose this type of reaction for studies aimed at elucidating the detailed paths of Pd(II) catalysis.

By analogy with the Wacker reaction, these reactions might be expected to proceed by addition of the Pd(II) and Y followed by elimination of Pd(II) and X, as shown in eq 17 and 18. One point on

$$CH_2{=}CHX + -PdY \rightarrow -Pd-CH_2CH\begin{smallmatrix}Y\\X\end{smallmatrix} \quad (17)$$

$$-Pd-CH_2CH\begin{smallmatrix}Y\\X\end{smallmatrix} \rightarrow CH_2{=}CHY + -Pd-X \quad (18)$$

which these studies would hopefully shed light is whether or not coordination of Y to Pd(II) is cequired for insertion across the olefinic double bond. Other questions which these studies would attempt to answer concern the stereochemistry of attack of various nucleophiles, Y, the need for coordination of organic substrates, and the role of coordination unsaturation. These questions are not only important to Pd(II) catalysis but are basic to homogeneous catalysis in general.[16]

All kinetic studies reported in this Account were carried out in acetic acid, and most involved palladium(II) chloride salts in LiCl-containing solutions. In order properly to interpret the kinetics, it is necessary to know the equilibria between Pd(II) and other species in solution. A molecular weight and spectral study[17] indicated the equilibria represented by eq 19 and 20 to be operative. K_3 has a value of 0.1 M^{-1}

$$Li_2Pd_2Cl_6 + 2LiCl \overset{K_3}{\rightleftharpoons} 2Li_2PdCl_4 \quad (19)$$

$$2LiCl \overset{K_D}{\rightleftharpoons} Li_2Cl_2 \quad (20)$$

and K_D is 2.6 M^{-1} at 25°. No complexing of LiOAc by Pd(II) could be detected in the chloride containing systems.

Vinyl Ester Exchange with Acetic Acid

The vinyl ester exchange reaction (eq 21), was first

$$CH_2{=}CHOOCR + HOAc \xrightleftharpoons{Pd(II)} CH_2{=}CHOAc + HOOCR \quad (21)$$

reported by Smidt and coworkers.[1] Their subsequent study of the mechanism indicated that the reaction proceeds by the route represented by eq 17 and 18.[18] The kinetics of the reaction was determined using either deuterated vinyl acetate[19] (R = CD_3) or vinyl propionate[20] (R = C_2H_5) as labels. Over a wide range of LiOAc and LiCl concentrations the rate expression is given by eq 22. The k_1' and k'' must cor-

$$\text{rate} = ([Li_2Pd_2Cl_6][\text{vinyl ester}]/[LiCl]) \times (k_1' + k_1''[LiOAc]) \quad (22)$$

respond to reaction with acetic acid solvent and acetate ion, respectively. Their relative values are such that at [LiOAc] = 0.1 M the k_1' path contributes only 10% of the total rate.

Two features of the kinetics deserve comment. First, the reaction is first order in dimer. If the first step of the reaction is formation of a Pd(II)–olefin π complex, as usually assumed for Pd(II)-catalyzed reactions, then the complex formation must be occurring *via* eq 23 rather than eq 24.[21]

$$Li_2Pd_2Cl_6 + C_2H_3OOCR \overset{K_4}{\rightleftharpoons} Li_2Pd_2Cl_5(C_2H_3OOCR) + LiCl \quad (23)$$

$$Li_2Pd_2Cl_6 + 2C_2H_3OOCR \overset{K_5}{\rightleftharpoons} 2LiPdCl_3(C_2H_3OOCR) \quad (24)$$

Second, the LiCl inhibition appears only to the first power. If coordination of both olefin and acetate were required, an $[LiCl]^2$ inhibition term would be expected in the rate expression. This result strongly

(12) J. Chatt, *Chem. Rev.*, **48**, 7 (1951).

(13) It should be pointed out that the kinetics do not absolutely require cis attack of Pd-OH. See P. M. Henry, *Advan. Chem. Ser.*, **No. 70**, 136 (1968).

(14) W. Kitching, Z. Rappoport, S. Winstein, and W. G. Young, *J. Amer. Chem. Soc.*, **88**, 2054 (1966).

(15) I. I. Moiseev, M. N. Vargaftik, S. V. Pestrikov, O. G. Levanda, T. N. Romanova, and Ya. K. Sirkin, *Dokl. Akad. Nauk SSSR*, **171**, 1365 (1969).

(16) J. P. Collman, *Accounts Chem. Res.* **1**, 136 (1968).

(17) P. M. Henry and O. W. Marks, *Inorg. Chem.*, **10**, 373 (1971).

(18) A. Sabel, J. Smidt, R. Jira, and H. Prigge, *Chem. Ber.*, **102**, 2939 (1969).

(19) P. M. Henry, *J. Amer. Chem. Soc.*, **93**, 3853 (1971).

(20) P. M. Henry, *J. Amer. Chem. Soc.*, **94**, 7316 (1972).

(21) Equation 16 would require a [vinyl ester]$^{1/2}$ term in the rate expression (eq 14).

suggests that acetate is attacking from outside the coordination sphere of Pd(II). This type of addition

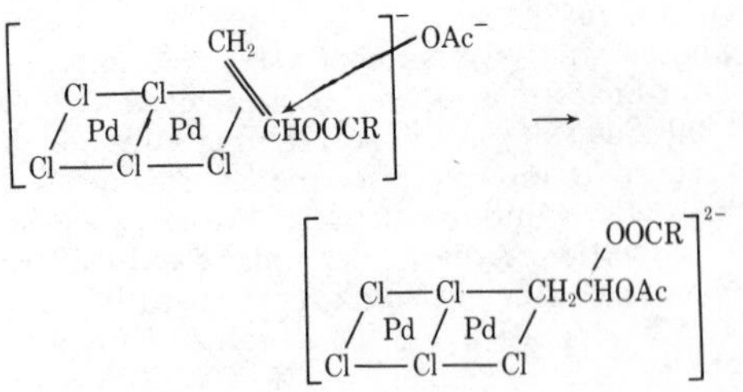

$$
\left[\begin{array}{c} \text{Cl—Cl—} \\ /\ \text{Pd}\ /\ \text{Pd}\ / \\ \text{Cl—Cl—Cl} \end{array} \begin{array}{c} \text{CH}_2 \\ \text{CHOOCR} \end{array} \right] \text{OAc}^- \longrightarrow
$$

$$
\left[\begin{array}{c} \text{OOCR} \\ \text{Cl—Cl—CH}_2\text{CHOAc} \\ /\ \text{Pd}\ /\ \text{Pd}\ / \\ \text{Cl—Cl—Cl} \end{array} \right]^{2-} \quad (25)
$$
$$
\mathbf{1}
$$

suggest trans stereochemistry, and indeed the products of oxidation of tetradeuteriocyclohexene are best explained by trans acetoxypalladation and cis palladium(II) hydride elimination.[22] The reaction scheme for formation of 3-cyclohexen-1-yl acetate is shown in eq 26.

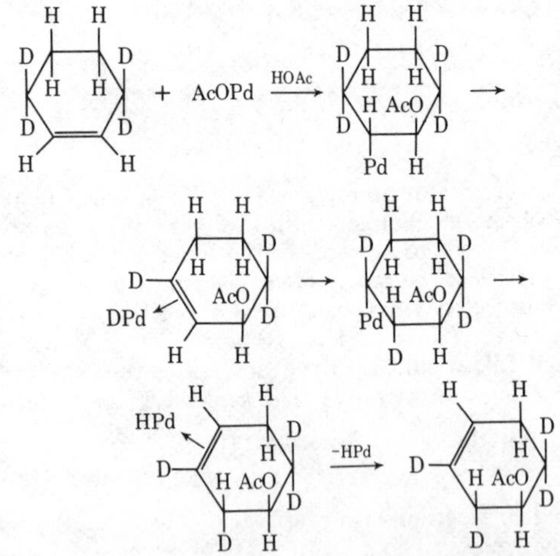

$$(26)$$

The most reasonable reaction sequence consistent with the rate expression is thus formation of a dimeric π complex *via* eq 23, followed by trans attack of acetate or acetic acid to give an acetoxypalladation adduct *via* eq 25. Reversal of eq 25 with ejection of $^-$OOCR instead of acetate completes exchange (eq 27).

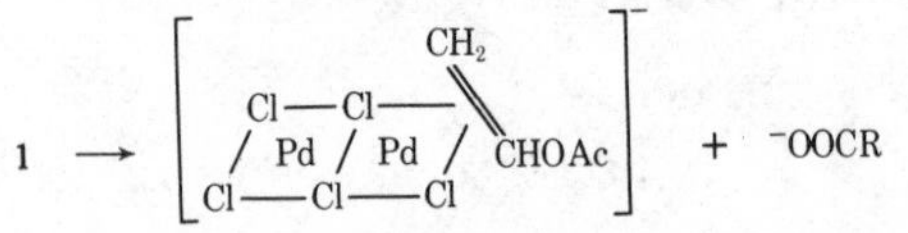

$$
\mathbf{1} \longrightarrow \left[\begin{array}{c} \text{CH}_2 \\ \text{Cl—Cl—} \\ /\ \text{Pd}\ /\ \text{Pd}\ /\ \text{CHOAc} \\ \text{Cl—Cl—Cl} \end{array} \right]^- + \ ^-\text{OOCR} \quad (27)
$$

To test this mechanism, the stereochemistry of the OAc_D exchange of *cis*- and *trans* 1-propen-1-yl acetate and propionate were determined. Exchange occurred only with isomerization.[19] As shown in eq 28 this result is consistent with the acetoxypalladation mechanism in which acetoxypalladation is stereospecific (A = addition, E = elimination). It is inconsistent with S$_N$2 attack of acetate on the carbon–oxygen bond, a mechanism which has been suggested for some Pd(II)-catalyzed oxidations, and which would predict retention of configuration.

(22) P. M. Henry and G. A. Ward, *J. Amer. Chem. Soc.*, **93**, 1494 (1971).

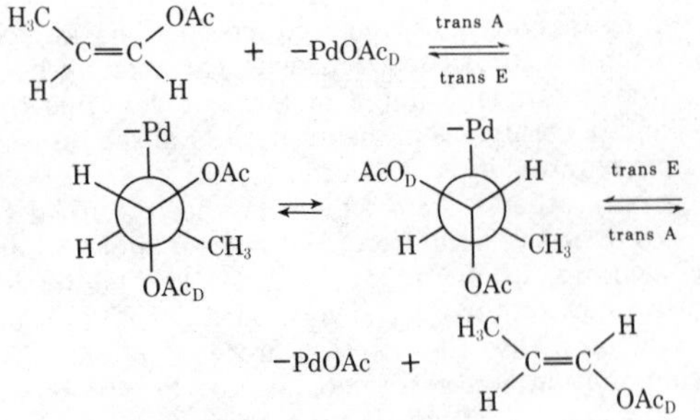

$$(28)$$

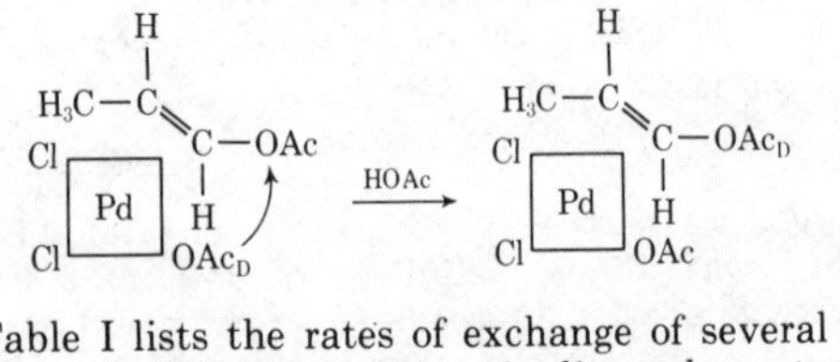

$$(29)$$

Table I lists the rates of exchange of several enol acetates.[19] Substitution on vinylic carbon strongly inhibits exchange. This result is consistent with the acetoxypalladation mechanism since addition of the elements of acetate and Pd(II) dimer across a double bond would have a large steric requirement.

Table I
Rates of Exchange of Various Enol Acetates at 25°

Enol acetate	$k, M^{-1} \sec^{-1}$
CH_2=CHOAc	2.0×10^{-2}
trans-CH$_3$CH=CHOAc	5.0×10^{-4}
cis-CH$_3$CH=C(CH$_3$)OAc	3.2×10^{-8}
1-Cyclopentenyl acetate	$<10^{-9}$

It is surprising at first glance that 1-acetoxy-1-cyclopentene does not exchange detectably since, on steric grounds, it would be expected to have approximately the same rate as 2-acetoxy-2-butene. However, as shown by eq 30, stereochemically pure acetoxypalladation (and deacetoxypalladation by the principle of microscopic reversibility) would not permit acetate exchange in this system.

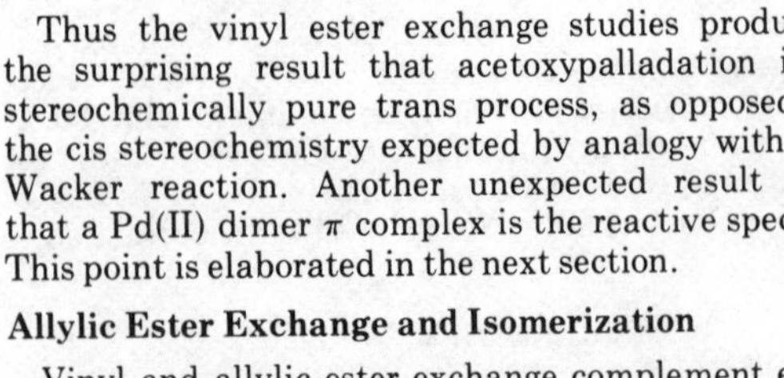

$$(30)$$

Thus the vinyl ester exchange studies produced the surprising result that acetoxypalladation is a stereochemically pure trans process, as opposed to the cis stereochemistry expected by analogy with the Wacker reaction. Another unexpected result was that a Pd(II) dimer π complex is the reactive species. This point is elaborated in the next section.

Allylic Ester Exchange and Isomerization

Vinyl and allylic ester exchange complement each other. Thus, unless optically active allylic esters are used, allylic ester exchange does not give information concerning stereochemistry of addition. However, exchange of unsymmetrical esters readily distinguishes between S_N2, π-allyl, and acetoxypalladation routes. The acetoxypalladation route would predict that exchange occurs only with isomerization of one allylic isomer into the other. For instance, crotyl propionate would exchange to give 2-buten-2-yl acetate (eq 31) while 3-buten-2-yl propionate would give crotyl

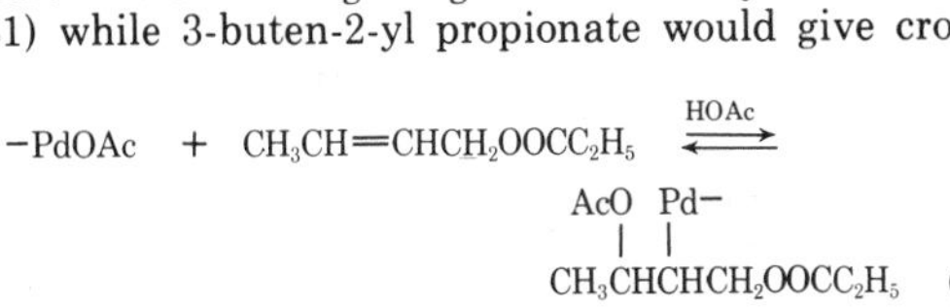

$$-PdOAc + CH_3CH=CHCH_2OOCC_2H_5 \overset{HOAc}{\rightleftharpoons}$$

$$\underset{CH_3CHCHCH_2OOCC_2H_5}{\overset{AcO\ \ Pd-}{\underset{|\ \ \ |}{}}} \quad (31)$$

$$\downarrow$$

$$\underset{CH_2CHCH=CH_2}{\overset{OAc}{\underset{|}{}}} + -Pd(OOCC_2H_5)$$

$$\underset{CH_3CHCH=CH_2}{\overset{OOCC_2H_5}{\underset{|}{}}} + -PdOAc \overset{HOAc}{\rightleftharpoons} \underset{CH_3CHCHCH_2OAc}{\overset{OOCC_2H_5}{\underset{|}{}}} \longrightarrow$$

$$\underset{Pd-}{\underset{|}{}}$$

$$CH_3CH=CHCH_2OAc + -Pd(OOCC_2H_5) \quad (32)$$

acetate (eq 32). The S_N2 and π-allyl routes would predict different product distributions.[23]

The kinetics of the exchange were first studied using allyl propionate as the organic substrate.[24] The rate expression at low allyl propionate concentration is given by eq 33. This rate expression is exactly

$$\text{rate} = \frac{[Li_2Pd_2Cl_6][\text{allyl propionate}]}{[LiCl]} \times$$

$$(k_2' + k_2''[LiOAc]) \quad (33)$$

analogous to that found for vinyl ester exchange, suggesting similar routes. The first step would be π-complex formation following by attack of acetate to give to the acetoxypalladation adduct, 3 (eq 34). Re-

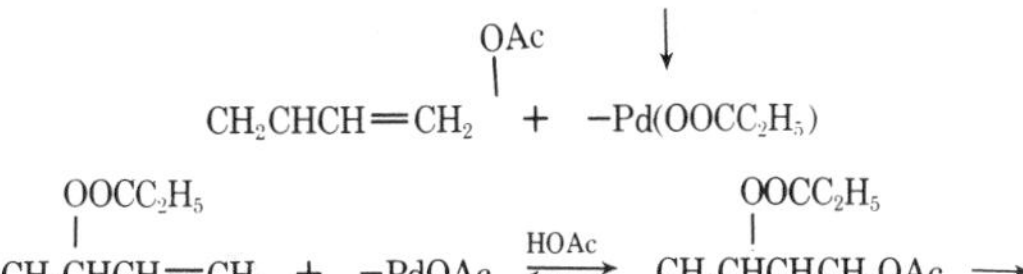

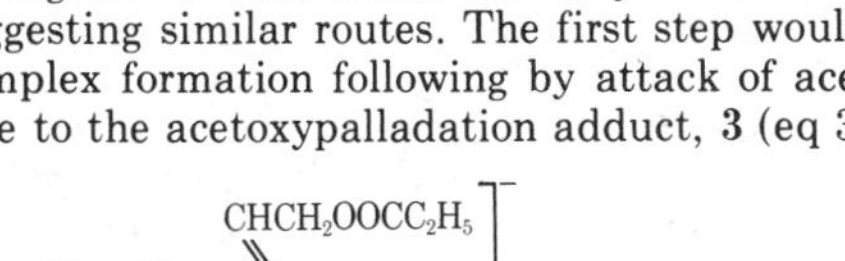

$$\left[\begin{array}{c} \overset{CHCH_2OOCC_2H_5}{\underset{\diagdown}{\parallel}} \\ \underset{Cl-Cl-Cl}{\overset{Cl-Cl}{\underset{/\ Pd\ /\ Pd\ /}{}}} CH_2 \end{array}\right] + LiOAc \overset{slow}{\longrightarrow}$$

2

$$\left[\begin{array}{c} \overset{CH_2OOCC_2H_5}{\underset{|}{}} \\ \overset{CH}{\underset{|}{}} \\ \underset{Cl-Cl-Cl}{\overset{Cl-Cl}{\underset{/\ Pd\ /\ Pd\ /}{}}} CH_2-OAc \end{array}\right]^{2-} \quad (34)$$

3

versal of this step, eliminating $OOCC_2H_5$ instead of OAc, would complete exchange. However, a complication, not present in the vinyl ester exchange, was noted. This complication was inhibition of rate by the allyl propionate itself. In addition the product, allyl acetate, was about as effective an inhibitor as allyl propionate. By analysis of the kinetic data it could be shown that the inhibition arose from π-complex formation to give an *unreactive* monomeric π complex according to eq 35 where K_6 has a value

$$Li_2Pd_2Cl_6 + 2 \text{ allyl ester} \overset{K_6}{\rightleftharpoons} 2LiPdCl_3(\text{allyl ester}) \quad (35)$$

(23) The S_N2 mechanism would predict no isomerization, while the π-alkyl route would require that both alkylic propionates give the same distribution of allylic acetates.

(24) P. M. Henry, *J. Amer. Chem. Soc.*, **94**, 1527 (1972).

of 0.25 M^{-1} for allyl propionate. This value of K_6 from kinetic measurements was confirmed by a spectral study. Since the kinetics require the dimeric π complex (see eq 23) to be the reactive species and this species could not be detected in appreciable concentration, the dimer π complex must be many times more reactive than the monomeric π complex. This difference is believed to result from electrostatic effects. In the rate-determining step of the reaction an acetate is attacking the π complex from outside the coordination sphere. The monomeric π complex would have considerably more negative charge on the Pd(II) containing the allyl ester than in the case of the dimeric π complex, **2**. The mutual repulsion of negative charge would tend to slow the rate of the monomeric π complex.

The exchange of the unsymmetrical esters, crotyl propionate and 3-buten-2-yl propionate, was now studied.[25] The reaction proved to be considerably more complicated than expected. By computer simulation of product distribution with time it could be shown that two separate reactions were taking place. One was the expected exchange reaction with a rate expression identical with that found for allyl propionate (eq 33). Furthermore it was shown that exchange occured only with isomerization of crotyl ester to 3-buten-2-ol ester and *vice versa*, a result expected for the acetoxypalladation mechanism (eq 31 and 32).

The unexpected reaction was isomerization without exchange (eq 36). It was found that, if crotyl

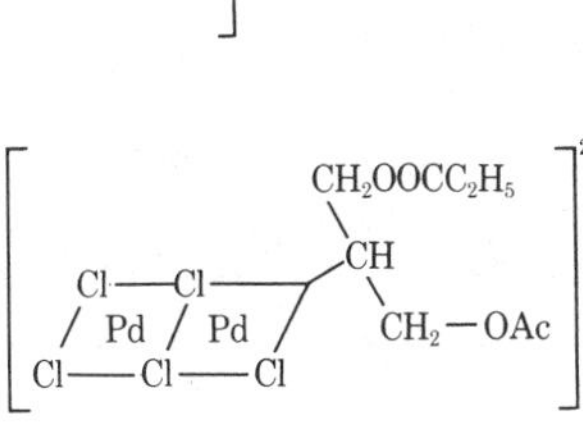

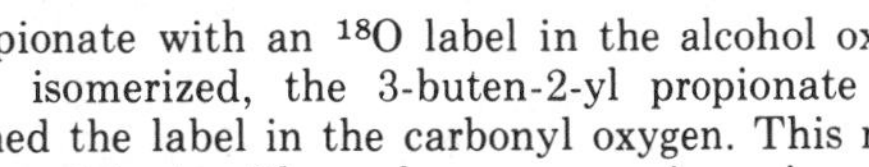

$$CH_3CH=CHCH_2OOCC_2H_5 \overset{HOAc}{\rightleftharpoons} \underset{CH_2=CHCHCH_3}{\overset{OOCC_2H_5}{\underset{|}{}}} \quad (36)$$

propionate with an ^{18}O label in the alcohol oxygen was isomerized, the 3-buten-2-yl propionate contained the label in the carbonyl oxygen. This result is consistent with exchange occurring *via* a 1,3-acetoxonium-type intermediate, **4**. Further evidence

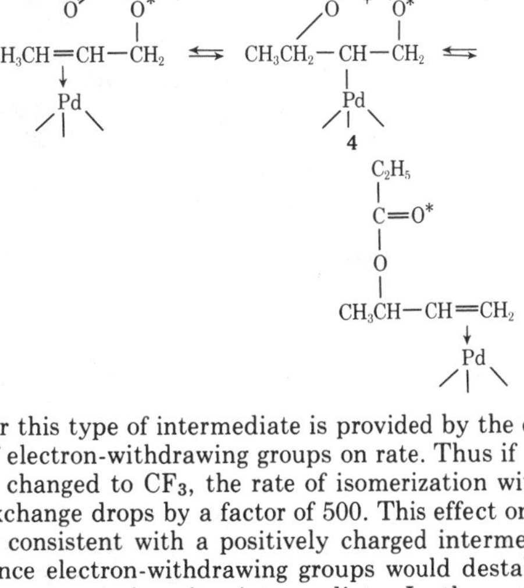

$$\underset{\substack{\downarrow \\ Pd \\ /\ |\ \diagdown}}{\overset{\substack{O \diagup C \diagdown O^* \\ | \\ CH_3CH=CH-CH_2}}{}} \rightleftharpoons \underset{\substack{\downarrow \\ Pd \\ /\ |\ \diagdown}}{\overset{\substack{O \diagup^+ \diagdown O^* \\ | \\ CH_3CH_2-CH-CH_2}}{}} \rightleftharpoons$$

4

$$\underset{\substack{\downarrow \\ Pd \\ /\ |\ \diagdown}}{\overset{\substack{C=O^* \\ | \\ O \\ | \\ CH_3CH-CH=CH_2}}{}} \quad (37)$$

for this type of intermediate is provided by the effect of electron-withdrawing groups on rate. Thus if C_2H_5 is changed to CF_3, the rate of isomerization without exchange drops by a factor of 500. This effect on rate is consistent with a positively charged intermediate since electron-withdrawing groups would destabilize the acetoxonium ion intermediate. In the author's

(25) P. M. Henry, *J. Amer. Chem. Soc.*, **94**, 5200 (1972).

knowledge this is the first example of an oxymetallation giving an unstable bridged intermediate such as **4**.

Vinyl Chloride Exchange with Radioactive Chloride

This exchange, the first to be studied which in-

$$CH_2{=}CHCl + LiCl^* \overset{Pd(II)}{\rightleftharpoons} CH_2{=}CHCl^* + LiCl \quad (38)$$

volved exchanging species other than acetate, provided some new surprises.[26] First the rate expression is extremely simple. Most important, there was no indication of the LiCl inhibition term which would

$$\text{rate} = k_3[Li_2Pd_2Cl_6][\text{vinyl chloride}] \quad (39)$$

be expected if π-complex formation were the initial

$$Li_2Pd_2Cl_6 + C_2H_3Cl \overset{K_7}{\rightleftharpoons} Li_2Pd_2Cl_5(C_2H_3Cl) + LiCl \quad (40)$$

step in the reaction. Second, the *cis*- and *trans*-1-chloropropenes isomerized into an equilibrium mixture much faster than they exchanged radioactive chloride from solution; therefore stereochemical infor-

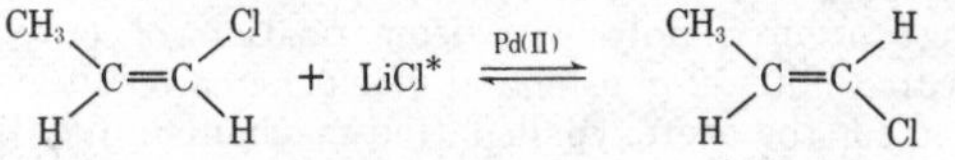 (41)

mation on the exchange reaction could not be obtained. Third, as shown by the data in Table II, 1-chlorocyclopentene exchanges at about the same rate as 2-chloro-2-butene.

Table II
Rate of Radioactive Exchange of Several Vinylic Chlorides at 25°

Vinylic chloride	k_3, M^{-1} sec^{-1}
$CH_2{=}CHCl$	10
$CH_3CH{=}CHCl$	0.23
$CH_3CH{=}C(CH_3)Cl$	0.049
1-Cyclopentenyl chloride	0.056

The fact that 1-chlorocyclopentene exchanges indicates that both cis and trans modes of chloropalladation are operative for, by the same arguments used for acetoxypalladation (see eq 30), if chloropalladation were stereochemically pure, cyclic vinylic chlorides should not exchange. Both modes of chloropalladation must have the rate expression of eq 39.

Since π-complex formation is almost certainly required for activation of vinylic chloride toward attack by chloride, the form of the rate equation most indicative of mechanism is eq 42, where all but the

$$\text{rate} = \frac{k_3[Li_2Pd_2Cl_6][\text{vinyl chloride}][LiCl]}{[LiCl]} \quad (42)$$

[LiCl] factor in the numerator is related to π-complex formation *via* eq 40.

The trans chloropalladation must occur by chloride attack from outside the coordination sphere, analogous to acetoxypalladation (eq 25). This type of attack would certainly cause the rate expression to have [LiCl] in the numerator. However the need for this [LiCl] factor in the rate expression for cis chloropalladation is not so obvious. Since cis chloropalladation must occur from the coordination sphere of the Pd(II), the [LiCl] factor most likely arises from the need to fill the incipient vacant coordination sphere on Pd(II). The scheme would be as given by eq 43 and 44. Exchange would be completed by de-

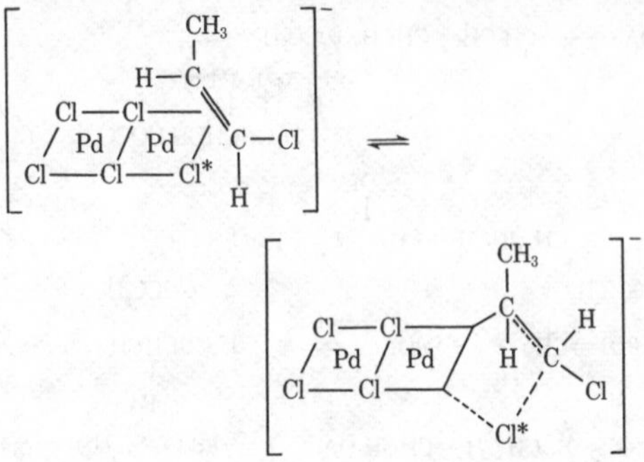 (43)

5

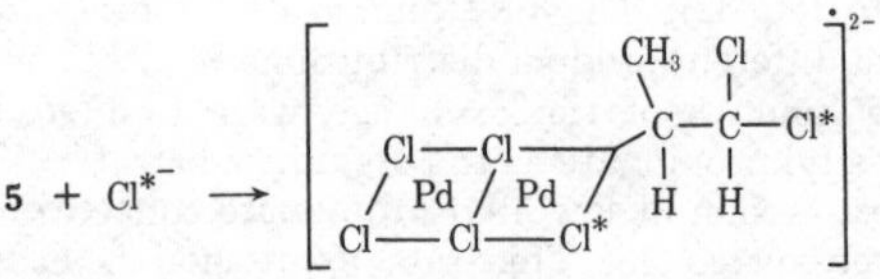 (44)

6

chloropalladation of nonradioactive chloride from **6**.

Allylic Trifluoroacetate Exchange with Chloride

Following the procedure used previously, the next logical exchange to study would be the allyl chloride radioactive chloride exchange. However, as this exchange would be expected to follow a mechanism analogous to the vinyl chloride–chloride exchange, it seemed more instructive to study the exchange of one type of functional group for another. The allylic trifluoroacetate exchange for chloride (eq 45), a reac-

$$CH_2{=}CHCH_2OOCCF_3 + LiCl \xrightarrow{Pd(II)} CH_2{=}CHCH_2Cl + LiOOCCF_3 \quad (45)$$

tion which goes to completion even at low chloride concentration, was chosen.[27]

The rate expression (eq 46) is analogous to that

$$\text{rate} = k_4[Li_2Pd_2Cl_6][C_3H_5O_2CCF_3] \quad (46)$$

found for vinylic chloride–chloride exchange, suggesting that the rate-determining step is chloropalladation. In support of the chloropalladation–detrifluoroacetoxypalladation mechanism are the re-

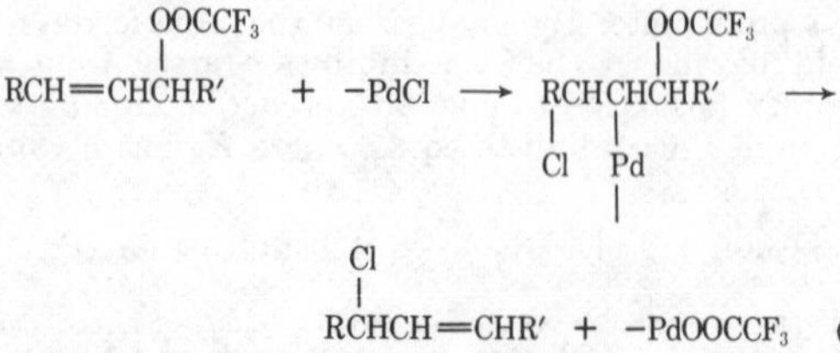 (47)

(26) P. M. Henry, *J. Org. Chem.*, 37, 2443 (1972).

(27) P. M. Henry, *Inorg. Chem.*, 11, 1876 (1972).

sults of exchange of crotyl and 3-buten-2-yl trifluoroacetate. Each exchanges only with isomerization, a result expected on the basis of this mechanism (R, R′ = CH_3 or H).

Vinyl Chloride Exchange with Acetate

To complete the cycle of acetate and chloride exchanges the exchange of vinyl chloride with acetate was studied.[28] This exchange has previously been studied by several workers[18,29-32] and it has been reported that, at least qualitatively, with *cis*- and *trans*-1-chloropropene exchange occurs with retention of configuration. Thus *cis*-1-chloropropene gives mainly *cis*-1-acetoxypropene. This result led to the suggestion[33] that exchange occurs by an acetoxypalladation–dechloropalladation mechanism (eq 17 and 18, X = Cl and Y = OAc), in which acetoxypalladation has the opposite stereochemistry from dechloropalladation.

The previous qualitive reports[18,32] of the stereochemistry of *cis*- and *trans*-1-chloropropene exchange was checked in a quantitative fashion in the present work. It was found that in fact the stereochemical results are not pure. Thus *cis*-1-chloropropene gives a mixture of *cis*- and *trans*-1-acetoxypropene, but the mixture is mainly cis (85%) with the remainder trans. *trans*-1-Chloropropene gave mainly *trans*-1-acetoxypropene.

Rate expression 48 is the first to exhibit a squared

$$\text{rate} = [Li_2Pd_2Cl_6][\text{vinyl chloride}]/[LiCl]^2 \times (k_5' + k_5''[LiOAc]) \quad (48)$$

inverse dependence on [LiCl]. The similarity of this rate expression to that for the Wacker reaction (eq 10) suggests that acetoxypalladation is occurring by attack of coordinated acetate on coordinated olefin.

$$Li_2Pd_2Cl_6 + C_2H_3Cl + LiOAc \rightleftarrows LiPd_2Cl_4(C_2H_3Cl)(OAc) + LiCl \quad (49)$$

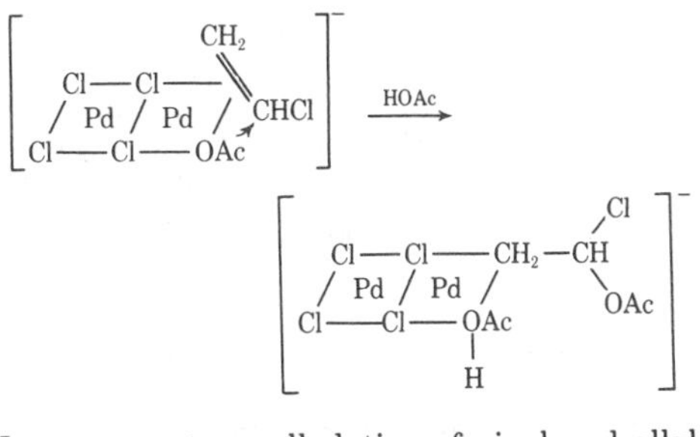
(50)

However, acetoxypalladation of vinyl and allylic esters does not occur by this route, and there is no reason to expect that vinyl chloride should react so differently from vinyl acetate.

In fact, a mechanism can be written which is entirely consistent with both previous and present kinetic and stereochemical results. The first step is, of course, π-complex formation *via* eq 40. The following steps are given by eq 51 and 52. The important point

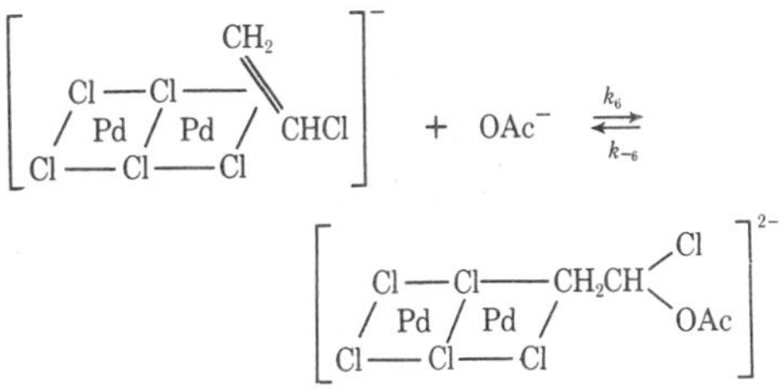
(51)

7

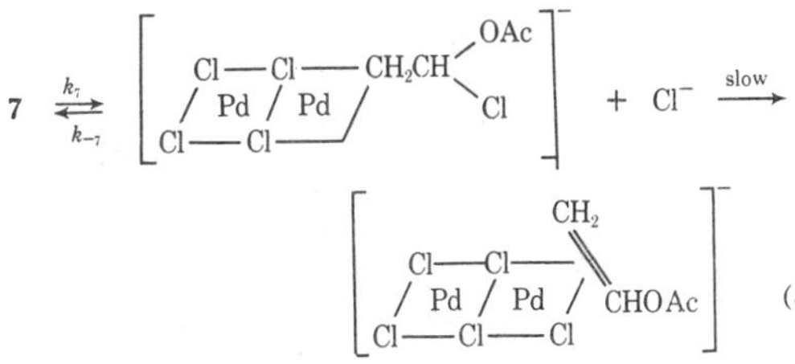
(52)

is that for dechloropalladation to occur (eq 52) a vacant coordination site must be present on the Pd(II). Thus LiCl inhibition would appear in both eq 40 and 52, accounting for the squared LiCl inhibition factor. The kinetics also require that eq 51 be reversible since eq 52 must be the rate-determining step.

The stereochemical results with *cis*- and *trans*-1-chloropropene indicate that not all dechloropalladation occurs by cis elimination. This result is consistent with the radioactive exchange studies on vinyl chlorides which indicate that chloropalladation is also not stereospecific. By the principle of microscopic reversibility, dechloropalladation should also be nonstereospecific.

Table III
Rates of Acetate Exchange of Several Vinylic Chlorides

Vinylic chloride	$k_1 \times 10^4$, sec^{-1}
CH_2=CHCl	1.94
trans-CH_3CH=CHCl	1.5×10^{-2}
CH_3CH=$C(CH_3)Cl$	2.2×10^{-4}
1-Cyclopentenyl chloride	2.6×10^{-3}

Table III lists the rate of exchange of several vinylic chlorides. As with previous exchanges methyl substitution on vinylic carbon strongly depresses rate, a result consistent with a large steric requirements for acetoxypalladation. The most interesting result, however, is the fact that 1-chlorocyclopentene reacts over ten times faster than 2-chloro-2-butene. As shown in Table I the rate of exchange of 1-acetoxycyclopentene was much slower than that of 2-acetoxy-2-butene, a result consistent with stereospecific acetoxypalladation (eq 30). In the case of radioactive chloride exchange, however, the rates for the cyclic and straight vinylic chlorides are about equal. Now 2-chloro-2-butene must exchange one-half the time it chloropalladates. 1-Chlorocyclopentene, on the other hand, exchanges less than one-half the times it chloropalladates because chloropalladation is mainly cis. Thus cis chloropalladation–cis dechloropalladation, which cannot give exchange, is preferred over cis

(28) P. M. Henry, *J. Amer. Chem. Soc.*, **94**, 7311 (1972).
(29) E. W. Stern, M. L. Spector, and H. P. Leftin, *J. Catal.*, **6**, 152 (1966).
(30) C. F. Kohll and R. Van Helden, *Recl. Trav. Chem. Pays-Bas*, **87**, 481 (1968).
(31) H. C. Volger, *Recl. Trav. Chem. Rays-Bas*, **87**, 501 (1968).
(32) E. W. Stern, *Catal. Rev.*, **1**, 73 (1967); see p 125.
(33) E. W. Stern and H. C. Volger, *Amer. Chem. Soc., Div. Petrol. Chem., Prepr.*, **14**, (4), F4 (1969).

chloropalladation–trans dechloropalladation or trans chloropalladation–cis dechloropalladation. The latter two routes give exchange. This analysis requires chloropalladation of 1-chlorocyclopentene to be faster than 2-chloro-2-butene. Finally, since acetoxypalladation has opposite stereochemistry from chloropalladation, the rates of exchange of the straight chain and cyclic vinylic chlorides should be close to the ratio of their rates of acetoxypalladation. As expected from the radioactive chloride results, 1-chlorocyclopentene exchanges chloride for acetate faster than 2-chloro-2-butene.

The first series of chloride and acetate exchanges in acetic acid catalyzed by Pd(II) in chloride-containing media present a consistent picture and elucidate some of the factors involved in acetoxypalladation and chloropalladation. Kinetics of exchange and other stereochemical evidence indicate acetoxypalladation to be a pure trans process while chloropalladation is mainly cis but occurs trans about 15% of the time. In retrospect this result is not surprising since in this system chloride is much more strongly complexed to Pd(II) than is acetate. Thus chloride is in position for a cis attack while acetate is not. Both chloride and acetate, however, are able to attack trans.

The important conclusion is that the stereochemistry of addition of metal ions and nucleophiles across double or triple bonds is not absolute for a given metal ion; rather, it depends on subtle factors such as coordination between nucleophile and the metal ion. Thus phenylpalladium, which is almost certainly a covalently bonded species, adds cis to cyclohexene.[34] On the other hand uncoordinated amines add trans to Pt(II)–monoolefin complexes.[35] Attack of nucleophiles on Pd(II)- and Pt(II)–diolefin complexes has also been found to be trans.[36-38] Among other factors, the fact that the diolefin occupies two coordination positions, thus preventing coordination of nucleophile, may be important.

Another well-studied metal ion addition to unsaturated double bonds is that of Hg(II) and nucleophiles. With simple olefins such as cyclohexene the addition of H_2O, acetate, and methanol is trans while addition to strained olefins such as norbornene is mainly cis[39,40] but depends somewhat on the olefin used.[41] However, dechloromercuration of olefins, which has not been studied to any extent, might well be cis. Thus exchange of cis and trans 1-chloropropene catalyzed by Hg(II) may well give stereochemical results similar to Pd(II).

Another result worthy of note is the need to fill an incipient coordination site on Pd(II) before cis chloropalladation can be achieved and conversely the need for an open coordination sphere on Pd(II) before dechloropalladation can be completed. The need for a vacant coordination site on a metal is a very important feature of homogeneous catalysis,[16] and the present results indicate this requirement to be very important in cis insertion processes.

(34) P. M. Henry and G. A. Ward, *J. Amer. Chem. Soc.*, **94**, 673 (1972).
(35) A. Panunzi, De Penzi, and G. Paiaro, *J. Amer. Chem. Soc.*, **92**, 3488 (1970).
(36) W. A. Whitla, H. M. Powell, and L. M. Venanzi, *Chem. Commun.*, 310 (1966).
(37) C. Panvatoni, G. Bombieri, E. Forsellini, B. Crosiani, and V. Belluco, *Chem. Commun.*, 187 (1969).
(38) J. K. Stille and R. A. Morgan, *J. Amer. Chem. Soc.*, **88**, 5135 (1966); **92**, 1274 (1970).
(39) T. G. Traylor and A. W. Baker, *Tetrahedron Lett.*, 14 (1959).
(40) M. M. Anderson and P. M. Henry, *Chem. Ind. (London)*, 2053 (1961).
(41) T. G. Traylor, *Accounts Chem. Res.*, **2**, 152 (1969).

Acid-Catalyzed Allylic Exchange

During the course of study of the effect of allylic propionate structure on rate of exchange with acetic acid, it was observed that 2-cyclohexene-1-yl propionate exchanged at a very slow rate even at 1.0 *M* LiOAc concentration. This rate was much slower than would be expected from steric effects. Even more surprising, the rate of exchange was faster when the LiOAc was omitted. Thus acetate has an inhibitory effect on this exchange as opposed to the catalytic effect on all the other acetate exchanges studied! This result suggested that the exchange is acid catalyzed. Acid catalysis was confirmed by runs at various [LiOAc] as well as by runs in which CF_3-COOH was added to the reaction mixture.[42] The rate expression for the exchange is given by eq 53

$$\text{rate} = ([Li_2Pd_2Cl_6][\text{allyl ester}](k_N + k_A[\text{acid}]) \qquad (53)$$

where k_N is the rate constant for the neutral reaction and k_A the rate constant for the acid-catalyzed reaction.

Two questions concerning this reaction must be answered. First, why is the cyclic allylic ester so much less reactive than straight allylic esters in the conventional acetate-catalyzed reaction; second, which is the mechanism of the acid-catalyzed reaction? The answer to the first question may lie in the conformational energies of the cyclohexene system, but the arguments are too complicated for a short review.

As to the mechanism of this new acid-catalyzed exchange, the rate expression does not contain the [LiCl] inhibition factor required for the formation of π complex (eq 54). As in previous cases where this

$$Li_2Pd_2Cl_6 + \text{allyl ester} \rightleftharpoons LiPd_2Cl_5(\text{allyl ester}) + LiCl \qquad (54)$$

factor did not appear, its absence is very likely due to cancellation by a [LiCl] factor in the denominator. The correct rate expression is that of eq 55.

$$\text{rate} = \frac{[Li_2Pd_2Cl_6][\text{allyl ester}][LiCl]}{[LiCl]}(k_N + k_A[\text{acid}]) \qquad (55)$$

Of the mechanisms which can be written for this exchange, one intriging possibility involves a Pd(IV) π-allyl. The acid catalysis would result from proto-

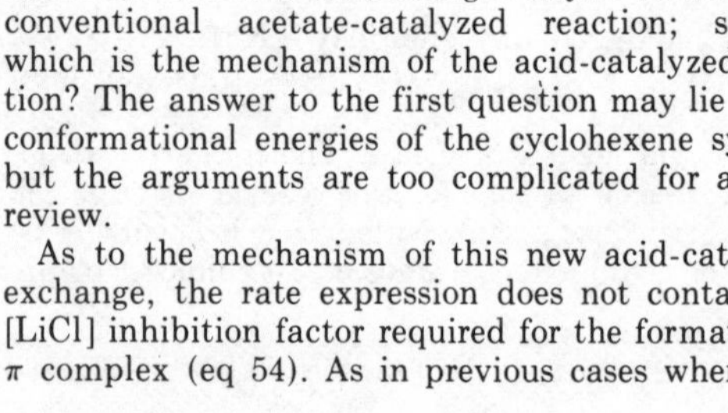

(56)

8

(42) F. T. T. Ng and P. M. Henry, *J. Org. Chem.*, **38**, 3338 (1973).

nation of the alcohol oxygen in the allylic ester, thus weakening the carbon–oxygen bond. The rate-determining step would be oxidative addition to give a Pd(IV)-π-allyl. The [LiCl] term in the numerator would result from the need to fill in the sixth coordi-

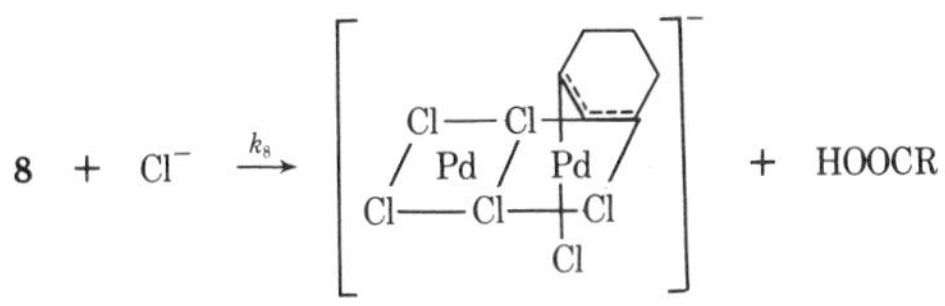
(57)

nation position on d^6 Pd(IV). Reversal of the oxidative addition would then give exchange.

Whatever the exact mechanism, this result indicates that the acetoxypalladation route is not the only means of allylic ester exchange. Further studies aimed at defining the exact mechanism of the acid-catalyzed reaction are planned. In particular, stereochemical studies should differentiate between the various possible reaction paths.

Cis–Trans Isomerization

In the discussion of radioactive chloride exchange of vinylic chlorides mention was made of a side reaction involving cis–trans isomerization without exchange with radioactive LiCl (eq 41).[26] *cis*- and *trans*-1-bromopropene were also isomerized without exchange of chloride for bromide.[43] Cis–trans isomerization without exchange was also observed in acetate-exchange studies with enol esters, although it was a less serious side reaction in this case. The isomerization was observed with the enol propionates

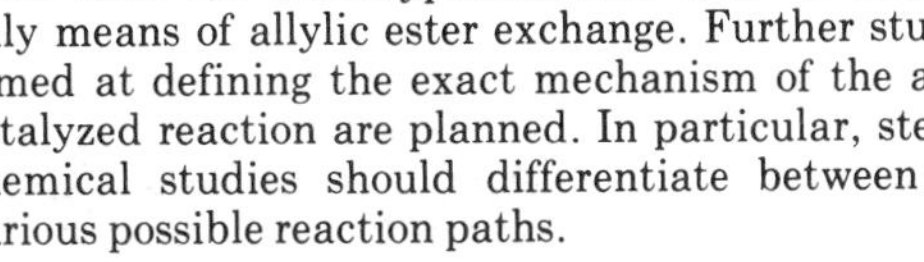
(58)

of propionaldehyde (R = CH_3, R′ = H), phenylacetaldehyde (R = C_6H_5, R′ = H) and 2-butanone (R = R′ = CH_3).

There have been many reports of double bond and cis–trans isomerizations catalyzed by noble metal salts. The mechanisms suggested for these reactions include intermolecular hydride transfer *via* metal hydrides, intramolecular hydride transfer *via* π-allyl hydrides, and reversible π-allyl complex formation. However, the isomerizations described in this work do not appear to proceed by any of these routes. Thus the π-allyl routes can be eliminated because the enol propionates of phenylacetaldehyde, which do not have allylic hydrogens, are isomerized.

The reactions also do not exhibit any of the features expected of hydride mechanisms such as exchange of hydrogens with deuterated solvent or deuterated olefins, acid catalysis, or double bond isomerization. In fact, if palladium(II) hydride is generat-

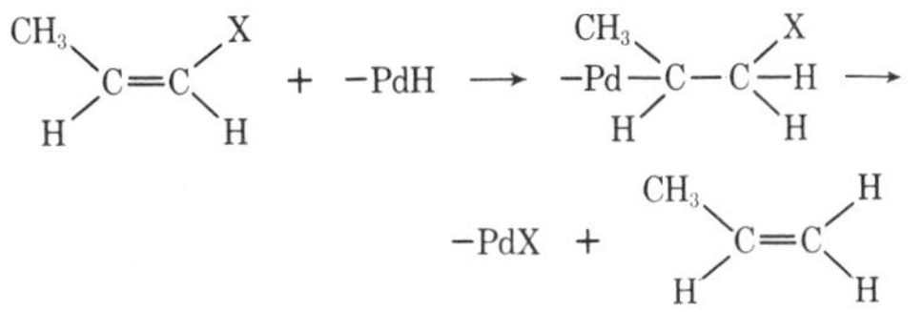
(59)

(43) P. M. Henry, *J. Amer. Chem. Soc.*, **93**, 3547 (1971).

ed *in situ*, isomerization is not observed. The only reaction is decomposition of the enol propionate or vinylic halide *via* eq 59 (X = $OOCC_2H_5$, Cl, or Br).

It thus appears that the mechanisms usually considered for Pd(II)-catalyzed isomerization are not operative in the present examples. The general rate expression for the isomerization is given in eq 60.

$$\text{rate} = (k_9[Li_2Pd_2Cl_6]/[LiCl] + k_{10}[Li_2Pd_2Cl_6] + k_{11}[Li_2Pd_2Cl_6]^{1/2})[\text{olefin}] \quad (60)$$

The enol propionate isomerization displays only the k_9 term,[20] the vinylic chloride isomerization has all three terms, while the vinylic bromide isomerization has the k_{10} and k_{11} terms.[44]

The k_9 term corresponds to π-complex formation *via* eq 23 while the k_{11} term[45] corresponds to π-complex formation *via* eq 24. Isomerization *via* the π complexes must be accomplished without the intervention of an external reagent. A definite mechanism cannot be proposed on the basis of the evidence to date, but an attractive one is a π–σ rearrangement to give a Pd(II)-bonded carbonium ion with sufficient lifetime for rotation (X = $OOCC_2H_5$, Cl, or Br).

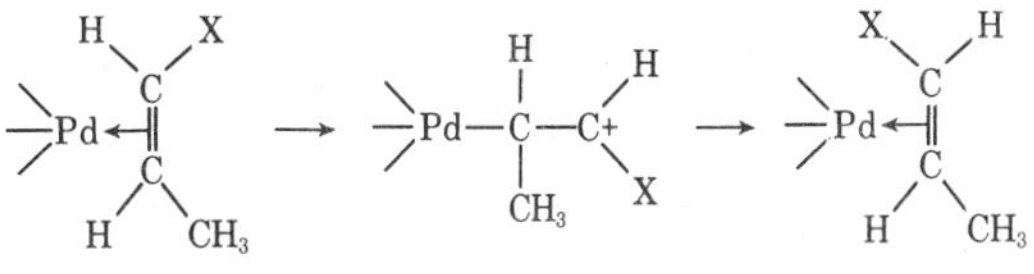
(61)

This mechanism will be tested by determining the effect of electron-releasing and -withdrawing substituents on the isomerization. A mechanism such as that represented by eq 61 would predict a large negative ρ.

The k_{10} term is of the same form as previously found for chloropalladation (eq 40), which suggests that isomerization may occur by nonstereospecific chloropalladation–dechloropalladation in the fashion reversed from that which gives exchange (A = addition, E = elimination).

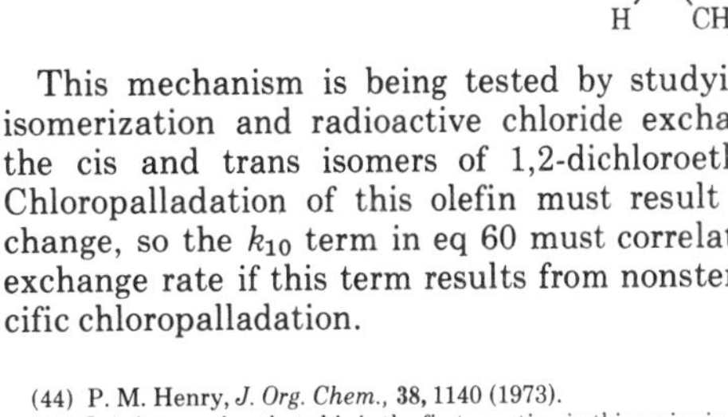
(62)

This mechanism is being tested by studying the isomerization and radioactive chloride exchange of the cis and trans isomers of 1,2-dichloroethylene. Chloropalladation of this olefin must result in exchange, so the k_{10} term in eq 60 must correlate with exchange rate if this term results from nonstereospecific chloropalladation.

(44) P. M. Henry, *J. Org. Chem.*, **38**, 1140 (1973).

(45) It is interesting that this is the first reaction in this series in which the monomeric π complex is a reactive species. The results advanced for the inactivity of this π complex in the exchange reactions are not valid in the case of isomerization since isomerization does not involve the attack of an external nucleophile.

Conclusions

As an example of the type of information which can be obtained in exchange studies, some representative results with just one system have been described. As well as defining the detailed mechanisms of acetoxypalladation and chloropalladation, these studies have revealed three new reactions, the allylic and cis–trans isomerizations and the acid-catalyzed exchange.

However, it is well to point out that, besides their own intrinsic interest, these studies should help in elucidating the mechanisms of other Pd(II)-catalyzed reactions. As an example, consider the oxidation of ethylene in HOAc. A general mechanism for vinyl ester formation consistent with other Pd(II) chemistry is given by eq 63–67 (X = OAc, Cl, etc.).

$$PdX_4^{2-} + C_2H_4 \underset{k_{-12}}{\overset{k_{12}}{\rightleftharpoons}} PdX_3(C_2H_4)^- + X^- \quad (63)$$

$$PdX_3(C_2H_4)^- + OAc^- \underset{k_{-13}}{\overset{k_{13}}{\rightleftharpoons}} X_3Pd-CH_2CH_2OAc \quad (64)$$

$$^{2-}X_3Pd-CH_2CH_2OAc \xrightarrow{k_{14}} {}^{2-}X_3PdH + CH_2{=}CHOAc \quad (65)$$

$$^{2-}X_3Pd-CH_2CH_2OAc \underset{k_{-15}}{\overset{k_{15}}{\rightleftharpoons}} {}^{-}X_2Pd-CH_2CH_2OAc + X^- \quad (66)$$

$$^{-}X_2Pd-CH_2CH_2OAc \xrightarrow{k_{16}} {}^{-}X_2PdH + CH_2{=}CHOAc \quad (67)$$

The last two equations represent the possibility that palladium(II) hydride elimination does not occur until a negative ligand is lost. This possibility must be considered since a vacant coordination position might well be a prerequisite for palladium(II) hydride elimination. This need for a vacant coordination site has precedent in the dechloropalladation step of vinyl chloride exchange with acetate.

If reaction 64 is the rate-determining step, of course the kinetics would give no information on eq 65–67. However if it were not, in principle we could determine if decomposition required the loss of X from the coordination sphere of Pd(II). In either case the kinetics would be difficult to interpret because the exact steps in arriving at $X_3PdCH_2CH_2OAc$ are not known. However, the acetoxypalladation reaction would be expected to be similar for ethylene and vinyl propionate. Furthermore, we know that the rate-determining step for exchange must be eq 25 since exchange must occur approximately one-half the time acetoxypalladation occurs.[46] Thus, by comparison of the rate expressions for exchange and oxidation, the rate-determining step for the latter should be more readily deduced.

Future Work

In regard to the $PdCl_2$–LiCl system in acetic acid the Pd(II)-catalyzed saponification of vinyl esters in wet acetic acid is being studied. A previous study[47] of this system indicated that a hydroxypalladation mechanism is operative. However, because the vari-

$$RCH{=}CHOAc + H_2O \xrightarrow[HOAc]{PdCl_2-LiCl} RCH_2CHO + HOAc \quad (68)$$

ous equilibria were not defined, the exact mode of hydroxypalladation was unclear. The present studies are aimed at elucidating the exact mechanism. This system is particularly complicated because water is not only a reactant but, in addition, it changes the solvent power of the acetic acid and thus affects the various equilibria in the system.

As indicated earlier, the cis–trans isomerization requires more study, as does the acid-catalyzed exchange. Work aimed at defining the generality of the allylic isomerization is planned: how many neighboring groups are capable of performing this isomerization and what other types of shifts, such as 1,2 shifts, take place?

Work has begun on a different system, the $Pd(OAc)_2$–NaOAc or –LiOAc system in HOAc. This chloride-free system is of particular interest because of two experimental observations. First, product distributions for oxidation of 1-olefins change with acetate concentration. For example, at low acetate concentration propylene gives isopropenyl acetate while at higher acetates the main product is allyl acetate.[48-51] Second, the rate of oxidation of ethylene has a complicated dependence on [NaOAc], first increasing with increasing [NaOAc] to about 0.3 *M* and then decreasing with further increase in [NaOAc].[15] Equilibrium and exchange studies should shed some light on these anomalies. Perhaps, as discussed earlier, the studies of the kinetics of oxidation and exchange will supplement each other.

Finally the work will be extended to other solvents such as methanol. Methanol lies between acetic acid and water in solvent characteristics and thus exchanges in it may be expected to exhibit intermediate mechanistic behavior.

Addendum (May 1976)

Recently the $Pd(OAc)_2$–NaOAc system in HOAc has been studied in some detail. This system was mentioned under *Future Work,* above. First, equilibria 69 and 70 were found to be operative in this system[52] (M = Na or Li):

$$Pd_3(OAc)_6 + 6MOAc \overset{K_{32}}{\rightleftharpoons} 3M_2Pd_2(OAc)_6 \quad (69)$$

$$M_2Pd_2(OAc)_6 + 2MOAc \overset{K_{21}}{\rightleftharpoons} 2M_2Pd(OAc)_4 \quad (70)$$

Unlike the chloride system, in which all equilibria are rapid, these equilibria took several hours to come to completion.

Next the π-complex equilibria between $Na_2Pd_2(OAc)_6$ and olefins was studied.[53] Again two equilibria (71 and 72) were detected (ol = olefin).

(46) The tendencies of OAc and OOCR to be lost from an intermediate such as I (eq 25) must be about equal.

(47) R. G. Schultz and P. R. Rony, *J. Catal.*, **16,** 133 (1970).

(48) D. Clark, P. Hayden, and R. D. Smith, *Discuss. Faraday Soc.* **1** (1968).

(49) R. Schultz and D. Gross, *Advan. Chem. Ser.*, **No. 70,** 97 (1968).

(50) T. Matsuda, T. Mitsuyasu, and Y. Nakamura, *Kōgyo Kagaku Zasshi,* **72,** 1751 (1969).

(51) J. E. McCaskie, Ph.D. Thesis, University of California, Los Angeles, Calif., 1971.

$$Na_2Pd_2(OAc)_6 + ol \underset{}{\overset{K'_{22}}{\rightleftharpoons}} NaPd_2(OAc)_5(ol) + NaOAc \quad (71)$$

$$Na_2Pd_2(OAc)_6 + 2ol \overset{K'_{21}}{\rightleftharpoons} 2NaPd(OAc)_3(ol) \quad (72)$$

In this case the first equilibrium is rapidly established while the second is attained only after several hours.

The exchange of vinyl propionate with acetic acid to give vinyl acetate was next studied over a range of [NaOAc] from zero to 1.0 *M*.[54] The rate was found first to increase with increasing [NaOAc], to reach a peak at 0.2 *M* [NaOAc], and then gradually to decrease with further increase in [NaOAc]. This behavior is consistent with the rate expression of eq. 73, if k_d is higher than k_t.

$$\text{rate} = (k_t[Pd_3(OAc)_6] + k_d[Na_2Pd_2(OAc)_6])[CH_2{=}CH{-}O_2CC_2H_5] \quad (73)$$

Thus the reason the rate attains a maximum at 0.2 *M* NaOAc is that practically the only species present is $Na_2Pd_2(OAc)_6$. Note that $Na_2Pd(OAc)_4$ is unreactive; the decrease in rate at higher [NaOAc] results from conversion of reactive $Na_2Pd_2(OAc)_6$ to unreactive $Na_2Pd(OAc)_4$. Also note that the rate expression (equation 73) does not explicitly contain a term in [NaOAc]. The role of the NaOAc is to convert one Pd(II) species to another.

The lack of an [NaOAc] term is interpreted in terms of equations 74 and 75; repression of equilibrium 74 by NaOAc is cancelled by its first order term involvement in equation 75.

$$Na_2Pd_2(OAc)_6 + C_2H_3O_2CC_2H_5 \rightleftharpoons NaPd_2(OAc)_5(C_2H_3O_2CC_2H_5 + NaOAc \quad (74)$$

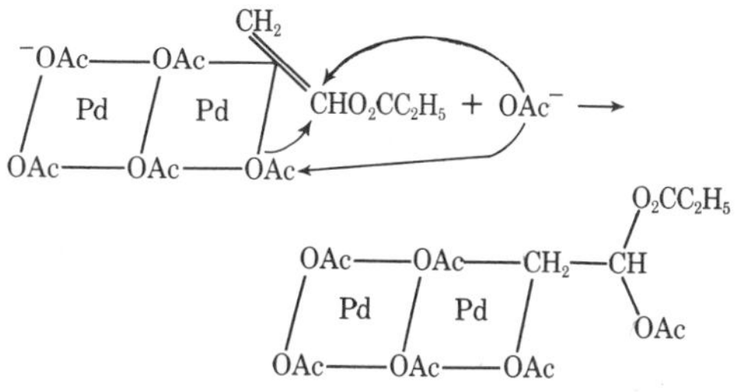
(75)

The attack of acetate in equation 75 is shown to occur from outside or inside the coordination sphere. Thus acetoxypalladation in this system should not be stereospecific. Acetoxypalladation in this system was shown to be non-stereospecific from the fact 1-cyclopenten-1-yl propionate exchanged rapidly (see equation 30 and the above discussion).

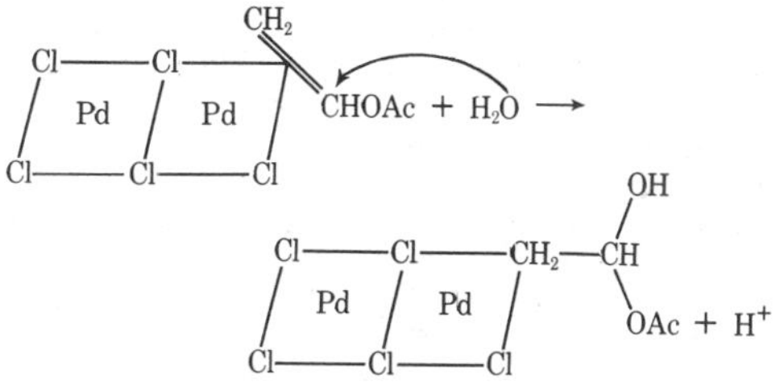
(76)

The saponification of vinyl esters in wet acetic acid was also mentioned under *Future Work*. This has been shown to proceed by trans attack of water on a dimeric π-complex (equation 76).[55] This indicates water can attack *cis* or *trans* depending on reaction conditions.

(52) R. N. Pandey and P. M. Henry, *Can. J. Chem.*, **52**, 1241 (1974).
(53) R. N. Pandey and P. M. Henry, *Can. J. Chem.*, **53**, 1833 (1975).
(54) R. N. Pandey and P. M. Henry, *Can. J. Chem.*, **53**, 2223 (1975).
(55) P. M. Henry, *J. Org. Chem.*, **38**, 2766 (1973).

Some Aspects of Organoplatinum Chemistry. Significance of Metal-Induced Carbonium Ions

Malcolm H. Chisholm and Howard C. Clark*

Department of Chemistry, University of Western Ontario, London, Ontario, Canada

Received September 25, 1972

Reprinted from Accounts of Chemical Research, *6*, *202* (*1973*)

Malcolm H. Chisholm received both his B.Sc. and his Ph.D. degrees from Queen Mary College of the University of London, the latter in 1969 with Professor D. C. Bradley. After 3 years as a postdoctoral fellow at the University of Western Ontario with Howard C. Clark, he was then appointed Assistant Professor of Chemistry at Princeton University.

Howard C. Clark is a native of New Zealand, having obtained his Ph.D. from the University of New Zealand in 1954. He also holds Ph.D. (1958) and Sc.D. (1972) degrees from Cambridge University. He joined the faculty at the University of British Columbia in 1957, and in 1965 was appointed Professor at the University of Western Ontario, where he is now Head of the Department of Chemistry. His research has included the study of transition-metal fluorides, Schiff's base complexes of transition metals, the stereochemistry of organotin compounds, insertion reactions of metal alkyl and metal hydrogen compounds, and the activation of unsaturated organic compounds by transition-metal catalysts.

The commercial importance of homogeneous and heterogeneous transition-metal-catalyzed reactions of olefins and acetylenes has generated tremendous interest in the activation of unsaturated molecules by transition-metal complexes. Noteworthy examples are (a) the hydroformylation of olefins to aldehydes, catalyzed by cobalt tetracarbonyl hydride,[1] (b) the hydrogenation of olefins catalyzed by $[P(C_6H_5)_3]_3RhCl$,[2] and (c) the oxidation of ethylene to acetal-

(1) D. Breslow and R. F. Heck, *Chem. Ind. (London)*, 467 (1960).

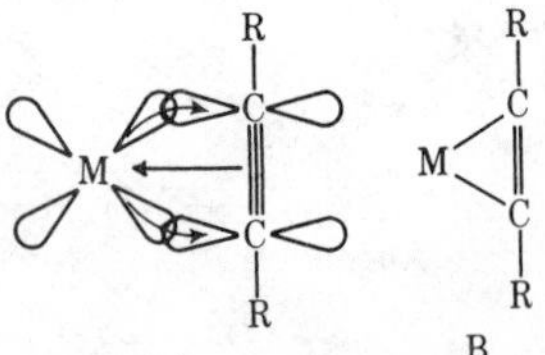

Figure 1. The Dewar–Chatt–Duncanson model of transition-metal–acetylene bonding.

dehyde in a continuous process catalyzed by $PdCl_2$–$CuCl_2$.[3]

Research into homogeneous catalysis has been stimulated by the belief that homogeneous catalysts can be "designed" which are more efficient than their heterogeneous counterparts. However, the role of a homogeneous catalyst is in many cases obscured by the very nature of the catalytic reaction: they are either very fast or very complex, usually both, and their reactive intermediates are normally inaccessible to direct or reliable observation.[4] The bonding in simple olefin and acetylene adducts with transition-metal complexes is therefore of particular interest since an understanding of the factors which lead to "stability" also illuminates the factors which produce "reactivity."

The original Dewar[5]-Chatt-Duncanson[6] model suggested that the bonding may be described as a combination of a σ interaction, olefin or acetylene π to metal "dsp," and a π interaction, metal "d" to olefin or acetylene π^* (Figure 1A) or in its extreme form as involving two metal-carbon σ bonds with complete rehybridization of the carbon atoms (Figure 1B).

For some complexes, *e.g.*, $RhCl[P(C_6H_5)_3]_2C_4F_6$, X-ray structural determinations have led to the conclusion that the bonding is that of A in Figure 1 while for others, *e.g.*, $Pt[P(C_6H_5)_3]_2C_4F_6$, bonding as in B is regarded as more appropriate. The concept of the synergic effect which is implicit in A is still accepted, although considerable debate has evolved concerning the relative magnitudes of σ and π interactions, the choice of metal orbitals employed in this bonding, and thus the coordination number and formal valency state of the metal.[4-14]

The sensitivity of σ- and π-bonding contributions to changes in the relative energies of the metal valence orbitals and the acetylene–olefin π and π^* molecular orbitals is clearly seen in the model of Maitlis,[8] *et al.* They consider that, for acetylene complexes, five situations could reasonably be postulated to occur, A → E, in which the energy of the metal valence electrons decreases relative to the acetylene π and π^* orbitals. Their case D was compared to the bonding described in A in Figure 1, and they predicted weak interaction between the metal and the acetylene, the bonding orbitals having mostly metal character and the antibonding orbital mostly acetylene. The pair of electrons originally on the acetylene is thus transferred to the metal.

If the relative magnitudes of the σ and π contributions differ, the polar nature of the metal–acetylene–olefin bond should facilitate reaction by an ionic mechanism. It has been suggested that fluoro olefins and fluoroacetylenes, which are strong π acids (*i.e.*, strong acceptors of metal d electrons into olefin or acetylene π^* orbitals) because of their electron-withdrawing substituents, react with low-valent electron-rich metal complexes *via* anionic or carbanion mechanisms.[15-18] Similarly, in the reactions of acetylenes with Pd(0)[19] and members of the cobalt triad,[20-24] which lead to benzenoid trimers and metallocyclopentadienes, the metal acts as a nucleophilic "catalyst." Indeed, this type of reaction is not limited to transition metals but is shared by other Lewis bases such as tertiary phosphines.[25]

On the other hand, olefin and acetylene complexes of higher valent and more electrophilic metals, in which bonding is dominated by the olefin–acetylene π to metal dsp contribution, have been predicted to be (i) unstable and (ii) susceptible to nucleophilic attack at the unsaturated carbon atoms.[8] It is this type of reaction which is discussed in detail in this Account, with special reference to the behavior of organoplatinum compounds. However, a third mode of reaction, involving free-radical formation, should also be recognized. Thus, the activation of unsaturated hydrocarbons may be compared with the activation of molecular hydrogen[26] or of alkyl halides[27] by transition-metal complexes and may occur by any one of these three mechanisms depending on the

(2) J. A. Osborn, F. H. Jardine, J. F. Young, and G. Wilkinson, *J. Chem. Soc. A*, 1711 (1966).
(3) J. Schmidt, W. Hafner, R. Jira, R. Sieber, J. Sedlmeier, and A. Sabel, *Angew. Chem.*, **74**, 93 (1962).
(4) L. Vaska, *Accounts Chem. Res.*, **1**, 335 (1968).
(5) M. J. S. Dewar, *Bull. Soc. Chim. Fr.*, **18**, C79 (1951).
(6) J. Chatt and L. A. Duncanson, *J. Chem. Soc.*, 2939 (1953).
(7) M. A. Bennett, Second Conference of the Coordination and Metal Organic Chemistry Division of the Royal Australian Chemical Institute, Monash, Australia, May 1968.
(8) E. O. Greaves, C. J. L. Lock, and P. M. Maitlis, *Can. J. Chem.*, **46**, 3879 (1968).
(9) F. R. Hartley, *Chem. Rev.*, **69**, 799 (1969), and references cited therein.
(10) J. H. Nelson, K. S. Wheelock, L. C. Cassachs, and H. B. Jonassen, *Chem. Commun.*, 1019 (1969).
(11) J. H. Nelson, K. S. Wheelock, L. C. Cassachs, and H. B. Jonassen, *J. Amer. Chem. Soc.*, **92**, 5110 (1970).
(12) J. H. Nelson and H. B. Jonassen, *Coord. Chem. Rev.*, **6**, 27 (1971).
(13) J. H. Nelson, K. S. Wheelock, L. C. Cassachs, and H. B. Jonassen, *J. Amer. Chem. Soc.*, **91**, 7005 (1969).
(14) V. Belluco, B. Crociani, R. Pietropaolo, and P. Uguaglia, *Inorg. Chim. Acta*, 19 (1969).
(15) R. Burt, M. Cooke, and M. Green, *J. Chem. Soc. A*, 2975 (1970).
(16) R. Burt, M. Cooke, and M. Green, *J. Chem. Soc. A*, 2971 (1970).
(17) T. Blackmore, M. I. Bruce, F. G. A. Stone, R. E. Davis, and A. Garza, *Chem. Commun.*, 852 (1971).
(18) H. D. Empsall, M. Green, and F. G. A. Stone, *J. Chem. Soc., Dalton Trans.*, 96 (1972).
(19) P. M. Maitlis, Plenary Lecture presented at the Fifth International Conference on Organometallic Chemistry, Moscow 1971.
(20) H. Yamazaki and M. Hagnitara, *J. Organometal. Chem.*, **17**, 22 (1967); **21**, 431 (1970).
(21) J. P. Collman, J. W. Kang, W. F. Little, and M. F. Sullivan, *Inorg. Chem.*, **7**, 1298 (1968).
(22) J. T. Mague and G. Wilkinson, *Inorg. Chem.*, **7**, 542 (1968).
(23) J. T. Mague, *J. Amer. Chem. Soc.*, **91**, 3983 (1969).
(24) S. McVey and P. M. Maitlis, *J. Organometal. Chem.*, **19**, 169 (1969).
(25) (a) M. A. Shaw, J. C. Tebby, R. S. Ward, and D. H. Williams, *J. Chem. Soc. C*, 2795 (1968); (b) N. E. Waite, J. C. Tebby, R. S. Ward, and D. H. Williams, *ibid.*, 1100 (1969); (c) E. M. Richards, J. C. Tebby, R. S. Wards, and D. H. Williams, *ibid.*, 1542 (1969); (d) N. E. Waite, J. C. Tebby, R. S. Ward, M. A. Shaw, and D. H. Williams, *ibid.*, 1620 (1971).
(26) (a) J. Halpern, *Annu. Rev. Phys. Chem.*, **70**, 207 (1968); (b) J. Halpern, *Chem. Eng. News*, **44**, 68 (Oct 31, 1966); (c) J. Halpern, *Advan. Chem. Ser.*, **No. 70**, 1 (1968).
(27) *E.g.*, for Ir(I): (a) P. B. Chock and J. Halpern, *J. Amer. Chem. Soc.*, **88**, 3511 (1966); (b) J. A. Labinger, R. J. Brans, D. Dolphin, and J. A. Osborne, *Chem. Commun.*, 612 (1970); (c) R. G. Pearson and W. R. Muir, *J. Amer. Chem. Soc.*, **92**, 5519 (1970); (d) J. S. Bradley, D. E. Connor, D. Dolphin, J. A. Labinger, and J. A. Osborne, *ibid.*, **94**, 4043 (1972).

particular system being examined and on the particular reaction conditions.

Certain aspects of the chemistry of organoplatinum compounds make them ideally suited to the consideration of these problems, namely[28] (a) the observance of well-defined oxidation states 0, +2, +4; (b) well-defined coordination numbers (three and four for Pt(0), four for Pt(2+) and six for Pt(4+)) in organoplatinum compounds; (c) their thermodynamic stability (and inertness to oxygen and water) and generally convenient rates of reaction; and (d) their suitability for the application of a variety of spectroscopic techniques.

In regard to points a and b, numerous illustrative compounds are known, *e.g.*, $[(C_6H_5)_3P]_2Pt(un)$ where un = olefin or acetylene,[29] $[(C_6H_5)_3P]_2Pt(CH_3)_2$, and $[(C_6H_5)_3P]_2Pt(CH_3)Cl$ and $[(C_6H_5)_3]_2Pt(CH_3)_nCl_{4-n}$ where n = 1–4.[29] While the assignment of the coordination number of platinum in $[(C_6H_5)_3P]_2Pt(un)$ is arbitrary, deviations from four-coordinate Pt(2+) and six-coordinate Pt(4+) are very rare, and in such cases the metal always attains an 18-electron valence shell *e.g.*, $Pt(SnCl_3)_5^-$,[30] $(\pi\text{-}C_5H_5)Pt[(C_6H_5)_3P]_2^+$,[31] $CF_3\text{-}[P(CH_3)_2C_6H_5]_2PtC_4(CH_3)_4^+$,[32] and $(\pi\text{-}C_5H_5)Pt(CH_3)_3$.[33] Their stability is indicated by the isolation of $[(CH_3)_3PtCl]_4$, which was one of the first organometallic compounds to be discovered;[34] this compound requires treatment with bromine to cleave the $Pt\text{-}CH_3$ bonds. Finally, examination of organoplatinum compounds and their reactions by the nmr method is particularly fruitful; the presence of ^{195}Pt (33% abundance, $I = \frac{1}{2}$) and hence the measurement of couplings between ^{195}Pt and other nuclei prove to be an invaluable tool in the understanding of trans influences[35] and chemical bonding.

Organoplatinum(2+) Cations

An unexpected product from the reaction of tetrafluoroethylene with $(R_3P)_2PtHCl$ in a glass container was[36,37] the cation $[(R_3P)_2PtCl(CO)]^+$, which is isoelectronic with the so-called Vaska's compound,[4,38] $(R_3P)_2IrCl(CO)$, and is similarly related to Wilkinson's[39] active catalyst, $(R_3P)_3RhCl$. This prompted us to prepare, and examine with respect to their catalytic activity, the related olefinic and acetylenic organoplatinum cations.

(28) F. A. Cotton and G. Wilkinson, "Advances in Inorganic Chemistry," 3rd ed, Wiley, New York, N. Y., 1972.

(29) (a) J. D. Ruddick and B. L. Shaw, *J. Chem. Soc. A*, 2801 (1969); (b) R. J. Cross, *Organometal. Chem. Rev.*, **2**, 97 (1967), and references therein.

(30) R. D. Cramer, R. V. Lindsey, C. T. Prewitt, and U. G. Stolberg, *J. Amer. Chem. Soc.*, **87**, 658 (1965).

(31) R. J. Cross and R. Wardle, *J. Chem. Soc. A*, 2000 (1971).

(32) M. H. Chisholm, H. C. Clark, D. B. Crump, and N. C. Payne, Submitted for publication.

(33) G. W. Adamson, J. C. J. Bart, and J. J. Daly, *J. Chem. Soc. A*, 2616 (1971).

(34) W. J. Pope and S. J. Peachey, *J. Chem. Soc.*, **95**, 571 (1909).

(35) (a) A. Pidcock, R. E. Richards, and L. M. Venanzi, *J. Chem. Soc. A*, 1707 (1966); (b) T. G. Appleton, H. C. Clark, and L. E. Manzer, *Coord. Chem. Rev.*, submitted for publication.

(36) H. C. Clark, P. W. R. Corfield, K. R. Dixon, and J. A. Ibers, *J. Amer. Chem. Soc.*, **89**, 3360 (1967).

(37) H. C. Clark, K. R. Dixon, and W. J. Jacobs, *J. Amer. Chem. Soc.*, **90**, 2259 (1968).

(38) (a) J. Halpern, *Accounts Chem. Res.*, **3**, 386 (1970); (b) J. P. Collman, *ibid.*, **1**, 136 (1968).

(39) G. Wilkinson, *et al.*, *J. Chem. Soc. A*, 1711, 1736 (1966); 1574 (1967); 1054 (1968).

The chloride ligand in *trans*-$PtCl(CH_3)L_2$ (where L = tertiary phosphine or arsine), I, is labile due to the high trans influence of the trans methyl group.[40,41] Hence the addition of a neutral donor ligand, L′, to I gives rise to equilibrium 1, for example when L′ = pyridine.[42] The addition of a silver salt of a nonpolarizable anion allows the isolation[40] of cationic platinum(2+) complexes for a variety of neutral ligands L′.

$$\underset{\text{I}}{trans\text{-}PtClCH_3L_2} + L' \underset{\text{or }(CH_3)_2CO}{\overset{CH_3OH}{\rightleftharpoons}} trans\text{-}[PtCH_3(L')L_2]^+Cl^- \quad (1)$$

Acetylenic Platinum(2+) Cations

With the hope of isolating methylplatinum acetylenic cations, *trans*-$[PtCH_3(RC{\equiv}CR')L_2]^+PF_6^-$ (II), we studied reaction 2. However, the products ob-

$$trans\text{-}PtCl(CH_3)L_2 + RC{\equiv}CR' + AgPF_6 \xrightarrow{\text{slovent}} \underset{\text{II}}{trans\text{-}[PtCH_3(RC{\equiv}CR')L_2]^+PF_6^-} + AgCl \quad (2)$$

tained from (2) showed a marked dependence on (i) the substituents of the acetylene, R and R′, (ii) the ligands on platinum, L, (iii) the solvent, and (iv) the reaction conditions.

The reaction of dialkyl- or diarylacetylenes in methanol or acetone did in many instances yield[42,43] II, while monoalkyl acetylenes, RC≡CH, in methanol or ethanol gave[42,44] cationic alkoxycarbene complexes (III) according to eq 3.

$$trans\text{-}PtCl(CH_3)L_2 + RC{\equiv}CH + AgPF_6 \xrightarrow{R'OH} \underset{\text{III}}{trans\text{-}[PtCH_3(RCH_2\ddot{C}OR')L_2]^+PF_6^-} + AgCl \quad (3)$$

Similarly, phenylacetylene gave[44] a benzylalkoxycarbene complex (III) when L = $P(CH_3)_2C_6H_5$, but when L = $As(CH_3)_3$ only acetylide formation occurred[42] as shown in (4). Reactions of monoalkyl acetylenes in aprotic polar solvents, such as tetrahydrofuran, also gave[45] acetylide formation as in (4).

$$trans\text{-}PtCl(CH_3)L_2 + PhC{\equiv}CH + AgPF_6 \xrightarrow[C_2H_5OH]{CH_3OH\text{ or}} \underset{\text{IV}}{[PtC{\equiv}CPh{\cdot}L_2]^+PF_6^-} + CH_4 + AgCl \quad (4)$$

$$L = As(CH_3)_3$$

Disubstituted acetylenes RC≡CR′ containing electron-withdrawing groups R and R′ produced[43] σ vinyl ether complexes in methanol (see 5), while in

$$trans\text{-}PtCl(CH_3)L_2 + RC{\equiv}CR + AgPF_6 \xrightarrow{CH_3OH} AgCl + CH_4 + \underset{\text{V}}{[PtCR{=}C(OCH_3)R{\cdot}L_2]^+PF_6^-} \quad (5)$$

aprotic polar solvents insertion into the methylplatinum bond occurred, [46] as in (6). In reactions 5 and 6 addition to the acetylenic triple bond leads to the

(40) H. C. Clark and J. D. Ruddick, *Inorg. Chem.*, **9**, 1226 (1970).

(41) F. Basolo, J. Chatt, H. B. Gray, R. G. Pearson, and B. L. Shaw, *J. Chem. Soc.*, 2207 (1961).

(42) M. H. Chisholm and H. C. Clark, *Chem. Commun.*, 763 (1970).

(43) M. H. Chisholm and H. C. Clark, *Inorg. Chem.*, **10**, 2557 (1971).

(44) M. H. Chisholm and H. C. Clark, *Inorg. Chem.*, **10**, 1711 (1971).

(45) M. H. Chisholm and H. C. Clark, unpublished results.

(46) M. H. Chisholm and H. C. Clark, *J. Amer. Chem. Soc.*, **94**, 1532 (1972).

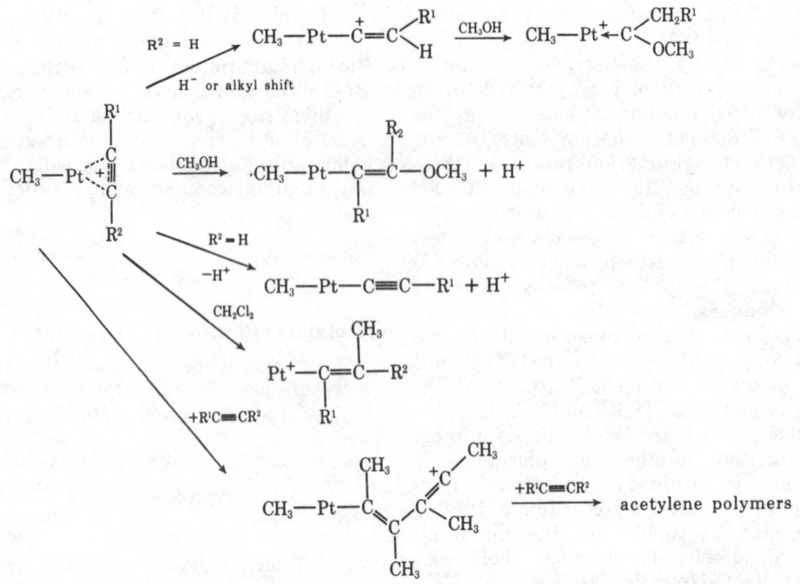

Figure 2.

$$trans\text{-}PtCl(CH_3)L_2 + RC{\equiv}CR + AgPF_6 \xrightarrow[\text{or THF}]{\text{acetone}} AgCl + [PtCR{=}C(CH_3)R{\cdot}L_2]^+PF_6^- \quad (6)$$

VI

stereospecific formation of trans and cis vinyl isomers respectively.[46]

The Carbonium Ion Model in Organoplatinum Chemistry

The products of reactions 3 to 6 are derived from the initial formation of cationic acetylenic platinum complexes (II) which show reactivity characteristic of carbonium ions (II′) and hence lead to products interpretable in terms of intramolecular rearrangements and/or nucleophilic addition.[46,47]

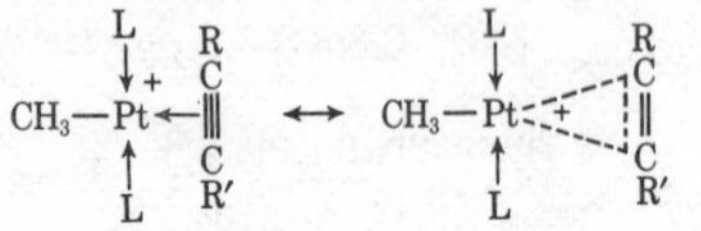

Reactions 2 to 6 led us to predict[46] that the carbonium ion reactivity of an acetylene $RC{\equiv}CR'$ coordinated to a platinum cation $[PtX{\cdot}Q_2]^+$ would be dependent on (i) the substituents on the acetylene, R and R′, (ii) the ligands on platinum, X and L, and (iii) the availability and nature of a nucleophile, which may be the solvent. Reactions 3 to 6 are explainable on this basis as shown in Figure 2. Thus, carbene formation (reaction 3) only occurs with terminal acetylenes in a protic solvent. In a polar but aprotic solvent, proton elimination or abstraction by solvent takes place to give acetylides. Moreover, the acidity of the acetylenic proton in the metal-induced carbonium ion is also influenced by the other ligands on platinum, as shown by the occurrence of both reactions 3 and 4 for phenylacetylene.[46]

(47) M. H. Chisholm, H. C. Clark, and D. H. Hunter, *Chem. Commun.*, 809 (1971).

Both cationic carbene complex formation and cationic acetylide complex formation prevent the isolation of cationic acetylenic complexes II, although such complexes have been isolated for dialkyl- or diarylacetylenes. The latter complexes are only slowly attacked by a nucleophilic protic solvent such as methanol to give methyl vinyl ether complexes (reaction 5), although on increasing the electrophilic character of the acetylene (*e.g.*, $HOCH_2C{\equiv}C\text{-}CH_2OH$, or $CH_3OOCC{\equiv}CCOOCH_3$), this reaction proceeds more rapidly. In the absence of a nucleophilic protic solvent, these electrophilic acetylenes react with the formation of insertion products (reaction 6). Indeed, in the absence of any nucleophile other than excess acetylene, polymerization is favored, as shown in Figure 2. Clearly, then, the choice of solvent and the ratio of reactants are critical in determining the course of these reactions.

Although many alkoxycarbene complexes of transition metals are now known,[48] reaction 3 is surprising and the mechanism deserves further comment. Initially we considered that the formation of the alkoxycarbene complexes III could be represented by (7), in which the cationic acetylenic complex II reacts with the solvent to produce a cationic vinyl ether complex, which then by hydride shift gives III.

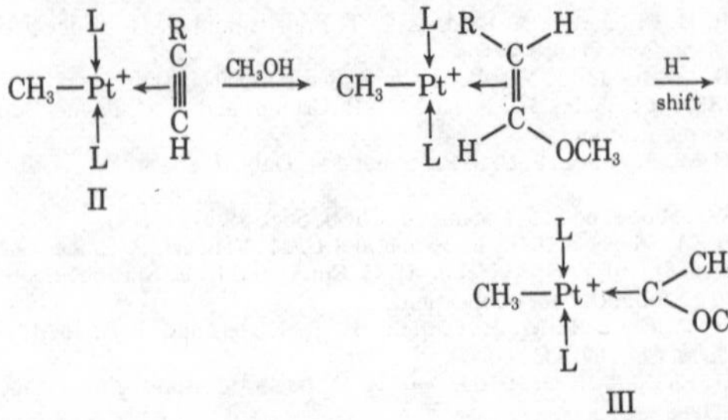

(7)

(48) (a) F. A. Cotton and C. M. Lukehart, *Progr. Inorg. Chem.*, in press; (b) D. J. Cardin, B. Cetinkaya, and M. F. Lappert, *Chem. Rev.*, 72, 545 (1972).

However, substitution of methyl (or ethyl) vinyl ether for the acetylene in reaction 3 led to the isolation of a stable methyl (or ethyl) vinyl ether cationic complex, which did not decompose in solution below 80°.[46] It should be noted that a carbonium ion mechanism involving H^- migration of a vinylic hydrogen is not favorable due to the orthogonality of the π cloud. This fact may be responsible for the inability of this π-coordinated vinyl ether to rearrange to give the alkoxycarbene ligand. Thus the formation of the platinum-stabilized carbonium ion, $CH_3PtC^+{=}CHR$ (VII) is essential for conversion to the alkoxycarbene ligand. The formation of VII can occur either by a hydride shift of II′ or by proton elimination (Figure 3). Deuterium-labeling studies[46] are consistent with the intramolecular mechanism involving hydride migration. Moreover, the formation[49] of Pt(4+) carbene complexes from cationic Pt(4+) acetylenic intermediates suggests that an oxidative addition–reductive elimination mechanism is not operative.

Other than the above reactions, many other aspects of organoplatinum chemistry clearly illustrate the generality of this carbonium ion model. For example, substitution of the methyl group attached to Pt in II by the more electron-withdrawing trifluoromethyl group increases the electrophilicity of the organoplatinum cation and hence increases the carbonium ion reactivity of the acetylene in *trans*-$[PtCF_3(RC{\equiv}CR)L_2]^+$. Although we have been unable to isolate such simple dialkylacetylene cations, they are reactive intermediates in the polymerization of acetylenes, a process which occurs much faster than for II in the absence of a polar protic solvent. A terminating step in the polymerization of dimethylacetylene is the formation[46,49] of the cationic tetramethylcyclobutadiene complex, $[PtCF_3\{C_4(CH_3)_4\}L_2]^+PF_6^-$, shown in reaction 8. Incidentally,

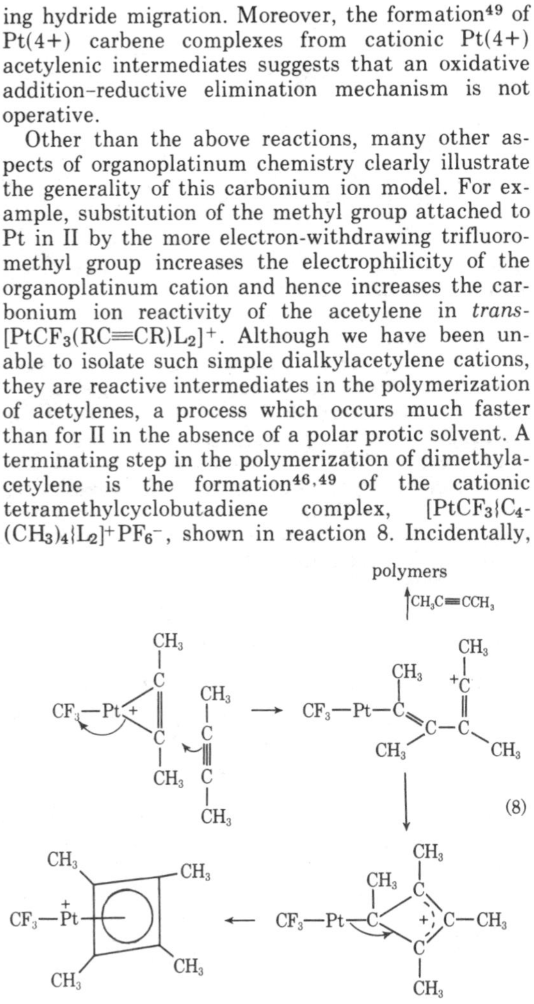

ligands L omitted for brevity

the geometry of this cation—pseudotetrahedral—is unusual for a Pt(2+) complex; moreover, its nmr parameters are consistent with fluxional behavior in solution.[32]

Another reaction in organoplatinum chemistry which is consistent with the carbonium ion model is

(49) M. H. Chisholm and H. C. Clark, *Chem. Commun.*, 1484 (1971).

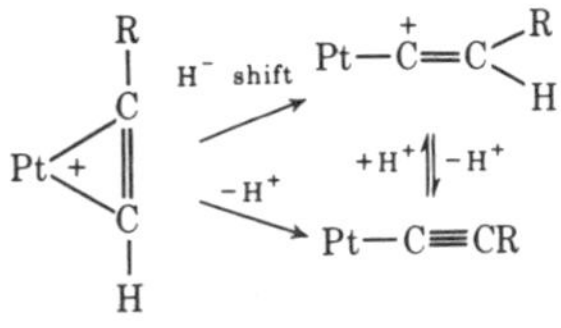

Figure 3.

the formation of an alkoxycarbeneplatinum(4+) cation *via* a cationic acetylenic Pt(4+) intermediate.[49]

$$[Pt(CH_3)_2CF_3(CH{\equiv}CCH_2CH_2OH)L_2]^+ \longrightarrow [Pt(CH_3)_2CF_3(\overline{CH_2CH_2O\ddot{C}}CH_2)L_2]^+$$

Data from ^{13}C nmr spectroscopy again illustrate the electron-deficient nature of the carbene carbon atom in these alkoxycarbene complexes.[50] By comparison with purely organic analogs, the ^{13}C shieldings of the carbene carbons are comparable to those of trialkylcarbonium ions.[51] Indeed, these platinum complexes could alternatively be considered as platinum-stabilized alkoxycarbonium ions, a nomenclature which more closely reflects their chemical behavior than does the name "alkoxycarbene." For example, carbenoid complexes undergo reactions with nucleophiles such as amines.[48] Likewise, we have found[45] that *trans*-$[PtCl\{\ddot{C}(OCH_3)CH_3\}L_2]^+$ reacts with pyridine to give neutral *trans*-$[PtCl(COCH_3)L_2]$, and the *N*-methylpyridine cation, a reaction which is the reverse of the well-known formation of cationic metal–carbene complexes from acyl derivatives.[48]

$$M{-}C(=O)R \xrightarrow{(CH_3)_3O^+BF_4^-} M^+{-}C(OCH_3)R$$

The role of platinum-induced carbonium ions is not limited to reactions described by eq 2; they are the reactive intermediates in the formation of neutral σ vinyl ether (eq 9), acyl (eq 10), and σ vinyl derivatives[52] (eq 11), shown below, which proceed *via* cationic acetylenic intermediates *trans*-$[PtCH_3(RC{\equiv}CR)L_2]^+Cl^-$.

$$trans\text{-}PtCl(CH_3)L_2 + RC{\equiv}CR + CH_3OH \longrightarrow trans\text{-}PtCl(CR{=}C(OCH_3)R)L_2 + CH_4 \quad (9)$$

$$trans\text{-}PtCl(CH_3)L_2 + RC{\equiv}CH + 2CH_3OH \longrightarrow trans\text{-}PtCl(COCH_2R)L_2 + CH_4 + (CH_3)_2O \quad (10)$$

$$trans\text{-}PtCl(CH_3)L_2 + RC{\equiv}CR + cat. \xrightarrow[solvent]{CH_2Cl_2} trans\text{-}Pt(CR{=}C(CH_3)R)\cdot Cl\cdot L_2 \quad (11)$$

cat. = trace of $[PtCH_3(acetone)Q_2]^+PF_6^-$

Thus, the formation of methyl vinyl ether complexes in (5) and (9) is stereospecific, giving only the trans vinylic (*i.e.*, trans with respect to C=C) isomer; it is thus analogous to the stereospecific bromination of an acetylene in methanol, which is traditionally regarded[53] as proceeding *via* a carbonium ion mechanism.

(50) M. H. Chisholm, H. C. Clark, L. E. Manzer, and J. B. Stothers, *Chem. Commun.*, 1627 (1971).

(51) J. B. Stothers, "Carbon-13 NMR Spectroscopy," Academic Press, New York, N. Y., 1972.

(52) M. H. Chisholm, H. C. Clark, and L. E. Manzer, *Inorg. Chem.*, **11**, 1269 (1972).

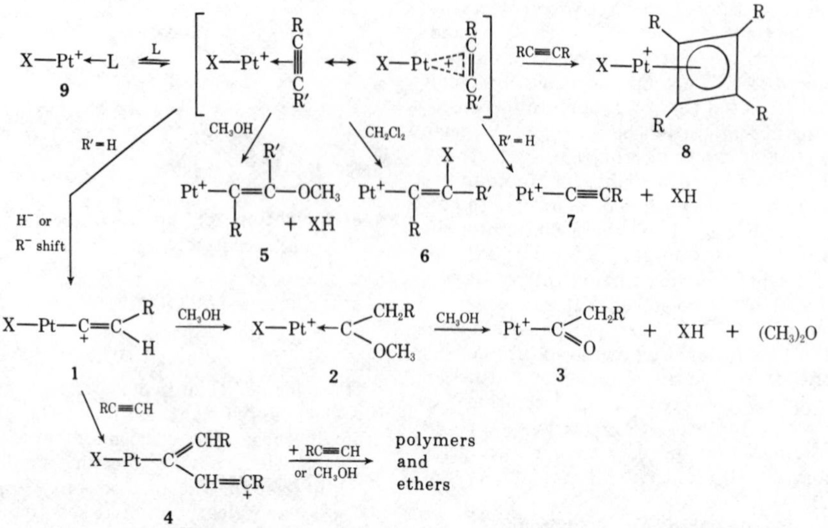

Figure 4. Some reactions of *trans*-$[PtX(RC\equiv CR')Q_2]^+Z^-$. Q = $P(CH_3)_2C_6H_5$ omitted from the figure for brevity. Z = PF_6 or Cl, except 8 = PF_6 only; X = CH_3 or CF_3 and R = alkyl or aryl for **1-4**. X = CH_3, R = R′ = alkyl, $COOCH_3$, COOH, CH_2OH for **5**; X = CH_3, R = R′ = $COOCH_3$, CF_3 for **6**; X = CH_3, R = phenyl or alkyl, solvent = THF or acetone for **7**; X = CF_3, R = CH_3 for **8**; X = CH_3 or CF_3, L = PR_3, py, CO, acetone, etc., for **9**.

Moreover, reactions 5 and 9 are analogous to a multitude of nucleophilic addition reactions of olefins in the presence of metal salts. The oxymercuration of simple unstrained olefins has long been known to be stereospecific in trans addition.[54] This was first predicted by Lucas, Hepner, and Winstein in 1939 by analogy to the behavior of the bromonium ion.[55] Similarly, platinum (II)- and palladium(II)-diolefin complexes of dicyclopentadiene, norbornadiene, and bicyclo[2.2.2]octadiene yield metal–carbon-bonded σ complexes with the methoxy group in the exo configuration on treatment with alkaline methanol.[56-60] Acyl formation, reaction 10, which occurs on prolonged contact with methanol, is considered to arise from further nucleophilic attack by the methanol on the electron-deficient carbene ligand,[46] *e.g.*, as in (12), and may be compared to the

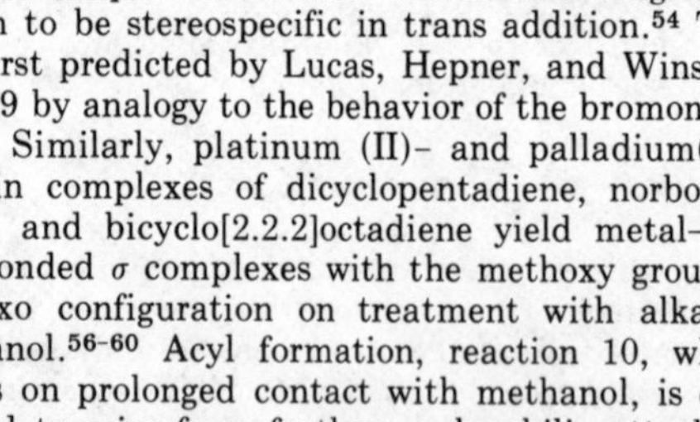

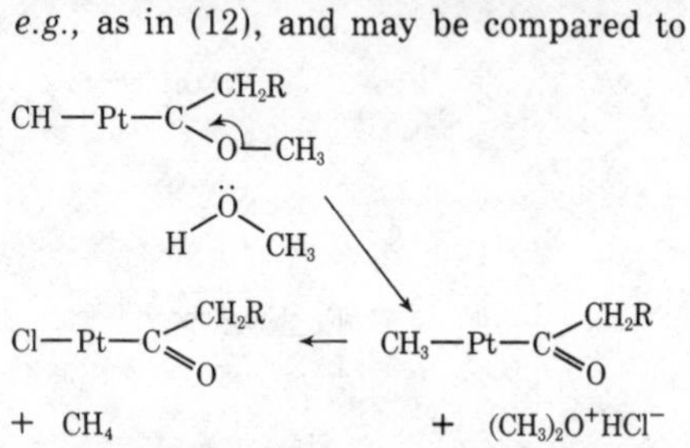

(12)

(53) (a) R. J. Morrison and R. M. Boyd, "Organic Chemistry," 2nd ed, Allyn and Bacon, Boston, Mass., 1966; (b) J. D. Roberts and M. Cassario, "Basic Principles of Organic Chemistry," W. A. Benjamin, New York, N. Y., 1965; (c) J. Much, "Advanced Organic Chemistry: Reaction, Mechanisms and Structure," McGraw-Hill, New York, N. Y., 1968.
(54) (a) W. Kitching, *Organometal. Chem. Rev., Sect. A*, 3 61 (1968); (b) J. Chatt, *Chem. Rev.*, **48**, 7 (1951).
(55) H. J. Lucas, R. F. Hepner, and S. Winstein, *J. Amer. Chem. Soc.*, **61**, 3102 (1939).
(56) J. K. Stille, R. A. Morgan, D. D. Whitehurst, and J. R. Doyle, *J. Amer. Chem. Soc.*, **87**, 3282 (1965).
(57) J. K. Stille and R. A. Morgan, *J. Amer. Chem. Soc.*, **88**, 5135 (1966).
(58) M. Green and R. I. Hancock, *J. Chem. Soc. A*, 2054 (1967).
(59) C. B. Anderson and B. J. Burreson, *J. Organometal. Chem.*, **7**, 181 (1967).
(60) C. B. Anderson and S. Winstein, *J. Org. Chem.*, **28**, 605 (1963).

reaction of the methoxymethyl carbene ligand and pyridine discussed earlier.

Reaction 11 provides an excellent illustration of the importance of the choice of solvent.[52] For R = CF_3, in a nonpolar solvent such as benzene, I gives[61] a relatively stable 1:1 adduct of known geometry[62] and ultimately *trans*-$PtCl[C(CF_3){=}C(CF_3)CH_3]L_2$, $Pt(C_4F_6)L_2$, and $PtCl_2(CH_3)_2L_2$.

In contrast, in dichloromethane solution, I plus hexafluorobut-2-yne, with a trace of *trans*-$[PtCH_3(acetone)L_2]^+$ as halide abstractor, rapidly gives[52] *only* the above vinylic product. This reaction may be represented by the catalyzed sequence 13.

(i) *trans*-$[PtCH_3(acetone)L_2]^+PF_6^- + CF_3C{\equiv}CCF_3 \xrightarrow{CH_2Cl_2} [L_2PtCCF_3{=}C(CH_3)CF_3]^+PF_6^- + \text{acetone}$

(ii) $[L_2PtCCF_3{=}C(CH_3)CF_3]^+PF_6^- +$ *trans*-$PtClCH_3L_2 \rightleftharpoons$ *cis*-$PtCl(CCF_3{=}C(CH_3)CF_3)L_2 + [PtCH_3Q_2]^+PF_6^-$

(iii) $[PtCH_3L_2]^+PF_6^- + CF_3C{\equiv}CCF_3 \longrightarrow [L_2PtCCF_3{=}C(CH_3)CF_3]^+PF_6^-$ (13)

Such activation of acetylenes as we have described above is not limited to methylplatinum compounds; it has also been observed[45,63,64] for a variety of other organoplatinum derivatives where the organic moiety bound to platinum is alkyl, phenyl, vinyl, or alkynyl, all of which exert a high trans influence. The reactions of acetylenes and their products are summarized in Figure 4. Nor is such activation limited to acetylenes. Other unsaturated ligands, un, in cationic complexes–intermediates $[PtX(un)L_2]^+$ are activated toward (i) insertion into the Pt–X bond, (ii) isomerization, and (iii) nucleophilic attack; un =

(61) H. C. Clark and R. J. Puddephatt, *Inorg. Chem.*, **10**, 17 (1971).
(62) B. W. Davies, N. C. Payne, and R. J. Puddephatt, *Can. J. Chem.*, **50**, 2276 (1972).
(63) H. C. Clark and R. J. Puddephatt, *Chem. Commun.*, 92 (1970); *Inorg. Chem.*, **9**, 2670 (1970).
(64) T. G. Appleton, M. H. Chisholm, H. C. Clark, and L. E. Manzer, *Inorg. Chem.*, **11**, 1786 (1972); *ibid.*, submitted for publication.

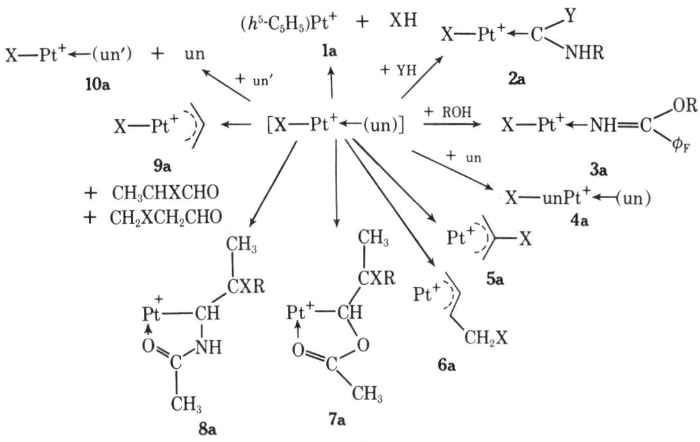

Figure 5. Some reactions of *trans*-$[PtX(un)Q_2]^+Z^-$. Q = PMe_2Ph $PMePh_2$, or PPh_3—omitted from figure for brevity. Z = PF_6, BF_4, ClO_4 or NO_3; **1a**, X = CH_3, un = cyclopentadiene; **2a**, X = Cl, un = RNC, Y = OCH_3, OC_2H_5, SR, NR_2 and NHR; **3a**, X = CH_3, CF_3, un = $\phi_F C{\equiv}N$ where ϕ_F is a perfluoroaryl group, R = CH_3 or C_2H_5; **4a**, X = H, un = $CH_2{=}CH_2$, $CH_3CH{=}CH_2$, $CH_3CH_2CH{=}CH_2$; **5a**, X = H, CH_3, un = $CH_2{=}C{=}CH_2$; **6a**, X = H, un = $CH_2{=}CH{-}CH{=}CH_2$; **7a**, X = H, un = $CH_2{=}CH{-}CH_2OCOCH_3$; **8a**, X = un = $CH_2{=}CHCH_2NHCO{-}CH_3$; **9a**, X = H, un = $(CH_2{=}CHCH_2)_2O$; **10a**, X = D, un = $CH_2{=}CHCH_2OCH_3$, $CH_2{=}CHCH_2OC_6H_5$, $CH_3CH_2\text{-}CH{=}CH_2$, un′ = $CH_3CH{=}CHOCH_3$, $CH_3CH{=}CHOC_6H_5$, and $CH_3CH{=}CHCH_3$, respectively, accompanied by D scrambling.

olefins, dienes, allenes, vinyl ethers, allyl alcohols, allylamines, allyl ethers, allyl esters, cyanides, and isocyanides.[47,64-76] These reactions are summarized in Figure 5.

Pt-C *vs.* Pt-H Insertion Reactions

The induced formation of cationic species in polar solvents can thus activate a wide variety of unsaturated compounds. Although the precise nature of the solvated cation and of the role of the solvent and counteranion are unknown, the attainment of the fourth coordination position trans to the ligand of high trans influence (*e.g.*, CH_3, H, C_6H_5, etc.) is clearly essential for activation. Thus, even in Pt-H or Pt-CH_3 insertion reactions, attainment of this position by un is a prerequisite. For example, the reaction of allene[47,75] with *trans*-$[PtCH_3(acetone)L_2]^+$ gives below 0° the *trans*-$[PtCH_3(\pi\text{-allene})L_2]^+$ cation, which at higher temperatures leads to $Pt(\pi\text{-}C_4H_7)L_2^+$, where $\pi\text{-}C_4H_7$ = π-2-methallyl. Similarly, the reaction of *trans*-$[PtH(acetone)L_2]^+$ with ethylene[74] at −78° gives *trans*-$[PtH(\pi\text{-}C_2H_4)L_2]^+$ which at higher temperature leads to the *trans*-$[Pt(C_2H_5)(\pi\text{-}C_2H_4)L_2]^+$ derivative. The mechanism by which such insertions proceed following attainment by un of that fourth coordination position is not clear, and the roles of the solvent and counteranion may well be critical.

It is instructive to compare the reactivities of $[CH_3PtL_2]^+$ and $[HPtL_2]^+$ cations. Firstly, the hydride cations $[HPtL_2]^+$ are the more reactive toward insertion; this may be partly attributable to the thermodynamic properties of the Pt-H bond compared with those of the Pt-C bond in either the $[CH_3PtL_2]^+$ cation or in the products. Secondly, the difference in reactivity may relate to the availability of alternative reaction mechanisms. Thus, reactions involving $[CH_3PtL_2]^+$ proceed in a Markovnikov manner by electrophilic attack of Pt^+: for example, as mentioned above, $Pt(\pi\text{-2-methallyl})L_2^+$ is formed from allene and *trans*-$[PtCH_3(acetone)L_2]^+$, whereas the analogous butadiene cation does not lead[75] to a π-allylic derivative by Pt-CH_3 insertion. In contrast, $[HPtL_2]^+$ can react either by Markovnikov or anti-Markovnikov mechanisms,[74] *i.e.*, Pt^+ or H^+ attack. A consequence of this apparent versatility is the formation[70,74] of π-allylic complexes from both allenes and 1,3-dienes with $[HPtL_2]^+$; moreover, this versatility is responsible for the characteristic isomerization and H-D exchange reactions of olefins with platinum hydrides.[77]

It is now generally accepted[78-80] that metal-carbon and metal-hydrogen insertion reactions form the basis of transition-metal-catalyzed polymerization and hydrogenation reactions of unsaturated hydrocarbons, respectively. The catalytic activity of d^8 transition metals has been attributed to their ability to expand their coordination spheres, and five-coordinate π complexes have been invoked[61,81-88] as the reactive intermediates in such reactions.

(65) H. C. Clark and L. E. Manzer, *Chem. Commun.*, 387 (1971).
(66) H. C. Clark and L. E. Manzer, *J. Organometal. Chem.*, **30**, C89 (1971).
(67) E. M. Badley, J. Chatt, R. L. Richards, and G. A. Sim, *Chem. Commun.*, 1322 (1969).
(68) E. M. Badley, J. Chatt, and R. L. Richards, *J. Chem. Soc. A*, 21 (1971).
(69) E. M. Badley, B. J. L. Kilby, and R. L. Richards, *J. Organometal. Chem.*, **27**, C37 (1971).
(70) A. J. Deeming, B. F. G. Johnson, and J. Lewis, *Chem. Commun.*, 1231 (1968).
(71) H. C. Clark and H. Kurosawa, *Chem. Commun.*, 957 (1971).
(72) H. C. Clark and L. E. Manzer, *Inorg. Chem.*, **10**, 2699 (1971).
(73) H. C. Clark and H. Kurosawa, *Chem. Commun.*, 150 (1970).
(74) H. C. Clark and H. Kurosawa, *Inorg. Chem.*, **11**, 1276 (1972).
(75) M. H. Chisholm and H. C. Clark, *Inorg. Chem.*, **12**, 991 (1973).
(76) H. C. Clark, *et al.*, unpublished results.
(77) H. C. Clark and H. Kurosawa, *Inorg. Chem.*, **12**, 357 (1973).
(78) M. L. H. Green "Organometallic Compounds," Vol. II, Methuen, London, 1968, p 312.
(79) L. Reich and A. Schindler, *Polym. Rev.*, **12**, 1 (1966).
(80) R. Cramer, *Accounts Chem. Res.*, **1**, 186 (1968).
(81) J. C. Bailar and H. A. Tayim, *J. Amer. Chem. Soc.*, **89**, 3420 (1967).
(82) J. C. Bailar and H. A. Tayin, *J. Amer. Chem Soc.*, **89**, 4330 (1967).
(83) W. H. Baddley and M. S. Frazer, *J. Amer. Chem. Soc.*, **91**, 3661 (1969).
(84) P. Ugualatti and W. H. Baddley, *J. Amer. Chem. Soc.*, **90**, 5446 (1968).
(85) G. W. Parshall and F. M. Jones, *J. Amer. Chem. Soc.*, **87**, 5356 (1965).
(86) H. C. Clark and W. S. Tsang, *J. Amer. Chem. Soc.*, **89**, 529 (1967).

While insertion may occur *via* the intramolecular rearrangement of such five-coordinate species, little direct evidence to establish this is yet available. Other possibilities, such as bimolecular or free-radical processes involving the five-coordinate species, may well need to be kept in mind. All of our present evidence would suggest that acetylenes are activated toward (i) rearrangements, (ii) Pt–C insertion, and (iii) nucleophilic attack, by coordination to a relatively electron-deficient metal atom as in $[PtX(RC{\equiv}CR')L_2]^+$. In contrast, coordination to a relatively electron-rich metal as in the five-coordinate π complex with I activates the acetylene to a much smaller extent and may indeed deactivate it toward such reactions.

Conclusions

In considering the generality of the metal-induced carbonium ion model, one must recognize that such reactions are not limited to those of cationic transition-metal complexes with unsaturated compounds. All that is required is that the metal be effectively acidic and capable of leading to a dipole-induced reaction. For example, the platinum–methyl insertion reactions with acetylenes, which proceed rapidly when the acetylene attains the fourth coordination position, may be compared with analogous insertion reactions involving aluminum trialkyls where kinetic data show[89] that the reactive species is $R_3Al(RC{\equiv}CR)$. On the other hand, formal positive charge on an organometallic cation does not always lead to acidic or electrophilic interaction with unsaturated compounds. For example, low-valent transition-metal cations may be relatively electron rich. The cationic Ir(+) acetylene complex $[Ir(CO)_2(PR_3)_2(H_3COOCC{\equiv}CCOOCH)]^+$[90] may be compared with the neutral Pt(2+) formed from I and the same acetylene; in both cases, the metal is acting as a strong nucleophile.

Certainly many organometallic- and transition-metal-catalyzed reactions can be considered to proceed *via* metal-induced or -stabilized carbonium ions. In particular, examples may be drawn from the following classes of reactions: metalation reactions,[54] proton-addition and hydride-abstraction reactions,[91] electrophilic substitution reactions of metallocenes.[92,93] Lewis acid cocatalyzed polymerization of olefins, dienes and acetylenes,[79,94] and metal-catalyzed σ rearrangements of strained alkanes.[95-97] Of these, the Ziegler–Natta process (involving a Lewis acid and an organohalo derivative of any early transition metal as cocatalysts) is of the greatest commercial significance, and its characteristic feature shows a close similarity to that of the organoplatinum systems discussed above. Both require the creation of a vacant and electrophilic coordination site,[98] and their differences result from the different thermodynamic properties of Ti–C and Pt–C σ bonds. Thus, many of the reactions of organoplatinum compounds described above serve as models for those which occur in more complex systems.

Addendum (May 1976)

Much information has been obtained recently on the detailed steps of Pt-H insertion with olefins and acetylenes. In the species *trans*-$PtHXL_2$ (where L is usually a tertiary phosphine) the nature of X is critically important in determining the insertion pathway,[99] and either four-coordinate (cationic) or five-coordinate (neutral)[100] intermediates may be involved. Where X is readily displaced so that the carbonium ion species *trans*-$[PtH(un)L_2]^+$ are involved, it is now well established that the insertion process involves the trans $\rightleftharpoons$ cis equilibrium.[101–103]

Carbonium ions are also accessible from both σ- and π-routes. Thus many reactions of organoplatinum(II) compounds containing σ-bonded unsaturated ligands should complement the reactions of π-bonded unsaturated hydrocarbons coordinated to organoplatinum(II) cations. Consistent with this view are the findings that platinum stabilized vinyl carbonium ions, $Pt\text{-}^+C{=}CHR$, act as alkoxycarbene ligand precursors in alcohols and may be formed from (i) cationic acetylenic intermediates (described above), (ii) protic acids and platinum(II) acetylides,[104] and (iii) solvolysis of α-chlorovinylplatinum(II) compounds.[105] *Trans*-$Pt(CCl{=}CH_2)_2(PMe_2Ph)_2$ is labile towards the stepwise reversible elimination of HCl to give *trans*-$Pt(CCl{=}CH_2)(C{\equiv}CH)(PMe_2Ph)_2$ and *trans*-$Pt(C{\equiv}CH)_2(PMe_2Ph)_2$, even in the solvents benzene and toluene.[106] With methanol, *trans*-$Pt(CCl{=}CH_2)_2(PMe_2Ph)_2$ reacts to give *trans*-$[Pt(CCl{=}CH_2)(C{<}^{OMe}_{CH_3})\text{-}(PMe_2Ph)_2]^+Cl^-$, which may be isolated as a hexafluorophosphate salt. Rather interestingly alcoholysis occurs only at one of the $Pt\text{-}CCl{=}CH_2$ groups,

(87) L. S. Meriwether, M. F. Leto, E. C. Coltump, and G. W. Kennerty, *J. Org. Chem.*, **27**, 3930 (1962).

(88) A. Furlini, I. Collamati, and G. Sartori, *J. Organometal. Chem.*, **17**, 463 (1969).

(89) P. E. M. Allen, A. E. Byers, and R. M. Lough, *J. Chem. Soc., Dalton Trans.*, 479, (1972).

(90) M. J. Church, M. J. Mays, R. M. F. Simson, and F. P. Stefanni, *J. Chem. Soc. A*, 2909 (1970).

(91) M. A. Haas, *Organometal. Chem. Rev., Sect. A*, **4**, 307 (1969).

(92) D. A. White, *Organometal. Chem. Rev., Sect. A*, **3**, 497 (1968).

(93) D. W. Slocum and C. R. Ernst, *Organometal. Chem. Rev., Sect. A*, **6**, 337 (1970).

(94) G. Lefebvre and Y. Chauvin in "Aspects of Homogeneous Catalysis I," R. Ugo, Ed., Carlo Manfredi, Milan, 1968.

(95) L. A. Paquette, *Accounts Chem. Res.*, **4**, 280 (1971).

(96) P. G. Gassman and T. J. Atkins, *J. Amer. Chem. Soc.*, **93**, 4597 (1971).

(97) J. E. Byrd, L. Cassar, P. E. Eaton, and J. Halpern, *Chem. Commun.*, 40 (1971).

(98) For example, in the $TiCl_4$–$(CH_3)_2AlCl$ system (which is a typical Ziegler–Natta catalyst) the active complex is CH_3TiCl_3–CH_3AlCl_2. On the basis of solvent studies, the following ionic solvated structure has been ascribed to the active species $[CH_3TiCl_2]^+[Al(CH_3)Cl_3]^-$, which allows formation of cationic olefinic intermediates (ref 92, p 130, and references therein).

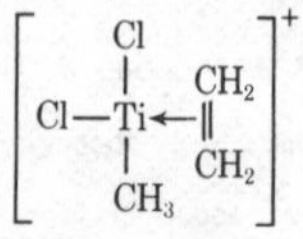

(99) H. C. Clark, C. R. Jablonski, and C. S. Wong, *Inorg. Chem.*, **14**, 1332 (1975).

(100) H. C. Clark, C. R. Jablonski, J. Halpern, A. Mantovani, and T. A. Weil, *Inorg. Chem.*, **13**, 1541 (1974).

(101) H. C. Clark and C. R. Jablonski, *Inorg. Chem.*, **13**, 2213 (1974).

(102) H. C. Clark and C. S. Wong, *J. Am. Chem. Soc.*, **96**, 7213 (1974).

(103) H. C. Clark and C. S. Wong, *J. Organomet. Chem.*, **92**, C31 (1975).

(104) M. H. Chisholm and D. A. Couch, *J.C.S. Chem. Commun.*, 42 (1974).

(105) R. A. Bell and M. H. Chisholm, *J.C.S. Chem. Commun.*, 818 (1974).

(106) R. A. Bell and M. H. Chisholm, *J.C.S. Chem. Commun.*, 200 (1976).

and when the reaction is carried out in MeOD, the vinylic protons of the Pt-CCl=CH_2 ligand in the cationic carbene complex are seen to not exchange with deuterons of the solvent.[107] The reactivity of the Pt-CCl=CH_2 group is thus dependent on the [L_2PtX] moiety to which it is bonded. X-ray studies show that *trans*-Pt(CCl=CH_2)$_2$(PMe_2Ph)$_2$ is a centrosymmetrical molecule. A long vinylic C–Cl bond distance, 1.809(6) Å, and a large Pt-C-C angle, 133–5°, provide structural evidence for the existence of the Pt-$^+$C=CH_2 moiety.[107]

The applicability of the metal-induced carbonium ion model to analogous systems of other transition metals has received attention recently. For example, methyl-iridium(III) and methyl-rhodium(III) cations[108] do not give isolable π-acetylene complexes, but in methanol, stable carbene complexes result, their formation presumably being due to methanol attack on a metal-induced carbonium ion.

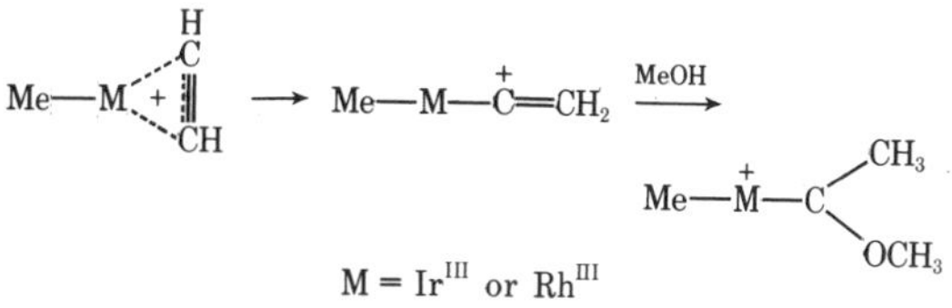

Also, detailed ^{13}C nmr studies[109] of cationic Pt^{II} carbene complexes provide data which are fully consistent with this carbonium ion model.

(107) R. A. Bell, M. H. Chisholm, and G. G. Christorph, *J. Am. Chem. Soc.*, **98**, 000 (1976).

(108) H. C. Clark and K. J. Reimer, *Inorg. Chem.*, **14**, 2133 (1975).

(109) M. H. Chisholm, H. C. Clark, J. E. H. Ward, and K. Yasufuku, *Inorg. Chem.*, **14**, 893 (1975).

Rapid Intramolecular Rearrangements in Pentacoordinate Transition Metal Compounds

John R. Shapley and John A. Osborn*

Department of Chemistry, Harvard University, Cambridge, Massachusetts 02138

Received November 14, 1972

Reprinted from ACCOUNTS OF CHEMICAL RESEARCH, *6, 305* (*1973*)

Pentacoordination is now found throughout the Periodic Table, but it occurs predominantly within two areas—in the compounds of group V (such as phosphorus(V)) and of the transition metals in a formal d^8 electronic configuration (*e.g.*, zerovalent iron).[1]

In group V, the ground-state geometry is almost exclusively trigonal bipyramidal (TBP) rather than square pyramidal (SP), and the relative energies of positional TBP isomers are determined by preference rules based on ligand electronegativity. Further, stereochemical nonrigidity is a characteristic and chemically important feature for which a general rearrangement mechanism has been established.[2]

In contrast, although many pentacoordinate transition metal complexes have been synthesized, and X-ray diffraction studies have established several solid-state structures, the factors determining the stability of geometric and positional isomers are not readily apparent. For d^8 complexes, most of the structures correspond closely to either TBP or, less commonly, SP geometry, although a few structures show quite large distortions from these idealized forms. In view of the close stability of SP and TBP forms in transition metal complexes, intramolecular rearrangement might be expected to occur readily. Indeed, at the outset of this work there were indications in the literature that this may well be so (*vide infra*), although no unambiguous example of such behavior had been reported.

In this Account we present the results of studies on a series of related complexes of d^8 electronic configuration, whose stereochemistries, dynamic behavior, and mechanism of rearrangement in solution have been investigated by nmr spectroscopy. From these studies some tentative generalizations have been drawn. However, pentacoordinate intermediates have also been invoked in ligand exchange reactions and catalytic processes involving d^8 metal complexes. Little consideration has been given to the detailed stereochemistry of such intermediates or to the implications of dynamic behavior occurring during their lifetime. Preliminary results involving such intermediates are also presented.

Background Observations.

Early ^{13}C nmr studies on iron pentacarbonyl found only one resonance line,[3] even though two signals (3:2 ratio) would be expected for the TBP ground-state geometry,[3] implying a rapid exchange of carbonyl ligands between nonequivalent sites. The first unambiguous evidence for a rapid intramolecular rearrangement in a d^8 pentacoordinate transition metal complex was reported in 1969 by Udovich and

John Osborn was born in Chislehurst, Kent, England, in 1939. He received his B.A. at Cambridge University and his Ph.D. at Imperial College, London, with Geoffrey Wilkinson. After a year as Imperial Chemical Industries Fellow at Imperial College, he was appointed Assistant Professor at Harvard University in 1967 and is presently Associate Professor. He is recipient of a Camille and Henry Dreyfus teaching-scholar award and an Alfred P. Sloan fellowship. His research interests are concerned with the synthesis, mechanistic studies, and reactivity of transition metal organometallics.

John R. Shapley received his B.S. from University of Kansas and his Ph.D. at Harvard University in 1971. He was a National Science Foundation postdoctoral fellow at Stanford University with James P. Collman, and is currently Assistant Professor at the University of Illinois at Champaign-Urbana.

(1) E. L. Muetterties and R. A. Schunn, *Quart. Rev., Chem., Soc.*, **20**, 245 (1966).

(2) *e.g.*, (a) G. M. Whitesides and H. L. Mitchell, *J. Amer. Chem. Soc.*, **91**, 5384 (1969); (b) F. H. Westheimer, *Accounts Chem. Res.*, **1**, 70 (1968); (c) R. R. Holmes, *ibid.*, **5**, 296 (1972).

(3) (a) F. A. Cotton, A. Danti, J. S. Waugh, and R. W. Fessenden, *J. Chem. Phys.*, **29**, 1427 (1958); (b) R. Bramley, B. N. Figgis, and R. S. Nyholm, *Trans. Faraday Soc.*, **58**, 1893 (1962).

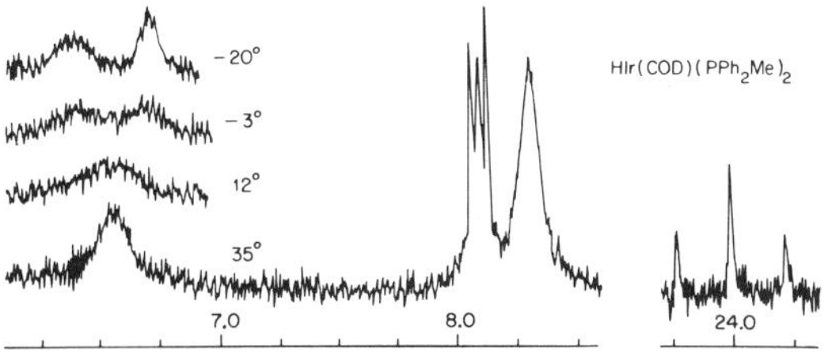

Figure 1.

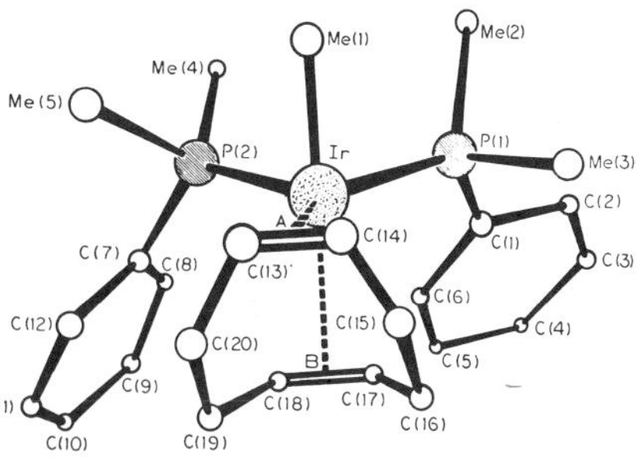

Figure 2.

Clark,[4] the variable low-temperature ^{19}F nmr spectrum of the compound $CF_3Co(CO)_3PF_3$ indicating two equilibrating isomers which exchange ligands between axial and equatorial sites. Contemporaneous studies on the compound[5] $HIr(CO)_2(P(C_6H_5)_3)_2$ and more recently on several $HM(PF_3)_4$ and related species[6] have provided further evidence that rapid isomer equilibration is a general phenomenon in such pentacoordinate complexes.

Diene Complexes of Pentacoordinate Ir(I) and Rh(I). Some Mechanistic Distinctions.

Concurrent with these latter reports,[7] we synthesized compounds of the type $RM(diene)L_2$ (M = Ir, Rh; R = H, CH_3, C_6H_5; L = tertiary phosphine or arsine; diene = 1,5-cyclooctadiene (COD) or norbornadiene (NBD)) and studied their ^1H nmr spectra over a wide temperature range. In each case, a low-temperature limiting spectrum was obtained which displayed features requiring the corresponding static structure to have the L ligands in symmetric positions, but the vinyl protons on the diene to be in two nonequivalent sites. The limiting spectrum measured at −20° for $HIr(COD)(PCH_3(C_6H_5)_2)_2$ in CH_2Cl_2 is shown in Figure 1; two separate vinyl resonances are observed for the 1,5-cyclooctadiene ligand (at τ 6.30 and 6.71), the hydride group attached to Ir appears as a triplet (τ 24.0) resulting from equal coupling to the two phosphorus nuclei, and only one methyl resonance (τ 8.08 virtually coupled) is observed for the two $PCH_3(C_6H_5)_2$ ligands. Moreover, only one ^{31}P resonance is observed for all compounds. The analogous NBD complexes of Rh(I) have equivalent methine protons which, by virtue of the symmetry of the NBD molecule, indicates that the nonequivalent vinyl protons are on the *different* double bonds. A TBP ground-state geometry is, therefore, indicated with the R group in one axial position, the phosphine ligands occupying two equatorial sites, and the diene spanning the remaining axial and equatorial sites (C_s symmetry). This is confirmed (Figure 2) by a single-crystal X-ray structural determination[7c] of $CH_3Ir(COD)(P(CH_3)_2C_6H_5)_2$. The only significant deviation from an idealized TBP structure is in the P–Ir–P angle which is 101.5° instead of the idealized 120°, undoubtedly because of interactions between the equatorial olefin and the phosphine ligands.

Variable-temperature nmr studies of these complexes show that two independent dynamic processes occur, the relative rates of which depend quite differently upon the composition of the complex. One process is that of ligand dissociation from the five-coordinate species, as evidenced by the collapse of coupling to the R group in the complexes involving phosphines and/or loss of the structure of the ligand methyl resonances. This process often occurs at distinctly higher temperatures (see Figure 3, +117° spectrum) which enables it to be observed separately from the second process; we defer discussion of such intermolecular processes until later. The second process causes substantial changes only in the diene resonances, most importantly, causing the coalescence of the separate vinyl signals which were observed in the limiting spectrum (*e.g.*, at 35° in Figure 1). This process is an intramolecular rearrangement which leads to interchange of the axial and equatorial double bonds found in the instantaneous structure.

The variable temperature nmr spectrum of $CH_3Ir(COD)(P(CH_3)_2C_6H_5)_2$ (as shown in Figure 3) provides the most information with regard to the mechanism of the rearrangement. In the limiting spectrum at −3° the appropriate signals for all the groups are observed. However, two distinct phosphine methyl resonances are observed as well as the separate vinyl proton signals. The two methyls on *each* dimethylphenylphosphine reside in different environments (*i.e.*, are diastereotopic) (Figure 2) and give rise to separate nmr signals. This is true even with rapid rotation around all bonds. However, should the two phosphine ligands interchange sites, coalescence of the separate methyl signals will occur. Inspection of Figure 3 (up to 87°), however, indicates that, when the temperature is raised into the range where coalescence of the vinyl proton signals occurs, the two phosphine methyl patterns remain distinct. Therefore, the intramolecular rearrangement we observe for this complex *interchanges only the nonequivalent double bonds and not the two phosphine ligands.* Similar behavior is observed for the other $P(CH_3)_2Ph$ complexes, and by analogy this conclusion is extended to the entire set of complexes.

What mechanistic information can be derived from these experimental observations? If we view the mechanism in a permutational (or topological) context,[8] we can see that several permutational pro-

(4) C. A. Udovich and R. J. Clark, *J. Amer. Chem. Soc.*, **91**, 526 (1969).
(5) G. Yagupsky and G. Wilkinson, *J. Chem. Soc. A*, 725 (1969).
(6) (a) P. Meakin, E. L. Muetterties, and J. P. Jesson, *J. Amer. Chem. Soc.*, **94**, 5271 (1972); (b) D. D. Titus, A. A. Orio, R. E. Marsh, and H. B. Gray, *Chem. Commun.*, 322 (1971).
(7) (a) J. R. Shapley and J. A. Osborn, *J. Amer. Chem. Soc.*, **92**, 6976 (1970); (b) D. P. Rice and J. A. Osborn, *J. Organometal. Chem.*, **30**, C84 (1971); (c) M. R. Churchill and S. A. Bezman, *Inorg. Chem.*, **11**, 2243 (1972).

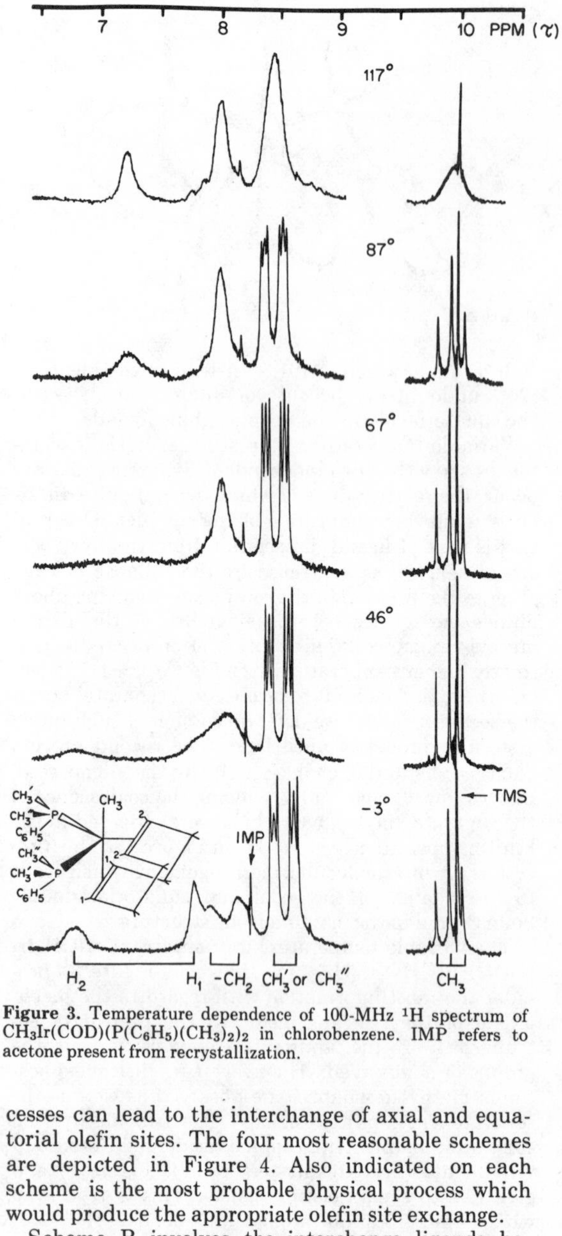

Figure 3. Temperature dependence of 100-MHz ^{1}H spectrum of $CH_3Ir(COD)(P(C_6H_5)(CH_3)_2)_2$ in chlorobenzene. IMP refers to acetone present from recrystallization.

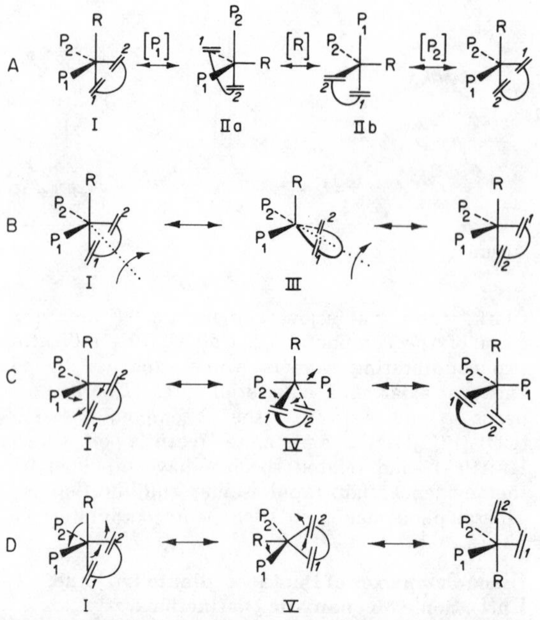

Figure 4. Mechanistic schemes to account for axial–equatorial equilibration of COD vinyl protons in the complexes $RIr(COD)P_2$.

cesses can lead to the interchange of axial and equatorial olefin sites. The four most reasonable schemes are depicted in Figure 4. Also indicated on each scheme is the most probable physical process which would produce the appropriate olefin site exchange.

Scheme B involves the interchange ligands between two sites within the trigonal-bipyramidal structure, one axial and one equatorial, without affecting remaining ligands. The actual physical motion depicted is a twist of the diene about a pseudotwofold axis perpendicular to the plane containing the double bonds. The intervening structure III might be either a transition state in which the relative disposition of R, P_1, and P_2 remain unchanged or an intermediate in which the ligands have relaxed somewhat to approximately SP geometry.

Each of the schemes C and D involves a permutation of three sites within the trigonal TBP structure. In C one axial and two equatorial ligands can be interchanged by a rotation about a pseudo-threefold axis, *via* IV as a transition state. Scheme D, on the other hand, couples both axial ligands with one equatorial ligand in a motion of the R group from one face to another of the pseudotetrahedral configuration of the remaining ligands.

Scheme A proceeds through two TBP intermediate structures, each of which is generated by permutation of four ligands within the structure preceding it. The specific mechanism for each step is the Berry or pseudorotation mechanism. Olefin site interchange can be accomplished by three reversible and sequential pseudorotations, with [P_1], [R], and [P_2] stepwise as pivots. This may seem a somewhat convoluted scheme at first, but two points may be noted: (i) the third pseudorotation about [P_2] is the reverse of the first about [P_1]; (ii) IIa and IIb are enantiomers and their interconversion might be expected to be facile. Olefin equilibration could also be achieved by a different Berry process, *e.g.*, configuration IV in C could be reached by a pseudorotation from I by using the equatorial olefin as a pivot. This scheme would involve the diene spanning equatorial sites in IV which would seem to be unfavorable energetically and, indeed, for reasons discussed below with respect to path C, this alternative Berry process cannot be operative here.

If these schemes are examined carefully, C and D, but *not* A and B, lead to interchange of the two ligands P_1 and P_2 while olefin site exchange takes place. The alternate Berry process involving IV as intermediate also causes P_1,P_2 interchange. Hence, only schemes A and B are in accord with the conclusions drawn from the spectral behavior of the PMe_2Ph compounds, providing, therefore, the first

(8) See, *e.g.*, (a) E. L. Muetterties, *J. Amer. Chem. Soc.*, **90**, 5097 (1968); (b) M. Gielen and N. Vanlautem, *Bull. Soc. Chim. Belg.*, **79**, 679 (1970); (c) J. R. Shapley, Ph.D. Thesis, Harvard University, 1971; (d) W. G. Klemperer, *J. Chem. Phys.*, **56**, 5478 (1972).

Table I

$RIr(COD)P_2$	ΔG_c^*, kcal/mol^{-1} [a]	
	R = CH_3	R = H
$P(CH_3)_2C_6H_5$	16.3	13.3
$P(CH_3)(C_6H_5)_2$	>16.9[a]	14.1
$P(C_6H_5)_3$	≫14.0[a]	17.7
$P(C_2H_5)(C_6H_5)_2$		17.6
$P(C_3H_7)(C_6H_5)_2$[c]		>20.6[b]
$P(C_6H_{11})(C_6H_5)_2$[c]		>20.1[b]

[a] ΔG_c^* value at coalescence temperature of vinyl resonances. [b] Lower limit since intermolecular exchange intervenes. [c] C_3H_7 = isopropyl; C_6H_{11} = cyclohexyl.

experimental evidence which conclusively eliminates some of the theoretically possible rearrangement mechanisms for pentacoordinate transition metal complexes. These results, however, do not permit a rigorous distinction between A and B.

Free energies of activation calculated from the nmr data obtained for the compounds $RIr(COD)L_2$ are presented in Table I. These numbers illustrate the dependence of a barrier to intramolecular rearrangements upon the composition of the complex and lead to several conclusions: (a) the barriers are medium-to-high within the range generally accessible by nmr techniques (*ca.* 5–20 kcal/mol); (b) the barrier is strongly dependent on the steric bulk of the phosphine ligands, *i.e.*, $P(C_6H_5)_2C_6H_{11}$ ~ $P(C_6H_5)_2C_3H_7$ > $P(C_6H_5)_3$ ~ $P(C_6H_5)_2C_2H_5$ > $P(C_6H_6)_2CH_3$ > $P(C_6H_5)(CH_3)_2$; (c) the barrier is somewhat higher for R = CH_3 than H. Similar trends are observed in the analogous Rh(I) complexes involving norbornadiene as ligand.[7b]

The dependence of the rearrangement barrier on the steric bulk of the ligands is readily interpreted in terms of scheme A. The conversion of I to IIa involves a greater amount of reorganization of the electronic and steric balance of the complex than does IIa to IIb. Hence it would be expected that I to IIa would be the rate-determining step of this sequence and, assuming the Hammond postulate, the transition state for this step will resemble the high-energy intermediate IIa. Proceeding from I to IIa leads to a compression of the angle P–Ir–P from the idealized 120 to 90°. This change should be more disadvantageous for large ligands, thus causing an increase in the barrier to rearrangement.

A similar explanation can also be advanced for B, where III might be expected to have a small (~90°) P–Ir–P angle.

Studies on Chelated Phosphine Complexes: Effect of Ring Strain.

In the complex $CH_3Ir(COD)(diphos)$ (diphos = 1,2-bis(diphenylphosphino)ethane), in which the phosphine ligands are linked, a sharp triplet pattern was observed for the methyl group, but no indication of vinyl proton nonequivalence was seen down to ~ −90°. The X-ray structure of this complex[9] showed the stereochemistry to be essentially that shown in Figure 2, the only significant difference being the *very small P–Ir–P angle of 84.9°*, presumably resulting from the geometric constraints produced by the chelating diphos ligand. This large compression of the P–Ir–P angle when compared to a nonchelating analog (*ca.* 17°) implies a high degree of strain within the five-membered chelate ring in the ground-state structure.

This observation, in conjunction with scheme A or B, provides an interpretation of our inability to achieve limiting spectra for the diphos complex. In A structures IIa and IIb would be expected to have a nearly ideal P–Ir–P angle of 90° which would mean the five-membered ring would be essentially *unstrained in these intermediates.*[10] Hence I is destabilized relative to IIa in the chelated molecules and the relief of ring strain afforded in IIa lowers the energy barrier for this rate-determining interconversion as compared to unstrained systems. Similarly, in B, III is expected to be less strained than I.

Study of the corresponding 1,3-bis(diphenylphosphino)propane complex, where the resulting six-membered ring would be less strained, lends credence to this interpretation. The limiting low-temperature spectrum is obtained at −45°, and the barrier to rearrangement is *ca.* 13.4 kcal/mol, a value greater than that estimated for the diphos complex but less than that for the unchelated analog. Further, the X-ray structure[11] of the 1,3-bis(diphenylphosphino)propane complex shows the P–Ir–P angle to be 93.4°, a value intermediate between those of the diphos and $P(CH_3)_2C_6H_5$ complexes. In the 1,4-(diphenylphosphino)butane complex the barrier is even higher (>16.5 kcal/mol), in agreement with the proposed mechanism.

If the chemical shift separation of the vinyl resonances in the diphos complex is assumed to be similar to that for the other chelating phosphines, an upper limit to the barrier for this complex can be estimated as <*ca.* 9 kcal/mol. Furthermore, as shown in Table II, ΔG_c^* for each of the chelating ligand complexes can be compared with that of an unchelated analog. These results establish that the chelated compounds have significantly lower barriers than their unchelated counterparts, the decrease amounting to as much as 7 kcal/mol.

Studies on Complexes with Tin Ligands: Effects of π Bonding

In view of the high degree of strain found for the ground-state isomer of $CH_3Ir(diphos)(COD)$, why is this positional isomer more stable than that with the methyl group equatorial and the chelating phosphine ligand comfortably spanning axial–equatorial sites (*e.g.*, IIa in Figure 4)? Might this not be illustrative of a site preference effect either of the strongly σ-bonding methyl group for the axial site or the more π-bonding phosphine ligand for the equatorial sites? If this were so, then the structure with the R group equatorial might be stabilized if the R group had *strong* π-bonding ability. Previous studies on $SnCl_3^-$ as a ligand had indicated it to be a very strong π-

(9) M. R. Churchill and S. A. Bezman, *Inorg. Chem.*, **12**, 260 (1973).

(10) The unstrained "bite" angle of diphos appears to be *ca.* 84°. For example, see J. A. McGinnety, N. C. Payne, and J. A. Ibers, *J. Amer. Chem. Soc.*, **91**, 6301 (1969).

(11) S. A. Bezman and M. R. Churchill, *Inorg. Chem.*, **12**, 531 (1973).

Table II

$CH_3Ir(COD)L_2$	ΔG_c^*, kcal/mol^{-1}	∠P-Ir-P, deg
L_2 = 1,2-bis(diphenylphosphino)ethane	<9.0[a]	84.9
L_2 = 1,2-bis(diphenylphosphino)propane	13.4	93.4
L_2 = 1,2-bis(diphenylphosphino)butane	>16.5	
L = $P(CH_3)(C_6H_5)_2$	>16.9	(101.5)[b]
L_2 = 1,2-bis(dimethylarsino)benzene	11.9	
L = $As(CH_3)_2C_6H_5$	>15.5	

[a] Estimated on basis that vinyl group separation is 120 Hz. [b] Value for $P(CH_3)_2C_6H_5$ complex.

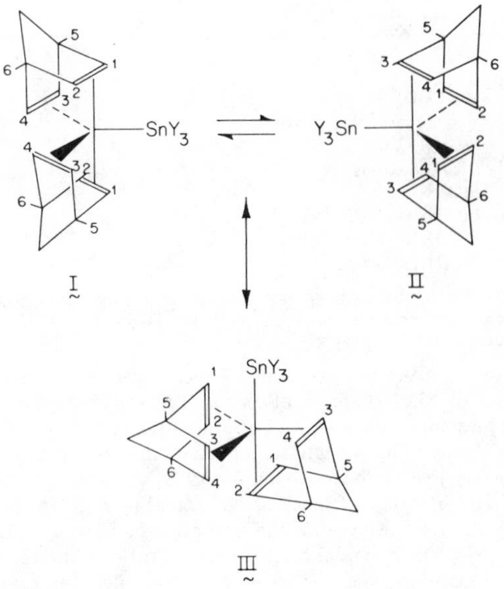

Figure 5.

acid ligand, and we noted that compounds of the type $Ir(diene)_2SnCl_3$ (diene = COD, NBD) had been isolated.[12,13] Moreover, the X-ray crystal structure[13] shows $Ir(COD)_2SnCl_3$ to be TBP, with $SnCl_3^-$ occupying an equatorial site and the diene ligands each spanning axial–equatorial positions. By displacement of one diene by two phosphine ligands, $Ir(diene)P_2SnCl_3$ complexes can be isolated which are analogous to the previously described iridium–alkyl complexes except that the σ-bonding alkyl has been replaced by the π-bonding $SnCl_3^-$ ligand.

The parent compound, $Ir(COD)_2SnCl_3$, shows only one vinyl resonance at 33°, instead of the four expected if the solid-state structure were retained and static in solution. Similarly eight compounds of the type $Ir(COD)L_2SnCl_3$ (L = phosphine, phosphonite, arsine) show ^1H nmr spectra much less complex than that expected for any individual static structure, but variable-temperature studies have failed as yet to achieve limiting spectra.

The corresponding NBD complexes, however, are more informative, since NBD will not span equatorial–equatorial sites readily and offers the methine resonances as a further structural probe. The ^1H nmr of $Ir(NBD)_2SnCl_3$ (33°) shows two vinyl and two methine resonances, consistent with a SP structure with $SnCl_3^-$ apical, but we prefer to assign this complex a TBP ground-state configuration as found for the corresponding COD complex. Rapid enantiomeric equilibration (I ⇌ II, Figure 5) using $SnCl_3^-$ as pivot in a Berry process would result in the observed spectrum, *i.e.*, equilibration of the vinyl protons 1 with 3, and 2 with 4, but *not* of the methine protons 5 and 6. In contrast, ten substituted derivatives of the type $Ir(NBD)L_2SnR_3$ (L = phosphine, phosphinite, or arsine; R = Cl or CH_3) display only *one* vinyl and *one* methine resonance at 33°. At −90°, however, the complexes $Ir(NBD)P_2SnCl_3$ (P = $P(CH_3)_2C_6H_5$, $P(CH_3)(C_6H_5)_2$, $P(C_6H_5)_3$, $P(C_4H_9)_3$) each show the *two* vinyl, *two* methine pattern. Further, the ^{31}P spectrum of the tri(*n*-butyl)phosphine species, $Ir(NBD)(P(C_4H_9)_3)_2SnCl_3$, at −80° is a sharp singlet. Hence, as before, $Ir(NBD)L_2SnCl_3$ are assigned as TBP[14] but undergo rapid equilibration between I and II (Figure 6) at −90°. At ambient temperatures these compounds must undergo a further rearrangement *via* a structure which causes coalescence of the pairs of vinyl and methine resonances. In the PMe_2Ph complex the diastereotopic methyl groups on the phosphine remain distinct during this process. The only intermediate consistent with these observations is III (Figure 6), with $SnCl_3$ axial and the two phosphine ligands in equatorial sites. The dynamic behavior observed for the system depicted in Figure 6 is therefore closely related to A in Figure 4 except *replacement of the alkyl (or hydride) ligand by $SnCl_3$ has altered the relative energies of the positional isomers.*

The calculated barriers (I + II → III) for the $Ir(NBD)P_2SnCl_3$ compounds are not greatly dependent on the steric bulk of the phosphine ligands, the spread of values (*ca.* 1 kcal/mol) being of the order of experimental error.[14] Interestingly, however, replacement of the monodentate phosphine ligands by the chelating diphos ligand causes an *increase* in the barrier by *ca.* 3 kcal/mol. This is precisely the reverse of that observed in the corresponding alkyl derivatives. However, inspection of path A (Figure 4) and Figure 6 shows that, whereas *with the chelating phosphine strain destabilization occurs in the ground state of the alkyl derivatives, it occurs in the transition state of these $SnCl_3^-$ complexes*, *i.e.*, the chelate spans the idealized 90° in the ground state of the $SnCl_3^-$ complexes, but on passing to the high-energy intermediate *III* the chelate must open to formally 120°. Strain thus develops in the *transition state*, thereby raising the barrier to intramolecular rearrangement. These observations can therefore be convincingly interpreted in terms of path A (the Berry mechanism). Similar arguments can be presented for path B, but they are intuitively much less satisfying.

(12) J. F. Young, R. D. Gillard, and G. Wilkinson, *J. Chem. Soc.*, 5176 (1964).
(13) P. Porta, H. M. Powell, R. J. Mawby, and L. M. Venanzi, *J. Chem. Soc. A*, 455 (1967).

(14) The X-ray crystal structure determined by Professor M. R. Churchill shows $Ir(NBD)(P(CH_3)_2Ph)_2SnCl_3$ to be a distorted TBP with $SnCl_3$ equatorial as depicted in idealized form in Figure 6. The distortion is toward the SP with $SnCl_3$ apical, which is the intermediate or transition state configuration for the interconversion I ↔ II. The low barrier observed for I ↔ II and the insensitivity of this barrier to steric factors can be seen to arise from this ground-state distortion.

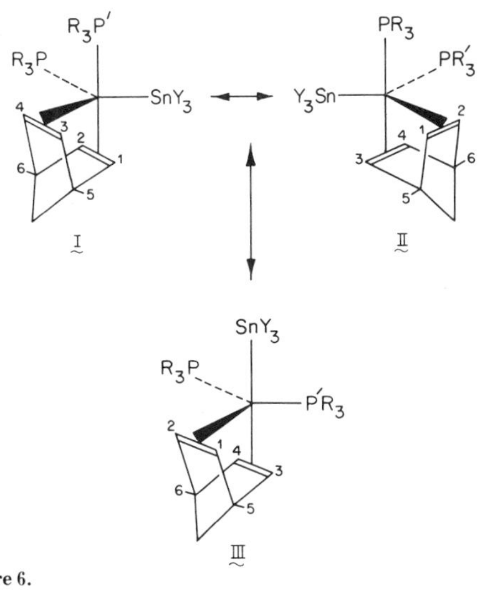

Figure 6.

On the assumption that these interpretations are correct, what predictions can be made? What if $SnCl_3^-$ were replaced by $Sn(CH_3)_3^-$ or PPh_3 in these complexes? Both these ligands are weaker π acceptors than $SnCl_3^-$, and thus isomers III (Figures 5 and 6) should be stabilized and isomers I and II destabilized by this substitution, and consequently the process I + II → III should be more facile. In accord with this prediction, in $Ir(NBD)_2Sn(CH_3)_3$ coalescence to the *one* vinyl, *one* methine pattern (I + II ↔ III) occurs at *ca.* 0°, whereas in the corresponding $SnCl_3^-$ complex the *two* vinyl, *two* methine pattern shows no sign of coalescence below +60°. Also, the compounds $Ir(NBD)_2P^+$ (P = $P(C_6H_5)_3$ or $P(CH_3)_2C_6H_5$) show closely similar behavior to the $Sn(CH_3)_3^-$ species, indicating that $Sn(CH_3)_3^-$ and these phosphine ligands have similar site preferences. Hence the analogous COD complexes would seem to display only one vinyl resonance even at low temperatures because isomer III (Figures 5, 6) is more readily accessible than for the NBD complexes, presumably because of the larger "bite angle" of COD.

Studies on $Ir(NBD)(COD)SnR_3$ (R = CH_3, Cl) are consistent with this viewpoint. The low-temperature spectrum (0°) (R = CH_3) shows two vinyl and two methine resonances of the NBD ligand, as well as two vinyl COD signals, consistent with a rapid equilibration of the type I ↔ II shown in Figure 5. At 90°, however, *both sets* of NBD resonances coalesce, but the COD vinyl sets *remain distinct.* Therefore, this process involves an isomer analogous to III (Figure 5), but with COD spanning the equatorial sites. The alternate isomer with NBD bridging two equatorial sites which would cause equilibration of the COD vinyl sets is clearly not readily accessible under these conditions.

Some Conclusions and Tentative Generalizations

Although to generalize on this restricted set of complexes may be premature, the following tentative proposals are offered as a basis for prediction and hence as a stimulus to further work.

1. Site Preferences. This property can be broadly related to the bonding properties of the ligands, those ligands which are strong π acceptors (*e.g.*, $SnCl_3^-$, CO) preferring equatorial sites in a TBP geometry, or, conversely, those which are strong σ donors (*e.g.*, H^-, CH_3^-) preferring axial sites. Ligands without starkly differentiated bonding properties may not exhibit strong positional preferences (*e.g.*, PR_3, $SnMe_3^-$). This observation is not unexpected since in TBP d^8 solid-state structures where steric effects are not significant the strongest π-bonding ligand is found in an equatorial site, *e.g.*, $Mn(CO)_4NO$. This qualification with regard to steric effects, however, is an important one, as we discuss next.

2. Ligand–Ligand Interactions. Steric interactions are not only dependent upon the steric bulk of the ligands involved but also on the relative positions within the coordination sphere, *i.e.*, for the TBP geometry the interactions decrease in the sequence equatorial–axial > equatorial–equatorial > axial–axial. Large ligand–ligand interactions may cause the stabilization of a positional isomer different from that predicted as the most stable by the site preference rule; *e.g.*, $RhH(CO)(PPh_3)_3$ is TBP with the hydride but also the carbonyl group axial. Site preference would direct the hydride axial and the carbonyl ligand equatorial. However, minimization of steric repulsion between the bulky phosphine ligands finally determines the most stable ground-state isomer, overriding the equatorial preference of the carbonyl ligand. In the analogous iridium(I) complex replacement of one triphenylphosphine by a carbonyl ligand to form $IrH(PPh_3)_2(CO)_2$ causes competition between site preference and steric effects, resulting in two isomers[5,6a] being populated in solution. However, in both isomers the positional preference of hydride for the axial site is evident, overriding purely steric considerations which would place the two phosphines in axial sites.

3. Ring Strain in Chelated Ligands It is clear that chelates which form four-, five-, or six-membered rings with the metal atom will generally prefer to bridge axial–equatorial sites rather than equatorial–equatorial. The diequatorial isomer therefore will be expected to be destabilized relative to the axial–equatorial isomer.

4. The Berry or Pseudorotation Mechanism The Berry mechanism appears to offer a consistent and intuitively appealing interpretation of the dynamic behavior observed for this series of complexes. However, let us examine in more detail what we imply by this statement.

For an intramolecular rearrangement it is important to maintain the distinction between the abstract (permutational) process defined by the initial and final configurations and any physical motion which relate the reactant and product states. The Berry mechanism as originally conceived was concerned with the internal rearrangement of pentacoordinate molecules of high symmetry (*e.g.*, D_{3h}, C_{2v}) in which a symmetric angular deformation of the ligands was proposed to cause axial–equatorial site interchange. Is the Berry mechanism applicable therefore to molecules of lower symmetry as are described here? Certainly when the axial (and/or equatorial) ligands differ greatly in properties (*e.g.*, bonding character), the bending motion transferring the axial pair of ligands to equatorial sites will not be symmetrical, and the physical motion that will connect such isomers is thus less well defined.

However, in all cases where discriminating experimental evidence is available,[2,7] the permutation process observed is that predicted by the Berry mechanism. It would appear justifiable, therefore, at this juncture to choose the Berry mechanism as a description for the stepwise connectivity of TBP isomers without the implication that the detailed physical motions involved need correspond *precisely* to the idealized model.

A permutationally equivalent description such as the turnstile mechanism[15] will also offer an adequate interpretation of the present data. Distinction between the Berry and turnstile mechanisms in pentacoordinate rearrangements rests on differences in the physical motions of the ligands and the nature (stereochemistry, energy) of the transition states. A compelling choice between the two mechanisms can only be based on an unambiguous experimental distinction (which has yet to be achieved) or on a sound theoretical basis. Nevertheless, we prefer to describe the rearrangement in terms of the Berry mechanism rather than the turnstile, since the former is more evident to manipulate conceptually and easier to illustrate in practice. Hence, if the Berry mechanism is used in this manner, and the relative stabilities of TBP configurations are related by the positional, steric, and strain rules mentioned above, an attractive and useful rationale of these (and other) experimental results is evident.

Some Applications of these Results

Other Fluxional Pentacoordinate TBP Molecules. Studies on $Fe(CO)_x(PF_3)_{5-x}$ (x = 0–5)[16] and $M(PF_3)_5$ (M = Fe, Rh, Os)[6a] show the barrier to rearrangement to be lower than in $RCo(PF_3)_{4-x}(CO)_x$ (R = H, CF_3)[4,17] where limiting spectra have been obtained. These observations can be explained qualitatively by the above proposals. Rearrangement of the molecules $RCoL_4$ of ground-state C_{3v} symmetry (R axial) must, if the Berry process is operative, proceed *via* an intermediate with the strongly σ-bonding R group at an unfavorable equatorial site. For ML_5 molecules no such unfavorable transition state or intermediate develops. In this context, it has been proposed from a study of pseudotetrahedral ML_4H molecules[6a] that hydrides may have access to a unique mechanism in which the hydride ligand may "hop" from one face to another of the pseudotetrahedral arrangement of the L ligands. For the complexes $IrH(diene)P_2$, however, *no single step nor any combination of single-step processes of the hop type* can account for our experimental observations, and application of the hop mechanism must be regarded with caution in TBP systems.

Exchange Reactions of Square d⁸ Complexes. The accepted mechanism for ligand substitution in a four-coordinate square-planar complex is an associative process proceeding through a TBP intermediate, in which the entering ligand occupies an equatorial site.[18] Since substitutions nearly always take place with retention of configuration (at least for Pt(II) complexes), the lifetime appears to be shorter than the time necessary for an intramolecular rearrangement. However, retention of configuration is not observed for the trialkylphosphine-catalyzed cis to trans isomerization of bis(trialkylphosphine)platinum dihalides. A mechanism involving intramolecular rearrangement of the five-coordinate intermediate can be proposed,[19] but an alternate process involving halide ionization may be operative.

We have investigated[20] the exchange of tertiary phosphine with the square complexes $[RPtP_3]^+$ (R = H, CH_3; P = tertiary phosphine) by nmr methods. The exchange process would be expected to involve the intermediate $[RPtP_4]^+$, stable hydride analogs of which undergo rapid internal rearrangement. However, in each case substitution at the phosphine cis to the hydride occurs faster than at the trans site and, furthermore, by a ratio which increases with increasing steric bulk of P. The five-coordinate intermediate of C_{3v} symmetry (R axial) which would be predicted as the most stable from the preference rules would lead directly to cis substitution. Trans substitution may arise either by intramolecular rearrangement of this intermediate *via* the less favorable isomer of C_{2v} symmetry (R equatorial) *or* by intermolecular exchange involving the direct formation of this latter isomer. In either case, this model would predict the energy of the C_{2v} structure to be increased (relative to the C_{3v} species) by increased steric interactions between the bulkier phosphine ligands resulting in the observed cis to trans relative rates of substitution. Hence, if intramolecular rearrangement is occurring the rate is considerably slower than intermolecular exchange.

However, in square complexes of the type M(diene)LCl (M = Rh, Ir; diene = COD, NBD; L = phosphine or arsine) the two distinct vinyl resonances coalesce into one signal on addition of free L.[21] This dynamic process is effectively a cis–trans isomerization, and, since ionization processes can be discounted, must occur *via* intramolecular rearrangement of the intermediate $M(diene)L_2Cl$.

The complexes $M(diene)L_2R$ (R = H, CH_3) show the effects of dissociation and exchange of L under certain conditions. When intramolecular rearrangement is slower than the intermolecular process, it is observed that the nonequivalent double bonds present in the stable five-coordinate complex are preserved in the four-coordinate intermediate.[7b] This intermediate must thus be square (and *not* tetrahedral), indicating that in the general equilibrium $M(diene)L_2X \rightleftharpoons M(diene)LX + L$ the idealized geometries involved are TBP and square, the relative energies of which are dependent on the nature of the ligands. Thus we have observed in this equilibrium

(15) I. Ugi, D. Marquarding, H. Klusacek, P. Gillepsie, and F. Ramirez, *Accounts Chem. Res.*, **4**, 288 (1971).

(16) C. A. Udovich, R. J. Clark, and H. Haas, *Inorg. Chem.*, **8**, 1066 (1969).

(17) C. A. Udovich, Ph.D. Thesis, Florida State University 1969.

(18) See, *e.g.*, C. H. Langford and H. B. Gray, "Ligand Substitution Processes," W. A. Benjamin, New York, N. Y., 1965.

(19) P. Haake and R. M. Pfeiffer, *J. Amer. Chem. Soc.*, **92**, 4996 (1970).

(20) D. P. Rice, Ph.D. Thesis, Harvard University, 1972.

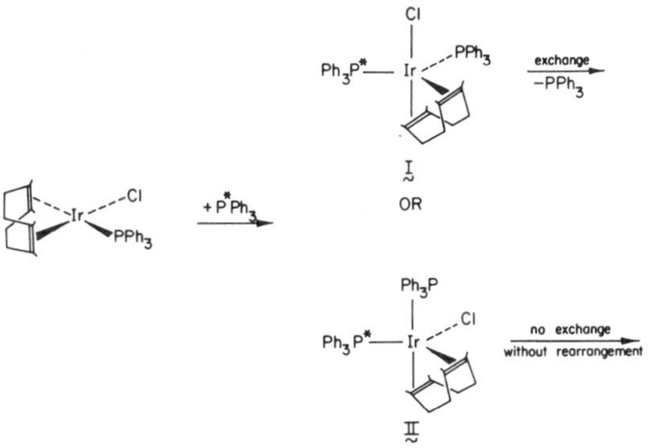

Figure 7. Intermediate TBP structures for the system Ir(COD)(PPh$_3$)Cl + PPh$_3$.

the pentacoordinate species is favored H > SnCl$_3$ > CH$_3$, and P(CH$_3$)$_2$C$_6$H$_5$ > P(CH$_3$)(C$_6$H$_5$)$_2$ > P(C$_3$H$_7$)(C$_6$H$_5$)$_2$ ~ P(C$_6$H$_{11}$)(C$_6$H$_5$)$_2$ > P(C$_6$H$_5$)$_3$. We therefore formulate the structure of the intermediate M(diene)L$_2$Cl in Figure 7 as TBP with either chloride at an axial (I) or equatorial site (II). If structure I were formed, exchange of L can proceed without axial–equatorial vinyl proton interchange, and the intramolecular and intermolecular processes are entirely separate. This is the case for M(diene)L$_2$R (R = H, CH$_3$). If *only* intermediate II were formed, *intermolecular exchange cannot take place without the occurrence of intramolecular rearrangement.* Indeed, under certain conditions, the intermolecular exchange rate may be determined by the intramolecular process. In this case the intra- and intermolecular rates are related, which is apparently the case for M(diene)L$_2$Cl where the rate of vinyl coalescence was taken as equal to the rate of intermolecular ligand exchange.[21] In general, however, the intra- and intermolecular exchange reaction must be regarded as totally separate processes.

Catalytic Processes. In several homogeneous catalytic reactions involving d^8 metal complexes, five-coordinate intermediates play a key role. For example, in the hydroformylation[22] of olefins with Co(I) complexes, at least four of the proposed intermediates are pentacoordinate, *e.g.*, for ethylene as substrate, HCo(CO)$_4$, HCo(CO)$_3$(C$_2$H$_4$), C$_2$H$_5$Co(CO)$_4$, and C$_2$H$_5$COCo(CO)$_4$. Clearly a complete understanding of the catalytic process requires details of the stereochemistry of such species and their configurational stability. In this connection, we have recently synthesized[23] the complexes HRu(NO)P$_3$ (P = tertiary phosphine) and examined their catalytic properties. The structure of HRu(NO)(P(C$_6$H$_5$)$_3$)$_3$ has been determined to be TBP both in solution and

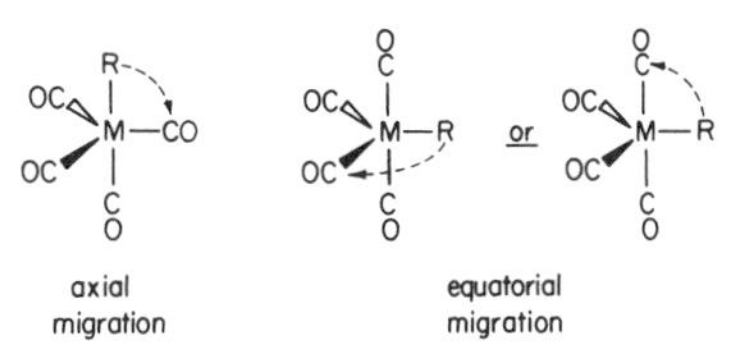

Figure 8.

solid state,[24] with hydride and nitrosyl ligands occupying axial sites. This complex in solution is found to catalyze very efficiently the isomerization of olefins; the complex with P(CH$_3$)(C$_6$H$_5$)$_2$ as ligand, however, is an ineffective catalyst. ^{1}H and ^{31}P nmr data reveal that this complex has a different configuration in solution, as might be expected from a competition of site preference and steric factors. The corresponding P(C$_3$H$_7$)(C$_6$H$_5$)$_2$ complex (C$_3$H$_7$ = isopropyl) in which both configurations are populated in solution is again catalytically active. It is thus tempting to associate catalytic activity with the presence of a *particular* configuration in solution.

A key step of several catalytic and stoichiometric reactions involving carbon monoxide is the migration of a bound alkyl group R to a carbonyl ligand, forming a transient four-coordinate d^8 acyl derivative, *i.e.*, RM(CO)$_4$ → RCOM(CO)$_3$. Such steps occur in the catalyzed hydroformylation of olefins[22] and in the aldehyde and ketone syntheses using Fe(CO)$_4^{2-}$ developed by Collman[25] and coworkers. Infrared evidence indicates that the pentacoordinate RM(CO)$_4$ species are TBP (C_{3v} symmetry, R axial), but we would expect rapid axial–equatorial carbonyl exchange *via* a TBP intermediate with R equatorial (C_{2v}). Hence the question of stereochemistry at the metal during the alkyl migration step may be posed. As is evident from Figure 8, migration of the alkyl from an axial site to an equatorial carbonyl produces approximately a tetrahedral ligand arrangement—which, in the spirit of a correlation diagram, is the excited state of the stable d^8 square-planar species. This latter, however, is readily generated by an alkyl migration from an equatorial site to an equatorial or axial carbonyl ligand in the C_{2v} species.[26] Hence a rapid intramolecular rearrangement of the ground-state C_{3v} structure to the reactive C_{2v} may be necessary for the migration reaction to occur readily and hence for the catalytic cycle to proceed.

We are grateful to our coworkers, S. A. Bezman, Dr. D. P. Rice, Dr. R. R. Schrock, and S. T. Wilson, for their contributions to this work. Particularly, we wish to thank Professor M. R. Churchill for his continuing interest in this project. Fellowships from National Science Foundation (J. R. S.) and The Henry and Camille Dreyfus Foundation (J. A. O.) are most gratefully acknowledged.

(21) K. Vrieze and P. W. N. M. van Leeuwen, *Progr. Inorg. Chem.*, **14**, 1 (1971), and references therein.

(22) *E.g.*, R. F. Heck, *Accounts Chem. Res.*, **2**, 10 (1969).

(23) S. T. Wilson and J. A. Osborn, *J. Amer. Chem. Soc.*, **93**, 3068 (1971).

(24) C. G. Pierpont, A. Pucci, and R. Eisenberg, *J. Amer. Chem. Soc.*, **93**, 3050 (1971).

(25) W. O. Siegel and J. P. Collman, *J. Amer. Chem. Soc.*, **94**, 2516 (1972), and references therein.

(26) It is clear that a square-planar species is *directly* produced by an equatorial–equatorial migration if *both* the R group and the reacting carbonyl are allowed to migrate toward each other.

Addendum (May 1976)

There has been continuing interest in the dynamic behavior of pentacoordinate complexes during the period since this Account was written. The following developments are particularly noteworthy.

A ^{31}P NMR study of $[Rh(P(OMe)_3)_5]BPh_4$ presented the first slow-exchange NMR spectrum of a ML_5 complex.[27] This has been followed by thorough studies of various cationic pentaphosphite complexes of cobalt and nickel triad metals.[28,29] Lineshape simulations have established a permutational exchange pathway consistent with the Berry mechanism and, for a particular metal, barriers to exchange depend on the bulk of the phosphite ligand. Although recent evidence[27,30] has continued to suggest a very low exchange barrier in $Fe(CO)_5$, two groups have observed that the requirement for olefin rotation in (olefin)$Fe(CO)_4$ derivatives to be coupled with axial-equatorial carbonyl exchange allows slow-exchange ^{13}C NMR spectra to be obtained.[31] Exchange barriers increase with the π-acceptor ability of the substituted olefin, presumably reflecting increased iron–olefin bond strength. The exchange patterns observed for isonitrile-substituted complexes are consistent with predictions of the Berry mechanism.[31b] Recently, the observation of separate axial and equatorial ^{13}C carbonyl resonances for several $X_3SnCo(CO)_4$ complexes has been reported.[32] The indication that axial-equatorial exchange is faster for X = Cl than for X = CH_3 is consistent with the site-preference/Berry mechanism model presented in this Account. Details of the crystal structure determination of $Ir(NBD)(PMe_2Ph)_2SnCl_3$ have been published.[33] The theoretical aspects of site preferences in both TBP and SP pentacoordinate molecules have been discussed.[34] The results of recent studies on the approximately square-pyramidal Fe(1,3-butadiene)$(CO)_3$ and related complexes[35] indicate relatively high barriers for apical–basal exchange. However, no electronic or steric ligand site-preference rules are yet apparent. Other workers[36] have also examined the question of intermolecular ligand exchange in $HPtL_3^+$ complexes vs. intramolecular exchange in an $HPtL_4^+$ intermediate. Extending the study to include palladium and nickel, they found k_{intra}/k_{inter} to increase in the order Pt < Pd < Ni. The general extent to which dynamic pentacoordinate intermediates are involved in ligand-catalyzed cis-trans isomerization of square planar complexes is not clear at present. Recent studies[37,38] have suggested both ionic and nonionic mechanisms, the predominance of one or the other being very dependent on the particular system studied.

(27) J. P. Jesson and P. Meakin, *J. Am. Chem. Soc.*, **95**, 1344 (1973).
(28) P. Meakin and J. P. Jesson, *J. Am. Chem. Soc.*, **95**, 7272 (1973).
(29) J. P. Jesson and P. Meakin, *J. Am. Chem. Soc.*, **96**, 5760 (1974).
(30) H. Mahnke, R. J. Clark, R. Rosanske, and R. K. Sheline, *J. Chem. Phys.*, **60**, 2997 (1974).
(31) (a) L. Kruczynski, L. K. K. LiShingman, and J. Takats, *J. Am. Chem. Soc.*, **96**, 4006 (1974). (b) S. T. Wilson, N. J. Coville, J. R. Shapley, and J. A. Osborn, *J. Am. Chem. Soc.*, **96**, 4038 (1974).
(32) D. L. Lichtenberger, D. R. Kidd, P. A. Loeffler, and T. L. Brown, *J. Am. Chem. Soc.*, **98**, 629 (1976).
(33) M. R. Churchill and K.-K. G. Lin, *J. Am. Chem. Soc.*, **96**, 76 (1974).
(34) A. R. Rossi and R. Hoffmann, *Inorg. Chem.*, **14**, 365 (1976).
(35) (a) L. Kruczynski and J. Takats, *J. Am. Chem. Soc.*, **96**, 932 (1974). (b) C. G. Kreiter, S. Stüber, and L. Wackerle, *J. Organomet. Chem.*, **66**, C49 (1974). (c) M. A. Bush and R. J. Clark, *Inorg. Chem.*, **14**, 226 (1975). (d) T. H. Whitesídes and R. A. Budnik, *Inorg. Chem.*, **14**, 226 (1975).
(36) P. Meakin, R. A. Schunn, and J. P. Jesson, *J. Am. Chem. Soc.*, **96**, 277 (1974).
(37) (a) D. G. Cooper and J. Powell, *J. Am. Chem. Soc.*, **95**, 1102 (1973). (b) D. G. Cooper and J. Powell, *Can. J. Chem.*, **51**, 1634 (1973). (c) J. Powell and D. G. Cooper, *J. Chem. Soc., Chem. Commun.*, 749 (1974).
(38) (a) D. A. Redfield, J. H. Nelson, R. A. Henry, D. W. Moore, and H. B. Jonassen, *J. Am. Chem. Soc.*, **96**, 6298 (1974). (b) D. A. Redfield and J. H. Nelson, *J. Am. Chem. Soc.*, **96**, 6219 (1974). (c) D. A. Redfield and J. H. Nelson, *Inorg. Nucl. Chem. Lett.*, **10**, 931 (1974).

Catalysis of Olefin and Carbon Monoxide Insertion Reactions

Gian Paolo Chiusoli

Centro Ricerche di Chimica Organica Montedison, Novara, Italy

Received February 7, 1973

Reprinted from ACCOUNTS OF CHEMICAL RESEARCH, **6**, 422 (*1973*)

During recent years insertion reactions on transition metal complexes have attracted the attention of organometallic chemists, both academic and industrial.[1-3] These reactions pose fundamental problems, and they have striking commercial potentialities.

Insertion reactions of allylnickel complexes have provided insights of particular value. Allylnickel complexes can be made by the oxidative addition of allyl halides ($CH_2{=}CHCH_2X$) to zerovalent nickel complexes, in the fashion of eq 1, in which L is a ligand such as CO or triphenylphosphine.[3]

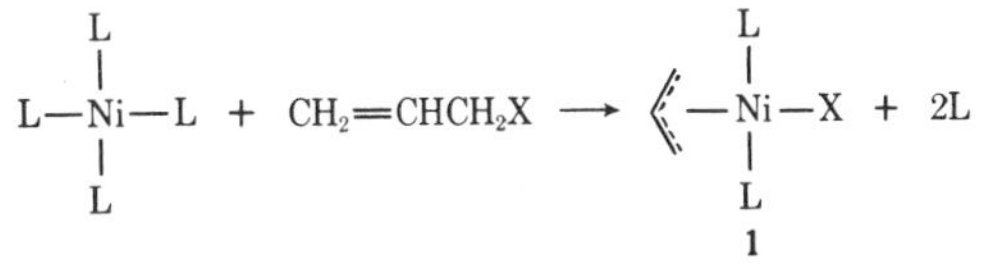

(1)

This reaction occurs at room temperature and atmospheric pressure, as do the further reactions to be described.

If the allylnickel complex obtained in eq 1 is exposed to CO in methanol solution, further reaction to form methyl vinylacetate occurs as represented by eq 2.[1-4]

$$\mathbf{1}\ (L = CO) + 3CO + CH_3OH \longrightarrow CH_2{=}CHCH_2COOCH_3 + Ni(CO)_4 + HX \qquad (2)$$

Other examples, also from our laboratory, are reported in eq 3 and 4. In reaction 3, acetylene inserts

$$\mathbf{1}\ (L = CO) + HC{\equiv}CH + 3CO + CH_3OH \longrightarrow CH_2{=}CHCH_2CH{=}CHCOOCH_3 + Ni(CO)_4 + HX \qquad (3)$$

$$\mathbf{1}\ (L = CO) + HC{\equiv}CH + 4CO + CH_3OH \longrightarrow$$

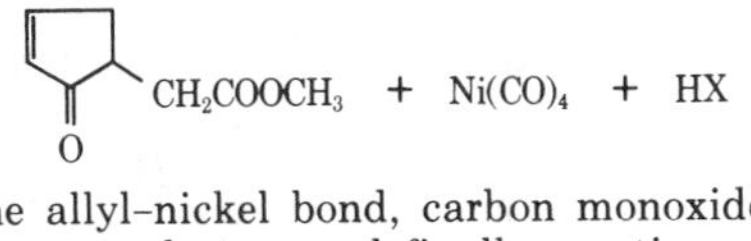

(4)

first into the allyl-nickel bond, carbon monoxide is inserted in a second step, and finally reaction with methanol cleaves the complex, forming methyl *cis*-2,5-hexadienoate. These reactions can be regarded as cooligomerizations involving multiple successive insertions.

We have shown that carbon monoxide or acetylene occupies a coordination site left free by the allyl group on passing from the π to the σ form[4] (eq 5 and 6). The final cleavage generating zerovalent nickel (reductive elimination) leads to methyl esters in the presence of methanol.

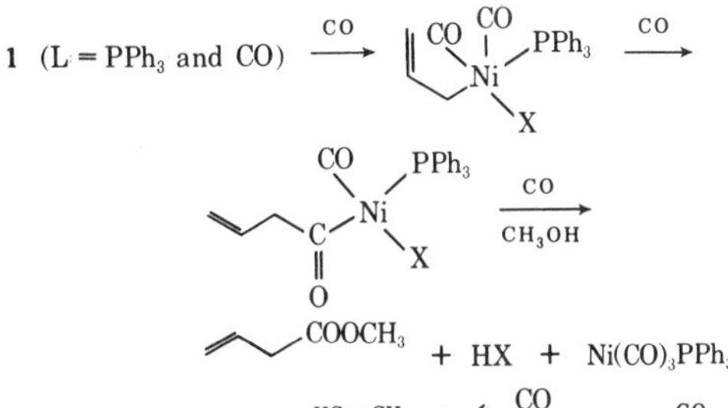

(5)

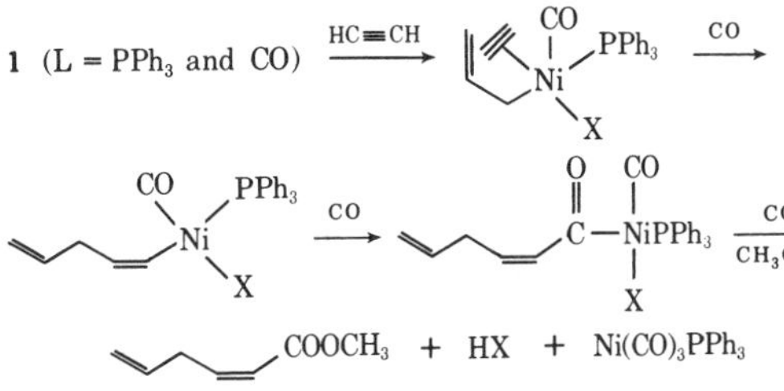

(6)

In the methanol cleavage step of reaction 6, the methanol competes for the acyl group with the terminal double bond of the coordinated chain. If the reaction is carried out with a low concentration of methanol in inert solvents, the attack of the terminal double bond on the nickel-bonded carbon predominates (reaction 4), and a cyclopentenone ring is formed. This step is followed by another carbon monoxide insertion before final attack by methanol. The ring-closure process can be schematized as in reaction 7.

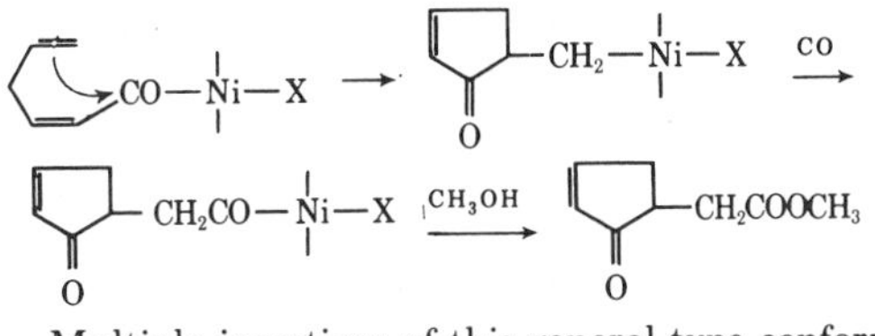

(7)

Multiple insertions of this general type conform to the pattern of Scheme I, in which M_1, M_2, etc., are inserting molecules such as CO and acetylene in the examples of eq 1–3. (For simplicity the insertion of a group of the same growing chain as in reaction 4 is not considered in Scheme I.)

A remarkable feature of these reactions is that, despite the presence of several kinds of insertable molecules in solution, one kind is chosen with high selectivity at each step of the addition process. The sequence of involvement of these molecules depends

Gian Paolo Chiusoli was born in Treviso, Italy, in 1923, and was graduated from the University of Padua. He is presently director of basic organic research in Montecatini Edison Research Center of Novara and Lecturer of Industrial Organic Chemistry in the University of Parma. His main research interests are organic syntheses *via* organometallic complexes, homogeneous catalysis, syntheses on ion pairs, and heterocyclic systems.

(1) G. P. Chiusoli, "XXIII Congress of Pure and Applied Chemistry, Special Lectures, Boston 1971," Vol. VI, Butterworths, London 1971; G. P. Chiusoli, *Gazz. Chim. Ital.*, **89**, 1332 (1959); G. P. Chiusoli and L. Cassar, *Angew. Chem. Int. Ed. Engl.*, **6**, 124 (1967).

(2) R. F. Heck, *Advan. Chem. Ser.*, **No. 49**, (1965); *Accounts Chem. Res.*, **2**, 10 (1969).

(3) P. Heimbach, P. W. Jolly, and G. Wilke, *Advan. Organometal. Chem.*, **8**, 29 (1970); E. O. Fischer and G. Bürger, *Z. Naturforsch. B*, **16**, 702 (1961).

(4) F. Guerrieri and G. P. Chiusoli, *J. Organometal. Chem.*, **15**, 209 (1968).

Scheme I

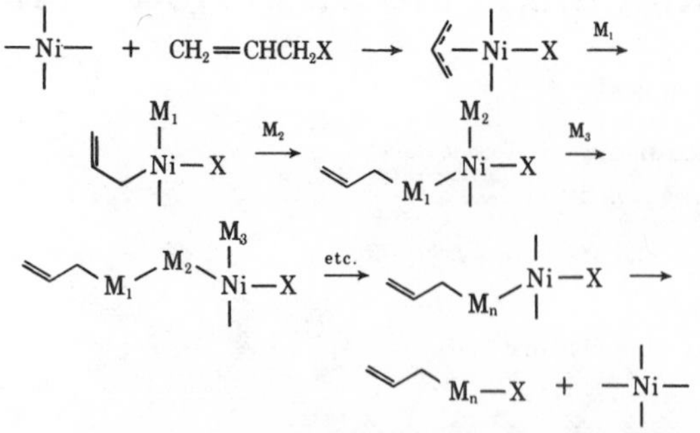

on their structures and on the reaction conditions. Another factor is the identity of the metal-bonded group. Allyl groups are mainly attacked by electrophiles and acyl groups by nucleophiles, while allyl groups show borderline behavior somewhat dependent on the substituents which the allyl group bears.

Other factors are the effective charge on the metal and the concentrations of the molecules to be inserted. Different molecules of appropriate electronic and geometric character can compete for insertion, and it should be possible to favor one of these by facilitating its coordination. There is thus the possibility of directing the process toward different structures by changing the building blocks.

Our efforts to determine what types of molecules possess the characteristics necessary for selective catalytic multistep reactions are the focus of this Account.

It turns out that these reactions require precise geometric arrangements. Steric factors play a very important role, and the synthesized molecules generally appear in one predominant stereochemical configuration.

Our first attempt to cause a double bond to react with an allyl group was carried out on olefins activated by electron-withdrawing substituents. Having in mind the tendency of the nickel-coordinated allyl group to react as a nucleophilic species and the known Michael-type reactions of carbanions with activated olefins, we expected this reaction to occur easily, provided that the activated olefin could take the same coordination site that inserting molecules occupy in the previously described reactions. Actually the insertion[5-7] goes very well in certain solvents, according to eq 8, written for the case of an allylnickel complex and methyl acrylate.

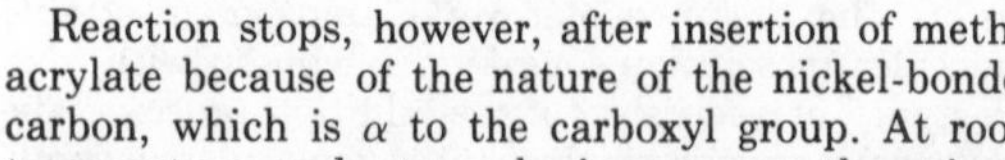

(8)

Reaction stops, however, after insertion of methyl acrylate because of the nature of the nickel-bonded carbon, which is α to the carboxyl group. At room temperature and atmospheric pressure there is no tendency to insert a molecule of carbon monoxide. A proton is taken up if the medium is protic, with formation of methyl hexenoate in very low yield, whereas hydrogen is lost from the carbon β to the carboxyl group in an aprotic medium. This reaction is likely to involve an intermediate nickel hydride which in part hydrogenates the α,β double bond first formed —incidentally, a trans double bond—with formation of some of the hexenoate as a secondary product.

An accurate balance of conditions is necessary to obtain this type of reaction. If a bis(π-allyl)nickel halide is caused to react in methanol, where it gives a monomeric cationic complex, the reaction with acrylate takes place only to a very slight extent. A possible interpretation is that the cationic character of the complex favors the coordination of methanol rather than the activated olefin, renders back-donation more difficult, and also favors nucleophilic attack of methanol on the allyl group. In benzene, where dissociation of the halide ion is hindered and formation of the monomeric species is effected by the activated olefin, the yield is quite satisfactory.

It is possible to achieve insertion in a protic medium, however, by adding a reducing agent such a powdered iron.[5,8] In this case the reactive species could possibly form according to eq 9.

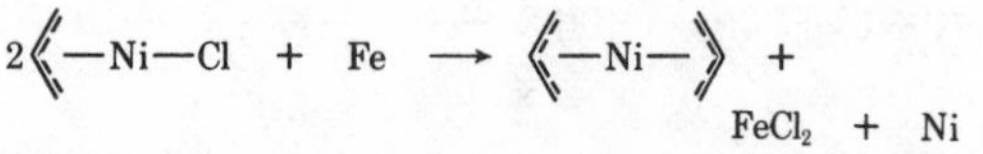

(9)

In fact, the reaction of bis(π-allyl)nickel, prepared by disproportionation of allylnickel chloride with pyridine, goes in an analogous way.

These experiments allowed us to form a picture of the influence of various factors and especially of the solvent on the ability of olefins to coordinate and to insert.

Other useful information came from the literature on reactions of nickel complexes, particularly from Wilke's group.[9] It has been reported, for example, that ethylene can be inserted into a bis(π-allyl) complex derived from two molecules of butadiene.

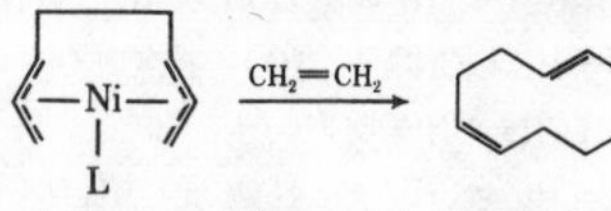

(10)

Olefin and Carbon Monoxide Insertion

We were interested, however, in a reaction including also carbonylation to carboxylic acids or their derivatives. In this field the closest example is the reaction of methallylnickel acetate with norbornene and carbon monoxide, reported by Porri and coworkers.[10] The reaction gives π-2-*exo*-methallyl-3-*exo*-carbomethoxynorbornane (reaction 11). Methallyl-

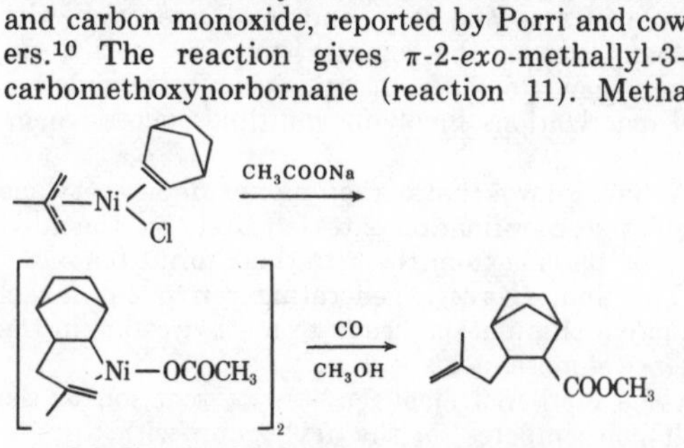

(11)

(5) G. P. Chiusoli, *Chim. Ind.* (*Milan*), **43,** 365 (1961).

(6) M. Dubini, F. Montino, and G. P. Chiusoli, *Chim. Ind.* (*Milan*), **47,** 839 (1965); M. Dubini and F. Montino, *J. Organometal. Chem.*, **6,** 188 (1966).

(7) G. P. Chiusoli and G. Bottaccio, *Chim. Ind.* (*Milan*), **44,** 131 (1962).

(8) G. P. Chiusoli and G. Cometti, unpublished results.

(9) G. Wilke, *et al.*, *Angew. Chem., Int. Ed. Engl.*, **2,** 105 (1963); **5,** 171 (1966); P. Heimbach and G. Wilke, *Justus Liebigs Ann. Chem.*, **727,** 183 (1969); M. F. Semmelhack, *Org. React.*, **19,** 115 (1972).

(10) M. C. Gallazzi, L. Porri, and G. Vitulli, Abstracts of papers presented at the Xth Congress of the Italian Chemical Society, Padua, June 17–21, 1968, Section XVI-2; M. C. Gallazzi, T. L. Hanlon, G. Vitulli, and L. Porri, *J. Organometal. Chem.*, **33,** C45 (1971).

nickel chloride is first allowed to react with norbornene, and then with sodium acetate to form a dimeric complex. Passage of carbon monoxide into a methanolic solution of the complex results in formation of the ester.

This reaction was carried out on a strained olefin, however, which is known to coordinate very well.

We tried to attack the problem of the direct insertion of simple double bonds, reasoning that the crucial point is to create conditions such that the double bond could win the competition with carbon monoxide for the appropriate coordination site.

The tendency to coordinate carbon monoxide is weakened by a low partial pressure of carbon monoxide, by higher temperatures, and by greater polarity of the medium, which favors formation of a cationic complex of nickel. The latter has a lower tendency to back-donation. Under these conditions, in fact, olefins react with the allyl group by insertion. Thus, ethylene under a pressure of 20–30 atm reacts at 40° with tetracarbonylnickel in a methanolic solution of allyl chloride (eq 12).[11] This reaction is catalytic and

$$CH_2{=}CHCH_2Cl + CH_2{=}CH_2 + CO + CH_3OH \xrightarrow{Ni} CH_2{=}CHCH_2CH_2CH_2COOCH_3 + HCl \quad (12)$$

parallels the analogous one with acetylene which gives methyl hexadienoate (eq 3). Yields vary from 40 to 60%. Several by-products are formed, including the coupling product of two allyl groups, 1,5-hexadiene, the carbonylation product of an allyl group, vinylacetic acid, and products deriving from the insertion of more than one molecule of ethylene and carbon monoxide such as $CH_2{=}CHCH_2CH_2CH_2CO$-$CH_2CH_2COOCH_3$, $CH_2{=}CHCH_2CH_2CH_2COCH_2$-$CH_2COCH_2CH_2COOCH_3$, and others, to be seen later.

With α-substituted olefins, the main product of the olefin insertion no longer was the open-chain acid or ester as in reaction 12, but a cyclopentanone derivative resulting from four successive insertions, double bond insertion, carbonylation, a new double bond insertion causing cyclization, and a new carbonylation, followed by final cleavage of the complex by water or alcohols. The process is depicted in eq 13, written for the case of crotyl chloride and 1-hexene. The reaction is catalytic. It takes place at atmospheric pressure and at a temperature of 40–70°, but gives rather poor yields (<5%).

$$CH_3CH{=}CHCH_2Cl + CH_2{=}CHCH_2CH_2CH_2CH_3 + 2CO + H_2O \xrightarrow{\text{Ni complex}}$$

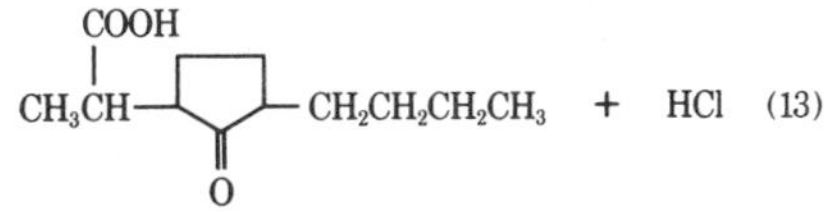

The tendency to give cyclopentanonic products and the low yield are both attributed to the slower hydrolysis of nickel-bonded acyl groups when substituents α to the acyl group are present. This enables the double bond of the coordinated chain to attack the acyl group in competition with the solvent. Simultaneously other reactions also occur. Thus among the products we find 3-pentenoic acid and higher acids. The origin of the latter becomes evident if one starts with allyl chloride. Even in the presence of 1-hexene the reaction takes another course, giving rise to a new cyclopentanonic structure, resulting from the insertion of 1,5-hexadiene, which is formed by coupling two allyl groups under the same reaction conditions.

$$CH_2{=}CHCH_2Cl + CH_2{=}CHCH_2CH_2CH{=}CH_2 + 2CO + H_2O \xrightarrow{\text{Ni complex}}$$

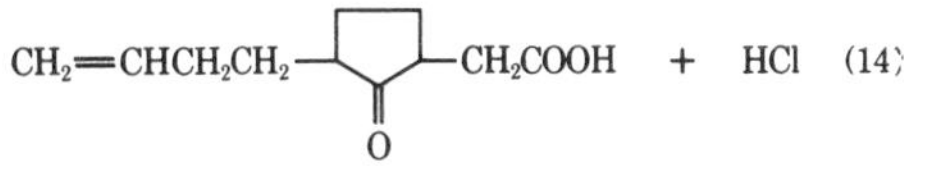

2

This reaction gives better yields (up to 40%) and is strongly favored by the polarity of the medium and by salts such as KPF_6, KF, and $SnCl_2$, which can enhance the cationic character of the catalytic complex.[11] Equation 15 shows our interpretation of the course of the reaction, exemplified for the case of allyl chloride and hexadiene.

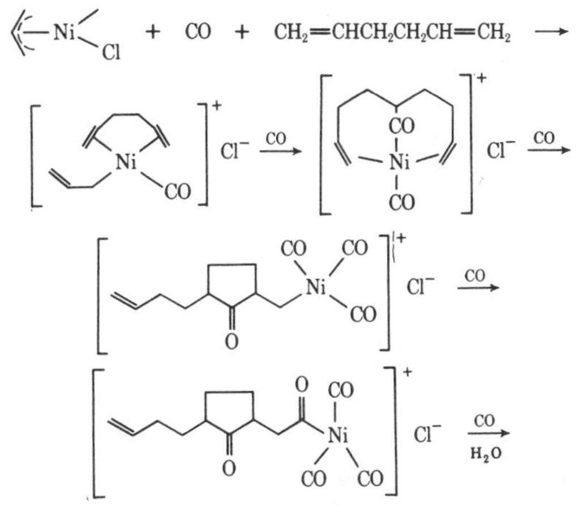

$$\mathbf{2} + Ni(CO)_4 + HCl \quad (15)$$

As in the preceding cases, the first step is the oxidative addition[12] of zerovalent nickel to allyl chloride. The resulting complex has cationic character in the solution used for the reaction. This enables hexadiene to coordinate as a chelating ring. One of the two double bonds should occupy a coordination site left free by the allyl group on its passage to the σ form. At this point the first insertion of carbon monoxide takes place with formation of an intermediate which has not been isolated. This intermediate should offer two pathways to cyclization, which are equal in the scheme but are different if a substituent is present on one of the two double bonds, as we shall see in a moment. The cyclization is followed by a new insertion of carbon monoxide and by final cleavage of the complex according to the scheme already shown in other cases. The product essentially is one of the two possible diastereomers. This means that also the cyclization is highly stereoselective.

The use of bis(π-allyl)nickel bromide, in place of allyl bromide and tetracarbonylnickel, with carbon monoxide and 1,5-hexadiene in methanol gives sup-

(11) G. P. Chiusoli and G. Cometti, *J. Chem. Soc., Chem. Commun.*, 1015 (1972); G. P. Chiusoli, G. Cometti, and V. Bellotti, *Gazz. Chim. Ital.*, in press.

(12) J. Halpern, *Accounts Chem. Res.* **3**, 386 (1970); J. P. Collman, *ibid.*, **1**, 137 (1968); A. J. Deeming and B. L. Shaw, *J. Chem. Soc. A*, 1562 (1969).

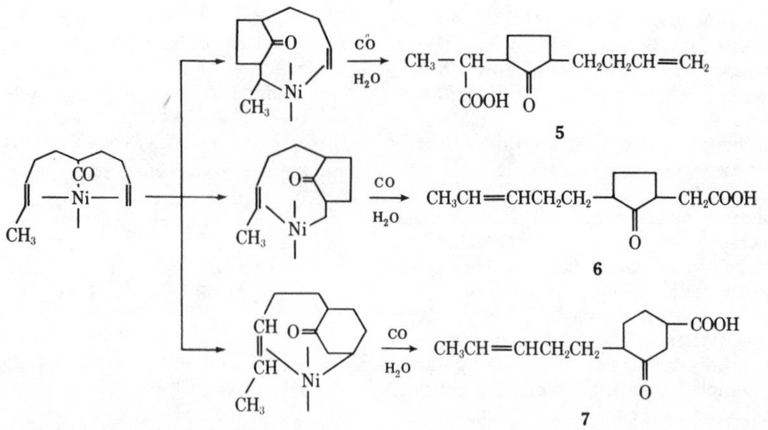

(16)

port to this interpretation.[11] Beside the product shown in eq 15, considerable amounts of a cyclohexanonic isomer (**3**) and of an open-chained compound corresponding to an intermediate stage in the addition process (**4**) are also obtained. The tendency to

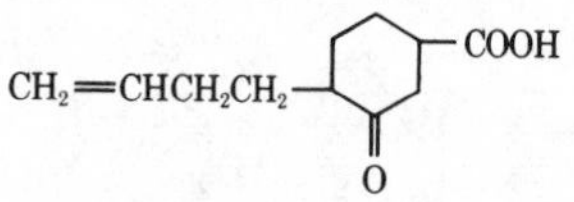

3

$CH_2{=}CHCH_2CH_2{-}CH(COOH){-}CH_2CH_2CH{=}CH_2$

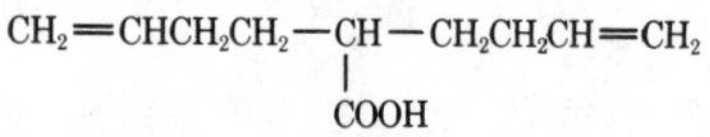

4

form the latter two products, whose presence is insignificant in the reaction with tetracarbonylnickel, is probably related to the lack of nickel-coordinated carbon monoxide groups. The complex would be less stable, and both penetration of the solvent into the coordination sphere (to cleave the complex before cyclization has occurred) and extension of the cyclizing chain to form a larger ring would be favored.

If the reaction is carried out using crotyl chloride instead of allyl chloride, the main reaction products are two, as already mentioned, corresponding to the possible pathways to cyclization (eq 16). There is also a third isomer, which corresponds to the formation of a cyclohexanonic ring. The three isomers have been obtained in the ratio **6**:**5**:**7** = 72:27:1 using KPF_6 as added salt. Compound **6** is a single diastereoisomer of the two possible. Compound **5** is a mixture of two diastereoisomers, one of which can be made to predominate (90:10) depending on reaction conditions. Thus cyclization appears to be stereoselective, whereas double bond insertion between two carbonyl groups is stereoselective to a lower extent.

The ratio between cyclopentanonic and cyclohexanonic structures indicates that the first are strongly preferred. The ratio between the two types of isomers deriving from ring closure at the allylic and at the second double bond of the diene, respectively, indicates that the terminally unsubstituted double bond is clearly preferred.

It is interesting to compare the amount of the product resulting from ring closure at the originally allylic double bond with that obtained with simple 1-olefins such 1-hexene, which in the absence of another double bond can only give the ring closure at the allylic double bond. In the absence of a chelation effect, one should obtain the same amount from one molecule of 1-hexene and half a molecule of 1,5-hexadiene (in a first approximation the conformation of the alkyl chain is neglected). To test this point we have caused a mixture of 1-hexene and 1,5-hexadiene in the mole ratio 2:1 to react with crotyl chloride. We have not considered the main product, which is the one derived from the ring closure at the second double bond of hexadiene, as mentioned before, and have only compared the other isomer with the product from 1-hexene. The first (eq 17a) is seven times the second (eq 17b). Chelation thus appears to play a very important role in determining the insertion of a double bond.

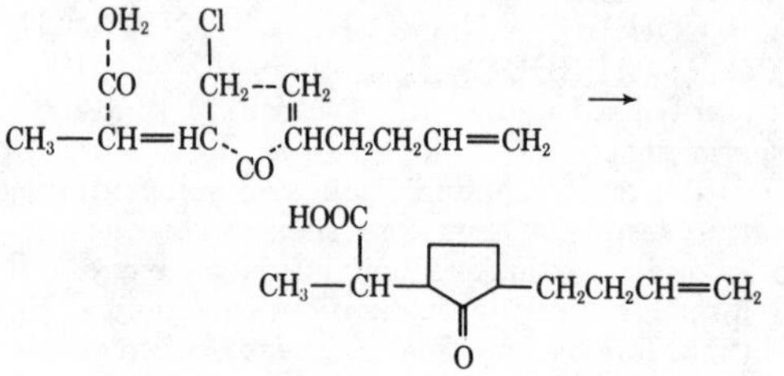

(17a)

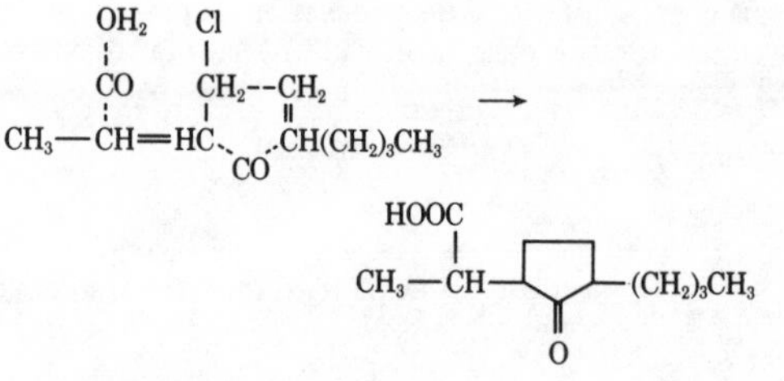

(17b)

Steric Effects

There are other requirements, however, that must be met to obtain a satisfactory reaction. At least one double bond has to be vinylic. If methallyl chloride, $CH_2{=}C(CH_3)CH_2Cl$, is caused to react with dimethallyl, $CH_2{=}C(CH_3)CH_2CH_2C(CH_3){=}CH_2$, in place of hexadiene, no insertion occurs under the same conditions—diluted solution of methanol or acetone-water, 45°, and atmospheric pressure—as for the reaction of hexadiene. The failure of dimethallyl to react should be attributed either to its lower ability to coordinate in comparison with 1,5-hexadiene or to steric hindrance to carbonylation.

An analogous effect is found in terminally substituted hexadienes. 2,6-Octadiene is practically inert under the conditions of reaction of hexadiene, thus revealing a sensitivity to the effect of substituents which is not equalled by any of the analogous reactions studied so far.

The failure of 2,6-octadiene to react is clearly due to the difficulty of the first attack of the allyl group on the double bond. In fact, a branched isomer of 2,6-octadiene, 3-methyl-1,5-heptadiene, reacts without difficulty, although to a lower extent than with 1,5-hexadiene.

As to the type of substituents on the allyl group, it should be noted that electron-withdrawing substituents cause cleavage of the nickel complex with formation of olefins according to a reaction we described some years ago.[7] Alkyl or aryl substituents can be present. Two types of products resulting from the two possible ways of cyclization are formed, depending on whether the substituents on the allyl group favor cyclization at the allylic double bond or not. For example, methallyl chloride gives the acid product of cyclization only at the side of the diene double bond (eq 18a) because cyclization at the side of the methallylic double bond to give a cyclopentanone ring (eq 18b) is even less favored than the analogous one on the crotyl group. On the other hand, cyclization to give a cyclohexanonic ring can occur, but in this case carbon monoxide has to attack a tertiary carbon atom. This attack being not favored, hydrogen elimination occurs with formation of a ketone (eq 19).

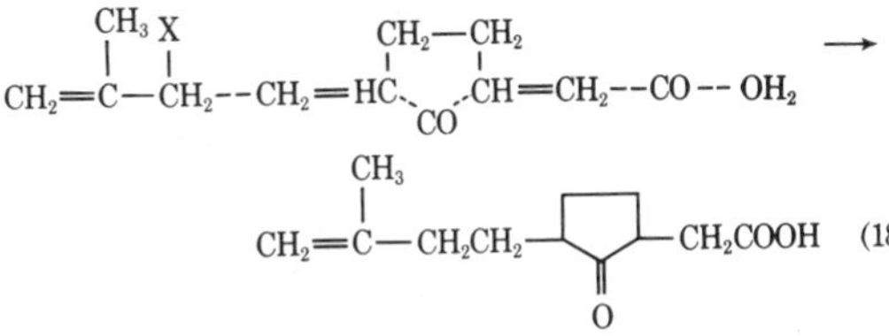

(18a)

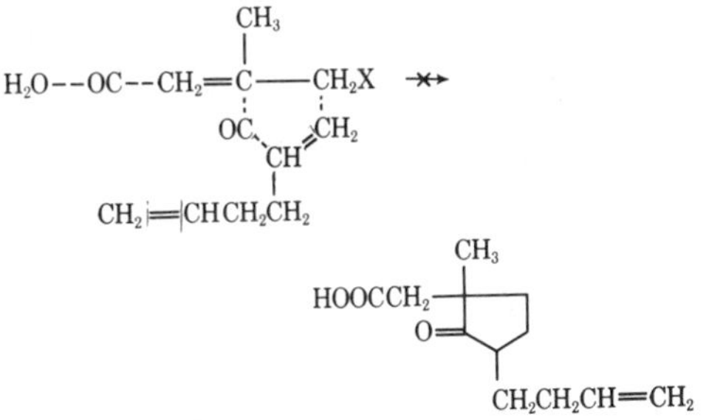

(18b)

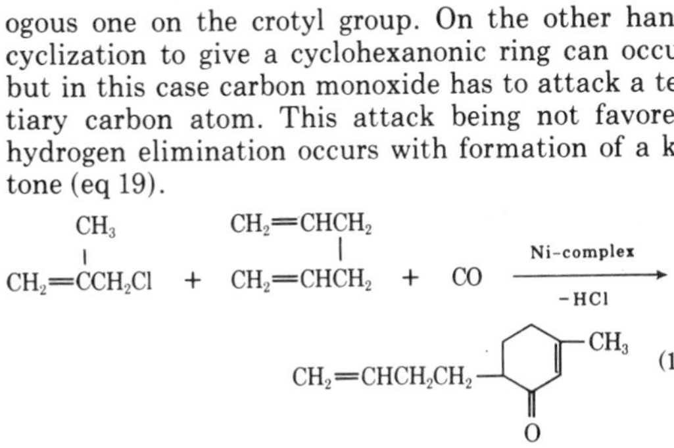

(19)

Geometry of the Chelating Molecule

Another important factor influencing the olefin insertion is the size of the chelating ring. To our surprise 1,4-pentadiene has not displayed a reactivity analogous to that of 1,5-hexadiene, although it does permit the closure of a five-membered ring. The only

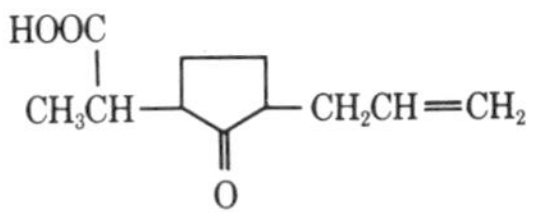

8

cyclopentanonic acid (**8**) derived from pentadiene and crotyl chloride, obtained to a limited extent, corresponds to the reaction of only one double bond of the diene. The poor yield may be due to the low chelating ability of the pentadiene system.

Most of the cyclopentanonic product consists of compounds resulting from reaction of the crotyl group with its branched dimer mentioned before, but the main carboxylic acid formed is 3-pentenoic acid, the direct carbonylation product of crotyl chloride. Compound **9**, derived from the insertion of one double bond of pentadiene without subsequent cyclization, is also formed.

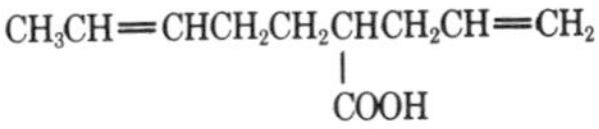

9

Cycloolefins like cyclooctene or cyclohexene do not react significantly, but *cis,cis*-1,5-cyclooctadiene easily gives rise to products derived from transannular cyclization or from cyclopentanone ring closure at the side of the originally allylic double bond. Thus crotyl chloride gives mainly **10** along with **11**. Both

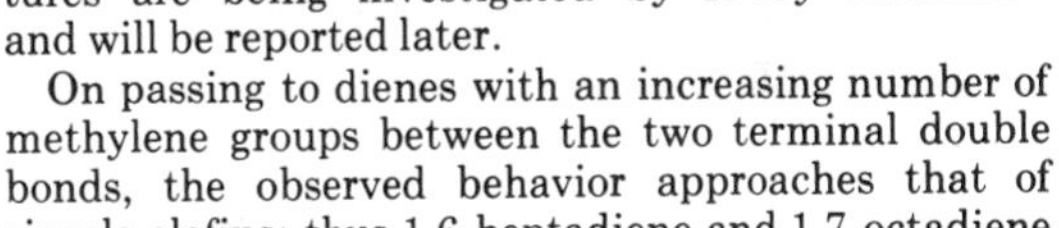

10 **11**

products are accompanied by minor amounts of diastereomers arising from dissymmetry at the tertiary carbon atom α to the carboxylic group. Their structures are being investigated by X-ray methods[13] and will be reported later.

On passing to dienes with an increasing number of methylene groups between the two terminal double bonds, the observed behavior approaches that of simple olefins; thus 1,6-heptadiene and 1,7-octadiene give low yields of compounds which result from a cyclization involving one double bond of the diene and one of the allyl group. The behavior of these two dienes is attributed to their low chelating ability. The size and geometry of the chelating system thus appear to have remarkable effects on reactivity.

Neutral Products from Stoichiometric Reactions

Up to now we have considered carboxylic acids or esters which constitute the main products of the catalytic synthesis. Neutral products are also found, corresponding to eq 20 and 21. These are stoichiometric syntheses which are responsible for destruc-

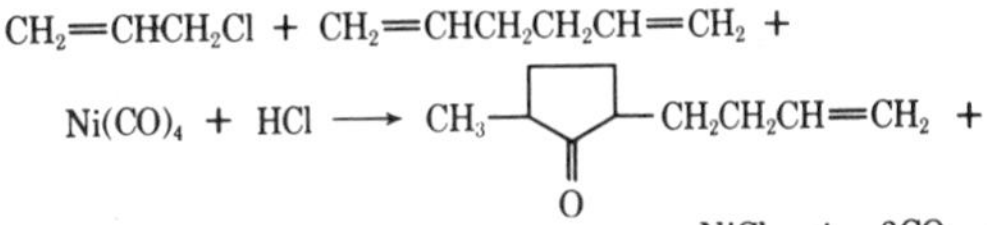

(20)

(13) *Cryst. Struct. Commun.*, **2**, 371 (1973); **3**, 495 (1974).

$$2CH_2\!=\!CHCH_2Cl + CH_2\!=\!CHCH_2CH_2CH\!=\!CH_2 + Ni(CO)_4 \longrightarrow$$

$$NiCl_2 + 2CO \quad (21)$$

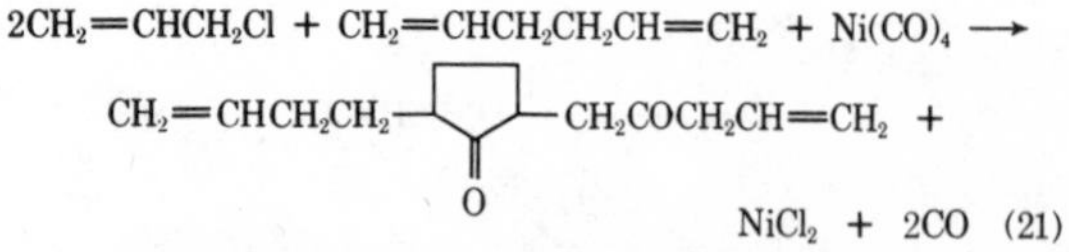

tion of the catalyst by transformation into a Ni^{II} species.

Double Cyclopentanone Ring Closure

It should be possible to continue the carbon monoxide and double bond insertion along a polyene chain, in which the hexadiene unit regularly repeats. This is the polybutadiene system. We have examined shorter chain compounds as models of more extended systems. *trans*-1,5,9-Decatriene gives reaction 22.

$$CH_3CH\!=\!CHCH_2Cl +$$
$$CH_2\!=\!CHCH_2CH_2CH\!=\!CHCH_2CH_2CH\!=\!CH_2 \xrightarrow[CO, H_2O]{Ni\ complex}$$

$$\quad (22)$$

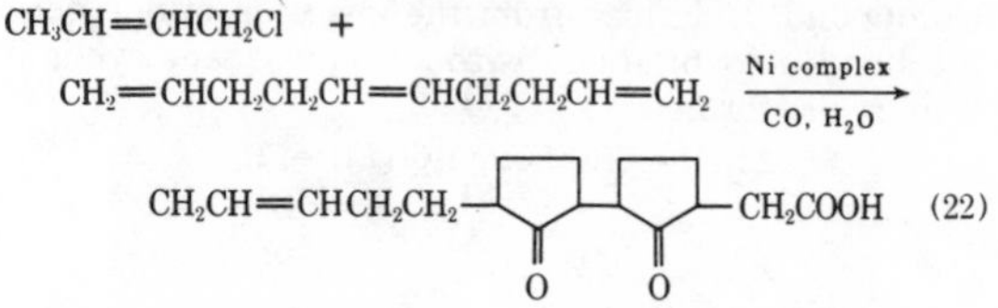

Part of the product consists of compounds with only one cyclopentanonic ring.

Insertion by Chelation of Unsaturated Acids

Another way in which we achieved insertion of a double bond takes advantage of a chelation effect.[14] For this purpose we used olefins containing salt-forming carboxylic groups at an appropriate distance from the double bond. This enabled the double bond to be coordinated as part of a chelating molecule. Suitable acids are the β,γ-unsaturated ones: α,β- or γ,δ-unsaturated acids are not effective. The synthesis is represented by eq 23, which is written for the

$$CH_2\!=\!CHCH_2Cl + CH_2\!=\!CHCH_2COONa + CO +$$
$$H_2O \xrightarrow{Ni(CO)_4} CH_2\!=\!CHCH_2CH_2CHCH_2COOH + NaCl \quad (23)$$
$$\qquad\qquad\qquad\qquad\qquad\quad |$$
$$\qquad\qquad\qquad\qquad\qquad COOH$$

case of allyl chloride and sodium vinylacetate. The reaction is carried out in an aqueous or alcoholic solution at a temperature of about 45° simply by mixing together the reagents under a carbon monoxide atmosphere.

Since under the same reaction conditions allyl chloride is carbonylated to vinylacetic acid, it is also possible to start with allyl chloride without adding sodium vinylacetate, provided that an acceptor of the liberated hydrogen chloride is present, such as magnesium oxide.

Another aspect to be noted is that the reaction is highly regiospecific, the products formed being those derived from the addition of the allyl group to the carbon atom in position 4 and of carbon monoxide to the carbon atom in position 3. We have interpreted this synthesis as in Scheme II.

Beside the oxidative addition of nickel to allyl halide, it involves coordination of the double bond of the transient nickel vinylacetate complex at a coordination site left free by the allyl group on passing

(14) G. P. Chiusoli, G. Cometti, and S. Merzoni, *Organometal. Chem. Syn.*, 1, 439 (1972).

Scheme II

$$CH_2\!=\!CHCH_2Cl + Ni(CO)_4 \longrightarrow$$

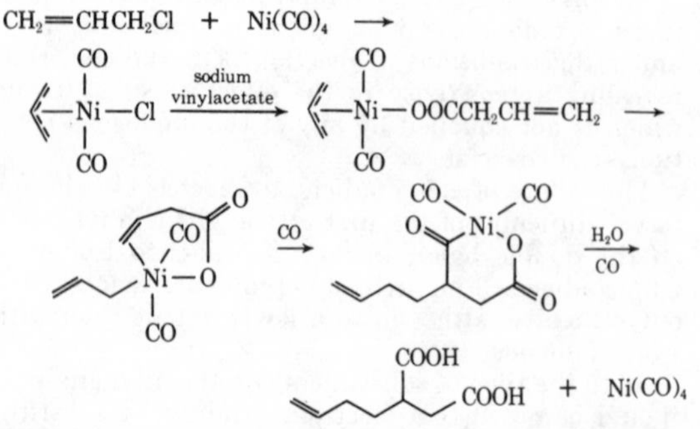

from the π to the σ form, insertion of the double bond and of carbon monoxide, and finally cleavage of the complex by water or alcohols. This interpretation is supported by the fact that reaction of allyl vinylacetate with tetracarbonylnickel gives the same butenylsuccinic acid.

Reaction 23 could also be directed to form butenylsuccinic anhydride. In this case only allyl chloride was used with magnesium oxide as neutralizing agent.

We could show that the anhydride is formed as a product of competition among all the nucleophilic species present in the reaction solution for the nickel-bonded acyl group. The acetate group, which is bonded to nickel, is in the best competitive site because it is coordinated adjacent to the acyl group.

The alkenylsuccinic synthesis allows the preparation of many acids and esters not easily accessible by other ways. In fact it is possible to vary the allyl halide on one side and the β,γ-unsaturated acid on the other, provided that steric effects do not hinder the reaction. These effects are, however, very important; a simple methyl group on the terminal carbon of vinylacetic acid decelerates the reaction remarkably.

Concluding Remarks

To sum up: in the course of stepwise additions on the metal the single blocks are selectively chosen according to the type of the C–metal bond and to the type of substrate. The reactions of the metal–carbon bond can be varied by using appropriate substituents and solvents in order to favor the insertion of one molecule or group in preference to others. The substrate to be inserted can also be favored by taking advantage of chelation effects, of the charge of the metal, and/or of steric and electronic effects caused by substituents. The rate of cleavage of the catalytic complex can also determine the course of a reaction.

From the synthetic point of view, the insertion reactions of olefins on nickel complexes afford a variety of acids and ketones simply by slightly changing the reaction conditions. Much remains to be done, however, toward understanding the catalytic process and the factors affecting it, before a better control of organic syntheses by stepwise additions on transition metal complexes is achieved.

I am indebted to my coworkers, whose dedication and experimental ability made possible the work on which this Account is based.

Addendum (May 1976)

3-Vinyl-1,5-hexadiene **12** with crotyl chloride and carbon monoxide, under the same conditions previously indicated for other dienes, gives mainly monocyclopentanonic products, but also about 15% of two bicyclic compounds **13** with condensed rings[15] was formed:

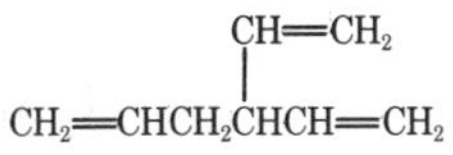

12

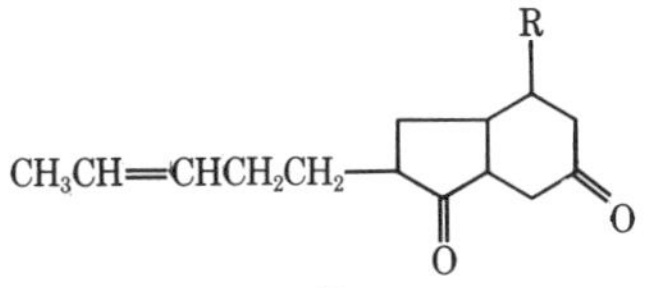

R = H, COOH

13

Reaction of a 1,4-pentadiene chain with carbon monoxide to give a cyclohexanone ring had not been observed previously in this type of synthesis. Apparently it is due to assistance of the adjacent 1,5-hexadiene chain.

Endo and *exo*-dicyclopentadiene reacted stereospecifically with crotyl chloride and carbon monoxide to give different products by selective reaction of the sole strained double bond.[16] Products were mainly polymeric ketones, but also monomeric acids were isolated. The x-ray structure of the acid **14** derived from the endo isomer shows that a cis,exo attack of the crotyl group and of carbon monoxide at the strained double bond has occurred. The cyclopentene double bond is localized in the position shown at least to the extent of 70%, the other 30% corresponding to the adjacent position:

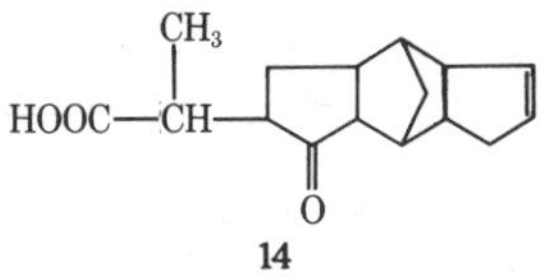

14

The polymeric material probably is a co-oligomerization product formed by further dicyclopentadiene and carbon monoxide insertions. Insertion of *endo*-dicyclopentadiene into an acylmanganese bond followed by carbonylation to form a complexed endo lactone ring has been recently described.[17]

Trans,trans-1,5,9,13-decatetratetraene **15** with methallyl chloride and carbon monoxide gave tricyclopentanone compound **16** with high selectivity, but with low conversion:

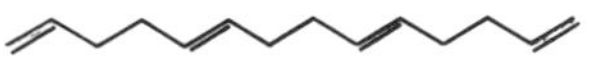

15

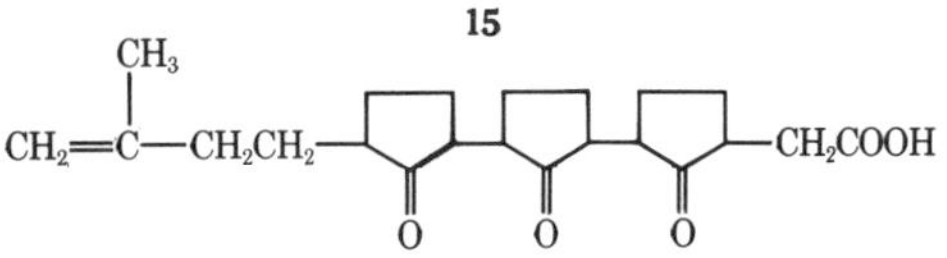

16

The x-ray structure shows that the hydrogen atoms α to the carbonyl groups are all placed on opposite sides in respect to the C-C(=O)-C planes. These are inclined by about 72° in respect to each other, so that the carbonyl groups describe a helix. This is a new type of stereospecific cycloco-oligomerization.[18]

All these reactions easily occur in the presence of carbon monoxide, which serves both to induce insertions (probably through pentacoordinated intermediates[4]) and to provide an acylnickel bond, subject to hydrolytic cleavage. The latter is an irreversible step, driving the preceding steps towards the final product. In this Account, for example, insertion of norbornene into an allylnickel bromide complex has been shown to occur in presence of carbon monoxide. In the absence of the latter, insertion is not observed. If, however, bromide is exchanged with acetate, insertion occurs to give a complex which reverts to the starting materials by replacing acetate with bromide.[19] Reversibility of insertion steps on metal complexes may prove to be a general phenomenon. Further discussions on insertion reactions are contained in recent books and reviews.[20]

(15) G. P. Chiusoli, G. Cometti, G. Sacchelli, and V. Bellotti, to be published.

(16) G. P. Chiusoli, g. Cometti, G. Sacchelli, V. Bellotti, G. D. Andreetti, G. Bocelli, and P. Sgarabotto, in press.

(17) B. L. Booth, M. Gardner, and R. N. Haszeldine, *J. Chem. Soc. Dalton Trans.*, 1856, 1863 (1975).

(18) G. P. Chiusoli, G. Cometti, G. Sacchelli, V. Bellotti, G. D. Andreetti, G. Bocelli, and P. Sgarabotto, in press.

(19) M. C. Gallazzi, L. Porri, and G. Vitulli, *J. Organomet. Chem.*, **97**, 131 (1975).

(20) (a) P. W. Jolly and G. Wilke, "Organic Chemistry of Nickel", Vol. 1 and 2, Academic Press, New York, 1974, 1975; (b) R. F. Heck, "Organotransition Metal Chemistry", Academic Press, New York, 1974; (c) J. Tsuji, "Organic Syntheses by Means of Transition Metal Complexes", Springer, Berlin, 1975; (d) R. Baker, *Chem. Rev.*, **73**, 487 (1973).

Homogeneous Catalytic Activation of C–H Bonds

George W. Parshall

Central Research Department, Experimental Station, E. I. du Pont de Nemours and Company, Wilmington, Delaware 19898

Received July 12, 1974

Reprinted from Accounts of Chemical Research, *8, 113 (1975)*

Very recently it has become apparent that some soluble transition-metal complexes can activate the C–H bonds of hydrocarbons in much the same way that the H–H bond of hydrogen is activated in catalytic hydrogenation reactions. This discovery points the way to a new type of homogeneous catalysis for reactions such as aromatic substitution[1] and opens some interesting opportunities for applications in synthesis.

The classical methods for substitution of arenes and alkanes involve attack of the hydrocarbon by a highly reactive reagent. The role of the catalyst has been to make the attacking reagent more reactive. In halogenation, for example, the catalyst generally promotes dissociation of the X–X bond, either by heterolytic cleavage to produce a highly reactive species with characteristics approaching those of X^+ or by homolytic cleavage to give a $X\cdot$ species. In either case, it is the halogen reagent that is activated rather than the hydrocarbon.

Recent work in several laboratories has shown that the C–H bonds in benzene are activated toward processes such as H–D exchange by a variety of transition-metal complexes including high-valent species such as $Ta^{V}H_3(C_5H_5)_2$, low-valent species like Ru^0-$(dmpe)_2$ (dmpe = $(CH_3)_2PCH_2CH_2P(CH_3)_2$), and electrophiles ($Pd^{2+}$ and Pt^{2+}). The activation is generally agreed to involve cleavage of a C–H bond by the metal and formation of a C–M bond. The C–M bond then undergoes reaction with a reagent such as D_2 or Cl_2. The process is similar to the metal-catalyzed activation of H_2 as illustrated by H–D exchange with a metal deuteride

$$\text{C-H} + \text{M-D} \rightleftharpoons \text{C-M}\!\begin{smallmatrix}\diagup \text{H}\\ \diagdown \text{D}\end{smallmatrix} \rightleftharpoons \text{C-D} + \text{M-H}$$

$$\text{H-H} + \text{M-D} \rightleftharpoons \text{H-M}\!\begin{smallmatrix}\diagup \text{H}\\ \diagdown \text{D}\end{smallmatrix} \rightleftharpoons \text{H-D} + \text{M-H}$$

George W. Parshall was born in Minnesota in 1929. He received his B.S. degree from the University of Minnesota and his Ph.D. from the University of Illinois. He is Research Supervisor in the Central Research Department, Du Pont Experimental Station, in Wilmington, Del. His major research interests are in homogeneous catalysis and in synthetic applications of organometallic compounds.

Although many of the studies of hydrocarbon activation involve the trivial reaction of hydrogen–deuterium exchange, more substantial synthetic applications are emerging. As described below, C–H activation can be utilized to introduce substituents such as $-Al(C_2H_5)_2$ directly onto an aromatic ring. Chelate effects can produce specific ortho substitution in aromatic rings, while enhanced steric effects can direct substituents into meta and para positions.

This Account relates several lines of research in our laboratories which bear on the problem of how a metal atom interacts with a C–H bond, either aromatic or aliphatic.

Background

To put the situation in its proper context, the activation of benzene C–H bonds by heterogeneous catalysts has been known for almost 40 years. Farkas and Farkas[2] discovered that a platinum film catalyzes the exchange between gaseous benzene and D_2 even at room temperature. Subsequent work has provided a reasonable picture of the mechanism.[3,4] As discussed later, there seems to be a close analogy between the processes occurring at the active sites of a metal surface and those occurring at the central metal atom in soluble complex.

The first observation of interaction between benzene and a soluble transition-metal compound was made by Chatt and Davidson.[5] They found that a reduction product of $RuCl_2(dmpe)_2$ reacts with benzene, naphthalene, and other aromatic hydrocarbons to give arylruthenium hydride complexes such as **1**.[6]

Catalytic activation of aromatic C–H bonds by soluble species was discovered by Garnett and cowork-

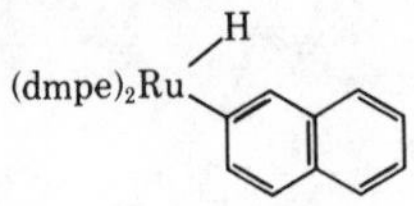

1

(1) G. W. Parshall, *Chem. Tech.*, **4**, 445 (1974).
(2) A. Farkas and L. Farkas, *Trans. Faraday Soc.*, **33**, 827 (1937).
(3) J. R. Anderson and C. Kemball, *Advan. Catal.*, **9**, 51 (1957).
(4) J. L. Garnett, *Catal. Rev.*, **5**, 229 (1971).
(5) J. Chatt and J. M. Davidson, *J. Chem. Soc.*, 843 (1965).
(6) S. D. Ibekwe, B. T. Kilbourn, U. A. Raeburn, and D. R. Russell, *J. Chem. Soc. A*, 1118 (1971).

ers.[4] Benzene and substituted benzenes react with D_2O in acetic acid in the presence of $[PtCl_4]^{2-}$ salts to produce deuterated benzenes. Although the catalyst appears to be soluble in the medium, the exchange resembles the heterogeneous catalytic exchange in two respects: (1) multiple exchange occurs, *i.e.* statistical analysis shows that more than one exchange occurs per catalytic event; (2) exchange also takes place in the side chains of alkylbenzenes. This observation of alkyl C–H exchange has led to the discovery of alkane activation by this catalytic system.[7,8]

Initial Studies

Much of the impetus for study of benzene C–H activation came from our studies of intramolecular aromatic substitution[9] in which a transition-metal atom reacts with an ortho aryl C–H bond of a donor ligand to form a metal–carbon bond. A typical example involves the interconversion of **2** and **3**. This transformation was first postulated[10] on the basis of selective ortho deuteration of the phenoxy groups of **2** by exposure to deuterium gas in solution at 100°. The ortho-bonded complex **3** has now been isolated.[11] Interestingly, its conversion to the Co–D analog of **2** by treatment with D_2 is accompanied by extensive ortho deuteration even under conditions in which **2** does not react with D_2 at a significant rate.

$$CoH[P(OC_6H_5)_3]_4 \rightleftharpoons$$

2

[structure: benzene ring with $Co[P(OC_6H_5)_3]_3$ and $O-P(OC_6H_5)_2$] $+ H_2$

3

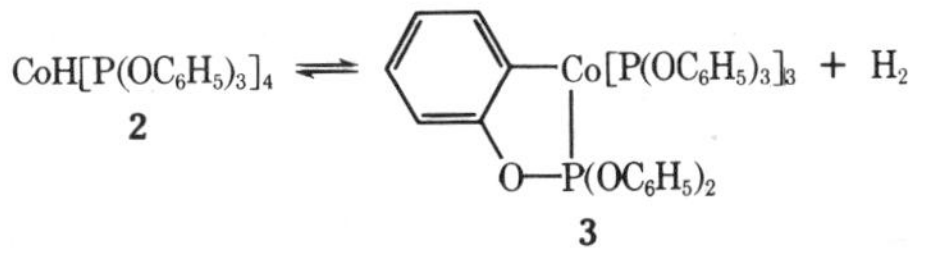

The occurrence of such intramolecular reactions raised the question as to whether or not similar reactions might not occur with benzene itself. Initial tests showed that two polyhydride complexes, $TaH_3(C_5H_5)_2$ and $IrH_5[P(CH_3)_3]_2$, do indeed catalyze the exchange of benzene with D_2. When benzene solutions of these complexes are heated with deuterium in a sealed tube, H_2 and HD appear in the gas phase. Deuterium is statistically exchanged with both benzene and ligand C–H and with M–H as well.[12]

Polyhydride Catalysts

Subsequent studies of the benzene–D_2 reaction have shown catalysis by polyhydride complexes of transition metals with odd numbers of valence electrons. Perhaps, because of the nature of our synthesis programs, greatest activity has been found for group V hydrides such as $NbH_3(C_5H_5)_2$[13] and $TaH_5(dmpe)_2$.[14] The latter has produced the fastest C_6H_6–D_2 exchange rates that we have yet observed with a soluble catalyst. Rhenium polyhydrides such as $ReH_5[P(C_6H_5)_3]_3$, which have been reported to ex-

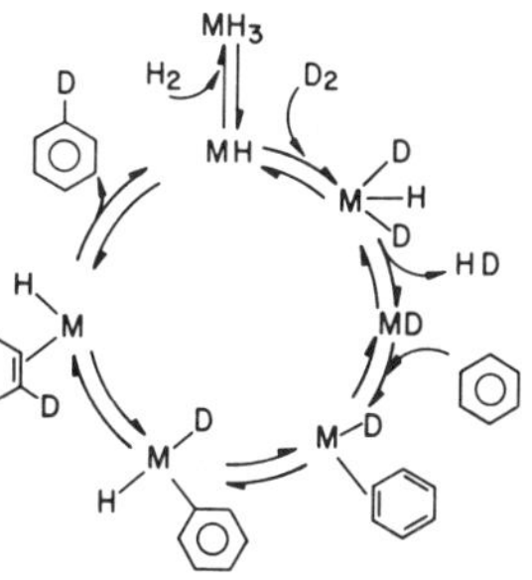

Figure 1. Proposed scheme for exchange between benzene and D_2 catalyzed by metal polyhydrides.

change Re–H with C_6D_6,[15] also catalyze exchange between benzene and deuterium. Complexes such as $[NbH(\mu\text{-}C_5H_4)(C_5H_5)]_2$ and $TaH(PR_3)(C_5H_5)_2$, which give rise to trihydrides on treatment with hydrogen, also catalyze exchange.

These metal hydride catalyzed exchanges have several features in common.[16,17] (1) Only aryl C–H bonds exchange. In contrast to Pt metal or $[PtCl_4]^{2-}$ catalysis, no deuterium enters the methyl groups of toluene or xylene. (2) Exchange occurs stepwise, one C–H bond at a time, again in contrast to the heterogeneous and $[PtCl_4]^{2-}$-catalyzed reactions. (3) Little deuteration occurs ortho to bulky groups such CH_3 or CF_3. (4) A deuterium isotope effect, k_H/k_D, of 2–3 suggests some degree of C–H bond breaking in the rate-determining step.

Reactions of substituted benzenes with deuterium in the presence of the polyhydride complexes are generally favored by electron-withdrawing substituents. The usual rate sequence is $C_6H_5F \sim C_6H_5CF_3 > C_6H_6 > C_6H_5OCH_3 \sim C_6H_5CH_3$ in competitive experiments, quite similar to that expected for a nucleophilic reaction. However, the range of rates is quite small, less than a factor of 10. In fact, with $TaH_3(C_5H_5)_2$ and $TaH_3(C_5H_4CH_3)_2$, there is no significant rate difference among the various monosubstituted benzenes. (*p*-Xylene is much slower, presumably because all the aryl hydrogens are ortho to methyl groups.)

The mechanism shown in Figure 1 has been proposed[16] to account for these results as well as other experimental data such as the position of deuterium substitution. A key intermediate in this scheme is a coordinatively unsaturated metal hydride formed by dissociation of H_2 from a trihydride species. (Other ligands have been neglected for simplicity and generality.) This monohydride (MH) is capable of exchange with D_2 as shown in the clockwise sequence and can coordinate a benzene molecule as illustrated with the analogous monodeuteride (MD). The coordinated benzene is proposed to transfer a hydrogen to the metal with simultaneous formation of a C–M bond, a formal oxidative addition to the metal atom. Reversal of the oxidative addition and dissociation of the deuterated benzene complete the catalytic cycle.

Although there is no direct experimental evidence for coordination of benzene to the metal in our system, sound precedent for both arene coordination

(7) M. B. Tyabin, A. E. Shilov, and A. A. Shteinman, *Dokl. Akad. Nauk SSSR*, **198**, 381 (1971).

(8) R. J. Hodges, D. E. Webster, and P. B. Wells, *J. Chem. Soc. A*, 3230 (1971).

(9) G. W. Parshall, *Accounts Chem. Res.*, **3**, 139 (1970).

(10) G. W. Parshall, W. H. Knoth, and R. A. Schunn, *J. Amer. Chem. Soc.*, **91**, 4990 (1969).

(11) L. W. Gosser, to be published in *Inorg. Chem.*

(12) E. K. Barefield, G. W. Parshall, and F. N. Tebbe, *J. Amer. Chem. Soc.*, **92**, 5234 (1970).

(13) F. N. Tebbe and G. W. Parshall, *J. Amer. Chem. Soc.*, **93**, 3793 (1971).

(14) F. N. Tebbe, *J. Amer. Chem. Soc.*, **95**, 5823 (1973).

(15) J. Chatt and R. S. Coffey, *J. Chem. Soc. A*, 1963 (1969).

(16) U. Klabunde and G. W. Parshall, *J. Amer. Chem. Soc.*, **94**, 9081 (1972).

(17) U. Klabunde, unpublished results.

and oxidative addition of a C–H bond exists in the reaction of benzene with a ruthenium(0) complex[5] described previously. More recently Green[18] and Brintzinger[19] have observed an irreversible addition of a benzene C–H bond to transient $M(C_5H_5)_2$ derivatives of molybdenum and tungsten, closely analogous to our proposed $MH(C_5H_5)_2$ catalysts based on niobium and tantalum.

The assumption of benzene coordination permits one to account for the small or nonexistent substituent effects observed in the exchange reactions of substituted benzenes if one assumes that oxidative addition and coordination are both rate-limiting factors and have opposing substituent effects. In the case of the niobium compounds which catalyze exchange preferentially with negatively substituted benzenes, oxidative addition is assumed to be rate controlling (consistent with the observed isotope effect, k_H/k_D = 2.3). Hence, for this series of catalysts, oxidative addition has some of the characteristics of a nucleophilic reaction.

Monohydride System. Recently it has been observed[17] that the monohydride, $RuHCl(PPh_3)_3$, catalyzes the exchange between benzene and deuterium gas under vigorous conditions. Although many aspects of this reaction remain unexplored, there are some interesting differences between this system and the polyhydride systems discussed above. When triphenylphosphine is added to the benzene–D_2 exchange reactions, it undergoes extensive deuterium substitution. With $RuHCl(PPh_3)_3$ as the catalyst, the deuterium appears exclusively in the ortho positions,[10] but with $TaH_5(dmpe)_2$, all the deuterium substitution occurs in the meta and para positions.[17] The difference seems to be that, with the ruthenium compound, exchange occurs by a conventional orthometalation mechanism while the tantalum compound effects exchange without coordination of the triphenylphosphine phosphorus to the metal. In effect, the latter catalyst treats triphenylphosphine as a $(C_6H_5)_2P$-substituted benzene in which $(C_6H_5)_2P$- is an inert but bulky substituent which directs exchange away from the ortho positions.

The differences between the two types of catalysts may possibly extend to the intimate interaction of the metal with the aryl C–H bond in the oxidative addition process. In the reaction of $(C_6H_5)_2P(CH_2)_nCH_3$ compounds with D_2 catalyzed by $RuHCl(PPh_3)_3$, alkyl as well as aryl C–H bonds undergo exchange. With the ruthenium catalyst, the methyl C–H bonds of $(C_6H_5)_2PCH_2CH_2CH_3$ are activated even at 20°. The exchange rates of methyl and ortho hydrogens (after statistical adjustment) are equivalent, and both exchanges show the same activation energy, *ca.* 6 kcal/mol, over a 140° range. Some deuteration of the methylene groups also occurs at temperatures above 100°, with α deuteration (E_{act} = *ca.* 13 kcal/mol) favored over β deuteration.

In the $RuHCl(PPh_3)_3$-catalyzed exchanges, oxidative addition to a coordinatively unsaturated ruthenium(II) species seems a likely mechanism for C–H bond activation. In comparing ortho aryl *vs.* γ methyl exchange it should be noted that the methyl group reaction involves an unstrained five-membered ring (A) while ortho exchange requires a relatively strained four-membered ring (B). For comparison of

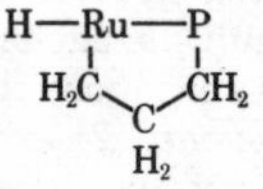

A

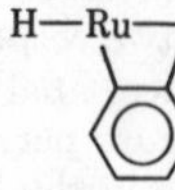

B

ring size effects, exchange of $(C_6H_5)_2POC_6H_5$ with deuterium was carried out in the presence of $RuHCl(PPh_3)_3$. Ortho deuteration of the phenoxy group (involving a five-membered ring intermediate) was at least 50 times as fast as in the P-bonded phenyl.

The ring size effect is similar to that reported[20] elsewhere for the deuteration of tri-*n*-propylphosphine in the complex

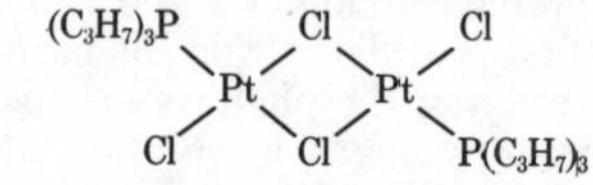

When this substance was treated with CH_3COOD–D_2O, deuterium was incorporated exclusively into the methyl groups, consistent with a five-membered ring intermediate as in A above. Such cyclic intermediates have been isolated in reactions of $PtCl_2(PR_3)_2$ complexes.[21] No α-CH_2 deuteration has been noted in the platinum complexes, although metalation of C–H bonds α to phosphorus has been observed in $Ru(dmpe)_2$[5,22] and in $IrH_5(PMe_3)_2$.[16]

In the γ deuteration catalyzed by the Ru and Pt complexes, precoordination of the γ carbon to the metal does not seem to be required for C–H bond cleavage. A reasonable transition state is represented by C. This configuration resembles the C–H/metal

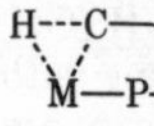

C

interaction detected in pyrazolylborate complexes of molybdenum by spectroscopic[23] and crystallographic[24,25] techniques.

Nonhydridic Catalysts

In the previous exchange studies from our laboratory, gaseous deuterium has been the source of labeled hydrogen and the catalysts have usually been metal hydrides. Now benzene activation has been detected with catalysts which contain no M–H bonds.

When a solution of ethylbis(π-cyclopentadienyl)-(ethylene)niobium (4) in benzene-d_6 is heated for extended periods, substantial amounts of deuterium (up to d_9) appear in the ethyl and ethylene ligands of the recovered complex.[26] This result is attributed to exchange between C_6D_6 and a transient species,

(18) M. L. H. Green and P. J. Knowles, *J. Chem. Soc. A*, 1508 (1971)
(19) K. L. T. Wong, J. L. Thomas, and H. H. Brintzinger, *J. Amer. Chem. Soc.*, **96**, 3694 (1974).
(20) C. Masters, *J. Chem. Soc., Chem. Commun.*, 191 (1973).
(21) A. J. Cheyney, B. E. Mann, B. L. Shaw, and R. M. Slade, *J. Chem. Soc. A*, 3833 (1971).
(22) F. A. Cotton, B. A. Frenz, and D. L. Hunter, *J. Chem. Soc., Chem. Commun.*, 755 (1974).
(23) S. Trofimenko, *Inorg. Chem.*, **9**, 2493 (1970).
(24) F. A. Cotton and V. W. Day, *J. Chem. Soc., Chem. Commun.*, **415** (1974).
(25) F. A. Cotton, T. LaCour, and A. G. Stanislowski, *J. Amer. Chem. Soc.*, **96**, 754 (1974).
(26) F. N. Tebbe, unpublished observations.

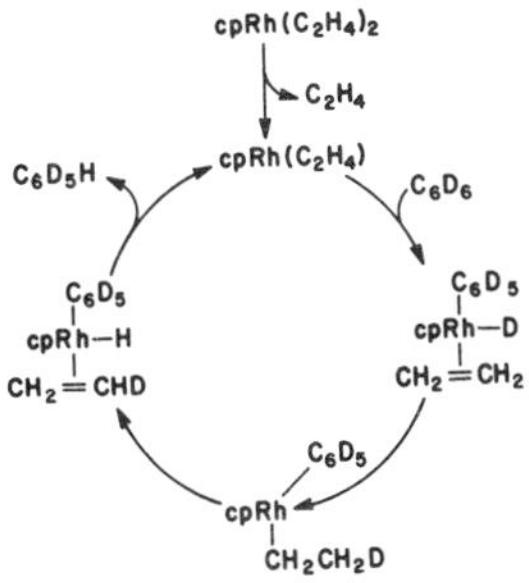

Figure 2. Mechanism for exchange between benzene-d_6 and coordinated ethylene.

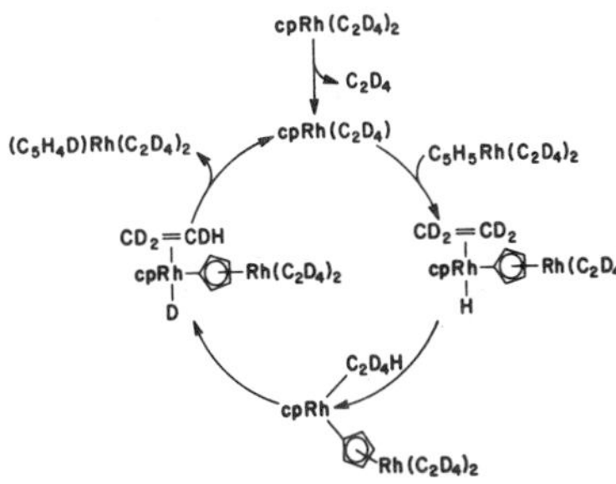

Figure 3. Proposed mechanism for H–D exchange between coordinated ethylene and cyclopentadienide ligands.

NbH(C_5H_5)$_2$ (**5**), by the mechanism of Figure 1. The monohydride (**5**), which has been proposed[13] as an intermediate in many reactions of NbH$_3$(C_5H_5)$_2$, may be formed by dissociation of two molecules of ethylene from **4**. [Dissociation of one ethylene to give NbH(C_2H_4)(C_5H_5)$_2$ has been observed.] If the ethylene dissociation is reversible, addition of C_2H_4 to NbD(C_5H_5)$_2$ provides a ready route to the observed incorporation of deuterium.

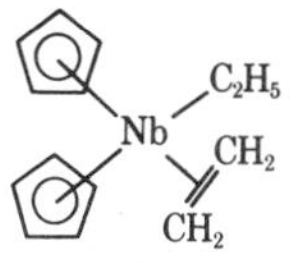

The path for C_6D_6–ligand exchange is less obvious for the rhodium complex **6** which reacts with benzene-d_6 at 130° to introduce deuterium into both the ethylene and cyclopentadienyl ligands.[27] Several other arenes, including pyridine-d_5, toluene-d_8, and nitrobenzene-d_5, also participate in this reaction as deuterium donors. The rate sequence is $C_6D_6 \sim C_5D_5N > C_6D_5NO_2 > C_6D_5CD_3$. With pyridine-$d_5$, deuterium enters the ethylene ligands, but little appears on the cyclopentadienyl ligands of the recovered complex.

Substitution of ^{1}H onto the deuteriobenzene is statistically random, consistent with one exchange per productive collision of arene and **6**. This behavior is in contrast to the multiple exchange encountered on the surface of metal films. The implication is that the arene activation is a homogeneous process. The solutions remain clear during the reaction. The absence of methyl C–D exchange in toluene-d_8 experiments also tends to exclude heterogeneous catalysis by metallic rhodium, which would be very active in this reaction.[28]

Another significant observation is that the exchange between C_6D_6 and **6** is repressed by the presence of C_2H_4 or C_2D_4. Similarly the intramolecular exchange between ethylenic and cyclopentadienyl C–H bonds in Rh(C_2D_4)$_2$(C_5H_5) is inhibited by C_2D_4. Significantly the exchange occurs only at temperatures above 115°, at which point ethylene dissociates from **6**.[29]

A mechanistic scheme to account for exchange between C_6D_6 and coordinated ethylene is shown in Figure 2. Ethylene dissociation from **6** generates a coordinatively unsaturated species which provides entry to the catalytic cycle. Oxidative addition[30] of a C–D bond to the coordinatively unsaturated Rh(C_2H_4)(C_5H_5) species generates a phenylrhodium deuteride. Addition of the Rh–D bond to ethylene followed by elimination of Rh–H accomplishes exchange by a well-established pathway.[31]

A strictly analogous scheme (Figure 3) can account for exchange of hydrogen and deuterium between the ethylene and cyclopentadienide ligands. In this speculative proposal, the cyclopentadienylrhodium complex is considered to behave as an arene. A cyclopentadienyl C–H bond of one molecule of complex is oxidatively added to the Rh(C_2H_4)(C_5H_5) species generated by dissociation of ethylene from a second. Thus, the apparently intramolecular exchange may occur by a bimolecular mechanism. This point has not been confirmed by kinetic studies, but intermolecular exchange between Rh(C_2D_4)$_2$(C_5H_5) and Rh(C_5H_5)(1,5-cyclooctadiene) has been observed, consistent with this mechanism.

A similar oxidative addition of a cyclopentadienyl C–H bond to the metal atom of a second molecule of complex is believed to occur in the dimerization of dicyclopentadienyl complexes of niobium and tantalum.[13,32]

$$\text{MH}_3(C_2H_5)_2 \underset{+H_2}{\overset{-H_2}{\rightleftharpoons}} (C_5H_5)(H)M \quad\text{—}\quad M(H)(C_5H_5)$$

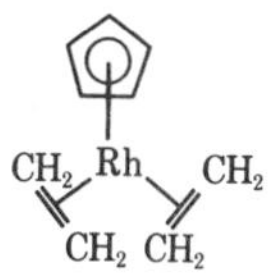

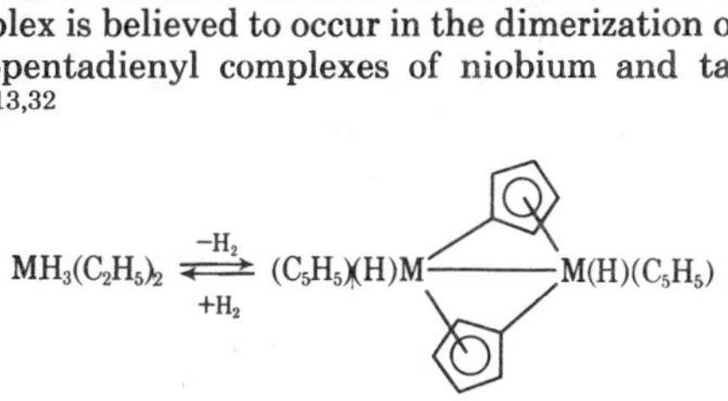

Such phenomena abound in the "titanocene" system[33] and considerably complicate the study of nitrogen fixation[34] by cyclopentadienyl complexes of titanium.

If the reaction schemes of Figures 2 and 3 are confirmed, a very satisfying unity of mechanism between the group V and group VIII metal catalyzed reactions will have been demonstrated. In addition, the mecha-

(27) L. P. Seiwell, *J. Amer. Chem. Soc.*, in press.

(28) C. Horrex, R. B. Moyes, and R. C. Squire, *Proc. Int. Conf. Catal., 4th* (1968).

(29) R. Cramer, *J. Amer. Chem. Soc.*, **94**, 5681 (1972).

(30) (a) J. P. Collman, *Accounts Chem. Res.*, **1**, 136 (1968); (b) J. Halpern, *ibid.*, **3**, 386 (1970).

(31) (a) R. Cramer, *Accounts Chem. Res.*, **1**, 186 (1968); (b) C. A. Tolman, *Chem. Soc. Rev.*, **1**, 337 (1972).

(32) L. J. Guggenberger, *Inorg. Chem.*, **12**, 294 (1973).

(33) F. N. Tebbe and L. J. Guggenberger, *J. Amer. Chem. Soc.*, **95**, 7870 (1973).

(34) (a) M. E. Volpin and V. B. Shur, *Organometal. React.*, **1**, 55 (1970); (b) E. E. van Tamelen, *Accounts Chem. Res.*, **3**, 361 (1970).

nism of Figure 2 permits catalytic exchange between different arenes *via* an ethylene ligand of the rhodium complex. Exchange between C_6D_6 and fluorobenzenes does indeed occur at 128° in the presence of **6**.[17]

Application to Synthesis

Although H–D exchange is a convenient tool for detecting potentially useful catalytic species, it is not generally a synthetically useful reaction in itself. An extension of the chemistry described above into synthetically useful substitution reactions has been developed by Dr. Tebbe.[26]

The reaction of $NbH_3(C_5H_5)_2$ (7) with triethylaluminum in benzene at elevated temperatures, in contrast to that at room temperature,[35] produces diethylphenylaluminum. This condensation reaction of

$$C_6H_6 + Al(C_2H_5)_3 \xrightarrow{7} C_6H_5Al(C_2H_5)_2 + C_2H_6$$

benzene and an alkylaluminum compound appears to be the first aromatic substitution other than HD exchange catalyzed by the transition-metal polyhydrides. Although it is possible that the catalysis occurs simply by a strong base mechanism,[36] it seems more likely to involve intermediates similar to those involved in H–D exchange.

The reaction of $NbH_3(C_5H_5)_2$ with $Al(C_2H_5)_3$ at room temperature[35] proceeds with evolution of ethane to give the hydride-bridged species **8**. Oxidative addition of a benzene C–H bond to **8** has the potential to generate **9**, which, in turn, can give $C_6H_5Al(C_2H_5)_2$ by a number of paths.

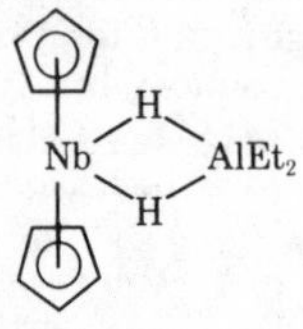

8

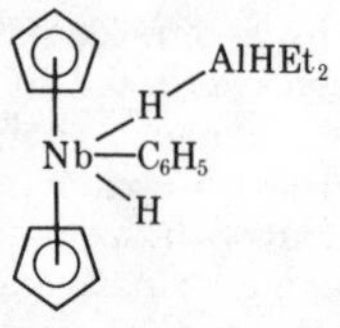

9

Relation to Heterogeneous Catalysis

The reactions of aromatic hydrocarbons described here, like the arene[4] and alkane[7,8] reactions catalyzed by $PtCl_4^{2-}$, show many similarities to those produced by metal films and supported metal catalysts. It is very tempting to propose that the intermediate species on a metal surface are similar to those involved in catalysis by soluble species. It is usually assumed that the three essential "sites" for exchange of arene C–H are disposed on three separate metal atoms as in D. In contrast, we propose that the exchange "sites" for a homogeneous catalyst are simply three adjacent orbitals on a single metal atom (E). In the metallic surface (D), the catalytically active metal atoms are bonded to "ligand" metal atoms. In

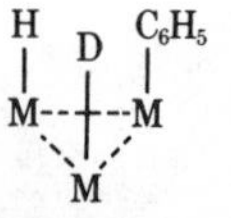

D

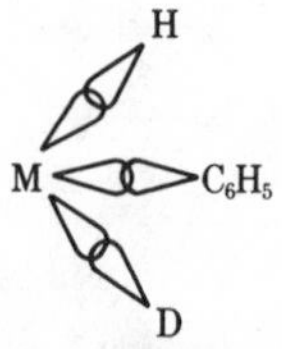

E

the soluble complex (E), hydrocarbon or tertiary phosphine ligands create the proper steric and electronic environment for catalytic activity.

One of the most interesting problems in aromatic hydrocarbon activation is study of the interrelation between the homogeneous and heterogeneous catalysts.[37] While it may be difficult to determine whether or not monoatomic sites analogous to E exist on a heterogeneous surface, the information may be quite rewarding. Similarly, development of soluble metal clusters with trimetallic sites like D may greatly increase the scope of hydrocarbon activation reactions accessible through homogeneous catalysis.

Saturated Hydrocarbons

One of the most dramatic developments in hydrocarbon activation has been the report by Shilov[7] that methane and ethane undergo exchange with D^+ in the presence of $PtCl_4^{2-}$. Given the observation that metal complexes can break C–H bonds in aliphatic ligands as described above, it is perhaps not surprising that alkanes also react. With ethane, exchange is proposed to proceed *via* ethyl and ethylene complexes.

$$[Pt—C_2H_5]^+ \underset{}{\overset{-H^+}{\rightleftharpoons}} Pt—\overset{CH_2}{\underset{CH_2}{\|}} \overset{+D^+}{\rightleftharpoons} [Pt—CH_2CH_2D]^+$$

An analogous mechanism for methane exchange requires a relatively unprecedented $Pt{=}CH_2$ intermediate in place of the ethylene complex. Support for the existence of such species now comes from the work of R. R. Schrock in our laboratories who has isolated and fully characterized a simple alkylidene complex.

$$(Me_3CCH_2)_3TaCl_2 \xrightarrow{2Me_3CCH_2Li} (Me_3CCH_2)_3Ta{=}CHCMe_3$$

Confirmation of the existence of such a multiply bonded species (frequently postulated in mechanisms of heterogeneous catalysis) helps to rationalize the phenomenon of multiple exchange encountered in Shilov's system and in CH_4–D_2 exchange catalyzed by metallic films. If two hydrogens are removed from CH_4 in its interaction with a metal such as Pt, products such as CH_2D_2 are not unexpected. Taken together, the results of Shilov and of Schrock support the proposal that homogeneous and heterogeneous C–H bond activation mehanisms are similar.

(35) F. N. Tebbe, *J. Amer. Chem. Soc.*, **95**, 5412 (1973).
(36) H. Lehmkuhl and R. Schaefer, U. S. Patent 3,341,562 (1967).

(37) G. W. Parshall, M. L. H. Green, *et al.*, in "Catalysis," F. Basolo and R. L. Burwell, Ed., Plenum Press, New York, N. Y., 1973.

Addendum (May 1976)

Activation by Proximity. Recent results suggest that almost any C-H bond can be activated by contact with an electron-rich transition metal center. The $PtCl_4^{2-}$ system remains unique in its efficient catalysis of HD exchange in alkanes, but there are now many examples of cleavage of alkyl C-H bonds in phosphine ligands. Several of these reactions involve bonds that are not activated by any obvious electronic factors. The only apparent activating factor is that the C-H bond may be close to a coordinatively unsaturated metal center. The selective HD exchange reactions of the γ-hydrogens of n-propylphosphine complexes cited above[17,20,38] are clear examples of this effect. As noted, the γ-positions in the alkyl chain are uniquely situated to form a strain-free five-membered metallocycle.

An interesting example of geometric activation appears in a biological system.[39] Substantial evidence indicates that rearrangements of the carbon skeletons of biological substrates (e.g. succinate $\rightleftharpoons$ methylmalonate) catalyzed by coenzyme B_{12} are initiated by hydrogen transfer from the substrate to a 5′-adenosyl group bound to cobalt. Both the forward and reverse reactions are proposed to occur through attack of a paramagnetic cobalt center on a CH_2 or CH_3 group. In this process, a major activating factor is the proximity of the C-H bond to the cobalt enforced by the geometry of the enzyme.

Arene Activation. As noted under *Background,* the first recognized arene C-H activation was the reaction of naphthalene with a zerovalent ruthenium complex (A), generated in situ, to give a β-naphthylruthenium hydride (B). Recent work in these laboratories[40] has shown that a similar reaction occurs in the analogous iron

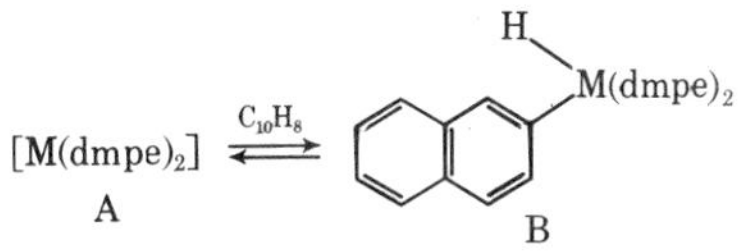

dmpe = $(CH_3)_2PCH_2CH_2P(CH_3)_2$

system when $FeCl_2(dmpe)_2$ is reduced with sodium and naphthalene. With iron, however, a ready equilibrium between A and B appears to exist. As a consequence, the cis and trans isomers of B equilibrate easily when M is iron. More importantly, other arenes rapidly exchange with the naphthyl iron hydride to form naphthalene and new $ArFeH(dmpe)_2$ complexes. Like the HD exchange reactions catalyzed by $NbH_3(C_5H_5)_2$, naphthalene displacement is favored by electron-withdrawing substituents on the arene reactant. The aryliron hydrides appear to be good models for study of the mechanism of arene C-H activation.

(38) A. A. Kiffen, C. Masters, and L. Raynand, *J. Chem. Soc. Dalton Trans.*, 853 (1975).

(39) (a) B. M. Babior, *Acc. Chem. Res.*, **8**, 376 (1975); (b) R. H. Abeles and D. Dolphin, *Acc. Chem. Res.*, **9**, 114 (1976); (c) R. Breslow and P. L. Khanna, *J. Am. Chem. Soc.*, **98**, 1297 (1976).

(40) S. D. Ittel, C. A. Tolman, A. D. English, and J. P. Jesson, *J. Am. Chem. Soc.*, in press.

Organoiron Complexes as Potential Reagents in Organic Synthesis

Myron Rosenblum

Department of Chemistry, Brandeis University, Waltham, Massachusetts 02154

Received September 4, 1973

Reprinted from ACCOUNTS OF CHEMICAL RESEARCH, *7, 122 (1974)*

Metal Assisted Cycloaddition Reactions

***monohapto*Allylmetal Complexes.** Three years ago our attention was drawn to the reactions of *monohapto*allyl–transition metal complexes[1] with SO_2.[2] These reactions generally led to the formation of metal allyl sulfones and were observed to occur with both retention and inversion of the allylic ligand (eq 1). The latter product was attributed to a concerted insertion process involving simultaneous interaction of SO_2 with the metal and the olefinic terminus (1).[2]

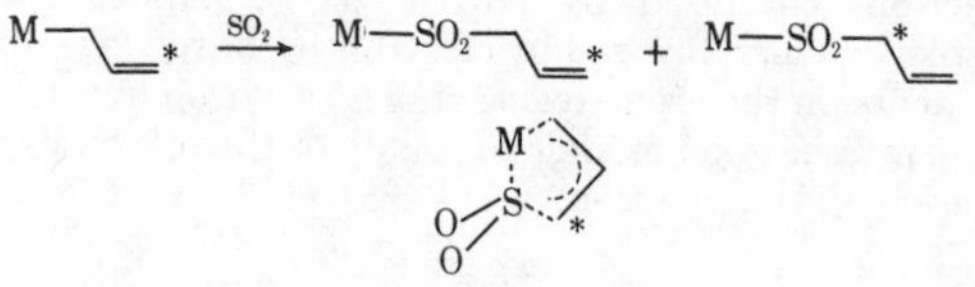

(1)

1

However, if SO_2 is viewed as an electrophile, it is apparent that the rearrangement reaction may alternatively be depicted as a two-step process involving initial formation of a dipolar ion (2, E = SO_2) and its subsequent collapse through displacement of the coordinated olefin by the anion (eq 2).

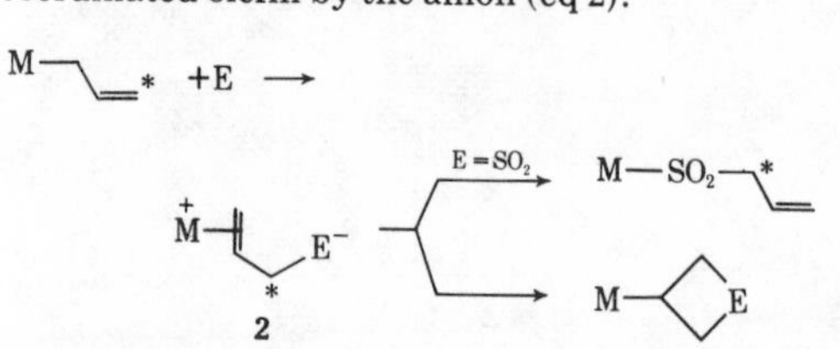

(2)

(3)

2

3

The first step of such a mechanism has ample analogy in the conversion of *monohapto*allyl–transition metal complexes to cationic olefin–metal complexes on protonation (eq 4)[3] while the second involving ligand exchange at a metal site (eq 5) is a process ubiquitous in transition metal chemistry.[4] More significantly, a nonconcerted mechanism exposes the possibility, not available to a one-step process, that the dipolar ion 2 may collapse through an internal cyclization reaction to give products of structure 3[5] (eq 3).

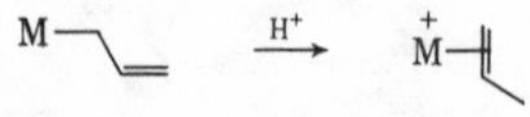

(4)

$$\mathrm{M{-}L + L' \rightleftarrows M{-}L' + L} \qquad (5)$$

While not observed with SO_2, the latter reaction path becomes the exclusive one with electrophilic olefins such as tetracyanoethylene (TCNE). This observation provided at once substantive evidence for a two-step process, a simple means for rationalizing the formation of similar products reported to be formed from *monohapto*propargylmetal complexes (*vide infra*), and a theoretical basis for extending these "metal-assisted cycloaddition reactions" to a larger variety of organometallic complexes and uncharged electrophiles.

Although these reactions appear to be quite general for *monohapto*allylmetal complexes, the most extensive investigations have been carried out with *monohapto*allyl derivatives of dicarbonyl*pentahapto*-cyclopentadienyliron (4). (In structure 4 and elsewhere, Fp stands for the h^5-$C_5H_5Fe(CO)_2$ moiety.) These were found to react within minutes, at room temperature, with TCNE, to give tetracyanocyclopentyl complexes of structure 5 in good yield.[5-7] The high reactivity of the double bond in these complexes no doubt reflects the capacity of the metal to

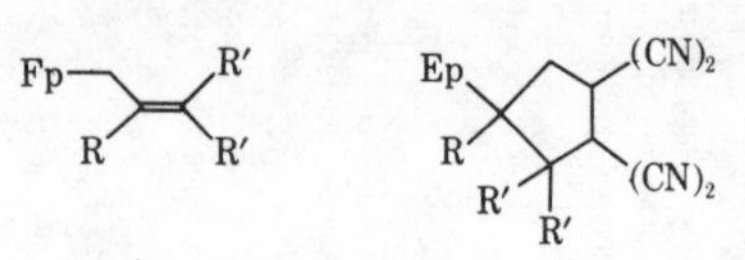

4　　5

a, R, R′ = H
b, R = Me; R′ = H
c, R = H; R′ = Me

Myron Rosenblum was born in New York City in 1925. He obtained his A.B. degree from Columbia University and his Ph.D. degree from Harvard University under R. B. Woodward. After spending 2 years at Columbia University in postdoctoral work with Gilbert Stork, he joined the Illinois Institute of Technology. He remained there until 1958 when he moved to Brandeis University, where he is Professor of Chemistry. His principal scientific interests are in organometallic chemistry.

(1) The nomenclature is that introduced by F. A. Cotton, *J. Amer. Chem. Soc.*, **90**, 6230 (1968), in which the h^n prefix (h^1, monohapto; h^2, dihapto, etc.) specifies the number of ligand atoms which are formally bonded to the metal.

(2) (a) F. A. Hartman, P. J. Pollick, R. L. Downs, and A. Wojcicki, *J. Amer. Chem. Soc.*, **89**, 2493 (1967); (b) F. A. Hartman and A. Wojcicki, *Inorg. Chim. Acta*, **2**, 289 (1968); (c) A. Wojcicki, *Accounts Chem. Res.*, **4**, 344 (1971).

(3) (a) M. L. H. Green and P. L. I. Nagy, *J. Chem. Soc.*, 189 (1963); (b) M. L. H. Green and A. N. Stear, *J. Organometal. Chem.*, **1**, 230 (1964); (c) M. Cousins and M. L. H. Green, *J. Chem. Soc.*, 889 (1963).

(4) It is not clear whether this step is best depicted as a concerted displacement reaction, or as one involving prior dissociation of the olefinic ligand and subsequent collapse of the resulting ion pair. Among six-coordinate complexes the latter mechanism is the prevalent path of substitution.

(5) W. P. Giering and M. Rosenblum, *J. Amer. Chem. Soc.*, **93**, 5299 (1971).

(6) S. R. Su and A. Wojcicki, *J. Organometal. Chem.*, **31**, C34 (1971).

(7) The closely related h^5-$C_5H_5M(CO)_3(h^1\text{-allyl})$ (M = Mo,[6] W[8]) as well as h^5-$C_5H_5Cr(NO)_2(h^1\text{-allyl})$[9] and ($h^1$-allylcobaloxime)–pyridine[8] also react with TCNE to give cycloaddition products. Qualitative observations suggest that the iron complex is the most reactive of these.[9]

(8) D. Wells, unpublished results.

(9) S. Raghu, unpublished results.

stabilize the positive charge in both the dipolar intermediate 2 and in the transition state leading to it.

Such (3 + 2) metal-assisted cycloaddition reactions may also be effected with cycloalkenyl derivatives. This is exemplified by the conversion of **6a–d** to the bicyclic derivatives **7a–d**.[10,11]

6 a–d, n = 1–4 **7**

Furthermore, other electrophilic olefins such as dichlorodicyanoquinone,[5] β,β-dicyano-*o*-chlorostyrene,[9,12] and dimethyl methylenemalonate[9] undergo (3 + 2) cycloaddition reactions with **4**, affording the adducts **8**,[13] **9**,[13] and **10**.

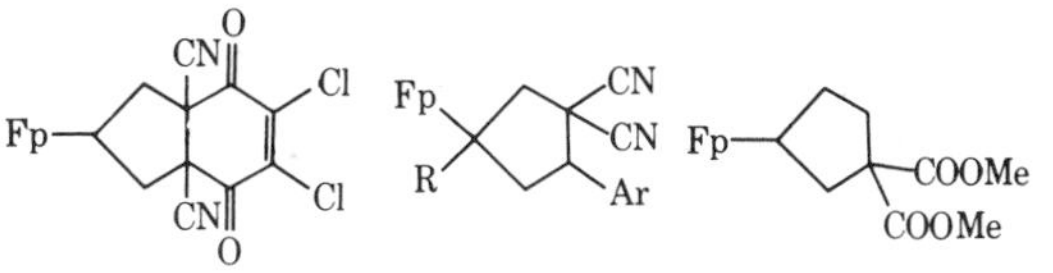

8 **9** **10**

Although neither alkyl nor phenyl isocyanate reacts with **4**, more reactive electrophiles such as toluenesulfonyl,[14] methoxysulfonyl,[14] and especially chlorosulfonyl isocyanate[14,15] enter smoothly into cycloaddition reactions with a number of (*monohapto*)allyliron complexes affording butyrolactam derivatives (**11**, **12**, and **13**).

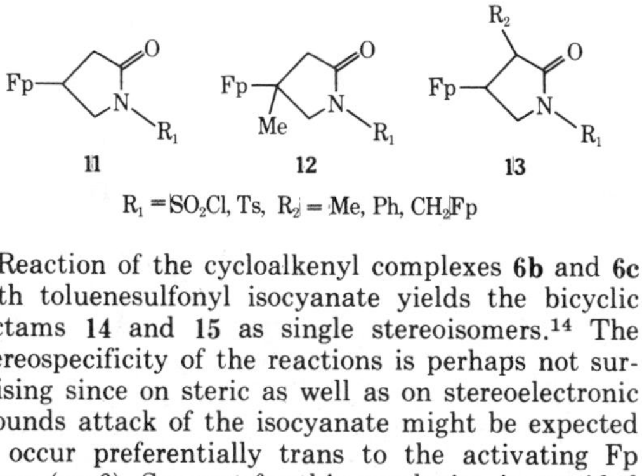

11 **12** **13**

R_1 = SO_2Cl, Ts, R_2 = Me, Ph, CH_2Fp

Reaction of the cycloalkenyl complexes **6b** and **6c** with toluenesulfonyl isocyanate yields the bicyclic lactams **14** and **15** as single stereoisomers.[14] The stereospecificity of the reactions is perhaps not surprising since on steric as well as on stereoelectronic grounds attack of the isocyanate might be expected to occur preferentially trans to the activating Fp group (eq 6). Support for this conclusion is provided by the nmr spectrum of **15** which shows triplet absorption for H-8 ($J_{1,8} = J_{5,8}$ = 5 Hz), consistent with an anti orientation for the Fp group.[16]

(10) A. Cutler, unpublished results.

(11) The reactions of substituted cyclopentenyldicarbonyl(*pentahapto*-cyclopentadienyl)iron complexes[10] and of **6** with isocyanates (*vide infra*) provide evidence for the stereochemical course of these reactions in which cycloaddition trans to the activating organometallic group is assumed.

(12) W. P. Giering, unpublished results.

(13) With the exception of **9** (R = Me) each adduct is obtained as a single stereoisomer, but their stereochemistry has not been established.

(14) W. P. Giering, S. Raghu, M. Rosenblum, A. Cutler, D. Ehntholt, and R. W. Fish, *J. Amer. Chem. Soc.*, **94**, 8251 (1972).

(15) Y. Yamamoto and A. Wojcicki, *Inorg. Nucl. Chem. Lett.*, 883 (1972); Y. Yamamoto and A. Wojcicki, *Inorg. Chem.*, **12**, 1779 (1973).

(16) E. Munck, C. S. Sodano, R. L. McLean, and T. H. Haskell, *J. Amer. Chem. Soc.*, **89**, 4158 (1967); C. W. Jefford, B. Waegell, and K. Ramey, *ibid.*, **87**, 2191 (1965); A. C. Oehlschlager and L. H. Zalkow, *J. Org. Chem.*, **30**, 4205 (1965); T. N. Margulis, L. Schiff, and M. Rosenblum, *J. Amer. Chem. Soc.*, **87**, 3269 (1965).

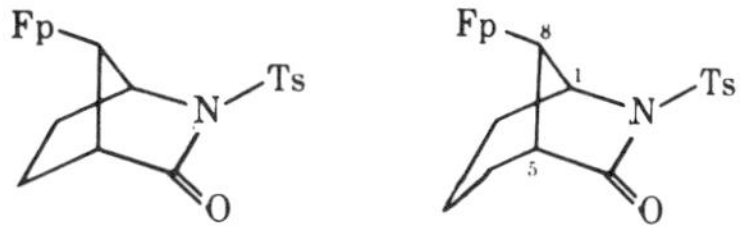

14 **15**

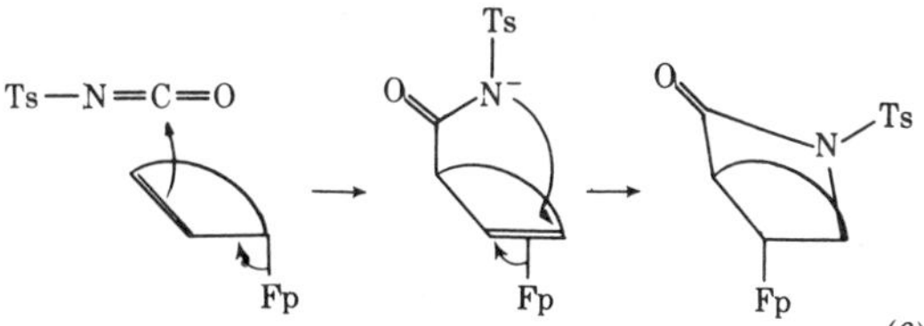

(6)

Preliminary experiments with *N*-sulfonylurethane[17] and sulfene have shown that these react readily with **4a** affording the isothiazoline dioxide **16**[14] and the sulfone **17**.[9]

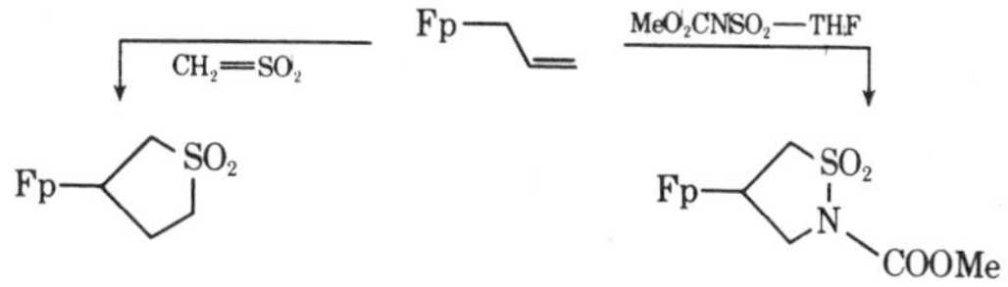

17 **16**

Synthesis of (*monohapto***Allyl**)dicarbonyl(*pentahapto***cyclopentadienyl**)**iron Complexes.** The parent complex was first prepared in 1963 by metalation of allyl chloride with the complex Fp anion.[3a] This method, which has been widely used for the preparation of many alkyl- and acyl-transition metal complexes, has been extended to the synthesis of a number of allyl–Fp and *monohapto*allyl–$M(CO)_5$ (M = Mn, Re)[2b] complexes substituted at C-1 and C-2 of the allyl chain. The parent cycloalkenyl complexes **6b–d** have also been prepared by this method in good yield from the corresponding cycloalkenyl chlorides.[10,14]

Deprotonation of the readily available cationic h^5-$C_5H_5Fe(CO)_2$(olefin) [(Fp(olefin))$^+$] complexes with tertiary amines occurs readily below room temperature and constitutes a second general route to *monohapto*allyl–Fp complexes. The deprotonation reaction appears to be highly stereospecific and to require the presence of a C–H bond trans to the metal–olefin bond. Thus, while **18a–c** are smoothly deprotonated, the cycloheptene complex **18d** is inert. Models show that, in contrast to the situation in the four-, five-, and six-membered-ring complexes, no allylic protons trans to the iron–olefin bond are available in the cycloheptene complex.[14]

A dramatic example of the stereospecificity of the deprotonation process is to be seen in the conversion of the cyclopentene complexes **19a,b** to **20a,b**, re-

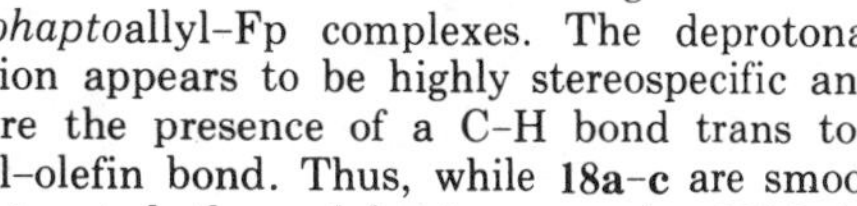

18 **19** **20**

a–d, n = 2–5

a, R = CN
b, R = SO_3H

(17) E. M. Burgess and W. M. Williams, *J. Amer. Chem. Soc.*, **94**, 4386 (1972).

spectively, notwithstanding the presence of the activating cyano and sulfonic acid groups.[14]

A number of 1-substituted allyl–Fp complexes have been prepared from the parent compound by the two-step process outlined below (eq 7), which can be carried out conveniently at low temperatures, without isolation of the intermediate cationic complex.

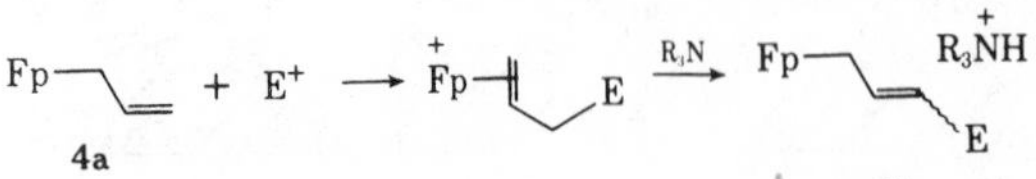

4a 21 (7)

The stereochemistry of the product **21** is highly dependent on the nature of E^+. Thus, acetylation with methyloxocarbonium tetrafluoroborate or alkylation with di- or trialkoxycarbonium ions followed by deprotonation with triethylamine gives exclusively the trans isomers **21** (E = Ac, $(RO)_2CR'$, $(RO)_3C$).[9,18] However, bromination with *N*-bromopyridinium bromide yielded only the cis bromo derivative (**21**, E = Br)[19] and alkylation with trimethyloxonium tetrafluoroborate gave an equal mixture of both cis and trans isomers (**21**, E = Me).[9] When **4a** is treated with SO_2 in the presence of $Me_3O^+BF_4^-$ the intermediate dipolar ion (**2**, E = SO_2, M = Fp) may be trapped by alkylation. Deprotonation yields the sulfone **21** (E = SO_2Me) as the trans isomer. With the exception of the bromo derivative, all of these 1-substituted allyl complexes yield normal cycloaddition products with TCNE.[9,20]

Reactions of (*monohapto*Propargyl)dicarbonyl(*pentahapto*cyclopentadienyl)iron Complexes. The general form of metal-assisted cycloaddition reactions of *monohapto*allylmetal complexes is readily extended to other systems, among them *monohapto*propargylmetal complexes.[21] These substances are available by metalation of propargyl halides or benzenesulfonates and their reaction with electrophiles may be depicted as proceeding through the formation and cyclization of a dipolar metal–allene complex (**22**;[5,22] eq 8).

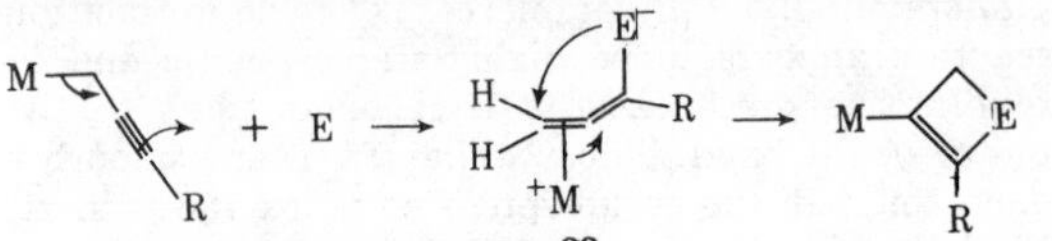

22 (8)

Among the heterocycles which have been prepared from propargyl–Fp complexes employing SO_2,[23] SO_3,[22] *N*-thionylaniline,[24] toluenesulfonyl isocyanate[14,15] and methyl *N*-sulfonylurethane[25a] are **23–27**.

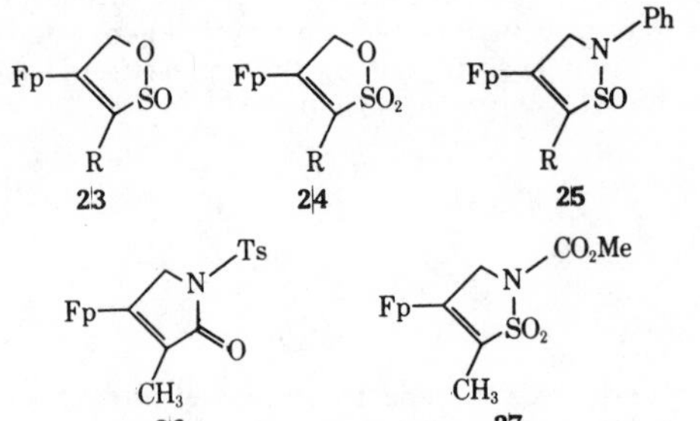

23 24 25 26 27

R = Me, Ph, CH_2Fp

Simple protonation of *monohapto*propargyl–Fp complexes allows the cationic Fp(allene) complex, postulated as an intermediate in the cycloaddition reactions, to be isolated.[25] The process is apparently highly stereospecific. Thus, protonation of **28a** affords the cis allene complex **29a** exclusively.[25a] This complex isomerizes by a first-order rate process to the more stable trans isomer **30**. The isomerization may be depicted as involving successive orthogonal migrations of the Fp group similar to those postulated to account for the averaging of methyl proton resonances in tetramethylalleneiron tetracarbonyl[26] (eq 9). The stereospecificity of the protonation reaction is best interpreted in terms of a concerted metal-assisted trans periplanar process (eq 9).

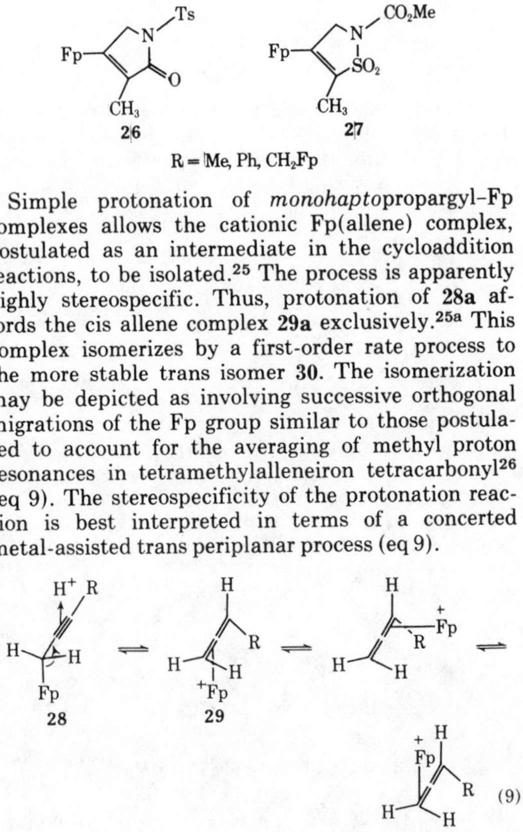

28 29 30 (9)

a, R = Me
b, R = Ph

In the reactions of the propargyl complexes with uncharged electrophiles it is undoubtedly this stereospecificity of electrophilic attack combined with distortion of the complexed allene ligand from linearity[27] (see **22**) which makes possible smooth conversion of the dipolar ion to cyclized product through trans addition of the anion to the coordinated double bond (eq 8).

Reactions of (*monohapto*Allenyl)dicarbonyl(*pentahapto*cyclopentadienyl)iron Complexes. Only the parent complex **31** is known, and this is readily prepared by the metalation of propargyl bromide or

(18) D. Ehntholt, unpublished results.

(19) K. Nicholas, unpublished results.

(20) The adduct derived from the (*monohapto*-1-bromoallyl)dicarbonyl(*pentahapto*cyclopentadienyl)iron complex (**21**, E = Br) is a β-bromoalkyldicarbonyl(*pentahapto*cyclopentadienyl)iron complex which apparently decomposes spontaneously by elimination of FpBr.

(21) In point of chronology the cycloaddition reactions of (*monohapto*propargyl)dicarbonyl(*pentahapto*cyclopentadienyl)iron complexes were reported before those of (*monohapto*allyl)dicarbonyl(*pentahapto*cyclopentadienyl)iron complexes, but the mechanism of these processes was apparently not recognized at the time. See, for example, ref 23.

(22) D. W. Lichtenberg and A. Wojcicki, *J. Organometal. Chem.*, **33**, C77 (1971).

(23) J. E. Thomasson, P. W. Robinson, D. A. Ross, and A. Wojcicki, *Inorg. Chem.*, **10**, 2130 (1971); M. Churchill, T. Wormald, D. A. Ross, J. E. Thomasson, and A. Wojcicki, *J. Amer. Chem. Soc.*, **92**, 1795 (1970).

(24) P. W. Robinson and A. Wojcicki, *Chem. Commun.*, 951 (1970).

(25) (a) S. Raghu and M. Rosenblum, *J. Amer. Chem. Soc.*, **95**, 3060 (1973); (b) D. W. Lichtenberg and A. Wojcicki, *ibid.*, **94**, 8271 (1972); (c) J. Benaim, J. Merour, and J. Roustan, *C. R. Acad. Sci., Ser. C*, **272**, 789 (1971).

(26) R. Ben-Shoshan and R. Pettit, *C. R. Acad. Sci., Ser. C*, **89**, 2231 (1967).

(27) T. G. Hewitt, K. Anzenhofer, and J. J. Deboer, *J. Organometal. Chem.*, **18**, P19 (1969); P. Racanelli, G. Paritini, A. Immirz, G. Allegra, and L. Povic, *Chem. Commun.*, 361 (1969); T. Kashiwagi, N. Yasuoka, N. Kasai, and M. Kukudo, *ibid.*, 317 (1969).

benzenesulfonate with the Fp anion.[28] While, in principle, electrophilic attack may occur at C-1 or C-3 of the allenic ligand, the latter site appears to be preferred. Protonation with $HPF_6 \cdot Et_2O$ at −20° gives the cationic acetylene complex **32** as a very air- and water-sensitive substance. With uncharged electrophiles such as TCNE, *β,β*-dicyano-*o*-chlorostyrene, and toluenesulfonyl isocyanate, the cycloaddition products (**33, 34**) are formed, the latter two defining the site of initial electrophilic attack at C-3 of the allenic chain. With *N*-carbomethoxysulfonylamine, the bidentate anion in the dipolar intermediate cyclizes preferentially through oxygen to give the oxathiazepine derivative **35**, the first member of this class of heterocycle.

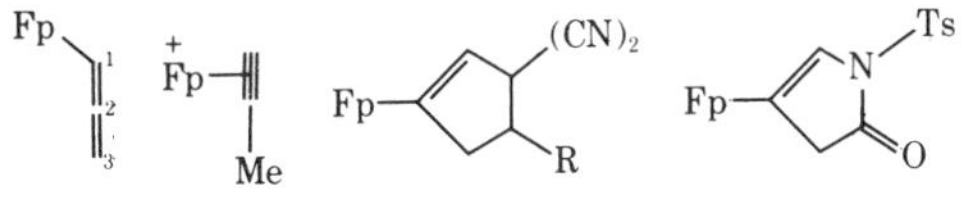

31 **32** **33** **34**

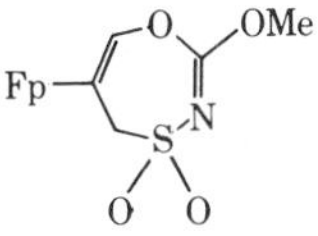

35

R = $(CN)_2$, *o*-C_6H_4Cl

Reactions of (*monohapto*Cyclopropylmethyl)dicarbonyl(*pentahapto*cyclopentadienyl)iron Complexes. Activation of metal-bonded ligands toward electrophilic attack is not confined to those having centers of unsaturation, but may also be observed with ligands possessing a strained single bond. The parent cyclopropylmethyl–Fp complex (**36**), prepared from cyclopropylmethyl tosylate and the complex anion, reacts with TCNE and with SO_2 to give **38** and **39**, respectively.[5] Each of these products is apparently derived from metal-assisted electrophilic attack on the cyclopropyl ring, followed by closure of the dipolar ion **37**, homologous to that generated with the allyl complex (eq 10).

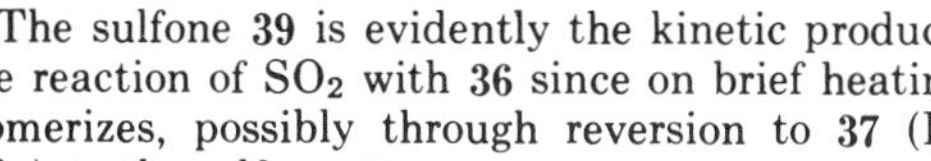

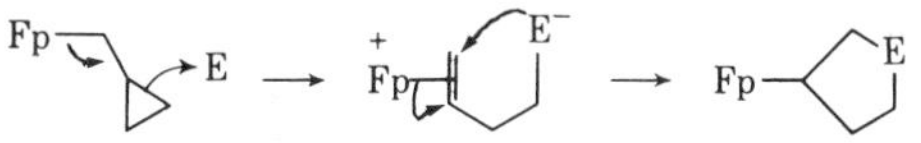

36 **37** (10)

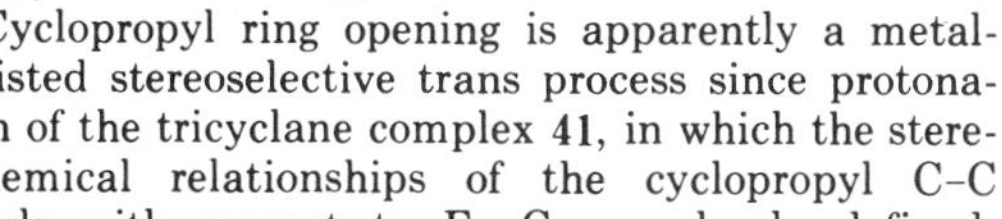

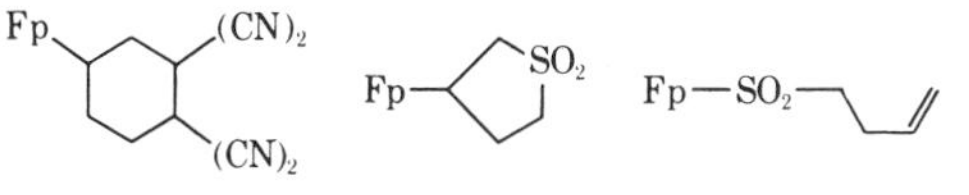

38 **39** **40**

The sulfone **39** is evidently the kinetic product in the reaction of SO_2 with **36** since on brief heating it isomerizes, possibly through reversion to **37** (E = SO_2), to the sulfone **40**.

Cyclopropyl ring opening is apparently a metal-assisted stereoselective trans process since protonation of the tricyclane complex **41**, in which the stereochemical relationships of the cyclopropyl C–C bonds with respect to Fp–C are clearly defined, yields only the exo norbornene complex **42** (60%), derived by cleavage of C-2–C-6 (eq 11). However, the endo isomer is unknown, and consequently its formation and decomposition in the course of the reaction cannot be excluded.

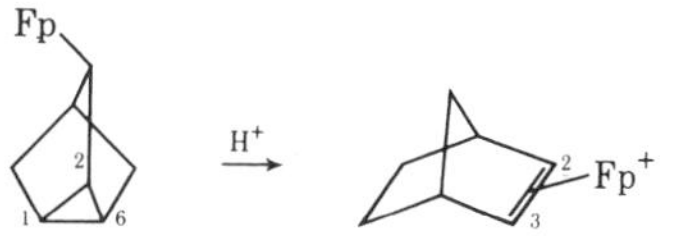

41 **42** (11)

Reactions of (*monohapto*Cyclopropyl)dicarbonyl(*pentahapto*cyclopentadienyl)iron. Two such substances, the parent compound **43** and the norcarane complex **44**, have been prepared either by metalation of cyclopropyl bromide with Fp^- or by alkylation of FpBr with cyclopropyllithium.[29] The reactions of these substances with electrophiles differ significantly from those of the isomeric *monohapto*allyl–Fp complexes.

43 **44**

Although protonation of **43** gives $Fp(propene)^+$, as does the isomeric *monohapto*allyl–Fp complex **4a**, the reaction evidently does not proceed through cleavage of C-2–C-3 and migration of the Fp group since the orthogonality of the Fp–C-1 and C-2–C-3 bonds precludes a concerted metal-assisted process. Thus, treatment of the 1-deuterio derivative (**43-*d***) with HBF_4 does not yield the 2-deuterio cation (**45b**) to be expected from a one-step synchronous reaction. The product is instead the 1-deuterio derivative, and more strikingly it is exclusively the cis isomer **45a**.

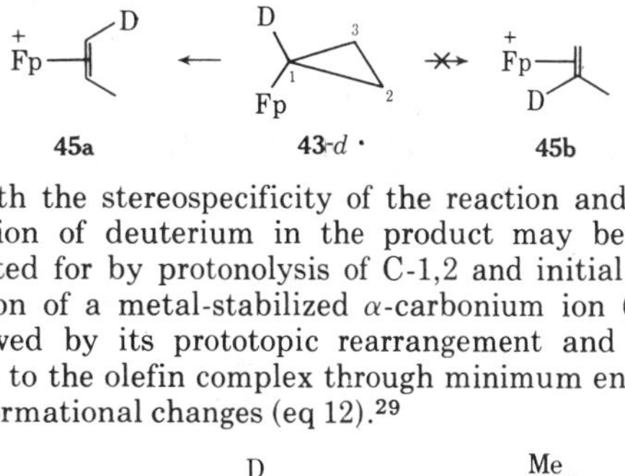

45a **43-*d*** **45b**

Both the stereospecificity of the reaction and the position of deuterium in the product may be accounted for by protonolysis of C-1,2 and initial formation of a metal-stabilized *α*-carbonium ion (**46**), followed by its prototopic rearrangement and collapse to the olefin complex through minimum energy conformational changes (eq 12).[29]

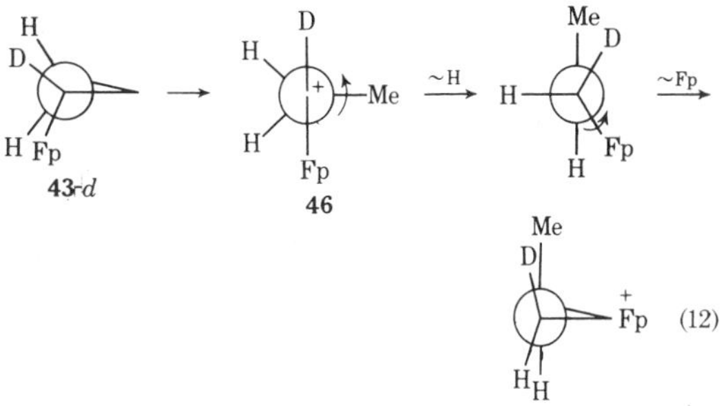

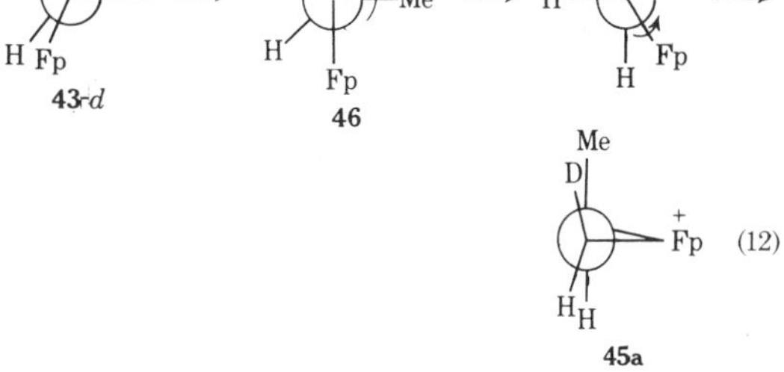

(12)

Metal-stabilized *α*-carbonium ions (**47**) are apparently also formed as intermediates in the reactions of the neutral electrophiles TCNE and SO_2 with **43**,

(28) J. Roustan and P. Cadiot, *C. R. Acad. Sci., Ser. C*, **268**, 734 (1969); P. W. Jolly and R. Pettit, *J. Organometal. Chem.*, **12**, 491 (1968); M. D. Johnson and C. Mayle, *Chem. Commun.*, 192 (1969).

(29) A. Cutler, R. W. Fish, W. P. Giering, and M. Rosenblum, *J. Amer. Chem. Soc.*, **94**, 4354 (1972).

which lead to the products **48** and **49**.[29] A small amount of the allyl sulfone **51** is also formed in the latter reaction through prototopic rearrangement of **47** (E = SO_2^-) to **50**, a process evidently competitive with the closure of **47** to the sultine **49**.

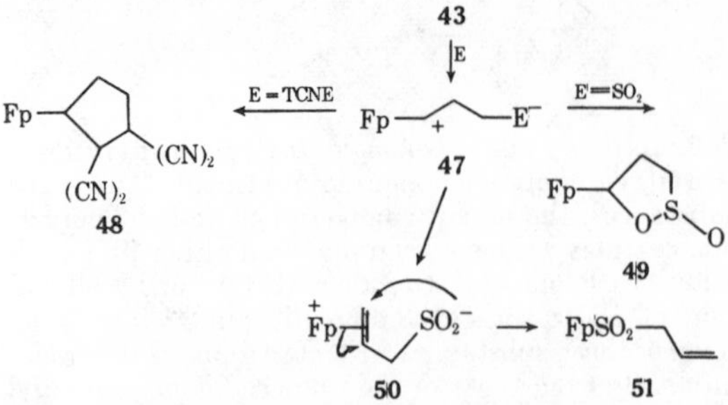

Reactions of Cationic Fp(olefin) Complexes

The addition of nucleophiles to metal-coordinated olefins and acetylenes is a reaction widely encountered in organometallic chemistry.[30] Since the second step of the cycloaddition reactions described above involves such a process, we were prompted to examine the chemistry of Fp(olefin) cations and in particular their reactions with nucleophiles.

Preparation of Fp(olefin) Cations. Several methods are available for the synthesis of these complexes. A number of these cations were first prepared by the reaction of dicarbonylcyclopentadienyliron bromide (FpBr) with simple olefins in the presence of Lewis acids,[31] by protonation of *monohapto*allyl-Fp complexes[3a] or by hydride abstraction from alkyl-Fp complexes.[32]

Two other methods have more recently augmented these. The first makes use of the Fp(isobutylene) cation **52**, which is readily prepared by protonation of methallyl-Fp **4b**. The reagent can be stored for prolonged periods at 0° without decomposition. When chlorocarbon solutions of this complex are heated briefly at 60° in the presence of an excess of olefin, exchange occurs leading to the formation of an Fp(olefin) cation (eq 13).[33]

(13)

52

The reaction is necessarily limited to the preparation of those olefin complexes which are thermally stable in solution under conditions used to effect exchange with the isobutylene complex.[34]

An alternative method which does not suffer from this limitation, and which may also be used for the preparation of olefin complexes having functional groups, makes use of epoxides as starting materials.[35] Treatment of these with the Fp anion affords alkoxy-Fp complexes, and these on protonation *in situ* and at low temperatures are converted to the corresponding Fp(olefin) cations. Equation 14 illustrates this transformation for ethylene oxide.

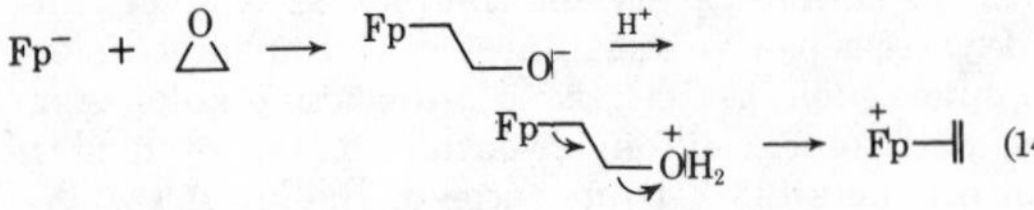

(14)

The reaction sequence has been applied successfully to a variety of cyclic and acyclic epoxides as well as to acrolein and crotonic ester epoxide and may also be used to prepare relatively unstable cations such as Fp(*trans*-stilbene)+. The transformation of epoxide to olefin complex proceeds with retention of configuration and constitutes an effective method for the stereospecific reduction of epoxides to olefins, since the cationic olefin complexes are readily decomposed on brief treatment with iodide, liberating the olefin.[35]

Nucleophilic Additions to Fp(olefin) Cations. Nucleophiles react with Fp(olefin) cations by one or more of three pathways. Direct addition of nucleophile to the complexed double bond is a very general process, but one which may be reversible with uncharged nucleophiles. Displacement of the olefin by the nucleophile as well as reductive processes resulting in the formation of dicarbonylcyclopentadienyliron dimer (Fp_2) often compete with the addition reaction. These latter two reactions generally become important and often predominant with reactive anionic nucleophiles, such as alkyl-Grignard or lithio reagents. However, lithium enolates, generated in THF solution with lithium bis(trimethylsilyl)amide,[36] add smoothly at −78° to a variety of acyclic and cyclic Fp(olefin) cations, affording neutral adducts such as **53** and **54**.[30,37]

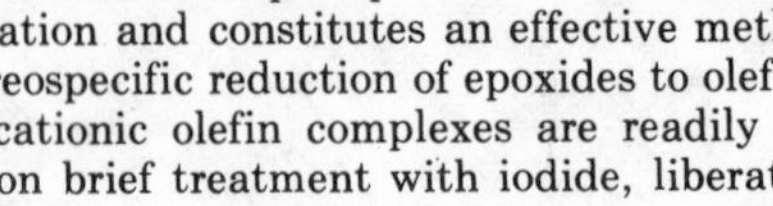

53　　**54**, $n = 2, 3$

R = $CH_2(COOMe)_2$; CH(COMe)COOEt; CMe(COOEt)$_2$

Dimethyl malonate adds to the Fp(propene) cation **55b** to give a 1:1 mixture of the adducts **56a** and **57a**, but the styrene complex **55c** reacts with greater regiospecificity, affording only **56b** in high yield.[38]

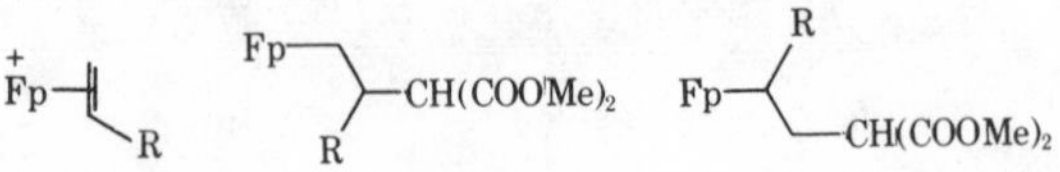

55a, R = H　　**56a**, R = Me　　**57a**, R = Me
b, R = Me　　**b**, R = Ph
c, R = Ph

Cyclohexanone pyrrolidine enamine reacts rapidly at 0° with cationic Fp(olefin) complexes, affording adducts such as **58** and **59** in high yield. With pro-

(30) For leading references see A. M. Rosan, M. Rosenblum, and J. Tancrede, *J. Amer. Chem. Soc.*, **95**, 3062 (1973).

(31) E. O. Fischer and K. Fichtel, *Chem. Ber.*, **94**, 1200 (1961); **95**, 2063 (1962).

(32) M. L. H. Green and P. L. I. Nagy, *J. Organometal. Chem.*, **1**, 58 (1963).

(33) W. P. Giering and M. Rosenblum, *Chem. Commun.*, 441 (1971).

(34) The mechanism of this exchange reaction is not known, but may involve dissociation of the isobutylene complex rather than displacement of one ligand by another.

(35) W. P. Giering, M. Rosenblum, and J. Tancrede, *J. Amer. Chem. Soc.*, **94**, 7170 (1972).

(36) M. W. Rathke, *J. Amer. Chem. Soc.*, **92**, 3222 (1970).

(37) The trans stereochemical assignments made for the cyclic adducts are well supported by the general course of additions of nucleophiles to metal coordinated olefins: A. Panunzi, A. De Renzi, and G. Parao, *J. Amer. Chem. Soc.*, **92**, 3489 (1970); J. K. Stille and R. H. Morgan, *ibid.*, **88**, 5135 (1966); M. Green and R. I. Hancock, *J. Chem. Soc. A*, 2054 (1967); C. B. Anderson and B. J. Burreson, *J. Organometal. Chem.*, **7**, 181 (1967); J. K. Stille and D. B. Fox, *Inorg. Chem. Lett.*, **5**, 157 (1969).

(38) A. M. Rosan, unpublished results.

pene and styrene complexes, the regiospecificity of these reactions parallels that observed with enolate anions. By contrast, the reactions of heteroatomic nucleophiles are highly regiospecific. The adducts **60** and **61** are the exclusive products obtained with these nucleophiles and the propene or styrene complexes.

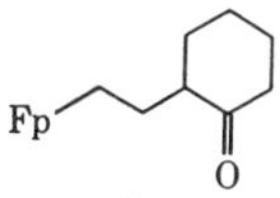

58

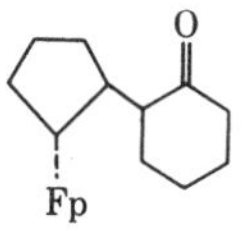

59

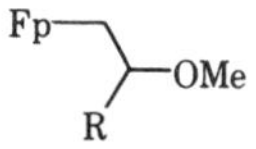

60, R = Me, Ph

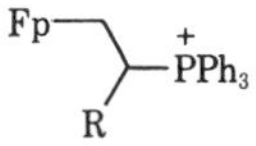

61, R = Me, Ph

Attempts to employ complexes **61** or the phosphonate **63**, derived from **62**, as Wittig reagents have thus far been unsuccessful, but the stabilized phosphorane, prepared from the adduct **64**, reacts with benzaldehyde to give **65**.[38]

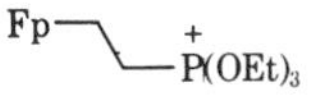

62

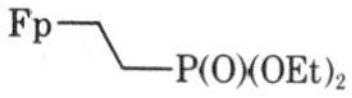

63

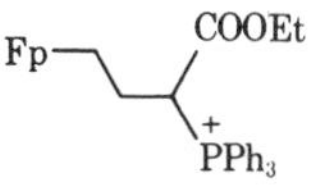

64

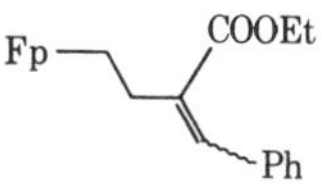

65

Organometallic Condensation Reactions. The combination of organometallic electrophile and nucleophile leads to a new condensation reaction. Thus, an equimolar mixture of **4a** and **55a** condenses at room temperature to give the dinuclear complex **66** in 60% yield.[30]

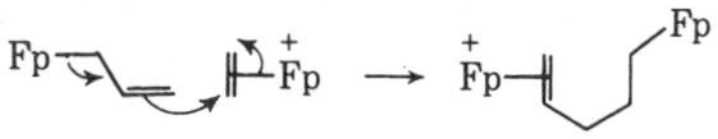

4a **55** **66** (15)

Similar condensations may be effected with **4a** or the cyclopentenyl complex **6a** as donor and the cationic ethylene, acrolein, or crotonic ester complexes as acceptors, yielding products such as **67–69**.

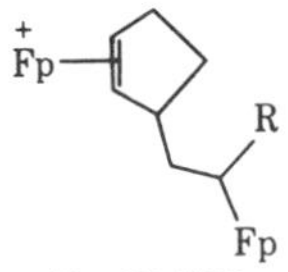

R = H, CHO

67

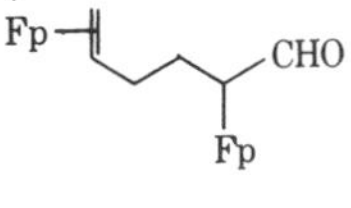

68

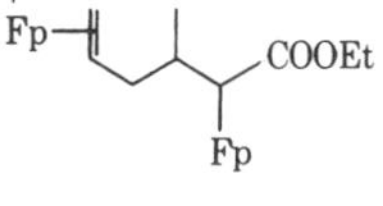

69

Each of the Fp groups in the dinuclear condensation products may be selectively removed. Thus, brief treatment of **66**, **67** (R = H), **68**, or **69** with sodium iodide in acetone solution at room temperature yields the mononuclear complexes **70–73**, while short contact of **66** with HCl at room temperature in methylene chloride solution affords **74**.

When the acceptor component is the Fp(1,3-butadiene) cation **75**, the condensation with allyl-Fp (**4a**) apparently proceeds through initial formation of **76**

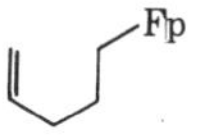

70

71

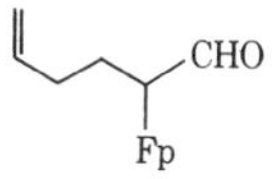

72

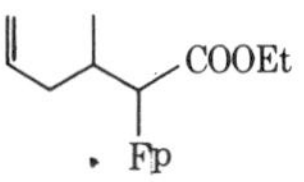

73

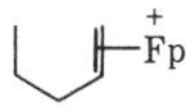

74

affording **78** and **79**.[39] These products were isolated as the neutral mononuclear complexes **81** and **82** in 40% yield after treatment of the reaction mixture with sodium iodide. The facile cyclization of **76** and **77** is not surprising since these intermediates contain both donor and acceptor components suitably disposed for intramolecular reaction.

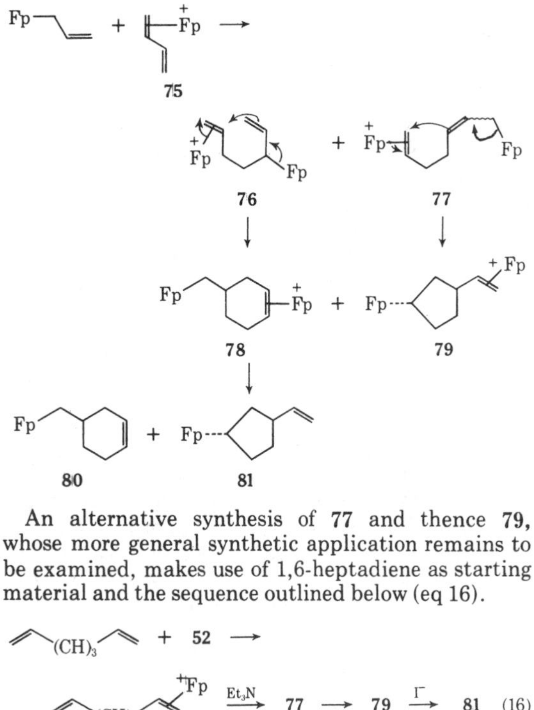

75 **76** **77** **78** **79** **80** **81**

An alternative synthesis of **77** and thence **79**, whose more general synthetic application remains to be examined, makes use of 1,6-heptadiene as starting material and the sequence outlined below (eq 16).

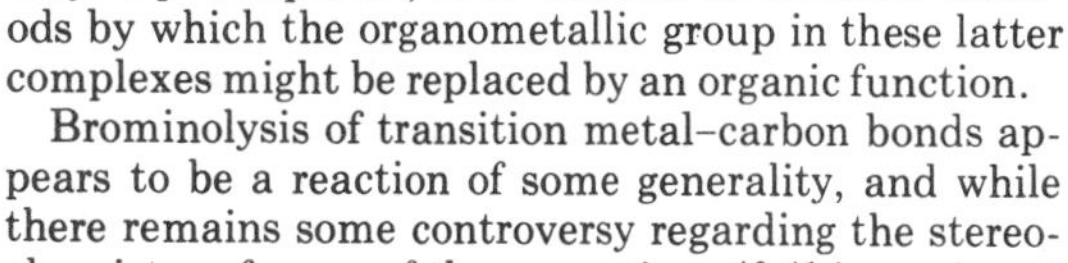

(16)

Demetalation Reactions

Since many of the reactions of *monohapto*allyl-Fp and Fp(olefin) complexes lead to the formation of alkyl-Fp complexes, it is natural to consider methods by which the organometallic group in these latter complexes might be replaced by an organic function.

Brominolysis of transition metal–carbon bonds appears to be a reaction of some generality, and while there remains some controversy regarding the stereochemistry of some of these reactions,[40,41] inversion of

(39) It is possible that the exclusive initial product is 76 and that 77 is formed from it by an allylic rearrangement.

(40) R. G. Pearson and W. R. Muir, *Inorg. Chem. Lett.*, **92**, 5519 (1970); R. W. Johnson and R. G. Pearson, *Chem. Commun.*, 986 (1970); J. A. Labinger, R. J. Baus, D. Dolphin, and J. H. Osborn, *ibid.*, 612 (1970); F. R. Jensen, V. Madan, and D. H. Buchanan, *J. Amer. Chem. Soc.*, **93**, 5283 (1971).

configuration in the reaction of bromine with a primary Fp-C bond appears well established.[42] It has recently been suggested that these and other electrophilic substitution reactions of metal-alkyl complexes proceed by nucleophilic attack on the oxidized complex (eq 17).[41]

$$M{-}R \xrightarrow{X_2} \overset{+\bullet}{M}{-}R + X^- \longrightarrow RX + M \qquad (17)$$

The brominolysis of Fp-C bonds is a very rapid reaction which may be carried out employing either bromine, pyridinium perbromide, or bromopyridinium bromide. So rapid is the reaction that it may be effected selectively at −78° in the presence of an olefin bond, as is illustrated by the conversion of **82** to **83**[19] (eq 18).

(18)

82 **83**

The conversion of transition metal-alkyl complexes to metal alkyl sulfone complexes by treatment with SO_2 has been examined for a number of complexes[43] and shown to proceed with inversion of configuration with FpCHDCHD(*t*-Bu).[42] By contrast, cleavage of the Fe-C bond in this latter complex with $HgCl_2$ proceeds with retention of configuration in the product, RHgCl.[42]

One-step replacement of the Fp group by either a carboxylic ester or acid function has more recently been achieved, through oxidation of alkyl-Fp complexes with either cupric chloride,[44] ferric chloride,[19] dichlorodicyanoquinone,[19] or ceric salts.[45] When the reaction is carried out in alcohol solution, the ester is obtained in good yield,[44,45] while in aqueous acetone the acid is formed.[19] It seems likely that the reaction proceeds through the oxidized form of the iron complex (**84**),[44,45] and that ligand transfer (R to carbonyl) in this intermediate is facilitated by the cationic charge which diminishes metal-carbonyl back-bonding and increases the positive charge at the carbonyl carbon atom. Nucleophilic displacement at the carbonyl group of the rearranged cation **85** is also likely to be facilitated by the positive charge on the metal (eq 19).

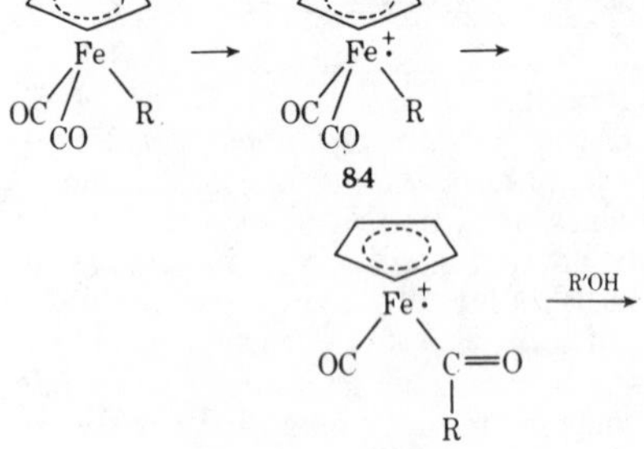

84 **85** (19)

The reaction has been applied successfully to both primary and secondary alkyl-Fp complexes and appears to be highly stereospecific. Thus, oxidative carboxylation of *cis*- and *trans*-4-methylcyclohexyl-Fp (**86** and **87**) with cupric chloride in ethanol solution gave the corresponding cis and trans ethyl esters, respectively (**88** and **89**).[44]

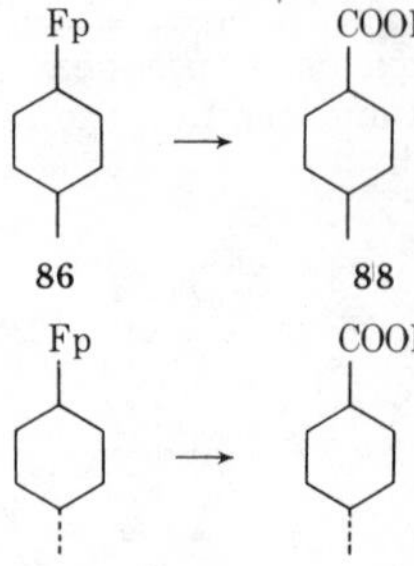

86 **88** **87** **89**

Much of the chemistry described in this Account represents the work of my research associates and students during the past 3 years. It is a pleasure to acknowledge my indebtedness to Dr. Warren P. Giering who began this research, and to Drs. Daniel Ehntholt, R. W. Fish, S. Raghu, Kenneth Nicholas, Darrell Wells, Messrs. Alan Cutler, Alan Rosan, and Jean Tancrede who have carried it forward so ably and with such spirit. The research has been supported by grants from the National Institutes of Health, by the National Science Foundation, and by the Army Research Office, Durham, which I gratefully acknowledge. This paper was written while I was on leave at the Hebrew University, Jerusalem, Israel. I wish to express my appreciation to the members of the Institute of Chemistry for their hospitality.

Addendum (May 1976)

Several further applications and explorations of the chemistry of (η^1-allyl) Fp and Fp (olefin) complexes have recently been reported. These include the use of the methyl vinyl ketone complex **90** as a component in regiospecific electroneutral and cationic Michael-type reactions:[46]

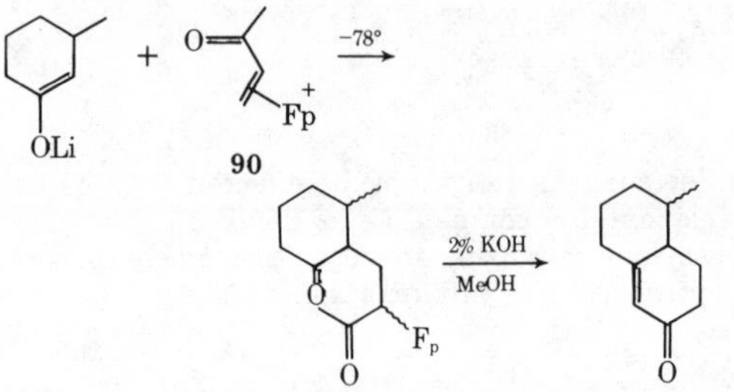

90 (20)

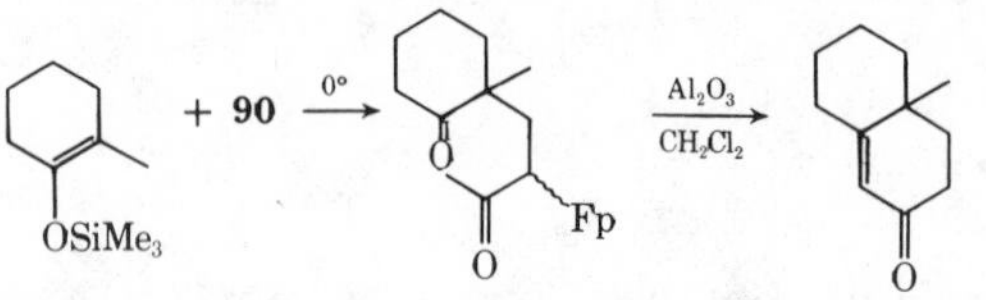

(21)

(41) S. N. Anderson, D. M. Ballard, J. Z. Chrzastowski, D. Dodd, and M. D. Johnson, *J. Chem. Soc., Chem. Commun.*, 685 (1972).

(42) G. M. Whitesides and D. J. Boschetto, *J. Amer. Chem. Soc.*, **93**, 1529 (1971).

(43) J. Bibler and A. Wojcicki, *J. Amer. Chem. Soc.*, **88**, 4862 (1966); A. Wojcicki and F. A. Hartman, *ibid.*, **88**, 844 (1966); F. A. Hartman and A. Wojcicki, *Inorg. Chem.*, **7**, 1504 (1968); M. Giaziani, J. Bibler, R. M. Montesano, and A. Wojcicki, *J. Organometal. Chem.*, **16**, 507 (1969).

(44) K. M. Nicholas and M. Rosenblum, *J. Amer. Chem. Soc.*, **95**, (1973).

As shown below, the preparation of hydroazulene complexes from tropyliumiron tricarbonyl (**91**) and η^1-allyl (Fp) (**92**) complexes has also been reported.[47]

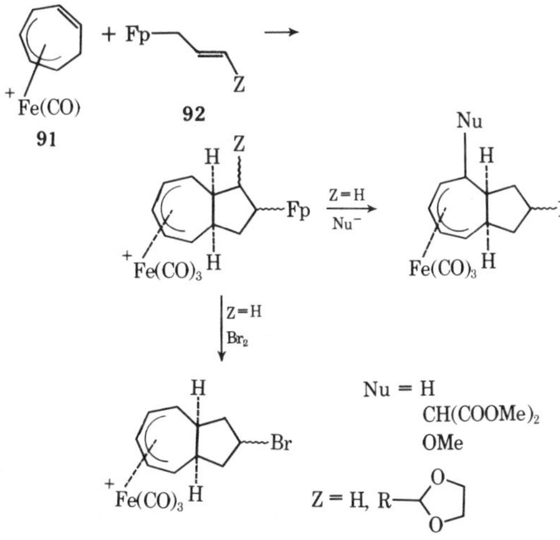

The sequence (eq. 14) by which epoxides may be transformed to Fp (olefin) cations and thence to free olefins with overall retention of configuration has now been shown to be applicable in deoxygenation of epoxides with *inversion* of stereochemistry.[48]

The regiospecificity and reversibility of the addition of several O, N, P, and S nucleophiles to monosubstituted Fp (olefin) cations has been examined.[49] A full paper describing the preparation of a number 1-substituted (η^1-allyl) Fp complexes (eq. 7) as appeared.[50] A full account of the preparation of (η^1-allyl) M complexes and their cycloaddition reactions has also been given.[51] A convenient synthesis of Fp(η^2-vinyl ether) and Fp(η^2-ketene acetal) cations has been described, and the structures of these cations are discussed.[52]

The use of the Fp cation as a protecting group for olefins has been reported.[53] Bromination, catalytic hydrogenation, and acetoxymercuration of unprotected centers of unsaturation may be carried out in the presence of the cationic Fp (olefin) function.

The preparation of the benzocyclobutadiene complex **93**,[54] the cyclobutadiene complex **94**,[55] and the dinuclear cyclobutadiene complex **95**[56] has been reported. Cation **96**[57] has been shown to be a useful hydride abstracting reagent.[58]

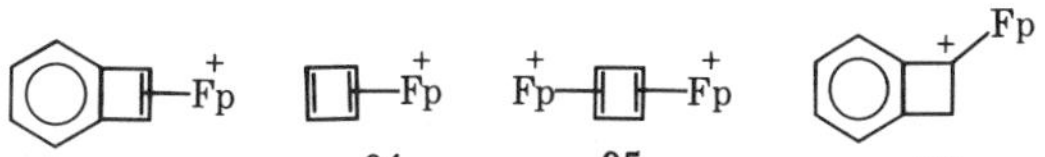

93 **94** **95** **96**

(45) S. N. Anderson, C. W. Fong, and M. D. Johnson, *J. Chem. Soc., Chem. Commun.*, 163 (1973).

(46) A. Rosan and M. Rosenblum, *J. Org. Chem.*, **40**, 3621 (1975).

(47) N. Genco, D. Marten, S. Raghu, and M. Rosenblum, *J. Am. Chem. Soc.*, **98**, 848 (1976).

(48) M. Rosenblum, M. R. Saidi, and M. Madhavarao, *Tetrahedron Lett.*, 4009 (1975).

(49) P. Lennon, M. Madhavarao, A. Rosan, and M. Rosenblum, *J. Organomet. Chem.*, **108**, 93 (1976).

(50) A. Cutler, D. Ehntholt, P. Lennon, K. M. Nicholas, D. F. Marten, M. Madhavarao, S. Raghu, A. Rosan, and M. Rosenblum, *J. Am. Chem. Soc.*, **97**, 3149 (1975).

(51) A. Cutler, D. Ehntholt, W. P. Giering, P. Lennon, S. Raghu, A. Rosan, M. Rosenblum, J. Tancrede, and D. Wells, *J. Am. Chem. Soc.*, **98**, 3495 (1976).

(52) A. Cutler, S. Raghu, and M. Rosenblum, *J. Organomet. Chem.*, **77**, 381 (1974).

(53) K. M. Nicholas, *J. Am. Chem. Soc.*, **97**, 3254 (1975).

(54) A. Sanders, C. V. Magatti, and W. P. Giering, *J. Am. Chem. Soc.*, **96**, 1610 (1974).

(55) A. Sanders and W. P. Giering, *J. Am. Chem. Soc.*, **97**, 919 (1975).

(56) A. Sanders and W. P. Giering, *J. Am. Chem. Soc.*, **96**, 5247 (1974).

(57) A. Sanders, L. Cohen, W. P. Giering, D. Kenedy, and C. V. Magatti, *J. Am. Chem. Soc.*, **95**, 5430 (1973).

(58) L. Cohen, W. P. Giering, D. Kenedy, C. V. Magatti, and A. Sanders, *J. Organomet. Chem.*, **65**, C57 (1974).

Approaches to the Synthesis of Pentalene *via* Metal Complexes

Selby A. R. Knox* and F. Gordon A. Stone*

Department of Inorganic Chemistry, The University, Bristol BS8 ITS, England

Received February 27, 1974

Reprinted from ACCOUNTS OF CHEMICAL RESEARCH, *7, 321* (*1974*)

Pentalene, which is predicted[1] to have in the ground state the polyolefin structure **1**, is unstable,

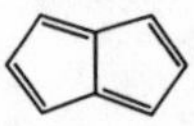

1

and attempts to prepare the compound by conventional organic synthetic procedures have failed. This is presumably due to its ready dimerization since 1-methylpentalene, identified from its ultraviolet and infrared spectra at −196°, was observed to dimerize at temperatures above −140°.[2] Moreover, although photolysis at −196° of the dimer of 1,3-dimethylpentalene led to spectroscopic identification of the monomer, quantitative regeneration of the dimer occurred on warming to 20°.[3] The presence of bulky substituents on the pentalene system appears to sterically hinder dimerization, and monomeric hexaphenyl-,[4] bis(1,3-dimethylamino)-,[5] and 1,3,5-tri-*tert*-butylpentalenes[6] exhibit fair stability.

Similar behavior is shown by cyclobutadiene, which is detectable only in a matrix at low temperatures.[7-10] As with pentalene, bulky substituents enhance stability, and tri-*tert*-butylcyclobutadiene can be prepared at −70°, although it decomposes on warming to room temperature.[11]

One of the most interesting aspects of the development of organometallic chemistry has been the discovery that transition metals can form stable complexes with organic molecules which are unstable under normal conditions.[12] A classic example is provided by the work of Pettit and his coworkers[13] on cyclobutadiene. Decades of attempts to synthesize this highly reactive hydrocarbon by classical methods were unsuccessful, but in 1965 a stable tricarbonyliron complex (**2**) was obtained by dehalogenation of 3,4-dichlorocyclobut-1-ene with $Fe_2(CO)_9$.[14]

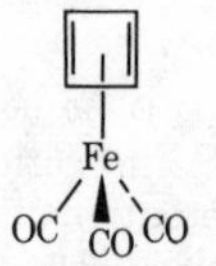

2

Subsequent release of cyclobutadiene from **2** with ceric ion illustrated the problem facing the early workers, for dimerization occurs at very low temperature. Considerable chemistry of cyclobutadiene is now known since **2** can be used as a source both of the free hydrocarbon and of substituted derivatives *via* electrophilic attack on the coordinated hydrocarbon.[13]

In view of the isolation of **2** it is not surprising that attention should have been given to the stabilization and characterization of pentalene through complexation with transition metals. Katz and coworkers[15]

Gordon Stone was born in Exeter, England, and received his Ph.D. degree from Cambridge University in 1952. After 2 years as a Fulbright Scholar at the University of Southern California, he spent several years at Harvard University, where his interests in organometallic chemistry began. He is now Professor of Inorganic Chemistry at Bristol University. His research has been marked by its breadth covering both the chemistry of the main-group elements and that of the transition metals.

Selby Knox was born in Newcastle-upon-Tyne, England, in 1944, and obtained B.Sc. and Ph.D. degrees from the University of Bristol. After postdoctoral work at the University of California, Los Angeles, with Professor Herbert D. Kaesz, he became a Lecturer in inorganic chemistry at Bristol University in 1972.

(1) N. C. Baird and R. M. West, *J. Amer. Chem. Soc.*, **93**, 3072 (1971).

(2) R. Block, R. A. Marty, and P. de Mayo, *J. Amer. Chem. Soc.*, **93**, 3071 (1971).

(3) K. Hafner, R. Dönges, E. Goedecke, and R. Kaiser, *Angew. Chem., Int. Ed. Engl.*, **12**, 337 (1973).

(4) E. L. Goff, *J. Amer. Chem. Soc.*, **84**, 3975 (1962).

(5) K. Hafner, K. F. Bangert, and V. Orfanos, *Angew. Chem., Int. Ed. Engl.*, **6**, 451 (1967).

(6) W. Weidemüller and U. Süss, *Angew. Chem., Int. Ed. Engl.*, **12**, 575 (1973).

(7) C. Y. Lin and A. Krantz, *J. Chem. Soc., Chem. Commun.*, 1111 (1972); A. Krantz, C. Y. Lin, and M. D. Newton, *J. Amer. Chem. Soc.*, **95**, 2744 (1973).

(8) S. Masamune, M. Suda, H. Ona, and L. M. Leichter, *J. Chem. Soc., Chem. Commun.*, 1268 (1972).

(9) G. Maier and B. Hoppe, *Tetrahedron Lett.*, 861 (1973); G. Maier and M. Schneider, *Angew. Chem., Int. Ed. Engl.*, **10**, 809 (1971); G. Maier, G. Fritschi, and B. Hoppe, *ibid.*, **9**, 529 (1970).

(10) O. L. Chapman, C. L. McIntosh, and J. Pacansky, *J. Amer. Chem. Soc.*, **95**, 614 (1973); O. L. Chapman, D. De La Cruz, R. Roth, and J. Pacansky, *ibid.*, **95**, 1337 (1973).

(11) S. Masamune, N. Nakamura, M. Suda, and H. Ona, *J. Amer. Chem. Soc.*, **95**, 8481 (1973).

(12) F. G. A. Stone, *Nature* (*London*), **232**, 534 (1971).

(13) R. Pettit, *Pure Appl. Chem.*, **17**, 253 (1968).

(14) L. Watts, J. D. Fitzpatrick, and R. Pettit, *J. Amer. Chem. Soc.*, **87**, 3253 (1965); J. D. Fitzpatrick, L. Watts, G. F. Emerson, and R. Pettit, *ibid.*, **87**, 3254 (1965).

(15) T. J. Katz and M. Rosenberger, *J. Amer. Chem. Soc.*, **84**, 865 (1962); T. J. Katz, M. Rosenberger, and R. K. O'Hara, *ibid.*, **86**, 249 (1964).

first synthesized the dianion 3, an aromatic ten-π-

3

electron system, from dihydropentalene. These investigators[16] treated various transition metal halides with the dianion in reactions paralleling the successful synthesis of *pentahapto*cyclopentadienyl complexes from metal salts and the cyclopentadienide ion. However, an X-ray crystallographic study[17] has shown that an iron complex, $Fe(C_8H_6)_2$, formed from ferrous chloride, is in actuality a derivative (4) of fer-

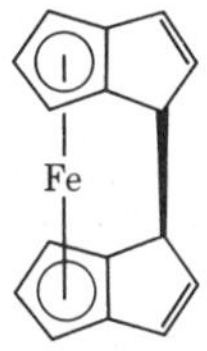

4

rocene, with a carbon–carbon bond joining the pair of C_5 rings not sandwiching the iron atom. Isoelectronic $[Co(C_8H_6)_2]^+$ probably has a similar structure. Binuclear metal complexes $Ni_2(C_8H_6)_2$ and $Co_2(C_8H_6)_2$ are obtained from $NiCl_2$ or $CoCl_2$ and 3, but in the absence of single-crystal X-ray crystallographic studies their nature is unclear. Reaction of the dianion with $[\eta^3\text{-}C_3H_5NiCl]_2$, however, affords[18] a product which has been shown[19] by single-crystal X-ray diffraction studies to contain an essentially planar pentalene ligand bridging two nickel atoms (**5a**). Although evidently diamagnetic, the molecule is overall two electrons short of satisfying the "18-electron" rule. In terms of the nickel–carbon (ring) bond lengths the structure is best represented by **5a**, although in a valence bond representation (**5b** and **5c**) the electron count is better revealed.

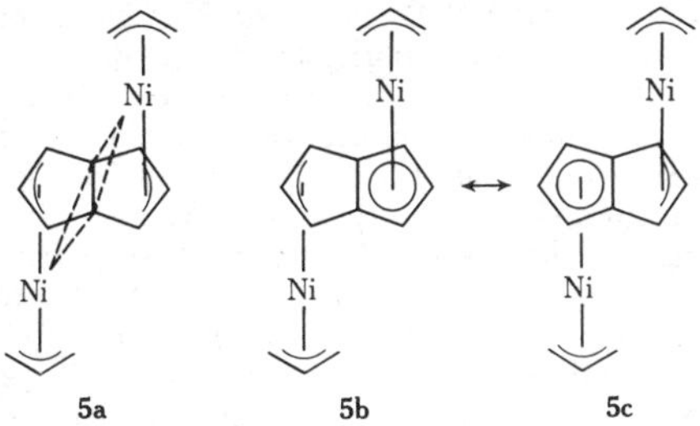

5a **5b** **5c**

Another approach to the isolation of pentalene as a metal complex involves dehydrogenation of dihydropentalenes with iron carbonyls. In this manner complexes $Fe_2(CO)_5(C_8H_5R)$ (6) have been obtained.[20] Compound 6 (R = H) has also been prepared by treating the dimer of pentalene with

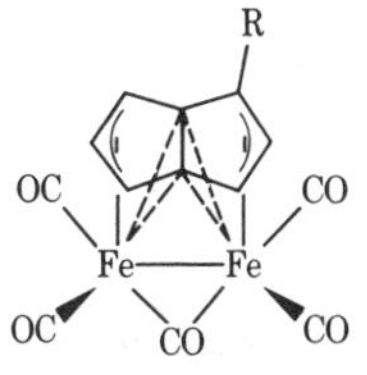

6, R = H, NMe$_2$, Ph

$Fe_2(CO)_9$.[21] The syntheses from dihydropentalenes again parallel cyclopentadienylmetal chemistry in that iron carbonyls dehydrogenate cyclopentadiene (C_5H_6) to give $[Fe(CO)_2(\eta^5\text{-}C_5H_5)]_2$.

In the light of our own work, described below, and in the absence of X-ray crystallographic studies on these molecules, it seems reasonable to formulate the metal-to-ring bonding in 6 as shown. It is important to note at this juncture that these syntheses of pentalene complexes involve organic precursors which already contain the relatively inaccessible bicyclo-[3.3.0] pentalene carbon skeleton.

Trimethylsilyl and Trimethylgermyl Derivatives of Ruthenium Carbonyl

Our own involvement in pentalene chemistry was in the first instance fortuitous. Following the discovery of convenient syntheses of dodecacarbonyltriruthenium in our laboratory[22] and elsewhere,[23] an apparently limitless area of organoruthenium chemistry has developed based on this carbonyl as a chemical reagent.[24] We have studied reactions of the carbonyl with trimethylsilane, -germane, and -stannane and characterized many polynuclear metal complexes containing ruthenium bonded to silicon, germanium, or tin.[25]

In the context of the "18-electron" rule, so commonly followed in metal carbonyl chemistry, the Me_3M ligands can be regarded formally as one-electron donors. Although complexes derived from the parent carbonyl $Ru_3(CO)_{12}$ with cyclic C_7 and C_8 olefins have been well studied, it seemed that related species formed from 7 or 8 would afford organoruthenium compounds of new structural types[26,27] because of different electronic requirements of ruthenium when coordinated with an MMe_3 ligand. In particular, compounds 7, at present somewhat rare ex-

(16) T. J. Katz and J. J. Mrowca, *J. Amer. Chem. Soc.*, **89**, 1105 (1967); T. J. Katz and N. Acton, *ibid.*, **94**, 3281 (1972); T. J. Katz, N. Acton, and J. McGinnis, *ibid*, **94**, 6205 (1972).

(17) M. R. Churchill and K.-K. G. Lin, *Inorg. Chem.*, **12**, 2274 (1973).

(18) A. Miyake and A. Kanai, *Angew. Chem., Int. Ed. Engl.*, **10**, 801 (1971).

(19) Y. Kitano, M. Kashiwagi, and Y. Kinoshita, *Bull. Chem. Soc. Jap.*, **46**, 723 (1973).

(20) D. F. Hunt and J. W. Russell, *J. Organomet. Chem.*, **104**, 373 (1976).

(21) W. Weidemüller and K. Hafner, *Angew. Chem., Int. Ed. Engl.*, **12**, 925 (1973).

(22) M. I. Bruce and F. G. A. Stone, *J. Chem. Soc. A*, 1238 (1967).

(23) F. Piacenti, P. Pino, M. Bianchi, G. Braca, and G. Sbrana, "Progress in Coordination Chemistry," M. Cais, Ed., Elsevier, Amsterdam, 1968, p 54.

(24) Those wishing to follow this rapidly developing area should refer to (a) M. A. Bennett, *Organometal. Chem.*, 1–3 (1972–1974); (b) J. A. McCleverty, *Annu. Survey 1972, J. Organometal. Chem.*, **68**, 423 (1974).

(25) These complexes included the first examples of covalent compounds containing six-membered heterocyclic metal rings. See S. A. R. Knox and F. G. A. Stone, *J. Chem. Soc. A*, 2874 (1971); A. Brookes, S. A. R. Knox, and F. G. A. Stone, *ibid.*, 3469 (1971); J. A. K. Howard and P. Woodward, *ibid.*, 3468 (1971).

(26) J. A. K. Howard, S. Kellet, and P. Woodward, *J. Chem. Soc. Dalton Trans.*, 2332 (1974).

(27) L. Vancea, R. K. Pomeroy, and W. A. G. Graham, *J. Am. Chem.Soc.*, **98**, 1407 (1976).

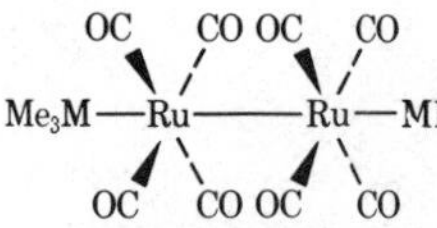

7, M = Si, Ge, Sn

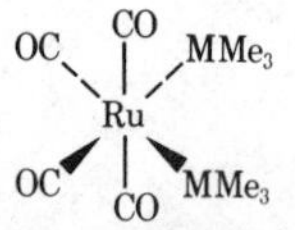

8, M = Si, Ge, Sn

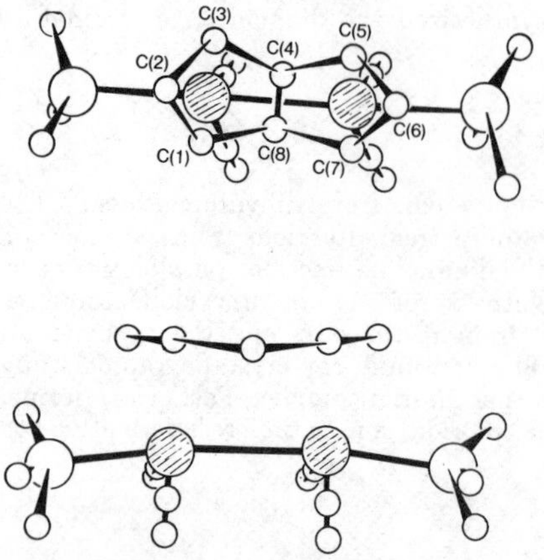

Figure 1. Molecular structure of the pentalene complex $[Ru(GeMe_3)(CO)_2]_2C_8H_6$:[31] large open circles, germanium.

amples of binuclear ruthenium species, should coordinate with cyclooctatetraene or cycloheptatriene following loss of carbon monoxide in such a manner that these ligands bridge the two ruthenium atoms. Compounds **8**, on the other hand, seemed likely precursors of mononuclear ruthenium complexes, as a result of direct replacement of carbon monoxide ligands. The possibility of insertion of C_8H_8 or C_7H_8 into the Me_3M–Ru bonds (*i.e.*, Me_3M group transfer from ruthenium to the coordinated hydrocarbon) had also to be considered in view of an interesting early observation of Gorsich.[28]

These admittedly naive ideas involving compounds **7** or **8** as precursors in organoruthenium chemistry were broadly supported by experiment.[29] However, a surprising capability of **7** and **8**, and of $Ru_3(CO)_{12}$ itself, to yield pentaleneruthenium complexes by dehydrogenative transannular ring closure of monocyclic C_8 olefins arose, and it is the study of this and related phenomena which we describe here.

Reactions of Cyclooctatetraenes

Both iron and ruthenium carbonyls are known to react with cyclooctatetraene (COT) to give several complexes whose study has done much to add to our knowledge of fluxional behavior.[30] For both metals, compounds of formula $(COT)M(CO)_3$, $(COT)M_2(CO)_6$, and $(COT)M_2(CO)_5$ have been thoroughly characterized. In addition, ruthenium forms a novel cluster complex of formula $(COT)_2Ru_3(CO)_4$ containing two cyclooctatetraene molecules. It was therefore with this hydrocarbon that we chose to initiate studies with **7** and **8**.

In refluxing hexane, complexes of formula $Ru(MMe_3)(CO)_2(C_8H_8MMe_3)$ were isolated (quantitatively for M = Si) from the reaction of COT with **8** (M = Si, Ge). An X-ray diffraction study completed on $Ru(SiMe_3)(CO)_2(C_8H_8SiMe_3)$ recently established structure **9**, the result of an MMe_3 migration to the C_8 ring and with one double bond not bonded to the ruthenium atom. These are fluxional molecules, with temperature-variable ^1H and ^{13}C nmr spectra (between limits of +100 and −60°) consistent with the oscillatory process **9a** ⇄ **9b** shown.

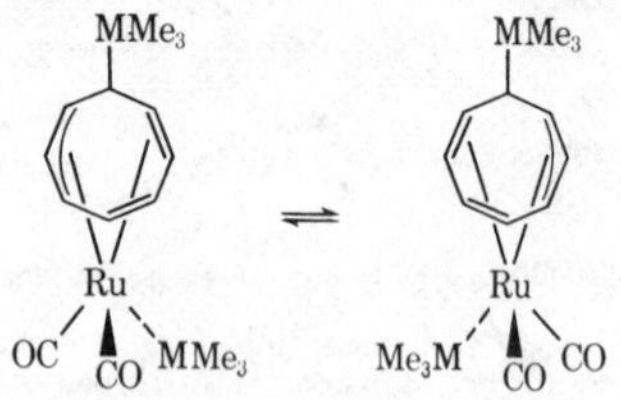

9a **9b**

If these complexes are refluxed in heptane, or if **8** and COT are heated in heptane or octane, crystalline compounds $Ru_2(MMe_3)_2(CO)_4(C_8H_6)$ (**10**) are

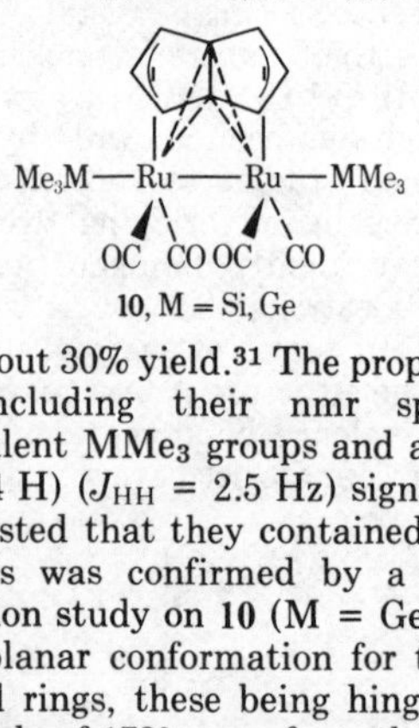

10, M = Si, Ge

obtained in about 30% yield.[31] The properties of these complexes, including their nmr spectra which showed equivalent MMe_3 groups and a triplet (2 H) and doublet (4 H) (J_{HH} = 2.5 Hz) signal for the ring protons, suggested that they contained pentalene as a ligand. This was confirmed by a single-crystal X-ray diffraction study on **10** (M = Ge),[31] which revealed a nonplanar conformation for the two fused five-membered rings, these being hinged to one another at an angle of 173° away from the Ru–Ru bond (Figure 1). There are two sets of Ru–C (ring) distances, with C(1)–C(3) and C(5)–C(7) on average 2.21 Å from their nearest Ru atom, while C(4) and C(8) are equidistant (2.53 Å) from each Ru atom. The bonding can therefore be approximately represented as involving two η^3-allyl units, each bonded to one ruthenium, coupled with a four-center four-electron interaction between the two ruthenium atoms and C(4) and C(8). A very related bonding scheme was earlier proposed for the cyclooctatetraene complex $Fe_2(CO)_5(COT)$.[30]

The role of the apparent intermediates **9** in the formation of the pentalene complexes **10** is as yet not understood. The process appears to occur with specific elimination of the MMe_3 group.

Although formation of **10** (M = Si) from $Ru(SiMe_3)_2(CO)_4$ occurs without by-products, the analo-

(28) R. D. Gorsich studied the reaction between $Mn(SnPh_3)(CO)_5$ and tetraphenylcyclopentadienone (*J. Amer. Chem. Soc.*, 84, 2486 (1962)) and obtained a compound formulated with a manganese–tin bond and a η^4-cyclopentadienone ligand. Subsequent study (R. D. Gorsich, *J. Organometal. Chem.*, 5, 105 (1966)) revealed that the complex was a substituted cyclopentadienyl derivative $Mn(C_5Ph_4OSnPh_3)(CO)_3$, *i.e.*, the Ph_3Sn group had transferred from the manganese atom to the organic moiety.

(29) The tin compounds **7** and **8** are less reactive and do not afford under similar conditions complexes analogous to those formed from the silicon and germanium compounds.

(30) For a review and summary of leading references, see F. A. Cotton, *Accounts Chem. Res.*, 1, 257 (1968).

(31) A. Brookes, J. A. K. Howard, S. A. R. Knox, F. G. A. Stone, and P. Woodward, *J. Chem. Soc., Chem. Commun.*, 587 (1973).

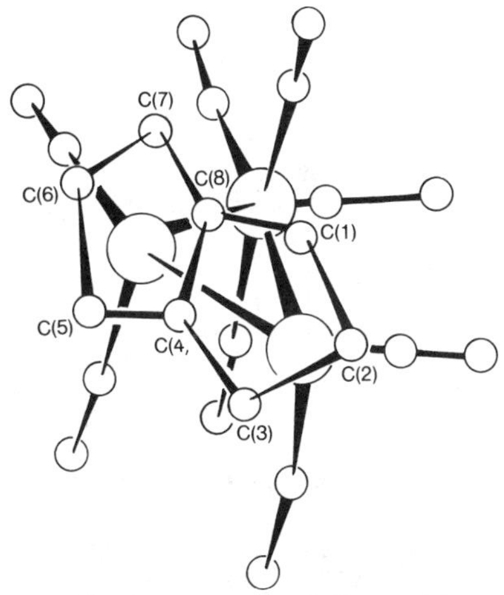

Figure 2. Molecular structure of the pentalene complex $Ru_3(CO)_8C_8H_6$.[34,36]

gous reaction between $Ru(GeMe_3)_2(CO)_4$ and COT, in addition to giving **10** (M = Ge), yields several other complexes, two of which have been identified as $Ru(GeMe_3)(CO)_2(C_8H_9)$ (**11**; M = Ge) and $Ru_2(GeMe_3)_2(CO)_4(C_8H_8)$ (**12**; M = Ge). The former is again

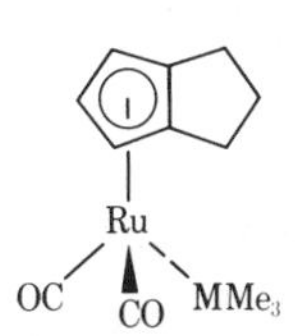

11, M = Si, Ge

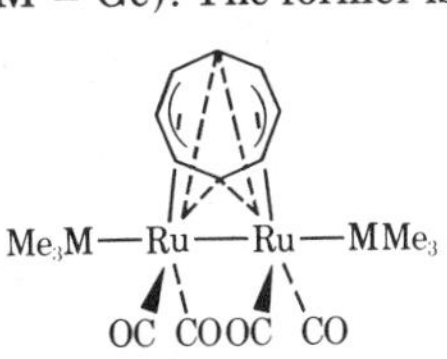

12, M = Si, Ge

the consequence of a ring-closure process, but here COT has abstracted a hydrogen from some source to give the tetrahydropentalenyl ligand,[32] a derivative of the cyclopentadienyl group.

Complexes **12** are of more interest. The mode of attachment of the C_8H_8 ring to the two metal atoms is likely to be similar to that in $Fe_2(CO)_5(COT)$, referred to above. As in **10**, the metal–ring bonding can be envisaged to involve two interannular η^3-allyl units and a four-center interaction. Because of this it was attractive to believe that **12** would lead to pentalene complexes **10** by loss of H from the central bridging carbons with concomitant C–C bond formation, but attempts to induce such a transformation have so far been unsuccessful. Complexes **12** can be prepared more easily by heating COT in hexane with **7**, compounds which already contain the $Ru_2(MMe_3)_2$ unit.

Formation of **10** by dehydrogenative transannular cyclization of COT prompted a more detailed study of reactions between cyclooctatetraene or its derivatives and ruthenium carbonyl complexes. In view of the inaccessibility of organic precursors containing a preformed pentalene C_8 skeleton, the development of syntheses of pentalene complexes from cyclooctatetraenes was attractive. Substituted cyclooctatetraenes can be fairly readily prepared, and if reactions of the type which gave **10** could be extended to these derivatives a variety of pentalene complexes might be identified. It was also necessary to establish whether pentaleneruthenium complexes could only be formed using the complexes **8** or whether ruthenium carbonyl itself could bring about similar dehydrogenation of COT. With this in mind, a reexamination [33] of the reaction of cyclooctatetraene with dodecacarbonyltriruthenium led to the discovery, albeit in very small yield, of two new trinuclear ruthenium complexes, in addition to the complexes previously obtained.[30] The formulas of the new crystalline compounds were $Ru_3(CO)_6(C_8H_9)_2$ and $Ru_3(CO)_8(C_8H_6)$.

The compound $Ru_3(CO)_8(C_8H_6)$ proved to be, as its formula suggested, a pentalene complex (**13**; R^1

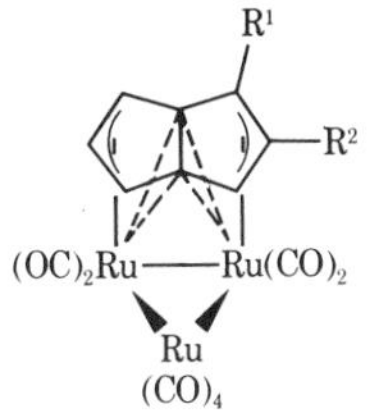

13, $R^1 = R^2 = H$; R^1 = Me or Ph, $R^2 = H$; $R^1 = H$, R^2 = Me or Ph

= R^2 = H). An X-ray diffraction study[34] revealed a nearly planar pentalene ligand, the angle between the planes of the two C_5 rings being 177°, with the rings again bending away from the molecular center (Figure 2). As in $Ru_2(GeMe_3)_2(CO)_4(C_8H_6)$ (**10**), the Ru–C (ring) distances fall into two groups, with C(1)–C(3) and C(5)–C(7) at a mean distance of 2.23 Å and the junction atoms C(4) and C(8) at 2.48 Å, so that the bonding can be similarly represented as in **10**. The most significant feature of the structure, however, is that the plane of the Ru_3 triangle makes an angle of 50° with the mean plane of the pentalene.

It was of considerable interest that the ^1H nmr spectrum of $Ru_3(CO)_8(C_8H_6)$ varied with temperature (Figure 3). The data indicate a degenerate rearrangement producing a time-averaged molecular plane of symmetry, and we favor the unique oscillatory rearrangement shown in Scheme I, compatible with the ground-state structure revealed by X-ray diffraction. One may note that such a process also requires the complementary "swinging" of the terminal CO ligands of the $Ru(CO)_2$ groups. The activation energy for the process is 12.8 ± 0.3 kcal/mol.

Scheme I

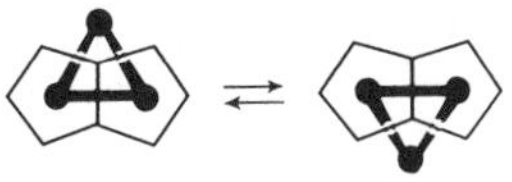

• = Ru, with CO groups omitted

The other new product of the reaction of $Ru_3(CO)_{12}$ and cyclooctatetraene, $Ru_3(CO)_6(C_8H_9)_2$, has recently been identified by an X-ray diffraction study[35] and is illustrated schematically as **14**. Again,

(32) Several years ago it was observed (T. H. Coffield, K. G. Ihrman, and W. Burns, *J. Amer. Chem. Soc.*, **82**, 4209 (1960)) that $Mn_2(CO)_{10}$ and COT reacted to give $Mn(CO)_3(C_8H_9)$.

(33) V. Riera, S.A. R. Knox, and F. G. A. Stone, to be published.

(34) J. A. K. Howard, S. A. R. Knox, V. Riera, F. G. A. Stone, and P. Woodward, *J. Chem. Soc., Chem. Commun.*, 452 (1974).

(35) R. Bau, B. Chaw-Kuo Chou, S. A. R. Knox, V. Riera, and F. G. A. Stone, *J. Organomet. Chem.*, **82**, C43 (1974).

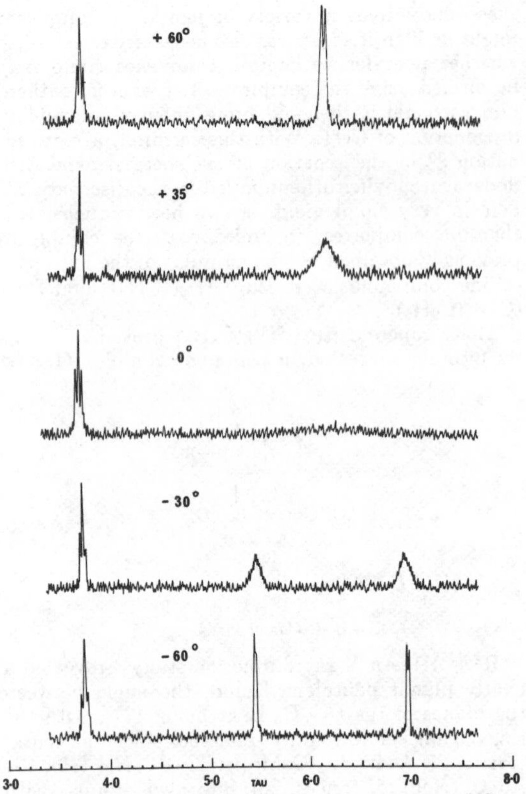

Figure 3. Variable-temperature ^1H nmr spectrum of $Ru_3(CO)_8(C_8H_6)$.

a ring closure is evident in the formation of the tetrahydropentalenyl ligand.

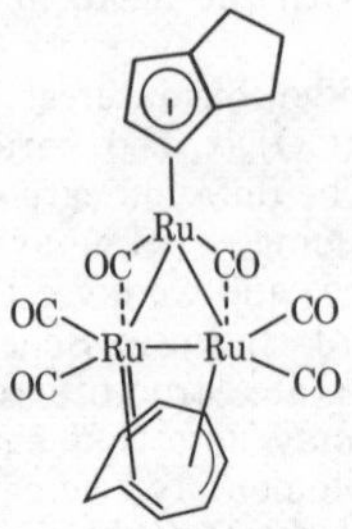

14

Phenyl- and methylcyclooctatetraenes react with **7** (M = Si) and with $Ru_3(CO)_{12}$ to give many organoruthenium complexes.[33] Several of these are of the pentalene type, and some are, as with COT, common to reactions involving both **7** (M = Si) and $Ru_3(CO)_{12}$. The former compound thus undergoes fission of its Ru–Si bonds, possibly with loss of Me_3SiH. Within the context of this Account four complexes are of interest. Each of **13** (R^1 = H, R^2 = Me or Ph; R^1 = Me or Ph, R^2 = H) was isolated, showing that during dehydrogenative ring closure of cyclooctatetraenes C_8H_7R, the position of central σ-bond formation is apparently not greatly influenced by the substituent. In fact, formation of 2-substituted pentalenes (2–7%) is marginally favored over 1-substituted pentalenes ($\sim$1%).

An interesting situation arises in that the 2-substituted complexes **13** (R^1 = H, R^2 = Me or Ph) are fluxional, while the 1-substituted isomers **13** (R^1 = Me or Ph, R^2 = H) are not. The two fluxional complexes appear from their nmr spectra to be undergoing a process analogous to that of $Ru_3(CO)_8(C_8H_6)$, with free energy of activation being essentially independent of the 2-substituent, at 12.8 ± 0.3 kcal/mol for each of the three.

A degenerate fluxional process of this type is not possible for complexes of the unsymmetrical 1-methyl- and 1-phenylpentalene, since the two components of the oscillation would be inequivalent. However, the possibility of a nondegenerate oscillation remains, but we have so far observed the population of only one form. The structure of $Ru_3(CO)_8[C_8H_4(SiMe_3)_2]$, to be discussed later, suggests that the favored species is that in which the $Ru(CO)_4$ group is furthest from the substituent.

A product isolated [33] in very low yield from the reaction between phenylcyclooctatetraene and $Ru_3(CO)_{12}$, which we initally believed to be a pentalene complex, has been characterized[36] by X-ray diffraction as a complex of bicyclooctatetraenyl, best represented by **15**. This complex has not been ob-

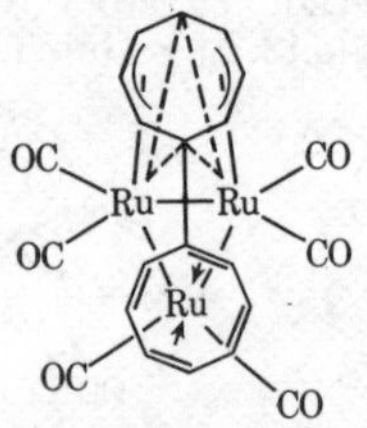

15

served in reactions involving COT, and may be formed *via* elimination of phenyl groups, since C–C bond formation between two rings has been observed in the reaction of cyclohexa-1,3-diene with $Ru_3(CO)_{12}$.[37] We cannot, however, discount the possibility that **15** arises from some bicyclooctatetraenyl impurity in the phenylcyclooctatetraene employed. The structure is unique with one COT ring in a "tub" conformation, coordinated as a 1,5-diene, an unprecedented mode for metals of the iron triad. The complex is formed readily from the direct reaction of $Ru_3(CO)_{12}$ and bicyclooctatetraenyl in refluxing toluene.[38]

Reactions of Cyclooctatrienes

In the context of the above results involving COT, and following an observation[39] (discussed below) that cycloocta-1,5-diene formed the tetrahydropentalenyl ligand in reactions with **7** and **8**, we initiated a study of reactions of cyclooctatrienes with ruthenium carbonyl complexes. We were especially hopeful that the yields of any pentalene complexes of the type **13** would exceed those obtained from cyclooctatetraenes.

Reduction of COT affords an isomeric mixture of

(36) J. A. K. Howard and P. Woodward, to be published.

(37) T. H. Whitesides and R. A. Budnik, *J. Chem. Soc., Chem. Commun.*, 87 (1973).

(38) J. D. Edwards, J. A. K. Howard, S. A. R. Knox, V. Riera, and P. Woodward, *J. Chem. Soc. Dalton Trans.*, 75 (1976).

(39) S. A. R. Knox, R. P. Phillips, and F. G. A. Stone, *J. Chem. Soc., Dalton Trans.*, 658 (1974).

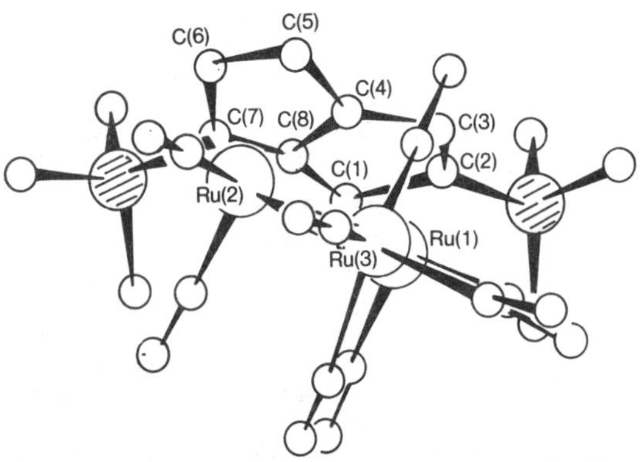

Figure 4. Molecular structure of the pentalene complex $Ru_3(CO)_8[C_8H_4(SiMe_3)_2]$.[36,43]

cycloocta-1,3,5-triene, cycloocta-1,3,6-triene, and bicyclo[4.2.0]octa-2,4-diene.[40] Under heptane reflux this mixture did not yield a pentalene complex upon reaction with **7** or **8**. The observed products were **11** and complexes $Ru(CO)_3(C_8H_{10})$ (**16**) and $Ru_2(CO)_6(C_8H_{10})$ (**17**) of the two tautomers of cy-

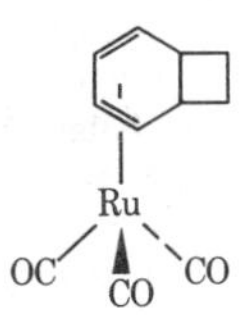

16

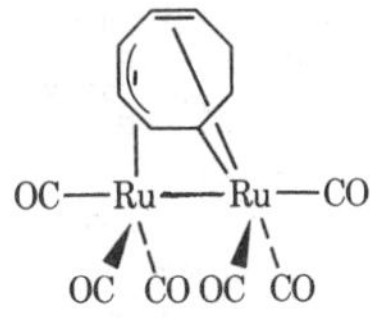

17

cloocta-1,3,6-triene. Not unexpectedly, both **16** and **17** are much more readily obtained from $Ru_3(CO)_{12}$ and the C_8H_{10} mixture.[41]

Since traces of $Ru_3(CO)_8(C_8H_6)$ were detected when the cyclooctatriene isomeric mixture was heated with $Ru_3(CO)_{12}$ in octane, it was decided to carry out studies on substituted cyclooctatrienes. The presence of bulky substituents appears to stabilize the pentalene system, as referred to earlier, and it seemed possible that pentalene complex formation might also be so favored. This work is at a preliminary stage, but the novel hydrocarbon 5,8-bis(trimethylsilyl)cycloocta-1,3,6-triene (**18**)[42] reacts with **7**

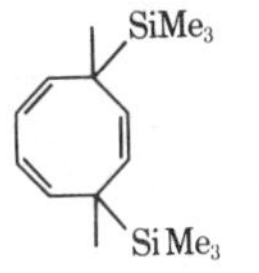

18

(M = Si) or $Ru_3(CO)_{12}$ to give a separable mixture of Me_3Si-containing analogs of **16** and **17**, and the bis(trimethylsilyl)pentalene complex $Ru_3(CO)_8[C_8H_4(SiMe_3)_2]$ (**19**)[43] in 10% yield, the structure of which has been established[36] by X-ray diffraction. The molecular configuration (Figure 4) is nearly identical with that of $Ru_3(CO)_8(C_8H_6)$: Ru(1)–Ru(3) = Ru(2)–Ru(3) = 2.806 (1) Å, Ru(1)–Ru(2) = 2.930 (1) Å; the mean Ru–C(1)–C(3) and Ru–C(5)–C(7) distance is 2.23 (1) Å and the mean Ru–C(4)–C(8) distance is 2.48 (1) Å. The C–C distances in the pentalene are all equal at 1.44 (1) Å mean, and the angle between the Ru_3 plane and the mean plane of the pentalene is again 50°. Like $Ru_3(CO)_8$ complexes of unsymmetrical 1-substituted pentalenes, the asymmetric pentalene confers nonfluxional character on **19**. Presumably for steric reasons the preferred

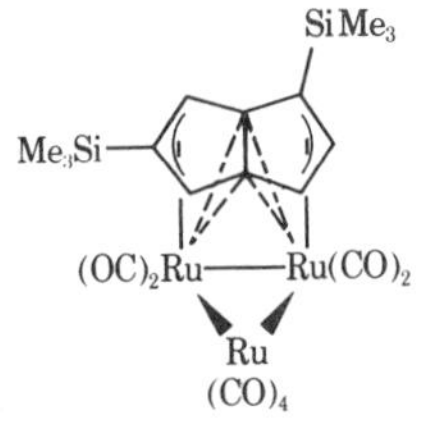

19

conformation is that in which the Me_3Si group on C(7) is furthest from Ru(3).

This route to substituted pentalene complexes is promising. The triene **20** reacts with $Ru_3(CO)_{12}$ to

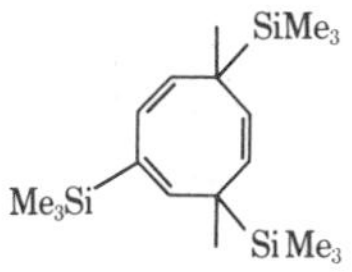

20

afford the 1,3,5-tris(trimethylsilyl)pentalene complex $Ru_3(CO)_8[C_8H_3(SiMe_3)_3]$ (**21**) in 15% yield.[43]

This symmetrically trisubstituted pentalene complex is fluxional, as expected, and able to undergo a degenerate oscillatory process similar to that invoked in Scheme I for $Ru_3(CO)_8(C_8H_6)$, so that at the high-temperature limit a time-averaged molecular plane of symmetry is generated (Figure 5). The free energy of activation for the oscillation (9.2 ± 0.2 kcal/mol) is appreciably lower than that for $Ru_3(CO)_8(C_8H_6)$ or 2-substituted pentalene complexes. A determination of the ground-state molecular structure of **21** would be of interest, in perhaps showing

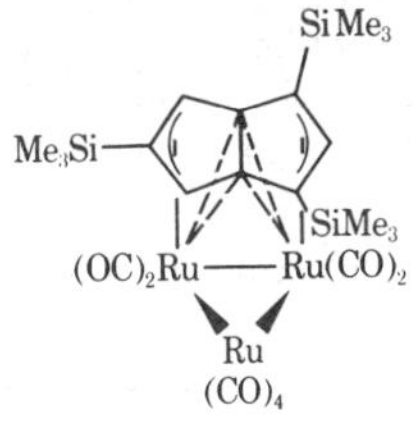

21

steric crowding by the $SiMe_3$ groups constraining the Ru_3 plane to make a larger angle than 50° with the pentalene plane, consistent with a lower energy oscillation.

Reactions of Cycloocta-1,5-diene

Cycloocta-1,5-diene does not yield a complex of pentalene upon reaction with either $Ru_3(CO)_{12}$, **7**, or **8**. A ring-closure process is in evidence, however, in that a major product from **7** or **8** is the tetrahydropentalenyl complex $Ru(MMe_3)(CO)_2(C_8H_9)$ (**11**).[39]

(40) A. C. Cope, A. C. Haven, F. L. Ramp, and E. R. Trumbull, *J. Amer. Chem. Soc.*, **74**, 4867 (1952).

(41) A. C. Szary, S. A. R. Knox, and F. G. A. Stone, *J. Chem. Soc., Dalton Trans.*, 66 (1974).

(42) J. M. Bellama and J. B. Davison, *J. Organomet. Chem.*, **86**, 69 (1975).

(43) J. A. K. Howard, S. A. R. Knox, F. G. A. Stone, A. C. Szary, and P. Woodward, *J. Chem. Soc. Chem. Commun.*, 788 (1974).

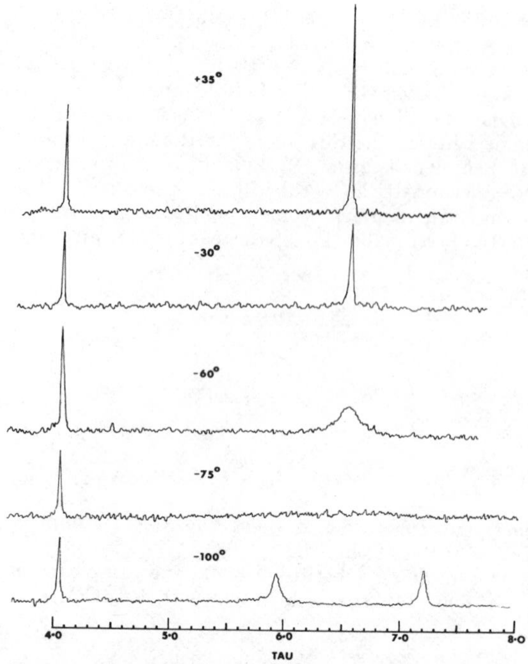

Figure 5. Variable-temperature ^{1}H nmr spectrum of $Ru_3(CO)_8[C_8H_3(SiMe_3)_3]$.

With 8 (M = Ge), additional products are $Ru(GeMe_3)_2(CO)_2(C_8H_{12})$ (22) and $Ru_2(GeMe_3)(\mu$-

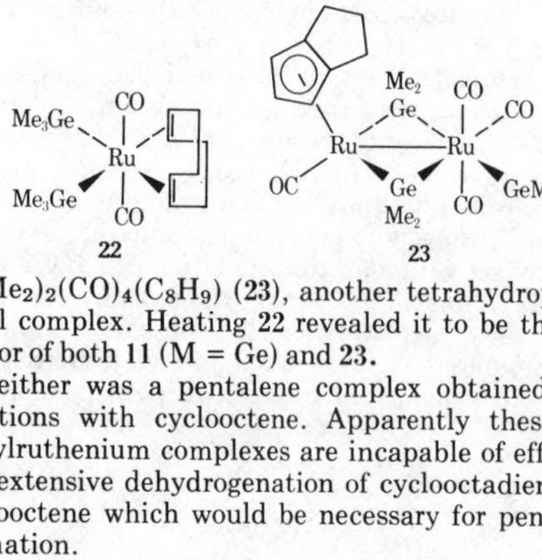

22 23

$GeMe_2)_2(CO)_4(C_8H_9)$ (23), another tetrahydropentalenyl complex. Heating 22 revealed it to be the precursor of both 11 (M = Ge) and 23.

Neither was a pentalene complex obtained from reactions with cyclooctene. Apparently these carbonylruthenium complexes are incapable of effecting the extensive dehydrogenation of cyclooctadiene and cyclooctene which would be necessary for pentalene formation.

Comments on Transannular Ring Closure

We have observed the capability of several carbonylruthenium complexes to effect transannular cyclization of cyclooctatetraenes, cyclooctatrienes, and cycloocta-1,5-diene to give pentalene and/or tetrahydropentalenyl complexes. Conversion of C_8H_8[32] or C_8H_{12}[44-46] to the bicyclic C_8H_9 tetrahydropentalenyl ligand has been observed previously, the former process involving addition of a hydrogen atom and the latter the removal of three hydrogen atoms per molecule of reactant hydrocarbon. The mechanisms of these reactions are not clear. Otsuka and Taketomi[45] found that dehydrogenative ring closure of the cycloocta-1,5-diene ligand in $Co(C_8H_{13})(C_8H_{12})$ is brought about by excess C_8H_{12} which is thereby converted to cyclooctene.

$$Co(C_8H_{13})(C_8H_{12}) + 2C_8H_{12} \longrightarrow Co(C_8H_9)(C_8H_{12}) + 2C_8H_{14}$$

The mechanism of pentalene complex formation by dehydrogenative ring closure is similarly obscure, and probably varies according to the carbonylruthenium complex employed. Complexes of pentalenes with the $Ru_3(CO)_8$ fragment can be prepared from either $Ru_3(CO)_{12}$ or **7**, so that the presence of MMe_3 ligands in the carbonylruthenium complex appears to be of no account. Formation of the pentalene complexes **10** from COT and **8**, however, appears to involve a crucial MMe_3 migration leading to **9** as an intermediate.

Such migrations are common in reactions of 7 and 8. Thus the bis(trimethylsilyl)pentalene complex **19** is the major product (30%) from trimethylsilylcyclooctatetraene and 7 (M = Si), while **20** on reaction with 7 (M = Si) affords **19** but no **21**.[43] Curiously, **18** and **17** (M = Si) gave **19** cleanly. This lability of MMe_3 groups inhibits mechanistic discussion of ring closure *via* 7 or 8. Ring closure of both cyclooctatrienes and -tetraenes induced by $Ru_3(CO)_{12}$ is, however, less complicated and allows scope for the study of reactions of various derivatives of these hydrocarbons in order to gain insight into the factors controlling the process.

It is noteworthy to record at this stage that we have observed migration of MMe_3 ligands from ruthenium to other hydrocarbons. The complexes $Ru(MMe_3)_2(CO)_4$ (**8**) are most adept in this respect, giving **24** and **25** in high yield with cycloheptatriene[47] and azulene,[48] respectively. In order to confirm the MMe_3 transfer implicit in the formulation **24**, a crystalline ring-substituted pentafluorophenyl derivative, **26**, was prepared from $7\text{-}C_6F_5C_7H_7$ and **8** (M = Si). The molecular structure was established by an X-ray diffraction study.[47] Surprisingly, an exo configuration was observed for the $SiMe_3$ group on the hydrocarbon and an endo configuration for the C_6F_5 group, although further experiments have suggested the migration is *intramolecular*.

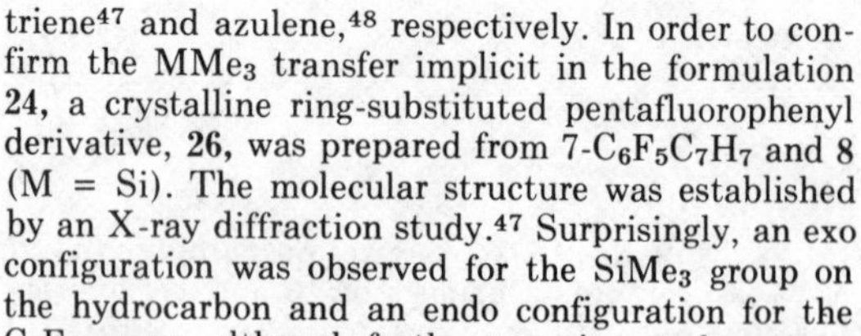

24 25 26

Conclusion

A major objective in preparing pentalene metal complexes must be a study of the chemistry of the pentalene system, either by its reactions while stabilized by coordination or by its release and subse-

(44) H. Lehmkuhl, W. Leuchte, and E. Janssen, *J. Organometal. Chem.*, **30**, 407 (1971).

(45) S. Otsuka and T. Taketomi, *J. Chem. Soc., Dalton Trans.*, 1879 (1972).

(46) K. K. Joshi, R. H. B. Mais, F. Nyman, P. G. Owston, and A. M. Wood, *J. Chem. Soc. A*, 318 (1968).

(47) See also A. Brookes, S. A. R. Knox, V. Riera, B. A. Sosinsky, and F. G. A. Stone, *J. Chem. Soc. Dalton Trans.*, 1641 (1975).

(48) S. A. R. Knox, B. A. Sosinsky, and F. G. A. Stone, *J. Chem. Soc. Dalton Trans.*, 1647 (1975).

quent trapping with various reagents. Thus far there have been no literature reports of experiments of this nature. However, it can be said that synthetic procedures are now at hand to make such studies imminent. Our own work has progressed to the point where pentalene complexes can be obtained in 30% yields, making 200–300-mg quantities readily available, and attention is now directed to such investigations. It should be appreciated, however, that with these complexes the pentalenes comprise a relatively small fraction of the total weight of complex, so that progress in this direction is perhaps likely to be slow.

We are indebted to our coworkers named in the references and to Drs. Judith Howard and Peter Woodward who, under the auspices of a Science Research Council Grant, have carried out several X-ray crystallographic studies crucial to the research. We also acknowledge the help of Professor R. Bau in determining the molecular structure of **14**.

Addendum (May 1976)

The most significant recent development has been the observation of a new mode of coordination of a pentalene ligand. Single crystal x-ray diffraction studies have established that the tri-substituted pentalene complex $Ru_3(CO)_8[C_8H_3(SiMe_3)_3]$ exists in two isomeric forms, which infrared and nmr spectroscopy show to be in equilibrium in solution.[49] One isomer (**21**) has the expected geometry based on earlier studies (Figures 2 and 4), with the pentalene bridging an edge of a triangle of ruthenium atoms, and similarly inclined at 51° to the Ru_3 plane. The other isomer has the new mode of coordination for a pentalene ligand, which lies almost parallel to and effectively spans the face of an Ru_3 triangle. The distribution of carbonyl groups is different from that in isomr **21**, with one uthenium atom carrying two CO groups and the others three each. When solutions of either isomer are allowed to stand at room temperature for several minutes, infrared and nmr signals caused by the other appear, and ultimately an equilibrium mixture develops with **21** (~90%) predominating. The change of equilibrium constant with temperature gives $\Delta H° = 0.9$ (±0.1) kcal/mol and $\Delta S° = 0.7$ (±0.3) e.u./mol.

(49) J. A. K. Howard, S. A. R. Knox, R. J. McKinney, R. F. D. Stansfield, F. G. A. Stone, and P. Woodward, *J. Chem. Soc. Chem. Commun.*, 557 (1976).

Study of the spectra of other $Ru_3(CO)_8$(pentalene) species shows that complexes of *symmetrical* pentalenes $Ru_3(CO)_8(2\text{-R-}C_8H_5)$ (R = H, Me, Ph, $SiMe_3$) are also predominantly edge bonded, but have a low equilibrium concentration of the face-bonded form.

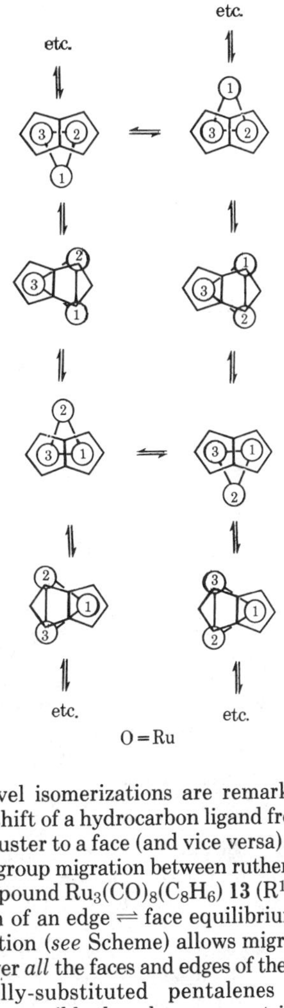

These novel isomerizations are remarkable in requiring the shift of a hydrocarbon ligand from the edge of a metal cluster to a face (and vice versa) with simultaneous CO group migration between ruthenium atoms. For the compound $Ru_3(CO)_8(C_8H_6)$ **13** ($R^1 = R^2 = H$), combination of an edge ⇌ face equilibrium and fluxional oscillation (*see* Scheme) allows migration of the pentalene over *all* the faces and edges of the cluster. For symmetrically-substituted pentalenes the same scrambling is possible though a more restricted motion is likely. It may be noted that the mobility of organic ligands on a metal surface may have features not unrelated to the behavior described here.

Disodium Tetracarbonylferrate—a Transition-Metal Analog of a Grignard Reagent

James P. Collman

Department of Chemistry, Stanford University, Stanford, California 94305

Received February 19, 1975

Reprinted from ACCOUNTS OF CHEMICAL RESEARCH, *8, 342 (1975)*

With the expectation that significant synthetic methodologies will emerge from organotransition-metal chemistry, organic chemists are becoming increasingly interested in this rapidly growing field. To date, few truly useful new organotransition reagents have been developed.[1] However as new reactions are discovered and the underlying reaction mechanisms are slowly clarified, practical applications become more likely. In this Account, I summarize our research on $Na_2Fe(CO)_4$ as a reagent for organic synthesis.

This work had its origin in 1970 when my postdoctoral associate, M. Cooke, was searching for a method of forming Ge–Ge bonds by coupling R_3GeCl with two electron reductants such as $Na_2Fe(CO)_4$. Being an organic chemist, Cooke tried methyl iodide in a model reaction. Treatment of CH_3I with $Na_2Fe(CO)_4$ followed by hydrolysis gave the characteristic odor of acetaldehyde. This lead was quickly developed into a general synthesis of homologous aldehydes.[2] Because of my past interest in oxidative addition,[3] reductive elimination,[3b] and migratory insertion,[3b] the potential of $Na_2Fe(CO)_4$ as a reagent for organic synthesis was evident, and the matter became vigorously pursued by my other students.

Synthesis of the Reagent. Our early experiments employed $Na_2Fe(CO)_4$ derived from $Fe(CO)_5$ and sodium–mercury amalgam (eq 1).[2] Because of the expense, difficulty in scale-up, and the presence of mercury salts and of colored polynuclear iron carbonyl impurities inherent in this procedure, we sought a better method for preparing $Na_2Fe(CO)_4$. Eventually we developed a very practical method[4,5] (eq 2)

$$\underset{\text{orange}}{Fe(CO)_5} \xrightarrow[\text{THF}]{\text{Na(Hg)}} \underset{\text{red-yellow}}{Na_2Fe(CO)_4} \quad (1)$$

$$\underset{\text{orange}}{Fe(CO)_5} + Na \xrightarrow[\underset{\text{deep blue}}{C_6H_5COC_6H_5}]{\text{dioxane, }100^\circ} \underset{\text{white}}{Na_2Fe(CO)_4 \cdot 1.5\text{dioxane}} \quad (2)$$

employing $Fe(CO)_5$, the least expensive iron carbonyl,[6] and metallic sodium, with an electron carrier (such as benzophenone ketyl) in an ethereal solvent under conditions where the sodium (mp 97.5°) is molten. At atmospheric pressure, boiling dioxane (bp 101°) is ideal, yielding a more soluble solvate (eq 2). This process is rapid, nearly quantitative, and easily scaled up. Present raw material costs in this preparation of $Na_2Fe(CO)_4$ depend substantially (~75%) on the current price of $Fe(CO)_5$. However, if a large-scale application for $Fe(CO)_5$ were developed,[7] raw material costs could drop below those of Grignard re-

Professor Collman's work covers a wide range of complex ion chemistry, from the classical coordination compounds to the newer organometallic combinations. He is a native of Nebraska, and studied at the University of Nebraska for his B.S. degree. Following receipt of the Ph.D. from the University of Illinois in 1958, he joined the faculty of University of North Carolina. In 1967, he moved to Stanford University, where he is professor of chemistry.

(1) Perhaps the most versatile and useful transition-metal reagents developed thus far are organocopper compounds: H. O. House, *Proc. Robert A. Welch Found. Conf. Chem. Res.*, **17**, 101 (1973).

(2) M. P. Cooke, *J. Am. Chem. Soc.*, **92**, 6080 (1970).

(3) (a) J. P. Collman and W. R. Roper, *J. Am. Chem. Soc.*, **87**, 4008 (1965); (b) J. P. Collman, *Acc. Chem. Res.*, **1**, 136 (1968); (c) J. P. Collman and W. R. Roper, *Adv. Organometal. Chem.*, **7**, 54 (1968).

(4) J. P. Collman and R. G. Komoto, U.S. Patent Application filed June 11, 1973.

(5) (a) J. P. Collman, R. G. Komoto, W. O. Siegl, S. R. Winter, and D. R. Clark, unpublished results; (b) S. R. Winter, Ph.D. Dissertation, Stanford University, 1973; (c) R. G. Komoto, Ph.D. Dissertation, Stanford University, 1974.

(6) In fact, $Fe(CO)_5$ is the least toxic and least expensive transition-metal carbonyl. The other iron carbonyls, $Fe_2(CO)_9$ and $Fe_3(CO)_{12}$, are derived from $Fe(CO)_5$ and thus are more expensive.

(7) For example, commercialization of a carbon monoxide process converting ilmenite to rutile would produce $Fe(CO)_5$ as a by-product: A. Vasnapu, B. C. Marek, and J. W. Jensen, Report of Investigations 7719, U.S. Department of the Interior, Bureau of the Mines, 1973.

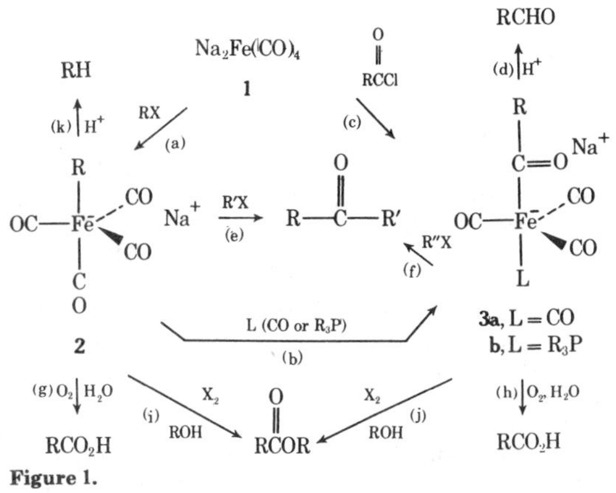

Figure 1.

agents, making $Na_2Fe(CO)_4$ competitive in fine chemicals manufacture. However, the extreme oxygen sensitivity of $Na_2Fe(CO)_4$ (spontaneously inflammable in air) will severely hinder its development on both industrial and laboratory scales.

Scope of Synthetic Applications. Conversions of aliphatic halides and sulfonates into aldehydes,[2] unsymmetric ketones,[8] carboxylic acids,[9] esters,[9] and amides[9] by means of $Na_2Fe(CO)_4$ (**1**) are outlined in Figure 1.[10] In a sense $Na_2Fe(CO)_4$ can be considered a transition-metal analog of a Grignard reagent. Principal advantages of these $Na_2Fe(CO)_4$ reactions are high yields, stereospecificity, and toleration of unmasked functional groups which would be attacked by the more reactive magnesium or lithium reagents. Limitations of $Na_2Fe(CO)_4$ derive from its basicity (pK_b about that of OH)[11] and the resulting tendency to cause eliminations. Thus for reaction a in Figure 1, tertiary substrates cannot be used, and secondary tosylates are preferred over secondary halides. Allylic halides cannot be employed since these afford stable 1,3-diene·$Fe(CO)_3$ complexes rather than the alkyliron(0) intermediate **2**. The preparation of aldehydes goes through the acyliron(0) intermediates **3**. Since pathway b, migratory insertion of the alkyliron(0) **2** to the acyliron(0) **3**, fails for alkyl groups bearing adjacent electronegative groups, the scope of aldehyde synthesis is limited to simple primary and secondary substrates. However, the acid chloride route (c) followed by (d) can also be employed.[12] Finally, alkylation of the alkyl or acyl intermediates **2** or **3** (steps e and f) affording ketones is restricted to reactive primary alkylating agents R′X, R″X (usually a primary iodide).

In spite of these restrictions a wide range of useful synthetic reactions can be carried out in high yield using $Na_2Fe(CO)_4$. Specific examples with yields are given in Figure 2. Reactions 3, 5, 6, and 9 illustrate

(8) J. P. Collman, S. R. Winter, and D. R. Clark, *J. Am. Chem. Soc.*, **94**, 1788 (1972).

(9) J. P. Collman, S. R. Winter, and R. G. Komoto, *J. Am. Chem. Soc.*, **95**, 249 (1973).

(10) The following abbreviations are used in this paper: L is a tertiary phosphine or CO, X_2 is Cl_2, Br_2, or I_2, THF is tetrahydrofuran, NMP is *N*-methylpyrrolidinone.

(11) P. Krumholtz and H. M. A. Stettiner, *J. Am. Chem. Soc.*, **71**, 3035 (1949).

(12) This route has been used for an aldehyde synthesis: Y. Watanabe, T. Mitsudo, M. Tanaka, K. Yamamoto, T. Okajima, and Y. Takegami, *Bull. Chem. Soc. Jpn.*, **44**, 2569 (1971).

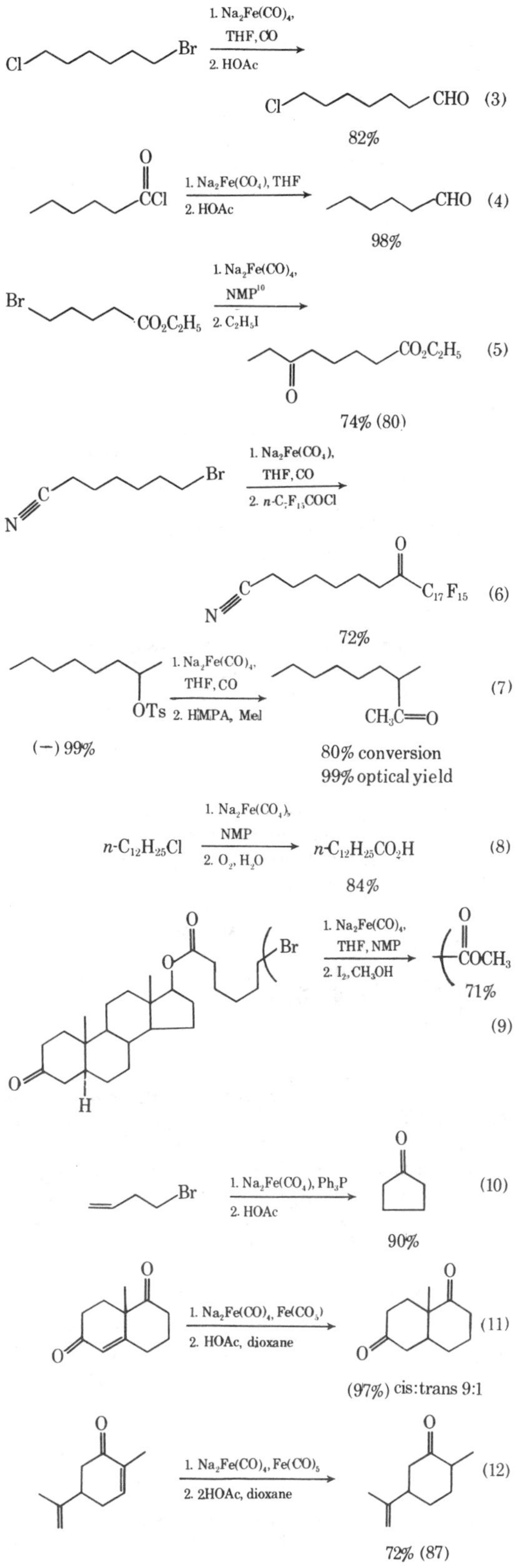

Figure 2. A survey of organic syntheses involving $Na_2Fe(CO)_4$.

the toleration of other functional groups. Comparison of eq 3 and 8 shows how the reactivity of the alkylation step a in Figure 1 may be controlled by the choice of solvent. A special case, affording hemifluorinated ketones[13] (eq 6), involves treating an alkyliron(0) intermediate (2) with a perfluoroacid chloride; however, this process is not effective with simple acid chlorides. The synthesis of cyclopentanone from 1-bromo-3-butene (eq 10) is also a special case limited to five- and six-ring ketones[5c,14] and is not illustrated in Figure 1. The mechanism of (10) seems to involve a variation of step b in which the olefin acts as an intramolecular ligand, L.

Perhaps the most important synthetic application of $Na_2Fe(CO)_4$ is the highly selective reduction of conjugated olefins illustrated in eq 11 and 12.[15] These reactions may involve $Na_2Fe_2(CO)_8$, prepared by reaction of $Na_2Fe(CO)_4$ with $Fe(CO)_5$ (eq 13) or generated in situ. Two equivalents of a mild acid are required. In terms of yield, stereoselectivity, and toleration of other functional groups (aldehyde, ketone, nitrile, unconjugated olefin, halides, and epoxides) this reagent[15] seems superior to other recently described reducing agents.[16]

$$Na_2Fe(CO)_4 + Fe(CO)_5 \longrightarrow Na_2Fe_2(CO)_8 + CO \quad (13)$$

The reagent $Na_2Fe(CO)_4$ is also useful for the synthesis of unusual inorganic substances. Two examples are shown in eq 14[17] and 15.[18] The latter is a complex reaction in which a Ge–Ge bond is cleaved.

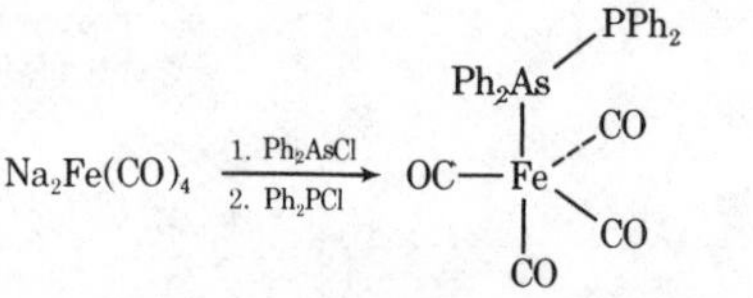

(14)

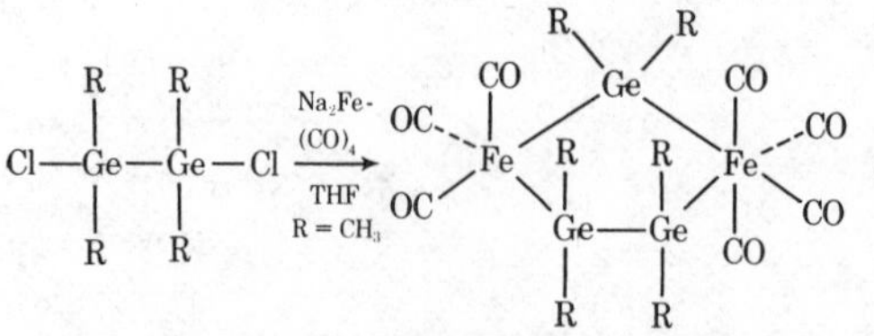

(15)

Mechanisms of Organic Syntheses Using $Na_2Fe(CO)_4$. Qualitative mechanistic patterns of organotransition-metal reactions[3b] were used as a guide for developing the synthetic applications of $Na_2Fe(CO)_4$. However, unexpected results such as striking solvent effects[19] prompted us to explore the reaction mechanisms in depth.[20] These studies, although still incomplete, provide a substantial understanding of the underlying mechanisms.

In Figure 1 pentacoordinate anionic alkyl- and acyliron(0) complexes (**2** and **3**) are shown as intermediates. In synthetic practice **2** and/or **3** are generated and used in situ. Several examples of these air-sensitive anionic iron(0) complexes have been isolated as air-stable crystalline $[(Ph_3P)_2N]^+$ salts,[21] characterized by elemental analyses and NMR and ir spectra, and demonstrated to undergo the individual steps b, d, e, f, g, and h illustrated in Figure 1.[5,22] The indicated trigonal-bipyramidal structures with bulky substituents in the apical position expected for five-coordinate d^8 complexes (C_{3v} symmetry) are consistent with the pattern of ν_{CO} frequencies exhibited by the isolated complexes. Proton and ^{13}C NMR spectra—especially for CO groups—suggest rapid equilibration of axial and equatorial groups, characteristic of many pentacoordinate complexes.

Let us next consider each reaction step and the evidence bearing on the corresponding mechanism. Reaction between an alkyl halide or sulfonate and $Na_2Fe(CO)_4$ forming a saturated d^8 complex **2** (step a) can be considered an oxidative addition[3] of a coordinatively saturated d^{10} complex, **1**, or equivalently as an SN2 attack at carbon by the nucleophilic reagent **1**. Substrate reactivities ($CH_3 > RCH_2 > RR'CH$, and RI > RBr > ROTs > RCl) resemble classic SN2 reactions.[20] Furthermore, the observed stereochemistry (overall inversion[8]—eq 7, Figure 2) is consistent with inversion[23] in step a, Figure 1, followed by retention[24] in the migratory insertion, step b.

The form of the rate law for the oxidative-addition step a is also consistent with an SN2 reaction when account is taken of the dominant role which ion pairing plays in this step. The rate of step a is dramatically increased by employing more polar solvents.[20,25] For example, addition of 10% NMP to a THF solution of $Na_2Fe(CO)_4$ increases the rate of step a 100-fold. Synthetic applications for this solvent-dependent reactivity are illustrated in Figure 2. In THF primary chlorides are very slow (eq 3), whereas in NMP these are useful substrates (eq 8). Solvent effects on the rate of the alkylation step a have been examined quantitatively for THF and NMP. These results clearly indicate *the more dissociated species are kinetically more active.*

For NMP solutions of $Na_2Fe(CO)_4$ freezing-point-depression studies indicate that the dominant species is a uni-unielectrolyte.[25] For reaction (a) under pseudo-first-order conditions in iron, the observed second-order rate constant increases with decreasing $[Fe]_T$ (total of $Na_2Fe(CO)_4$ concentration). The ob-

(13) J. P. Collman and N. W. Hoffman, *J. Am. Chem. Soc.*, **95**, 2689 (1973).
(14) (a) J. Y. Merom, J. L. Roustan, C. Charrier, and *J. Organometal. Chem.*, **51**, C24 (1973); (b) J. P. Collman, M. P. Coo, J. N. Cawse, and ke R. G. Komoto, unpublished results.
(15) J. P. Collman, R. G. Komoto, R. Wahren, and P. L. Matlock, unpublished results.
(16) (a) R. Noyori, I. Umeda, and T. Ishigami, *J. Org. Chem.*, **37**, 1542 (1972); (b) R. K. Boeckman and R. Michalak, *J. Am. Chem. Soc.*, **96**, 1623 (1974); (c) S. Masamune, G. S. Bates, and P. E. Georghiou, *ibid.*, **96**, 3686 (1974).
(17) J. P. Collman, R. G. Homoto, and W. O. Siegl, *J. Am. Chem. Soc.*, **94**, 5905 (1972).
(18) J. P. Collman and J. K. Hoyano, unpublished results.
(19) J. P. Collman, J. N. Cawse, and J. I. Brauman, *J. Am. Chem. Soc.*, **94**, 5905 (1972).
(20) J. P. Collman, J. I. Brauman, R. G. Finke, and J. N. Cawse, unpublished results.
(21) This ion is often useful in forming kinetically stable crystalline salts of oxygen-sensitive anions: J. K. Ruff and W. J. S. Chlientz, *Inorg. Syn.*, in press.
(22) W. O. Siegl and J. P. Collman, *J. Am. Chem. Soc.*, **94**, 2516 (1972).
(23) Other oxidative additions found to go by inversion at carbon include: (a) P. K. Wong, K. S. Y. Lau, and J. K. Stille, *J. Am. Chem. Soc.*, **96**, 3956 (1974), and references therein; (b) G. M. Whitesides and D. J. Boschetto, *ibid.*, **91**, 4313 (1969).
(24) All alkyl–acyl migratory insertions which have been studied with chiral centers proceed with retention at that center—see, for example, ref 23b.
(25) (a) J. N. Cawse, Ph.D. Dissertation, Stanford University, 1973. (b) In NMP the solvent-separated ion pair may be the kinetically dominant species.

Table I
Effect of Solvents on Reaction Rates

Cation in THF	Alkyl to acyl[a] (b), Ph_2PMe, 0°	Alkyl ketone[a] reaction (e), CH_3I, 0°	Acyl ketone[a] reaction (f), CH_3I, 25°	$HFe(CO)_4^-$ [a–c] reaction, C_2H_5I, 35°
Li^+	6×10^{-2}	1×10^{-1}	1×10^{-3}	1×10^{-3}
$(Ph_3P)_2N^+$	$<2 \times 10^{-5}$	2×10^{-1}	1×10^{-1}	2×10^{-3}
Na^+	1×10^{-2}	1×10^{-1}	1×10^{-3}	1×10^{-3}
Na^+, 1% NMP	3×10^{-3}	1×10^{-1}	3×10^{-3}	
Na^+, 10% NMP	1×10^{-4}	1×10^{-1}	3×10^{-2}	

[a] Second-order rate constants, k_2, in M^{-1} sec^{-1} [b] The product of this reaction is ethane. [c] These values are imprecise because triple ions have not been taken into account. However the trends with respect to gegenion are probably correct.

served rate data can be accommodated by a single dissociative equilibrium (eq 16) with $K_D \approx 0.2\ M$. A

$$Na_2Fe(CO)_4 \overset{K_D}{\rightleftharpoons} Na^+ + \{NaFe(CO)_4^-\} \qquad (16)$$

conductometric titration of $Na_2Fe(CO)_4$ in NMP using cryptand as the titrant shows two equivalence points corresponding to the titration of the first and second sodium ion.[20] These conductivity data are consistent with the kinetic results. From the ratio of conductivities at the initial (no cryptand present) and the first equivalent point, K_D (eq 16) is calculated to be $K_D \leq 0.3\ M$.

An estimate of the dissociation constant for the second sodium ion is $\sim 10^{-3}\ M$, which is consistent with the notion that the free ion $Fe(CO)_4^{2-}$ is not kinetically important—even in NMP. This conclusion is also supported by a small (50%) common ion rate depression for step a in the presence of an 8-fold excess of $NaBPh_4$ over $Na_2Fe(CO)_4$ in NMP.[20]

In THF $Na_2Fe(CO)_4$ is much less dissociated and correspondingly less soluble than in NMP. Changes of conductivity of $Na_2Fe(CO)_4$ in THF as a function of concentration give an estimate of $K_D \approx 10^{-5}\ M$ (eq 16).[20] This is consistent with the failure of dicyclohexyl-18-crown-6 crown ether to produce a breakpoint in the conductometric titration in THF since in this solvent the sodium binding constants of this crown ether and $[NaFe(CO)_4]^-$ seem to be of comparable magnitude. Addition of 1 equiv of crown ether to $Na_2Fe(CO)_4$ in THF increases the rate of reaction a with both alkyl chlorides and bromides by $\sim$60-fold—again indicating a greater reactivity for the species $[NaFe(CO)_4]^-$. For such a small K_D the observed second-order rate constant should be proportional to $[Fe]_T^{-1/2}$. This relationship was found to be followed within experimental error. Activation energies were determined for reaction between $Na_2Fe(CO)_4$ and an alkyl bromide in THF. The large negative entropy of activation and small enthalpy ($\Delta S^\ddagger = -40 \pm 5$ eu; $\Delta H^\ddagger = 7.2 \pm 0.3$ kcal mol^{-1}) are similar to those found for other bimolecular oxidative-addition reactions involving transition metals and alkyl halides.

Thus the *20,000-fold difference in reactivity* of $Na_2Fe(CO)_4$ in NMP compared with THF[20,25a] can be accounted for by differences in the extent of dissociation of the first Na^+ in the two solvents with the assumption that $NaFe(CO)_4^-$ is the kinetically dominant species.[25b] Comparison of the reactivity of $Na_2Fe(CO)_4$ with those for other transition-metal nucleophiles can be made by using the Pearson logarithmic nucleophilicity parameter.[26a] The estimated value,[26b] 16.7, for $Na_2Fe(CO)_4$ in NMP is as high as any other nucleophile on record.

Ion-pairing effects[26c] are also important in the migratory insertion reaction converting alkyliron(0) complexes **2** into corresponding acyl complexes—step b, Figure 1. In our early studies we were surprised to find this reaction to be slower in better solvating media such as HMPA or NMP compared with THF.[19] Thus step b is dramatically retarded by adding a few percent NMP to THF solutions of **2** (Table I). Furthermore, the rate of migratory insertion shows marked dependence on the nature of the gegencation: $Li^+ > Na^+ \gg [(Ph_3P)_2N]^+$ (Table I).

Thus unusual behavior can be quantitatively accounted for by assuming that *the tight ion pair,* $NaRFe(CO)_4$, *is the kinetically active species*—more so even than a solvent-separated ion pair by a factor of $>10^2$. Further evidence for ion pairing comes from a crown ether conductometric titration of $NaRFe(CO)_4$ (**2**) which shows a sharp break at 1:1 crown ether to $NaRFe(CO)_4$ in THF. One equivalent of crown ether results in a 150-fold decrease in the rate of reaction b in THF.

The rate law for these migratory insertions is first order with respect to the tight ion pair of the acyl complex of **2** and first order with respect to ligand L.[20,25] The concentration dependence for the conductivity of the acyl salt **2** shows the presence of triple ions.[25b] The rate dependence of the migratory insertion reaction upon the total iron concentration has a nearly zero slope, but this can be qualitatively accommodated by taking into account ion-pairing equilibria which include the formation of triple ions.[26c]

$$[Na[RFe(CO)_4]Na]^+ + [RFe(CO)_4][Na][RFe(CO)_4]^-$$

triple ions

$$\updownarrow$$

$$\underset{\text{tight ion pair}}{\{NaRFe(CO)_4\}} \overset{K_1}{\rightleftharpoons} \underset{\text{solvent-separated ion pair}}{Na^+||RFe(CO)_4^-} \overset{K_2}{\rightleftharpoons} Na^+ + RFe(CO)_4^- \qquad (17)$$

The observed second-order rate law for reaction b is consistent either with a concerted mechanism or one involving formation of a hypothetical four-coordinate, unsaturated intermediate, **4** (Figure 3), under conditions where $k_1 \gg [L]k_2$. Because of prior evi-

(26) (a) R. G. Pearson, H. Sobel, and J. Songstad, *J. Am. Chem. Soc.*, **90**, 319 (1968). (b) This parameter was estimated from the rates of *n*-alkyl chloride in NMP, and then correcting for the reactivity ratios for methyl vs. *n*-alkyl and for chloride vs. iodide. We consider such parameters to have very limited value. (c) A general reference on ion-pairing effects is: "Ions and Ion Pairs in Organic Reactions", Vol. 2, M. Szwarc, Ed., Wiley-Interscience, New York, N.Y., 1974.

$$\{NaRFe(CO)_4\} \underset{k_{-1}}{\overset{k_1}{\rightleftharpoons}} \{NaRCFe(CO)_3\} \overset{k_2}{\underset{L}{\rightarrow}} \{RCFe(CO)_3L\}$$

2   **4**   **3**

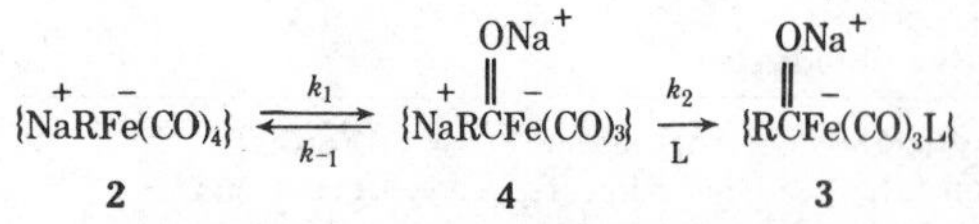

Figure 3. Rate $= k_1 k_2 [2][L]/(k_{-1} + [L]k_2)$. If $k_{-1} \gg [L]k_2$, rate $= (k_1/k_{-1})k_2[2][L]$. { } denotes tight ion pair.

R = H, C_2H_5 **7**

Na$^+$ δ 279 δ 362 δ 254

[(Ph$_3$P)$_2$N]$^+$ δ 261

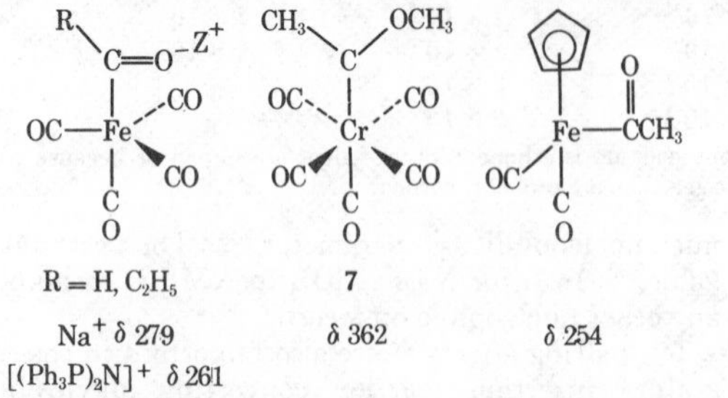

Figure 4. ^{13}C chemical shifts for acyl and carbene carbon atoms in ppm relative to TMS.

dence for a two-step path in alkyl–acyl migratory insertions,[27] we prefer this mechanism in the present case. Furthermore, the near-zero activation entropy ($\Delta S^{\ddagger} = -2$ eu) found for reaction b is inconsistent with a concerted mechanism. Nevertheless we have so far not been able to find conditions under which the rate law for (b) deviates from second-order behavior.

In THF the acyliron(0) complex **3** also forms a strong ion pair with Na$^+$ (or Li$^+$), as shown by conductometric titration (Figure 3), and the cation dependence of the acyl ir stretching mode. By using this change in the infrared spectrum, we were able to show that the acyl complex forms a more stable ion pair, **6**, than that of the alkyl complex, **5** (eq 18). The

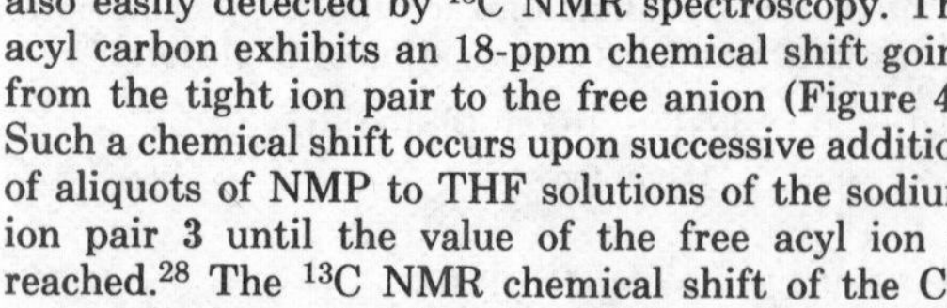

$$[(Ph_3P)_2N]^+[C_2H_5\overset{O}{\overset{\|}{C}}Fe(CO)_4]^- + \{Na^+C_2H_5Fe(CO)_4^-\} \overset{K}{\underset{\frac{K}{100}>}{\rightleftharpoons}}$$

$$\{C_2H_5\overset{ONa^+}{\overset{\|}{C}}Fe(CO)_4\} + [(Ph_3P)_2N]^+[C_2H_5Fe(CO)_4]^- \quad (18)$$

6

formation of the sodium–acyliron tight ion pair **6** is also easily detected by ^{13}C NMR spectroscopy. The acyl carbon exhibits an 18-ppm chemical shift going from the tight ion pair to the free anion (Figure 4). Such a chemical shift occurs upon successive addition of aliquots of NMP to THF solutions of the sodium ion pair **3** until the value of the free acyl ion is reached.[28] The ^{13}C NMR chemical shift of the CO groups (a sharp singlet) is insensitive to ion pairing, suggesting that the cation is associated with the acyl oxygen in **6**. The acyl tight ion pair **6** may be considered structurally related to carbene complexes such as **7** (Figure 4) which exhibit a characteristically low-field ^{13}C NMR signal for the carbene carbon.[29]

The reverse of step b which would involve conver-

(27) (a) I. S. Butler, F. Basolo, and R. G. Pearson, *Inorg. Chem.*, **6**, 2074 (1967); (b) R. W. Glyde and R. J. Mawby, *Inorg. Chim. Acta*, **4**, 331 (1970); **5**, 317 (1971); *Inorg. Chem.*, **10**, 854 (1971).

(28) This ^{13}C chemical shift vs. NMP "titration curve" exhibits a breakpoint at 2.5 NMP per Na$^+$.

(29) C. G. Kreiter and V. Formacek, *Angew. Chem., Int. Ed. Engl.*, **11**, 141 (1972); G. M. Bodner, S. B. Kahl, K. Bork, B. N. Storhoff, J. E. Wuller, and L. J. Todd, *Inorg. Chem.*, **12**, 1071 (1973).

$$RCFe(CO)_4^- + R'X \text{ (or } H^+\text{)} \longrightarrow \text{[9]}$$

$$\downarrow (C_2H_5)_3O^+BF_4^-$$

10 $\xrightarrow{\ \ \ \ }$ RCR' or RCH

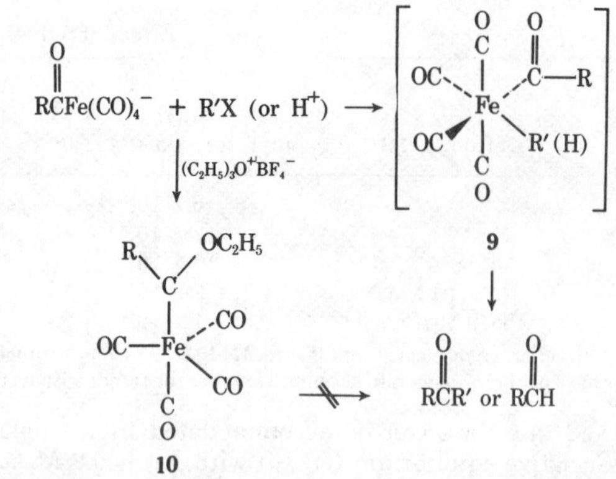

Figure 5.

sion of **3** to **4** in Figure 3 is very slow, as shown by the failure of the acyl complex **3** to exchange with ^{13}CO under ambient conditions. This kinetic stability of acyl complexes was employed to prepare the thermodynamically unstable formyl complex **8** (eq 19)[30] by a variation of step c (Figure 1). Hitherto no formyl complex had been characterized. Apparently the failure of the hydride NaHFe(CO)$_4$ to form the formyl complex under 1000 psi CO is due to an unfavorable equilibrium (K_1/K_{-1} very small in Figure 3, R = H). A similar explanation can be advanced for the failure of electronegatively substituted alkyl groups (such as **11**, eq 20) to migrate, thus limiting the scope of the sequence ((a), (b), (d)) forming aldehydes.

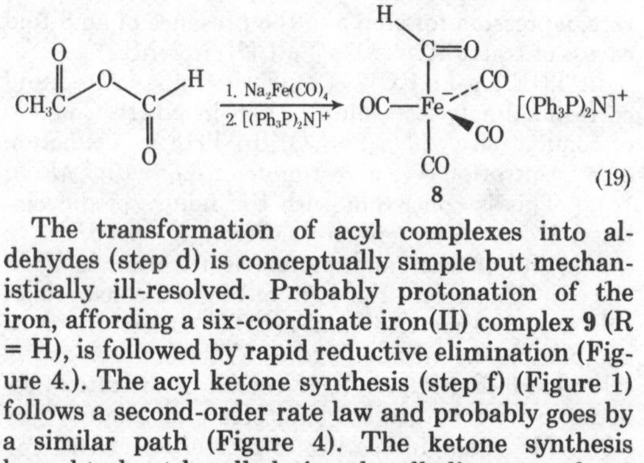

$$CH_3\overset{O}{\overset{\|}{C}}\overset{O}{\underset{\|}{O}}\overset{O}{\overset{\|}{C}}H \xrightarrow[2.\ [(Ph_3P)_2N]^+]{1.\ Na_2Fe(CO)_4} \text{[8]} [(Ph_3P)_2N]^+ \quad (19)$$

The transformation of acyl complexes into aldehydes (step d) is conceptually simple but mechanistically ill-resolved. Probably protonation of the iron, affording a six-coordinate iron(II) complex **9** (R = H), is followed by rapid reductive elimination (Figure 4.). The acyl ketone synthesis (step f) (Figure 1) follows a second-order rate law and probably goes by a similar path (Figure 4). The ketone synthesis brought about by alkylating the alkyliron complex **2** (step e, Figure 1) is also a second-order reaction (100 times faster than step f; see Table I).[20] This "alkyl ketone synthesis" is more complex and must involve at least two intermediates. Alkyl–alkyl hydrocarbon coupling has not been detected.[31] That step e, Figure 1, does not involve the unsaturated acyl intermediate **4** (Figure 4) is certain because (e) does not show parallel kinetic responses due to the ion-pairing effects described above for the alkyl–acyl migration (see Table I). Intermediates such as **9** have not been detected. That alkylation at oxygen affording an oxycarbene intermediate, **10**, is not involved in step f is clear since **10**[32] was found to be quite stable (Figure

(30) J. P. Collman and S. R. Winter, *J. Am. Chem. Soc.*, **95**, 4089 (1973).

(31) The reductive elimination of two saturated alkyl groups forming an alkane would be a most useful synthetic procedure; however, this seems to be an unfavorable process.

(32) (a) N. W. Hoffman, Ph.D. Dissertation, Stanford University, 1973; (b) H. L. Conder and M. Y. Darensbourg, *J. Organometal. Chem.*, **67**, 93 (1974).

5). Similar, stable acyloxy carbene complexes are formed by reaction of the anionic acyl complexes **3** with acid chlorides blocking a potential synthesis of α-diketones.[33]

The mechanisms for paths g, h, i, and j (Figure 1) have not been studied. It seems probable that these involve oxidation of **2** or **3**, affording very reactive iron(III) acyl complexes. Oxidative enhancement of migratory insertion and solvolysis of oxidized acyls have precedents.[34]

Mechanisms for the reducing reactions (eq 11 and 12, Figure 2) are poorly understood. The coordinatively saturated hydride slowly adds to α,β-unsaturated esters, affording the kinetically determined product **11** (eq 20) which is reduced by protonation (step k, Figure 1),[5] but the first step in eq 20 is too slow to account for the rate of reductions in reactions such as eq 11 and 12 (Figure 2).

$$CH_2{=}CHCO_2CH_3 \xrightarrow[2.\,[(Ph_3P)_2N]^+]{1.\,NaDFe(CO)_4} [DCH_2CH[Fe(CO)_4]CO_2CH_3]^-[(Ph_3P)_2N]^+ \quad (20)$$

11

We are still investigating mechanisms for the reduction reactions (eq 11, 12) and oxidatively induced migration.[34] These may bring to a close our studies of this reagent.

This Account is based on the experimental and intellectual efforts of my students and associates: J. Cawse, D. R. Clark, M. P. Cooke, R. Finke, N. W. Hoffman, R. G. Komoto, P. L. Matlock, W. O. Siegl, R. Wahren, and S. R. Winter; support was provided by the National Science Foundation (MPS70-01722-A03) and the Center for Materials Research (N00014-67-A-0112-0056 and DAHC15-73-G15). I am also indebted to my colleague J. I. Brauman for collaboration in the mechanistic studies discussed herein.

Addendum (May 1976)

Since the original publication of this Account, four salient developments have occurred.

X-ray crystallographic structure determinations of $Na_2Fe(CO)_4 \cdot 1.5$ dioxane,[35] $K_2Fe(CO)_4$,[36] and (222-cryptate·Na)$_2$Fe(CO)$_4$[36] have been completed. The free $Fe(CO)_4^{2-}$ unit in (222-cryptate·Na)$_2$Fe(CO)$_4$ has T_d symmetry, C–Fe–C angle 109°. However two types of interactions between Na^+ and $Fe(CO)_4^{2-}$ in $Na_2Fe(CO)_4 \cdot 1.5$ dioxane significantly distort the $Fe(CO)_4^{2-}$ unit. Notably, long range Na^+ . . . C and Na^+ . . . -Fe interactions (2.95 and 3.09 Å, respectively) are probably responsible for the distortion of one C–Fe–C angle to 129.7°. Shorter range Na^+ . . . O interactions also occur at 2.32 Å. In $K_2Fe(CO)_4$, again there is an association of the alkali metal ions with the C–Fe–C region of $Fe(CO)_4^{2-}$. The K . . . Fe distance (3.62 Å) is significantly longer than the Na . . . Fe distance, suggesting a weaker M^+ . . . Fe interaction. This is consistent with the lower degree of angular distortion in $K_2Fe(CO)_4$ (C–Fe–C = 121.0°).

An x-ray crystallographic structure determination of $[(Ph_3P)_2N]^+[HFe_2(CO)_8]^-$ has appeared.[37] The $HFe_2(CO)_8^-$ unit is isostructural with $Fe_2(CO)_9$, showing two bridging carbonyl groups and suggesting a symmetrically bridging hydride.

A complete mechanistic study of the selective reduction of the double bond of α,β-unsaturated carbonyl compounds by $NaHFe_2(CO)_8$ in THF has been completed.[38] A preliminary report has appeared.[39] Several unusual mechanistic features were found, including: (a) an associative mechanism for $RCH{=}CHCO_2R'$ coordination to $NaHFe_2(CO)_8$; (b) reversible, regiospecific migratory insertion, $Na^+[HFe_2(CO)_8(RCH{=}CHCO_2R')] \rightleftharpoons Na^+[RCH_2CH(Fe_2(CO)_8)CO_2R']^-$; and (c) rate-determining iron–iron bond cleavage, $Na^+[RCH_2CH(Fe_2(CO)_8)CO_2R']^- \rightarrow Na^+[RCH_2CHC(Fe(CO)_4)CO_2R']^- + Fe(CO)_4$.

Mechanistic studies of oxidative-addition to $Na_2Fe(CO)_4$, the migratory insertion reaction, and the ketone reaction (steps a, b, and f, respectively, Figure 1) are now complete and have been submitted for publication.[20]

The assistance of Richard G. Finke in updating the material in this Account is gratefully acknowledged.

(33) Low yields of α-diketone have been found after destructive distillation mixtures formed from acid chlorides and the acyl complex: Y. Sawa, M. Ryang, and S. Tsutsumi, *J. Org. Chem.*, **35**, 4183 (1970).

(34) (a) K. M. Nicholas and M. Rosenblum, *J. Am. Chem. Soc.*, **95**, 4449 (1973); (b) M. Rosenblum, *Acc. Chem. Res.*, **7**, 122 (1974).

(35) H. B. Chin and R. Bau, *J. Am. Chem. Soc.*, **98**, 2434 (1976).

(36) R. Bau, R. G. Teller, H. Chin, R. G. Finke, and J. P. Collman, unpublished results.

(37) H. P. Chin, Ph.D. Thesis, University of Southern California, 1975.

(38) J. P. Collman, R. G. Finke, P. L. Matlock, and J. I. Brauman, to be submitted for publication.

(39) J. P. Collman, R. G. Finke, P. L. Matlock, and J. I. Brauman, *J. Am. Chem. Soc.*, **98**, 4085 (1976).

Organic Chemistry of Metal Vapors

Kenneth J. Klabunde

Department of Chemistry, University of North Dakota, Grand Forks, North Dakota 58202

Received September 27, 1974

Reprinted from ACCOUNTS OF CHEMICAL RESEARCH, *8, 393 (1975)*

An intense interest in the synthetic uses of high-temperature species at low temperature has developed in recent years. Metal atoms (vapors)[1] are among the most easily produced high-temperature species, and their synthetic chemistry is interesting and useful. This Account deals with the organic chemistry of these metal vapors, especially transition-metal vapors.

The chemistry of high-temperature species is an old and fruitful field. Most studies of high-temperature species have been carried out at low temperature (e.g., 77 K). Reasons for low temperature are twofold: (1) reaction rates are slowed down so that intermediate reactive species can be observed spectroscopically (i.e., matrix isolation spectroscopy); (2) usually the high-temperature species must be generated in a vacuum system, and in order to maintain low vapor pressures in the system all incoming reactants must be immediately condensed.

Many reactive particles can be considered to come under the heading of high-temperature species. Hydrogen atoms,[2] organic radicals,[3] alkali metal atoms,[2,4] divalent carbon,[5] boron,[6] silicon,[7] free atoms of the transition elements,[8] and others have been produced. However, even though there have been many excellent flow system and matrix isolation studies of such species,[2,8] only in very recent years have chemists really started using these materials on a synthetic basis, that is, as reagents to make new, interesting, and usually isolable compounds. In this regard, Timms[9] has written an excellent review in an attempt to orientate the whole field of synthesis employing high-temperature species at low temperature. Many species are discussed in that review of 1972, which mainly covered inorganic reactions. Since that time, however, much work has been done, most of it concerned with organic reactions of metal atoms.

Metal Vaporizations

Using metal vapors as chemical reagents sounds exotic. However, for many years industrial concerns have vaporized metals for deposition as thin films on materials ranging from plastics to clean metal films, and the technology needed even for carrying out industrial scale continuous vaporizations is available. Only in very recent years have chemists begun to apply these vaporization methods in chemical reactions.

Most metals vaporize as mainly monatomic species.[1] Thus, the metallic elements may now be studied in their atomic state, and we can begin to fill a void concerning their chemistry. That is, much is known about the chemistry of metals through their compounds, but very little is known about the chemistry of the elements themselves. *The vaporization method yields these elements in reactive high chemical potential states, a condition which allows each element to have a rich and varied chemistry of its own at low temperature.*

Pioneering Work

Skell and Wescott[10] must be given much of the credit for bringing things to the point where synthesis can be carried out using high-temperature species. They devised a carbon vapor reactor for study of the reactions of carbon species, C_1, C_2, and C_3, with organic compounds at 77 K (−196°C).[11] Carbon was vaporized by arcing carbon electrodes in a vacuum chamber. The resulting carbon vapor was codeposited with organic substrates on the liquid-nitrogen-cooled walls of the reactor. Today, the same basic reactor design is used for most metal-vapor chemistry.

At the same time that the Skell and Wescott work

Kenneth J. Klabunde, born in 1943 in Madison, Wisconsin, received his B.A. from Augustana College and Ph.D. from the University of Iowa (with D. J. Burton). He went on to do postdoctoral work at The Pennsylvania State University (with P. S. Skell), and in 1970 was appointed Assistant Professor at the University of North Dakota, where he is now Associate Professor. His research interests are both organic and inorganic, including (1) reactive atoms and molecules as synthetic reagents, (2) organometallic and fluorocarbon compounds, and (3) catalysis in fuel conversion processes.

(1) B. Siegel, *Q. Rev. Chem. Soc.*, **19,** 77 (1965).
(2) E. W. R. Steacie, "Atomic and Free Radical Reactions", 2nd ed, Reinhold, New York, N.Y., 1954; R. J. Cvetanovic, *Adv. Photochem.*, **1,** 115 (1963).
(3) A. M. Brass and H. P. Broida, "Formation and Trapping of Free Radicals", Academic Press, New York, N.Y., 1960.
(4) Von B. Mile, *Angew. Chem.*, **80,** (13), 519 (1968).
(5) W. Kirmse, "Carbene Chemistry", Academic Press, New York, N.Y., 1971; M. Jones, Jr., and R. A. Moss, Ed., "Carbenes", Vol. 1, Wiley, New York, N.Y., 1973.
(6) P. L. Timms, *Chem. Commun.*, 258 (1968).
(7) P. S. Skell and P. W. Owen, *J. Am. Chem. Soc.*, **94,** 5434 (1972); also cf. *ibid.*, **89,** 3933 (1967).
(8) D. M. Mann and H. P. Broida, *J. Chem. Phys.*, **55,** 84 (1971); M. Poliakoff and J. J. Turner, *J. Chem. Soc., Perkin Trans. 2,* 70, 93 (1974); D. W. Green and D. M. Gruen, *J. Chem. Phys.*, **60,** 1797 (1974); G. A. Ozin and A. V. Voet, *Acc. Chem. Res.*, **6,** 313 (1973).
(9) P. L. Timms, *Adv. Inorg. Radiochem.*, **14,** 121 (1972).
(10) P. S. Skell, L. D. Wescott, J. P. Golstein, and R. R. Engel, *J. Am. Chem. Soc.*, **87,** 2829 (1965).
(11) P. S. Skell, J. J. Havel, and M. J. McGlinchey, *Acc. Chem. Res.*, **6,** 97 (1973).

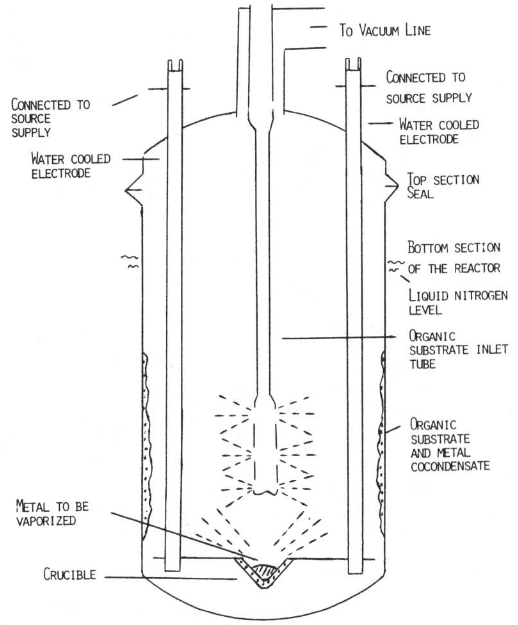

Figure 1.

appeared, Timms, Kents, Ehlert, and Margrave published their first detailed work on the low-temperature chemistry of high-temperature-generated SiF_2. They used techniques similar to those of Skell and Wescott.[12]

Timms was the first to apply these techniques to metals.[13] He has been concerned with inorganic as well as organic substrates, and has modified and improved the method to the point where it is now a viable synthetic technique.[9]

The amounts of metals that can be vaporized in current reactors are substantial, usually 0.5–2 g, sometimes up to 50 g, and reactors capable of handling kilograms of metals may be feasible.[14]

Apparatus, Reaction Conditions, and Limitations

We have designed an apparatus that is simple to assemble and operate. A description of it was published previously,[15,16] and the theory and operation of it are explained in Timms' article[9] and by Figure 1. A simple metal atom reactor of this type, suitable for undergraduate laboratory work as well as for research, will soon be marketed by Kontes–Martin Glass Co.

In this apparatus metals are vaporized using resistive heating. The metal vapor deposits on the inside walls of the reactor while vapors of organic substrate are directed through a shower head and codeposited in the same area. Low-temperature walls are required from a practical viewpoint since, in order to maintain good vacuum and not allow the substrate to come into the zone of the high-temperature source, the reactants must immediately be condensed. This must be a temperature where the vapor pressure of the reactants is $\lesssim 10^{-5}$ Torr. Experimentally, this is not difficult to achieve since liquid nitrogen cooling is satisfactory. However, higher temperatures are sometimes desirable in terms of increased reaction rates between the substrate and high-temperature species A competing low activation energy process is always repolymerization of the vaporized species, which is often a serious problem, one that will be discussed in more detail later.

Similar codeposition setups can be used employing metal vaporizations with electron guns,[7,9,17] arcs,[18] and lasers.[19] Metals Mg, Ca, Ti, V, Cr, Mn, Fe, Co, Ni, Pd, Pt, Cu, Ag, Zn, and Cd have all been vaporized using resistive heating and our simple apparatus. Also, by resistive heating others have vaporized Mo, W, Au, Ge, and Sn as well. Thus, we are not limited in vaporization techniques. Instead, the main disadvantage in using a vapor deposition reactor such as the one described is the substrate vaporization. It must enter the reactor as a vapor, which limits its boiling range to about −80 to +200°C at atmospheric pressure. Substrates boiling below that (N_2, CO, $CH_2{=}CH_2$) cause high vapor pressures in the reactor, and substances boiling above that cause difficulty in just passing them into the reactor even under good vacuum. Refinements of the apparatus utilizing heated inlet passages or liquid inlets with "flash pans" extend the range somewhat. However, in order to study solid or liquid substrates dissolved in solutions, Timms[20,21] and Green[17] have cleverly devised rotating solution reactors where the metal vapors proceed upward into a cold film of solution on the upper inside of the reactor, or are sprayed onto the walls as a solution by a rapidly rotating disk. The vaporization source is stationary, but the reactor revolves in a cold bath exactly like a rotary evaporator apparatus. The low-temperature bath is cold enough to keep the vapor pressure of the solvent and substrate low enough so that the metal can be vaporized.[20,21] Alkane solvents are best since they appear not to interact with metal atoms.

We have carried out solution-phase metal atom reactions using the reactor shown in Figure 1.[15,16] However, special vaporization sources were employed so that the metal vapor was directed downward into a cold, stirred solution of the desired reagents. This method does appear to have wide applicability, and will certainly find more and more use in our laboratories.

Organic Reactions

The reactivity of each metal–substrate pair depends on a number of complicated factors. Among these are acid–base properties, electronic spin state of the metal atom, availability of orbitals for π com-

(12) P. L. Timms, R. A. Kent, T. C. Ehlert, and J. L. Margrave, *J. Am. Chem. Soc.*, **87**, 2824 (1965).
(13) P. L. Timms, *Chem. Commun.*, 1525 (1968).
(14) W. Reichelt, Symposium "Metal Atoms in Chemical Synthesis," Merchsche Gesellshaft fur Kunst and Wissenshaft e.v., Darmstadt, West Germany, May 1974, *Angew. Chem.*, **87**, 239 (1975); *Angew. Chem., Int. Ed. Engl.*, **14**, 218 (1975).
(15) K. J. Klabunde, *Angew. Chem.*, **87**, 309 (1975); *Angew. Chem. Int. Ed. Engl.*, **14**, 287 (1975).
(16) K. J. Klabunde and H. F. Efner, *Inorg. Chem.*, **14**, 789 (1975).
(17) F. W. S. Benfield, M. L. H. Green, J. S. Ogden, and D. Young, *J. Chem. Soc., Chem. Commun.*, 866 (1973).
(18) P. S. Skell and J. E. Girard, *J. Am. Chem. Soc.*, **94**, 5518 (1972); also private communications with P. S. Skell and J. E. Girard.
(19) E. Koerner von Gustorf, O. Jaenicke, and O. E. Polansky, *Angew. Chem., Int. Ed. Engl.*, **11**, 532 (1972).
(20) P. L. Timms, *Chem. Eng. News*, **52**, 23 (April 29, 1974); R. MacKenzie and P. L. Timms, *J. Chem. Soc., Chem. Commun.*, 650 (1974).
(21) P. L. Timms, *Angew. Chem.*, **87**, 295 (1975); *Angew. Chem., Int. Ed. Engl.*, **14**, 273 (1975).

plexation, and, very crudely, the energy input (heat of vaporization) to the metal. Probably most important are the acid–base properties and the π complexation (very low activation energy processes, which is needed for very low temperature reactions). Indeed, one of the most efficient reactions we have found is Ca atoms (strong base) with perfluoro olefins (capable of accepting bases and possessing π systems).[22] Furthermore, it would be expected that metals with low ionization potentials would interact efficiently with electron-acceptor systems. Also, weak bonds (e.g., C–I) should be susceptible to radical-type processes with metal atoms, and metals with readily available d orbitals, such as the transition metals, should complex well with organic compounds possessing nonbonding electrons or π electrons.

Intuitively, it would be expected that electron-transfer processes (charge transfer), radical abstraction processes, and σ-donor and π-donor complexations should take place when organic molecules are allowed to cocondense with metal atoms. However, always in competition with these processes is the repolymerization of the metal atoms. Since this repolymerization is of obvious importance (with regard to metal film formations and to efficiency of metal atom–substrate reactions), a great deal of research regarding this process is warranted. As yet, little definitive work has been done.

Regarding the dimerization of metal atoms in a low-temperature matrix Hanlan and Ozin have found the dimers to be readily formed and reactive with CO to form unusual metal carbonyl dimers (by matrix isolation spectroscopy).[23] From qualitative observations in our laboratory we can say that, unless nonbonding or π electrons are available in the chemical substrate being deposited with the metal atoms (so that low-temperature complex formation can occur), the metal atoms will rapidly repolymerize at 77 K. Thus, alkanes have little effectiveness at preventing metal atom recombination at 77 K. Perfluoroalkanes behave similarly to alkanes in this respect, which would indicate that nonbonding electrons that are tightly held (nonbasic) are not effective in complexation with metal atoms.

Organohalides with Metal Atoms. The formation of Grignard reagents is a famous reaction of a metal with organohalides. Skell and Girard[18] were able to produce the long-sought-after nonsolvated Grignard reagents simply by cocondensing Mg atoms with R–X. The formation of these RMgX reagents did not take place immediately upon −196°C cocondensation. Apparently first a black RX–Mg complex was formed that rearranged to RMgX on matrix warm-up. Although not yet investigated in much detail, Skell and Girard's nonsolvated Grignard reagents exhibited unusual properties. Thus, acetone was not added by the nonsolvated reagents—only proton abstraction took place. Also, crotonaldehyde added the Grignard 1,2 rather than 1,4 as normally.[18]

Skell and Girard have made the only comparisons of thermally vaporized metal vs. arc vaporized metal. Thermally (resistively heated) vaporized Mg yields atoms in a ground singlet state which readily form charge-transfer complexes at low temperature with alkyl halides and ammonia. However, arc vaporized Mg (brilliant purple arc) yields high concentrations of excited triplet state Mg atoms which react in radical-type processes.[18]

(22) K. J. Klabunde, J. Y. F. Low, and M. S. Key, *J. Fluorine Chem.*, **2,** 207 (1972).
(23) L. A. Hanlan and G. A. Ozin, *J. Am. Chem. Soc.*, **96,** 6324 (1974).

Timms reported very early on the reactions of Cu, Ag, and Au atoms with alkyl halides.[9,13] These metals, especially Cu, acted as halogen atom abstraction reagents. Alkyl radicals were formed which coupled and disproportionated. In some situations these types of reactions can be synthetically useful, i.e., B_2Cl_4 can be produced in 10-g batches.[24]

$$2\text{Cu atoms} + BCl_3 \longrightarrow B_2Cl_4 + 2CuCl$$

The radical-abstraction work is somewhat reminiscent of Skell's early work on the vapor-phase reactions of Na–K vapor with organohalides.[25] Here, however, high temperatures were employed, and some Wurtz-type syntheses were quite successful.

Allyl halides react with metal atoms. Piper and Timms[26] cocondensed Ni atoms with allyl chloride and allyl bromide to yield the π-allyl dimers. The re-

$$\text{Ni atoms} + CH_2{=}CH{\cdot}CH_2X \longrightarrow (\eta^3\text{-}C_3H_5NiX)_2$$

actions were efficient, and this serves as a good method for producing the complexes. Platinum atoms[27] reacted with allyl halides in the same fashion, as did palladium atoms[28] (under some conditions even palladium metal itself reacts with allyl halides).

Most of our metal atom research has been concerned with organohalide reactions. First, we reported on the production of nonsolvated perfluoroorganozinc halides.[29] Only iodides reacted with the

$$\text{Zn atoms} + R_f\text{–I} \longrightarrow R_f\text{–Zn–I}$$

readily formed zinc vapor. It was interesting to find that these nonsolvated R_fZnI compounds were reactive and unstable even at relatively low temperatures, which is not true for the solvated analogs.[30] In the case of CF_3ZnI, decomposition to yield CF_2 readily took place.

Calcium atoms reacted very efficiently with perfluoro olefins and hexafluorobenzene to yield C–F insertion.[22] This is the only C–F metal atom insertion reported. Perfluoro alkanes did not react, but perfluoro unsaturated systems were quite capable of accepting the strong-base calcium atom.

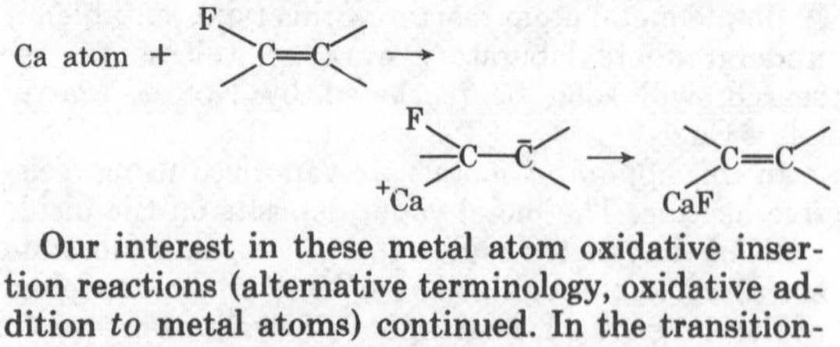

Our interest in these metal atom oxidative insertion reactions (alternative terminology, oxidative addition *to* metal atoms) continued. In the transition-metal series many fascinating species can be envisioned by carrying out these insertions. For example,

(24) P. L. Timms, *J. Chem. Soc. A*, 830 (1972).
(25) P. S. Skell and R. G. Deorr, *J. Am. Chem. Soc.*, **89,** 4688 (1967).
(26) M. J. Piper and P. L. Timms, *J. Chem. Soc., Chem. Commun.*, 50 (1972).
(27) P. S. Skell and J. J. Havel, *J. Am. Chem. Soc.*, **93,** 6687 (1971).
(28) H. F. Efner, unpublished results from this laboratory.
(29) K. J. Klabunde, M. S. Key, and J. Y. F. Low, *J. Am. Chem. Soc.*, **94,** 999 (1972).
(30) R. D. Chambers, W. K. R. Musgrave, and J. Savory, *J. Chem. Soc.*, 1993 (1962).

the formation of RNiX, RPdX, and RPtX seemed interesting since RPdX species (no other ligands) have been proposed as reaction intermediates in a number of important reactions,[31] such as palladium-promoted arylations, alkylations, etc. Similarly, RCOPdCl has been discussed as a likely short-lived species in palladium-catalyzed CO insertion reactions, olefin carbonylations, and reaction mechanisms relating to the Rosenmund reduction.[31,32]

Using the metal atom technique we succeeded in synthesizing some RMX and RCOMX (M = Ni, Pd, Pt) species.[15,33-35] Although we expected high reactivity and thermal instabilities for many of these, the low temperature and inert conditions afforded by this technique served to stabilize the species. So after completion of the codeposition of M atoms and R–X, the reaction matrix could be warmed to some desired temperature (e.g., −78°) and then a secondary substrate added to react with or trap the RMX species.

$$\text{M atom} + \text{R–X} \xrightarrow{\text{PEt}_3} \begin{array}{c} \text{PEt}_3 \\ | \\ \text{RMX} \\ | \\ \text{PEt}_3 \end{array}$$

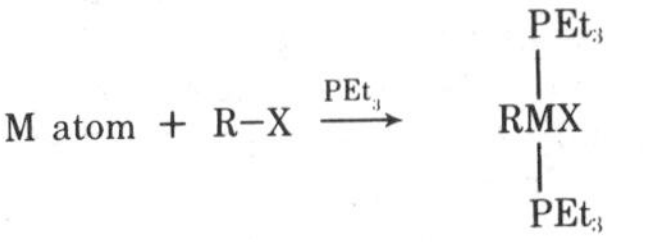

stable adduct

For nickel, we found that *alkyl* halide reactions only yielded NiX_2 and products resulting from R· coupling and disproportionation reactions. No direct evidence for RNiX was obtained, although it is possibly an intermediate en route to products. Similar results were obtained for C_6H_5NiX. However, in the case of R_fNiX and C_6F_5NiX, low-temperature trapping studies were successful, and trapping experiments at different temperatures showed that C_6F_5NiCl was formed at very low temperature (< −130°C), and decomposed about −80°C.[15]

For palladium, no direct evidence for RPdX was obtainable when R = CH_3, C_2H_5. Products attributable to R· formation were found, although these products could also be explained in terms of RPdX reactions.[34] Palladium atoms with C_6H_5Br yielded C_6H_5PdBr. This species was found to be unstable on warming above ca. −100°C.[34] However, palladium atoms with C_6F_5X yielded C_6F_5PdX, found to be stable at room temperature and only slightly sensitive to air or water. We expected the C_6F_5 moiety to lend stability to the RMX species, but were surprised to find such good stability for this formally coordinatively unsaturated compound. This material can be isolated as a red-brown powder, is soluble in organic solvents (dimeric and trimeric), and adds ligands instantaneously to form the bis adducts (trans).[36] Perfluoroalkylpalladium halides are also stable compounds according to our results. Thus, CF_3Br, CF_3I,

$$C_6F_5PdBr + L \longrightarrow \begin{array}{c} L \\ | \\ C_6F_5PdBr \\ | \\ L \end{array}$$

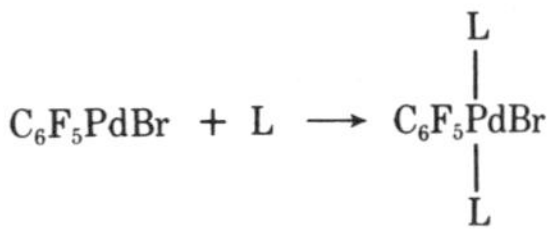

L = NR_3, SR_2, PR_3, AsR_3, NH_3, etc.

C_2F_5I, and n-C_3F_7I yield R_fPdX compounds that are soluble in organic solvents and only moderately air and temperature sensitive.

One nonfluorinated RPdX compound has been found to be stable at room temperature; that is $C_6H_5CH_2PdCl$ formed efficiently from Pd atoms and $C_6H_5CH_2Cl$. This material is a rust-colored solid soluble in some organic solvents and air sensitive. It decomposes in solution at about 40°C, and at 100°C in the solid state. Spectral evidence (NMR, uv) tends to indicate allyl-type bonding with the arene system.[37,38] On incremental addition of Et_3P to this compound (I), a π- to σ-rearrangement is observed first (loss of allyl type bonding), and then a second mole of Et_3P is taken up to form the stable bisphosphine adduct.[37] The chemistry of $C_6H_5CH_2PdCl$ is

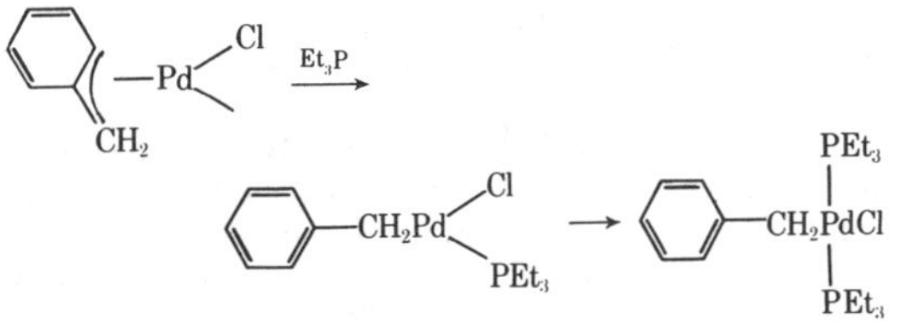

quite interesting, and corresponds to that which would be predicted on the basis of Heck's proposal of $C_6H_5CH_2PdCl$ as a catalysis intermediate.[31b] Other similar compounds that have good stability are $C_6H_5CH(CF_3)PdCl$ and various Me-substituted systems. In some of the substituted systems, low-temperature NMR studies indicate that preferred conformers can be frozen out.

Acyl halides also oxidatively add to palladium atoms. In all examples studied the resultant product readily liberated CO. In carrying out low-temperature Et_3P trapping studies and product analyses, we were able to learn a lot about what R groups stabilized the intermediates, their thermal stabilities, and modes of decomposition. In the case of R = n-C_3F_7 the $R_fCOPdCl$ could be trapped at −78°C, but when R = CF_3 some $R_fCOPdCl$ and R_fPdCl were trapped, and when R_f = C_6F_5 only $R_fCOPdCl$ was trapped.

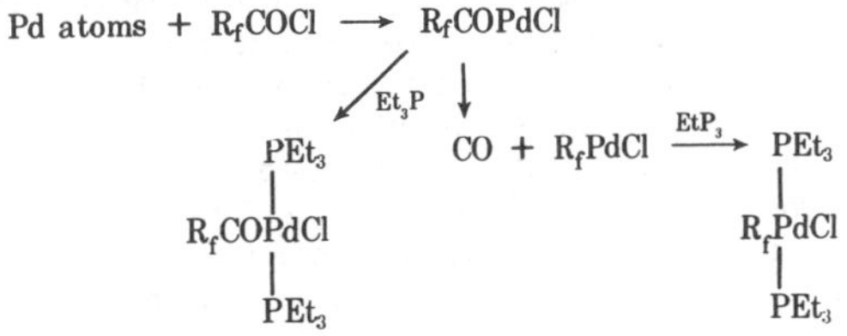

Platinum atoms reacted with C_6F_5Br and $C_6H_5CH_2Cl$ to yield stable RPdX compounds. In the case of C_6F_5PtBr, addition of Et_3P yielded both the cis and trans bisphosphine adducts (in the Pd case

(31) (a) R. F. Heck, *J. Am. Chem. Soc.*, **90**, 5518, 5526, 5531, 5535, 5546 (1968); P. M. Henry, *Tetrahedron Lett.*, 2285 (1968); T. Hosokawa, C. Calvo, H. B. Lee, and P. M. Maitlis, *J. Am. Chem. Soc.*, **95**, 4914, 4924 (1973); (b) R. F. Heck and J. P. Nolley, Jr., *J. Org. Chem.*, **37**, 2320 (1972); H. A. Dieck and R. F. Heck, *J. Am. Chem. Soc.*, **96**, 1133 (1974).

(32) J. Tsuji, M. Morikawa, and J. Kiji, *J. Am. Chem. Soc.*, **86**, 4851 (1964); *Tetrahedron Lett.*, 1437 (1963); J. Tsuji, K. Ohno, and T. Kajimoto, *ibid.*, 4565 (1965).

(33) K. J. Klabunde and J. Y. F. Low, *J. Organomet. Chem.*, **51**, C33 (1973).

(34) K. J. Klabunde and J. Y. F. Low, *J. Am. Chem. Soc.*, **96**, 7674 (1974).

(35) K. J. Klabunde, J. Y. F. Low, and H. F. Efner, *J. Am. Chem. Soc.*, **96**, 1984 (1974).

(36) B. B. Anderson, unpublished results from this laboratory.

(37) J. S. Roberts, unpublished results from this laboratory; J. S. Roberts and K. J. Klabunde, *J. Organomet. Chem.*, **85**, C13 (1974).

(38) R. B. King and A. Fronzaglia, *J. Am. Chem. Soc.*, **88**, 709 (1966); F. A. Cotton and M. D. LaPrade, *ibid.*, **90**, 5418 (1968).

only trans complexes have been found).

Thus, using the metal atom technique we have been able to produce some interesting RMX species that have not as yet been available by other techniques. Information on stabilities and modes of decomposition has been obtained. The most interesting facet of this work is that some of these RMX species are much more stable than any one would have anticipated. More structural and chemical properties need determination, and other metals are being investigated (such as Co, Fe, Mn, and Ti).

Some general observations we have made in these studies have been useful: (1) efficiency of oxidative addition to metal atoms follows the order C–I > C–Br > C–Cl; (2) aryl and benzyl halides react much more efficiently than alkyl halides. Product yields, matrix color changes, and previous knowledge of arene–metal complexes tend to indicate that π-arene–metal complexes first form before oxidative insertion occurs. This tends to "cage or preserve" the metal atoms and to prevent them from polymerization with themselves in the matrix. The π complexation would be a very low activation energy process, and then on matrix warm-up, probably during color change, the higher activation energy C–X insertion must occur.

Arenes (π Complexes with Metal Atoms). One of the most important developments in the metal atom field has been synthesis of bis(arene) sandwich compounds.[39] This was a needed breakthrough in

$$\text{Cr atoms} + C_6H_6 \longrightarrow (C_6H_6)_2Cr$$

terms of speed and convenience in synthesis, and in expanding the scope of arene sandwich compounds available (conventional reducing Friedel–Crafts methods are quite limited as to what substituents can be tolerated on the arene ring). Now a host of variously substituted bis(arene)chromium(0) compounds have been made simply by condensing Cr atoms with the appropriate arene ligand.[16,39–43]

Bis(arene) Cr(0) complexes

C_6H_5X	o, m, p-C_6H_4XY
$X = F, Cl, CF_3,$	$X = F, Cl, CF_3, CH_3,$
$C(O)OCH_3, CH_3,$	$CH(CH_3)_2$
$C_2H_5, CH(CH_3)_2$	$Y = F, Cl, CF_3, CH_3, CH(CH_3)_2$

It is most interesting that these normally air-sensitive complexes are stabilized to air oxidation by electronegative substituents.[16,41,42] We have been most concerned with the F- and CF_3-substituted systems[15,16,44] and found, for the *m*- and *p*-$(CF_3)_2C_6H_4$ ligands, that the sandwich chromium complexes were formed in good yield and were indefinitely stable to air at room temperature, and even for short periods of time at 200°C. Probably the first step in the oxidative decomposition of these compounds is the one-electron oxidation of chromium(0) to chromium(1+), which would be disfavored by the presence of electronegative substituents, and this is dramatically demonstrated by studying polarographic one-electron oxidations (collaborative work with P. Treichel and G. Essenmacher). Furthermore, recent X-ray crystallographic evidence indicates that F and CF_3 have very significant effects on the structures of these sandwich compounds of V as well as of Cr (collaborative work with L. Radonovich and C. Zuerner).

In related spectroscopic investigations, McGlinchey and Tan[43] have recently made some careful ^{19}F NMR studies on a series of substituted bis(arene)-chromium(0) complexes and conclude that the overall electron-withdrawing effect of a π-bonded Cr on each ring is similar to the effect of four ring fluorine substituents (on a noncomplexed benzene ring). They point out the interesting possibility of carrying out nucleophilic substitution of fluoride on $(C_6H_4F_2)_2Cr$ as readily as can be done with hexafluorobenzene, and find methoxide effective in this respect.

Although halogen substitution directly on the aromatic ring (F, Cl), lends some added air stability, it also imparts explosive character to some of these sandwich compounds. In the case of hexafluorobenzene as a ligand, the complexes were so sensitive and explosive that a pure compound has not been isolable from the C_6F_6–Cr system (similar results have been obtained with the C_6F_6–V[44] and C_6F_6–Ni[44] systems). With fluorinated benzenes only two fluorine substituents can be tolerated. Higher fluorine content (three–six fluorines) results in unstable, very explosive complexes.[45] Endothermic compounds can be obtained utilizing the metal atom technique, and we should be wary of all halogenated π complexes because of the potentially highly exothermic M–X bond formation that can occur on decomposition. So far no problems have been encountered with CF_3-substituted systems, however.

Silvon, Van Dam, and Skell have been able to prepare bis(arene) complexes of Mo and W.[46] These

$(C_6H_5X)_2Mo$	$(C_6H_5X)_2W$	$(o\text{-}(CH_3)_2C_6H_4)_2W$
$X = H, CH_3, F, Cl, OCH_3,$	$X = H, CH_3, OCH_3, F$	
$N(CH_3)_2, COOCH_3$		

metals have very high heats of vaporization, but amounts up to about 0.5 g can be sublimed by careful resistive heating of wires of Mo or W.

Benfield, Green, Ogden, and Young have employed an electron gun in a rotating reactor for vaporizing Mo and Ti.[17] These workers were also able to synthesize $(C_6H_6)_2Mo(0)$, as well as a very intriguing sandwich compound of Ti: $(C_6H_6)_2Ti$. This material must be isolated with great care because of its sensitivity and possible autocatalytic decomposition (probably induced by the presence of Ti metal film). It is the first true example of a Ti(0) sandwich complex and is red-orange and diamagnetic.

We have synthesized a series of bis(arene)vanadium(0) complexes by vaporizing V from resistively

$(C_6H_5X)_2V$	$(1,4\text{-}C_6H_4F_2)_2V$
$X = H, F, Cl, CF_3$	

heated tungsten boats.[15,16,44] These compounds are much more sensitive to air than the analogous Cr compounds. However, again electron-withdrawing

(39) P. L. Timms, *Chem. Commun.*, 1033 (1969).
(40) P. L. Timms, *J. Chem. Educ.*, **49**, 782 (1972).
(41) R. Middleton, J. R. Hull, S. R. Simpson, C. H. Tomlinson, and P. L. Timms, *J. Chem. Soc., Dalton Trans.*, 120 (1973).
(42) P. S. Skell, D. L. Williams-Smith, and M. J. McGlinchey, *J. Am. Chem. Soc.*, **95**, 3337 (1973).
(43) M. J. McGlinchey and T. S. Tan, *Can. J. Chem.*, **52**, 2439 (1974).
(44) K. J. Klabunde and H. F. Efner, *J. Fluorine Chem.*, **4**, 115 (1974).
(45) Private communications with P. L. Timms.
(46) M. P. Silvon, E. M. Van Dam, and P. S. Skell, *J. Am. Chem. Soc.*, **96**, 1945 (1974).

groups have a stabilizing effect, CF_3 being the best substituent to use.

Mixed cyclic complexes are also available utilizing metal atoms. As evidence of this, Van Dam and Skell reported that W atoms cocondensed with arene–cyclopentadiene mixtures yielded mixed sandwich complexes with W–H bonding:[47] $[(\eta^5\text{-}C_5H_5)(\eta^6\text{-}C_6H_6)WH]$.

Evidence for unstable arene π complexes has also been found. The first such observation was from an iron atom–benzene cocondensation yielding a very explosive unstable material.[39] Iron atoms with toluene formed a complex that could be warmed to −94° and then allowed to react with PF_3 to yield bis(trifluorophosphine)tolueneiron(0).[48] A number of unstable arene complexes were found when a series of transition metals (V, Cr, Mn, Fe, Co, Ni, and Pd) were cocondensed with hexafluorobenzene and benzene.[15,44]

By condensing substrate mixtures with metal vapors, Timms and coworkers[41] have prepared $Cr(arene)(PF_3)_3$ complexes (arenes = benzene, hexafluorobenzene, cumene, and mesitylene). Similarly, iron vapor condensed with a mixture of benzene and PF_3 yielded $C_6H_6Fe(PF_3)_2$. The $C_6F_6(PF_3)_3$ was formed in low yield, but is very stable, which is interesting in light of the fact that $(C_6F_6)_2Cr$ is not stable and is very explosive.

Diene–Metal Atom Reactions. Timms demonstrated that cyclopentadiene was reactive with metal atoms.[9,39] Ferrocene could be produced with hydrogen gas as a by-product. Similarly, chromocene[9,42]

$$Fe^0 + C_5H_6 \longrightarrow (C_5H_5)_2Fe + H_2$$
$$(C_5H_5)_2Cr$$
$$C_5H_6CoC_5H_5$$
$$C_5H_7NiC_5H_5$$

could be synthesized, but in the cases of Co and Ni hydrogen transfers took place, yielding the compounds shown.[9] No hydrogen was evolved in the Ni reaction, and trihapto as well pentahapto bonding to nickel resulted.

Recently Mo and W atoms have been condensed with cyclopentadiene. In these cases hydrogen was not evolved, and $(C_5H_5)_2MH_2$ derivatives were formed.[47,49] By extrapolation, then, it is possible that Cr atoms plus cyclopentadiene initially yield the dihydride derivative $(C_5H_5)_2CrH_2$ which decomposes to yield chromocene and hydrogen. Iron may go through a similar pathway. However, studies on mechanisms are needed and would be of great interest since labile M–H bonds may be involved.

1,5-Cyclooctadiene also interacts well with transition-metal atoms. The known bis(1,5-cyclooctadiene)platinum(0) complex was synthesized employing Pt atoms.[27] Also, MacKenzie and Timms recently disclosed the preparation of the very unusual compound bis(1,5-cyclooctadiene)iron(0) by reaction of Fe atoms with a *solution* of the diene in methylcyclohexane solvent.[20,21] This compound is formally coordinatively unsaturated, similar in that sense to the RPdX compounds previously discussed.

Skell and coworkers[42,50–52] synthesized some extremely interesting 1,3-butadiene complexes of transition metals. Some reactions and products are shown.

$$Mo^0 + 1{,}3\text{-}C_4H_6 \longrightarrow (1{,}3\text{-}C_4H_6)_3Mo$$

$$W^0 + 1{,}3\text{-}C_4H_6 \longrightarrow (1{,}3\text{-}C_4H_6)_3W$$

$$Fe^0 + 1{,}3\text{-}C_4H_6 \longrightarrow (1{,}3\text{-}C_4H_6)_nFe \xrightarrow{CO} (1{,}3\text{-}C_4H_6)_2FeCO$$

$$Cr^0 + 1{,}3\text{-}C_4H_6 + PF_3 \longrightarrow 1{,}3\text{-}C_4H_6\ Cr(PF_3)_4$$

$$Ni^0 + 1{,}3\text{-}C_4H_6 \longrightarrow 1{,}3\text{-}C_4H_6\text{–Ni polymer}$$

The most interesting are tris(1,3-butadiene)molybdenum and -tungsten.[51] These are air and thermally stable materials. Recent X-ray studies showed that the diene molecules are wrapped around the metal atoms so that all 12 carbons are equidistant from the metal.[53] These materials represent rare examples of compounds where Mo and W are zerovalent without CO ligand stabilization.

Koerner von Gustorf and coworkers have also interacted butadiene with iron atoms as well as with Cr atoms.[19] Subsequent reaction of the diene–Cr complex with CO yielded a butadienechromium carbonyl complex of limited stability. This complex had been

$$Cr^0 + 1{,}3\text{-}C_4H_6 \longrightarrow (1{,}3\text{-}C_4H_6)_nCr \xrightarrow{CO} 1{,}3\text{-}C_4H_6Cr(CO)_4$$

proposed earlier as a catalysis intermediate, and this work elucidated some of the properties of this labile species. In these metal atom studies, laser heating was employed for the metal vaporizations, and large-scale reactions could readily be carried out.[19] This method would appear to have considerable promise for doing large-scale vaporizations of refractory metals, and perhaps for the generation and study of electronically excited metal atoms.

Olefins with Metal Atoms. Propene has been cocondensed with metals Al, Co, Dy, Er, Ni, Pd, Pt, and Zr in initial orientating experiments in an attempt to demonstrate differences and similarities between the metals.[54] Organometallic products were not isolated and characterized, but rather were decomposed with D_2O to mark C–M bonds. Aluminum atoms with propene yielded mainly 1,2-dialuminoalkanes (σ bonding),[55] whereas Ni and Pt yielded mainly π complexes (unlabeled propene released on D_2O addition).[27,42,54] In the aluminum work some 1,4-dialumino-C_6 products were also formed[55] Matrix isolation spectroscopy work in conjuction with product analysis indicated that the dialuminoalkanes were indeed formed from aluminum atoms at −196°C (not on warming of the matrix or from active aluminum surfaces or clusters).[55]

Indirect evidence has been obtained for the formation of π-allylnickel hydride from the reaction of Ni

(47) E. M. Van Dam, W. N. Brent, M. P. Silvon, and P. S. Skell, *J. Am. Chem. Soc.*, **97**, 465 (1975).
(48) D. L. Williams-Smith, L. R. Wolf, and P. S. Skell, *J. Am. Chem. Soc.*, **96**, 4042 (1974).
(49) M. J. D'Amiello and E. K. Barefield, *J. Organomet. Chem.*, **76**, C50 (1974).
(50) P. S. Skell, J. J. Havel, D. L. Williams-Smith, and J. J. McGlinchey, *J. Chem. Soc., Chem. Commun.*, 1098 (1972).
(51) P. S. Skell, E. M. Van Dam, and M. P. Silvon, *J. Am. Chem. Soc.*, **96**, 626 (1974).
(52) D. L. Williams-Smith, L. R. Wolf, and P. S. Skell, *J. Am. Chem. Soc.*, **94**, 4042 (1972).
(53) M. Yevitz and P. S. Skell, International Union of Chrystallography Intercongress Symposium on Intra- and Intermolecular Forces, August 14–16, 1974, The Pennsylvania State University, Series 2, Vol. 2, paper E9.
(54) P. S. Skell, *Proc. Int. Congr. Pure Appl. Chem.*, **23** (4), 215 (1971).
(55) P. S. Skell and L. R. Wolf, *J. Am. Chem. Soc.*, **94**, 7919 (1972).

atoms with propene.[42] Thus, the condensate from propene and Ni atoms yields both nondeuterated propene and monodeuteriopropene upon D_2O hydrolyis. In further support of the π-allyl Ni–H interme-

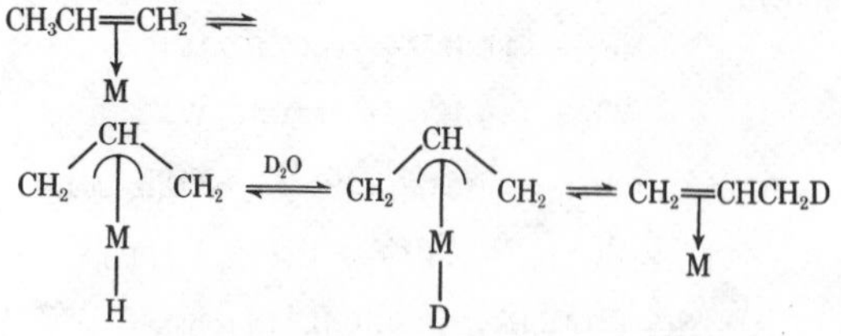

diate, cocondensation of C_3H_6 plus C_3D_6 with Ni atoms leads to H–D scrambled products.[50]

We have found that perfluoro olefins form complexes with Ni, Pd, and Pt atoms at low temperature.[35] On warm-up the complexed olefin is released unchanged. The (olefin)$_3$Pd complex shown below reacted with triethylphosphine[35] or with pyridine[28] to form the corresponding (olefin)PdL_2 species in high yields. Since the olefin becomes σ bonded in the complexes, this reaction sequence is an example of oxidative addition of a C–C bond to a palladium atom.[35]

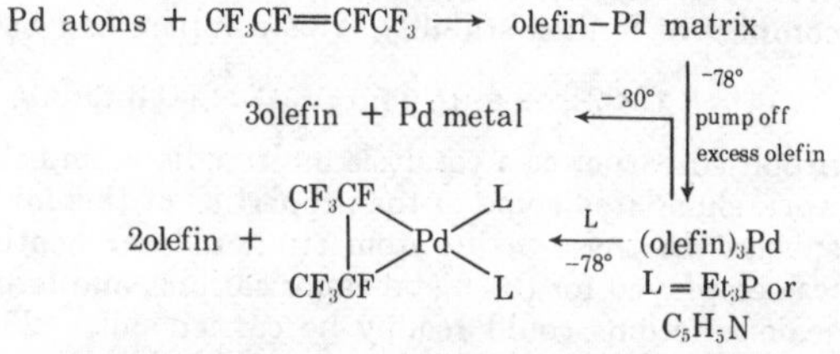

Acetylenes, Trienes, and Tetraenes with Metal Atoms. Bis(cycloheptatriene)chromium(0) has been synthesized using Cr atoms plus cycloheptatriene.[42] With Mo and W atoms, however, hydrogen transfers took place, and the interesting η^7,η^5 systems were obtained.[47]

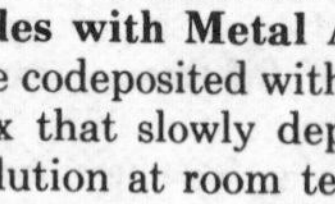

Some acetylenes were trimerized to benzenes by chromium vapor, but no bis(arene)chromium(0) complexes could be isolated from the reaction mixtures.[42] The absence of the bis(arene) sandwich complexes indicates that probably only one arene ring was generated and bonded to chromium at any time.

Acid Anhydrides with Metal Atoms. Hexafluoroacetic anhydride codeposited with palladium atoms yielded a complex that slowly deposited palladium metal while in solution at room temperature.[35] The structure of this material has not been determined as yet, but on addition of Et_3P to the complex, *cis*-bis(triethylphosphine)perfluorodiacetatopalladium(II)($(Et_3P)_2Pd(OCOCF_3)_2$) was formed. Formally, this is an example of the C–O oxidative insertion by palladium. However, it is likely that the initial complex is zerovalent in palladium.

Active Metal Slurries Formed from Metal Atom–Solvent Cocondensations. Currently, there is great interest in the production and chemistry of high-surface-area clean-metal slurries.[56] We have found metal vapors to be useful in this regard. Thus, Mg vapor cocondensed with tetrahydrofuran (THF) followed by warming yielded a black Mg–THF slurry that was extremely reactive for Grignard reagent preparations.[57]

An active Ni–THF slurry can be produced similarly. This slurry, as is the case with Mg, has very finely divided metal present and can actually be manipulated by syringe. This Ni slurry is quite reactive with organohalides, and, in cases such as benzyl chloride and iodobenzene, can be used as coupling reagent under very mild conditions.[58] In attempts to learn more about the physical and structural aspects of the metal powder resulting from Ni–THF condeposition followed by pump-off of excess THF, we have utilized electron-scanning microscopy and X-ray powder techniques as well as chemical reactions. These studies have shown the material is actually a nickel(0) etherate (Ni:THF $\simeq$ 5:1), has essentially no crystalline character, and is in the form of small globules in the 0.5-μm range.[28,58] On pyrolysis this material decomposes with the formation of Ni–Ni bonds, leading to large Ni crystals.

Currently we are studying other metal atom–solvent systems. One aspect of the work is to produce "solvated metal atoms", and polyethers and arenes appear to be promising substrates from preliminary data. We find that Ni atoms can be solvated by toluene at low temperature and have used this Ni–toluene solution for the low-temperature deposition of small nickel crystallites on alumina. These materials, as well as the metal slurries that can be made, should have many uses in chemical synthesis and catalysis. It is encouraging that the metal–ether slurries show good storage properties (if kept under a rigorous argon atmosphere).

General Observations

The metal atom method gives us a wide scope in the synthesis of new organometallic compounds, some of them very "high-energy materials". Uses of these new materials will very likely be closely related to homogeneous catalysis because they often are "coordinatively unsaturated". Active-metal slurries formed this way also have promise, particularly for use in low-temperature organometallic synthesis schemes.

Useful aspects of the technique are: (1) reactors are becoming quite simplified and larger scale, (2) bulk metals are cheap chemical reagents, and (3) the method can serve as a quick indicator of what organometallic compounds have good stability, and then more conventional methods can be devised for synthesizing these materials.

Financial support from the National Science Foundation (GP-42376) and also the outstanding work of Dr. James Y. F. Low, Howard F. Efner, John S. Roberts, Bruce B. Anderson, and Thomas O. Murdock are appreciated.

(56) R. D. Rieke and L. C. Chao, *Synth. React. Inorg. Met.-Org. Chem.*, **4**, 101 (1974); R. D. Rieke and P. M. Hudnall, *J. Am. Chem. Soc.*, **94**, 7178 (1972); R. D. Rieke, S. J. Uhm, and P. M. Hudnall, *J. Chem. Soc., Chem. Commun.*, 269 (1973); R. D. Rieke and S. E. Bales, *ibid.*, 879 (1973); R. D. Rieke and S. E. Bales, *J. Am. Chem. Soc.*, **96**, 1775 (1974); C. A. Brown and V. K. Ahuja, *J. Org. Chem.*, **38**, 2226 (1973).

(57) K. J. Klabunde, H. F. Efner, L. Satek, and W. Donley, *J. Organomet. Chem.*, **71**, 309 (1974).

(58) T. O. Murdock, unpublished observations from this laboratory.

Addendum (May 1976)

Organohalides with Metal Atoms. In recent work some R_2M and RM compounds have been prepared by reaction of metal vapors with organic halogen compounds. One example is the reaction of C_6F_5Br with cobalt vapor.[59] A disproportionation yielding (*bis*)-pentafluorophenyl(cobalt)II (1) and cobalt dibromide occurred. An interesting feature of 1 is that it bonds toluene in π-fashion, yielding what appears to be the first example of a (π-arene)MR_2 complex (2) (as shown by an x-ray structure determination).[59] This work points out the advantage of synthesizing these reactive "coordinatively unsaturated" organometallics by the metal atom method in the absence of donor solvents, since "$(C_6F_5)_2Co$" had previously been reported but as an ether solvate not showing the interesting reactivity of 1.[60]

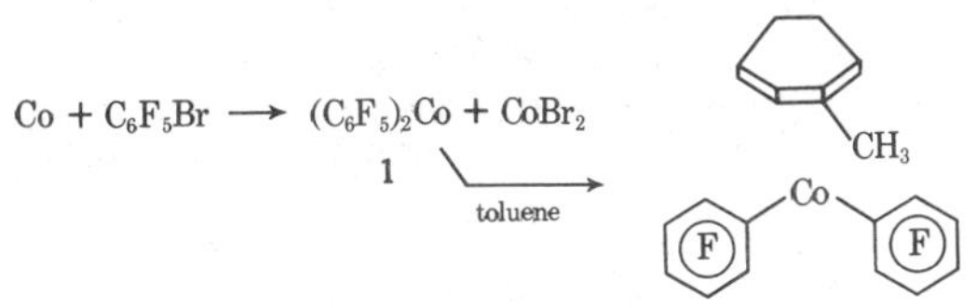

$$Co + C_6F_5Br \longrightarrow (C_6F_5)_2Co + CoBr_2$$

Silver atoms have been codeposited with perfluoroalkyl iodides yielding R_fAg compounds.[61] In this way the previously unknown and relatively unstable n-C_3F_7Ag and CF_3Ag were prepared. Also, a series of experiments on preparation of $(CF_3)_2CFAg$ (a known compound[62]) showed (1) that high R_fI/Ag ratios favor R_fAg formation, and thus that R_fI- - -Ag complex formation is probably occurring at the $-196°C$ codeposition temperature, and (2) that the R_fAg compounds are much more stable solvated with CH_3CN than unsolvated.

$$Ag + R_fI \rightarrow AgI + R_nAg$$

Arenes (π-Complexes with Metal Atoms). In further work on the synthesis of *bis*(arene) Cr and V complexes by the metal atom method,[63–65] a 2,6-dimethylpyridine complex with Cr has been prepared and its structure determined.[65] A systematic x-ray structural investigation[66] of substituted *bis*(arene) Cr and V complexes made by the metal atom method has shown, notably, that in $(1,4-C_6H_4F_2)_2V$, a small boat deformation (C–F carbons up out of plane) is present and that the ring carbons are eclipsed (d_2 symmetry).[66] Additional *bis*(arene) Ti(O) complexes have also been prepared;[67] they can not be made by any other methods.

Olefins with Metal Atoms. Syntheses of *tris*(ethylene) Ni and Pd have been successful with use of metal atoms. The method lends itself well to production and low temperature isolation of such labile species. *Bis*(cyclooctadiene) Pd and *bis*(norbornadiene) Pd have been prepared.[68]

Tetraenes with Metal Atoms. Codeposition of Nd atoms with cyclooctatetraene (COT) followed by workup in THF yielded a new type of asymmetric complex $[Nd(COT)(THF)_2][Nd(COT)_2]$ which exists as an anion–cation pair. The presumed species present before THF addition is $Nd_2(COT)_3$.[69] Butadiene oligomerization catalyzed by metal atoms has been shown to be a highly complex process.[70]

Solvated Metal Atoms and Active Metal Slurries. Storeable active metal slurries can be prepared by condensing metal vapors with appropriate solvents. Slurries of Zn, Cd, and Sn in polar and non-polar solvents react with alkyl halides to form organometallics under mild conditions.[71,72] Never before have direct RX–Cd reactions been carried out in solvents other than hexamethylphosphortriamide.

Solvated Ni atoms have been described.[73] These are low temperature toluene–Ni or THF–Ni adducts. In the case of toluene, a low temperature Ni atom-toluene solution has been found to be very reactive with other donor ligands.[73] Removal of the weakly complexing solvent results in the formation of high surface area active metal powders, each of differing appearance and reactivity when a different solvent is involved (with the same metal). X-ray powder studies, scanning electron microscope studies, and some chemical studies were reported for these powders from Ni–hexane, Ni–toluene and Ni–THF.[73]

(59) B. B. Anderson, C. Behrens, L. Radonovich, and K. J. Klabunde, *J. Am. Chem. Soc.*, in press.

(60) C. F. Smith and C. Tamborski, *J. Organomet. Chem.*, **32**, 257 (1971).

(61) K. J. Klabunde, *J. Fluorine Chem.*, **7**, 95 (1976).

(62) W. T. Miller and R. J. Burnard, *J. Am. Chem. Soc.*, **90**, 7367 (1968.

(63) M. J. McGlinchey and T. S. Tan, *J. Am. Chem. Soc.*, **98**, 2271 (1976).

(64) V. Graves and J. J. Lagowski, *Inorg. Chem.*, **15**, 577 (1976).

(65) L. H. Simons, P. E. Riley, R. E. Davis, and J. J. Lagowski, *J. Am. Chem. Soc.*, **98**, 1044 (1976).

(66) L. J. Radonovich, C. Zuerner, H. F. Efner, and K. J. Klabunde, *Inorg. Chem.*, in press.

(67) M. T. Anthony, M. L. H. Green, and D. Young, *J. Chem. Soc., Dalton Trans.*, 1419 (1975).

(68) R. Alkins, R. MacKenzie, P. L. Timms, and T. W. Turney, *J. Chem. Soc., Chem. Commun.*, (18), 764 (1975).

(69) S. R. Ely, T. E. Hopkins, and C. W. DeKock, *J. Am. Chem. Soc.*, **98**, 1624 (1976).

(70) V. M. Akhmedov, M. T. Anthony, M. L. H. Green, and D. Young, *J. Am. Chem. Soc.*, 1412 (1975).

(71) T. O. Murdock and K. J. Klabunde, *J. Org. Chem.*, **41**, 1076 (1976).

(72) T. O. Murdock, unpublished results.

(73) K. J. Klabunde, H. F. Efner, T. O. Murdock, and R. Ropple, *J. Am. Chem. Soc.*, **98**, 1021 (1976).

Iron–Sulfur Coordination Compounds and Proteins

Stephen J. Lippard

Department of Chemistry, Columbia University, New York, New York 10027

Received August 28, 1972

Reprinted from ACCOUNTS OF CHEMICAL RESEARCH, *6,* 282 *(1973)*

Although powerful oxidizing and reducing agents exist among transition metal–aquo complexes (Fe(III)|Fe(II), 0.77 V; Cr(III)|Cr(II), −0.41 V), these simple ions are unavailable as biological redox catalysts because of extensive hydrolysis in the physiological pH range. Transition metal ions are found in the active sites of many redox metalloproteins, however.[1] Here the metal is generally coordinated to donor atoms supplied by the rich variety of amino acid functional groups. Occasionally a special ligand such as the heme group serves to bind the metal to the protein.

The isolation of highly purified bacterial and plant ferredoxins nearly a decade ago[2] led to intensive research activity on these and related iron–sulfur proteins (Table I).[3] X-Ray structural studies of rubredoxin,[4] a bacterial ferredoxin,[5] and the high-potential iron protein (HiPIP) from *Chromatium*[6] (Figure 1) established that iron is coordinated to the sulfur atoms of cysteine and, except for rubredoxin, to a biologically unique form of sulfur that can be released as hydrogen sulfide upon mild acidification. The geometry of the iron–sulfur core in the plant ferredoxins, adrenodoxin, and related Fe_2S_2 proteins has not yet been directly established, but a host of indirect studies[7] have converged on the structure shown in Figure 2a as the most likely candidate. An alternative structure has been proposed[8] for adrenodoxin (Figure 2b) and will be discussed below.

Biological interest in the iron–sulfur proteins has centered around their widespread occurrence in redox systems, fulfilling such diverse functions as nitrogen fixation, steroid hydroxylation, and photosynthesis. The proteins appear to function as high- or low-potential "wires" in electron-transport chains. Perhaps the most interesting are the ferredoxins, which have some of the lowest reduction potentials (Table I) in biology.[2] The low value for the ferredoxin reduction potential compared to the Fe(III)|Fe(II) couple is not simply the result of sulfur *vs.* oxygen coordination, since rubredoxin has a potential of −57 mV. Even more striking is the difference of 730 mV between the values for the HiPIP and bacterial ferredoxins (Table I) despite the fact that the Fe_4S_4 cores in both proteins are geometrically quite similar.[9]

Apart from the biological interest in these proteins, they have attracted the attention of the coordination chemist. One reason is a purely synthetic one, for the preparation and characterization of iron–sulfur coordination compounds using only biologically significant ligands[10] have, until recently, remained elusive. The chief difficulties are the oxidation of sulfide or mercaptide ligands by iron(III) salts, the tendency for iron(II) complexs to form as insoluble, intractable polymers, and the oxygen sensitivity of the iron–sulfur compounds obtained.[11] The fact that iron–sulfur chromophores (also oxygen sensitive[12]) of this kind are found in the proteins[13] has thus stimulated considerable synthetic activity by coordination chemists.

Coordination chemists have also been interested in the low reduction potentials exhibited by certain classes of iron–sulfur proteins and the dependence of the reduction potential on the degree of polymerization and the local protein environment (Table I). Thus a second objective has been to examine the ef-

Professor Lippard's research activities include synthetic and physical studies of redox metalloproteins and related coordination compounds. Impetus for his recent slant to more biological areas was provided by a recent sabbatical leave spent as a Guggenheim Fellow in the laboratory of Professor Bo Malmström in Göteborg, Sweden. Professor Lippard's training was in inorganic chemistry at MIT, where he received his Ph.D. in 1965 with F. A. Cotton. He is now Professor of Chemistry at Columbia.

(1) For a comprehensive review see L. E. Bennett, *Progr. Inorg. Chem.*, **18,** 1 (1973).

(2) (a) L. E. Mortenson, R. C. Valentine, and J. E. Carnahan, *Biochem. Biophys. Res. Commun.*, **7,** 448 (1962); (b) K. Tagawa and D. I. Arnon, *Nature (London)*, **195,** 537 (1962).

(3) For reviews, see (a) R. Malkin and J. C. Rabinowitz, *Annu. Rev. Biochem.*, **36,** 113 (1967); (b) T. Kimura, *Struct. Bonding (Berlin)*, **5,** 1 (1968); (c) D. O. Hall and M. C. W. Evans, *Nature (London)*, **223,** 1342 (1969); (d) J. C. M. Tsibris and R. W. Woody, *Coord. Chem. Rev.*, **5,** 417 (1970).

(4) K. D. Watenpaugh, L. C. Sieker, J. R. Herriott, and L. H. Jensen: (a) Abstracts, American Crystallographic Association Meeting, Iowa State University, Ames, Iowa, 1971, p 52; (b) *Cold Spring Harbor Symp. Quant. Biol.*, **36,** 359 (1971).

(5) L. C. Sieker, E. Adman, and L. H. Jensen, *Nature (London)*, **235,** 40 (1972).

(6) (a) G. Strahs and J. Kraut, *J. Mol. Biol.*, **35,** 503 (1968); (b) C. W. Carter, Jr., S. T. Freer, Ng. H. Xuong, R. A. Alden, and J. Kraut, *Cold Spring Harbor Symp. Quant. Biol.*, **36,** 381 (1971).

(7) (a) W. R. Dunham, G. Palmer, R. H. Sands, and A. J. Bearden, *Biochim. Biophys. Acta*, **253,** 373 (1971), and references cited therein; (b) M. Poe, W. D. Phillips, J. D. Glickson, C. C. McDonald, and A. San Pietro, *Proc. Nat. Acad. Sci. U. S.*, **68,** 68 (1971); (c) C. E. Johnson, R. Commack, K. K. Rao, and D. O. Hall, *Biochem. Biophys. Res. Commun.*, **43,** 564 (1971).

(8) T. Kimura, Y. Nagata, and J. Tsurugi, *J. Biol. Chem.*, **246,** 5140 (1971).

(9) C. W. Carter, Jr., J. Kraut, S. T. Freer, R. A. Alden, L. C. Sieker, E. Adman, and L. H. Jensen, *Proc. Nat. Acad. Sci. U. S.*, **69,** 3526 (1972).

(10) The term "biologically significant ligands" will be used in the present context to mean, broadly, ligands with donor functions similar to those found in the naturally occuring amino acids and, in a more restricted sense, ligands containing the $-CH_2S^-$ group. The sulfide ion, S^{2-} (*cf.* Table I and Figures 1 and 2a) and the Cys-S-S⁻ moiety (Figure 2b) are also included.

(11) (a) D. Coucouvanis, S. J. Lippard, B. G. Segal, and J. A. Zubieta, *Proc. Int. Conf. Coord. Chem., 12th,* 190 (1969); (b) D. Coucouvanis and S. J. Lippard, unpublished results.

(12) D. Petering, J. A. Fee, and G. Palmer, *J. Biol. Chem.*, **246,** 643 (1971).

(13) The existence of the iron–sulfur chromophores in the proteins has on occasion led to an advancement in coordination chemistry. For example, the first evaluation of the crystal field splitting parameter *Dq* for mercaptide sulfur came through the analysis of the optical spectrum of rubredoxin (W. A. Eaton and W. Lovenberg, *J. Amer. Chem. Soc.*, **92,** 7195 (1970)), there being no simple metal–sulfur complex from which this information could be extracted.

Table I
Physical and Chemical Properties of Iron–Sulfur Proteins[a]

Protein	Source	E_0', mV	Moles/mole of protein: Iron	Sulfide	Cysteine	n	Mol wt
Ferredoxin[b]	Bacteria	−500 (−400)	4(8)	4(8)	4(8)	1(2)	6,000
Ferredoxin	Plants	−430	2	2	5	1	12,000
Adrenodoxin	Animals	−270[c]	2	2	4	1	12,000
Putidaredoxin	Bacteria	−235	2	2		1	12,000
Rubredoxin	Bacteria	−57	1	0	4	1	6,000
High-potential iron proteins	Bacteria	+330	4	4	4	1	9,500

[a] Data are obtained from ref 2. Abbreviations: *n*, number of electrons involved in redox process. [b] Numbers in parentheses refer to the 8Fe–8S *Clostridium* and other proteins in which there are two Fe_4S_4 units (see ref 9). [c] J. J. Huang and T. Kimura, *Biochemistry*, 12, 406 (1973).

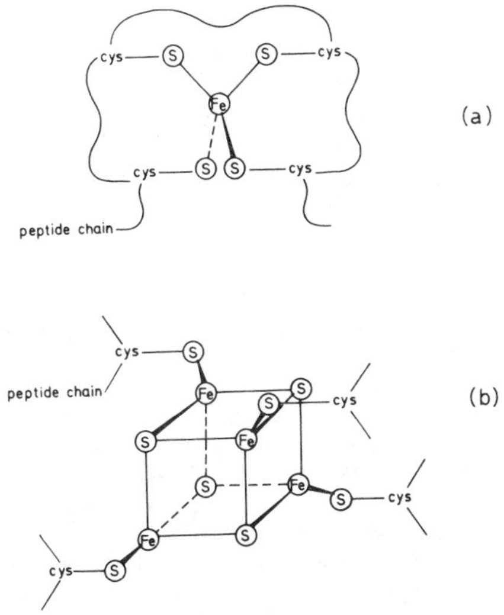

Figure 1. Structures of the iron–sulfur chromophores in (a) *C. pasteurianum* rubredoxin and (b) high-potential iron protein (HiPIP) from *Chromatium* and *C. pasteurianum* ferredoxin. The drawings are idealizations (see ref 4–6 and 9 for details).

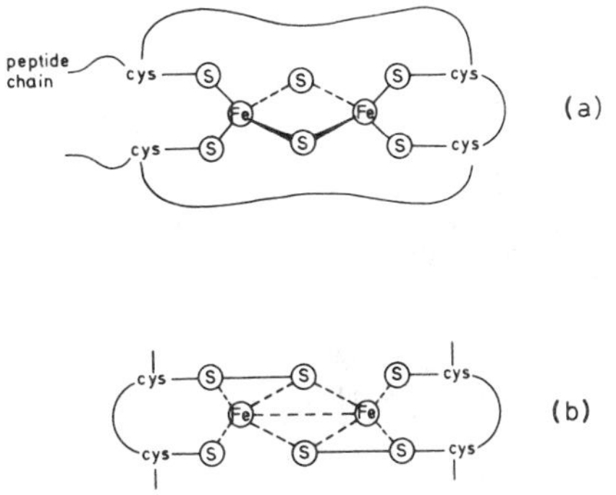

Figure 2. (a) Model for the iron–sulfur core in Fe_2S_2 proteins;[7] (b) structure proposed[8] for the iron–sulfur redox center in adrenodoxin. In the oxidized form, (a) has two antiferromagnetically coupled high-spin iron(III) centers which upon reduction are converted to an iron(II)–iron(III) pair, also antiferromagnetically coupled.[7]

fect of factors such as extent of polymerization, choice of ligand (*e.g.*, RCH_2S^- *vs.* S^{2-}), solvent polarity, overall charge, and steric strain[14] on the redox properties of iron–sulfur complexes. Here again imaginative synthetic chemistry is required to reproduce the subtle environmental factors of a protein sheath using relatively simple ligand molecules.

In this Account, several experimental studies of iron–sulfur coordination compounds, mostly performed in this laboratory, are examined. These investigations have generated specific suggestions concerning the structural and redox properties of the proteins. We caution at the outset that only direct studies of the protein systems themselves can prove or disprove the validity of the ideas put forth. While such work is in progress both in our laboratory and elsewhere, it lies outside the scope of this Account.

The Fe–S–S–C Linkage

Before X-ray diffraction results were available for any of the iron–sulfur proteins, it was proposed, based on phenylmercuric acetate titrations of dihydroorotate dehydrogenase,[15] that the labile sulfur (released as H_2S upon acidification) might be in combination with cysteinyl sulfur in the form of a persulfide unit, I.

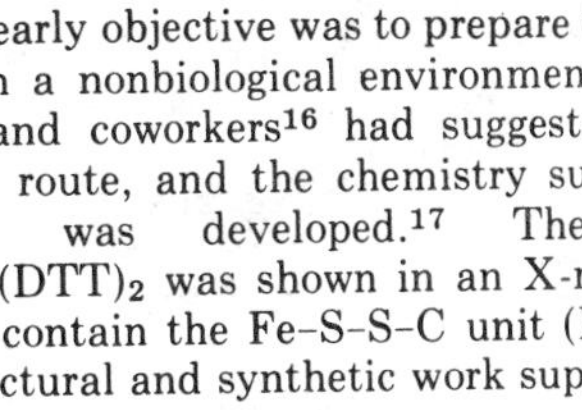

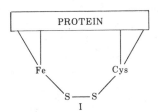

I

Thus an early objective was to prepare the Fe–S–S–R moiety in a nonbiological environment. Studies by Fackler and coworkers[16] had suggested a possible synthetic route, and the chemistry summarized by eq 1–3 was developed.[17] The compound $Fe(TTD)(DTT)_2$ was shown in an X-ray diffraction study to contain the Fe–S–S–C unit (Figure 3).[17,18] This structural and synthetic work supplied little bi-

(14) The "unusual" stereochemical properties of metal ion cores in redox proteins have long been recognized: B. G. Malmström, *Pure Appl. Chem.*, **24**, 393 (1970), and references cited therein.

(15) R. W. Miller and V. Massey, *J. Biol. Chem.*, **240**, 1453 (1965).

(16) J. P. Fackler, D. Coucouvanis, J. A. Fetchin, and W. C. Seidel, *J. Amer. Chem. Soc.*, **90**, 2784 (1968), and references cited therein.

(17) D. Coucouvanis and S. J. Lippard, *J. Amer. Chem. Soc.*, **90**, 3281 (1968). Abbreviations: TTD = thio-*p*-toluoyl disulfide, $CH_3C_6H_4CS_3^-$; DTT = dithio-*p*-toluate, $CH_3C_6H_4CS_2^-$.

(18) D. Coucouvanis and S. J. Lippard, *J. Amer. Chem. Soc.*, **91**, 307 (1969).

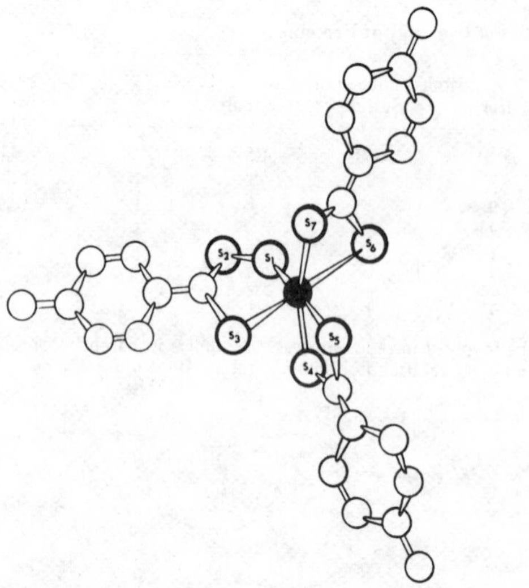

Figure 3. Molecular structure of $Fe(TTD)(DTT)_2$. The iron atom is at the center and carbon atoms are not labeled (reproduced from ref 18).

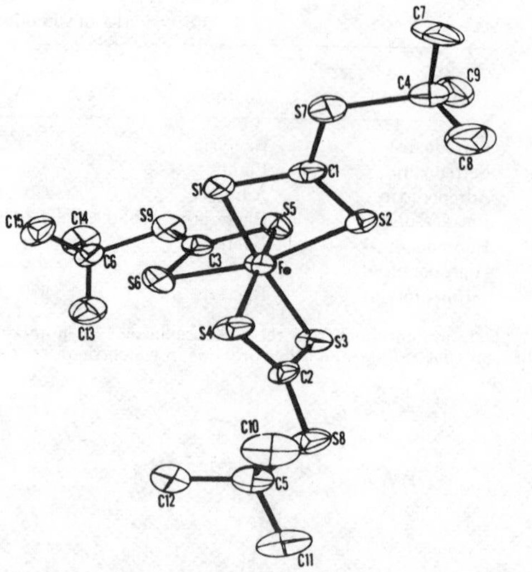

Figure 4. Molecular structure of $Fe(S_2CSR)_3$, R = *t*-butyl, showing the 50% probability thermal ellipsoids (reproduced from ref 23).

$$1.5Zn(TTD)_2 + FeCl_3 \rightleftharpoons Fe(TTD)_2(DTT) + \text{other products} \quad (1)$$

$$Fe(TTD)_2(DTT) + (C_6H_5)_3P \rightleftharpoons Fe(TTD)(DTT)_2 + (C_6H_5)_3PS \quad (2)$$

$$Fe(TTD)(DTT)_2 + (C_6H_5)_3P \rightleftharpoons Fe(DTT)_3 + (C_6H_5)_3PS \quad (3)$$

ological insight, however; it served merely to provide a small molecule prototype in the event that such a unit might be found in proteins.

Studies by Kimura, Nagata, and Tsurugi[8] revived interest in this chemistry and in the possible existence of an Fe-S-S-C linkage for at least some classes of iron–sulfur proteins. Extending reactions 2 and 3 to the protein adrenodoxin, these workers were able to titrate quantitatively the labile sulfur with triphenylphosphine to form triphenylphosphine sulfide. Neither S^{2-} nor RS^- will undergo such a redox reaction. Kimura, *et al.*, proposed the structure shown in Figure 2b for the iron–sulfur core in adrenodoxin.

This structure is not compatible with sulfhydryl titration data[19] on adrenodoxin, however (see ref 7a for discussion of related proteins). Moreover, since the work was performed in 33% ethanol, the protein active site may have been denatured, with oxidation of S^{2-} and Cys-S^- to form Cys-S-S-S-Cys. This species could then react with triphenylphosphine. The formation of trisulfides has been proposed to account for the oxygen sensitivity of spinach ferredoxin and putidaredoxin in the presence of denaturants.[12] On the other hand, organic solvents appear to have only minimal effect on the stability of oxidized adrenodoxin;[20] the abstraction of sulfur by triphenylphosphine also occurs under anaerobic conditions.[8] High-resolution X-ray diffraction studies of the protein would resolve this point. In the persulfide structure (Figure 2b), the closest S–S distance would be ~2.0–2.1 Å, compared to ~2.8 Å or more for the structure shown in Figure 2a.

Should a persulfide structure, or some variation thereof,[21] exist for adrenodoxin or any of the other iron–sulfur proteins, the electron-transport process would have to be reevaluated in terms of a non-metal-based redox reaction, *e.g.*, a sulfide ⇌ persulfide equilibrium. A persulfide structure may well occur in dihydroorotate dehydrogenase, being compatible with the sulfhydryl titer of the protein.[15] With the possible exception of adrenodoxin, however, there is no compelling reason to doubt the structure proposed[7] in Figure 2a for the Fe_2S_2 proteins listed in Table I. This structure will be assumed in the ensuing discussion.

Mercaptide-Bridged Thioxanthate Dimers of Iron(III)

As indicated in Table I, the reduction potentials of the Fe_2S_2 proteins are ~0.2–0.4 V lower than for rubredoxin. It was therefore of interest to obtain mono- and binuclear iron complexes with similar sulfur donor atom sets and to compare their redox properties.

Carbon disulfide elimination from the tris(*n*-alkyl thioxanthato)iron(III) complexes produced binuclear iron(III) complexes, eq 4–6.[22] In the case where R =

$$Na^+RS^- + CS_2 \underset{CS_2}{\overset{THF}{\rightleftharpoons}} Na^+(RSCS_2^-)\ \text{(yellow solution)} \quad (4)$$

$$FeCl_3(aq) + \text{yellow solution} \longrightarrow \underset{\text{unstable}}{Fe(S_2CSR)_3} + NaCl \quad (5)$$

$$2Fe(S_2CSR)_3 \longrightarrow [Fe(SR)(S_2CSR)_2]_2 + 2CS_2 \quad (6)$$

$$R = C_2H_5,\ n\text{-}C_3H_7,\ n\text{-}C_4H_9,\ C_6H_5CH_2$$

(19) T. Kimura, *Struct. Bonding (Berlin)*, **5**, 1 (1968).
(20) T. Kimura, *Biochem. Biophys. Res. Commun.*, **43**, 1145 (1971).

(21) (a) E. Bayer, H. Eckstein, H. Hagenmaier, D. Josef, J. Koch, P. Krauss, A. Röder, and P. Schretzmann, *Eur. J. Biochem.*, **8**, 33 (1969); (b) G. T. Kubas, T. G. Spiro, and A. Terzis, *J. Amer. Chem. Soc.*, **95**, 273 (1973).
(22) D. Coucouvanis, S. J. Lippard, and J. A. Zubieta, *J. Amer. Chem. Soc.*, **91**, 761 (1969); **92**, 3342 (1970).

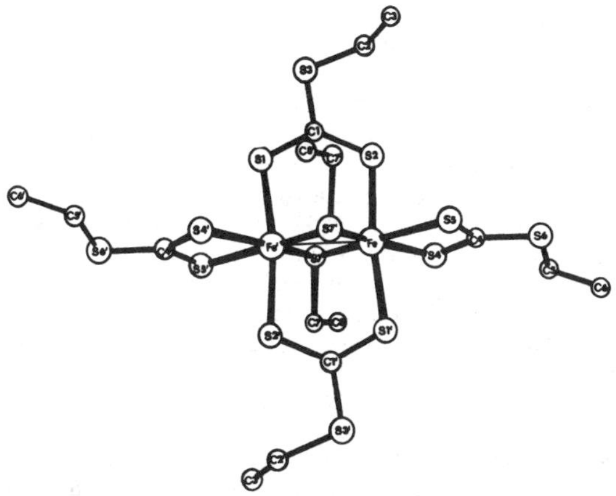

Figure 5. Molecular structure of $[Fe(SR)(S_2CSR)_2]_2$, R = ethyl (reproduced from ref 24).

t-C_4H_9, monomeric $Fe(S_2CSR)_3$ was isolated under the conditions employed. The solid-state structures (the solution geometries are similar[22]) of both monomeric[23] and dimeric (R = ethyl)[24] products are shown in Figures 4 and 5, respectively. In both cases the coordination geometry of the iron(III) atom is a distorted octahedron. The dimer has an iron–iron bond as determined from geometric and magnetic criteria[24] and electronic structural considerations.[22] A qualitative bonding scheme is shown in Figure 6a.[22,25]

Although the $Fe(S_2CSR)_3$ and $[Fe(SR)(S_2CSR)_2]_2$ complexes differ from rubredoxin and the Fe_2S_2 proteins in several respects (the iron is six-coordinate and low-spin, the dimers do not contain labile sulfur, the thioxanthate ligands contain unsaturated C⋯S donor atoms), it was of interest to compare their electrochemical properties since they afforded an opportunity to study the redox potentials of mono- and binuclear iron(III) complexes with nearly identical sulfur donor ligands. Voltammetric studies in dichloromethane solution (Figure 7) established that the half-wave reduction potentials of the dimers (irreversible, two-electron reduction) were ~0.3 V more negative than that of the monomer (reversible, one-electron process).[22] This result parallels the relative potentials of the mononuclear and binuclear iron–sulfur proteins (Table I), although the correlation may well be coincidental. As shown in Figure 7, electrons added to $[Fe(SR)(S_2CSR)_2]_2$, R = n-propyl, on the cathodic sweep are removed at a more positive potential, close to that observed for the reduction of the *tert*-butyl monomer. This behavior has been ascribed to a disruption of the metal–metal bonding interaction, possibly accompanied by a structural rearrangement.[22,25]

Since the hypothetical (short-lived?) $[Fe(SR)(S_2CSR)_2]_2^{2-}$ dianions generated electrochemically were not isolated, synthetic and structural studies of the isoelectronic cobalt(III) compounds

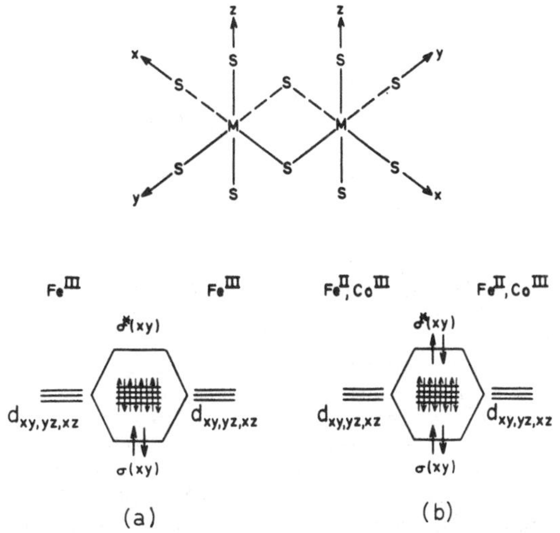

Figure 6. Qualitative bonding scheme for (a) $[Fe(SR)(S_2CSR)_2]_2$ and (b) $[Co(SR)(S_2CSR)_2]_2$ complexes showing the orbitals involved in the metal–metal interaction. For discussion see ref 22 and 25.

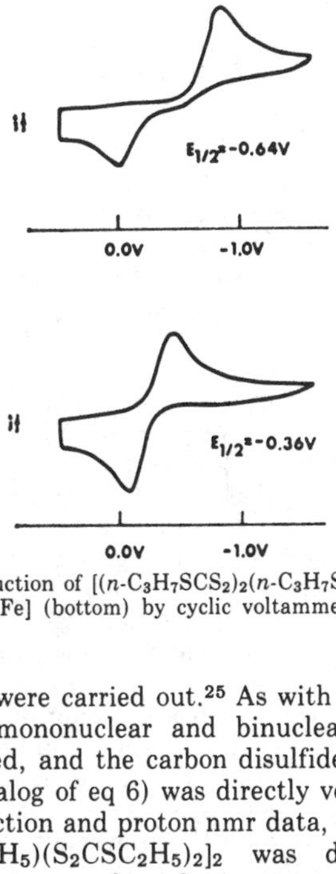

Figure 7. Reduction of $[(n\text{-}C_3H_7SCS_2)_2(n\text{-}C_3H_7S)Fe]_2$ (top) and $[(t\text{-}C_4H_9SCS_2)_3Fe]$ (bottom) by cyclic voltammetry (reproduced from ref 22).

(Figure 6b) were carried out.[25] As with the iron system, both mononuclear and binuclear complexes were obtained, and the carbon disulfide elimination reaction (analog of eq 6) was directly verified. From X-ray diffraction and proton nmr data, the structure of $[Co(SC_2H_5)(S_2CSC_2H_5)_2]_2$ was determined.[25] Whereas the iron analog adopts structure II, the cobalt derivative was found to have structure III.

From the X-ray results it was clear that the cobalt dimer has no metal–metal bond, as anticipated from electronic structural considerations (Figure 6).[26] Because of the long Co⋯Co distance of 3.32 Å (compared to 2.62 Å for the iron dimer), the thioxanthate ligands are no longer able to bridge across the top and bottom of the M_2S_2 rhombus (structure II).[25]

(23) D. F. Lewis, S. J. Lippard, and J. A. Zubieta, *Inorg. Chem.*, **11**, 823 (1972).
(24) D. Coucouvanis, S. J. Lippard, and J. A. Zubieta, *Inorg. Chem.*, **9**, 2775 (1970).
(25) D. F. Lewis, S. J. Lippard, and J. A. Zubieta, *J. Amer. Chem. Soc.*, **94**, 1563 (1972).
(26) See L. F. Dahl, E. R. de Gil, and R. D. Feltham, *J. Amer. Chem. Soc.*, **91**, 1653 (1969), for a general treatment of the stereochemical consequences of metal–metal bonding in ligand-bridged binuclear complexes.

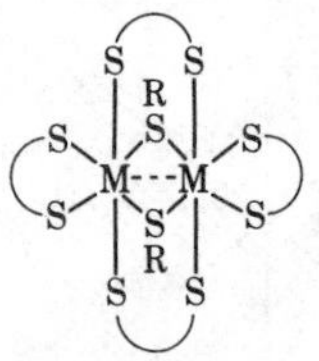

II

III

The occurrence of structure III for $[Co(SC_2H_5)(S_2CSC_2H_5)_2]_2$ provides some support, albeit indirect, for the idea that a structural rearrangement such as II → III occurs in the electrochemical reduction of the iron(III) analog. The slowness of this rearrangement compared to the electron-transfer step could then account for the observed irreversibility of the electrode process. This analysis presumes the $\sigma^*(xy)$ orbital (Figure 6) to be populated in the reduction of the $[Fe(SR)(S_2CSR)_2]_2$ compounds. The high energy of this orbital and the presence of the bridging thioxanthate ligands render this an energetically unfavorable process, thus accounting for the low reduction potential of the dimers compared to monomeric $Fe(S_2CSR)_3$.

A Rationale for the Low and Varying Reduction Potentials of the Fe_2S_2 Proteins

The results just discussed, together with studies carried out chiefly by Dahl and coworkers,[26,27] demonstrate that the metal–metal bond order in a bridged M_2X_2 dimer can have a profound influence on its geometry. The redox behavior of the system also appears to be sensitive to geometric rearrangements or reactions[28] accompanying the transfer of electrons into or out of the orbitals involved in metal–metal bonding.

These observations suggest[25] the following rationale for the low and varying reduction potentials of the Fe_2S_2 proteins. If the geometries of the Fe_2S_2(S-Cys)$_4$ centers in these proteins are constrained by the surrounding polypeptide backbone to be the same in both oxidized and reduced forms, addition of an electron cannot be accompanied by a geometric adjustment. Assuming that the electronic structure of the Fe_2S_2(S-Cys)$_4$ center would require such an adjustment upon reduction (*vide infra*), but that the constraints of the polypeptide chain do not allow it, the protein in its reduced form would be a good electron donor. Changes in the amino acid composition among various classes of Fe_2S_2 proteins could monitor the constraint on the iron–sulfur redox center, producing the observed variations in the reduction potentials (Table I). Besides the reduction potential, other properties of the reduced protein would be affected, for example, the unusual epr spectra observed for the reduced plant ferredoxins.[2] The "unique" properties of metalloproteins have previously been attributed to the presence of highly specific ligand geometries available in the macromolecule but not in simple ligand systems.[29] In a sense, the present hypothesis is an extension of these ideas.

(27) N. G. Connelly and L. F. Dahl, *J. Amer. Chem. Soc.*, **92,** 7472 (1970), and references cited therein.

(28) See, for example, J. A. Ferguson and T. J. Meyer, *Inorg. Chem.*, **11,** 631 (1972).

Before discussing possible experimental tests of the above rationale, let us consider in somewhat more detail exactly how the model might be applied to the Fe_2S_2 proteins. To recapitulate, the major features are that (1) the electron enters a relatively high energy state and (2) the protein does not structurally rearrange to accommodate the extra electron. In the case of $[Fe(SR)(S_2CSR)_2]_2$, the high-energy state was the $\sigma^*(xy)$ orbital (Figure 6). For the Fe_2S_2 proteins, however, several lines of evidence indicate the presence of antiferromagnetically coupled, high-spin, tetrahedral iron(III) centers which upon reduction are converted to one iron(II) and one iron(III) center, also antiferromagnetically coupled.[7] The estimated J values[7a] show the interaction to be considerably weaker than expected for an iron–iron single bond. Bearden and Dunham assign[30] the electron in the reduced protein to the d_{z^2} orbital of the iron(II) ion. Since this orbital is directed toward the half-filled d_{z^2} orbital of the neighboring iron(III) center (the z axis is taken along the Fe···Fe vector), a nonbonded electron repulsive term would render this a high-energy state compared to the oxidized protein.[30] Failure of the Fe_2S_2 chromophore to relax this state energetically by making a geometric adjustment would result in a low reduction potential, as suggested above. The effect is like that of a metal–metal bonded system, only weaker. It is noteworthy that the reduction of the distorted tetrahedral iron(III) center to iron(II) in rubredoxin also involves the d_{z^2} orbital ($e^2t_2^3 \rightarrow e^3t_2^3$). Here there is no neighboring iron atom, however, and the reduction potential is ~0.35 V more positive. Since the electron does not enter a σ-antibonding (t_2) orbital there is no reason to expect a gross geometric change, and none is experimentally observed.[4]

It is prudent to underscore the speculative nature of the foregoing analysis. There are, to be sure, alternative explanations for the various reduction potentials (see, for example, ref 1). Yet the rationale does present a working hypothesis which, like any other, must stand or fall based on experimental studies. The most revealing of these would be to obtain high-resolution X-ray data for both oxidized and reduced forms of an Fe_2S_2 protein. Equally important would be to examine the effect on the reduction potential of chemical modifications of the redox centers of the proteins, and studies of this kind are in progress. Finally, investigating the effect of steric constraints on the electronic properties of simple iron–sulfur coordination compounds would provide further information on which to assess the rationale for the redox behavior of the proteins. Some preliminary work of this nature is outlined below.

Sterically Constrained Complexes with Biologically Significant Ligands

The dithio acid, persulfide, and thioxanthate complexes of iron(III) described above do not qualify as biologically significant in the sense defined here,[10]

(29) (a) B. G. Malmström and T. Vänngård, *J. Mol. Biol.*, **2,** 118 (1960); (b) B. L. Vallee and R. J. P. Williams, *Proc. Nat. Acad. Sci. U. S.*, **59,** 498 (1968).

(30) A. J. Bearden and W. R. Dunham, *Struct. Bonding (Berlin)*, **8,** 1 (1970).

although they have influenced our thinking about the iron–sulfur proteins. A similar remark applies to iron–sulfur coordination compounds of a wide variety prepared and characterized in other laboratories. Two notable examples are $(h^5\text{-}C_5H_5)_4Fe_4S_4$[31] and $Fe[S_2(PR_2)_2N]_2$,[32] the structures of which closely approximate the iron–sulfur cores in the bacterial ferredoxins (and HiPIP) and rubredoxin, respectively.

Lately, several iron–sulfur complexes with biologically significant ligands have been reported to be "identified" in solution by optical spectroscopy.[33] These compounds have varying degrees of stability and, in some cases, their electronic spectra bear a striking resemblance to those of certain iron–sulfur proteins. A major shortcoming of those studies, however, is their failure to produce crystalline products suitable for chemical and structural analysis. Since little is known about the geometries of the complexes formed in solution, new information concerning the structural possibilities for the iron–sulfur proteins has not been provided. This is not to minimize the value of the work in demonstrating that the optical properties of the proteins are congruent with data obtained on iron–sulfur chromophores constituted from biologically significant ligands.

Recently we have synthesized and characterized crystalline iron–sulfur complexes using ligands of type IV.[34] These ligands cannot afford an exact

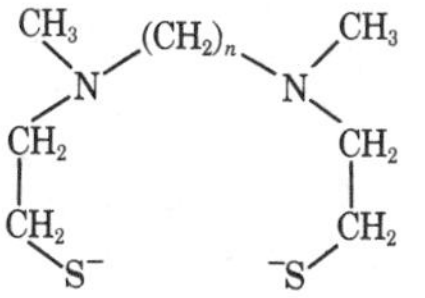

IVa, $n = 2$
b, $n = 3$

match to the coordination environment of the iron–sulfur proteins since labile sulfur is not available and nitrogen donor atoms almost certainly will coordinate, an unlikely possibility, although not yet entirely eliminated, for the Fe_2S_2 proteins. Nevertheless, the ligands were designed for several specific reasons. First, they do provide the desired $-CH_2S^-$ function. Second, extensive polymerization was expected to be suppressed by the multidentate character of the ligand. Third, chemical control of steric strain is possible through variation of n.

In the case where $n = 1$ or 2, a study of models showed that a severely strained complex would result from coordinating IV to a single metal ion with a pseudotetrahedral geometry. For $n = 3$, the strain is relieved. Finally, since primary mercaptide sulfur

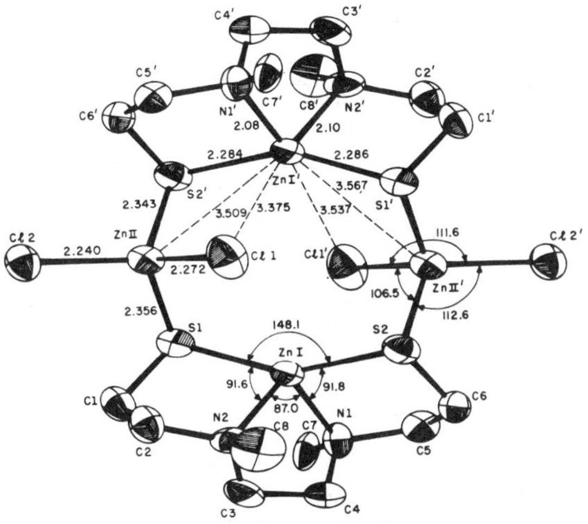

Figure 8. The tetranuclear cluster in $[Zn_2Cl_2L]_2 \cdot 2H_2O$ (reproduced from ref 36).

atoms have a known tendency to form three bonds and bridge metal atoms,[35] the ligands were expected to provide low molecular weight oligomeric compounds. Since in the absence of additional ligands the coordination number of the metal would be limited to four (monomer, V) or five (dimer, VI), iron–

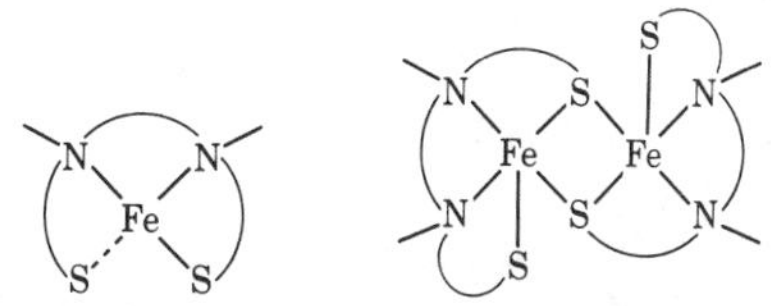

V VI

sulfur complexes with a resultant weak ligand field might obtain. These would be of obvious interest since high-spin iron(III) centers are known to occur in the proteins. In brief, then, the employment of multidentate ligands IV was based upon a desire for control and variation of the stereochemical properties of the donor atom set at the expense of exact duplication of the protein ligands.

To study its coordination properties, ligand IVa was allowed to react with zinc(II) chloride to form an air-stable, crystalline salt, the structure of which is shown in Figure 8.[36] Two relevant features of this structure are: (1) the tetrahedral geometry of type I zinc atoms is strongly distorted and (2) a tetranuclear array of metals with bridging mercaptide atoms is formed. Both of these results were anticipated for ligand IVa, as mentioned above.

The synthesis of the iron(II) complexes of IVa and IVb, FeL and FeL′, respectively, was achieved by reaction of the appropriate ligand in excess with ferric acetylacetonate.[37] Red-brown crystals of both compounds were obtained and shown by X-ray diffraction to have structures similar to VI.[37] The iron(II) atoms in the binuclear complexes are in a dis-

(31) (a) R. A. Schunn, C. J. Fritchie, and C. T. Prewitt, *Inorg. Chem.*, **5**, 892 (1966); (b) C. H. Wei, G. R. Wilkes, P. M. Treichel, and L. F. Dahl, *ibid.*, **5**, 900 (1966).

(32) (a) A. Davison and E. S. Switkes, *Inorg. Chem.*, **10**, 837 (1971); (b) A. Davison and D. L. Reger, *ibid.*, **10**, 1967 (1971); (c) M. R. Churchill and J. Wormald, *ibid.*, **10**, 1778 (1971).

(33) (a) A. Ali, F. Fahrenholz, J. C. Garing, and B. Weinstein, *J. Amer. Chem. Soc.*, **94**, 2556 (1972); (b) Y. Sugiura and H. Tanaka, *Biochem. Biophys. Res. Commun.*, **46**, 335 (1972); (c) L. G. Stadtherr and R. B. Martin, *Inorg. Chem.*, **11**, 92 (1972); (d) S. A. Grachev, L. I. Shchelkunova, Yu. A. Makashev, and F. Ya Kul'ba, *Zh. Neorg. Khim*, **16**, 198 (1971); (e) C. S. Yang and F. M. Huennekens, *Biochemistry*, **9**, 2127 (1970); (f) A. Tomita, H. Hirai, and S. Makishima, *Inorg. Chem.*, **7**, 760 (1968); **6**, 1746 (1967).

(34) W. J. Hu, K. D. Karlin, D. Barton, and S. J. Lippard, *Proc. Int. Conf. Coord. Chem.*, *14th*, 598 (1972).

(35) S. E. Livingstone, *Quart. Rev., Chem. Soc.*, **19**, 386 (1965).

(36) W. J. Hu, D. Barton, and S. J. Lippard, *J. Amer. Chem. Soc*, **95**, 1170 (1973).

(37) W. J. Hu, K. D. Karlin, and S. J. Lippard, unpublished results.

torted trigonal-bipyramidal environment. Although the bonded iron–donor atom distances remain essentially constant for the two molecules, there are severe angular distortions in $(FeL)_2$. As a consequence, the nonbonded Fe···Fe distance is reduced to 3.205 (7) Å from the value of 3.371 (2) Å in relatively unstrained $(FeL')_2$.

The structural results demonstrate the effect of external ligand constraints (such as a protein might produce) on the geometry of the Fe_2S_2 bridging system. If nonbonded electron interactions do affect the reduction potentials in such systems, as suggested above, then it would be interesting to examine the redox properties of $(FeL)_2$ and $(FeL')_2$. Unfortunately, the compounds decompose or are insoluble in most solvents tried to date. A study of the dependence of the magnetic susceptibilities of the solid complexes over the range $80 < T < 400$ K, however, shows them to have measurably different electronic structures. Both contain antiferromagnetically coupled, high-spin iron(II) atoms, but the room temperature moment and Neél temperature for $(FeL')_2$ are 4.2 BM and 160 K, respectively, while the corresponding values for $(FeL)_2$ are 3.4 BM and ~350 K. Thus, the structural and magnetic properties of Fe_2S_2 dimers in a weak field environment using biologically significant donor ligands are sensitive to steric strain supplied by the ligand. Further studies are in progress.

Summary and Overview

In a review of the bioinorganic chemistry of vitamin B_{12} and related compounds, it was suggested that the biological studies inspired more advances in the coordination chemistry of cobalt than *vice versa.*[38] A similar comment is applicable, at least in part, to the iron–sulfur systems discussed here. The synthetic goal of producing crystalline and well-characterized iron–sulfur complexes with only biologically significant ligands has been realized, not only in the preparation of $(FeL)_2$ but also in recent work from the laboratory of Holm.[39] These latter compounds, which have been structurally characterized by Ibers,[39,40] provide the closest simulation yet to the actual iron–sulfur centers in both the Fe_4S_4 and Fe_2S_2 protein classes. The extent to which any of the above preparative achievements will enhance our understanding of how the proteins function remains an open question. It appears that a good beginning has been made, however.

Reasoning by admittedly speculative analogy to the iron(III) and cobalt(III) thioxanthate systems, a rationale for the low and varying reduction potentials of the Fe_2S_2 proteins has been proposed. New iron–sulfur complexes with sterically constraining polydentate ligands have been synthesized to provide some experimental criteria on which to base further assessment of this rationale. Besides primary mercaptides, the ligands chosen contain nitrogen donor atoms most likely not available to the iron in the proteins. Nevertheless, they may be more suited to the purpose of correlating redox behavior with steric strain than all sulfur donor ligands in which a strained configuration is lacking. There seems to be no compelling reason to strive for perfect duplication of the protein active site environment (*e.g.*, through the use of polypeptide ligands) to simulate or attempt to understand its properties. Indeed, no molecule smaller than the protein itself is likely to be capable of displaying *all* its relevant physical and chemical properties.

The discussion in this Account has focused primarily on rubredoxin and the Fe_2S_2 proteins, in which the oxidation states of the metal and ligand atoms are known. An important piece of information which is lacking for the Fe_4S_4 proteins is the overall charge on the tetranuclear cluster. Thus, differences in the average valence state of the iron atoms in the bacterial ferredoxins and the HiPIP molecules could account for their different reduction potentials.[39a] A further question is whether the electron in the reduced Fe_4S_4 cluster is distributed over all four iron atoms or whether, as in the Fe_2S_2 proteins, it is more highly localized.

In the Fe_4S_4 proteins, the choice of S^{2-} has a natural explanation in its role as a triply bridging ligand, for which cysteine sulfur would be somewhat inferior.[41] In the Fe_2S_2 proteins, however, it would appear that cysteine sulfur could replace the labile sulfide as the bridging atoms. It is therefore important that the possibility of a Fe–S–S–C linkage, discussed earlier, or Fe–S–S–Fe units[21] receive a critical evaluation. Again the difference in overall charge on the iron–sulfur core, for S^{2-} *vs.* RSS^- or RCH_2S^- donor ligands, may be an important factor.

Besides the relatively well-characterized iron–sulfur proteins listed in Table I, numerous others constitute essential parts of biological redox systems.[1,3d] For example, epr studies of mitochondrial and submitochondrial particles provide strong evidence for iron–sulfur redox cores,[42] but little detailed information about purified protein materials is yet available. These proteins offer potential challenges for future work and will possibly reveal a new relevance of iron–sulfur coordination compounds already well understood.

The experimental studies carried out in the author's laboratory were made possible by generous support from the National Institutes of Health under Grant GM-16449 and by the fruitful labors of the several coworkers cited in individual references. This article was written during a sabbatical leave, supported in part by the John Simon Guggenheim Foundation, at the Chalmers Institute of Technology in Göteborg, Sweden. The hospitality and stimulation of Professor Bo Malmström are gratefully acknowledged.

(38) D. L. Brown, *Progr. Inorg. Chem.*, **18**, 177 (1973).

(39) (a) T. Herskovitz, B. A. Averill, R. H. Holm, and J. A. Ibers, *Proc. Int. Conf. Coord. Chem., 14th*, 1 (1972); (b) T. Herskovitz, B. A. Averill, R. H. Holm, J. A. Ibers, W. D. Phillips, and J. F. Weiher, *Proc. Nat. Acad. Sci. U. S.*, **69**, 2437 (1972); (c) R. H. Holm, private communication.

(40) M. R. Snow and J. A. Ibers, *Inorg. Chem.*, **12**, 249 (1973).

(41) Primary mercaptides can utilize all valence electrons to coordinate three metal atoms (W. Harrison, W. C. Marsh, and J. Trotter, *J. Chem. Soc., Dalton Trans.*, 1009 (1972), and references cited therein), but this is rare.

(42) (a) N. R. Orme-Johnson, W. H. Orme-Johnson, R. E. Hansen, H. Beinert, and Y. Hatefi, *Biochem. Biophys. Res. Commun.*, **44**, 446 (1971); (b) T. Ohnishi, D. F. Wilson, T. Asakura, and B. Chance, *ibid.*, **46**, 1631 (1972).

Addendum (May 1976)

Detailed magnetic studies of the binuclear, sulfur-bridged complexes $(FeL)_2$ and $(FeL')_2$ have now been completed.[43] A theory based on the hypothesis of orbitally dependent exchange was derived to explain the susceptibility vs. temperature curves for these two molecules. The two iron(II) atoms 3.21 Å apart in $(FeL)_2$ engage in significant antiferromagnetic exchange ($|J| \simeq 50\ cm^{-1}$) by direct overlap of 3d orbitals directed along the iron–iron internuclear axis. This direct exchange is reduced by at least 50% in the less sterically constrained molecule $(FeL')_2$, where the Fe–Fe distance is 0.16 Å longer. The magnetic exchange interaction is thus seen to be extremely sensitive to the iron–iron distance, a result that might be useful in understanding the detailed magnetic properties of the iron–sulfur proteins.

The complex $(FeL)_2$ reacts reversibly with carbon monoxide and with RNC, R = CH_3 or *tert*-C_4H_9, to form the bis-adducts.[44] Infrared spectral results suggest formation of the *cis* isomers. The similarity between $FeL(CO)_2$ and $Fe(cys)_2(CO)_2$,[45] cys = cysteinate, may be noted.

The crystal structure of the green complex FeL(NO) shows it to be a pentacoordinate monomer with slightly distorted tetragonal pyramidal geometry.[46] The nitrosyl ligand is apical, and the Fe–N–O bond angle is 157°. This extent of bending is exactly that expected for a $[MNO]^7$ complex.[47] The NO stretching frequency in FeL(NO) occurs as a strong, broad band at 1650 cm^{-1} in the infrared spectrum, and the epr spectrum at 233° K in acetonitrile consists of a triplet with g_{av} = 2.043, $A_{iso}(^{14}N)$ = 12.0 G. Very similar values were observed for FeL′((NO) (ν_{NO} = 1630 cm^{-1}, broad; g = 2.041; A = 12.0 G).

The reaction of nitrosonium salts with $(FeL)_2$or $(CoL)_2$ affords the $[(ML)_2\text{-}NO]^+$ cations in high yield.[48,49] X-ray structural studies show the products to consist of two distorted octahedra sharing a common face comprised of two bridging thiolate sulfur atoms and a symmetrically bridging nitrosyl ligand. The binuclear iron complex achieves an 18-electron configuration by forming an iron–iron bond (d_{Fe-Fe}, 2.468 Å) that is lacking in the cobalt analog (d_{Co-Co}, 2.770 Å). These structural results, together with infrared and Mössbauer studies of the adducts, are best interpreted according to eq. (7) involving oxidative addition of nitrosonium ion to the bimetallic iron(II) or cobalt(II) center.

$$(M^{II}L)_2 + NO^+ \longrightarrow [(M^{III}L)_2(NO^-)]^+ \qquad (7)$$

Apart from the synthetic and structural results obtained in our own laboratory and just summarized, Holm, Ibers, and co-workers have synthesized and structurally characterized $Fe(SR)_4^-$, $Fe_2S_2(SR)_4^{2-}$, and $Fe_4S_4(SR)_4^{2-}$ complexes that have provided substantial insight into the stereochemical and physical properties of iron–sulfur centers in the proteins.[50]

(43) A. P. Ginsberg, M. E. Lines, K. D. Karlin, S. J. Lippard, and F. J. DiSalvo, *J. Am. Chem. Soc.*, in press.
(44) K. D. Karlin and S. J. Lippard, *J. Am. Chem. Soc.*, in press.
(45) (a) M. Schubert, *J. Am. Chem. Soc.*, **54**, 4077 (1932); *ibid.*, **55**, 4563 (1933) (b) J. H. Wang, A. Nakahara, and E. B. Fleischer, *J. Am. Chem. Soc.*, **80**, 1109 (1958); (c) A. Tomita, H. Hirai, and S. Makishima, *Inorg. Chem.*, **6**, 1746 (1967); *ibid.*, **7**, 760 (1968).
(46) K. D. Karlin, H. N. Rabinowitz, D. L. Lewis, and S. J. Lippard, to be submitted for publication.
(47) (a) J. H. Enemark and R. D. Feltham, *Coord. Chem. Rev.*, **13**, 339 (1974); (b) R. Hoffman, M. M. L. Chen, M. Elian, A. R. Rossi, and D. M. P. Mingos, *Inorg. Chem.*, **13**, 2666 (1974), and references cited therein.
(48) K. D. Karlin, D. L. Lewis, H. N. Rabinowitz, and S. J. Lippard, *J. Am. Chem. Soc.*, **96**, 6519 (1974).
(49) H. N. Rabinowitz, K. D. Karlin, and S. J. Lippard, submitted for publication.
(50) R. H. Holm, *Endeavour*, **34**, 38 (1975), and references cited therein.

The Biochemical Significance of Porphyrin π Cation Radicals

David Dolphin*

Department of Chemistry, Harvard University, Cambridge, Massachusetts 02138

Ronald H. Felton

School of Chemistry, Georgia Institute of Technology, Atlanta, Georgia 30332

Received September 22, 1972

Reprinted from ACCOUNTS OF CHEMICAL RESEARCH, *7, 26 (1974)*

Porphine (**1**), an aromatic tetrapyrrolic macro-

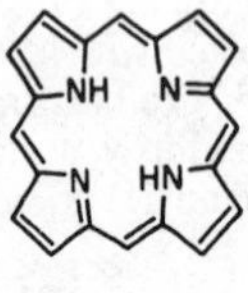

1

cycle,[1] serves as the nucleus for all the naturally occurring porphyrins and related systems. Metalloporphyrins are readily prepared by metal ion displacement of the central nitrogen protons, and all of these systems are characterized by an intense absorption band in the uv (Soret band) and weaker absorptions (ϵ ~20,000) in the visible and near ir.

Solid-state structure determinations[2] of porphyrins and metalloporphyrins show that the molecule is not totally rigid and may show ruffling of the skeleton by alternate upward and downward displacement of the pyrrole rings from a planar arrangement. Nonetheless, in an environment free from crystal forces, spectroscopic evidence is consistent with a general planar description of the porphyrin. In metalloporphyrins the metal may be found[2,3] either in or out of this plane, depending upon the metal and its axial ligands.

Interest in porphyrin chemistry is generated by the presence of these species and closely related substances in biological redox processes; of particular importance is their appearance and function in biological oxidations.

The biosphere depends upon photosynthesis to carry out the conversion of photonic into chemical energy as well as to maintain an oxidizing atmosphere for catabolism (conversion of organic compounds to CO_2) which provides the energy necessary to drive endergonic biochemical processes. The classical reaction of photosynthesis requiring chlorophyll a (**2**) involves CO_2 fixation, *viz.*

$$CO_2 + H_2O \xrightarrow[\text{"chlorophyll"}]{h\nu} [CH_2O] + O_2 \qquad (1)$$

David Dolphin received both his B.Sc. and Ph.D. degrees from the University of Nottingham. In the fall of 1965 he moved to Harvard where he spent a year as a postdoctoral fellow with R. B. Woodward. He then spent 7 years on the faculty at Harvard and in January 1974 joined the faculty at the University of British Columbia. The structure, synthesis, chemistry and biochemistry of porphyrins, vitamin B_{12}, and related macrocycles are the main focus of his research.

Ronald Felton received his B.S. from M.I.T. and Ph.D. in Chemical Physics at Harvard University. He joined the faculty at Georgia Tech in 1968 where he is now an Associate Professor. Professor Felton is interested in the statistical theory of electronic energies as well as the physical properties of porphyrin ions.

where $[CH_2O]$ is a carbohydrate and water serves as the oxidizable hydrogen donor. A similar reaction occurs in certain photosynthetic bacteria which utilize bacteriochlorophyll (**3**) and hydrogen donors other than water.

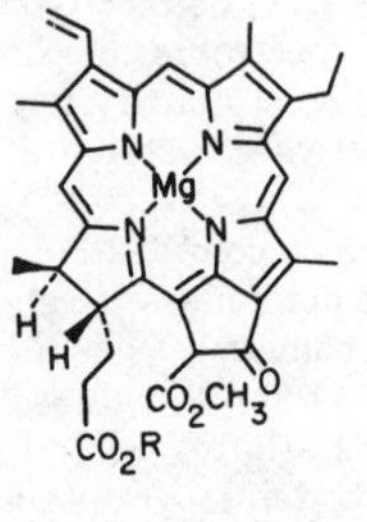

2

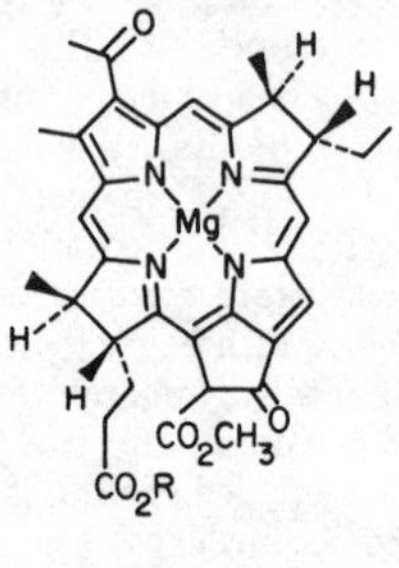

3

The photosynthetic process divides naturally into light-driven primary reactions[4,5] and the subsequent CO_2 reduction steps studied by Calvin, Bassham, and coworkers.[6] Under the title of primary reactions is included the conversion of light energy into reductants and oxidants and, in the case of green plants and blue-green algae, the chemical and physical states involved in the oxygen-evolving apparatus.

Heme protein participation[7] in oxygen transport, peroxide reduction and disproportionation, the mitochondrial electron transport chain, and drug metabolism (cytochrome P450) stresses the biological importance and diverse roles of the iron porphyrins. Iron protoporphyrin IX (**4**) (heme) is the prosthetic group of hemoglobin, myoglobin, catalase, peroxi-

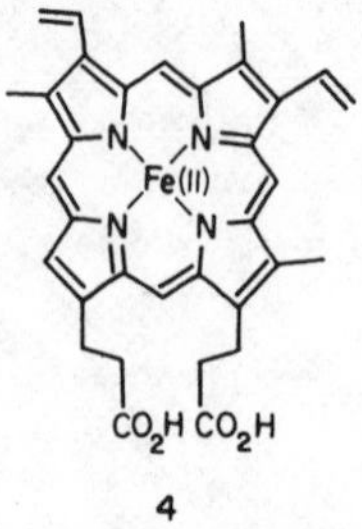

4

(1) J. E. Falk, *Porphyrins Metalloporphyrins,* (1964).
(2) E. B. Fleischer, *Accounts Chem. Res.,* **3,** 105 (1970).
(3) J. L. Hoard, *Science,* **174,** 1295 (1971).
(4) N. I. Bishop, *Annu. Rev. Biochem.,* **40,** 197 (1971).
(5) R. K. Clayton, *Annu. Rev. Biophys. Bioeng.,* **2,** 131 (1973).
(6) M. Calvin and J. A. Bassham, "The Photosynthesis of Carbon Compounds," W. A. Benjamin, New York, N. Y., 1962.
(7) H. R. Mahler and E. H. Cordes, "Biological Chemistry," 2nd ed., Harper and Row, New York, N. Y., 1971, Chapter 15.

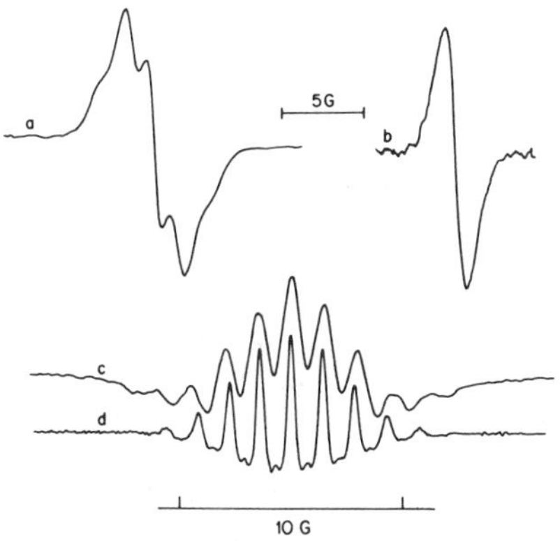

Figure 1. First derivative epr spectra of (a) $[Mg^{II}OEP]\cdot^{+}ClO_4^-$ and (b) $[Mg^{II}OEP\text{-}meso\text{-}d_4]\cdot^{+}ClO_4^-$ in methanol at −50°. Second derivative epr spectra of (c) $[Mg^{II}TPP]\cdot^{+}ClO_4^-$ and $[Mg^{II}TPP\text{-}d_{20}]\cdot^{+}ClO_4^-$ (perdeuterated phenyl groups) in chloroform.

dase, and many of the cytochromes. The respiratory pigment hemoglobin contains four heme prosthetic groups and is distributed in red blood cells; myoglobin is a monomer found in muscle cells. Both pigments reversibly bind oxygen for use in cellular catabolism. Hydroperoxidases are hemiproteins (iron is present as Fe^{3+} in the resting enzyme) which serve to catalyze the reaction

$$2H_2O_2 \xrightarrow{\text{catalase}} 2H_2O + O_2 \qquad (2)$$

in the case of catalase or a peroxidative reaction

$$ROOH + H_2A \xrightarrow{\text{peroxidase}} ROH + H_2O + A \qquad (3)$$

The reaction pathways by which these enzymes act are complex, and we will have occasion later to remark on the structure of some intermediates observed during their catalytic cycles.

Considerable emphasis (arising historically from studies on the heme function in hemoglobin and the cytochromes) has been placed upon the role of the central metal ion. However, the ubiquity of metalloporphyrins in nature and the many varied functions they perform suggest that, besides modifying metal redox potentials or binding the metal at an appropriate site in the protein, the porphyrin ring itself possesses properties necessary for proper biological function. In this Account the results of porphyrin oxidations will be discussed in relation to their biological function and significance.

Oxidation of Model Porphyrins

Polarographic studies[8] have shown that a variety of porphyrins and metalloporphyrins undergo two successive one-electron oxidations. The synthetic pigments most thoroughly investigated are metallooctaethylporphyrins (**5**) (MOEP) and metallotetraphenylporphyrins (**6**) (MTPP). Porphyrins containing magnesium and zinc may be readily oxidized by iodine or bromine[9-11] in CH_2Cl_2, $CHCl_3$, and

(8) A. Stanienda and G. Biebl, *Z. Phys. Chem.* (*Frankfurt am Main*), **52,** 254 (1967); A. Stanienda, *Z. Phys. Chem.* (*Leipzig*), **229,** 257 (1965).

(9) J.-H. Fuhrhop and D. Mauzerall, *J. Amer. Chem. Soc.*, **90,** 3875 (1968).

(10) R. H. Felton, D. Dolphin, D. C. Borg, and J. Fajer, *J. Amer. Chem. Soc.*, **91,** 196 (1969).

(11) J.-H. Fuhrhop and D. Mauzerall, *J. Amer. Chem. Soc.*, **91,** 4174 (1969).

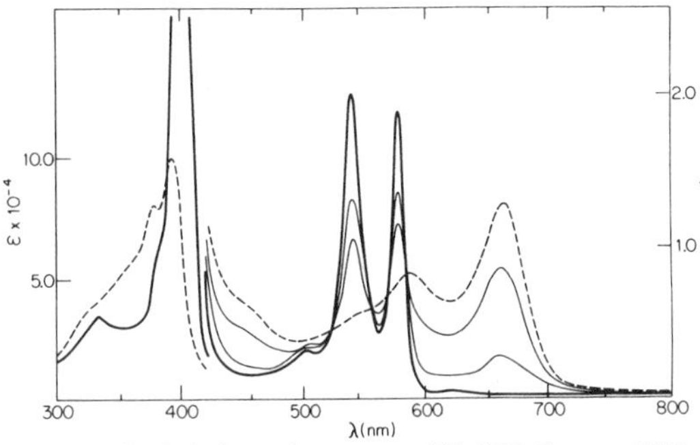

Figure 2. Optical absorption spectra of MgOEP (heavy solid line) and $[Mg^{II}OEP]\cdot^{+}ClO_4^-$ (dashed line) in methylene dichloride.

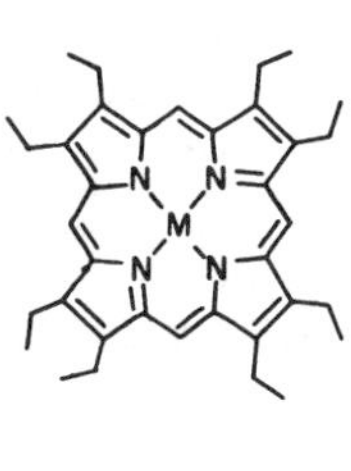

5

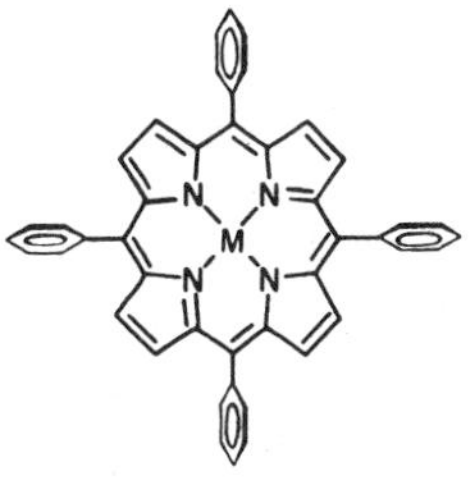

6

CH_3OH, by electrolysis in CH_2Cl_2 or butyronitrile,[12] or by photochemical oxidation.[13] Stoichiometric titrations with chemical oxidants or coulometry during electrolysis agree with the one-electron process observed polarographically, and analyses[12,14] of the epr hyperfine structure (hfs) of the singly oxidized species $[Zn^{II}TPP]\cdot^{+}$, $[Mg^{II}TPP]\cdot^{+}$, and $[Mg^{II}OEP]\cdot^{+}$ have clearly shown that the radicals are formed by electron abstraction from porphyrin orbitals. However, when the epr spectra of $[Mg^{II}OEP]\cdot^{+}$ and $[Mg^{II}TPP]\cdot^{+}$ were compared (Figures 1a and 1b), striking differences were noted. Thus $[Mg^{II}TPP]\cdot^{+}$ shows hfs from four equivalent nitrogens and from phenyl protons, while $[Mg^{II}OEP]\cdot^{+}$ exhibits a five-line spectrum due to four equivalent meso protons, but shows no coupling to the nitrogens (assignments were verified by deuterium substitution (Figures 1a and 1b)). Hence the epr spectra, while establishing the oxidized products as π cation radicals, lead to the question: why should spin density distributions be different in these radicals? Moreover, one must ask why are the optical spectra (Figures 2 and 3) of these apparently similar species so different?[15]

Theoretical considerations suggest an answer to these questions. The theory of porphyrin optical spectra developed by Gouterman and coworkers[18] indicates that the highest filled orbitals, a_{1u} and a_{2u},

(12) J. Fajer, D. C. Borg, A. Forman, D. Dolphin, and R. H. Felton, *J. Amer. Chem. Soc.*, **92,** 3451 (1970).

(13) A. N. Sidorov, V. Ye Kholmogorov, R. P. Yestigneyeva, and G. N. Kol'tsova, *Biofizika,* **12,** 143 (1968).

(14) J. Fajer, D. C. Borg, A. Forman, R. H. Felton, L. Vegh, and D. Dolphin, *Ann. N. Y. Acad. Sci.*, **206,** 349 (1973).

(15) The pronounced 660-nm absorption in $Mg^{II}OEP\cdot^{+}$ led to a preliminary identification[9] of this species as a metallophlorin[16,17] radical formed by nucleophilic attack at a meso carbon; however, the epr data obviate this structure.

(16) R. B. Woodward, *Ind. Chim. Belge,* **11,** 1293 (1962).

(17) D. Mauzerall, *J. Amer. Chem. Soc.*, **84,** 2437 (1962).

(18) M. P. Gouterman, *J. Mol. Spectrosc.*, **6,** 138 (1961); C. Weiss, H. Kobayashi, and M. P. Gouterman, *ibid.*, **16,** 415 (1965).

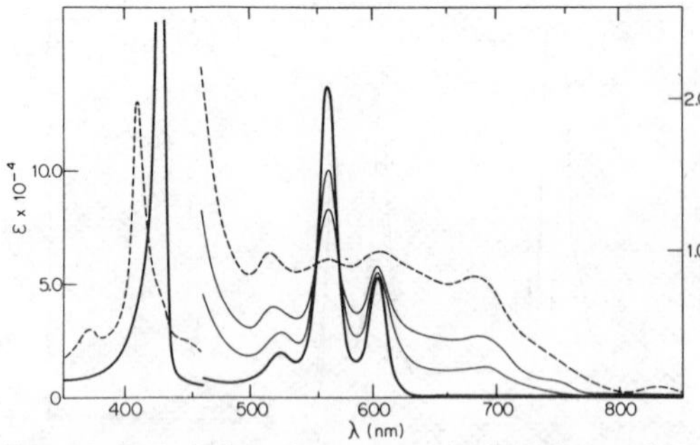

Figure 3. Optical absorption spectra of MgTPP (heavy solid line) and $[Mg^{II}TPP]\cdot^{+}ClO_4^-$ (dashed line) in methylene dichloride.

are almost degenerate in the unoxidized species. Therefore a π cation radical produced by electron abstraction from one or the other of these orbitals would lead to either a $^2A_{2u}$ or $^2A_{1u}$ ground state. This qualitative consideration is borne out by the results of open-shell calculations which predict an energy gap of 2000 to 3000 cm^{-1} between these two states.[12] In the $^2A_{2u}$ state, spin density appears on the meso carbon and nitrogen atoms, while the $^2A_{1u}$ state has low spin density on these atoms. In the latter state the spin density is primarily confined to α-pyrrolic carbon atoms. Predicted values for the nitrogen and hydrogen hyperfine coupling constants agree with those observed in $[Zn^{II}TPP]\cdot^{+}$. The small energy gap implies that peripheral substituents on the porphyrin ring, the metal, or its axial ligands might be expected to influence which ground state is lowest in energy. The epr data cited above show this to be the case; $[Mg^{II}OEP]\cdot^{+}$ has a $^2A_{1u}$ ground electronic state while $[Mg^{II}TPP]\cdot^{+}$ has the spin distribution of a $^2A_{2u}$ ground state.

Electronic transitions in the doublet manifold will differ as one or the other state is the ground state of the system. We have used optical spectra, particularly in the region 500–700 nm, to assign ground states to the π cation radicals. Examples[11,19,20] of species with A_{1u} spectra are $[Mg^{II}OEP]\cdot^{+}ClO_4^-$ and Br^-, $[Zn^{II}OEP]\cdot^{+}ClO_4^-$ and Br^-, and $[Ni^{II}TPP]\cdot^{+}ClO_4$ and Cl^-; A_{2u} spectra are displayed *inter alia* by $[Zn^{II}TPP]\cdot^{+}ClO_4^-$ and Br^-, $[Ni^{II}OEP]\cdot^{+}ClO_4^-$, $[Cu^{II}TPP]\cdot^{+}ClO_4^-$, and $[Cu^{II}OEP]\cdot^{+}ClO_4^-$. The last two examples indicate the subtlety of which interaction or set of interactions serves to select a specific ground state. Both oxidized copper porphyrins, though differing in peripheral substituents, display spectra characteristic of electron abstraction from an a_{2u} orbital.[8]

Important evidence in favor of the two-state theory was obtained with $Co^{II}OEP$.[21] Chemical oxidation with bromine proceeded by successive reversible one-electron oxidations, *viz.*

$$Co^{II}OEP \xrightarrow{\frac{1}{2}Br_2} [Co^{III}OEP]^{+}Br^{-} \xrightarrow{\frac{1}{2}Br_2} [Co^{III}OEP]^{2+}, 2Br^{-} \quad (4)$$

The resulting optical spectrum was typical of a $^2A_{1u}$ state. Electrochemical preparation of the π cation radical with tetrapropylammonium perchlorate as supporting electrolyte yielded $[Co^{III}OEP]\cdot^{2+}$-$2ClO_4^-$, whose spectrum was that of a typical $^2A_{2u}$ ground state. When the dibromide salt was treated with a slight excess of silver perchlorate, the bromide ligands were removed and the optical spectrum changed to that of the perchlorate salt (Figure 5a). This dramatic and reversible change in spectra as the axial ligands were changed is best explained by a switching of ground states.

The further electrochemical oxidation of the π cation radicals at potentials about 0.3 V above those required for the first oxidation brought about a second reversible one-electron oxidation to give a π dication.[12] During these oxidations the epr signal of the π cation radical decreased and disappeared when the oxidation was complete, and simultaneously the optical spectra changed to give an almost featureless spectrum in the visible and a single absorption in the uv region. Whereas the π cation radicals are relatively stable species which we routinely recrystallized from nucleophilic solvents such as a methanol, the π dications are powerful electrophiles which react readily with a variety of nucleophiles such as methanol, water, acetate, and halides to give initially an isoporphyrin (**7**; X = original nucleophile). In the

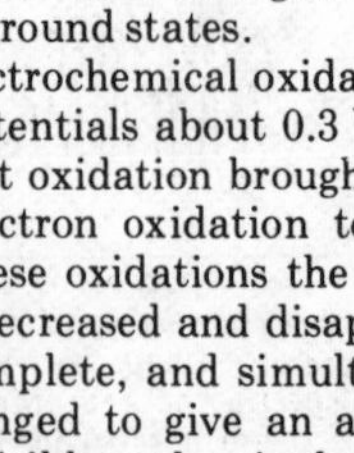

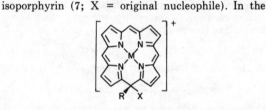

7

case of the *meso*-tetraphenyl derivatives (**7**; R = phenyl), these isoporphyrins are stable and have been isolated and characterized[22] as the perchlorate salts. Their optical spectra are characterized by a strong absorption band at 850–900 nm. With the octaethyl derivatives, however, the intermediate isoporphyrin (**7**; R = H) can lose a proton to give the neutral meso-substituted metalloporphyrin.[20] This series of reactions presents a convenient synthetic route to a variety of meso-substituted porphyrins and represents the first example of nucleophilic substitution at the periphery of the porphyrin nucleus.

The Occurrence of Metalloporphyrin π Cation Radicals in Nature

Photosynthesis. The ease of formation and stability of the metalloporphyrin π cation radicals led to the possibility that such species might occur in nature. The search began in the photosynthetic systems where chlorophyll a (Chl, **2**), a magnesium dihydroporphyrin (chlorine), and bacteriochlorophyll (BChl, **3**), a magnesium tetrahydroporphyrin (bacteriochlorine), play principal roles in the initial light-induced step of photosynthesis.

In the chloroplasts of plants, the majority of chlorophyll molecules have optical absorptions similar to the cell-free pigment. These function as antennae to

(19) A. Wolberg and J. Manassen, *J. Amer. Chem. Soc.*, **92**, 3982 (1970).
(20) D. Dolphin, Z. Muljiani, K. Rousseau, D. C. Borg, J. Fajer, and R. H. Felton, *Ann. N. Y. Acad. Sci.*, **206**, 177 (1973).
(21) D. Dolphin, A. Forman, D. C. Borg, J. Fajer, and R. H. Felton, *Proc. Nat. Acad. Sci. U. S.*, **68**, 614 (1971).
(22) D. Dolphin, R. H. Felton, D. C. Borg, and J. Fajer, *J. Amer. Chem. Soc.*, **92**, 743 (1970).

harvest and transport light energy to specific Chl a molecules that absorb to the red (at approximately 700 nm) of the antennae molecules. This species is known as pigment 700 (P700), and the differences in principal absorption maxima of P700 and cell-free Chl a (677 nm) arise from the local environment of P700 in the photosynthetic unit. Available evidence suggests that P700 is Chl a dimer, perhaps stabilized by a bridging water molecule.[23] A similar situation arises in photosynthetic bacterial chromatophores with BChl or chlorobium chlorophyll serving as antennae that transport energy to P865, the specialized[24] BChl a molecule in the reaction center. Upon illumination, *in vivo* bleaching of P700 or P865 and the concurrent appearance of epr signals have been interpreted as photooxidation of these species.[25-30]

The oxidation of cell-free chlorophyll a by methanolic ferricyanide produced an epr signal which mimicked those seen in photosynthesizing chloroplasts, and it was these observations that suggested that the photochemical step in photosynthesis was in fact a photooxidation of chlorophyll a. However, the site from which the electron was removed during this oxidation and the structure and fate of the remaining oxidized species have been the subjects of much speculation.

The electrochemical oxidation of chlorophyll a in methylene dichloride proceeded with the removal of one electron per molecule and gave a yellow solution (Figure 4a) which, upon electrochemical reduction, regenerated the initial chlorophyll a. Electrophoresis of this oxidized species showed it to be a cation and the radical character of the oxidized species was evident from its epr spectrum, which consisted of a single narrow line (9 G wide, peak to peak) at g = 2.0026. This epr signal closely resembled that reported for photosynthetic systems and suggested a similarity between the electrochemically oxidized chlorophyll a and the photooxidized P700 of nature. This similarity was further exemplified by the close resemblance of the difference spectra (Figure 4b) of chlorophyll a and electrochemically[31] generated Chl a$\cdot^+$ with that of dark and photosynthesizing chloroplasts.[32] The only major difference between these two spectra was the bathochromic shift of the change at 700 nm compared to that of 677 for cell-free chlorophyll a, a shift which reflects the special environment of P700 in the photosynthetic unit. The spectral analogies between the electrochemically generated Chl a$\cdot^+$ and photooxidized P700 strongly support

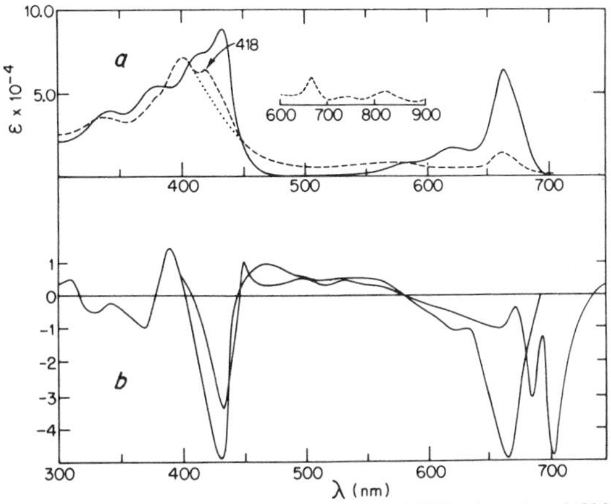

Figure 4. (a) Optical absorption spectra of Chl a (——) and Chl a$\cdot^+$ (---) in methylene dichloride. A small amount of decomposition product (λ_{max} 418 nm) appears during the electrochemical oxidation. (b) The optical difference spectrum[32] of Chl a and Chl a$\cdot^+$ (calculated from a) and the difference spectrum of photosynthesizing chloroplasts. The observation of an absorption band in the near infrared of the chlorophyll π cation radical (---) suggested that such an absorption should be observed in the near-infrared difference spectrum of photosynthesizing chloroplasts; such an observation[32] has been recently made.

the identification of Chl a$\cdot^+$ as the chlorine moiety of oxidized P700.

A characterization of Chl a$\cdot^+$ was then needed because of its role in photosynthetic processes. While there was general agreement that bleached P700 was a positive ion of chlorophyll, no evidence was adduced which demonstrated that this species was specifically a π cation radical. Fortunately this identification can be established by comparisons between Chl a$\cdot^+$ and oxidized metalloporphyrins and chlorines.

Two reversible one-electron waves ($E_{1/2}(1)$ and $E_{1/2}(2)$) are observed in the polarographic oxidation of chlorophyll a and b, pheophytin a and b (magnesium-free chlorophylls), and zinc *meso*-tetraphenylchlorine.[8,31] The difference in half-wave potentials, $\Delta = E_{1/2}(2) - E_{1/2}(1)$, is 0.27 (±0.03) V for these chlorines. For a series of metalloporphyrins Δ = 0.30 (±0.03) V and for metalloetioporphyrins[8] Δ = 0.35 (±0.1) V. If this difference is examined rather than the individual oxidation potentials, then the influence of the metal is minimized. Moreover, Δ is constant when the sites of the two reversible one-electron abstractions are the same, and the prior unambiguous demonstration of π cation radical formation in metalloporphyrins demonstrates that chlorophyll a is similarly oxidized. Other possibilities such as oxidation of an unsaturated side chain or of the isocyclic ring of chlorophyll are ruled out by this argument and the favorable spectral comparisons between Chl a$\cdot^+$ and the π cation radical of zinc tetraphenylchlorine.

Similar comparisons[26,33] between the properties of BChl$\cdot^+$, produced by iodine oxidation or electrolysis, and those of bleached P865 attest that P865$\cdot^+$ contains a π cation radical of BChl.

Since a dimer or aggregate structure for P700 or P865 is favored by epr and circular dichroism[34,35]

(23) J. R. Norris, R. A. Uphaus, H. L. Crespi, and J. J. Katz, *Proc. Nat. Acad. Sci. U. S.* **68**, 625 (1971).
(24) J. R. Norris, M. E. Dryan, and J. J. Katz, *J. Amer. Chem. Soc.*, **95**, 1680 (1973). The possibility that a moiety of the dimer is bacteriophaeophytin (Mg-free bacteriochlorophyll) is not excluded by these authors' results.
(25) D. H. Kohl, in "Biological Applications of Electron Spin Resonance," H. M. Swartz, J. R. Bolton, and D. C. Borg, Ed., Wiley-Interscience, New York, N. Y., 1972, p 213.
(26) J. D. McElroy, G. Feher, and D. Mauzerall, *Biochim. Biophys. Acta*, **267**, 363 (1972).
(27) W. W. Parson, *Biochim. Biophys. Acta*, **153**, 248 (1968).
(28) P. A. Loach and K. Walsh, *Biochemistry*, **8**, 1908 (1969).
(29) J. R. Bolton, R. K. Clayton, and D. W. Reed, *Photochem. Photobiol.*, **9**, 209 (1969).
(30) J. T. Warden and J. R. Bolton, *J. Amer. Chem. Soc.*, **94**, 4351 (1972).
(31) D. C. Borg, J. Fajer, R. H. Felton, and D. Dolphin, *Proc. Nat. Acad. Sci. U. S.*, **67**, 813 (1970).
(32) T. Hiyama and B. Ke, *Biochim. Biophys. Acta*, **267**, 160 (1972).

(33) P. A. Loach, R. A. Bambara, and F. J. Ryan, *Photochem. Photobiol.*, **13**, 247 (1971).

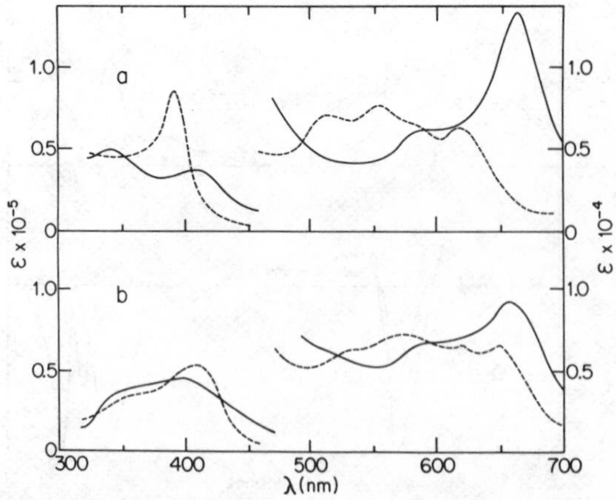

Figure 5. Comparison of the optical absorption spectra of (a) $[Co^{III}OEP]\cdot^{2+}2Br^-$ (——) and $[Co^{III}OEP]\cdot^{2+}2ClO_4^-$ (---); (b) Cat I (——) and HRP I (---). Catalase is tetrameric and the ϵ above is per hemin.

measurements, it is natural to inquire if the Chl a or BChl moiety might serve as the primary acceptor for the photochemically ejected electron and yield the anion radical of Chl or BChl, *e.g.*

$$(BChl)_2 \xrightarrow{h\nu} BChl\cdot^+ + BChl\cdot^- \quad (5)$$

Vacuum electrolytic reduction of BChl dissolved in dimethylformamide affords a facile route to the anion radical; comparison[36] of its optical absorption spectrum (maxima at 990 and 920 nm) and epr spectrum (g = 2.0028) with the spectra of the presumed primary acceptor[37-39] (no absorption in the ir and g = 2.005) distinguishes between these species. However, the photochemical reaction leading to charge separation might involve $BChl\cdot^-$ as a transient intermediate, observable, perhaps, by near-ir flash spectroscopic studies of reaction center preparations.

Catalase and Peroxidase. When nature uses chlorophyll as a source of electrons it is not too surprising that it is the dihydroporphyrin ring rather than the divalent magnesium ion which supplies them. With iron porphyrins, however, it has generally been assumed that the iron atom itself is the entity which undergoes the redox reaction, and in the cytochromes, which function *via* an Fe(II) ⇌ Fe(III) couple, there is no doubt that it is the metal which is the *eventual* site of electron capture or release.

There are two closely related series of iron porphyrin containing enzymes, the catalases[40] (Cat) and the peroxidases[41] (which are typified by horseradish peroxidase (HRP)). The resting enzymes both contain trivalent iron and are oxidized by hydrogen peroxide. The first intermediate observed spectrophotometrically during this oxidation is the so-called primary compound (Cat I or HRP I) which has two electrons less than the parent ferrihemoprotein. A one-electron reduction of the green primary compound forms the brown-red secondary compound (Cat II or HRP II). While the first step in the catalytic cycle of these two enzymes is the same, *i.e.*, a two-electron oxidation, by hydrogen peroxide, to their primary compounds, the two enzymes then perform different functions. Cat I oxidizes a second molecule of hydrogen peroxide to molecular oxygen and is itself reduced back to the ferrihemoprotein, while HRP I reacts with a hydrogen donor AH_2 to give a free radical and the secondary compound of the enzyme HRP II (eq 6) which can in turn oxidize a second donor molecule with the formation of the ferrihemoprotein (eq 7).

$$\text{HRP I} + AH_2 \longrightarrow \text{HRP II} + AH\cdot + H^+ \quad (6)$$

$$\text{HRP II} + AH_2 \longrightarrow \text{HRP} + AH\cdot + H^+ \quad (7)$$

There are a number of ways in which one might describe the electronic distribution in the primary complexes. It was originally suggested[42] that the primary compounds were complexes between hydrogen peroxide and the ferric iron, *i.e.*, $Fe^{III}OOH$, but this concept was challenged[40,43] when it was found that nonperoxidatic oxidants could give rise to the primary compounds. It was then suggested that the two electrons could be removed from the ferric iron to give pentavalent iron,[44] or by some combination of oxidations of the iron, the porphyrin, or the protein.[45-47] But such proposals were considered unlikely since an oxidation of either the porphyrin ring or the protein would give rise to free radicals which were considered to be too unstable and could not account for the stability exhibited by the primary complexes. To the contrary, it is our contention that the primary complexes of both catalase and horseradish peroxidase are porphyrin π cation radicals. We saw earlier that the π cation radical $[CO^{III}OEP]\cdot^{2+}$ could, as a function of its axial ligands, be made to occupy either of two ground states. Compare now the optical spectra of these two ground states with the optical spectra of the primary compounds of catalase and horseradish peroxidase (Figure 5). The similarity strongly suggests that Cat I is a π cation radical which exists in a $^2A_{1u}$ ground state and that HRP I is likewise a π cation radical with a $^2A_{2u}$ ground state.

If porphyrin ring oxidation accounts for one oxidizing equivalent, then from where in the primary complexes is the other electron removed? A further oxidation of the ring would result in a π dication which can be excluded since its optical spectrum[12] is so different from that of Cat I or HRP I. Oxidation of the protein would result in the formation of an organic free radical which should be detectable by epr, but so far has not been observed. A suggested ferric

(34) E. A. Dratz, A. J. Schultz, and K. Sauer, *Brookhaven Symp. Biol.*, **19**, 303 (1967).
(35) K. Sauer, in "Methods in Enzymology," Vol. 24B, A. San Pietro, Ed., Academic Press, New York, N. Y., 1972.
(36) J. Fajer, D. C. Borg, A. Forman, D. Dolphin, and R. H. Felton, *J. Amer. Chem. Soc.*, **95**, 2739 (1973).
(37) P. A. Loach and R. L. Hall, *Proc. Nat. Acad. Sci. U. S.*, **68**, 1010 (1971).
(38) R. K. Clayton, *Proc. Nat. Acad. Sci. U. S.*, **69**, 44 (1972).
(39) G. Feher, M. Y. Okamura, and J. D. McElroy, *Biochim. Biophys. Acta*, **267**, 222 (1972).
(40) P. Nicholls and G. R. Schonbaum, in "The Enzymes," 2nd ed., P. D. Boyer, H. Lardy, and K. Myrback, Ed., Academic Press, New York, N. Y. 1963, Chapter 6.
(41) K. G. Paul, in "The Enzymes," 2nd ed., P. D. Boyer, H. Lardy, and K. Myrback, Ed., Academic Press New York, N. Y., 1963, Chapter 7.
(42) B. Chance, *Nature (London)*, **161**, 914 (1948).
(43) P. George, *J. Biol. Chem.*, **201**, 413 (1953).
(44) M. E. Winfield, in "Haematin Enzymes," J. E. Falk, R. Lemberg, and R. K. Morton, Ed., Pergamon Press, Elmsford, N. Y., 1961, p 245.
(45) A. S. Brill and R. J. P. Williams, *Biochem. J.*, **78**, 253 (1961).
(46) J. Peisach, W. E. Blumberg, B. A. Wittenberg, and J. B. Wittenberg, *J. Biol. Chem.*, **243**, 1871 (1968).
(47) A. S. Brill, *Compre. Biochem.*, **14**, 447 (1966).

isoporphyrin structure[48] also may be excluded, since authentic species have recently been prepared;[49] their electronic absorption spectra show characteristic bands at 870–900 nm (ϵ ~14,000) which are absent in Cat I or HRP I.

These considerations require that the oxidation of the ferrihemoproteins be accounted for by the loss of one electron from the metal (to give Fe(IV)) and one from the ring (to give a π cation radical). The overall electronic structure of the primary complexes can then be represented as

$$[Fe^{IV}HRP]^{\cdot 3+} \quad \text{and} \quad [Fe^{IV}Cat]^{\cdot 3+} \qquad (8)$$

The secondary compounds, prepared by a one-electron reduction of the primary compounds, display an optical spectrum typical of a metalloporphyrin. This suggests that the electron adds to the ring and thus the secondary compounds are derivatives of tetravalent iron porphyrins, a view consonant with the original hypothesis of George.[43] Supporting evidence comes from Mossbauer spectroscopy[50,51] which demonstrates that the electronic structure of iron in the primary and secondary compounds is the same but differs from that found in the parent ferrihemoprotein.

The existence of tetravalent iron in metalloporphyrins is controversial, but the preparation and characterization[52,53] of singly oxidized ferric porphyrins support this view. Electrochemical oxidation with simultaneous coulometry of $Fe^{III}TPP^{+}Cl^{-}$, $Fe^{III}OEP^{+}Cl^{-}$, and the μ-oxo dimers, $(Fe^{III}TPP)_2O$ and $(Fe^{III}OEP)_2O$, afforded stable products. Proton magnetic resonance spectra of the oxidized porphyrins were interpreted as evidence against a π cation radical formulation, since both the magnitudes of the paramagnetic shifts and the line widths are inconsistent with appreciable spin density at the porphyrin periphery. This conclusion is supported by extended Hückel calculations[54] of the high-spin iron monomer which predict that the highest filled MO is composed of approximately equal amounts of $d_{x^2-y^2}$ and nitrogen σ orbitals. Similarly, μ-oxo dimer orbitals likely to be involved in the oxidation contain negligible contributions from porphyrin π atomic orbitals. The observation of an integer spin ($S = 2$) in $[Fe^{IV}TPP]^{2+}$ and $[Fe^{IV}OEP]^{2+}$ agrees well with a similar result in the two-electron oxidation of cytochrome c peroxidase (with ethyl hydrogen peroxide), where Mossbauer spectra suggest that the iron in the oxidized enzymes is characterized by an integer spin.[55]

It is interesting to speculate as to why these two similar enzymes function so differently, for while catalase decomposes hydrogen peroxide by a diffusion-controlled rate, horseradish peroxidase shows no catalytic activity of this type and is stable in the presence of excess hydrogen peroxide.[40] Moreover, while HRP I reacts with hydrogen donor substrates to give free radicals,[41] catalase decomposes hydrogen peroxide by a nonradical pathway.[56] We suggest that the two ground states and the different electronic distributions associated with them account for this difference in reactivity.

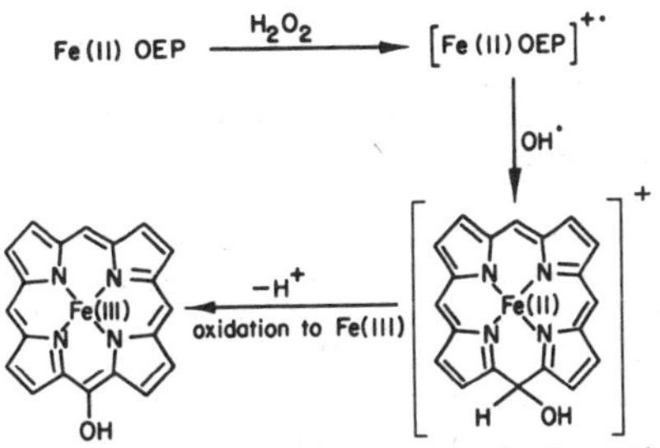

Figure 6. Mechanism proposed to account for the oxidation of ferrous octaethylporphyrin to ferric *meso*-hydroxyoctaethylporphyrin.

Cytochromes and Heme Catabolism. The catabolism of heme to bile pigments proceeds through the intermediacy of *meso*-hydroxyhemin.[57] Ferric *meso*-hydroxyoctaethylporphyrin was formed when ferrous, but not ferric, octaethylporphyrin was treated with hydrogen peroxide in pyridine.[58] This unusual observation was explained by assuming an analogy to Fenton's reagent, where ferrous iron catalyzes the conversion of hydrogen peroxide to a hydroxyl radical. By analogy it was assumed that the $Fe^{II}OEP$ and hydrogen peroxide would react to give $Fe^{III}OEP$ and an hydroxyl radical, and that these would react in turn to give ferric *meso*-hydroxyoctaethylporphyrin. But meso substitution of $Fe^{III}OEP$ with either hydroxyl or benzoyloxy radicals does not occur even though metalloporphyrins such as Zn- and MgOEP react very rapidly with benzoyl peroxide to give *meso*-benzoyloxyporphyrins.[20] The mechanism by which the last reaction proceeds may, however, throw some light on the meso hydroxylation of $Fe^{II}OEP$.

When $Mg^{II}TPP$ is treated with a slight excess of benzoyl peroxide it is oxidized to the corresponding π cation radical $[Mg^{II}TPP]^{\cdot +}$, and in the presence of excess benzoyl peroxide this π cation radical reacts to give *meso*-benzoyloxymagnesium tetraphenylisoporphyrin. When the same reaction is carried out with $Mg^{II}OEP$, the π cation radical is once again formed, and this reacts to give the *meso*-benzoyloxyisoporphyrin, which reacts further to give the corresponding meso-substituted MgOEP. If one applies this same mechanism to the meso hydroxylation of $Fe^{II}OEP$, the steps could be represented as follows (Figure 6). This mechanism requires that the one-electron oxidation of a ferrous porphyrin initially generates the corresponding ferrous porphyrin π cation radical and not the ferric porphyrin, and such an

(48) D. H. Busch, K. Farmery, V. Goedken, V. Katovic, A. C. Melnyk, C. R. Sperati, and N. Tokel, *Advan. Chem. Ser.*, **100**, 44 (1971).
(49) J. A. Guzinski and R. H. Felton, *J. Chem. Soc., Chem. Commun.*, 714 (1973).
(50) T. H. Moss, A. Ehrenberg, and A. J. Bearden, *Biochemistry*, **8**, 4159 1969).
(51) Y. Maeda and Y. Morita, in "Structure and Function of Cytochromes," K. Okunuki, M. D. Kamen, and I. Sekuzu, Ed., University of Tokyo Press, 1968, p 523.
(52) R. H. Felton, G. S. Owen, D. Dolphin, A. Forman, D. Borg, and J. Fajer, *Ann. N. Y. Acad. Sci.*, **206**, 504 (1973).
(53) R. H. Felton, G. S. Owen, D. Dolphin, and J. Fajer, *J. Amer. Chem. Soc.* **93**, 6332 (1971).
(54) M. Zerner, M. P. Gouterman, and H. Kobayashi, *Theor. Chim. Acta.*, **6**, 363 (1966).
(55) G. Lang, *Quart. Rev. Biophys.*, **3**, 1 (1970).
(56) R. K. Bonnichsen, B. Chance, and H. Theorell, *Acta Chem. Scand.*, **1**, 685 (1947).
(57) C. O'Heocha, in "Porphyrins and Related Compounds," T. W. Goodwin, Ed., Academic Press, New York, N. Y., 1968.
(58) R. Bonnett and M. J. Dimsdale, *Tetrahedron Lett.*, 731 (1968); R. Bonnett and M. J. Dimsdale, *J. Chem. Soc.*, 2540 (1972).

hypothesis explains the lack of reaction of the ferric porphyrin toward hydrogen peroxide.

If the oxidation of the ferrous porphyrin can give a ferrous porphyrin π cation radical, then this offers an attractive explanation to account for the redox properties of the cytochromes, such as cytochrome c. Crystallographic structure determinations have shown that the iron atoms of cytochrome c[59] and cytochrome b_5[60] are so well "protected" by the porphyrin and protein that *in vivo* electron carriers cannot approach the iron directly. How then do the cytochromes function when the iron atom is so inaccessible? Suggestions have been advanced[61,62] that the initial attack in the electron-transfer process occurs at the porphyrin, a segment of which is exposed to the exterior of the protein sheath. Complete removal of an electron would give the ferrocytochrome π cation. Rapid internal transfer could then occur between the ring and metal to give the ferrihemoprotein.

$$\mathrm{Fe^{II}Cyt\ c} \xrightarrow{-e^-} [\mathrm{Fe^{II}Cyt\ c}]^{\cdot +} \xrightarrow[\text{transfer}]{\text{internal}} \mathrm{Fe^{III}Cyt\ c} \quad (9)$$

Analogies are known for such internal transfers. Thus the one-electron oxidation of $\mathrm{Pb^{II}OEP}$ gives $[\mathrm{Pb^{II}OEP}]^{\cdot +}$, and the second one-electron oxidation of this species gives $\mathrm{Pb^{IV}OEP}$.[63,64] One can envisage this second oxidation as an oxidation of the metal or of the ring, but in either case an internal redistribution of electrons must take place to give the tetravalent lead porphyrin. Similar internal conversion of this type has been observed in cobalt complexes of porphyrin-like macrocycles[48,65] and during the oxidation of $\mathrm{Pd^{II}TPP}$ to $\mathrm{Pd^{IV}TPP}$.[66]

$$\mathrm{Pb^{II}OEP} \xrightarrow{-e^-} [\mathrm{Pb^{II}OEP}]^{\cdot +} \begin{cases} \xrightarrow{-e^-} [\mathrm{Pb^{III}OEP}]^{\cdot +} \\ \xrightarrow{-e^-} [\mathrm{Pb^{II}OEP}]^{2+} \end{cases} \xrightarrow{\text{internal transfer}} \mathrm{Pb^{IV}OEP}$$

One may ask if systems such as a ferrous π cation radical and a ferric porphyrin (eq 9) differ only in the way in which their electronic structures are written are they not simply both resonance forms of the same molecule, and if so why consider them as two distinct species? It is clear, however, in the cases of the cobalt and lead species described above, that while the two partners in the reaction are isoelectronic there is apparently an energy barrier between their interconversion. We envisage that such a barrier may be the result of either a conformation change or change in ligation of the metal atom occurring during the internal transfer.

The role of ferrocytochrome π cation in the electron-transfer mechanism is speculative; indeed, reaction of ferricytochrome with electrons or hydrogen atoms produced by pulse radiolysis has been interpreted[67,68] as evidence that transfer occurs *via* the protein residues.[69] However, the mechanism indicated in eq 9 might be tested by pulse radiolysis or flash photolysis of the protein-free heme. At present, no mechanistic assignment can be made with surety.

Conclusion

Metalloporphyrins, chlorines, and bacteriochlorines undergo reversible one-electron oxidations to give π cations. It has been shown by spectral comparisons that the *in vivo* bleaching of the photosynthetic systems P700 and P865 corresponds to the formation of a π cation radical of chlorophyll or bacteriochlorophyll. Moreover, the spectral properties of the enzymatically active primary complexes of both catalase and horseradish peroxidase suggest that they are π cation radicals containing tetravalent iron, a characterization which is supported by the preparation of authentic iron(IV) porphyrins. The formation of π cation radicals when zinc or magnesium porphyrins are treated with organic peroxides, as well as the difference in reactivity of the ferrous and ferric complexes under these conditions, leads to the hypothesis that electron transport involving the cytochromes takes place *via* ring oxidation followed by rapid internal conversion.

Attention has begun to focus on the influence of axial ligands on the course as well as the rate[70] of electron transfer. The possibility of these acting to switch the electron transfer from metal to ring is realized in ruthenium complexes[71] and is likely important in hematin enzyme function.

We wish to thank all of our collaborators who have been involved in the work described here, especially our colleagues Drs. D. C. Borg and J. Fajer at Brookhaven. The work performed in our laboratories was supported by the National Institutes of Health (AM 14343 and AM 14344) and the National Science Foundation (GP-16761 and GP-17061). Aspects of the work described have been performed at Brookhaven with the support of the U. S. Atomic Energy Commission.

(59) T. Takano, R. Swanson, O. B. Kallai, and R. E. Dickerson, *Symp. Quant. Biol.*, 397 (1971).
(60) F. S. Mathews, M. Levine, and P. Argos, *J. Mol. Biol.*, **64**, 449 (1972).
(61) C. E. Castro, *J. Theor. Biol.*, **33**, 475 (1971).
(62) N. Sutin and A. Forman, *J. Amer. Chem. Soc.*, **93**, 5274 (1971); N. Sutin, *Chem Brit.*, **8**, 148 (1972).
(63) D. G. Whitten, T. J. Meyer, F. R. Hopf, J. A. Ferguson, and G. Brown, *Ann. N. Y. Acad. Sci.*, **206**, 516 (1973).
(64) M. P. Gouterman, P. Smith, and D. Dolphin, unpublished results.
(65) N. S. Hush and I. S. Woolsey, *J. Amer. Chem. Soc.*, **94**, 4107 (1972).
(66) L. Vegh and D. Dolphin, unpublished results.
(67) N. N. Lichtin, A. Shafferman, and G. Stein, *Science*, **179**, 680 (1973).
(68) N. Lichtin, J. Ogdan, and G. Stein, *Isr. J. Chem.*, **9**, 579 (1971).
(69) R. E. Dickerson, T. Takano, D. Eisenberg, O. B. Kallai, and L. Samson, "Proceedings of the Wenner-Gren Symposium on Oxidative Enzymes, Aug 1970," Wenner-Gren Foundation, Stockholm, Sweden, 1971.
(70) K. M. Kadish and D. G. Davis, *Ann. N. Y. Acad. Sci.*, **206**, 495 (1973).
(71) G. M. Brown, F. R. Hopf, J. A. Ferguson, T. J. Meyer, and D. G. Whitten, *J. Amer. Chem. Soc.*, **95**, 5939 (1973).

Addendum (May 1976)

Since the publication of this Account we have demonstrated[72] an internal electron transfer of the type proposed in eq. 9. Thus the one-electron oxidation of $Ni^{II}TPP$ generated the corresponding π-cation radical $[Ni^{II}TPP]\cdot^{+}$. At room temperature this cation radical exhibited an absorption spectrum characteristic of the $^2A_{1u}$ ground state, and an EPR spectrum with g = 2.0041 with a peak-to-peak width of 48.2 G. However upon cooling a CH_2Cl_2 solution of $[Ni^{II}TPP]\cdot^{+}ClO_4^{-}$ to 77 K, the green, room temperature solution turned orange-red. These optical absorption changes were paralleled by the behavior of the EPR spectra. At 77 K the free radical signal was replaced by an EPR spectrum with $g_\perp = 2.286$ and $g_\parallel = 2.086$. These values are characteristic of the low-spin d^7 Ni(III) complexes reported by Busch[73] and by Wolberg and Manassen.[74] Additionally, the electronic absorption spectrum at 110 K displayed the Soret (418 nm) and visible (526 nm) bands of a typical metalloporphyrin in which bonding ligand π orbitals are filled.

The reversible, temperature-dependent transformation can be described as:

$$[Ni^{II}TPP]\cdot^{+}, ClO_4^{-} \underset{\Delta}{\overset{-\Delta}{\rightleftharpoons}} [Ni^{III}TPP]^{+}, ClO_4^{-}$$

This intramolecular electron transfer thus establishes a precedent for the reversible electron route postulated in the cytochromes and encourages us to seek the same phenomenon in iron porphyrins.

(72) D. Dolphin, T. Niem, R. H. Felton, and I. Fujita, *J. Am. Chem. Soc.*, **97**, 5288 (1975).

(73) F. V. Lovecchio, E. S. Gore, and D. H. Busch, *J. Am. Chem. Soc.*, **96**, 3109 (1974).

(74) A. Wolberg and J. Manassen, *Inorg. Chem.*, **9**, 2365 (1970).

Novel Metalloporphyrins—Syntheses and Implications

David Ostfeld*

Department of Chemistry, Seton Hall University, South Orange, New Jersey 07079

Minoru Tsutsui*

Department of Chemistry, Texas A&M University, College Station, Texas 77843

Received May 7, 1973

Reprinted from ACCOUNTS OF CHEMICAL RESEARCH, *7, 52 (1974)*

The last few years have seen substantial growth in metalloporphyrin research. Certainly most of this stems from interest in the biological systems to which these compounds are related. However metalloporphyrins are also studied for other reasons. These include the search for new semiconductors,[1] superconductors,[2] anticancer drugs,[3] and catalysts.[4] Several porphyrin-related compounds, particularly the phthalocyanines,[5] have proved useful as dyes. Even without their biological and industrial implications the properties of metalloporphyrins would be studied for their purely theoretical importance. Thus, even to the nonspecialist, a knowledge of the more recent or unusual synthetic metalloporphyrins could be useful.

The porphyrins are compounds formed by adding substituents to the nucleus of porphine (1). The nat-

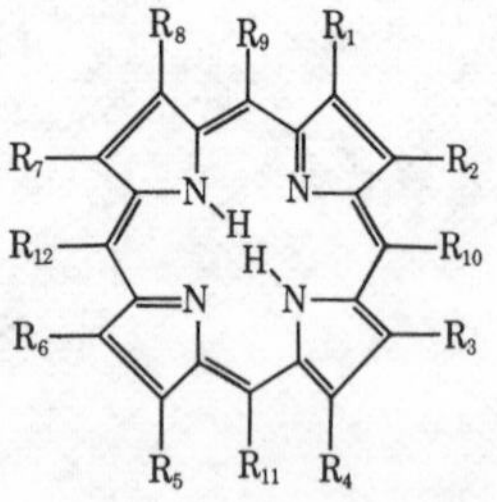

1

urally occurring porphyrins are generally formed by adding substituents to positions 1–8 and are named according to the number and type of substituents. The roman numerals after the name indicate the pattern of substitution. Of the large number of possible arrangements, few have been found in nature. Table I lists several porphyrins, both natural and synthetic, along with their trivial names. Certain common porphyrins will hereafter be referred to by abbreviations listed in Table I.

Upon removal of the pyrrole protons, porphyrins readily complex with a variety of metals. While free porphyrins are biologically unimportant, metalloporphyrins are widely found in nature. It is with certain of these metal complexes that this Account deals.

Several biologically important porphyrin-containing species are listed in Table II. Strictly speaking, chlorophyll, cobalamin, and the *d* cytochromes are not porphyrins (2–4), but the similarity between

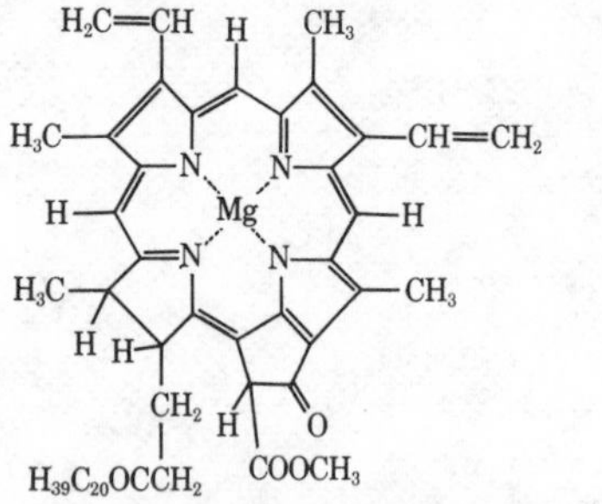

2, tetrapyrrole complex from chlorophyll a

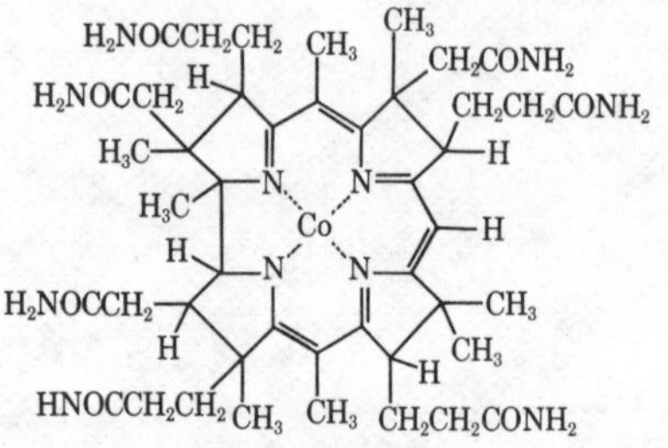

3, tetrapyrrole complex from cobalamin

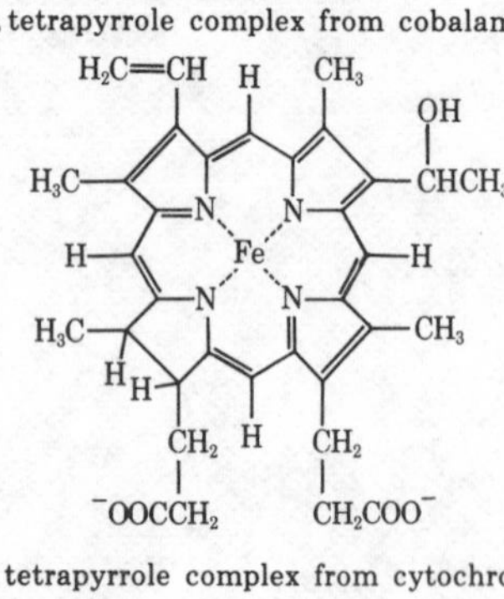

4, tetrapyrrole complex from cytochrome *d*

David Ostfeld was born in Chicago in 1942. He earned his B.S. in Chemistry from the University of Illinois and his Ph.D. from Cornell University. Three years of postdoctoral work, including a year with Professor Tsutsui, were devoted to studying metalloporphyrin chemistry. Currently he is Assistant Professor of Chemistry at Seton Hall University.

Minoru Tsutsui, a native of Japan, received a B.A. degree from Gifu University in Japan, M.S. degrees from both Tokyo University of Literature and Science and Yale University, and a Ph.D. degree from Yale in 1954. He has been on the staff at Tokyo University, Monsanto Chemical Company, and New York University. Since 1968 he has been Professor of Chemistry at Texas A&M University. His research interests include organotransition metal and organolanthanide chemistry and the chemistry of novel metalloporphyrins.

(1) M. Tsutsui, work in progress.
(2) A. D. Adler, *J. Polym. Sci., Part C*, **29**, 73 (1970).
(3) J. P. Macquet and T. Theophanides, *Can. J. Chem.*, **51**, 219 (1972).
(4) E. B. Fleischer and M. Krishnamurthy, *J. Amer. Chem. Soc.*, **94**, 1382 (1972).
(5) D. Patterson, "Pigments, an Introduction to Their Physical Chemistry," Elsevier, New York, N. Y., 1967, p 44.

Table I
Some Common Porphyrins and Their Trivial Names

Porphyrin	Position on ring[a]								
	1	2	3	4	5	6	7	8	9–12
Proto-porphyrin IX	M	V	M	V	M	P	P	M	H
Deutero-porphyrin IX	M	H	M	H	M	P	P	M	H
Hemato-porphyrin IX	M	B	M	B	M	P	P	M	H
Meso-porphyrin IX[b]	M	E	M	E	M	P	P	M	H
Phylloporphyrin	M	E	M	E	M	H	P	M	CH_3 + 3 H
Tetraphenyl-porphine[c]	H	H	H	H	H	H	H	H	C_6H_5
Octaethyl-porphine[d]	E	E	E	E	E	E	E	E	H

[a] A, CH_2COOH; B, $CHOHCH_3$; E, CH_2CH_3; M, CH_3; P, CH_2CH_2COOH; V, $CHCH_2$. [b] H_2MP. [c] H_2TPP. [d] H_2OEP.

Table II
Some Biological Compounds Containing Porphyrin-like Ligands

Compound	Metal	Compound	Metal
Hemoglobin	Fe	Peroxidase	Fe
Myoglobin	Fe	Oxidase	Fe
Cytochrome	Fe	Cobalamin (vitamin B_{12})	Co
Catalase	Fe	Chlorophyll	Mg

these heterocycles and the porphyrins is obvious. Biosynthesis of certain chlorophylls is even known to involve magnesium protoporphyrin as an intermediate.[6] It is therefore generally acceptable to consider model studies on porphyrins as applicable to cobalamin, the chlorophylls, and other nonporphyrin macrocycles.

The study of porphyrin chemistry can be said[7] to have begun in 1880 with Hoppe-Seyler's isolation of hematoporphyrin from hemin (5) and phylloporphy-

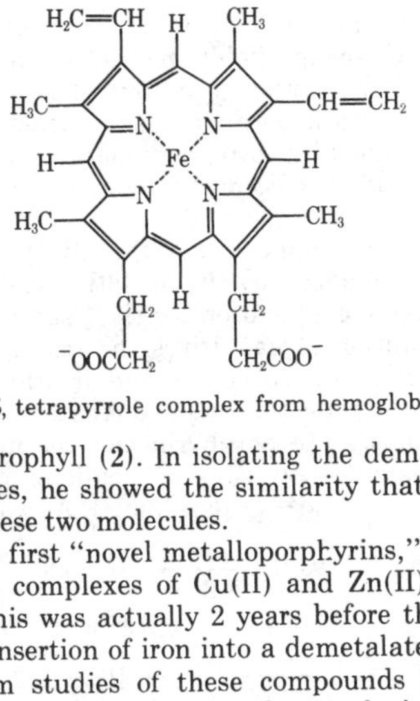

5, tetrapyrrole complex from hemoglobin

rin from chlorophyll (2). In isolating the dematalated macrocycles, he showed the similarity that existed between these two molecules.

In 1902 the first "novel metalloporphyrins," mesoporphyrin IX complexes of Cu(II) and Zn(II), were prepared.[8] This was actually 2 years before the first successful reinsertion of iron into a demetalated porphyrin.[9] From studies of these compounds it was found that mesoporphyrin complexes of zinc and copper, metals lacking a common 3+ state, contain no chlorine. This showed the chlorine in [porphyrin–iron(III) chloride] to be bound to the metal and not to the porphyrin.

Synthetic work continued so that, by 1964, Falk[10] was able to list porphyrin complexes for 28 metallic elements. Figure 1 contrasts the metals for which porphyrin complexes were listed by Falk with those for which such complexes have subsequently been prepared. References are given to the syntheses of the new compounds.[11-30]

The characteristics which porphyrins share with other multidentate nitrogen-donor ligands need not be discussed here. What must be emphasized are those characteristics unique to the porphyrin ligand.

Porphyrins, in common with other macrocyclic ligands, have a central hole of essentially fixed size. In certain complexes the metal is unable to fit into this hole and, as has been shown by Hoard,[31] lies out of the porphyrin plane. In hemoglobin, for example, the iron atom in the oxygenated form lies roughly in plane, while the metal lies out of plane in the deoxy form. The transition between these states is thought to be responsible for the cooperative nature of oxygen binding in hemoglobin.[32]

Also characteristic of complexes of porphyrins and other macrocycles is the necessity of simultaneously breaking all metal–ligand bonds to remove the ligand. Using several macrocyclic ligands, Busch and coworkers demonstrated[33] that, consequently, the

(6) G. S. Marks, "Heme and Chlorophyll," Van Nostrand, Princeton, N. J., 1969, p 140.
(7) H. Fischer and H. Orth, "Die Chemie des Pyrrols," Vol. 2, Part 1, Akademie Verlagsgesellschaft, Leipzig, 1937, p 158.
(8) J. Zaleski, *Z. Physiol. Chem.*, **37**, 54 (1902).
(9) P. P. Laidlaw, *J. Physiol.*, **31**, 464 (1904).
(10) J. E. Falk, "Porphyrins and Metalloporphyrins," Elsevier, New York, N. Y., 1964, p 134.

(11) D. B. Boyland and M. Calvin, *J. Amer. Chem. Soc.*, **89**, 5472 (1967).
(12) A. Treibs, *Justus Liebigs Ann. Chem.*, **728**, 115 (1969).
(13) R. S. Becker and J. B. Allison, *J. Phys. Chem.*, **67**, 2669 (1963).
(14) A. D. Adler, F. R. Longo, F. Kampas, and J. Kim, *J. Inorg. Nucl. Chem.*, **32**, 2443 (1970).
(15) J. W. Buchler, G. Eikelmann, J. Puppe, K. Rohbock, H. H. Schneehage, and D. Weck, *Justus Liebigs Ann. Chem.*, **745**, 135 (1971).
(16) M. Tsutsui, R. A. Velapoldi, K. Suzuki, and T. Koyano, *Angew. Chem., Int. Ed. Engl.*, **7**, 891 (1966).
(17) M. Tsutsui, M. Ichikawa, F. Vohwinkel, and K. Suzuki, *J. Amer. Chem. Soc.*, **88**, 854 (1966).
(18) J. W. Buchler, L. Puppe, K. Rohbock, and H. H. Schneehage, *Ann. N. Y. Acad. Sci.*, **206**, 116 (1973).
(19) J. W. Buchler and K. Rohbock, *Inorg. Nucl. Chem. Lett.*, **8**, 1073 (1972).
(20) E. B. Fleischer and T. S. Srivastava, *Inorg. Chim. Acta*, **5**, 151 (1972).
(21) M. Tsutsui, D. Ostfeld, and L. M. Hoffman, *J. Amer. Chem. Soc.*, **93**, 1820 (1971).
(22) B. C. Chow and I. A. Cohen, *Bioinorg. Chem.*, **1**, 57 (1971).
(23) N. Sadasivan and E. B. Fleischer, *J. Inorg. Nucl. Chem.*, **30**, 591 (1968).
(24) E. B. Fleischer and D. Lavallee, *J. Amer. Chem. Soc.*, **89**, 7132 (1967).
(25) Z. Yoshida, H. Ogoshi, T. Omura, E. Watanabe, and T. Kurosaki, *Tetrahedron Lett.*, 1077 (1972).
(26) H. Ogoshi, T. Omura, and Z. Yoshida, *J. Amer. Chem. Soc.*, **95**, 1668 (1973).
(27) B. R. James and D. V. Stynes, *J. Amer. Chem. Soc.*, **94**, 6225 (1972).
(28) D. Ostfeld, M. Tsutsui, C. P. Hrung, and D. C. Conway, *J. Amer. Chem. Soc.*, **93**, 2548 (1971).
(29) E. B. Fleischer and A. Laszlo, *Inorg. Nucl. Chem. Lett.*, **5**, 373 (1969).
(30) M. Tsutsui and C. P. Hrung, *Chem. Lett.*, 941 (1973).
(31) J. L. Hoard, *Science*, **174**, 1295 (1971), and references contained therein.
(32) M. F. Perutz and L. F. TenEyck, *Cold Stream Harbor Symp. Quant. Biol.*, **36**, 295 (1971).
(33) D. H. Busch, K. Farmery, V. Goedken, V. Katovic, A. C. Melnyk, C. R. Sperati, and N. Tokel, *Advan. Chem. Ser.*, **No. 100**, 44 (1971).

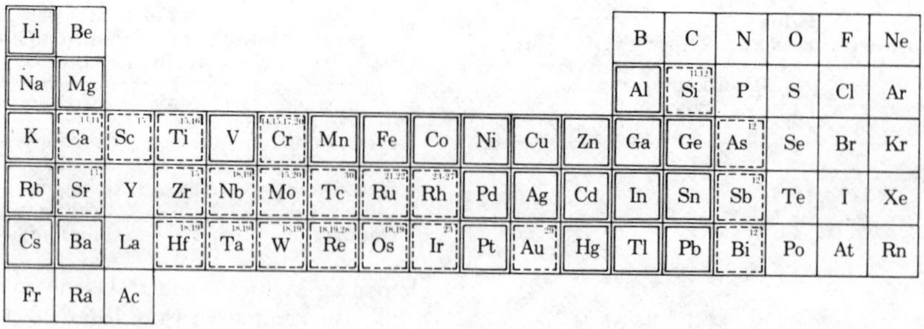

Figure 1. Metals for which porphyrin complexes have been made: □, porphyrins mentioned by Falk;[10] ⬚, more recent metalloporphyrins.

dissociation of a metal–macrocycle complex is about 10^6 times slower than for a similar noncyclic ligand.

Porphyrins and certain other unsaturated ligands (*e.g.*, bipyridyl) are distinguished by their delocalized π systems. Overlap between ligand π orbitals and metal orbitals of proper symmetry produces a moderately high ligand field strength.[34] Also, by back-accepting π electron density from complexed metals, the ligands facilitate the reduction of complexed metal to a low oxidation state.[27,35,36] It should be mentioned, however, that not all unusual porphyrin oxidation states are so easily explained. For example, the most stable silver porphyrin has the metal in a 2+ and not a 1+ state.[37] No explanation for this has yet been proposed.

Synthesis

The first of the new synthetic methods to be discovered was the insertion of a metal from a carbonyl complex.[17,38] Instead of simple ligand exchange with the porphyrin dianion, the porphyrin is oxidized by the reduction of the porphyrin pyrrole protons. For example, letting Por stand for porphyrin

$$Cr(CO)_6 + PorH_2 \xrightarrow[\text{decalin}]{185°} [Por{-}Cr^{II}] + 6CO\uparrow + H_2\uparrow$$

The mechanism for this class of reaction has never been studied, but it is assumed that the evolution of hydrogen and carbon monoxide drives the equilibrium toward metal insertion. By inference from the rhenium complexes (discussed later) it can be surmised that the metal is oxidized in a series of one-electron steps, each of which is accompanied by the reduction of a pyrrole proton.

$$M(0) + PorH_2 \longrightarrow [HPor{-}M^{I}] + H_2\uparrow$$

$$[HPor{-}M^{I}] \longrightarrow [Por{-}M^{II}] + H_2\uparrow$$

The carbonyl method has been used to synthesize previously unknown complexes of chromium,[17,38] molybdenum,[20] ruthenium,[21,22] rhodium,[23] iridium,[23] rhenium,[28] and technetium.[30] Metalloporphyrins containing some of these metals [chromium,[14,15] molybdenum,[15] rhodium,[39] and rhenium[19]] have subsequently been prepared by other methods, but the complexes formed from carbonyls are still unique. In some cases [ruthenium(II), iridium(I), rhenium(I), and technetium(I)] a carbonyl ligand is retained by the metal. Also low oxidation states are generally obtained from metal carbonyls. Several novel metal(I) porphyrin complexes which have been prepared using metal carbonyls (discussed later) are certainly unique to this method.

A similar method involves the oxidation of hydride ions from a metal hydride complex. For example[40]

$$AlH_3 + PorH_2 \xrightarrow[\text{THF}]{65°} \xrightarrow{+H_2O} [Por{-}AlOH] + 3H_2\uparrow$$

The small number of metals for which hydrides are available has probably restricted the use of this method. One other example of a metal hydride reacting with a porphyrin is the reaction of sodium borohydride with tetraphenylporphine. The product, which has never been characterized, may be a boron porphyrin.[41]

It does not appear that the utility of these reactions is enhanced by the oxidation and reduction. A more likely explanation for their success is that: (1) the evolution of gas drives the reaction to completion; (2) metal carbonyls and hydrides dissolve in the same organic solvents as do the porphyrins.

Previously the porphyrin and the metal salt had been dissolved by using an acidic or basic medium (*e.g.*, acetic acid[42] or pyridine[12]). As was pointed out

$$Fe(C_2H_3O_2)_2 + PorH_2 \xrightarrow[HC_2H_3O_2]{118°} \xrightarrow{+O_2,\ Cl^-} [Por{-}Fe^{III}Cl] + 2HC_2H_3O_2$$

$$AlCl_3 + PorH_2 \xrightarrow[\text{pyridine}]{80°} \xrightarrow[HC_2H_3O_2]{} [Por{-}Al(C_2H_3O_2)] + 3HCl\uparrow$$

by Adler,[14] these reactions have some serious flaws. In an acidic medium the porphyrin exists primarily in the unreactive protonated form; in basic media the reactivity of the metal is decreased by complexation with hydroxide, pyridine, etc.

The solubility problem was circumvented by Tsutsui through the use of an organometallic compound

(34) Good discussions of the electronic structure of metalloporphyrins are given by: M. Zerner, M. Gouterman, and H. Kobayashi, *Theoret. Chem. Acta*, **6**, 363 (1966); H. Kobayashi and Y. Yanagawa, *Bull. Chem. Soc. Jap.*, **45**, 450 (1972).

(35) D. A. Clarke, D. Dolphin, R. Grigg, A. W. Johnson, and H. A. Pinnock, *J. Chem. Soc. C*, 881 (1968).

(36) I. A. Cohen, D. Ostfeld, and B. Lichtenstein, *J. Amer. Chem. Soc.*, **94**, 4522 (1972).

(37) G. D. Dorough, J. R. Miller, and F. M. Huennekens, *J. Amer. Chem. Soc.*, **73**, 4315 (1951).

(38) M. Tsutsui, R. A. Velapoldi, K. Suzuki, F. Vohwinkel, M. Ichikawa, and T. Koyano, *J. Amer. Chem. Soc.*, **91**, 6262 (1969).

(39) L. K. Hanson, M. Gouterman, and J. C. Hanson, *J. Amer. Chem. Soc.*, **95**, 4822 (1973).

(40) H. H. Inhoffen and J. W. Buchler, *Tetrahedron Lett.*, 2057 (1968).

(41) P. S. Clezy and J. Barrett, *Biochem. J.*, **78**, 798 (1961).

(42) D. W. Thomas and A. E. Martell, *J. Amer. Chem. Soc.*, **81**, 5111 (1959).

as a metal source. Diphenyltitanium was found to react with mesoporphyrin IX dimethyl ester to give the titanyl porphyrin.[16] The organometallic com-

$$(C_6H_5)_2Ti + PorH_2 \xrightarrow[\text{mentylene}]{240°} \xrightarrow{+O_2} [Por{-}Ti{=}O] + 2C_6H_6$$

pound used here is soluble in nonpolar organic solvents, and the stable organic moiety formed in the reaction helps drive the equilibrium in the desired direction. Presumably air oxidizes the titanium to the 4+ state. Other organometallics could be used in a similar manner. Indeed, the use of organometallics is probably one of the most powerful methods available for metal insertion. However, the inconvenience involved in choosing and obtaining organometallic derivatives and the availability of other methods have prevented greater use of organometallics as starting materials.

Buchler and coworkers[15] found metal acetylacetonates both readily available and reasonably soluble in organic solvents. They studied the reaction of octaethylporphine with a number of metal acetylacetonates, using melts of phenol, quinoline, and imidazole as solvents.

$$M(acac)_n + PorH_2 \longrightarrow [Por{-}M(acac)_{n-2}] + 2Hacac$$

Besides a number of previously known metalloporphyrins, new complexes of scandium and zirconium were prepared. Of their reactions the only ones that failed were attempted syntheses of cerium and thorium porphyrins.

Instead of supplying the metal in the form of a complex, Adler and coworkers solved the solubility problem by using *N,N*-dimethylformamide (DMF) as a solvent.[14] DMF, due to its high dielectric constant, can dissolve both a porphyrin and a metal salt. While one cannot exclude the possibility of another form of solvent assistance, such as a stabilized transition state, no evidence of this has been found. Adler obtained improved yields of many known compounds by using DMF. This method may also be useful for the synthesis of new compounds. Among the compounds thus synthesized was the once-elusive chromium porphyrin. Due to the simplicity of

$$4PorH_2 + 4CrCl_2 + O_2 \xrightarrow[\text{DMF}]{153°} 4[Por{-}Cr^{III}Cl] + 4HCl\uparrow + 2H_2O$$

this method and the ready availability of starting materials, it is not surprising to see compounds which were once made by other methods now being prepared in DMF.[39]

The use of polar reaction media had not been previously unknown. It is presumably due to their lower dielectric constants that solvents such as acetone, dioxane, and ethanol[43] were not so successful as DMF.

Buchler[18] further extended the synthesis of metalloporphyrins in media of high dielectric constant. Using a phenol melt and appropriate metal halides or oxides, he prepared complexes of scandium, tantalum, tungsten, osmium, and rhenium. In benzonitrile, porphyrins and metal halides reacted to give an equally novel array of metalloporphyrins, including

(43) Reference 5, p 135.

those of chromium, molybdenum, tungsten, and niobium. For example

$$PorH_2 + H_2WO_4 \xrightarrow[C_6H_5OH]{220°} [Por{-}W{=}O(OC_6H_5)]$$

$$PorH_2 + MoCl_2 \xrightarrow[C_6H_5CN]{195°} [Por{-}Mo{=}O]_2O$$

A tungsten complex is particularly noteworthy, since attempted syntheses using tungsten hexacarbonyl had been unsuccessful.[44]

Except for the lanthanides and actinides, at least one porphyrin complex has been made for virtually every metallic element. However, due to the importance of oxidation state and axial ligation it would be incorrect to assume that all porphyrin complexes of potential interest are known. What might be stated is that any reasonable porphyrin complex desired by the modern chemist should be attainable through the powerful synthetic methods now available.

Bridged and Metal-Metal-Bonded Species

Among the reasons for interest in bridged and metal-metal-bonded complexes has been the desire to prepare model compounds for studying the role of heme iron in mitochondrial electron transfer.[45] In mitochondria, the cellular bodies in which oxidation occurs, heme proteins called cytochromes are crucial links in the electron-transfer chain. A typical chain, as found in beef heart, is[46]

$$e^- \longrightarrow \text{cyt } b \longrightarrow \text{cyt } c_1 \longrightarrow \text{cyt } c \longrightarrow \boxed{\begin{matrix}\text{Cu}\\ \text{cyt } a_3\\ \text{cyt } a\end{matrix}} \longrightarrow O_2$$

The classes of cytochromes (*a, b, c,* etc.) differ in their spectra, oxidation potentials, porphyrin substitution patterns, and attached proteins. However all (except the *d* cytochromes) are iron porphyrins.

While electron transfer *via* the porphyrin ring is a possibility,[47] work has concentrated on transfer through the axial sites.[48] Cytochrome *b* is known to coordinate two imidazoles in its axial position,[49] while the *c* cytochromes bind to an imidazole and a methionine sulfur.[50] It has been theorized that an electron leaving cytochrome *c* goes by way of the methionine sulfur.[51] Imidazole, with its double bonds, could also be a suitable bridge for electron transfer.

Ostfeld and Cohen[52] have studied the polymer $[\text{-(FeTPP)-(imidazolate)-}]_n$ (**6**). Magnetic studies indicate spin coupling through the imidazolate ligands, thus demonstrating the ability of this ligand to serve as an electron bridge when complexed to an iron porphyrin.

(44) T. S. Srivastava and M. Tsutsui, unpublished observations.

(45) W. S. Caughey, J. L. Davies, W. H. Fuchsman, and S. McCoy in "Structure and Function of Cytochromes," K. Okunuki, M. D. Kamen, and I. Sekuzu, Ed., University Park Press, Baltimore, Md., 1968, p 20.

(46) H. R. Mahler and E. H. Cordes, "Biological Chemistry," Harper and Row, New York, N. Y., 1971, p 683.

(47) C. E. Castro and H. F. Davis, *J. Amer. Chem. Soc.*, **91**, 5405 (1969).

(48) I. A. Cohen, C. Jung, and T. Governo, *J. Amer. Chem. Soc.*, **94**, 3003 (1972).

(49) F. S. Matthews, P. Argos, and M. Levine, *Cold Stream Harbor Symp. Quant. Biol.*, **36**, 387 (1971).

(50) R. E. Dickerson, T. Takano, D. Eisenberg, O. B. Kallai, L. Samson, A. Cooper, and E. Margoliash, *J. Biol. Chem.*, **246**, 1511 (1971).

(51) T. Takano, R. Swanson, O. B. Kallai, and R. E. Dickerson, *Cold Stream Harbor Symp. Quant. Biol.*, **36**, 397 (1971).

(52) D. Ostfeld and I. A. Cohen, submitted for publication.

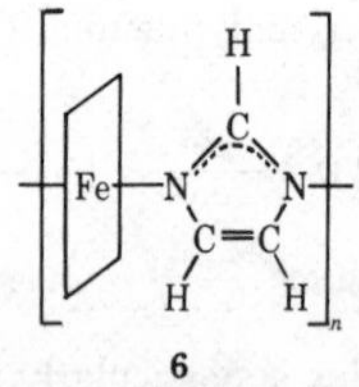

6

Compared to the *b* and *c* cytochromes, considerably less is known about the structure of cytochrome oxidase (the boxed species in the electron-transfer scheme shown above). Even the order of and connection between the copper and the *a* cytochromes are unclear. For this reason a variety of model compounds can be considered relevant to the cytochrome oxidase system.[53]

The best known example of metal–metal interaction in metalloporphyrins is the oxo-bridged "hematin" dimer **7**, only recently recognized to be a dimer and not a monomeric hydroxide.[54] The strong antiferromagnetic coupling[55] between the two iron(III) atoms shows the ability of this bridge to conduct electrons. Caughey[53] has proposed the possibility of iron–iron or iron–copper oxo bridges in cytochrome oxidase. Similar oxygen-bridged dimers have also been prepared for porphyrins of aluminum(III)[40] and scandium(III).[15] In porphyrin complexes of niobium, molybdenum, tungsten, and rhenium, two O=M(V) groups are joined by an oxygen bridge (**8**).[19]

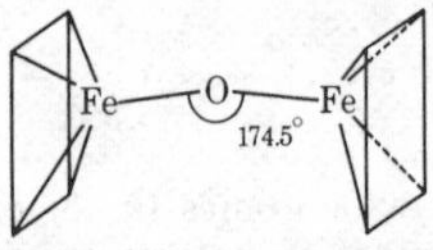

7

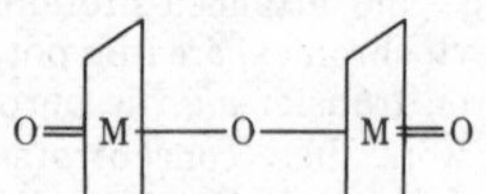

8

Like the iron dimer, the scandium dimer forms instead of the nonexistent monomeric hydroxide.[15] The porphyrin hydroxide of aluminum(III), however, does exist and is dimerized by heating under vacuum.[40] This leads to the possibility that other 3+

$$[\mathrm{Por{-}Al^{III}(OH)}] \xrightarrow[\text{vacuum}]{\Delta} [\mathrm{Por{-}Al^{III}{-}O{-}Al^{III}{-}Por}]$$

metals [*e.g.*, Co(III), Mn(III), Ga(III)] can form oxo-bridged porphyrin dimers upon removal of water from the hydroxide. An oxo-bridged dimer of manganese(III) phthalocyanine is, in fact, already known.[56]

Another recently prepared bridging system is a trimer involving two iron(III) porphyrin azides and an iron(II) porphyrin.[57] The magnetic moment of the ferric iron in this compound is 5.0 μ_B independent of temperature. This indicates the presence of four unpaired electrons instead of the five expected for high-spin iron(III). That pairing occurs between only two of the electrons can be explained by the occurrence of a single orbital available for electron occupancy across the length of the molecule.

A good case for metal–metal bonding has been made by Whitten and coworkers[58] with the photochemical dimerization of ruthenium porphyrin. Upon irradiating a pyridine solution of ruthenium carbonyl octaethylporphine, two molecules of carbon monoxide are expelled.

$$2[\mathrm{(OEP{-}Ru)(py)(CO)}] \xrightarrow{h\nu} \mathrm{py(OEP{-}Ru){-}(Ru{-}OEP)py} + 2\mathrm{CO}\uparrow$$

This reaction does not occur when the porphyrin is tetraphenylporphine.[58] Presumably the TPP phenyl rings, which are known to lie almost perpendicular to the porphyrin plane,[59] interact such that two porphyrin molecules cannot approach sufficiently close for a metal–metal bond to form. Study of this compound by single-crystal X-ray diffraction would be desirable.

Metal–metal interaction also occurs in solid chromium(II) mesoporphyrin IX dimethyl ester. The magnetic moment of 2.84 μ_B[17] for each chromium atom is substantially less than the four-spin value of 4.90 μ_B expected[60] for square-planar chromium(II). However solutions of the chromium complex do not show a reduced moment, and a value of 5.19 μ_B is observed.[17] Apparently the close approach of chromium atoms in the solid phase causes a sufficient rise in the energy of the d_z orbitals to bring about electron pairing. As in the case of ruthenium porphyrin dimerization, tetraphenylporphine seems to inhibit metal–metal bonding. For solid Cr^{II}TPP a magnetic moment of 4.9 μ_B has been observed.[61] Metal–metal interaction may be a general phenomenon for paramagnetic metalloporphyrins containing no axial ligands, since a somewhat low magnetic moment has also been found for Rh^{II}TPP.[27]

The only known metal–metal bond in which a metalloporphyrin bonds to a nonporphyrin metal is an adduct of trimethyltin and iron tetraphenylporphine.[62] Mössbauer and far-ir data indicate that this air-sensitive, diamagnetic compound contains a metal–metal bond between iron(III) and tin(IV).

Unusual Geometries

Recent synthetic work can be used to show the geometric limits within which metalloporphyrin formation is possible. For example, the absence of porphyrin complexes for the lanthanide and actinide elements seems to indicate that metals above a certain size cannot be incorporated into a porphyrin ring. This is consistent with Hoard's finding[31] that iron in certain electronic states will not fit into a metalloporphyrin, but instead sits slightly out of the porphyrin plane.

Additional geometries which metalloporphyrins can assume are illustrated by two rhenium porphy-

(53) W. S. Caughey, *Advan. Chem. Ser.*, **No. 100,** 248 (1971).
(54) I. A. Cohen, *J. Amer. Chem. Soc.*, **91,** 1980 (1969).
(55) T. H. Moss, H. R. Lilienthal, G. Moleski, G. A. Smythe, M. C. McDaniel, and W. S. Caughey, *J. Chem. Soc., Chem. Commun.*, 263 (1972).
(56) E. B. Fleischer, *Accounts Chem. Res.*, **3,** 105 (1970).
(57) I. A. Cohen, *Ann. N. Y. Acad. Sci.*, **206,** 453 (1973).
(58) W. Sovocol, F. R. Hopf, and D. G. Whitten, *J. Amer. Chem. Soc.*, **94,** 4350 (1972).
(59) E. B. Fleischer, C. K. Miller, and L. E. Webb, *J. Amer. Chem. Soc.*, **86,** 2342 (1964).
(60) B. N. Figgis and J. Lewis, *Progr. Inorg. Chem.*, **6,** 134 (1964).
(61) N. J. Gogan and Z. U. Siddiqui, *Can. J. Chem.*, **50,** 720 (1972).
(62) I. A. Cohen and B. Lichtenstein, to be published.

rins, μ-(Por)[$Re(CO)_3$]$_2$ and (PorH)$Re(CO)_3$.[28,63] The manner in which the metal atoms were bonded to the porphyrin ring was not initially clear, and several speculative structures were based[63] on the almost identical ir spectra given by the mono- and dirhenium complexes. A single-crystal X-ray diffraction analysis has given structure **9** for the dirhenium complex;[64] presumably the monorhenium complex is similar (**10**).

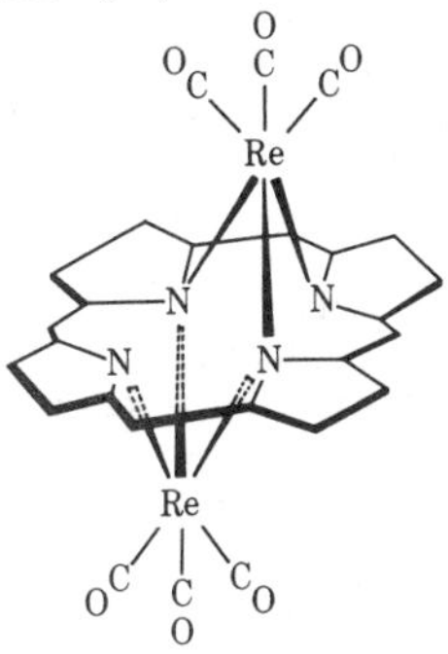

9

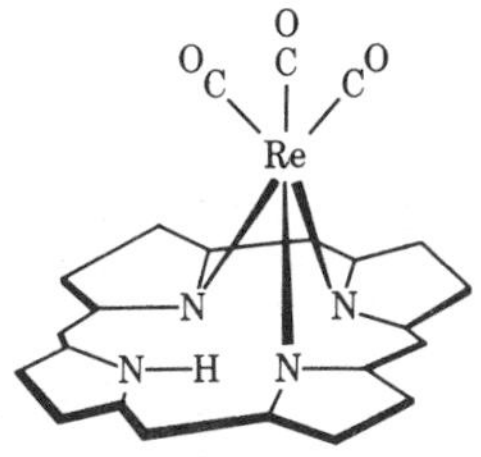

10

In the reported structure each rhenium atom is bonded to three of the nitrogen atoms in a distorted porphyrin ring. The metal–metal distance, though somewhat long for bonding,[64] is short enough that some interaction is a reasonable possibility.

These complexes can be used as models for two proposed metal-insertion intermediates. The monorhenium complex **10** resembles Fleischer's proposed "sitting atop complex" (**11**),[65] while the dirhenium complex **9** has been proposed by Hambright[66] as a model for another possible intermediate (**12**). The latter comparison may no longer apply, since the X-ray structure of **9** shows that both metal atoms lie off of the S_2 axis normal to the prophyrin plane. Alternately it might be concluded that intermediate **12** also has its two metal atoms located nonaxially.

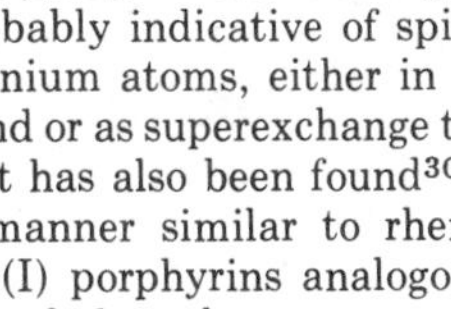

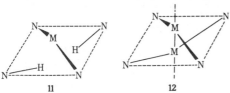

11

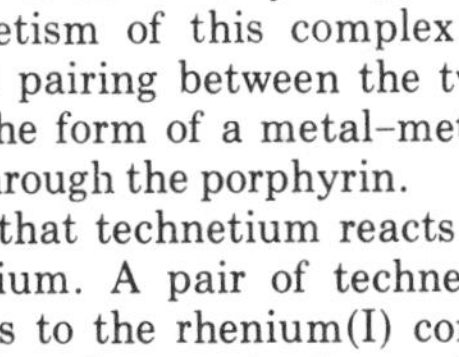

12

When $Re(CO)_5Br$ is used instead of $Re_2(CO)_{10}$, a compound is obtained of stoichiometry Re_2(Por)-$(CO)_4Br_2$.[67] The diamagnetism of this complex is probably indicative of spin pairing between the two rhenium atoms, either in the form of a metal–metal bond or as superexchange through the porphyrin.

It has also been found[30] that technetium reacts in a manner similar to rhenium. A pair of technetium(I) porphyrins analogous to the rhenium(I) compounds have been prepared, as has a mixed complex μ-(Por)[$Re(CO)_3$][$Tc(CO)_3$].[68]

(63) D. Ostfeld, M. Tsutsui, C. P. Hrung, and D. C. Conway, *J. Coord. Chem.*, **2**, 101 (1972).
(64) D. Cullen, E. Meyer, T. S. Srivastava, and M. Tsutsui, *J. Amer. Chem. Soc.*, **94**, 7603 (1972).
(65) E. B. Fleischer, E. I. Choi, P. Hambright, and A. Stone, *Inorg. Chem.*, **3**, 1284 (1964).
(66) P. Hambright, *J. Chem. Soc., Chem. Commun.*, 13 (1972).
(67) M. Tsutsui and C. P. Hrung, to be published.
(68) M. Tsutsui and C. P. Hrung, *J. Amer. Chem. Soc.*, **95**, 5777 (1973).

Heating a solution of monotechnetium complex causes disproportionation.[67] Disproportionation does

$$2[(PorH)Tc(CO)_3] \xrightarrow[\text{decalin}]{193^\circ} PorH_2 + Por[Tc(CO)_3]_2$$

not occur with the monorhenium complex. A solution containing both uncomplexed porphyrin and ditechnetium complex shows no sign of the reverse reaction upon being heated. Apparently the monotechnetium complex, and by inference the monorhenium complex, are unstable. Only great kinetic inertness allows these novel mono complexes to be isolated.

Yoshida and coworkers[26] have used the carbonyl method to prepare a complex containing two rhodium atoms per porphyrin. It is thought that both rhodium atoms lie on the same side of the porphyrin plane and bond to each other through chloride bridges. However an X-ray structure determination will be necessary before this compound can be further discussed.

Further use of "novel" synthetic metalloporphyrins is expected in elucidating the geometries which axial ligands can assume. One approach would be to examine the metalloporphyrin complexes of metals known to exhibit coordination numbers greater than six. Scandium porphyrin forms complexes in which axially coordinated acetate or acetylacetonate acts as a bidentate ligand.[15] Considering the size and electronic structure of the metal and the steric requirement of the ligands, it is to be expected that the scandium is situated above the porphyrin plane. Zirconium(IV)[15] and hafnium(IV)[18,19] form porphyrin complexes each containing two bidentate acetate ligands. It has been proposed[15,19] that the metal atoms are also out of the porphyrin plane, thus coordinating both acetates on the same side of the porphyrin (**13**). This has been confirmed[69] by single crystal X-ray diffraction.

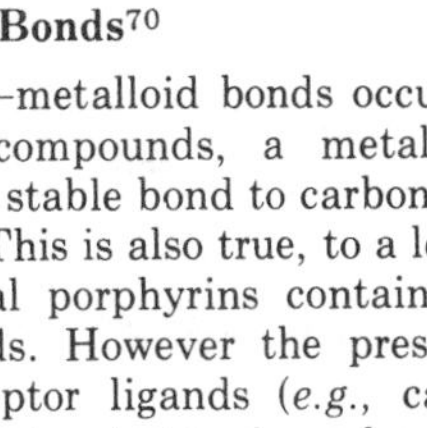

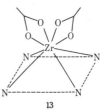

13

Metal–Carbon Bonds[70]

Since carbon–metalloid bonds occur commonly in nonporphyrin compounds, a metalloid porphyrin complex with a stable bond to carbon is not germane to this article. This is also true, to a lesser extent, for transition metal porphyrins containing cyanide or carbonyl ligands. However the presence of strong, neutral, π-acceptor ligands (*e.g.*, carbonyls, phosphines, and unsaturated hydrocarbons) is frequently required for transition metals to bond to alkyl, aryl, or acyl groups.[71] This type of organometallic bond is also formed in the presence of porphyrin or porphyrin-like ligands.

(69) J. L. Hoard, private communication.
(70) A more complete discussion of this topic is given by M. Tsutsui, *Proc. N. Y. Acad. Sci.*, in press.
(71) M. L. H. Green, "Organometallic Compounds," Vol. 2, Methuen, London, 1968, p 203.

Alkyl and acyl complexes of cobalt porphyrins have been formed by oxidative addition to the appropriate Co(I) complex.[35] A second method, using a

$$[Por-Co^{I}]^{-} + RX \xrightarrow{\text{CH}_3\text{OCH}_2\text{CH}_2\text{OCH}_3} [Por-Co^{III}R] + X^{-}$$

Grignard reagent, has been used to prepare alkyl and aryl cobalt porphyrin complexes by a nonoxidative route.[35] The two alkyl iron porphyrins reported[35] as

$$[Por-Co^{III}(py)(Br)] + RMgBr \longrightarrow$$
$$[Por-Co^{III}R] + MgBr_2 + py$$

forming from iron(III) porphyrin and a Grignard reagent are less well characterized, the presence of alkyl ligands being based entirely on C, H, and N analyses.

An oxidative addition has also been used to prepare methyl rhodium porphyrin.[27] In this case the methyl group comes from an *N*-methylporphyrin (in a presumably intramolecular transfer) rather than from an alkyl halide. In addition, rhodium porphyrins have been reported to form acyl complexes, again by an intramolecular route.[72,73]

$$H^{+}[Por-Rh^{I}] \longrightarrow [Por-Rh^{III}] + HN(CH_3)_2$$

$$\overset{|}{\underset{CH_3CN(CH_3)_2}{\overset{O}{\|}}} \qquad \overset{|}{\underset{}{CH_3C=O}}$$

$$[Por-Rh^{III}(Cl)(CO)] \longrightarrow [Por-Rh^{III}Cl]$$
$$\overset{|}{O=\overset{}{C}OCH_2CH_3}$$

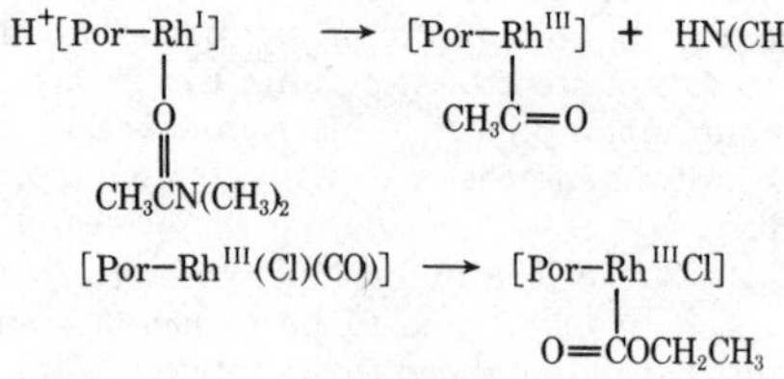

The significance of alkyls of cobalt and rhodium porphyrins lies in their similarity to vitamin B_{12}, another compound forming metal–alkyl bonds from an M(I) species.[74] This resemblance is shared by several bidentate ligands,[75] pairs of which can assume a square-planar coordination geometry, and by a number of less well known macrocycles.[33]

Schrauzer[75] has ascribed the stability of these alkyls to the strong, planar ligand field of the porphyrin or porphyrin-like ligand. Consistent with this theory, Busch[33] found cobalt alkyl bonds to be stable only in the presence of macrocycles with greater than a certain minimum ligand field strength. Clearly nothing in these arguments should limit organometallic porphyrins to those of cobalt and rhodium. Nor is it obvious that rhodium porphyrins cannot undergo intermolecular addition and cobalt porphyrins, intramolecular addition. More work will be necessary to discover the scope of carbon–metal bonding in metalloporphyrins.

The authors are indebted to their coworkers, whose names appear in the literature cited. This research has been supported in part by grants from the National Science Foundation and the Office of Naval Research.

Addendum (May 1976)

The rapid progress in the subject area made during the recent years is well described in the following articles.

(72) B. R. James and D. V. Stynes, *J. Chem. Soc., Chem. Commun.*, 1261 (1972).

(73) I. A. Cohen and B. C. Chow, *Inorg. Chem.*, **13**, 488 (1974).

(74) H. A. O. Hill, J. M. Pratt, and R. J. P. Williams, *Chem. Brit.*, **5**, 156 (1969).

(75) G. N. Schrauzer, *Accounts Chem. Res.*, **1**, 97 (1968).

(76) M. Tsutsui, C. P. Hrung, D. Ostfeld, T. S. Srivastava, D. L. Cullen, and E. F. Meyer, Jr., Unusual Metalloporphyrin Complexes of Rhenium and Technetium, *J. Am. Chem. Soc.*, **97**, 3952 (1975).

(77) M. Tsutsui and G. A. Taylor, Out-of-Plane Metalloporphyrins, *J. Chem. Educ.*, **52**, 7156 (1975).

(78) M. Tsutsui and G. A. Taylor, Metalloporphyrins with Unusual Geometrics, in "Porphyrins and Metalloporphyrins," K. Smith, Ed., Elsevier, Amsterdam, 1975, pp. 299–313.

Resonance Raman Spectroscopy: a New Structure Probe for Biological Chromophores

Thomas G. Spiro

Department of Chemistry, Princeton University, Princeton, New Jersey 08540

Received April 1, 1974

Reprinted from ACCOUNTS OF CHEMICAL RESEARCH, *7, 339 (1974)*

The central challenge of modern biochemistry is to elucidate biological function in terms of molecular structure. A considerable catalog of information on biomolecular architecture is already available from X-ray diffraction studies on crystalline or partially ordered materials. Numerous spectroscopic methods have been introduced to monitor structural features and detect changes which accompany biological function. Among them, vibrational spectroscopy offers high promise, since vibrational frequencies, available from Raman or infrared spectra, are sensitive to geometric and bonding arrangements of localized groups of atoms in a molecule.

The study of vibrational spectra has played a leading role in structural investigations of small molecules, and a substantial body of systematic knowlege has been formed.[1] Application to biological materials is beset with difficulties, however. Water, the ubiquitous biological medium, is an excellent absorber of infrared radiation, leaving only restricted "windows" for infrared spectroscopy. These can be somewhat extended by using D_2O as well as H_2O solutions.[2] Raman spectroscopy does not suffer as much from this limitation, since water is a poor Raman scatterer. Lasers now provide the high light power density required for Raman spectroscopy and allow examination of minute quantities (microliters) of material. The chief obstacle encountered with biological materials, however, is their complexity. A molecule containing N atoms has $3N - 6$ ($3N - 5$ for linear molecules) normal modes of vibration. The macromolecules of biology contain thousands of atoms and have far too many vibrational frequencies to be resolved, let alone assigned in a normal Raman or infrared spectrum. Fortunately these frequencies tend to group themselves into more-or-less discrete bands, which can be identified with certain classes of structure. These bands can then be used to monitor changes in gross conformation, and this technique has been fruitfully applied to proteins, nucleic acids, and lipids.[3,4]

If one is interested in structural features of a specific site of biological function within a macromolecule, then the myriad vibrations of the whole molecule are a serious interference. What is needed is a selective technique that samples only the vibrations of the atoms in the vicinity of the site. This can be provided by *resonance* Raman spectroscopy, if the atoms in the site give rise to an isolated electronic absorption band. A normal Raman spectrum is obtained by illumination of the sample in a transparent region of its spectrum. In resonance Raman spectroscopy, the illumination is within an absorption band. Most of the Raman bands are attenuated by the ab-

Thomas G. Spiro received his B.S. degree from UCLA in 1956 and the Ph.D. from MIT in 1960, with Professor David N. Hume. A postdoctoral year at the University of Copenhagen, with Professor Carl J. Ballhausen, was followed by a year as research chemist at the California Research Corp., LaHabra, Calif., and another postdoctoral year at the Royal Institute of Technology, Stockholm, with the late Professor Lars Gunnar Sillén. In 1963 he joined the faculty at Princeton University where he is now Professor of Chemistry. His research program has dealt with structural inorganic chemistry and the role of metal ions in biology, and now focuses on applications of Raman spectroscopy to biological systems.

(1) E. B. Wilson, J. C. Decius, and P. C. Cross, "Molecular Vibrations," McGraw-Hill New York, N. Y., 1955.

(2) G. J. Thomas, Jr., in "Physical Techniques in Biochemical Research," 2nd ed, Vol. 1A, A. Weissberger, Ed., Academic Press, New York, N. Y., 1971, Chapter 4.

(3) J. L. Koenig, *J. Polym. Sci., Part D*, 6, 59 (1972).

(4) T. G. Spiro in "Chemical and Biochemical Applications of Lasers," C. B. Moore, Ed., Academic Press, New York, N. Y., 1974, Chapter 2.

sorption, but some bands may be greatly enhanced. This effect is due to a coupling of electronic and vibrational transitions, and the vibrational modes which are subject to enhancement are localized on the grouping of atoms which give rise to the electronic transition, *i.e.*, the chromophore. Resonance Raman spectroscopy therefore provides a means of monitoring vibrational frequencies of a chromophore, independent of its (nonabsorbing) matrix. Since biological chromophores (hemes, flavins, metal ions, etc.) are usually at sites of biological function, the technique offers high promise as a new probe of biological structure.

While its application to biology is new, the resonance Raman effect has long been familiar to physicists. Resonance enhancement was implicit in the treatment of Raman scattering theory given by Van Vleck[5] and by Placzek,[6] some 40 years ago, and has been the subject of several experimental studies in the interval.[7-9] Only recently, however, have multifrequency, and now tunable, lasers become available, with which one can deliberately tune in to the absorption bands afforded by biological chromophores. This Account provides a brief introduction to the principles of the technique and to some recent applications, particularly to heme proteins, which have proven to be fertile subjects for study.

Characteristics of Resonance Raman Scattering

The Scattering Process. Raman spectroscopy involves a light-scattering experiment in which the frequency of the scattered light is analyzed. Most of the scattered light emerges at the incident frequency (Rayleigh scattering), but occasionally a photon scatters inelastically from a molecule and is shifted from its original frequency by a quantum of energy corresponding to a molecular transition of the sample. The transition may be translational (*e.g.*, lattice modes of a crystal), rotational, vibrational, or electronic in nature. All of these processes have been detected in Raman scattering, but we are here concerned with the vibrational effect, a schematic representation of which is given in Figure 1. A photon can lose energy by raising a molecule to an excited vibrational state, or gain energy by inducing the reverse process, producing Stokes or anti-Stokes lines, respectively, in the Raman spectrum. Since the fraction of molecules occupying excited states decreases exponentially with increasing energy, the intensity of anti-Stokes lines falls off rapidly with increasing frequency shift.

Resonance enhancement of Raman bands comes into play when the energy of the incident light approaches that of an electronic transition. If the photon is actually absorbed in the electronic transition and then reemitted, the process is called fluorescence. The conceptual distinction between scattering and fluorescence under resonance conditions (having to do with the lifetime of the photon–molecule complex[10]) is subtle, but for practical purposes it is easy to distinguish the two processes by varying the exciting frequency. For Raman scattering the frequency shifts are independent of the exciting frequency, while for fluorescence, the absolute frequency remains constant. Moreover in liquids, fluorescence almost always occurs in a broad spectral band, because of molecular perturbations during the lifetime of the excited state while Raman bands are sharp. For this reason fluorescence can easily obscure resonance Raman spectra.

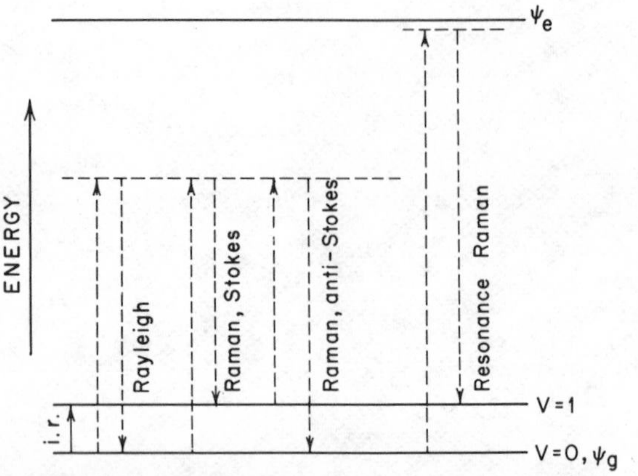

Figure 1. Energy level scheme illustrating the vibrational Raman scattering process.

A promising approach to reducing such interference is to time-resolve the emitted light, using pulsed laser sources. Because of the short lifetime of many fluorescence processes, however, there are practical difficulties in devising an effective "fluorescence filter."[11] Fortunately many biological chromophores have low fluorescence yields, thanks to effective quenching mechanisms.

Raman Intensities. The total intensity of a Raman line for randomly oriented molecules is given by

$$I = \frac{2^7\pi^5}{3^2c^4} I_0 \nu_s^{\,4} \sum_{ij} |\alpha_{ij}|^2 \quad (1)$$

where I_0 is the intensity of the incident light, ν_s is the frequency of the scattered light, and α_{ij} is an element of the scattering tensor.[12] This tensor is viewed classically as the molecular polarizability (Rayleigh scattering) or its derivative with respect to the nuclear displacements (Raman scattering). Its quantum mechanical expression, through second-order perturbation theory, is given by the Kramers–Heisenberg–Dirac dispersion equation[12]

$$(\alpha_{ij})_{mn} = \frac{1}{h}\sum_{e}\left[\frac{(M_j)_{me}(M_i)_{en}}{\nu_e - \nu_o} + \frac{(M_i)_{me}(M_j)_{en}}{\nu_e + \nu_s}\right] \quad (2)$$

Here m and n are the initial and final states of the molecule, while e is an excited state, and the summation is over all excited states. The quantities $(M_j)_{me}$ and $(M_i)_{en}$ are electric dipole transition moments, along the directions *j* and *i*, from m to e and

(5) J. H. Van Vleck, *Proc. Nat. Acad. Sci. U. S.*, **15**, 754 (1929).
(6) G. Placzek, "Rayleigh and Raman Scattering," UCRL Translation No. 526L from "Handbuch der Radiologie," Vol. 2, E. Marx, Ed., Leipzig, Akademische Verlagsgesellschaft VI, Leipzig, p 209.
(7) J. Behringer in "Raman Spectroscopy," Vol. 1, H. A. Szymanski, Ed., Plenum Press, New York, N. Y., 1967, p 168.
(8) W. Holzer, W. F. Murphy, and H. J. Bernstein, *J. Chem. Phys.*, **52**, 399 (1970).
(9) R. J. Gillespie and M. J. Morton, *J. Mol. Spectrosc.*, **30**, 78 (1969).
(10) P. F. Williams, D. L. Rousseau, and S. H. Dworetsky, *Phys. Rev. Lett.*, **32**, 196 (1974).
(11) R. P. Van Duyne, D. L. Jeannaire, and D. F. Shriver, *Anal. Chem.*, **46**, 213 (1974).
(12) J. Tang and A. C. Albrecht in "Raman Spectroscopy," Vol. 2, H. A. Szymanski, Ed., Plenum Press, New York, N. Y., 1970, Chapter 2.

from e to n, while ν_e is the frequency of the transition from m to e, and ν_0 and ν_s are the frequencies of the incident and scattered photons. In the nonresonance region $\nu_0 \ll \nu_e$, and α_{ij} is independent of the exciting frequency. As ν_0 approaches ν_e, however, α_{ij} is subject to preresonance enhancement through the first term on the right-hand side of eq 2. When $\nu_e - \nu_0$ becomes very small (it is prevented from reaching zero by inclusion of a damping constant to allow for a finite electronic line width), then one element in the summation, corresponding to the resonant electronic transition, dominates all others, assuming that the transition moments are sizable. When $\nu_e \gg \nu_0$ the resonance enhancement again decreases.

Equation 2 is unspecific about the initial and final states of the molecule and therefore gives no information as to which vibrations are subject to resonance enhancement. To examine this question it is customary to use the adiabatic approximation (separability of electronic and vibrational wave functions).[13] One can then expand the electronic wave function in a Taylor series of the nuclear displacements, using the Herzberg–Teller formalism.[12] When this is done, it becomes apparent that the vibrations which experience resonance enhancement are similar to those which (in the excited states) lend intensity to the electronic absorption spectrum.[12] Consequently resonance Raman spectroscopy can be considered as a form of high-resolution vibronic spectroscopy.

An alternative approach is to apply third-order time-dependent perturbation theory to the scattering process.[14] The results are similar to those stemming from the Herzberg–Teller approach, and the most convenient formulation[15] is

$$\alpha_{ij} = \frac{1}{h}\sum_{e,f}\left[\frac{(M_j)_{ge}h_{ef}{}^{\Delta\nu}(M_i)_{fg}}{(\nu_e - \nu_o)(\nu_f - \nu_s)} + \frac{(M_i)_{ge}h_{ef}{}^{\Delta\nu}(M_j)_{fg}}{(\nu_e + \nu_s)(\nu_f + \nu_o)}\right] \quad (3)$$

where the summation is over all excited electronic states e and f, taken in pairs, and g is the electronic ground state. Here $h_{ef}{}^{\Delta\nu}$ is a vibronic coupling matrix element, connecting states e and f by a particular vibration $\Delta\nu$. Some terms which are expected to be of lesser magnitude have been omitted from the full equation[14] for α_{ij}.

Two kinds of resonance process are expected to be important, corresponding to the conditions e = f and e ≠ f in eq 3, which Albrecht and Hutley[16] have called *A* and *B* terms, respectively. *A* terms involve vibrational interaction with a single excited electronic state, through the Frank-Condon overlaps. Only for totally symmetric vibrations are these overlaps nonzero for both the initial and final states of the molecule. *B* terms involve vibronic mixing of two excited states, e and f. The active vibrations may have any symmetry which is contained in the direct product of the two electronic transition representations. Moreover, *A* and *B* terms may be distinguished by the excitation frequency dependence of the preresonance enhancement.[16]

(13) Nonadiabatic effects may be nonnegligible; however, *cf.* J. M. Friedman and R. M. Hochstrasser, *Chem. Phys.*, **1**, 457 (1973).

(14) W. L. Peticolas, L. A. Nafie, P. Stein, and B. Fanconi, *J. Chem. Phys.*, **52**, 1576 (1970).

(15) D. W. Collins, D. B. Fitchen, and A. Lewis, *J. Chem. Phys.*, **59**, 5714 (1973).

(16) A. C. Albrecht and M. C. Hutley, *J. Chem. Phys.*, **55**, 4438 (1971).

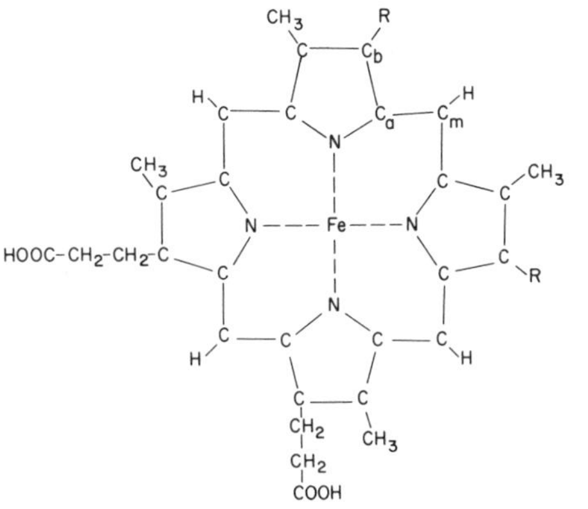

Figure 2. Structure of heme, indicating pyrrole substituents which occur in hemoglobin and cytochrome *c* (from ref 17).

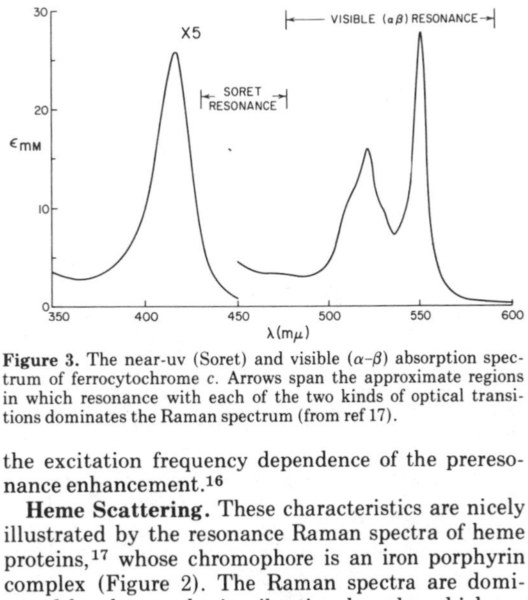

Figure 3. The near-uv (Soret) and visible (α–β) absorption spectrum of ferrocytochrome *c*. Arrows span the approximate regions in which resonance with each of the two kinds of optical transitions dominates the Raman spectrum (from ref 17).

Heme Scattering. These characteristics are nicely illustrated by the resonance Raman spectra of heme proteins,[17] whose chromophore is an iron porphyrin complex (Figure 2). The Raman spectra are dominated by the porphyrin vibrational modes which are enhanced by resonance with the allowed electronic transitions in the visible and near-ultraviolet region. These are π–π^* transitions,[18] polarized in the porphyrin plane, and the strong Raman bands are at frequency shifts (1000–1700 cm^{-1}) appropriate for in-plane stretching vibrations of the ring. Within this set different bands are enhanced depending on whether the excitation wavelength lies in the vicinity of the visible (α and β) or near-ultraviolet (Soret) bands. Figure 3 shows a typical heme absorption spectrum, that of reduced cytochrome *c*, while Figure 4 illustrates the alteration in the Raman spectra of oxy- and deoxyhemoglobin on changing the excitation wavelength.

(17) T. G. Spiro and T. C. Strekas, *J. Amer. Chem. Soc.*, **96**, 338 (1974), and references cited therein.

(18) M. Gouterman, *J. Chem. Phys.*, **30**, 1139 (1959); *J. Mol. Spectrosc.*, **6**, 138 (1961).

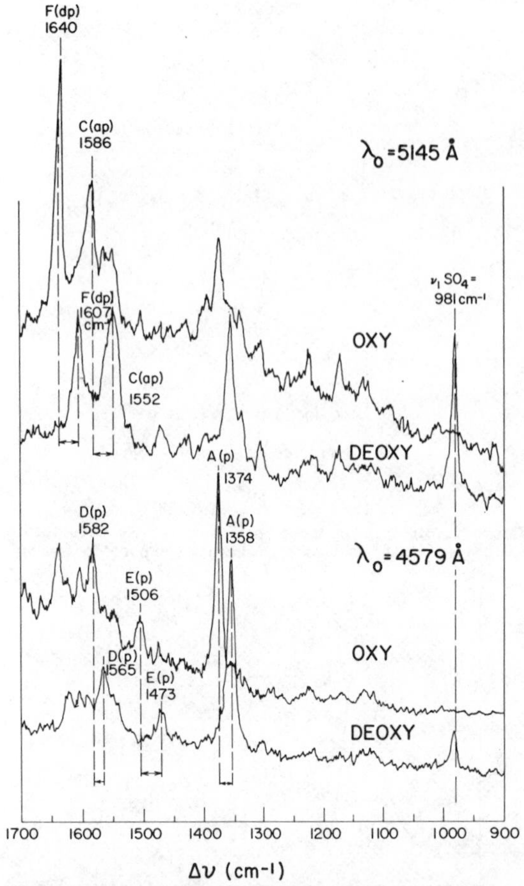

Figure 4. Resonance Raman spectra of oxy- and deoxyhemoglobin in the α–β (λ_0 5145 Å) and Soret (λ_0 4519 Å) scattering regions. The solutions were 0.68 and 0.34 mM in heme for oxy- and deoxyhemoglobin, respectively, and the latter contained 0.4 M $(NH_4)_2SO_4$, the $\nu_1(SO_4^{2-})$ band (981 cm^{-1}) of which is indicated. Frequency shifts for corresponding bands are marked by the arrows (from ref 17).

The modes which are in resonance with the intense Soret band are all of type A, and are totally symmetric.[19,20] No A terms are seen in the α–β region, presumably because the absorption (transition moment) is much lower in the α than in the Soret band. A rich variety of vibronic, type B, modes are observed,[21] however. The β band is known to be a vibronic side band on the α band, the result of vibronic mixing between the α and Soret transitions.[18] Both are of E_u symmetry, under the effective D_{4h} point group of the chromophore, and the allowed symmetry of the vibronically active modes is $E_u \times E_u = A_{1g} + B_{1g} + B_{2g} + A_{2g}$. The A_{1g} modes, however, have been shown to be ineffective in vibronic mixing.[22] Indeed all of the bands which are in-resonance with the α–β bands are either depolarized (B_{1g} or B_{2g}) or are anomalously polarized, the latter having greater intensity in the perpendicular than in the parallel scattering component. Anomalously polarized bands have contributions from A_{2g} modes, which have antisymmetric scattering tensors, *i.e.*, $\alpha_{xy} = -\alpha_{yx}$. Observation of these bands[21] in hemoglobin and cytochrome *c* was the first experimental confirmation of antisymmetric vibrational scattering, which is forbidden in the nonresonance region, although the phenomenon was predicted 40 years ago by Placzek.[6] Recent experiments[23,24] have also confirmed Placzek's suggestion,[6] recently reemphasized by McClain,[25] that antisymmetric scattering can be measured directly using circularly polarized as well as linearly polarized illumination.

Equation 3 predicts resonance maxima at both the incident and the scattered frequency. This prediction has been confirmed for heme proteins for the visible electronic transition. The plot of intensity *vs.* excitation wavelength, called the excitation profile, peaks at the center of the α band,[26] *i.e.*, the position of resonance of the incident frequency, $\nu_e = \nu_o$. As the excitation frequency is increased into the region of the β band, the Raman intensity increases again and reaches a second maximum, which, however, lies at increasingly higher frequency with increasing vibrational frequency of the individual Raman bands.[21] The intensity maxima are at positions corresponding to the sum of the electronic and vibrational frequencies, $\nu_e + \Delta\nu$. This is just the position of resonance of the scattered frequency, $\nu_e = \nu_s = \nu_o - \Delta\nu$.

Applications

Heme Structure. The dominant features of the resonance Raman spectra of heme proteins are porphyrin ring modes in the 1000–1700-cm^{-1} region.[17] Vibrations involving the iron atom are unfortunately not strongly enhanced, presumably because the resonant π–π^* transitions are largely localized on the porphyrin ring.[18] Assignment of the numerous ring modes is difficult, since they contain contributions from many internal coordinates. They can be readily catalogued, however, *via* their three different states of polarization, and overlapping bands can often be resolved by varying the excitation wavelength to take advantage of their differential enhancement. Consequently corresponding modes can reliably be correlated from one heme derivative to another.[17]

Figure 5 shows a frequency correlation diagram for a series of heme protein derivatives, chosen for their well-characterized spin (high spin or low spin) and oxidation (Fe(II) or Fe(III)) states. The hemoglobin spectra display more bands than do those of cytochrome *c*, probably because of the two peripheral vinyl substituents on protoporphyrin IX, known to be conjugated with the ring, which are replaced in cytochrome *c* by thioether links to the protein (Figure 2). The vinyl groups may themselves be responsible for one or two high-frequency vibrations, and can also induce Raman activity into infrared-active modes (E_u) since their asymmetric disposition destroys the center of symmetry of the chromophore. Most of the frequencies remain nearly constant for all the derivatives, but a few display substantial

(19) T. C. Strekas and T. G. Spiro, *J. Raman Spectrosc.*, **1**, 197 (1973).
(20) L. A. Nafie, M. Pézolet, and W. L. Peticolas, *Chem. Phys. Lett.*, **20**, 563 (1973).
(21) T. G. Spiro and T. C. Strekas, *Proc. Nat. Acad. Sci. U. S.*, **69**, 2622 (1972).
(22) M. H. Perrin, M. Gouterman, and C. L. Perrin, *J. Chem. Phys.*, **50**, 4137 (1969).
(23) M. Pézolet, L. A. Nafie, and W. L. Peticolas, *J. Raman Spectrosc.*, **1**, 455 (1973).
(24) J. Nestor and T. G. Spiro, *J. Raman Spectrosc.*, **1**, 539 (1973).
(25) W. M. McClain, *J. Chem. Phys.*, **55**, 2789 (1971).
(26) T. C. Strekas and T. G. Spiro, *J. Raman Spectrosc.*, **1**, 387 (1973).

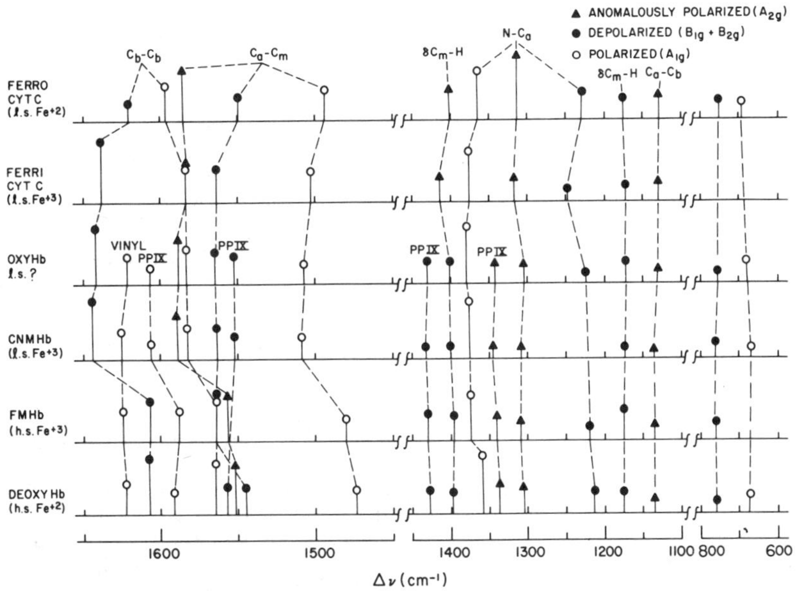

Figure 5. Correlation diagram for resonance Raman bands of hemoglobin and cytochrome *c*. Spin and oxidation states of the various species are indicated. See text for discussion of the oxyhemoglobin oxidation state. The lengths of the solid lines are roughly proportional to the observed relative intensities at 5145 Å excitation (4965 Å for FMHb) for anomalously polarized (▲) and depolarized (○) bands, and at 4579 Å for polarized (○) bands. Suggested assignments (approximate) to various heme internal coordinates are indicated at the top. The bands marked PPIX (protoporphyrin) and VINYL are observed for hemoglobin derivatives only (from ref 17).

shifts. Conversion from Fe(III) to Fe(II) without change in spin state produces a slight but systematic lowering of the ring frequencies (compare ferri- with ferrocytochrome *c* (low spin), and fluoromethemoglobin with deoxyhemoglobin (high spin)), suggesting a concomitant weakening of the bonds in the ring. Such weakening may be attributable to population of π^* antibonding ring orbitals *via* back-donation of electrons from the iron d_π orbitals, which should be greater for Fe(II) than for Fe(III).[17]

Conversion from low- to high-spin states, without change in oxidation state, produces pronounced frequency lowering for some of the Raman bands (compare cyano (low spin) with fluoromethemoglobin (high spin) (Fe(III)) and ferrocytochrome *c* (low spin) with deoxyhemoglobin (high spin) (Fe(II)). These shifts, which are much larger than those accompanying oxidation-state change, can be related to the alterations in heme structure which are known to accompany spin-state changes.[27] In low-spin hemes, the iron atom, whether Fe(II) or Fe(III), lies in the plane of the four pyrrole nitrogen atoms, but high-spin Fe(II) or Fe(III), in which the antibonding d orbitals are half-occupied, are too large to fit into the central porphyrin cavity. In high-spin hemes the iron atom moves out of the porphyrin plane, and the pyrrole groups tilt, with the nitrogen atoms pointed at the iron atoms, producing a doming of the entire heme group.[27] It is no doubt this doming which produces the observed frequency lowering of the ring modes, through changes both in kinematic coupling and in the force field.

Electron Distribution in Oxy- and Carbonmonoxyhemoglobin. While deoxyhemoglobin contains high-spin Fe(II), oxyhemoglobin is diamagnetic. Over the years a considerable controversy has developed over whether the iron–oxygen complex should be thought of as $Fe^{2+}O_2$ (low spin)[28,29] or as $Fe^{3+}O_2^-$ (low spin, with exchange coupling between unpaired electrons on Fe^{3+} and O_2^-),[30] or even as $Fe^{4+}O_2^{2-}$.[31] It is therefore of some interest that, on the basis of its resonance Raman frequencies, oxyhemoglobin classifies unambiguously as low-spin Fe(III).[17,32] Figure 5 shows that all of its frequencies are nearly the same as those of cyanomethemoglobin. Consistent with this, the O–O stretching frequency, recently located in the infrared spectrum of packed red cells,[33] is at a frequency, 1104 cm^{-1}, characteristic of superoxide.

On the other hand the resonance Raman frequencies of carbonmonoxyhemoglobin are found to be the same as those of O_2Hb, including the band at 1377 cm^{-1}, which is a reliable oxidation-state marker (see Figure 6).[17] Yet the formulation $Fe^{3+}CO^-$ is uncongenial, as the CO^- radical would be highly unstable. Relief of charge on a metal center *via* back-donation to π-acceptor ligands such as CO and O_2 is a familiar concept in inorganic chemistry, however. Indeed, the lowering of the stretching frequency[34] of carbon monoxide on binding to hemoglobin is comparable to that found in transition-metal carbonyls. One need not insist that oxyhemoglobin contains superoxide radical to allow that substantial back-donation to the bound O_2 (or CO) molecule occurs, leaving the iron atom with about the same charge as in low-spin Fe(III) hemes.

(27) J. L. Hoard, *Science*, **174**, 1295 (1971).

(28) L. Pauling, *Nature (London)*, **203**, 182 (1964).
(29) J. S. Griffith, *Proc. Roy. Soc., Ser. A*, **235**, 23 (1956).
(30) J. Weiss, *Nature (London)*, **203**, 83 (1964).
(31) H. B. Gray, *Advan. Chem. Ser.*, **No. 100**, 365 (1971).
(32) T. Yamamoto, G. Palmer, D. Gill, I. T. Salmeen, and L. Rimai, *J. Biol. Chem.*, **248**, 5211 (1973).
(33) C. H. Barlow, J. C. Maxwell, W. J. Wallace, and W. S. Caughey, *Biochem. Biophys. Res. Commun.*, **55**, 91 (1973).
(34) J. O. Alben and W. S. Caughey, *Biochemistry*, **7**, 175 (1968).

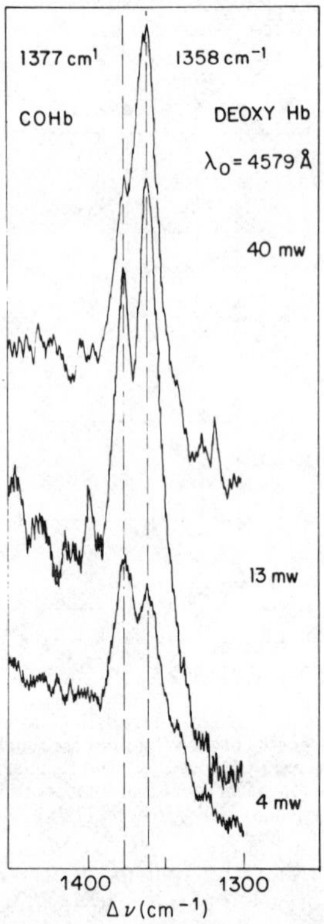

Figure 6. Portion of the Raman spectrum of carbonmonoxyhemoglobin (1377-cm^{-1} peak), showing reversible photodissociation to deoxyhemoglobin (1358-cm^{-1} peak). The sample, 0.66 m*M* COHb with sodium dithionite added, was placed under argon and run in a (sealed) rotating cell, at ~2000 rpm. The changing peak heights correspond to changing ratios of COHb and deoxyHb at the indicated power levels (measured at the sample) of 4579-Å Ar^+ laser radiation. The spectra were recorded in order of *decreasing* power levels, and the absorption spectrum of the solution at the end of the experiment corresponded to that of pure COHb. Photodissociation is less pronounced with longer wavelength excitation (from ref 17).

Other Applications. The number of resonance Raman applications to biological systems is increasing rapidly, and space precludes more than a listing of some of the chromophores for which spectra have recently been reported. Cobalt-substituted hemoglobin gives Raman spectra[35] nearly identical with those of protein-free cobalt porphyrins, suggesting very little doming of the heme group, in contrast to native hemoglobin. Raman spectra of oxidized cytochrome *c′* are indicative of a heme structure intermediate between those of low- and high-spin hemes,[36] as are those of horseradish peroxidase.[37,38] Cytochrome oxidase spectra[20,39] give some indication of inequivalence of heme a and a_3.[39] Chlorophyll[40,41] and vitamin B_{12}[42,43] give spectra similar to those of hemes. Carotenes give intense resonance scattering,[44,45] readily observable *in situ* in plant tissues.[46] Rhodopsin may involve a charge-transfer complex between retinal and aromatic protein side chains, as evidenced by resonance enhancement of both retinal and aromatic amino acid vibrations upon excitation in the rhodopsin absorption band.[47] Transferrin spectra show enhancement of tyrosine modes,[48-50] confirming tyrosine coordination of the bound Fe^{3+}. Resonance Raman spectra of the iron-sulfur proteins rubredoxin[51,52] and adrenodoxin[53] show Fe–S stretching modes, while those of "blue" copper proteins show both Cu–S and Cu–N (or Cu–O) stretching modes.[54] Oxygen bound to hemerythrin[55] and hemocyanin[56] has been shown by the resonance Raman technique to be peroxide-like; an O_2^{2-} group apparently bridges two Fe^{3+} or Cu^{2+} ions, respectively, in the oxyproteins. It is possible to introduce an artificial chromophore, *i.e.*, a "resonance Raman label," into biological sites. Thus the structural consequences of binding methyl orange to bovine serum albumin,[57] and of binding chromophoric haptens to their antibodies[58] have been explored with resonance Raman spectroscopy.

Most of the studies cited in this account have appeared within the last 2 years. It is clear that applications of resonance Raman spectroscopy to biological systems, still in their infancy, are growing rapidly. It should be stressed, however, that the technique is limited to systems with intense absorption at wavelengths corresponding to available laser sources, which currently operate in the red or blue regions of the spectrum. This limitation is being rapidly overcome by the development of lasers with increasingly wide tuning ranges. It will soon be possible to tune Raman spectrometers through the ultraviolet region, where nucleotides and aromatic amino acids absorb strongly. The scope of resonance Raman studies will then become wide indeed.

The author is indebted to his coworkers, whose names appear in the literature cited. This research has been supported by grants from the National Science Foundation and from the National Institutes of Health.

(35) W. H. Woodruff, T. G. Spiro, and T. Yonetani, *Proc. Nat. Acad. Sci. U. S.*, **71**, 1065 (1974).
(36) T. G. Spiro and T. C. Strekas, *Biochim. Biophys. Acta*, **351**, 237 (1974).
(37) G. Rakshit and T. G. Spiro, submitted for publication.
(38) T. M. Loehr and J. S. Loehr, *Biochem. Biophys. Res. Commun.*, **55**, 218 (1973).
(39) I. Salmeen, L. Rimai, D. Gill, T. Yamamoto, G. Palmer, C. R. Hartzell, and H. Beinert, *Biochem. Biophys. Res. Commun.*, **52**, 1100 (1973).
(40) M. Lutz, *C. R. Acad. Sci., Ser. B*, **275**, 497 (1972).
(41) M. Lutz and J. Breton, *Biochem. Biophys. Res. Commun.*, **53**, 413 (1973).
(42) E. Mayer, D. J. Gardiner, and R. E. Hester, *Biochim. Biophys. Acta*, **297**, 568 (1973).
(43) W. Wozniak and T. G. Spiro, *J. Amer. Chem. Soc.*, **95**, 3402 (1973).
(44) J. Behringer and J. Brandmüller, *Ann. Phys. (Leipzig)*, **4**, 234 (1959).
(45) L. Rimai, R. G. Kilponen, and D. Gill, *J. Amer. Chem. Soc.*, **92**, 3829 (1970).
(46) D. Gill, R. G. Kilponen, and L. Rimai, *Nature (London)*, **227**, 743 (1970).
(47) A. Lewis, *J. Raman Spectrosc.*, **1**, 473 (1973).
(48) Y. Tomimatsu, S. Kint, and J. R. Scherer, *Biochem. Biophys. Res. Commun.*, **54**, 1067 (1973).
(49) P. Carey and N. M. Young, *Can. J. Biochem.*, **52**, 273 (1974).
(50) V. Miskowski, B. P. Gaber, and T. G. Spiro, *J. Amer. Chem. Soc.*, in press.
(51) T. V. Long and T. M. Loehr, *J. Amer. Chem. Soc.*, **92**, 6384 (1970).
(52) T. V. Long, T. M. Loehr, J. R. Allkins, and W. Lovenberg, *J. Amer. Chem. Soc.*, **93**, 1810 (1971).
(53) S.-P. W. Tang, T. G. Spiro, K. Mukai, and T. Kimura, *Biochem. Biophys. Res. Commun.*, **53**, 869 (1973).
(54) V. Miskowski, S.-P. W. Tang, T. H. Moss, E. R. Shapiro, and T. G. Spiro, submitted for publication.
(55) J. B. R. Dunn, D. F. Shriver, and I. M. Klotz, *Proc. Nat. Acad. Sci. U. S.*, **70**, 2582 (1973).
(56) J. S. Loehr, T. B. Freeman, and T. M. Loehr, *Biochem. Biophys. Res. Commun.*, **56**, 510 (1974).
(57) P. R. Carey, H. Schneider, and H. J. Bernstein, *Biochem. Biophys. Res. Commun.*, **47**, 588 (1972).
(58) P. R. Carey, A. Froese, and H. Schneider, *Biochemistry*, **12**, 2198 (1973).

Addendum (May 1976)

A more up-to-date and detailed review of applications of resonance Raman studies to biological systems is contained in Chapter 3 of "Advances in Infrared and Raman Spectroscopy," Volume 1, R. J. H. Clark and R. E. Hester, Ed., Heyden and Son, Ltd., New York, 1975, by T. G. Spiro and T. M. Loehr. Chapter 4 of the same volume by R. J. H. Clark discusses application to inorganic molecules and ions.

Mechanism of Cobalamin-Dependent Rearrangements

Bernard M. Babior

Department of Medicine, Tufts–New England Medical Center Hospital, Boston, Massachusetts 02111

Received March 7, 1975

Reprinted from ACCOUNTS OF CHEMICAL RESEARCH, *8, 376 (1975)*

The corrins are an extensive family of compounds among which are included a unique group of cobalt-containing organometallic reagents of great biological significance. These compounds were found at the end of a long search for the agent that cured pernicious anemia, a disease which, before the discovery of corrins, was almost always fatal.

Bernard M. Babior received an M.D. from the University of California School of Medicine, San Francisco, in 1959, and a Ph.D. in biochemistry from Harvard University in 1965. After postdoctoral training he joined the faculty of Harvard Medical School. In 1972 he moved to Tufts Medical School where he is currently Associate Professor of Medicine. He studies the biochemistry of corrinoids and the physiology and biochemistry of white blood cells.

In 1922, Minot and Murphy found that pernicious anemia could be cured by liver fed in very large quantities.[1] Twenty-six years later, the isolation of the therapeutic principle in liver was announced independently by Smith[2] and by Folkers.[3] The substance they isolated was a beautiful red crystalline compound that prevented pernicious anemia at doses of only 1–2 μg per day. Over the next several years partial structures of this compound, which was given the names vitamin B_{12} and cobalamin, were de-

(1) G. R. Minot and W. P. Murphy, *J. Am. Med. Assoc.*, **87,** 470 (1926).
(2) E. L. Smith and L. F. J. Parker, *Biochem. J.*, **43,** VIII (1948).
(3) E. L. Rickes, N. G. Brink, F. R. Koniuszy, T. R. Wood, and K. Folkers, *Science*, **107,** 396 (1948).

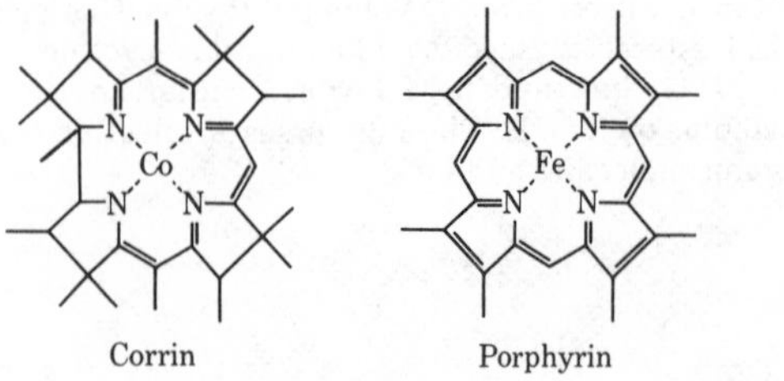

Figure 1. Corrin and porphyrin. The lines extending from the periphery of the macrocycles represent alkyl or acyl substituent groups. Strictly speaking, the terms "porphyrin" and "corrin" refer to the metal-free macrocycles. For purposes of emphasis, however, the structures are shown containing metals usually associated with them under biological conditions.

scribed, and cobalt was identified as a constituent. The complete structure was not unravelled until 1958, however, when X-ray crystallography in Hodgkin's laboratory[4] revealed the compound to be a tetrapyrrole macrocycle structurally related to the then well-known porphyrins. The cobalt atom lay in the center of the macrocycle, coordinated to the four pyrrole nitrogens.

While the early chemical studies were being carried out, investigations were simultaneously under way to find a specific biological function for the corrinoids. Such a function was first demonstrated by Barker and his associates, who found a corrinoid requirement for an enzyme-catalyzed interconversion of glutamic and β-methylaspartic acids.[5] This reaction turned out to be the prototype for the cobalamin-dependent rearrangements, which are the subject of this Account.

At the time of Barker's discovery, structural studies had indicated that cyanide was coordinated to the corrinoid cobalt atom in one of the two positions not occupied by the pyrrole nitrogens of the macrocycle. Barker's investigations, however, revealed that in the biologically active corrinoid the cyanide was replaced by an adenine nucleoside fragment. X-Ray crystallography, again performed in Hodgkin's laboratory, showed this fragment to be the 5′-deoxyadenosyl residue, attached to the corrinoid by a σ bond between its 5′-carbon atom and the cobalt.[6]

Co-Adenosylcorrinoids serve as cofactors in two types of reactions: rearrangements of a characteristic type, and the reduction of ribonucleotides to deoxyribonucleotides by certain single-celled organisms.[7] The discussion to follow will be concerned primarily with the mechanism of action of adenosylcorrinoids, as illustrated by their role in the corrinoid-dependent rearrangements. In the discussion, adenosylcobalamin will be chosen as representative, since most of the biochemical studies have been carried out with this compound.

Chemistry of Cobalamin

Structure. The basic structural unit of the corrins is the tetrapyrrole macrocycle. In this respect corrins

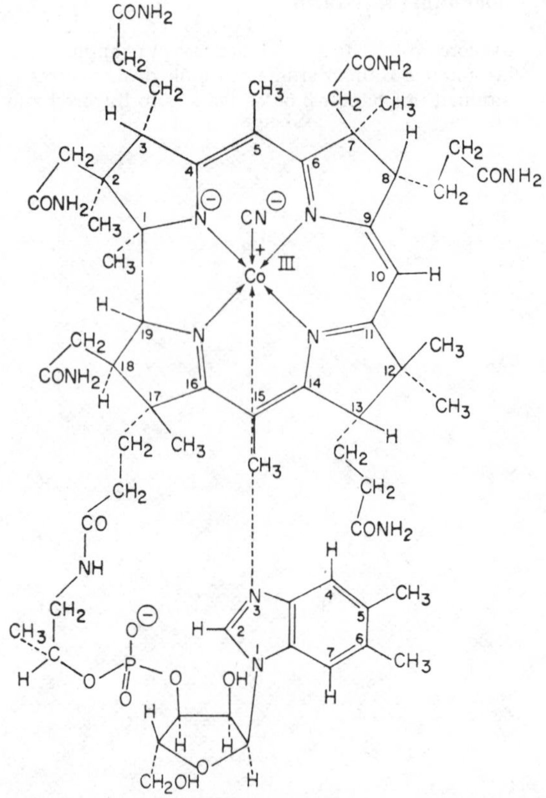

Figure 2. Cyanocobalamin (from J. M. Pratt, "Inorganic Chemistry of Vitamin B_{12}" Academic Press, New York, N.Y., 1972, p 2; reproduced with permission of the author and Academic Press).

resemble porphyrins. A closer look, however, reveals that the differences between corrins and porphyrins are as important as the similarities (Figure 1). (a) The biologically important porphyrins contain iron or magnesium; the corrins, cobalt. (b) The porphyrins possess four methine (methylidyne) bridges, one between each adjacent pair of pyrrole rings. In the corrins, one of the methine bridges is missing, so that two of the pyrrole rings are joined directly by a bond between their α carbons. (c) The porphyrins are fully unsaturated and aromatic, while the corrins are extensively reduced. The unsaturated portion of the corrins consists of a chain of six double bonds involving the nitrogens and α carbons of the pyrrole rings which partially encircles the cobalt.

What distinguishes cobalamins from other corrins[8] is the pattern of substitution of the carboxyl groups at the periphery of the ring (Figure 2). All seven are amidated, six with NH_3 and the seventh with (R)-2-aminopropanol. Attached to the propanolamine through a phosphodiester linkage is an unusual ribonucleoside which in cobalamins contains 5,6-dimethylbenzimidazole as the heterocyclic base. At neutral pH this base is coordinated to the α position of the cobalt through its free nitrogen atom. Under sufficiently acid conditions, however, the base will protonate and leave the cobalt coordination sphere, to be replaced by water or another suitable ligand.

(4) D. C. Hodgkin, *Science*, **150**, 979 (1965).
(5) H. A. Barker, H. Weissbach, and R. D. Smyth, *Proc. Natl. Acad. Sci. U.S.A.*, **44**, 1093 (1958).
(6) P. G. Lenhert and D. C. Hodgkin, *Nature (London)*, **192**, 937 (1961).
(7) F. K. Gleason and H. P. C. Hogenkamp, *Biochim. Biophys. Acta*, **277**, 466 (1972).

(8) IUPAC–IUB Commission on Biochemical Nomenclature, *Biochemistry*, **13**, 1555 (1974).

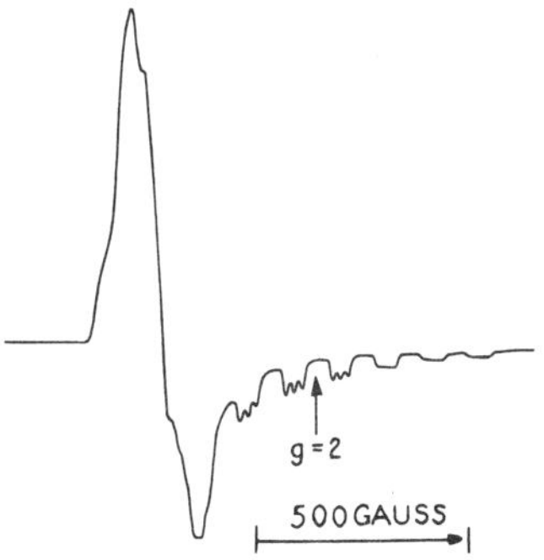

Figure 3. ESR spectrum of cob(II)alamin[12] (reproduced with permission of the author and The American Society of Biological Chemists, Inc.)

The cobalamins themselves vary in the ligand coordinated to the cobalt in the β position. This ligand may be a nucleophile, or it may be an alkyl group connected to cobalamin by a carbon–cobalt σ bond. Though a great many cobalamins with varying Coβ ligands have been prepared and studied, only four are regularly isolated from living systems: hydroxocobalamin,[9] cyanocobalamin, and the *Co*-alkyl derivatives, methyl- and adenosylcobalamin. The two naturally occurring alkyl derivatives are both active as coenzymes, though the reactions for which methylcobalamin is required are different than those requiring adenosylcobalamin; neither hydroxocobalamin nor cyanocobalamin have any cofactor activity.

Reactions. The portion of the cobalamin molecule of major biochemical interest is the cobalt atom and its β ligand. The following brief discussion will deal with aspects of the chemistry of this region of the molecule that are particularly pertinent to the role of the cofactor in the catalysis of adenosylcobalamin-dependent rearrangements.

Redox Properties of Cobalamin. The naturally occurring cobalamins discussed above all contain trivalent cobalt.[10] Under appropriate conditions the metal can be reduced to a lower valence state. Reduction by one electron produces cob(II)alamin, a paramagnetic compound containing low-spin divalent cobalt.[11] Cob(II)alamin gives a characteristic ESR signal[12] showing hyperfine coupling between the unpaired electron and the cobalt nucleus ($I = 7/2$) (Figure 3). It is indefinitely stable in aqueous solution under anaerobic conditions, but oxidizes slowly back to hydroxocobalamin in air.[13]

A further one-electron reduction yields cob(I)alamin, containing monovalent cobalt.[11a] Unlike cob(II)alamin, which is relatively stable, cob(I)alamin is extremely reactive. It is readily oxidized by agents as weak as H^+, so that it is unstable in aqueous solution,[14] giving up an electron to form cob(II)alamin with the evolution of H_2 gas. It is also extremely nucleophilic,[15] a property that is illustrated by its use in the synthesis of alkyl cobalamins from alkyl halides:

$$\text{cob(I)alamin} + CH_3I \longrightarrow \text{methylcobalamin} + I^-$$

Alkylcobalamins. Alkylcobalamins are compounds containing an alkyl group covalently attached to the metal by a carbon–cobalt σ bond. They are usually prepared by reacting cob(I)alamin with an organic compound substituted with a nucleophilic leaving group. The formation of methylcobalamin from cob(I)alamin and methyl iodide, illustrated above, is an example. Co(II)- and Co(III)-containing corrinoids have also been used on occasion for the alkylation reaction.[16,17]

The outstanding chemical property of alkylcobalamins is their susceptibility to photolysis. Illumination of a solution of alkylcobalamin leads to homolysis of the carbon–cobalt bond[18] (but see ref 19). Subsequent reactions of the immediate products of photolysis with other components of the reaction mixture determine the identity of the final products, which consequently vary from case to case.

$$\text{alkylcobalamin} \xrightarrow{h\nu} R\cdot + \text{cob(II)alamin} \longrightarrow \text{products}$$

In general, alkylcobalamins are stable to air and water. Certain of them, however, are susceptible to acid- or base-catalyzed heterolysis of the carbon–cobalt bond. In particular, adenosylcobalamin is cleaved by both acid and base.[20] In acid, the products are hydroxocobalamin, adenine, and 2,3-dihydroxy-4-pentenal,[20a] while base heterolysis yields cob(I)alamin and 4′,5′-anhydroadenosine.[20b]

Although the heterolytic reactions of adenosylcobalamin have been well characterized in chemical systems, the evidence obtained to date in enzymatic systems points to homolytic cleavage as the biologically significant process.

Adenosylcobalamin-Dependent Rearrangements

The rearrangements catalyzed by adenosylcobalamin-dependent enzymes all involve the migration of a hydrogen from one carbon atom to an adjacent one in exchange for a group X that migrates in the opposite direction. Nine such rearrangements have so far been discovered[21] (Table I). In three of these rear-

(9) Known as aquacobalamin when the Coβ-hydroxo group is protonated. In this paper, hydroxocobalamin will be used to refer to both forms.

(10) G. Boehm, A. Faessler, and G. Rittmayer, *Z. Naturforsch., Teil B*, **9**, 509 (1954).

(11) (a) B. Jaselkis and H. Diehl, *J. Am. Chem. Soc.*, **76**, 4354 (1954); (b) S. A. Cockle, H. A. O. Hill, J. M. Pratt, and R. J. P. Williams, *Biochim. Biophys. Acta*, **177**, 686 (1969).

(12) J. H. Bayston, F. D. Looney, L. R. Pilbrow, and M. E. Winfield, *Biochemistry*, **9**, 2164 (1970).

(13) J. M. Pratt, *J. Chem. Soc.*, 5154 (1964).

(14) S. L. Tackett, J. W. Collat, and J. C. Abbott, *Biochemistry*, **2**, 919 (1963).

(15) (a) A. W. Johnson, L. Mervyn, N. Shaw, and E. L. Smith, *J. Chem. Soc.*, 4146 (1963); (b) G. N. Schrauzer and E. J. Deutsch, *J. Am. Chem. Soc.*, **91**, 3341 (1969).

(16) G. N. Schrauzer, J. W. Sibert, and R. J. Windgassen, *J. Am. Chem. Soc.*, **90**, 6681 (1968).

(17) F. Wagner and K. Bernhauer, *Ann. N.Y. Acad. Sci.*, **112**, 580 (1964).

(18) (a) R. T. Taylor, L. Smucker, M. L. Hanna, and J. Gill, *Arch. Biochem. Biophys.*, **156**, 521 (1973); (b) H. P. C. Hogenkamp, *Biochemistry*, **5**, 417 (1966); (c) K. N. Joblin, A. W. Johnson, M. F. Lappert, and B. K. Nicholson, *J. Chem. Soc., Chem. Commun.*, 441 (1975).

(19) Under certain conditions, corrins (and related cobalt complexes) in which the metal is alkylated with a group bearing a hydrogen on the β carbon will photolyze to cob(I)alamin and a terminal olefin. This process has been explained in terms of concerted photoelimination (K. N. V. Duong, A. Ahond, C. Merienne, and A. Gaudemer, *J. Organomet. Chem.*, **55**, 375 (1973)), but other work suggests a two-step mechanism in which homolysis by light is followed by the transfer of a β hydrogen from the alkyl radical to cob(II)alamin to yield the final products (R. H. Yamada, S. Shimizu, and S. Fukui, *Biochim. Biophys. Acta*, **124**, 197 (1966)).

(20) (a) H. P. C. Hogenkamp and H. A. Barker, *J. Biol. Chem.*, **236**, 3079 (1961); (b) G. N. Schrauzer and J. W. Sibert, *J. Am. Chem. Soc.*, **92**, 1022 (1970).

(21) (a) H. A. Barker, V. Rooze, F. Suzuki, and A. A. Iodice, *J. Biol. Chem.*, **239**, 3260 (1964); (b) H. G. Wood, R. W. Kellermeyer, R. Stjernholm,

Table I
Adenosylcobalamin-Dependent Rearrangements

Enzyme	Reaction	Group X
Glutamate mutase[21a]	L-Glutamate ⇌ L-*threo*-β-methylaspartate	–CH(NH_2)COOH
Methylmalonyl-CoA mutase[21b]	L-Methylmalonyl-CoA ⇌ Succinyl-CoA	–COSCoA
2-Methyleneglutarate mutase[21c]	2-Methyleneglutarate ⇌ Methyl itaconate	–C(=CH_2)COOH
Diol dehydrase[21d]	Ethylene glycol ⟶ acetaldehyde	–OH
	1,2-Propanediol ⟶ propionaldehyde	
Glycerol dehydrase[21e]	Glycerol ⟶ β-hydroxypropionaldehyde	–OH
Ethanolamine ammonia-lyase[21f,g]	Ethanolamine ⟶ acetaldehyde + NH_4^+	–NH_2
	2-Aminopropanol ⇌ propionaldehyde + NH_4^+	
β-Lysine mutase[21h]	L-β-Lysine (L-3,6-diaminohexanoic acid) ⇌ 3,5-diaminohexanoic acid	–NH_2
α-Lysine mutase[21i]	D-Lysine ⇌ 2,5-diaminohexanoic acid	–NH_2
Ornithine mutase[21j]	D-Ornithine ⇌ 2,4-diaminopentanoic acid	–NH_2

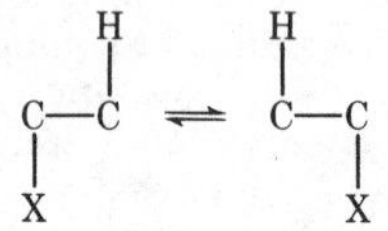

rangements, the migrating group is a bulky alkyl or acyl residue. In the remaining six, the reaction involves the migration of an electronegative group, either OH or NH_2.

These reactions have been studied from a number of points of view. Particular attention has been paid to their stereochemical features and to the structures of the enzymes responsible for their catalysis. These topics are beyond the scope of the present Account, which will be restricted to a discussion of the mechanism of the rearrangements.

In considering the mechanism of adenosylcobalamin-dependent rearrangements, it is useful to separate the hydrogen-transfer step from the group X migration. In the following discussion, the mechanism of hydrogen transfer will be treated in some detail, since this step is reasonably well understood. Group X migration, about which much less is known, will be discussed primarily in terms of model reactions, since there is little pertinent enzymological data regarding this step.

Hydrogen Transfer. In the initial studies on the mechanism of hydrogen transfer, the fate of the migrating hydrogen atom was examined using deuterated substrate. These studies showed that all of the label appeared in the expected location in the product, a result taken to indicate that the migration of hydrogen was intramolecular. In fact, this was an erroneous conclusion, and the experiment that revealed the error provided the key to the mechanism of hydrogen transfer. This experiment,[22] performed in Abeles' laboratory, took advantage of the fact that diol dehydrase converts both ethylene glycol and propylene glycol to the respective aldehydes. When the enzyme was allowed to act on a mixture of unlabeled ethylene glycol and [1,1-^{2}H]propylene glycol, deuterated acetaldehyde was obtained. This could only have arisen by the transfer of a deuterium from one molecule to another. Thus, it was shown that hydrogen transfer was not necessarily *intra*molecular, as previously thought, but could occur by an *inter*molecular process as well.

From this observation, Abeles postulated that the enzyme–adenosylcobalamin complex serves as an intermediate hydrogen carrier, first accepting a hydrogen from the substrate and then, in a subsequent step, giving a hydrogen back to the product. Experiments with tritiated substrate confirmed this formulation, showing that the migrating hydrogen atom was transferred from the substrate to the Coβ-linked carbon atom of the cofactor[23] (the C-5′ atom of the cobalt-linked adenosyl group; henceforth this will be referred to as the C-5′ position). Subsequent investi-

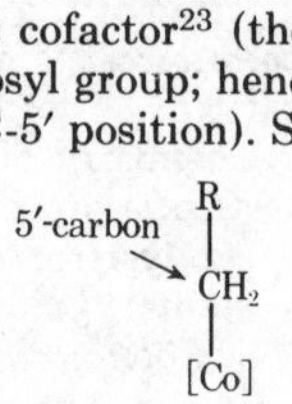

gations showed that other adenosylcobalamin-dependent rearrangements involve a similar transfer of hydrogen.[24]

A remarkable feature of the transfer of hydrogen to and from adenosylcobalamin is that both of the C-5′ positions participate in this reaction.[23] This is surprising, because enzymes are ordinarily able to distinguish between the *pro-R* and *pro-S* positions of meso carbon atoms such as the C-5′ carbon of adenosylcobalamin. Considerable experimental evidence indicates that this unusual labeling pattern results from a process whereby the two C-5′ hydrogen atoms of the cofactor and the migrating hydrogen of the substrate all become equivalent. This process is postulated to involve cleavage of the carbon–cobalt bond of adenosylcobalamin followed by the reversible transfer of hydrogen from substrate to the C-5′ carbon to form 5′-deoxyadenosine (I). Rotation of the methyl group would lead almost instantly to equilibration of the three hydrogens.

and S. H. G. Allen, *Ann. N.Y. Acad. Sci.*, **112**, 661 (1964); (c) H. F. Kung, S. Cederbaum, L. Tsai, and T. C. Stadtman, *Proc. Natl. Acad. Sci. U.S.A.*, **65**, 978 (1970); (d) H. A. Lee, Jr., and R. H. Abeles, *J. Biol. Chem.*, **238**, 2367 (1963); (e) K. L. Smiley and M. Sobolov, *Ann. N.Y. Acad. Sci.*, **112**, 706 (1964); (f) B. H. Kaplan and E. R. Stadtman, *J. Biol. Chem.*, **243**, 1787 (1968); (g) T. J. Carty, B. M. Babior, and R. H. Abeles, *ibid.*, **249**, 1683 (1974); (h) E. E. Dekker and H. A. Barker, *ibid.*, **243**, 3232 (1968); (i) C. G. D. Morley and T. C. Stadtman, *Biochemistry*, **10**, 2325 (1971); (j) Y. Tsuda and H. C. Friedmann, *J. Biol. Chem.*, **245**, 5914 (1970).

(22) R. H. Abeles and B. Zagalak, *J. Biol. Chem.*, **241**, 1245 (1966).

(23) P. A. Frey, M. R. Essenberg, and R. H. Abeles, *J. Biol. Chem.*, **242**, 5369 (1967).

(24) (a) R. L. Switzer, B. G. Baltimore, and H. A. Barker, *J. Biol. Chem.*, **244**, 5268 (1969); (b) B. M. Babior, *Biochim. Biophys. Acta*, **167**, 456 (1968); (c) W. W. Miller and J. H. Richards, *J. Am. Chem. Soc.*, **91**, 1498 (1969); (d) H. F. Kung and L. Tsai, *J. Biol. Chem.*, **246**, 6436 (1971); (e) J. Rétey, F. Kunz, T. C. Stadtman, and D. Arigoni, *Experientia*, **25**, 801 (1969).

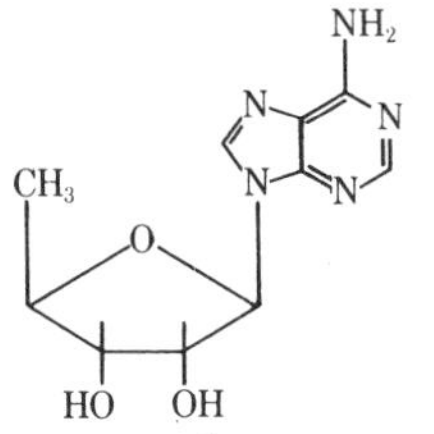

I

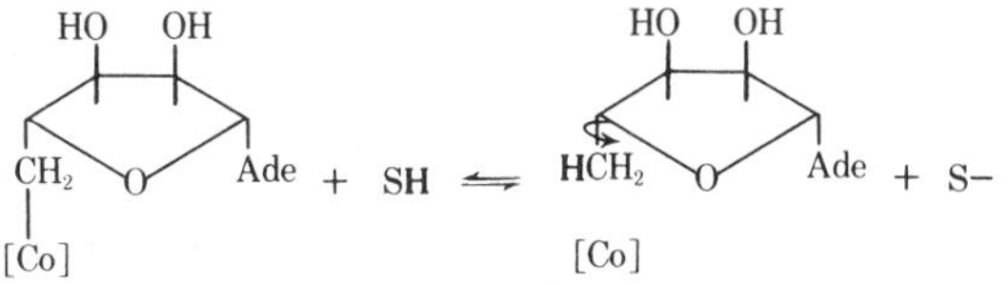

5′-Deoxyadenosine. The involvement of 5′-deoxyadenosine in adenosylcobalamin-dependent rearrangements was first proposed by Ingraham in 1964.[25] Its actual formation was shown 2 years later by Wagner et al.,[26] who found that, when diol dehydrase and adenosylcobalamin were incubated with glycolaldehyde, the cofactor was converted to 5′-deoxyadenosine and a cobalamin derivative that was not identified at the time. The extra hydrogen on 5′-deoxyadenosine was supplied by glycolaldehyde, which was oxidized to glyoxal in the reaction.

Similar reactions have since been demonstrated with several enzymes, using a variety of substrate analogs[27] (Table II). These reactions have in common the features that the *Coβ*-adenosyl residue of the cofactor is converted to 5′-deoxyadenosine and that (where investigated) the third hydrogen of the methyl group is derived from the analog. It is clear that the ability to form 5′-deoxyadenosine from adenosylcobalamin is a rather general property of enzymes catalyzing adenosylcobalamin-dependent rearrangements. The difficulty in interpreting these results in terms of a catalytic role for 5′-deoxyadenosine is that all the reactions listed in Table II are irreversible: 5′-deoxyadenosine is produced at the cost of the destruction of the cofactor. Unless it were possible to show that 5′-deoxyadenosine formation is reversible, it could be argued that this compound is not a catalytic intermediate, but is merely the product of an abortive side reaction leading to the inactivation of the enzyme–cofactor complex. Attempts to demonstrate reversibility by showing exchange of 5′-deoxyadenosine into adenosylcobalamin have failed uniformly, indicating that a reversible process involving the formation *and release* of 5′-deoxyadenosine does not take place. The experiments described below, however, have demonstrated reversible formation of 5′-deoxyadenosine by ethanolamine ammonia-lyase without release from the enzyme.

When ethanolamine ammonia-lyase was denatured during incubations with substrates (i.e., while engaged in catalysis), some of the cofactor was released as 5′-deoxyadenosine.[28] This finding was similar to results obtained with substrate analogs. With true substrates, however, 5′-deoxyadenosine formation was reversible. The first evidence for this was the finding that the fraction of total cofactor recovered as 5′-deoxyadenosine varied with the denaturing agent (Table III). Such variability could be seen if a reversible interconversion were occurring between two forms of enzyme–cofactor complex, one form carrying the cofactor as adenosylcobalamin and the other as 5′-deoxyadenosine plus a corrin fragment.

Table II
5′-Deoxyadenosine Formation in the Presence of Substrate Analogs

Enzyme	Analog promoting 5′-deoxyadenosine formation
Diol dehydrase	Glycolaldehyde[26]
	Chloroacetaldehyde[27a]
Methylmalonyl-CoA mutase	Malonyl-CoA[27b]
β-Lysine mutase	L-3,5-Diaminohexanoate[27c,a]
Ethanolamine ammonia-lyase	Ethylene glycol[27d]
	Acetaldehyde[27e,a]

[a] Product of the catalytic reaction.

Table III
Fraction of Cofactor in the Form of 5′-Deoxyadenosine as a Function of the Denaturing Agent

Denaturing agent	Fraction as 5′-deoxyadenosine, %
Ethanolamine as Substrate	
Hot 1-propanol	3.6 ± 0.4[a]
Tetrahydrofuran	1.3 ± 0.2
Propanolamine as Substrate	
Trichloroacetic acid	86.9 ± 1.3
Heat	89.3 ± 5.3
Ethanol	12.7 ± 3.2

[a] Mean ± standard error.

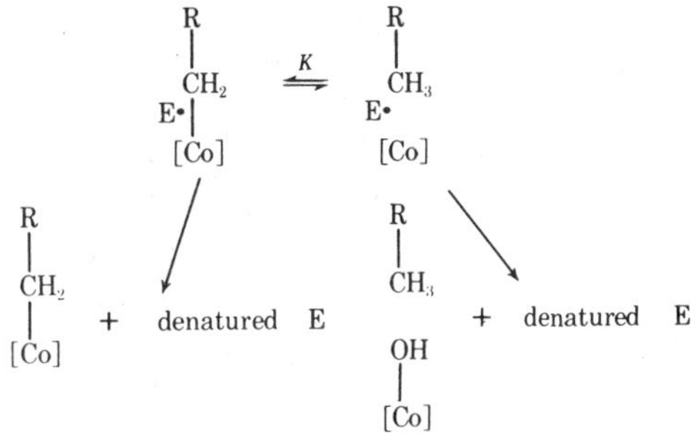

During catalysis, the relative amounts of the two forms would depend on the equilibrium constant *K*. Upon denaturation, however, differences in the susceptibility of the two forms to a given denaturing agent could pull the equilibrium to the left or right during denaturation. If the relative susceptibilities varied from agent to agent, the extent of cleavage in

(25) L. L. Ingraham, *Ann. N.Y. Acad. Sci.*, **112,** 713 (1964).
(26) O. W. Wagner, H. A. Lee, Jr., P. A. Frey, and R. H. Abeles, *J. Biol. Chem.*, **241,** 1751 (1966).
(27) (a) T. H. Finlay, J. Valinsky, K. Sato, and R. H. Abeles, *J. Biol. Chem.*, **247,** 4197 (1972); (b) B. M. Babior, A. D. Woodams, and J. D. Brodie, *ibid.*, **248,** 1445 (1973); (c) J. J. Baker, C. Van der Drift, and T. C. Stadtman, *Biochemistry,* **12,** 1054 (1973); (d) B. M. Babior, *J. Biol. Chem.*, **245,** 1755 (1970); (e) T. J. Carty, B. M. Babior, and R. H. Abeles, *J. Biol. Chem.*, **246,** 6313 (1971).
(28) (a) B. M. Babior, *J. Biol. Chem.*, **245,** 6125 (1970); (b) B. M. Babior, T. J. Carty, and R. H. Abeles, *ibid.*, **249,** 1689 (1974).

Table IV
Reversible Formation of Adenosylcobalamin

	Distribution of isotope, cpm × 10^{-4}			
	Propanol-amine only		Propanol-amine followed by ethanol-amine	
Compound	^{3}H	^{14}C	^{3}H	^{14}C
5′-Deoxyadenosine	19.4	3.0	0.7	0.4
Adenosylcobalamin	2.0	0.6	0.6	3.1
Acetaldehyde			19.1	
Total	21.4	3.6	21.4	3.5

an incubation terminated with one agent would differ from that seen in a similar incubation terminated with another.

Assuming the above explanation to be correct, the results presented in Table III suggest that there is much less 5′-deoxyadenosine at the active site during steady-state deamination of ethanolamine than during propanolamine deamination. The best evidence for reversibility was based on this observation and consists of the finding that the addition of ethanolamine to a reaction mixture originally containing enzyme, adenosylcobalamin, and propanolamine led to a reversal of the high 5′-deoxyadenosine/adenosylcobalamin ratio prevailing during propanolamine deamination.[28b]

In the experiment, [1-^{3}H]propanolamine was incubated with [^{14}C]adenosylcobalamin and enzyme. Analysis of ^{14}C distribution in a reaction terminated at 15 sec with trichloroacetic acid showed that 80% of the cofactor had been converted to 5′-deoxyadenosine (Table IV). Moreover, both 5′-deoxyadenosine and intact adenosylcobalamin were labeled with tritium, which had been transferred to the cofactor from the substrate. The ratio of tritium specific activities in 5′-deoxyadenosine and adenosylcobalamin was that expected if the two were in equilibrium.

In an otherwise identical reaction mixture to which unlabeled ethanolamine was added at 15 sec followed by trichloroacetic acid 1 min later, the isotope distribution was very different. Analysis of ^{14}C showed that 5′-deoxyadenosine present at the time the ethanolamine was added had been converted almost completely back to adenosylcobalamin. Moreover, the tritium that was present in both adenosylcobalamin and 5′-deoxyadenosine at 15 sec was now to be found in the acetaldehyde formed by the enzyme-catalyzed deamination of unlabelled ethanolamine. It is apparent that the 5′-deoxyadenosine present on the enzyme before the addition of ethanolamine must have been converted back to catalytically competent adenosylcobalamin.

These findings show that an adenosylcobalamin-dependent enzyme can catalyze the reversible cleavage of the cofactor at the carbon–cobalt bond and provide strong, though not absolutely conclusive, evidence that 5′-deoxyadenosine is one of the products of cleavage. The residue of uncertainty regarding this point relates to the possibility that the third hydrogen atom involved in the equilibrium between the two C-5′ hydrogens and the substrate hydrogen may not be connected to the C-5′ carbon in the active enzyme but may only attach to it during denaturation. This possibility notwithstanding, reversible formation of 5′-deoxyadenosine by carbon–cobalt bond cleavage and hydrogen abstraction remains the best explanation for hydrogen transfer in adenosylcobalamin-dependent rearrangements.

Species with Unpaired Electrons. The foregoing indicates that cleavage of the carbon–cobalt bond occurs during adenosylcobalamin-dependent rearrangements. This bond can be broken in four ways: heterolytically with carbonium ion formation; heterolytically with carbanion formation; heterolytically with olefin formation;[19,20b] and homolytically. Evidence obtained by ESR spectroscopy indicates that the mechanism of these rearrangements involves homolysis of the carbon–cobalt bond with the participation of species with unpaired electrons.

Of the enzymes listed in Table I, ESR signals have been reported with three: ethanolamine ammonia-lyase, glycerol dehydrase, and diol dehydrase.[29] With all three, the signals are seen when the enzyme–cofactor complexes are frozen while engaged in catalysis (Figure 4). With diol dehydrase, signals also appear when the enzyme is incubated with cofactor in the presence of analogs that promote cleavage of the carbon–cobalt bond.[29c] Signals are seen with ethanolamine ammonia-lyase plus cofactor even in the absence of substrate or a substrate analog.[29a]

In all cases, the ESR signals consist of two components: a broad low-field peak representing enzyme-bound cob(II)alamin; and one or two narrower peaks situated upfield from the cob(II)alamin peak. The species responsible for the upfield peaks have been the subject of controversy, but in two instances they have been unequivocally identified as radicals derived from substrate or a substrate analog.[29b,30] The species represented by the high-field signal appearing during the ethanolamine ammonia-lyase catalyzed deamination of propanolamine[29] (Figure 4) was identified by ESR spectrometry of isotopically labeled substrates (Figure 5). Spectra with unlabeled and [2-^{2}H]propanolamine are essentially the same. With [1,1-^{2}H]propanolamine, however, the signal is several gauss narrower than control, while substantial broadening is seen with [1-^{13}C]propanolamine. These findings identify the species in question as the 1-hydroxy-2-aminopropyl radical. Similarly, an enzyme-bound radical derived from the substrate analog was identified as one of the products of the reaction between diol dehydrase and chloroacetaldehyde.[30]

Are these ESR-absorbing species relevant to the catalytic mechanism? Two arguments have been offered supporting their participation in catalysis. The first is that the paramagnetic species account for a major fraction (5–65%, depending on the enzyme and

(29) (a) B. M. Babior, T. H. Moss, and D. C. Gould, *J. Biol. Chem.*, **247**, 4389 (1972); (b) B. M. Babior, T. H. Moss, W. H. Orme-Johnson, and H. Beinert, *ibid.*, **249**, 4537 (1974); (c) T.H. Finlay, J. Valinsky, A. S. Mildvan, and R. H. Abeles, *ibid.*, **248**, 1285 (1973); (d) S. A. Cockle, H. A. O. Hill, R. J. P. Williams, S. P. Davies, and M. A. Foster, *J. Am. Chem. Soc.*, **94**, 275 (1972).

(30) J. E. Valinsky, R. H. Abeles, and A. S. Mildvan, *J. Biol. Chem.*, **249**, 2751 (1974).

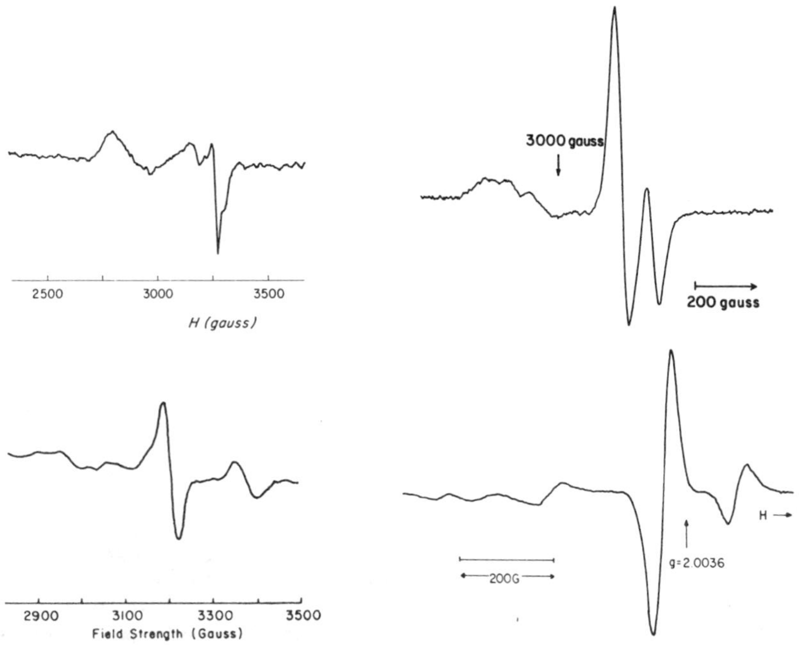

Figure 4. ESR spectra of enzymes frozen while engaged in catalysis. Top left: ethanolamine ammonia-lyase, ethanolamine.[29a] Top right: ethanolamine ammonia-lyase, propanolamine.[29b] Bottom left: diol dehydrase, propanediol.[29c] Bottom right: glycerol dehydrase, glycerol[29d] (reproduced with permission of the author and The American Society of Biological Chemists, Inc.).

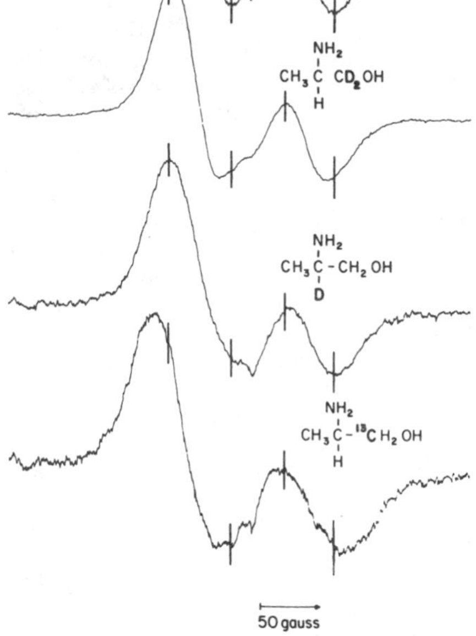

Figure 5. Identification of 1-hydroxy-2-aminopropyl as the free radical formed during the ethanolamine ammonia-lyase catalyzed deamination of propanolamine[29b] (reproduced with permission of the author and The American Society of Biological Chemists, Inc.).

substrate) of the active sites present in a reaction mixture.[29] It seems unreasonable to propose that species representing such a large proportion of the total enzyme are not involved in catalysis. Even stronger arguments are provided by experiments using rapid freezing techniques. For two reactions—the ethanolamine ammonia-lyase catalyzed deamination of L-2-aminopropanol[29b] and the dehydration of L-propanediol by diol dehydrase[31]—such experiments have shown that the rates of appearance of the ESR signals exceed the catalytic turnover numbers. The formation of the paramagnetic species in these reactions is thus a kinetically competent process, a result strongly suggesting a catalytic role for these species.

Most of the free-radical signals consist of two peaks, a feature which is not completely understood. Since the peaks do not separate when the microwave frequency is increased,[29b–d] they represent a single absorption split by an interaction with a nearby paramagnetic species with spin $\frac{1}{2}$. One group has proposed that the nearby magnet is the unpaired electron of enzyme-bound cob(II)alamin, while another has proposed a hydrogen nucleus as the interacting species. The question is unresolved at present.

The current view of the mechanism of hydrogen transfer is

$$\begin{array}{ccc}
R & R & R \\
| & | & | \\
CH_2 & \dot{C}H_2 & \dot{C}H_3 \\
| & \rightleftharpoons \quad + SH \quad \rightleftharpoons & + S\cdot \\
[Co] & [\dot{C}o] & [\dot{C}o]
\end{array}$$

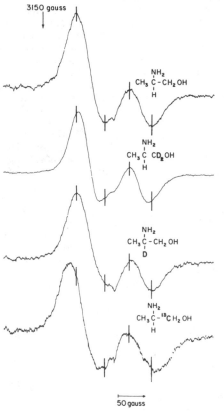

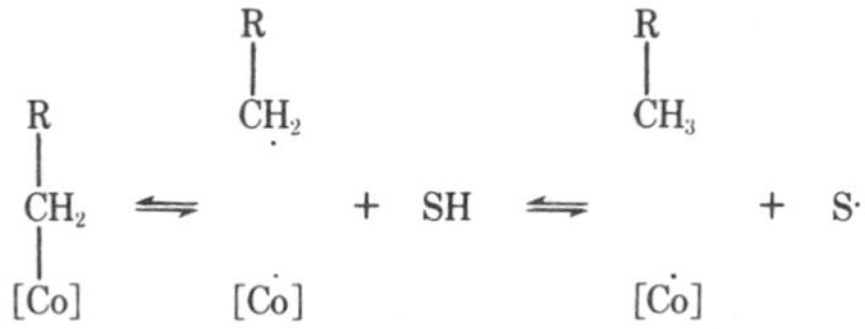

(31) J. E. Valinsky, R. H. Abeles, and J. A. Fee, *J. Am. Chem. Soc.*, **96**, 4709 (1974).

Table V
Model Reactions for Group X Migration

Reaction	Class	Ref
HOCH$_2$CHCH$_3$ $\xrightarrow{\text{H}^+}$ CH$_2$CHOHCH$_3$ |(Co) |(Co)	$\sigma \rightleftharpoons \pi$ rearrangement	35a
CH$_2$CH$_2$OH $\xrightarrow{\text{OH}^-}$ (Co)$^-$ + CH$_3$CHO |(Co)	Hydride shift (nucleophilic attack)	16
(cyclopentanone CH$_2^-$ / COOEt) → (cyclohexanone COOEt)	Carbanion rearrangement	35b
C$_6$H$_5$C(Me)$_2$CH$_2$Cl → C$_6$H$_5$C(Me)$_2$CH$_2$· → ·C(Me)$_2$CH$_2$C$_6$H$_5$ → HC(Me)$_2$CH$_2$C$_6$H$_5$	Radical rearrangement	35c
CH$_3$CH—CH$_2$ → CH$_3$CH—CHOH → CH$_3$CH$_2$CH(OH)$_2$ |OH |OH +|O|H	Carbonium ion rearrangement	23

Homolysis of the carbon–cobalt bond of adenosylcobalamin is followed by abstraction of a substrate hydrogen by the resulting adenosyl radical, forming 5′-deoxyadenosine and the substrate radical. After rearrangement the process is reversed, the product radical removing a hydrogen atom from 5′-deoxyadenosine to re-form the adenosyl radical. This then rejoins the enzyme-bound cob(II)alamin to reconstitute the original cofactor.

An Alternative Mechanism. Based largely on model reactions and on studies of the effect of N$_2$O on enzyme-catalyzed reactions that are open to varying interpretations, Schrauzer has proposed a mechanism for adenosylcobalamin-dependent rearrangements involving heterolysis of the carbon–cobalt bond with the participation of cob(I)alamin and 4′,5′-anhydroadenosine as catalytic intermediates. This proposal, which has been the subject of some dispute among various groups working in the field, is discussed at length in papers from Schrauzer's[20b,32] and Abeles'[27e,33] laboratories. Interested readers are referred there for details.

Group X Migration. Unlike hydrogen transfer, the mechanism of group X migration is not understood at all. The enzymological experiments that have been performed with a view to understanding this process have consisted mostly of attempts to demonstrate enzyme-catalyzed exchange of postulated intermediates into substrate or product.[34] Like the previously discussed attempts to show exchange of 5′-deoxyadenosine into adenosylcobalamin, these have all failed. The only positive finding that might be related to the mechanism of group migration is the observation by Morley and Stadtman that α-D-lysine

mutase, a pyridoxal phosphate and adenosylcobalamin requiring enzyme that converts α-D-lysine to 2,5-diaminohexanoic acid, catalyzes a pyridoxal phosphate dependent exchange of hydrogen between solvent and C-6 of the substrate.[21i] This suggests Schiff base formation between pyridoxal phosphate and the ω-amino group of D-lysine. How this relates to the rearrangement, though, is not clear.

Lacking enzymological data, investigators have attempted to explain the mechanism of group X migration in terms of model reactions. A partial list of these reactions is presented in Table V. This list, by no means exhaustive, contains representatives of several reaction types. Each of these reactions may be defended as the model that truly describes group X migration, since each involves a group migration, and in the absence of further enzymological data none can be ruled out. This proliferation of model reactions seems to suggest that, while the models are of considerable intrinsic interest and great heuristic value, elucidation of the mechanism of enzyme-catalyzed group X migration will require studies with enzymes.

Conclusions

Adenosylcobalamin is one of a group of closely related biological organometallic reagents all of which are *Co*-alkyl derivatives of corrinoid compounds. It is required as the cofactor for certain enzyme-catalyzed rearrangements in which a hydrogen atom migrates from one carbon atom to an adjacent one in exchange for a group X that migrates in the opposite direction. The nature of group X is highly variable: it may be a complex alkyl (e.g., –CH(NH$_2$)COOH)) or acyl (–COSCoA) group, or a conventional nucleophilic leaving group (–OH or –NH$_2$).

The mechanism of hydrogen transfer in these reactions is reasonably well understood. An enzyme-catalyzed homolysis of the carbon–cobalt bond of the cofactor leads to the appearance at the active site of cob(II)alamin and a free radical, 5′-deoxyadenos-5′-

(32) (a) G. N. Schrauzer and R. J. Windgassen, *J. Am. Chem. Soc.*, **89**, 143 (1967); (b) G. N. Schrauzer, R. J. Holland, and J. A. Seck, *ibid.*, **93**, 1503 (1971); (c) G. N. Schrauzer, J. A. Seck, and R. J. Holland, *Z. Naturforsch., Teil C*, **28**, 1 (1973).

(33) (a) M. K. Essenberg, P. A. Frey, and R. H. Abeles, *J. Am. Chem. Soc.*, **93**, 1242 (1971); (b) P. A. Frey, M. K. Essenberg, R. H. Abeles, and S. S. Kerwar, *ibid.*, **92**, 4488 (1970).

(34) (a) H. A. Barker, R. D. Smyth, E. J. Wawszkiewicz, M. N. Lee, and R. M. Wilson, *Arch. Biochem. Biophys.*, **78**, 468 (1958); (b) R. Swick, *Proc. Natl. Acad. Sci. U.S.A.*, **48**, 288 (1962); (c) M. Sprecher, M. J. Clark, and D. B. Sprinson, *J. Biol. Chem.*, **241**, 872 (1966); (d) H. F. Kung and T. C. Stadtman, *ibid.*, **246**, 3378 (1971); (e) R. C. Bray and T. C. Stadtman, *ibid.*, **243**, 381 (1968).

(35) (a) K. L. Brown and L. L. Ingraham, *J. Am. Chem. Soc.*, **96**, 7681 (1974); (b) J. N. Lowe and L. L. Ingraham, *ibid.*, **93**, 3801 (1971); (c) H. Eggerer, P. Overath, F. Lynen, and E. R. Stadtman, *ibid.*, **82**, 2643 (1960).

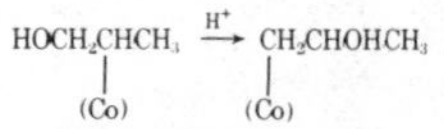

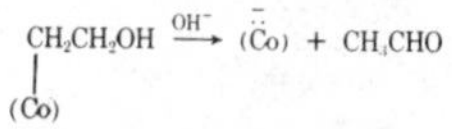

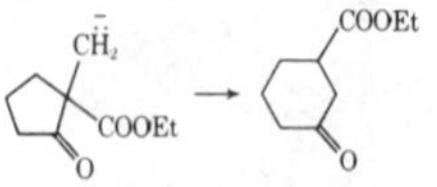

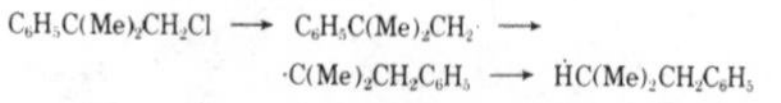

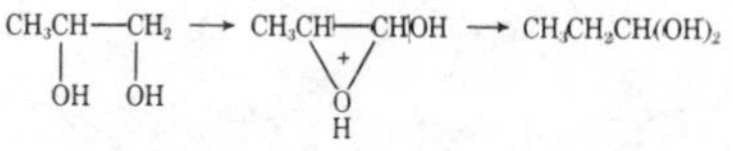

yl. The radical then abstracts the migrating hydrogen atom from the enzyme-bound substrate molecule, forming 5′-deoxyadenosine and the substrate radical. After migration of group X, the rearranged radical retrieves a hydrogen from 5′-deoxyadenosine to form the product, with the reappearance of the 5′-deoxyadenos-5′-yl radical. This then recombines with cob(II)alamin to regenerate the cofactor and complete the catalytic cycle.

Though hydrogen transfer is understood, at least in outline, little is known about the mechanism of group X migration. Enzymological experiments have thus far provided almost no information concerning this process. Conversely, experiments with chemical systems have provided an abundance of results. These have been interpreted in terms of a large number of reaction mechanisms, each of which represents an attractive model for group X migration and among which the lack of biochemical data makes it impossible to choose. Future progress in adenosylcobalamin-dependent rearrangements will undoubtedly yield a clarification of the mechanism of the migration step. With the rapid advances that have taken place in recent years in enzyme mechanism studies, it is likely that a deeper understanding of these rearrangements, including an explanation for the group X migration step, is not too far away.

Much of the research described in this Account was conducted during the tenure of a Research Career Development Award (AM 11950) from the National Institutes of Health. Recent work has been supported by a grant (AM 16589) from the National Institute of Arthritis, Metabolism and Digestive Diseases.

Addendum (May 1976)

Recent experiments have shed light on two areas of uncertainty mentioned in the foregoing Account. The first has to do with the question of whether the third hydrogen atom involved in the equilibrium between the two C-5′ hydrogens and the substrate hydrogen is attached to the C-5′ carbon while the enzyme is engaged in catalysis—i.e., whether 5′-deoxyadenosine is ever actually present at the active site of the working enzyme. Evidence concerning this point was obtained by determining the origin of the third methyl hydrogen in the 5′-deoxyadenosine formed during the ethanolamine ammonia-lyase catalyzed deamination of L-2-aminopropanol. If the active site of the working enzyme contained an adenosyl fragment with only two C-5′ hydrogen atoms, then 5′-deoxyadenosine would have to arise when the adenosyl fragment, dislodged by the disruption of the active site, abstracted a nearby hydrogen in a nonspecific reaction. Under these circumstances, the third C-5′ hydrogen would almost certainly be derived from the solvent or the protein rather than the substrate, since the substrate would leave the vicinity of the adenosyl fragment during the denaturation reaction and thus would not be available as a hydrogen donor. On the other hand, if the 5′-deoxyadenosine isolated from the reaction mixture were released as such from the active site, then the third C-5′ hydrogen should come from the substrate—specifically, from the carbinol group of the amino alcohol.

Investigation of these alternatives with the ethanolamine ammonia-lyase system, using [1,1-D_2] propanolamine and [5,5-D_2] adenosylcobalamin as substrate and cofactor respectively, showed that the third C-5′ hydrogen of 5′-deoxyadenosine was derived from the substrate. This result provides strong support for the notion that 5′-deoxyadenosine is a true catalytic intermediate.[36]

Quantum mechanical calculations have provided information concerning the origin of the line shapes of the ESR spectra of adenosylcobalamin-dependent enzymes which are engaged in catalysis. Spectra were calculated for a model system consisting of two unpaired electrons, one with the properties of a free radical electron (isotropic g-tensor, no interaction with the nucleus) and the other with the properties of the unpaired electron on cob(II)alamin (anisotropic g-tensor and a hyperfine interaction with a nucleus of spin $\frac{7}{2}$), coupled by a weak exchange interaction. By varying only the magnitude of the exchange coupling, it was possible to calculate from this model system a series of spectra which agreed remarkably well with spectra observed with ethanolamine ammonia-lyase, diol dehydrase, and glycerol dehydrase. These calculations indicate that the configurations of the ESR spectrum of adenosylcobalamin-dependent enzymes can be explained by an interaction between the free radical electron and the electron on the nearby cob(II)alamin molecule.[37]

(36) K. Sato, J. C. Orr, B. M. Babior, and R. H. Abeles, *J. Biol. Chem.*, **251**, 3734 (1976).

(37) K. L. Schepler, W. R. Dunham, R. H. Sands, J. A. Fee, and R. H. Abeles, *Biochim. Biophys. Acta*, **397**, 510 (1975).

Synthetic Oxygen Carriers of Biological Interest

Fred Basolo,* Brian M. Hoffman,* and James A. Ibers*

Department of Chemistry, Northwestern University, Evanston, Illinois 60201

Received April 2, 1975

Reprinted from ACCOUNTS OF CHEMICAL RESEARCH, *8, 384 (1975)*

Certain metal–ligand combinations provide the delicate balance required to form a 1:1 dioxygen complex without the metal (M) and/or the ligand (L) being irreversibly oxidized. These systems, which are

$$M(L)_n + O_2 \rightleftharpoons M(L)_n(O_2) \quad (1)$$

called oxygen carriers, have long been recognized in vitally important biological processes, and as early as 1852 Fremy[1] reported that solid cobalt ammine salts absorb oxygen from air and release it again when dissolved in water. However, the significance of synthetic oxygen carriers was not apparent until 1938.

Since the initial discovery by Tsumaki[2] that Schiff base chelates of cobalt(II) are oxygen carriers, there has been a continued interest in this property of such metal complexes. Much of the research effort has been devoted toward the more fundamental aspects of the problem, such as the metal–ligand properties of a good oxygen carrier, the nature of the metal–dioxygen bonding, and the structure. However, considerable research has also been devoted to possible applications of these oxygen carriers. For example, Calvin and his students[3] extensively investigated the absorption–desorption of molecular oxygen on various solid cobalt(II)–Schiff-base chelates. These complexes were studied for the purpose of isolating pure oxygen from air, and oxygen so produced was used for several months aboard a destroyer tender for welding and cutting.[4a] Other applications of synthetic oxygen carriers include their use as catalysts for reactions of molecular oxygen.[4b] It is even possible that they may eventually find use in artificial blood.[5]

Twelve years ago, Vaska[6] reported that $IrCl(CO)(P(C_6H_5)_3)_2$ is an oxygen carrier, and the structure of its oxygen complex was determined.[7] Other low-valent transition-metal compounds of this type have since been found to be oxygen carriers and/or homogeneous catalysts for certain reactions of dioxygen.[8] Structural studies on these complexes have consistently revealed a triangular MO_2 arrangement with each oxygen atom equidistant from the metal and with the O–O bond lengthened over that in dioxygen. Because of structural differences[9] from

Fred Basolo, a native of Illinois, obtained his Ph.D. degree in 1943 at the University of Illinois. His first position was with Rohm and Haas, and he joined the faculty at Northwestern University in 1946. His research interest is in transition-metal chemistry. Professor Basolo is the 1975 recipient of the ACS Award for Distinguished Service in the Advancement of Inorganic Chemistry sponsored by Mallinckrodt, and this Account is based in part on his Award address.

Brian Hoffman, born in Chicago, obtained his Ph.D. degree from the California Institute of Technology in 1966. After a postdoctoral year in the Biology Department at Massachusetts Institute of Technology, he joined the faculty at Northwestern University. The study of transition ions in biological systems is one aspect of Professor Hoffman's research interests.

James A. Ibers was born in Los Angeles, and received his B.S. and Ph.D. degrees (1954) from the California Institute of Technology. After a postdoctoral year in Australia, he spent 6 years as a chemist with Shell Development Company and 3 years as a chemist at Brookhaven National Laboratory. Professor Ibers joined the faculty at Northwestern in 1965. His research interests are in syntheses and structures of transition-metal and organometallic complexes.

(1) E. Fremy, *Justus Liebigs Ann. Chem.*, **83**, 227, 289 (1852).

(2) T. Tsumaki, *Bull. Chem. Soc. Jpn.*, **13**, 252 (1938).

(3) See, for example, A. E. Martell and M. Calvin, "Chemistry of the Metal Chelate Compounds", Prentice-Hall, Englewood Cliffs, N.J., 1952, pp 336–357.

(4) (a) R. F. Stewart, P. A. Estep, and J. J. S. Sebastian, *U.S. Bur. Mines, Inf. Circ.*, **No. 7906** (1959); (b) A. E. Martell and M. M. Taqui Khan, *Inorg. Biochem.*, **2**, 654 (1973).

(5) R. P. Geyer, *Bull. Parenter. Drug Assoc.*, **28** (2), 88 (1974); J. E. Baldwin, *Newsweek*, 91 (March 24, 1975).

(6) L. Vaska, *Science*, **140**, 809 (1963).

(7) J. A. Ibers and S. J. La Placa, *Science*, **145**, 920 (1964).

(8) J. P. Collman, *Acc. Chem. Res.*, **1**, 136 (1968); J. Valentine, *Chem. Rev.*, **73**, 235 (1973).

(9) J. L. Hoard, "Structural Chemistry and Molecular Biology", A. Rich and N. Davidson, Ed., W. H. Freeman, San Francisco, Calif., 1968, p 573.

natural oxygen carriers, it became increasingly apparent that the study of low-valent second- and third-row transition-metal oxygen complexes, though interesting from many points of view, probably lacked significant biological relevance. These oxygen carriers will not be discussed in this Account.

Research on synthetic oxygen carriers at Northwestern University was largely prompted by the isolation and characterization of the monoadduct of dioxygen with cobalt Schiff-base (SB) complexes, Co(SB)(O_2). Work was then extended to studies of the reversible reaction of molecular oxygen with cobalt(II) porphyrins, cobalt(II) hemoglobin, cobalt(II) myoglobin, and other metal-substituted hemoglobins as well. More recently this work has been extended to the interaction of O_2 with Fe, Mn, and Cr synthetic systems.

Cobalt(II)–Schiff-Base Complexes

Calvin and coworkers[3] conducted an exhaustive study of the oxygen-carrying properties of solid cobalt(II)–Schiff-base complexes. Floriani and Calderazzo[10] studied the reaction of oxygen with *N,N'*-ethylenebis(salicylideniminato)cobalt(II), (Co(salen)) and some of its derivatives in various solvents. Except for 3-methoxysalen the products were the usual diamagnetic peroxo-bridged species of the type (B)(salen)Co^{III}–(O_2^{2-})–Co^{III}(salen)(B), where B stands for a nitrogenous base, such as pyridine. Most cobalt(II) complexes react with oxygen to yield such species, and these may also be readily oxidized to the corresponding superoxo-bridged complexes, Co(III)–(O_2^-)–Co(III). Much more significant was their observation that Co(3-methoxysalen) in pyridine (py) solution reacts with oxygen to yield the monomeric complex Co(3-methoxysalen)(py)(O_2). This was the first report of the isolation of 1:1 dioxygen–cobalt complex, and it represents a significant development because oxyhemoglobin is a 1:1 dioxygen–iron complex.

Our independent experiments,[11] carried out concurrently with those of Floriani and Calderazzo,[10] made use of *N,N'*-ethylenebis(acetylacetoniminato)cobalt(II) (Co(acacen)), which is more soluble than

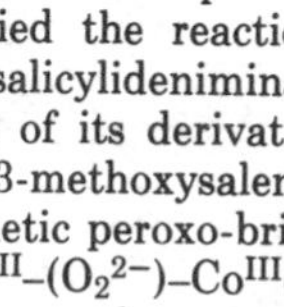

Co(acacen)

Co(salen). The reaction of Co(acacen) in a coordinating solvent (e.g., dimethylformamide, DMF) or in a noncoordinating solvent (e.g., toluene) with added base (e.g., py) at room temperature results in a nonstoichiometric, continuous slow uptake of oxygen over a period of days (Figure 1). Under these conditions, it appears the cobalt(II) is catalyzing the oxidation of the organic ligand. However, at temperatures near 0° or below there is a rapid and reversible uptake of dioxygen, corresponding to the formation of the 1:1 Co(O_2) complex (Figure 1). Various experimental techniques were then used to establish the

(10) C. Floriani and F. Calderazzo, *J. Chem. Soc. A*, 946 (1969).

(11) A. L. Crumbliss and F. Basolo, *Science*, **164**, 1168 (1969); *J. Am. Chem. Soc.*, **92**, 55 (1970).

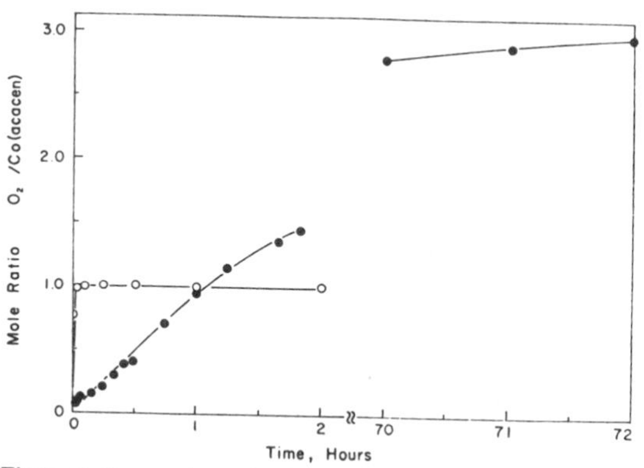

Figure 1. Oxygen absorption by DMF solutions of Co(acacen): ○, data collected at 6°C and 760 mmHg of O_2 pressure; ●, data collected at 25°C and 480 mmHg of O_2 pressure (from ref 11).

$$\text{Co(acacen)(B)} + O_2 \rightleftharpoons \text{Co(acacen)(B)}(O_2) \qquad (2)$$

presence of these 1:1 complexes (eq 2) in solution and in the solid state. The solid complexes have a very strong ir band in the region of 1140 cm^{-1} (which disappears when O_2 is removed and reappears when O_2 is returned). This we attribute to the O–O stretching vibration.

EPR spectra[12] of these systems in both liquid and frozen solution clearly demonstrate that the oxygen complexes are monomeric. These spectra exhibit interaction with but one ^{59}Co nucleus, in contrast with the two interacting ^{59}Co nuclei of μ-superoxo dimers.[13] The EPR technique has been used extensively to establish the presence of monomeric complexes of dioxygen in numerous cobalt systems.[14]

The spin Hamiltonian parameters for these monomeric oxygen complexes permit definite conclusions to be drawn about their electronic and geometric structures.[12] Our analysis shows that there is an ~90% transfer of spin density from Co(II) to O_2 upon complex formation. This result, in conjunction with the ir measurements, is compelling evidence for charge transfer upon complex formation. Thus the 1:1 cobalt dioxygen complexes should be *formally* described[12,14f] as Co(III)–O_2^-. Analysis[12] of magnetic resonance and magnetic moment data also permitted discrimination between the three plausible geometries of the Co(III)–O_2^- linkage. We concluded that

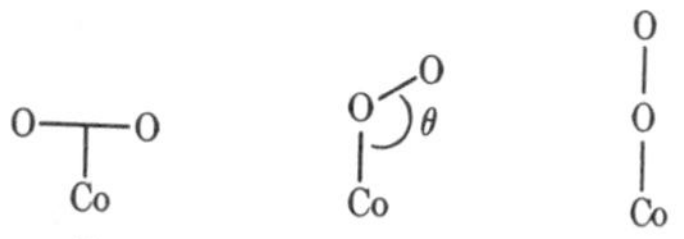

I II III

only nonsymmetric bonding, as in structure II, is consistent with experiment, with the simplest model involving an angle θ of 120°.

(12) B. M. Hoffman, D. L. Diemente, and F. Basolo, *J. Am. Chem. Soc.*, **92**, 61 (1970).

(13) M. Mori, J. A. Weil, and J. K. Kinnaird, *J. Phys. Chem.*, **71**, 103 (1967).

(14) (a) H. Kon and N. Sharpless, *Spectrosc. Lett.*, **1**, 45 (1969); (b) J. H. Bayston, N. King, F. D. Looney, and M. E. Winfield, *J. Am. Chem. Soc.*, **91**, 2775 (1969); (c) F. A. Walker, *ibid.*, **92**, 4235 (1970); (d) S. A. Cockle, H. O. Hill, and R. J. P. Williams, *Inorg. Nucl. Chem. Lett.*, **6**, 131 (1970); (e) D. L. Diemente, B. M. Hoffman, and F. Basolo, *Chem. Commun.*, 467 (1970); (f) B. M. Hoffman, T. Szymanski, and F. Basolo, *J. Am. Chem. Soc.*, **97**, 673 (1975).

These conclusions about the electronic and geometric properties, made solely on the basis of our spectroscopic data, have recently been confirmed. Rodley and Robinson[15] have carried out an X-ray crystallographic study of a $Co(SB)(py)(O_2)$ complex. They find that the nonsymmetric structure (II) is indeed correct, with $\theta = 125°$. Further, the O–O distance of 1.2 Å is different from that of the O_2^{2-} ion in K_2O_2 (1.49 Å). In addition, EPR studies[16] using ^{17}O-enriched dioxygen show that nearly 100% of the spin density is indeed located on dioxygen, and that the two oxygen atoms are nonequivalent as required by structure II.

Cobalt(II)–Porphyrin Complexes

With the discovery of the 1:1 $Co(O_2)$–Schiff-base complexes it was apparent that parallel studies on cobalt–porphyrin complexes were in order, since metalloporphyrins are utilized in biological systems. Although an extensive literature already existed on metalloporphyrin chemistry, essentially all of that chemistry had been carried out in aqueous systems.

By 1970, when this work was initiated, the significant result of Perutz[17] that the dioxygen molecule travels through a hydrophobic pocket to reach the coordination site on the Fe atom in hemoglobin was well known. Consequently we reasoned that one might simulate this hydrophobic environment on the protein by abandoning aqueous systems and working in aprotic solvents. Most of our early studies[18] were carried out in toluene which offered the advantage of reasonable spectral clarity down to about −70°. In order to stay as close as possible to biologically relevant systems, we chose to use the porphyrin protoporphyrin IX dimethyl ester (Figure 2a), which differs from the porphyrin in the heme group of hemoglobin through esterification of the two acid groups. The most common synthetic porphyrin, tetraphenylporphyrin (TPP) (Figure 2b), is less closely related to the natural porphyrins. Its popularity arises from its ease of synthesis and purification.

Initial experiments,[18,19] indicated that, if cobalt(II) protoporphyrin IX dimethyl ester (Co(PIXDME)) is dissolved in toluene and a base is added, the equilibrium is strongly to the right.[20] At room temperature in the presence of dioxygen no apparent reaction occurs, but if the solution is cooled and exposed to dioxygen reaction 4 takes place. These reactions are

$$\text{Co(PIXDME)} + \text{B} \rightleftharpoons \text{Co(PIXDME)(B)} \quad (3)$$

$$\text{Co(PIXDME)(B)} + O_2 \rightleftharpoons \text{Co(PIXDME)(B)}(O_2) \quad (4)$$

monitored readily by either EPR[19] or visible[18] spectroscopy, with the latter technique offering some advantages for quantitative measurements.

The EPR parameters[21] of $\text{Co(PIXDME)(B)}(O_2)$ are not qualitatively different from those of the $Co(SB)(B)(O_2)$ complexes. Thus, the formal description of the metal–dioxygen linkage as Co^{III}–O_2^- seems to be largely independent of the equatorial ligand.

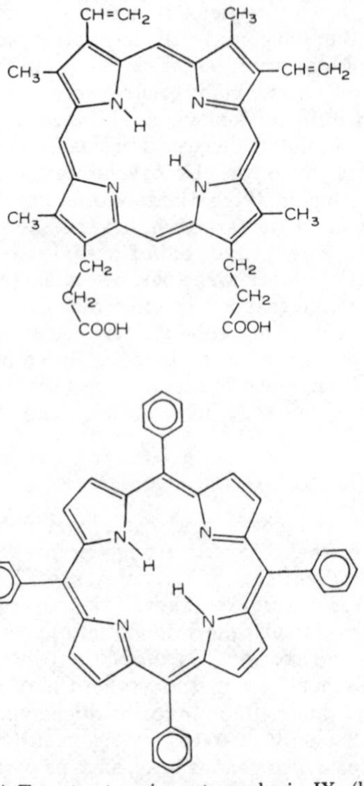

Figure 2. (a) Top structure is protoporphyrin IX. (b) Bottom structure is 5,10,15,20-tetraphenylporphyrin.

Cobalt(II) and Other Metal Hemoglobins and Myoglobins

The successful preparation of Co–O_2 complexes suggested to us the possibility of introducing a functional paramagnetic probe into hemoglobin (Hb) and myoglobin (Mb) by removing the heme group and reconstituting the protein with a cobalt(II) porphyrin to form an oxygen-carrying cobalt metalloprotein. The study of such metal-substituted proteins would permit us to observe directly the influence of the protein environment upon the metalloporphyrin by comparing the oxygen-binding and spectroscopic properties with those of Co(PIXDME) in solution. Also, variation of the stereochemical properties of the functioning metal atom binding site of hemoglobin would allow us to examine the involvement of the heme in the conformational changes which constitute hemoglobin allosteric, or cooperative, ligand binding (see below).

We thus prepared globins from hemoglobin and myoglobin and reconstituted them with cobalt(II) protoporphyrin IX (Co(PIX)).[21] These *coboglobins*, the cobalt analogs of hemoglobin (CoHb) and myoglobin (CoMb), indeed function as oxygen-carrying proteins.[21,22] EPR studies showed that the cobalt

(15) G. A. Rodley and W. T. Robinson, *Nature (London)*, **235**, 438 (1972).
(16) (a) E. F. Vansant and J. H. Lunsford, *Adv. Chem. Ser.*, **No. 121**, 441 (1973); (b) E. Melamud, B. L. Silver, and Z. Dori, *J. Am. Chem. Soc.*, **96**, 4689 (1974).
(17) M. F. Perutz, "The Harvey Lecture", Academic Press, New York-London, 1969, p 213.
(18) H. C. Stynes and J. A. Ibers, *J. Am. Chem. Soc.*, **94**, 1559 (1972).
(19) F. A. Walker, *J. Am. Chem. Soc.*, **92**, 4235 (1970).
(20) D. V. Stynes, H. C. Stynes, B. R. James, and J. A. Ibers, *J. Am. Chem. Soc.*, **95**, 1796 (1973).
(21) B. M. Hoffman and D. H. Petering, *Proc. Natl. Acad. Sci. U.S.A.*, **67**, 637 (1970).
(22) B. M. Hoffman, C. A. Spilburg, and D. H. Petering, *Cold Spring Harbor Symp. Quant. Biol.*, **36**, 343 (1971).

atom in deoxy-CoMb and CoHb is five-coordinate and, as is true for the iron in natural proteins, is coordinated to the imine nitrogen of the proximal histidine residue.[21] Further, Co(PIX) has the same orientation within the protein as does the heme.[23] Comparison with the spectrum of the monoimidazole complex of Co(PIXDME) in organic solvents shows that the five-coordinate Co(II) of CoHb exhibits no significant perturbations which are caused by its incorporation into the protein environment.[21] Exposing CoHb (CoMb) to oxygen gives oxy-CoHb (oxy-CoMb).[21–24] EPR studies again show that the Co–O_2 linkage within the protein is not significantly different from that of an oxygenated cobalt porphyrin in solution.[21]

Comparisons of the properties of the naturally occurring heme group both in and out of a protein environment were previously impossible, since heme cannot ordinarily be prepared in a five-coordinate state and heme in solution is oxidized to hemin upon exposure to oxygen. We reported[25] the first such comparison when we used Mössbauer emission spectroscopy to study heme complexes produced in frozen solution by nuclear decay from the isomorphous Co-labeled compound. The Mössbauer parameters of the daughter compound of ^{57}Co(PIX)(py)(O_2) agree well with the values for oxyhemoglobin,[26] suggesting that the electronic structure of the heme iron in oxy-Hb is not measurably influenced by the protein. This parallels the conclusions from EPR studies of oxy-CoHb.[21] Thus, the stabilization of the heme group in hemoglobin is probably not ascribable to protein-induced changes of the electronic state or iron in oxyheme.

From discussions of the implications of CoHb linkage properties, we realized that investigation of other metal-substituted hemoglobins (e.g., MnHb, ZnHb, and ZnMb[27]) could offer additional valuable insights into the nature of hemoglobin cooperativity. Such substitutions have provided new physical and chemical probes with which we can observe the influence of a protein environment on the physical and chemical properties of a prosthetic group. EPR studies of the paramagnetic (Co(II), Co(III)–O_2^-,[21,23] and Mn(II)[28]) metal-substituted proteins primarily give information about the influences of the protein environment on the metal center. However, EPR studies of zinc porphyrins and of ZnHb and ZnMb in the photoexcited triplet state provide a probe to examine the interactions between the porphyrin macrocycle itself and the protein environment.[29]

We have found[29] that the porphyrin–protein interactions in ZnHb differ from those in ZnMb; chain differences within hemoglobin are also observed. Stresses in hemoglobin which, for example, tend to stabilize a quasi-D_{2d} type of ruffling of the macrocycle could account for the observations. Indeed, it may be that the presence of such stresses is responsible for the unique properties of deoxy-Hb, and that changes in the ruffling of the porphyrin are an important component in the mechanism of cooperative ligand binding. (See section on hemoglobin cooperativity). Despite the fact that MnHb cannot reversibly bind molecular oxygen, there is a high degree of parallelism between the molecular properties of MnHb and those of Hb.[30–32] In particular, the redox reactions of MnHb are important to an understanding of hemoglobin cooperativity (see below). Preliminary X-ray studies show that Mn^{III}Hb resembles Fe^{III}Hb very closely.[33] Since Mn^{III}Hb undergoes a change in quaternary structure upon reduction, just as does Fe^{III}Hb, MnHb retains both the major functional and structural properties of Hb.

Thermodynamics of Dioxygen Binding to Cobalt(II) Complexes

Long before EPR experiments permitted the formulation $Co^{III}(O_2^-)$ for these complexes, it was suggested that the ease of oxygen uptake by metal complexes should depend on the ease of oxidation of the metal in a given complex.[34] This implies that metals which are readily oxidized (e.g., Fe(II)) react with oxygen irreversibly, metals difficult to oxidize (e.g., Ni(II)) do not react with oxygen, whereas metals between the two extremes (e.g., Co(II)) may react with oxygen reversibly. Furthermore, reversible dioxygen uptake ability of a complex depends not just on the metal but also on the ligand.

Equilibrium constants, K_{O_2}, have been determined for the reaction of dioxygen with cobalt(II) in many different ligand environments (eq 5). For a given co-

$$\mathrm{Co(L)(B)} + \mathrm{O_2} \overset{K_{O_2}}{\rightleftharpoons} \mathrm{Co(L)(B)(O_2)} \qquad (5)$$

balt(II)–Schiff-base chelate, Co(L), a linear correlation was found[35] between K_{O_2} and the ease of oxidation of cobalt(II) to cobalt(III) as B varies (Figure 3). There are enough data to show that the protonic base strength of the axial base does not correlate with oxygen uptake. For example, the values of K_{O_2} for changes in axial base decrease in the order *n*-butylamine > 1-methylimidazole > piperidine, although the pK_a's of these protonated bases are 10.6, 7.25, and 11.3, respectively. That 1-methylimidazole is much more effective in promoting complex formation than is expected from its protonic basicity may arise from its good π-electron-donating property.

This same type of behavior had been observed earlier[20,36,37] for cobalt porphyrins. Figure 4 shows a plot

(23) (a) J. C. W. Chien, and L. C. Dickinson, *Proc. Natl. Acad. Sci. U.S.A.*, **69**, 2787 (19?3); (b) L. C. Dickinson and J. C. W. Chien, *J. Biol. Chem.*, **248**, 5005 (1973).

(24) T. Yonetani, H. Yamamoto, and G. V. Woodrow, *J. Biol. Chem.*, **249**, 682, 691 (1974).

(25) L. Marchant, M. Sharrock, B. M. Hoffman, and E. Munck, *Proc. Natl. Acad. Sci. U.S.A.*, **69**, 2396 (1972).

(26) G. Lang and W. Marshall, *Proc. Phys. Soc., London*, **87**, 3 (1966).

(27) (a) S. F. Andres and M. Z. Atassi, *Biochemistry*, **9**, 2268 (1970); (b) J. J. Leonard, T. Yonetani, and J. B. Calles, *ibid.*, **13**, 1461 (1974).

(28) T. Yonetani, H. Yamamoto, J. E. Erman, J. S. Leigh, and G. H. Reed, *J. Biol. Chem.*, **247**, 2447 (1972).

(29) B. M. Hoffman, *J. Am. Chem. Soc.*, **97**, 1688 (1975).

(30) C. Bull, R. G. Fisher, and B. M. Hoffman, *Biochem. Biophys. Res. Commun.*, **59**, 140 (1974).

(31) Q. H. Gibson, B. M. Hoffman, R. H. Crepeau, S. J. Edelstein, and C. Bull, *Biochem. Biophys. Res. Commun.*, **59**, 146 (1974).

(32) B. M. Hoffman, Q. H. Gibson, C. Bull, R. H. Crepeau, S. J. Edelstein, R. G. Fisher, and M. J. McDonald, *Ann. N.Y. Acad. Sci.*, **244**, 174 (1975).

(33) K. Moffat, R. S. Loe, and B. M. Hoffman, *J. Am. Chem. Soc.*, **96**, 5259 (1974).

(34) L. H. Vogt, Jr., H. M. Faigenbaum, and S. E. Wiberley, *Chem. Rev.*, **63**, 269 (1963).

(35) M. J. Carter, L. M. Engelhardt, D. P. Rillema, and F. Basolo, *J. Chem. Soc., Chem. Commun.*, 810 (1973); M. J. Carter, D. P. Rillema, and F. Basolo, *J. Am. Chem. Soc.*, **96**, 392 (1974).

(36) D. V. Stynes, H. C. Stynes, J. A. Ibers, and B. R. James, *J. Am. Chem. Soc.*, **95**, 1142 (1973).

(37) J. A. Ibers, D. V. Stynes, H. C. Stynes, and B. R. James, *J. Am. Chem. Soc.*, **96**, 1358 (1974).

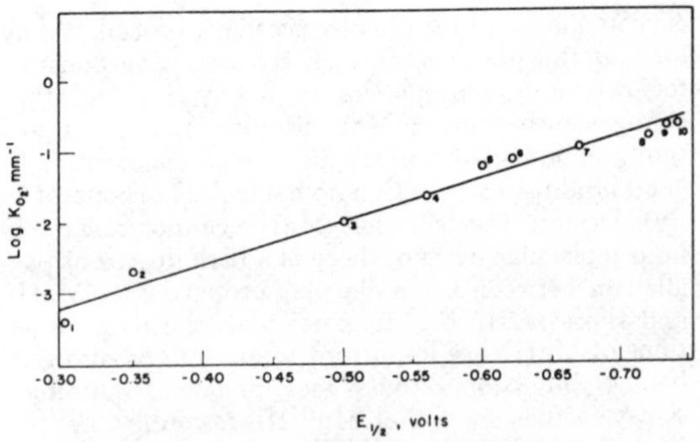

Figure 3. Comparison of oxygen uptake (log K_{O_2}) at −21°C for Co(benacen)B to the polarographic half-wave potentials ($E_{1/2}$) for $Co^{II \to III}$(benacen)B_2: 1,PPh_3; 2, CNpy; 3, py; 4, 3,4-lutidine; 5, piperidine; 6, *sec*-$BuNH_2$; 7, 5-Cl-N-MeIm; 8, N-MeIm; 9, *i*-$BuNH_2$; 10, *n*-$BuNH_2$. benacen = benzoylacetylacetonate ion, PPh_3 = triphenylphosphine, py = pyridine, and Bu = butyl (from ref 35).

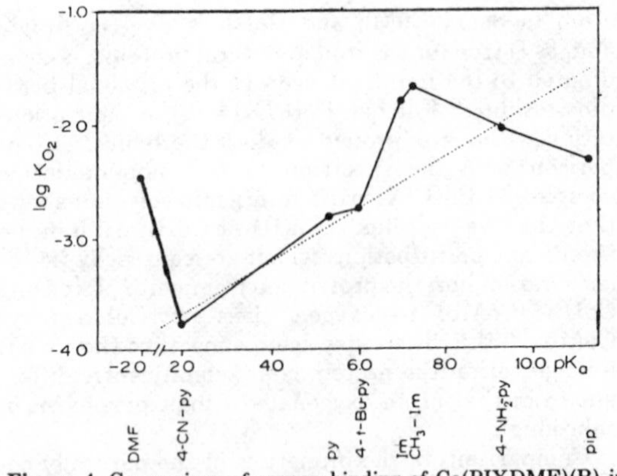

Figure 4. Comparison of oxygen binding of Co(PIXDME)(B) in toluene at −45°C as a function of the pK_a of the base, B. The dashed line shows the correlation of log K_{O_2} with pK_a for the pyridine systems (from ref 20).

of log K_{O_2} vs. pK_a for different axial bases on Co(PIXDME) in toluene solution. Neither for these equilibrium constants nor for the derived thermodynamic quantities, ΔH, ΔG, or ΔS, is there a simple, general relationship that is related to the pK_a's. However, as the dashed line indicates, there is a reasonably linear correlation of the equilibrium constant with the pK_a's of the substituted pyridines. Tetraphenylporphyrin and other synthetic porphyrins yield approximately the same results.[38]

We attribute such differences to the π-donor properties of these ligands. In the Pauling model[39] the bonding of an oxygen atom to a metal complex may be explained in terms of the σ donation from a nonbonding sp^2 lone pair on oxygen to the d_{z^2} orbital on the metal, accompanied presumably by synergistic π back-bonding from the filled d_{xz} (or d_{yz}) orbitals into the empty π^* orbitals of oxygen. Since ligands coordinated trans to oxygen will compete for π electron density on the metal, the binding of O_2 to Co(P)(B), where P is any porphyrin, will be sensitive to the π-donating ability or π-accepting ability of the axial ligand B. Good π acceptors will decrease π electron density on the metal, resulting in a poorer oxygen carrier, whereas good π donors will promote oxygenation by increasing the electron density available for back-bonding. Imidazoles are much better π donors than pyridines; the strong π-donor properties of DMF have also been demonstrated. On the other hand, the log K_{O_2} for piperidine is much less than that expected from basicity arguments. It is possible here that steric effects play a role through the interactions of the ring protons with parts of the porphyrin core, preventing close approach of the base to the metal. We also found[35,40] that a change from the essentially nonpolar solvent toluene to the polar solvent dimethylformamide (DMF) has a pronounced effect on the equilibrium of eq 5, leading to much greater degree of oxygenation. We ascribe this large solvent effect to the stabilization of the polar Co(III)–O_2^- species in the more polar solvents. Although cobalt porphyrins are reversibly oxygenated at low temperatures in aprotic solvents, it is known[36] that in the presence of a protonic base irreversible oxidation occurs.

For changes in the quadridentate equatorial ligand, keeping the same axial base, there is also a linear correlation[35] between the log K_{O_2} and the $E_{1/2}$ values for various cobalt(II) complexes. In general, cobalt(II) porphyrins have a smaller tendency to form dioxygen complexes than do cobalt(II)–Schiff-base chelates. This is perhaps the result of the porphyrin ligand having a greater tendency to delocalize π electron density away from cobalt atom than does the less aromatic Schiff-base ligand. Representative thermodynamic data for dioxygen uptake by cobalt(II) complexes of Schiff bases and of porphyrins are given in Table I. Also shown in this table for comparison are corresponding values for CoMb and Mb.

The protein environment, although producing no major effect on the electronic structure of the oxygen adduct, does produce a large effect on its stability.[22,41] We can directly observe the influence of the globin (apoprotein) upon the reactivity of the metalloporphyrin (prosthetic group) by comparing oxygen

Table I
Representative Thermodynamic Data for Reversible Oxygen Binding to Cobalt(II) Complexes, to Myoglobin, and to a Modified Silica Gel

System[a]	log K_{O_2}, mm^{-1}	ΔH, kcal/mol	ΔS, eu	$P_{1/2}$, Torr[e]	Ref
Co(benacen)-(MeIm)[b]	−2.98	−17.5	−73.5	955	35
Co(PIXDME)-(MeIm)[b]	−4.10	−11.8	−59	12,500	20
CoMb[c] (sperm whale)	−1.65	−13.3	−53	45	41
Mb[c] (sperm whale)	−0.3	−18.1	−60	0.50	*f*
(IPG)Fe(TPP)[d]				900	72

[a] Data in the table are from literature values extrapolated to 20°C. [b] Solvent is toluene. [c] Solvent is water, phosphate buffer at pH 7. [d] Solid–gas phase. [e] $P_{1/2}$ is the pressure at which one-half of the complex is present in the oxygenated form. [f] M. H. Keyes, M. Falley, and R. Lumry, *J. Am. Chem. Soc.*, **93**, 2038 (1971).

(38) F. A. Walker, *J. Am. Chem. Soc.*, **95**, 1150, 1154 (1973).
(39) L. Pauling, *Nature (London)*, **203**, 182 (1964).
(40) H. C. Stynes and J. A. Ibers, *J. Am. Chem. Soc.*, **94**, 5125 (1972).
(41) C. A. Spilburg, B. M. Hoffman, and D. H. Petering, *J. Biol. Chem.*, **247**, 4219 (1972).

binding of the five-coordinate Co(PIXDME)(MeIm) complex with that of CoMb. For example, at 25°, the incorporation of Co(PIX) into globins of whale, horse, or harbor-seal myoglobins increases the K_{O_2} by a factor of ~300. Through the use of a formal thermochemical cycle oxygen binding to protein and to solution porphyrin can be related, and the protein contribution to the thermodynamics of dioxygen binding to the metalloprotein can be determined. We find that the protein provides a significantly favorable (positive) entropic contribution to O_2 binding.[22,41] The enhancement of oxygen binding by horse CoMb is entirely entropic. Binding to the harbor-seal CoMb is actually disfavored enthalpically compared with the free porphyrin, but this is compensated by a further favorable change in the entropy. The whale CoMb does exhibit a rough balance between favorable entropy and enthalpy contributions. Thus, by combining the results for model compound and protein, we see how these three myoglobins of identical function and similar structure achieve similar O_2 binding properties near physiological temperatures through a different balance of enthalpic and entropic factors.[42]

The molecular basis of the more favorable entropy of dioxygen binding in the protein environment has been discussed,[41] and at least in part is related to the effects of polar solvents on binding to Co(PIX) described above. Formation of the strongly dipolar Co(III)–O_2^- linkage from a neutral cobalt(II) porphyrin and a neutral oxygen molecule causes solvent reorganization which accommodated the charge separation and thus stabilizes the metal–dioxygen bond. This reorganization would appear thermodynamically as a negative contribution to the entropy of oxygen binding to solution porphyrin. If the hydrophobic protein crevice is already organized to accommodate a polar metal–oxygen group, the negative entropy associated with the formation of the crevice would appear as part of the entropy of formation of the metalloprotein itself, not as a contribution to the entropy change upon oxygen binding.

Hemoglobin Cooperativity

A chief source of the ongoing interest in ligand binding to hemoglobin rests in the following allosteric properties which result from its tetrameric nature.[43] The dioxygen binding curve is "cooperative" (autocatalytic), not hyperbolic. The degree of cooperative binding can be expressed by the Hill constant, n, which has the value $n \approx 2.8$. For comparison, independent O_2 binding by the four hemes of Hb would require $n = 1.0$, whereas all-or-none binding to the four hemes would require $n = 4$. Hemoglobin exhibits two other so-called linkage properties: first, the binding of O_2 is pH dependent, although there is no ionizable group at the heme (Bohr effect); second, the oxygen affinity is altered by the binding of an organic phosphate to Hb, although the binding site is far removed from the hemes (phosphate effect).

The ligated and unligated forms of Hb, respectively denoted as R and T, differ both in the conformation of the individual chains (tertiary structure) and in the relative orientation of the chains (quaternary structure). The linkage effects exhibited by Hb arise from the reversible transition between these two forms. Perutz and Ten Eyck[44] concluded that ligand binding can be accompanied by Bohr and phosphate effects and the normal quaternary structure transition, even though cooperativity ($n > 1$) is weak or absent; thus, these linkage effects are diagnostic of a quaternary structure change. The degree of cooperativity, as measured by the n value, appears to reflect the details of the ligation process and thus the nature and type of intermediates which occur as the four hemes of Hb react.

In constructing a detailed scheme for the ligation process of Hb, a question can be formulated as to the nature of an "allosteric trigger",[44] the mechanism by which the ligation (or oxidation) of five-coordinate, iron(II) heme induces conformational changes within the protein. Our preparation of cobalt- and manganese-substituted hemoglobins has provided a unique opportunity to help answer this question. We have found that CoHb retains the full Bohr and phosphate effects upon oxygen binding,[22,45] and that the binding of O_2 is cooperative.[21] The Hill constant is at least 2.3[23,24] and probably is as great as 2.5,[46] whereas that of Hb is 2.8. Thus, following Perutz's arguments,[44] CoHb must undergo a quaternary structure change upon oxygenation, and the stereochemical changes that occur as the five-coordinate, low-spin cobalt(II) protoporphyrin IX of CoHb binds oxygen must be compatible with the mechanism that triggers this quaternary structure transition.

This phenomenon of "heme–heme interaction" or cooperativity must have its basis at the molecule level, even though the Fe atoms are separated by ≈25 Å. The problem has been the subject of considerable speculation that is not appropriate to review here. Suffice it to say that the "trigger mechanism" proposed by Perutz, which is based on ideas of Williams[47] and Hoard,[48] may be illustrated in the top of Figure 5. In deoxy-Hb the high-spin Fe(II) atom is about 0.8 Å above the mean plane of the porphyrin. Upon oxygenation the Fe atom, which is variously

Table II
Effect of pH on Oxygen Binding to CoHb and Hb[a]

	DPG,[d] mM	log $P_{1/2}$ pH 7.3[b]	log $P_{1/2}$ pH 9.4[c]	Δ log $P_{1/2}$
CoHb	0.2	2.04	0.96	1.08
	0	1.56	0.96	0.60
Hb	0.2	0.63	−0.47	1.10
	0	0.21	−0.47	0.68

[a] From ref 45. [b] 0.05 M Bis-Tris; 4°C. [c] 0.05 M Tris; 4°C. [d] DPG, diphosphoglycerate.

(42) R. Lumry and S. Rajender, *Biopolymers*, **9**, 1125 (1970).
(43) E. Antonini and M. Brunori, "Hemoglobin and Myoglobin in their Reactions with Ligands", North-Holland Publishing Co., Amsterdam, 1971.
(44) M. F. Perutz and C. F. Ten Eyck, *Cold Spring Harbor Symp. Quant. Biol.*, **36**, 315 (1971).
(45) G. C. Hsu, C. A. Spilburg, C. Bull, and B. M. Hoffman, *Proc. Natl. Acad. Sci. U.S.A.*, **69**, 2127 (1972).
(46) B. M. Hoffman and C. Bull, unpublished.
(47) R. J. P. Williams, *Fed. Proc., Fed. Am. Soc. Exp. Biol.*, **20** (Suppl. 10), 5 (1961).
(48) J. L. Hoard in "Hemes and Hemoproteins", B. Chance, R. W. Estabrook, and T. Yonetani, Ed., Academic Press, New York, N.Y., 1966, p 9.

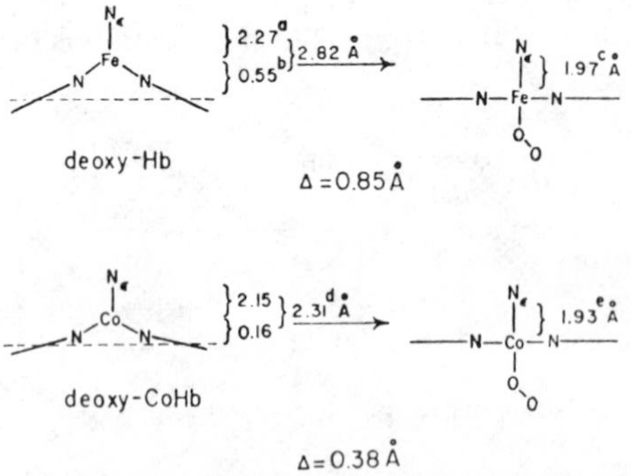

Figure 5. Illustration of the calculation of the total movement of the proximal histidine in Hb (top) and CoHb (bottom). The displacements are taken relative to the 24-atom porphyrin core (from ref 52).

thought to be Fe(II) or Fe(III), is now low spin, and it is believed to move into the mean plane of the porphyrin. Perutz proposes that the high-spin Fe(II) ion is held in a stressed five-coordinate state by the protein and that this stress or tension is released upon oxygenation. The resultant movement triggers a relaxation throughout a subunit which is ultimately transmitted to salt bridges.

However, the cobalt atom of CoHb is low spin even in the deoxy state.[21] Since low-spin Co(II) exhibits a smaller covalent radius than does high-spin Fe(II), we[21,22,45] predicted that the displacement of the Co(II) atom in CoHb would be relatively small compared with that of the Fe(II). Although CoHb has not been crystallized, our studies[49–52] of the stereochemistry of cobalt porphyrins have given considerable evidence regarding this prediction and have permitted reliable estimates of the possible motion to be expected in CoHb. As the model for deoxy-CoHb we have studied the structure of Co(OEP)(MeIm)[52] (OEP = octaethylporphyrin; MeIm = methylimidazole). On the basis of spectroscopic data similar to that discussed above, it is clear[21] that the cobalt in oxy-CoHb is low-spin six-coordinate Co(III) and thus a possible model is $Co(TPP)(Im)_2{}^+$ [51] (Im = imidazole). Whereas the proposed movement of the N atom of the proximal histidine group in Hb is about 0.85 Å on oxygenation, an upper limit of about 0.38 Å can be placed on the similar movement in CoHb. Since CoHb nevertheless displays cooperativity, it does not appear that cooperativity depends primarily upon the spin state and stereochemistry of the metal in the deoxy protein.[46,52] We thus concluded[22,52] that there are inherent shortcomings, or at least unresolved difficulties, in the trigger mechanism of Perutz.

A counterproposal has been advanced[53] in which it is suggested that in deoxy-CoHb tension within the protein increases the Co–histidine bond length and deforms the porphyrin in such a way that the net motion of the proximal histidine upon ligation is made equal to that in Hb. Direct evidence against this tension model in CoHb comes from resonance Raman studies of Hb and CoHb[54] and from EPR experiments of CoHb and model systems.[55,56] Since none of these experiments gives indications of tension, there appears to be no convincing evidence that the globin exerts a force which greatly increases the out-of-plane displacement of the Co atom or stretches the Co–N (of imidazole) bond in CoHb.

A second mechanism for cooperative effects in Hb is Hopfield's "linearly distributed energy model", in which the free energy of cooperativity is stored as small strains in many bonds.[57] Although attractive, this model may not account for the fact that deoxy-CoHb has the T quaternary structure, whereas met-Hb has the R structure since the metalloporphyrin geometries should be approximately the same in both molecules. Furthermore, the application of the model employs the specific value $n = 2.3$ for CoHb to relate Co(PIX) structures to CoHb function. However, as noted above, the Hill constant for CoHb of 2.3 cannot be considered as firm, and is probably not even the correct lower limit.

The implications of CoHb cooperativity for the nature of the Hb allosteric mechanism can be widened by using other metal-substituted hemoglobins. Met-Hb is in the high-spin, aqua form at low pH, and undergoes an ionization to the largely low-spin hydroxo form at p$K \sim 8$. This transition has been suggested to contribute to an observed increase of the n value for the redox reaction with pH.[44] We have observed that MnHb also exhibits an increase from $n = 1.4$ at pH 6.5 to $n = 2$ at pH 9, mimicking the behavior of Hb.[32,46] This observation is particularly significant, for Mn^{III}Hb does *not* exhibit a heme-linked ionization in the pH range of 5–10, and, therefore, Mn^{III}Hb does not display a "met-aqua" to "met-hydroxo," high–low spin transition.[32] Prompted by these results, we have directly addressed the question of the coupling of spin-state and cooperative redox behavior by examining the Hb redox equilibrium in the presence of F^- and of N_3^-.[58] In this way the oxidized heme is either constrained to remain high spin through binding of F^- or low spin through binding of N_3^-. The results are completely consistent with those for MnHb.

These experiments prove that the pH-induced changes in redox cooperativity do not arise from the pH dependence of the oxidized-heme spin state. The changes thus appear to reside in a pH dependence of the allosteric equilibrium; in addition, chain nonequivalence which depresses n at low pH[59] may diminish with increasing pH. On the other hand, a value of n of 2.2 appears to be the maximum obtainable with the high-spin forms, whereas azidomet-Hb exhibits an n of 2.7.[58] Whether this modest increase results from an effect of spin-state change on the position of the proximal histidine and to other ligand-induced changes in heme conformation[29] or whether

(49) J. A. Ibers, J. W. Lauher, and R. G. Little, *Acta Crystallogr., Sect. B*, **30**, 268 (1974).
(50) R. G. Little and J. A. Ibers, *J. Am. Chem. Soc.*, **96**, 4440 (1974).
(51) J. W. Lauher and J. A. Ibers, *J. Am. Chem. Soc.*, **96**, 4447 (1974).
(52) R. G. Little and J. A. Ibers, *J. Am. Chem. Soc.*, **96**, 4452 (1974).
(53) J. L. Hoard and W. R. Scheidt, *Proc. Natl. Acad. Sci. U.S.A.*, **70**, 3919 (1973); **71**, 1578 (1974).
(54) W. H. Woodruff, T. G. Spiro, and T. Yonetani, *Proc. Natl. Acad. Sci. U.S.A.*, **71**, 1065 (1974).
(55) L. H. Marchant, Ph.D. Dissertation, Northwestern University, 1973.
(56) R. G. Little, B. M. Hoffman, and J. A. Ibers, *Bioinorg. Chem.*, **3**, 207 (1974).
(57) J. J. Hopfield, *J. Mol. Biol.*, **77**, 207 (1973).
(58) C. Bull and B. M. Hoffman, *Proc. Natl. Acad. Sci. U.S.A.*, in press.
(59) C. P. Hensley, S. J. Edelstein, P. C. Wharton, and Q. H. Gibson, *J. Biol. Chem.*, **250**, 952, (1975).

it arises from an interaction of the N_3^- ligand with amino acid residues on the distal side of the heme and to an abolition of chain differences in the R state cannot be inferred from present data. However, since the results for coboglobin[21,22,45,50,56] and other studies[59,60] of Hb redox reactions do not support a simple coupling between spin state, the position of the heme-linked histidine, and cooperativity, the idea of an "allosteric trigger" thus seems to be either inappropriate or at least limited in its applicability.

Iron(II) Porphyrin Complexes

Research on synthetic oxygen carriers has very recently experienced the significant experimental development that iron(II) porphyrins can react reversibly with dioxygen (eq 6). One of the major difficul-

$$Fe(P)(B)_2 + O_2 \rightleftharpoons Fe(P)(B)(O_2) + B \qquad (6)$$

ties encountered in attempts to obtain oxygen carriers of iron(II) complexes is the large driving force toward the irreversible formation of the stable μ-oxo dimer (eq 7). Considerable recent work has been

$$Fe^{II} + O_2 \rightleftharpoons Fe(O_2) \xrightarrow{Fe^{II}} Fe^{III}-O-Fe^{III} \qquad (7)$$

aimed at overcoming this problem, and three approaches have been successful: (1) *steric*—in such a fashion that dimerization is inhibited; (2) *low temperature*—in order that reactions leading to dimerization are very slow; and (3) *rigid surfaces*—attachment of the iron complex on the surface in a manner that dimerization is prevented. The elegant steric approaches[61-64] were not done in our laboratories and are not discussed here. However, the low-temperature and the rigid-surface approaches which we have used are discussed.

One of the first synthetic iron(II) oxygen carriers[65] was an iron porphyrin with a built-in imidazole. The fact that it is sterically capable of forming a μ-oxo dimer and yet reversibly oxygenates at $-45°$ suggested to us that the low temperature and not the neighboring group effect of an attached imidazole was responsible. Indeed the simple ferrous porphyrins, $Fe(TPP)(B)_2$, in methylene chloride solution at $-79°$ are excellent oxygen carriers.[66] Similar observations were made independently in other laboratories using different iron(II) porphyrins.[67] The spectral changes show that $Fe(TPP)(py)_2$ is oxygenated in the presence of dioxygen and deoxygenated when the solution is purged with dinitrogen. Several oxygenation-deoxygenation cycles are possible at $-79°$ with a minimum of irreversible oxidation occurring, and at this temperature a solution of the complex in one atmosphere of oxygen is stable for at least 1 day. The

(60) T. Yonetani, H. Yamamoto, F. Kayne, and G. Woodruff, *Biochem. Soc. Trans.*, **2**, 44 (1973).

(61) J. E. Baldwin and J. Huff, *J. Am. Chem. Soc.*, **95**, 5757 (1973).

(62) J. P. Collman, R. R. Gagne, T. R. Halbert, J. C. Marchon, and C. A. Reed, *J. Am. Chem. Soc.*, **95**, 7863 (1973).

(63) J. P. Collman, R. R. Gagne, C. A. Reed, W. T. Robinson, and G. A. Rodley, *Proc. Natl. Acad. Sci. U.S.A.*, **71**, 1326 (1974).

(64) J. Almog, J. E. Baldwin, and J. Huff, *J. Am. Chem. Soc.*, **97**, 227 (1975).

(65) C. K. Chang and T. G. Traylor, *J. Am. Chem. Soc.*, **95**, 5810 (1973).

(66) D. L. Anderson, C. J. Weschler, and F. Basolo, *J. Am. Chem. Soc.*, **96**, 5599 (1974).

(67) G. C. Wagner and R. J. Kassner, *J. Am. Chem. Soc.*, **96**, 5593 (1974); W. S. Brinigar and C. K. Chang, *ibid.*, **96**, 5595 (1974); J. Almog, J. E. Baldwin, R. L. Dyer, J. Huff, and C. J. Wilkerson, *ibid.*, **96**, 5600 (1974).

Table III
Comparisons between Kinetic and Thermodynamic Parameters for Reactions of O_2 and CO with a Simple Ferrous Porphyrin and with Heme Proteins

Ferrous system[a]	k_3/k_2	k_{-3}/k_{-2}	K_{CO}/K_{O_2}
Fe(TPP)(MeIm)	0.031	$<1.1 \times 10^{-3}$	>30
Myoglobin[b] (horse)	0.016	1.08×10^{-3}	14.4
Myoglobin[c] (aplysia)	0.033	0.29×10^{-3}	114
Isolated α chains[d,e]	0.071	0.52×10^{-3}	133
Isolated β chains[d,e]	0.030	0.17×10^{-3}	180

[a] The values for the heme proteins are at 20°C in aqueous solution (pH 7.0–7.4), whereas for Fe(TPP)(MeIm) the conditions are $-79°$C and methylene chloride solution (from ref 69). [b] G. A. Millikan, *Proc. Roy. Soc., Ser. B*, **120**, 366 (1936). [c] B. A. Wittenberg, M. Brunori, E. Antonini, J. B. Wittenberg, and J. Wyman, *Arch. Biochem. Biophys.*, **111**, 576 (1965). [d] M. Brunori, R. W. Noble, E. Antonini, and J. Wyman, *J. Biol. Chem.*, **241**, 5238 (1966). [e] R. W. Noble, Q. H. Gibson, M. Brunori, E. Antonini, and J. Wyman, *J. Biol. Chem.*, **244**, 3905 (1969).

stoichiometry of the reaction corresponds to 1 mol of dioxygen per mol of iron (eq 8).

$$Fe(TPP)(B)_2 + O_2 \rightleftharpoons Fe(TPP)(B)(O_2) + B \qquad (8)$$

Under 1 atm of oxygen, complete complex formation occurs in methylene chloride; only a small amount of complex is formed in toluene, whereas the degree of complex formation in ethyl ether lies between these extremes. Similarly, dioxygen complexes of cobalt are stabilized by polar solvents. The behavior in polar solvents supports formulation of the iron complex as $Fe^{III}(O_2^-)$. This formulation is in accord with the ir spectrum[68] of HbO_2.

Kinetic and equilibrium studies have been made on the reactions of $Fe(TPP)(B)_2$ with oxygen and with carbon monoxide in methylene chloride at $-79°$. The results of this investigation[69] are in accord with a rate-determining dissociation process (eq 9), followed by the rapid reaction of the five-coordinated intermediate, $Fe(TPP)(B)$, with either oxygen (eq 10) or carbon monoxide (eq 11).

$$Fe(TPP)(B)_2 \underset{k_{-1}}{\overset{k_1}{\rightleftharpoons}} Fe(TPP)(B) + B \qquad (9)$$

$$Fe(TPP)(B) + O_2 \underset{k_{-2}}{\overset{k_2}{\rightleftharpoons}} Fe(TPP)(B)(O_2) \qquad (10)$$

$$Fe(TPP)(B) + CO \underset{k_{-3}}{\overset{k_3}{\rightleftharpoons}} Fe(TPP)(B)(CO) \qquad (11)$$

Some of the data obtained are given in Table III, along with corresponding data for Mb and for the isolated α and β chains of Hb. The simple five-coordinate ferrous porphyrin Fe(TPP)(MeIm) displays the same trends in reacting with dioxygen and carbon monoxide as do the more complicated heme proteins. This suggests that the simple system is a satisfactory model for the active site of heme proteins. Although the carbon monoxide complex is more stable than the corresponding oxygen complex, dioxygen reacts more rapidly than does carbon monoxide with the five-coordinate iron species. Nitric oxide, which contains

(68) C. H. Barlow, J. C. Maxwell, W. J. Wallace, and W. S. Caughey, *Biochem. Biophys. Res. Commun.*, **55**, 91 (1973).

(69) C. J. Weschler, D. L. Anderson, and F. Basolo, *J. Chem. Soc., Chem. Commun.*, 757 (1974); *J. Am. Chem. Soc.*, in press.

an unpaired electron, also reacts with heme proteins faster[43] than does CO. Thermodynamically Fe(porphyrin)(B)(NO) is more stable than Fe(porphyrin)(B)(CO), and this stability is augmented in the protein.[70]

The ratio k_2/k_{-1} (see eq 9 and 10) is approximately equal to 1 for different axial bases, which shows that there is no large kinetic preference of iron porphyrin for oxygen compared with nitrogenous bases. This illustrates the subtlety of the heme environment in the oxygen-carrying proteins. The distal imidazole can stabilize the oxygen complex through hydrogen bonding, but geometric constraints fortunately prevent its coordination to the iron center. The values of k_2/k_{-1} show that if the distal imidazole could bind to heme, it would seriously compete with oxygen for the available site, and the protein would cease to function as an effective oxygen carrier.

The third successful method of obtaining an oxygen carrying iron complex is to attach it to the surface of a solid so that two iron atoms cannot approach each other to form a dimer. In a classic experiment Wang[71] reported the first synthetic Fe(II) oxygen carrier. He showed that oxygen binds reversibly to 1-(2-phenylethyl)imidazole heme diethyl ester embedded in a matrix of an amorphous mixture of polystyrene and 1-(2-phenylethyl)imidazole. We have found[72] that the attachment of Fe(TPP) to a rigid modified silica gel support also inhibits dimerization and produces an efficient oxygen carrier. Collman and Reed[73] reported that cross-linked polystyrene, containing imidazole ligands coordinated to Fe(TPP), was not a rigid enough support to prevent dimerization upon treatment with oxygen.

The modified silica gel employed is one prepared earlier for other purposes[74] and contains 3-imidazolylpropyl (IPG) groups bonded to surface atoms of silicon. This then reacts with Fe(TPP)(B_2) to coordinate the iron(II) porphyrin to imidazole groups attached to the silica surface. Heating (IPG)-Fe(TPP)(B) in flowing helium removes the axial base (pyridine or piperidine) and generates the active coordination site, (IPG)Fe(TPP) (see eq 12), resembling that of the five-coordinated iron(II) porphyrin in hemoglobin and myoglobin. The open coordination site can then reversibly bind dioxygen.

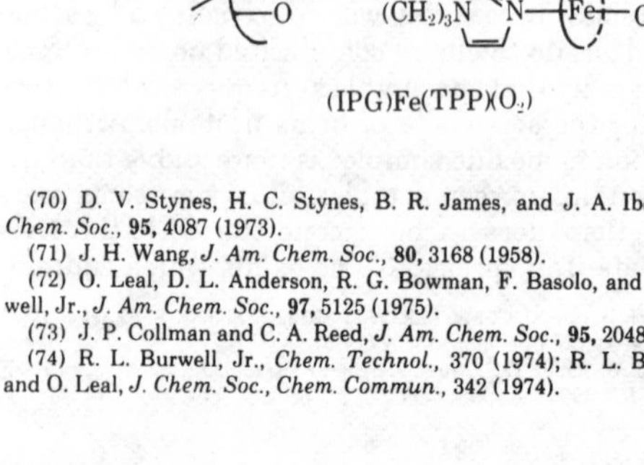

(12)

The chemisorption of oxygen is weak. It is irreversible at −127° and has a $P_{1/2}$ of 0.4 Torr at −78°, and a $P_{1/2}$ of 230 Torr at 0°. Data for human myoglobin[43] extrapolated to 0° give a $P_{1/2}$ of 0.14 Torr. Successful O_2 binding on the modified silica gel, however, confirms the utility of the rigid-surface approach.

Conclusion

During the last 5 years research on synthetic oxygen carriers in our laboratories has been devoted almost entirely to systems related to the natural oxygen carriers. The discovery of 1:1 O_2–Co(Schiff base) complexes, followed by similar observations on 1:1 O_2–Co(porphyrin) complexes, and the reconstitution of hemoglobin with Co in place of Fe have afforded an excellent opportunity to attack some of the central problems in biological oxygen carriers. For example, various experiments enabled us to assess the role of the globin in the uptake of oxygen and other small molecules by biological molecules. Moreover, the combination of results from the model cobalt porphyrin systems and cobalt-substituted hemoglobin has enabled us to test critically current theories for the allosteric properties of hemoglobin. More recently with the discovery of stable, reversibly oxygenated iron porphyrin systems we anticipate an even greater opportunity to test critically the role of the globin in hemoglobin chemistry. The current efforts to study hemoglobin substituted with other metals, such as Mn and Zn, and to study analogous porphyrin systems has already begun to provide further insight into the biological systems. Extension of the techniques used to obtain oxygen carriers with Fe complexes to corresponding Cr, Mn, and V systems has now produced new synthetic oxygen carriers.[75]

We thank all of our coworkers whose names appear in the references and we gratefully acknowledge that without their efforts the research on synthetic oxygen carriers done at Northwestern University would not have been accomplished. This research was in part financially supported by National Institutes of Health Grants HL 13412, HL 13531, and HL 13157, by National Science Foundation Grants GP-27626, MPS 71-02605 and BMS-00478, by the Petroleum Research Fund, administered by the American Chemical Society, and by the Materials Research Center, Northwestern University.

Addendum (May 1976)

An apparent contradiction[76] to our picture of the bonding in a reversible (Co–O_2) or (Fe–O_2) linkage has now been removed. Recently Collman et al.[77] corrected their erroneous assignment of a 1385 cm^{-1} band to the O–O stretching vibration in an iron–picket fence dioxygen adduct and have found the true vibration at 1160 cm^{-1}, using the techniques of differential infrared spectroscopy applied so successfully by Barlow et al.[68]

(70) D. V. Stynes, H. C. Stynes, B. R. James, and J. A. Ibers, *J. Am. Chem. Soc.*, **95**, 4087 (1973).

(71) J. H. Wang, *J. Am. Chem. Soc.*, **80**, 3168 (1958).

(72) O. Leal, D. L. Anderson, R. G. Bowman, F. Basolo, and R. L. Burwell, Jr., *J. Am. Chem. Soc.*, **97**, 5125 (1975).

(73) J. P. Collman and C. A. Reed, *J. Am. Chem. Soc.*, **95**, 2048 (1973).

(74) R. L. Burwell, Jr., *Chem. Technol.*, 370 (1974); R. L. Burwell, Jr., and O. Leal, *J. Chem. Soc., Chem. Commun.*, 342 (1974).

(75) C. J. Weschler, B. M. Hoffman, and F. Basolo, *J. Am. Chem. Soc.*, **97**, 5278 (1975).

(76) J. P. Collman, R. R. Gagne, H. B. Gray, and J. W. Hare, *J. Am. Chem. Soc.*, **96**, 6523 (1974).

(77) J. P. Collman, J. I. Brauman, T. R. Halbert, and K. S. Suslick, private communication.

to Hb and Mb. Collman et al.[77] now completely agree with our earlier views that the correct formulations of these various complexes are Fe(III)–O_2^- and Co(III)–O_2^- in the appropriate model and biological systems.

A detailed optical and epr study of Mn(TPP)(O_2)[75] has now been completed.[78] Dioxygen binds to Mn(TPP)(py) with replacement of pyridine to form a five-coordinate complex which, surprisingly, exhibits an $S = 3/2$ state. The epr results indicate extensive Mn $\rightarrow$ O_2 charge transfer, the appropriateness of an Mn(IV)(O_2^{-2}) valency formalism, and suggest the symmetric, "edge-on" mode of binding. Such binding has been shown by x-ray diffraction to occur in Ti(OEP)(O_2).[79]

Reed[80] recently reported the preparation of Cr(II)(TPP)(py) and the observation that the solid reacts with dioxygen to form Cr(TPP)(py)(O_2). This compound exhibits an infrared band of medium intensity at 1142 cm^{-1}, which is assigned to the O–O stretching vibration. Again it appears that, as in the earlier Co(III)–O_2^- systems,[12] the chromium complex has a bent Cr–O_2 structure and a Cr(III)–O_2^- formulation. Of interest is the observation that the chromium complex is not an oxygen carrier, consistent with the fact that superoxide ion is not a good enough reducing agent to reduce Cr(III) to Cr(II).

Finally, the arguments presented in the present Account concerning the Hb allosteric trigger have received increased support from diverse sources. Apparent evidence for tension on the heme group in Hb cited by Hoard and Scheidt,[53] namely the Fe-mean plane distance of 0.75 Å in Hb vs. that of 0.55 Å in the model system Fe(TPP)(2-MeIm), has recently been cancelled by the new refinement of the Hb structure. The Fe-mean plane distance is now thought to be ~0.60 Å,[81] a value indistinguishable from that in the model. In recent x-ray absorption fine-structure (EXAFS)[82] and resonance Raman[83,84] studies, no evidence could be found for deformation of either an unliganded or a liganded heme when the Hb conformation is shifted between the T to R states. In short, no evidence could be found for the proposed tension.

(78) B. M. Hoffman, C. J. Weschler, and F. Basolo, *J. Am. Chem. Soc.*, **98** (Sept. 1976).

(79) R. Guilard, M. Fontesse, P. Fournari, C. Lecomte, and J. Protas, *J. C. S. Chem. Commun.*, 161 (1976).

(80) S. Yee, J. Wong, J. Kouba, B. Gonzalez, and C. A. Reed, Centennial ACS Meeting, New York, April, 1976, Inorg 51.

(81) G. Fermi, *J. Mol. Biol.*, **97**, 237 (1975).

(82) P. Eisenberger, R. G. Shulman, G. S. Brown, and S. Ogawa, *Proc. Natl. Acad. Sci.*, **73**, 491 (1976).

(83) T. G. Spiro and J. M. Burke, *J. Am. Chem. Soc.*, in press.

(84) D. M. Scholler, B. M. Hoffman, and D. F. Shriver, *J. Am. Chem. Soc.*, in press.

Index

D

E

K

L

M

Q

R

S

T

Q

R

S

T